Marine Sponges: Chemicobiological and Biomedical Applications

Ramjee Pallela • Hermann Ehrlich
Editors

Marine Sponges: Chemicobiological and Biomedical Applications

Springer

Editors
Ramjee Pallela
IKP Knowledge Park
Genome Valley, Turkapally
Hyderabad
Telangana
India

Hermann Ehrlich
Institute of Experimental Physics
TU Bergakademie Freiberg
Freiberg
Sachsen
Germany

ISBN 978-81-322-2792-2 ISBN 978-81-322-2794-6 (eBook)
DOI 10.1007/978-81-322-2794-6

Library of Congress Control Number: 2016947950

Printed on acid-free paper

This Springer imprint is published by Springer Nature
The registered company is Springer (India) Pvt. Ltd.

Preface

Sponges (phylum Porifera) are probably one of the earliest branching animals and their fossil record dates back to the Precambrian. These simply organized, exclusively aquatic and almost sessile organisms habituate in both fresh and marine waters. According to the modern view, the divisions of sponges include four classes: Demospongiae (horn sponges), Hexactinellida (glass sponges), Calcarea (calcareous sponges) and Homoscleromorpha (for details see World Porifera database. Available: http://www.marinespecies.org/). Over million years of evolution, sponges developed a unique strategy for survival via their attachment to hard substrates, and consequently, being open to all possible predators including microorganisms and metazoans. They contain mechanically stable, usually mineralized skeletons, and possess the ability to synthesize secondary metabolites with antiviral, antibacterial, antifungal and cytotoxic activities.

Biomineralogy of diverse sponges skeletal structures like biosilica- and calcite-based spicules is well investigated and represents a source for bioinspired materials science today. The organic chemistry of the same structures is also of immense scientific interest due to the presence of very special biopolymers including collagens, chitin, keratin-like sponging, as well as a broad variety of proteins with enzymatic activities. Some of them, for example, silicateins, are involved in biosilicification. Although there has been tremendous activity within numerous scientific groups to obtain better understanding of the structural biology of the biomineralization phenomenon in sponges, a detailed knowledge of the genetic, biochemical and molecular mechanisms involved is still lacking. Even recent discoveries on highly hydroxylated collagen in anchoring spicules of glass sponges, and chitin within skeletal fibrous networks of some keratose demosponges, produced more open questions than answers. Both the mechanical and optical properties of selected spicular structures observed in some sponges have attracted the attention of experts from such scientific fields as biomimetics, bionics and biophotonics. Spongin- and chitin-based three-dimensional scaffolds isolated from demosponges possess high biomimetic potential for biomedicine and tissue engineering.

Due to their excellently developed strategy of chemical defense, sponges (mostly demosponges) are important within the fields of marine biotechnology and marine pharmacology. Moreover, they are considered as the best resources for highly active therapeutic molecules amongst other marine invertebrates. Several biological compounds isolated from marine sponges have been further explored for pharmaceutical drug development. Ara-C is the first isolated compound for cancer treatment that is used widely. Usually, sponges synthesize different substances with antibiotic activity simultaneously by themselves (Verongida sponges), or due to their symbiotic microorganisms. Consequently, they represent intriguing examples for experts which are involved in research on multidrug-resistant organisms.

Various experts in marine sponge related area globally have contributed valuable chapters in the present book. We strongly believe that this would provide enough insights into the amazing world of marine sponges to the readers. The present book is an attempt to compile the novel information available on sponge metabolites, including low molecular weight bioactive compounds, and structural biopolymers and their isolation techniques as well as the biomedical applications at the same place. It is an essential reading for the novice and expert in the field of marine biotechnology and pharmacology, natural product researchers and chemists, industrialists as well as students.

Hyderabad, India Ramjee Pallela

Contents

1 **Introduction to the Global Scenario of Marine Sponge Research** 1
P.V. Bramhachari, Hermann Ehrlich, and Ramjee Pallela

2 **Global Constraints, Prospects, and Perspectives of Marine Sponge Research** 25
Baboucarr Lowe, Jayachandran Venkatesan, Hermann Ehrlich, and Se-Kwon Kim

3 **Chemical Ecology of Marine Sponges** 37
Narsinh L. Thakur and Anshika Singh

4 **Bioeroding Sponges in Aquaculture Systems** 53
P. Sunil Kumar

5 **Marine Sponge-Associated Actinobacteria and Their Biological Properties** 57
Panchanathan Manivasagan and Se-Kwon Kim

6 **Novel Insights on the Symbiotic Interactions of Marine Sponge-Associated Microorganisms: Marine Microbial Biotechnology Perspective** 69
P.V. Bramhachari, Satish Mutyala, Ira Bhatnagar, and Ramjee Pallela

7 **Remarks on the Chemo Biological Applications of Marine Sponges** 97
P. Sunil Kumar

8 **Sponges as Biomonitors of Metal Toxicity in the Aquatic Systems** 105
Koigoora Srikanth and Janapala Venkateswara Rao

9 **Biologically Active Metabolites from Sponges and Their Activities** 115
A. Shanmugam and S. Vairamani

10 **Bioactive Potential of Sponge Secondary Metabolites** 143
Irudayaraj Rajendran

11 **Typification of Chemical Compounds of Marine Sponge Metabolites** 167
Irudayaraj Rajendran

12 **Bioactive Alkaloids from Marine Sponges** 257
Keisham S. Singh and Mahesh S. Majik

13 **Proteoglycans from Marine Sponges and Their Biomedical Applications** 287
Ramachandran Karthik and Ramachandran Saravanan

14 **Marine Sponge-Derived Antiangiogenic Compounds for Cancer Therapeutics** 305
Kalimuthu Senthilkumar, Govindan Ramajayam, Jayachandran Venkatesan, Se-Kwon Kim, and Byeong-Cheol Ahn

15 **Chronicles of Sponge Biomaterials: The Saga in Biomedicine** 315
Ira Bhatnagar, Ramjee Pallela, P.V. Bramhachari, and Kranti Kiran Reddy Ealla

16 **Biomedical Potential of Marine Sponges** 329
Sushrut Sharma, Renesha Srivastava, Ananya Srivastava, Pawan Kumar Maurya, and Pranjal Chandra

17 **Sponge Biomass for the Development of Biomedical Products and Their Applications** 341
Naveen Kumar Mekala, Rama Raju Baadhe, and Sreenivasa Rao Parcha

18 **Marine Sponges as Future Biomedical Models** 349
Jayachandran Venkatesan, Sukumaran Anil, Elna P. Chalisserry, and Se-Kwon Kim

19 **Chemistry and Biology of Marine Sponge Collagens** 359
Kota Sobha and Devarai Santhosh Kumar

20 **Biomedical Applications of Marine Sponge Collagens** 373
Ramjee Pallela, Hermann Ehrlich, and Ira Bhatnagar

About the Editors

Dr. Ramjee Pallela, Ph.D., M.B.A. is currently working as a chief manager at IKP Knowledge Park (IKP), Hyderabad, India. His career at IKP has involved igniting scientific brains through meetings and personal interactions, inspiring proposals for start-up grants and mediating the process with Biotechnology Industry Research Assistance Council (BIRAC) through IKP. Before his engagement with IKP, Dr. Pallela served as research scientist at the International Centre for Genetic Engineering and Biotechnology (ICGEB), New Delhi, India. With a Ph.D. from Osmania University [CSIR-IICT as pedestal workstation], India, Dr. Pallela has engaged into biomedical research during his postdoctoral stay at South Korea spanning from 2009 to 2012. He started working towards transplanting his scientific acumen into management roles with a P.G. diploma in patent law from the National Academy of Legal Studies and Research (NALSAR) University of Law, Hyderabad, and with an M.B.A. degree from Pondicherry University, India. Accolades for his scientific excellence are marked with more than 40 research publications in varied national and international journals/books, which are majorly related to toxicology, marine sponge biology, and development of novel biopolymer scaffolds and their applications in bone tissue engineering.

Hermann Ehrlich, Prof. Dr. rer. nat. habil (1957) after the defense of the Ph.D. thesis (1984) served as a postdoctoral researcher at Max Bergmann Center of Biomaterials and Institute of Materials Science in Dresden. After successful habilitation in 2011 at Christian-Albrechts University in Kiel, he holds a W3 Heisenberg full professor position at the Institute of Experimental Physics at the TU Bergakademie Freiberg. His research is focused on marine biomaterials, biominerals, biocomposites, and biomimetics. Using biochemical, cellular, molecular, and analytical approaches, he and his coworkers, for the first time, discovered and characterized chitin and novel hydroxylated collagen in the skeletal formations of marine sponges. During last 10 years, he has published over 70 peer-reviewed articles, 8 book chapters, and 2 monographs and is additionally holding 4 patents.

He represented numerous invited and keynote lectures at Columbia University, at Yale University, and at the Massachusetts Institute of Technology as well. Recently, H. Ehrlich has been nominated for the Gottfried Wilhelm Leibniz Prize 2015.

Introduction to the Global Scenario of Marine Sponge Research

1

P.V. Bramhachari, Hermann Ehrlich, and Ramjee Pallela

Abstract

Spongology has grown into a discipline attracting a progressively growing population of hundreds of scientists across the world. Several marine sponges harbor dense and diverse microbial communities of huge ecological and biotechnological significance. Sponges represent an evolutionarily divergent group of species with widespread physiological and ecological traits. They also host complex communities of microbial symbionts and thus are ideal model to test functional equivalence and evolutionary convergence that exists in complex symbiont communities across phylogenetically divergent hosts. This review highlighted the largest part of promising research domains in sponge diversity, taxonomy, ecology, cell culture, metagenomics, drug discovery, marine natural products and applications of sponges in biomaterials, tissue engineering, regenerative medicine, and advanced methodologies used for the bioprospection of marine microorganisms. Genome, transcriptome, and metagenome analysis has revealed extraordinary insights into the sponge symbiotic functions, its ecological role, and biotechnological significance. Recent developments in metagenomics also provided novel avenues in sponge metabolite production. This review has covered the recent findings regarding dynamics of sponges, and several interesting research areas, that we believe are deserving of increased attention.

P.V. Bramhachari (✉)
Department of Biotechnology & Botany, Krishna University, Machilipatnam 521 001, Andhra Pradesh, India
e-mail: veerabramha@gmail.com

H. Ehrlich
Institute of Experimental Physics, TU Bergakademie Freiberg, Leipziger 23, Freiberg 09599, Germany
e-mail: Herman.Ehrlich@physik.tu-freiberg.de

R. Pallela
IKP Knowledge Park, Genome Valley, Turkapally, Hyderabad 500078, Telangana, India
e-mail: rpallela@gmail.com

R. Pallela, H. Ehrlich (eds.), *Marine Sponges: Chemicobiological and Biomedical Applications*,
DOI 10.1007/978-81-322-2794-6_1

Keywords
Sponges • Symbionts • Metagenomics • Drug discovery • Biomaterials and tissue engineering

1.1 Introduction

Marine sponges signify noteworthy constituents of benthic communities all through the world, in terms of both biomass and their prospective to influence benthic or pelagic processes (Dayton 1974, 1989; Gili and Coma 1998; Maldonado et al. 2005). Sponges are among the oldest of the multicellular metazoan animals and possess reasonably petite in the way of differentiation and coordination of tissues (Simpson 1984). They are sessile, filter-feeding organisms which, notwithstanding a simple body plan, are remarkably efficient at obtaining food from the contiguous water (Reiswig 1971; Vogel 1977; Pile et al. 1996). Spongology, the study of all aspects of the sponge biology, ecology, taxonomy, and chemistry of sponges, has full fledged into a discipline attracting a progressively increasing population of hundreds of scientists worldwide, numerous of whom devote a lifetime career to the study of this group. Apart from nurturing academic interest, sponges play an important role in human health as producers of chemical compounds with useful pharmaceutical properties, including antitumor, anti-infective, and anti-inflammatory properties (Pomponi 2006). Despite their evolutionary divergence, sponges have maintained many common physiological characters and ecological roles, including the filter feeding of planktonic microorganisms and particulate matter (Taylor et al. 2007). Interestingly, the sponges congregate complex communities of microbial symbionts, and extensive research over a decade documented the phylogenetic diversity and biogeography of sponge-associated microorganisms (Schmitt et al. 2012). Sponges have been the primary focus of much recent interest due to two key factors: (i) they form close associations with an extensive variety of diverse microorganisms, and (ii) most importantly, they are an affluent source of biologically active secondary metabolites. Marine sponges lack neurons or any other kind of nervous tissue. Instead, they have an exceptional body plan, characterized by different reproductive modes as well as cellular totipotency and mobility balancing the lack of true tissues and organs (Brusca and Brusca 1990). Apparently, the analysis of sponge genomes and transcriptomes has revealed a complex variety of signaling molecules and proteins necessary for a postsynaptic scaffold (Sakarya et al. 2007; Conaco et al. 2012). The conformity of physiological evidence strongly suggests that glutamatergic signaling occurs in sponges (Elliott and Leys 2007, 2010) and acts analogous to that seen in other metazoans may be used to coordinate the sponge behavior. Whereas the sensory organs are well known from ctenophores, in sponges an exceptional mechanism for transducing sensory information from the environment has as yet remained largely unknown. The sponge structural integrity is conferred upon by siliceous or calcareous spicules (Simpson 1984), and these skeletal components are the basis for much of sponge biology and taxonomy. A wide range of spicule types are secreted, many of which are characteristic of particular taxa (Hooper and van Soest 2002). Collagenous tissues, such as spongin, also play a pivotal role in providing structural support and, together with spicules, allow the development of very large individuals, such as those found among many tropical species.

The existence of these putative symbionts alongside bacterium-digesting archaeocytes to some extent is paradoxical and implies either recognition of different microbial types by the sponge cells or defensive of symbiont cells to prevent consumption (Wilkinson et al. 1984). Many marine sponges are associated with dense and phylogenetically diverse microbial consortia including bacteria, archaea, and single-celled eukaryotes (fungi and microalgae) that can account for nearly half of the animal's biomass

(Hentschel et al. 2006; Taylor et al. 2007). The dominant symbiotic bacterial taxa in marine sponges are *Proteobacteria*, *Actinobacteria*, *Chloroflexi*, *Firmicutes*, *Acidobacteria*, and *Cyanobacteria* (Fieseler et al. 2004; Scheuermayer et al. 2006; Taylor et al. 2007). For instance, photosynthetically fixed carbon from cyanobacterial symbionts provides >50 % of the energy requirements of certain tropical sponges (Wilkinson 1983), while other microorganisms may contribute to host defense by means of producing biologically active metabolites (Unson et al. 1994; Schmidt et al. 2000). A wide range of chemical and functional diversity has been observed among bioactive compounds such as polyketides, alkaloids, fatty acids, peptides, and terpenes (Thomas et al. 2010a, b). Molecular tools are increasingly being used to address ecological questions, to increase our understanding of marine ecosystems, and to aid in their management and conservation (Féral 2002). Genetic variation in populations can have ecologically important effects and is an important component of biodiversity (Hughes et al. 2008). When there is genetic variability for traits that affect fitness, the genetic diversity of a population as a whole can be important for its resilience and capacity to adapt to adverse changing conditions. This becomes particularly vital in the face of climate change, where the capacity for a population to evolve in response to elevated temperatures may be influenced by its levels of genetic variation (Hughes et al. 2003).

Marine sponges are regarded as the most relevant reservoir of biologically active metabolites in the seas (Piel 2004), with more than 280 new structures reported in 2010 and similar numbers in previous years (Blunt et al. 2012). With growing evidence, microbial symbionts rather than the host might in fact produce several of the documented sponge-derived bioactive compounds (Piel 2004; Hentschel et al. 2012). Polyketides and nonribosomal peptides are often evoked as examples of metabolites found in sponges with a likely bacterial symbiont origin. These molecules, synthesized by large multifunctional enzymes called polyketide synthases (PKS) and nonribosomal peptide synthetases (NRPS), comprise substance classes that are typical for microorganisms (Piel 2004; Fisch et al. 2009). They possess intricate and diverse structures that display a wide range of relevant pharmaceutical bioactivities including antitumoral, antifungal, and antiparasitic (Staunton and Weissman 2001; Hochmuth and Piel 2009). Cultivation-independent approaches such as metagenomics and single-cell genomics have been of utmost relevance for the discovery of novel biosynthetic gene clusters – including PKS and NPRS operons – from recalcitrant or hard-to-cultivate sponge symbionts (Piel et al. 2004; Fisch et al. 2009; Bayer et al. 2013). Marine sponges with an exceptionally rich chemistry have been the source of several bioactive secondary metabolites. So far, more than 5,300 different products have been isolated from sponges and their associated microorganisms (Laport et al. 2009). Blunt et al. (2010) in a Natural Product Report review described 287 new compounds from marine sponges isolated in 2009. Noteworthy examples of bioactive secondary metabolites isolated from marine sponges are hemiasterlin and discodermolide. Hemiasterlin is a cytotoxic tripeptide originally isolated from the marine sponge *Hemiasterella minor*, currently in Phase I trial (Talpir et al. 1994). The polyketide natural product discodermolide, isolated from the marine sponge *Discodermia dissoluta*, has potent cytotoxicity to human and murine cell lines (Kingston et al. 2011). Certain marine sponges have been recognized as potentially rich sources of various bioactive compounds. According to the database, around 319 compounds have been reported from the genus *Xestospongia*, 244 compounds from the genus *Theonella*, 222 compounds from the genus *Halichondria*, and 118 metabolites from the genus *Aplysina*, among other sponges (Hu et al. 2011). In principle, sponges have great potential for cell culture because of the presence of totipotent stem cells (i.e., archaeocytes) and because cells can be easily dissociated from its tissue, due to their loosely cellular organization. However, despite efforts by several research groups, a continuous sponge cell line has not yet been developed, and the number of primary sponge cell cultures developed is very inadequate (Rinkevich 2005; Pomponi 2006; Schippers et al. 2012). This escalating and diverse research interests have greatly enhanced our knowledge of sponge biology, and yet, as obvious throughout this review article, recent advances on the sponges have been highlighted with updated knowledge of these enigmatic sponge associations. Here we aim to provide a comprehensive review of the current knowledge of the ecology and biotechnological potential of sponges.

1.2 Microbial Diversity of Marine Sponges Producing Bioactive Compounds

Many marine sponges are associated with dense and phylogenetically diverse microbial consortia including bacteria, archaea, and single-celled eukaryotes (fungi and microalgae) that can account for nearly half of the animal's biomass (Hentschel et al. 2006; Taylor et al. 2007). Sponges are filter feeders capable of processing enormous volumes of seawater, providing a rich source of microorganisms. So far, more than 25 bacterial phyla have been reported from sponges (Schmitt et al. 2012; Webster and Taylor 2012). The dominant bacterial taxa in marine sponges are *Proteobacteria*, *Actinobacteria*, *Chloroflexi*, *Firmicutes*, *Acidobacteria*, and *Cyanobacteria* (Fieseler et al. 2004; Scheuermayer et al. 2006; Taylor et al. 2007). Marine microorganisms are well known for being capable of producing bioactive secondary metabolites. A wide range of chemical and functional diversity has been observed among bioactive compounds such as polyketides, alkaloids, fatty acids, peptides, and terpenes (Thomas et al. 2010a, b). Most of the compounds isolated from marine microorganisms have shown biological properties such as antimicrobial, antitumor, and anticancer activities. The phylum *Actinobacteria* dominates in the production of therapeutic compounds followed by *Proteobacteria*. Among fungi, members of the *Ascomycota* are predominant producers of bioactive molecules, and members of *Deuteromycota* are also a potential group for exhibiting bioactivity (Thomas et al. 2010a, b).

The actinomycetes are of particular relevance due to their unmatched capacity to produce novel and bioactive secondary metabolites. About 7,000 compounds have been isolated from this bacterial taxon alone (Jensen et al. 2005). The anticancer compounds salinosporamide and sporolide from the actinomycete *Salinispora tropica* (Fenical et al. 2009), as well as the antitumor antibiotic marinomycin from the obligately marine genus *Marinispora* (Kwon et al. 2006), are vital examples of metabolites from marine actinomycetes. Other interesting taxa associated with marine sponges are bacteria of the order *Sphingomonadales*, which are yellow-pigmented, gram-negative, rod-shaped bacteria that contain glycosphingolipids (GSLs) in their cell envelope and were first described by Yabuuchi et al. (1990). Glycosphingolipids are a class of compounds shown to be the potent stimulators of natural killer T cells (Sriram et al. 2005; Mattner et al. 2005). Long et al. (2007) also reported the synthesis and evaluation of stimulatory properties of the GSL-1 to GSL-4 series of glycosphingolipids isolated from the *Sphingomonadaceae* family. GSL-1 was found to be a potent NKT cell stimulator. GSL-4, a metabolite isolated from a *Sphingomonas* strain, has been previously found to have NKT cell stimulatory properties (Sriram et al. 2005). Interestingly, Laroche et al. (2007) suggested that glycolipids of the marine sponge *Plakortis simplex* are produced by microbial symbionts rather than by the sponge itself. Sphingomonads also produce other types of secondary metabolites, for example, the diketopiperazine glionitrin B was reported to produce using a microbial coculture of the bacterium *Sphingomonas* sp. KMK-001 and the fungus *Aspergillus fumigatus* KMC-901 (Park et al. 2011).

Other interesting taxa of marine organisms associated with marine sponges that have a pharmacological significance are *Cyanobacteria*, e.g., *Lyngbya*, *Oscillatoria*, *Symploca*, *Calothrix*, *Leptolyngbya*, *Dichothrix*, *Geitlerinema*, *Schizothrix*, *Aphanothece*, *Blennothrix*, and *Synechocystis* (Nagarajan et al. 2011). *Cyanobacteria* are known to produce diverse structural classes of metabolites. Malyngamide

H is an ichthyotoxic amide isolated from the marine cyanobacterium *Lyngbya majuscula* (Orjala et al. 1995). Isomalyngamide A, a fatty acid amide isolated from the Taiwanese *Lyngbya majuscula*, was found to have therapeutic potential against tumor cell migration (Chang et al. 2011). It is known, for example, that the occurrence of scytonemins, which are metabolites composed of either an aminocyclohexenone or an aminocyclohexenimine ring, containing amino acid or amino alcohol substituents, is restricted to *Cyanobacteria* or cyanobacterial lichens (Klisch and Hader 2008). Thus, it is more likely that scytonemins are appealing molecules due to their pharmacological potential as modulator of cell cycle control and inflammation (Stevenson et al. 2002). According to Rateb et al. (2011), more than 1,000 compounds have been isolated from marine fungi. For example, 14 anthracenedione derivatives were separated from the secondary metabolites of the mangrove endophytic fungus *Halorosellinia* sp. and exhibited potent anticancer activity (Zhang et al. 2010).

1.3 Role of Molecular Markers in Deciphering the Sponge Diversity

The study of intraspecific genetic diversity and its distribution, ideally, genetic diversity, would be measured across functionally significant gene loci to show ecologically and evolutionary relevant levels of diversity. However, identifying loci that are under selection, and determining the strength of that selection, is extremely challenging and is beyond the scope of most studies interested in genetic variation and its distribution. The use of genetic markers to detect cryptic species and formulate phylogenetic hypotheses has revolutionized systematics and taxonomy in the last 30 years. From the early studies with allozymes (Thorpe and Sole-Cava 1994) to the recent analyses of DNA sequences (Avise 2004), molecular systematics has mostly corroborated classical taxonomy.

1.3.1 Allozymes

In early studies, allozymes (polymorphic enzymatic proteins) were frequently used as population markers for sponges, including studies on the genetic diversity of populations (van Oppen et al. 2002). Allozymes have many practical benefits, as they are relatively simple to use, cost-effective, give quick results, and can be used where there is no prior knowledge of the genome. They also tend to show high intraspecies variability in sponges (Uriz and Turon 2012).

1.3.2 Mitochondrial DNA (mtDNA)

Sequences from the mitochondrial genome are often used as molecular markers to investigate phylogeny or population genetic structure. The cytochrome c oxidase subunit 1 (COI) has been used to infer sponge phylogeny (Erpenbeck et al. 2007) and is currently being used in the Sponge Barcoding Project (Vargas et al. 2012). However, in sponges, mtDNA shows an unusually slow rate of evolution and often has low variability within species (Duran et al. 2004a, b; Hoshino et al. 2008), and as a result, its use in intraspecies work is generally discouraged (van Oppen et al. 2002; Uriz and Turon 2012). Despite this, there are some instances where levels of intraspecific variability have been higher: DeBiasse et al. (2010) used the COI region to determine population structure in *Callyspongia vaginalis*, and Rua et al. (2011) found other mtDNA regions with higher nucleotide diversity than the COI region in a number of Demosponge species.

1.3.3 Microsatellites

Microsatellites have been used in variety of ways to study sponge populations. Population structure has been studied in multiple species at a range of spatial scales. Blanquer and Uriz (2010) detected genetic structure within populations, between populations in the same region, and between regions in *Scopalina lophyropoda*, using seven microsatellite loci. Duran et al. (2004a, b) also found genetic structure between and within sites at a large geographic scale in *Crambe crambe* using six microsatellites (this is in contrast to the low levels of structure found in the same species using mtDNA (Duran et al. 2004a, b)). Population structure at a small spatial scale has also been detected using microsatellites in *C. crambe* (Calderón et al. 2007) and *S. lophyropoda* (Blanquer et al. 2009). Microsatellites have also been used to study population genetic diversity of *Spongia officinalis* in the Mediterranean (Dailianis et al. 2011), to assess temporal as well as spatial structure in *Paraleucilla magna* populations in the northeast Iberian Peninsula (Guardiola et al. 2012), and to detect intraorganism genetic heterogeneity in *S. lophyropoda* (Blanquer and Uriz 2011). This range of studies shows the utility of microsatellites and their effectiveness in a range of species.

1.3.4 Amplified Fragment Length Polymorphism (AFLP)

It is possible to obtain contaminant-free DNA, for example, Lopez et al. (2002) characterized AFLP markers in the sponge *Axinella corrugate* from DNA from a cell culture of the sponge cells – but these methods are time-consuming and practically constraining, making markers with species-specific primers far easier to use when multiple samples are to be processed. In addition, AFLPs are dominant markers and therefore cannot distinguish between heterozygote and homozygote states, meaning that heterozygosity in a population cannot be measured (Mueller and Wolfenbarger 1999).

1.3.5 Nuclear DNA

Regions of nuclear DNA have been useful in sponge population genetic and phylogeographic studies. The ITS regions are popular and widely used, with PCR primers available for conserved regions which flank more variable regions (Uriz and Turon 2012). Although these markers have proved useful in studies on phylogeny and phylogeography (van Oppen et al. 2002; Uriz and Turon 2012), there are some disadvantages to their use at the intraspecies population genetics level.

1.3.6 DNA Barcoding

DNA barcoding provides exciting new means for quick species identification and discovery. The use of DNA signature sequences (DNA barcodes) in sponge taxonomy, supplementing conventional morphological characters, will revolutionize future ways in which we conduct taxonomic research to define and describe species. There is a huge gap in taxonomy to be filled, and crystallizing the current knowledge in a rapidly changing field, like sponge taxonomy, is bound to be a step backward. A good example of how the use of multiple datasets has helped understanding taxonomic relationships in sponges can be found in a research report of Erpenbeck et al. (2006a). In sponges, there are quite few works using COI sequences for species-level taxonomy (Schroder et al. 2003; Duran et al. 2004a; Nichols and Barnes 2005; Wörheide 2006), but it appears that the barcoding region of COI may be too conserved in sponges (Wörheide et al. 2004; Erpenbeck et al. 2006b). For example, several species of *Chondrilla* that could be identified through allozymes, ribosomal sequences, and conventional taxonomy would all be clustered into a single species.

1.4 Sponge Molecular Microbiology

In recent years, with the advent of better molecular tools, the sponge microbiology has attained a new shape and unlocked the bacterial treasure residing within the sponge hosts (Nocker et al. 2007). For instance, sponge microbial population study utilized the advanced microbial ecology tools like PCR-DGGE (polymerase chain reaction-denaturing gradient gel electrophoresis; (Webster et al. 2004; Wichels et al. 2006; Hardoim et al. 2009; Radwan et al. 2010), RFLP (restriction fragment length polymorphism), 16S rRNA gene clone libraries (Hill et al. 2005), fluorescent in situ hybridization (FISH) techniques (Bruck et al. 2008; Schmitt et al. 2012), and more recently the next-generation sequencing techniques (Webster et al. 2010; Schmitt et al. 2012). The main advantage of next-generation sequencing (NGS) technique is the massive parallel pyrosequencing of multiple samples using multiplex approach (using DNA barcodes) generating hundreds of thousands of sequence reads (Siqueira et al. 2012) and overcomes the shortcomings in the traditional techniques.

1.5 Sponge Metagenomics

Metagenomics is the genetic analysis of a complex microbial mixture that can be used to analyze sponge microbial associations. The power of this technique was shown by Venter's group, who cloned all genes from 1 m^3 of seawater (Venter et al. 2004). Knowing the genes involved in the biosynthesis of bioactive compounds would enable their expression in a suitable host organism and the production of the bioactive compound more efficiently and with higher yield. Although this sounds simple, in reality it is exceptionally complex. First, because the bioactive compounds are complex, their metabolic pathways are also complex and are likely to involve a large number of genes, all of which need to be identified. This has been achieved by Haygood and colleagues for bryostatin from bryozoan-associated microorganisms (Hildebrand et al. 2004) and for sponge-associated microorganisms by Piel et al. (2004). In both cases, genes involved in the polyketide synthesis pathways were isolated. These examples emphasize that the approach is perhaps promising and might be extended to other bioactive compounds. Metagenomics revealed to be a very powerful tool for the exploitation of bioactive compounds from marine bacterial communities, since it is extremely hard to isolate and cultivate symbiotic bacteria of marine macroorganisms, e.g., sponges that has been recently indicated as promising source of novel compounds, in particular as anticancer, by a large body of literature (Schirmer et al. 2005; Kennedy et al. 2007). A first success of this approach was in 2002 by using beetles (Piel 2002), and it gave the input to perform metagenomics on the marine sponges. The first work employing this strategy on sponges dates back to 2004 and were performed again by Piel et al. (2004). They isolated and identified several putative PKS clusters from a highly complex metagenome of the marine sponge *Theonella swinhoei*. The total DNA was extracted and cloned in cosmids and the library was screened by using appropriate PCR primers. In an effort to better elucidate the genes involved in secondary metabolite biosynthesis, Fieseler et al. (2006) identified the polyketide synthase (PKS type I) genes in a metagenomic library from over 20 marine sponges from the Mediterranean, the Pacific, and the Caribbean for a bioprospecting effort. These genes are accountable for synthesis of many novel pharmacologically active metabolites obtained from the marine environment. Approximately 90,000 cosmid clones representing 3.2 Gb of DNA from the Pacific sponge *Theonella swinhoei* and a library of 30,000 fosmid clones from *Aplysina aerophoba* were screened for PKS genes by PCR.

After learning more about the diversity of sponge microbial symbionts, the function evaluation of the microbial symbionts represents the frontier and hot issue of sponge symbioses. Investigations on single strain, functional gene, and genome have suggested the functions of symbiotic microbes in sponges, such as producing bioactive compounds, nitrogen cycling, and carbon fixation (Hallam et al. 2006; Siegl et al. 2011; Hentschel et al. 2012; Kamke et al. 2013). Modern omics provides a promising strategy for understanding the metabolic diversity of the sponge symbionts. In 2010, Thomas et al. (2010a, b) first explored the functional genomic signature of bacteria associated with the sponge *Cymbastela concentrica* by shotgun sequencing. Thereafter, Liu et al. (2012) analyzed the bacterial functional proteins in the sponge *Cymbastela concentrica* using metaproteogenomic technique. Recently, Fan et al. (2012) investigated the metabolisms of the bacterial communities of six sponges using metagenomics and suggested the functional equivalence and evolutionary convergence in complex microbial communities of sponge symbionts. Omics investigations revealed previously unknown diversity and functions of sponge symbionts (Thomas et al. 2010a, b); Liu et al. 2012; Fan et al. 2012; Trindade-Silva et al. 2012), but to date, only bacterial community reports of shallow-water sponges were documented.

1.6 Functional Metagenomics of Sponge and Its Symbionts

Marine symbioses could significantly influence the ecology, physiology, and evolution of partners, for example, symbioses between the marine invertebrates and bacteria may explain the high biomasses observed in the environs of deep-sea hydrothermal vents and around cold seeps (Duperron et al. 2009). Sponge-microbe symbioses have been suggested by the presence of a core microbial community and sponge-specific microbial lineages as well as the microbial vertical transmission (Taylor et al. 2007; Schmitt et al. 2012; Simister et al. 2012). In particular, the adhesins, adhesion-related proteins, ankyrin repeat proteins (ARPs), tetratricopeptide repeat domain-encoding proteins (TPRs), and transposable insertion elements observed recently in sponge metagenome and sponge bacterial genome suggest a close association of bacterial symbionts with their sponge host (Hallam et al. 2006; Siegl et al. 2011; Thomas et al. 2010a, b; Hentschel et al. 2012; Kamke et al. 2013; Liu et al. 2012; Fan et al. 2012; Trindade-Silva et al. 2012).

Investigations on single strain, functional gene, and genome have suggested the functions of symbiotic microbes in sponges, such as producing bioactive compounds, nitrogen cycling, and carbon fixation (Hallam et al. 2006; Hentschel et al. 2012; Kamke et al. 2013; Siegl et al. 2011). Modern omics provides a promising strategy for understanding the metabolic diversity of the sponge symbionts. In 2010, Thomas et al. first explored the functional genomic signature of bacteria associated with the sponge *Cymbastela concentrica* by shotgun sequencing (Thomas et al. 2010a, b). Thereafter, Liu et al. analyzed the bacterial functional proteins in the sponge *Cymbastela concentrica* using metaproteogenomic technique (Liu et al. 2012). Recently, Fan and colleagues investigated the metabolisms of bacterial communities of six sponges using metagenomics and suggested the functional equivalence and evolutionary convergence in complex microbial communities of sponge symbionts (Fan et al. 2012). Omics investigations revealed previously unknown diversity and functions of sponge symbionts (Thomas et al. 2010a, b; Liu et al. 2012; Fan et al. 2012; Trindade-Silva et al. 2012). Hoffmann et al. (2009) revealed a complex nitrogen cycle in sponge *Geodia barrette*; however, only nitrification, denitrification, and anammox processes were also included. In the metagenomic investigation of six species of shallow-water sponges by Fan et al. (2012), only denitrification and ammonia oxidization were analyzed. In this study, nitrogen fixation, assimilation, DNRA, ammonia oxidization, and complete denitrification were suggested in the deep-sea sponge *Lamellomorpha* sp. Siegl et al. (2011) suggested the assimilation of NH_3 by *Poribacteria* through the single-cell genomics). It is well known that different ankyrin repeat proteins (ARPs) and tetratricopeptide repeat proteins (TPRs) are often found in facultative or obligate symbionts that could modulate the host's behavior (Pan et al. 2008). Meanwhile, it is known that transcriptases play an important role in genetic exchange and rearrangement and consequently facilitate the evolutionary adaptation of microbial populations to specific niches (Moliner et al. 2010).

1.7 Sponges Possess a Repertoire of Transient Receptor Potential Channels

Sponges (Porifera), one of the earliest evolving phyla, lack conventional muscles and nerves and yet sense and respond to changes in their fluid environment. We demonstrated here the presence of nonmotile cilia in sponges and studied their role as flow sensors. An analysis of sponge transcriptomes shows the presence of several transient receptor potential (TRP) channels including PKD channels known to be involved in sensing changes in flow in other animals. Whereas sensory organs are well known from ctenophores, in sponges the mechanism for transducing sensory information from the environment has yet remained unknown. By using an emergent model system, the freshwater sponge, Ludeman et al. (2014) investigated the ultrastructure and physiology of the cilia and also studied its molecular evolution of sensory channels of the TRP channel family in Porifera. Interestingly a 700aa homolog of pkd2 (Type II TRP) was identified in *Corticium candelabrum* (Homoscleromorpha), and a 178aa sequence of a pkd2 (Type II TRP) gene was found in the freshwater Spongilla lacustris (Demospongiae) (Ludeman et al. 2014). Interestingly, this group also established an evidence of a 978aa sequence of a Type II TRP (ML) in *Sycon coactum* (Calcarea), and several sequences with similarity to various Type I TRP channels were found in all four Porifera classes. These candidates were included in an alignment containing more than 100 representatives for all the TRP families across bilaterians. Analysis of sponge genomes and transcriptomes also revealed a complex assortment of signaling molecules and proteins necessary for a postsynaptic scaffold (Sakarya et al. 2007; Conaco et al. 2012). Together with this physiological evidences, glutamate, GABA, and NO in coordinating behavior and glutamatergic signaling were also shown to occur in the sponge *Ephydatia muelleri* (Demospongiae) (Elliott and Leys 2007, 2010), and this suggests that a signaling system similar to that seen in other metazoans may be used to coordinate sponge behavior.

1.8 Cryptochrome-Based Photoreceptor System in Sponges

Sponges respond to external light or mechanical signals with contractile or metabolic reactions and are devoid of any nervous or muscular system. Furthermore, the elements of a photoreception and phototransduction system exist in demosponges. Recently, a cryptochrome-based photoreceptor system has been discovered in the demosponge. The hypothesis in sponges that the siliceous skeleton acts as a substitution for the lack of a nervous system and allows light signals to be transmitted through its glass fiber network is supported by the research findings that the spicules are efficient light waveguides and, secondly, sponges have the enzymatic machinery for the generation of light. It is interesting to note that, much before understanding light signal recognition in sponges, on a molecular level, the first neuronal receptor was cloned in the demosponge *Geodia cydonium* (Perović et al. 1999). The metabotropic glutamate/GABA-like receptor has been found to undergo sensitization to the excitatory amino acid glutamate, resulting in an increase in the intracellular calcium concentration (Ca^{2+}). As a first molecule involved in light recognition in sponges, the cryptochrome has been cloned and functionally analyzed in the demosponge Suberites domuncula (Müller et al. 2010). Searches in sequence databases, including expressed sequence tags (ESTs) from *S. domuncula* (SpongeBase 2010) or genomic tags from the demosponge *Amphimedon queenslandica* (Srivastava et al. 2010), revealed that the opsin-based light sensory apparatus is missing in sponges, even though the covalently bound cofactor retinal is synthesized in *S. domuncula* (Müller et al. 2011a, b). Likewise, the master control gene, *Pax6*, for eye development in Bilateria (Gehring and Seimiya 2010) has not been explored in sponges so far.

Consecutively, Müller et al. (2010) proposed that, in sponges, the cryptochrome represents the major photoreceptive system, a finding that has been corroborated recently in the sponge *A. queenslandica* (Rivera et al. 2012). The experimental data gathered indicate that it is blue light that is most sensitively perceived by the cryptochrome system; this light spectral range is generated by the sponge luciferase system (Müller et al. 2009a, b) and also exists in the marine twilight zone, where sponges exist. The bioluminescence emission spectrum of the

S. domuncula luciferase (at pH 8.0) ranges between 480 and 620 nm. The spicules from siliceous sponges allow the transmission of light within the wavelength range from 600 to 1,300 nm (Müller et al. 2006); hence, the proposed coupling of luciferase-generated light to the spicules occurs within the white light spectrum. The expression of the cryptochrome gene is correlated with the light-dark cycle and showed maximum efficiency during the light phase (Müller et al. 2010). In the *S. domuncula* and also the *A. queenslandica* systems (Müller et al. 2010; Rivera et al. 2012), cryptochrome, with its flavin-based cofactor, is coupled to the siliceous spicular system. In *S. domuncula*, the skeletal elements, the monaxonial tylostyles, comprise dimensions of about 200 μm in length and 5–10 μm in diameter. These siliceous spicules have been proven to act as light waveguides (Cattaneo-Vietti et al. 1996; Aizenberg et al. 2005; Müller et al. 2006), allowing blue light to pass through. Müller et al. (2013) have identified and cloned two additional potential molecules of the sponge cryptochrome photoreception system, the guanine nucleotide-binding protein β-subunit, related to β-transducin, and the nitric oxide synthase (NOS)-interacting protein in *Suberites domuncula*. Interestingly, the cryptochrome and NOSIP are light-inducible genes. Apparently the modern studies showed that NOS inhibitor L-NMMA impairs both morphogenesis and motility of the sponge cells. The same research group has also provided new evidence that, in *S. domuncula*, the cryptochrome system is coupled not only to the NOS pathway but also to G protein-coupled signal transduction. Transducin is the linker molecule between the photochemical reaction, *cis-trans* retinal isomerization by light, and the downstream signaling cascade Müller et al. (2013).

1.9 Impacts of Marine Natural Products on Drug Discovery

Around half of the drugs approved for clinical use between 1981 and 2006 are natural products or analogs inspired by them (Newman and Cragg 2007). Despite this successful record, pharmaceutical companies have sought drug candidates not in natural products but in libraries of synthetic compounds (Paterson and Anderson 2005; Li and Vederas 2009). Natural products often have complicated structures with lots of chiral centers, which do not only make it difficult to determine the structures but also impede in supply and manufacturing by chemical synthesis. Nevertheless, recent improvements in analysis technologies and synthesis techniques have overcome the obstacles (Gerwick and Moore 2012). With easier access to natural products, drug discovery scientists have reassessed the structural and functional diversity. Furthermore the natural products in particular from marine organisms are in their "renaissance" in drug discovery (Paterson and Anderson 2005). Numerous drugs originated from marine natural products have been approved in the past few years (Gerwick and Moore 2012), for example, ziconotide, trabectedin, and eribulin. Ziconotide (also known as w-conotoxin), a peptide originally discovered in a tropical cone snail *Conus magus*, was the first marine-derived compound approved in the USA in 2004 for the management of severe and chronic pain (Olivera 2000). Ziconotide potently inhibits the transmission of nerve signals by blocking the N-type voltage-sensitive calcium channels specifically (Olivera et al. 1987). Trabectedin is a marine natural product isolated from the tunicate *Ecteinascidia turbinate* (Rinehart et al. 1993). Trabectedin is the first marine anticancer drug, which were approved in the European Union in 2007 for the treatment of soft tissue sarcoma (Verweij 2009). The mode of action has been deliberated to be covalent modification of DNA, resulting in cell apoptosis. Halichondrin B, isolated from the marine sponge *Halichondria okadai*, (Hirata and Uemura 1986) was structurally simplified and pharmaceutically optimized to be eribulin (Yu et al. 2013). Eribulin mesylate was approved by the US Food and Drug Administration (FDA) in 2010 for the treatment of metastatic breast cancer (Gerwick and Moore 2012). Eribulin was shown to trigger apoptosis of cancer cells by irreversible disruption of microtubules.

1.9.1 Sponge Natural Products Targeting Tumor-Associated Enzymes

Recent advances in molecular biology have revealed the pathways and gene expression underlying carcinogenesis and cancer phenotype. The knowledge have developed a new field of cancer therapy termed molecular-targeted therapy (Rosa et al. 2008). Molecular-targeted therapy refers to a type of medication by attacking specific molecular targets such as enzymes or receptors which are overexpressed or highly activated only in tumor cells (Sledge 2005). Molecular-targeted therapy is, thus, expected to be more effective and less harmful than other therapies using cytotoxic agents which can cause deleterious effects on normal cells (Rosa et al. 2008). However, the efficacy of the screenings (Swinney and Anthony 2011), together with the impacts of marine natural products on drug discovery, suggests a high possibility that target-based screenings using extracts of marine organisms will lead to discoveries of novel drug candidates, e.g., two molecular targets associated with tumorigenesis, tumor invasion, and metastasis: cathepsin B and histone deacetylase 1 (HDAC1). Cathepsin B occupies a central node of the proteolytic signal amplification network in mammals. This lysosomal cysteine protease has been documented to be highly upregulated in tumor cells and to play a pivotal role in tumor invasion and metastasis (Buck et al. 1992). Acetylation and deacetylation of histone proteins are regulated by histone acetyltransferases (HATs) and histone deacetylases (HDACs), respectively (Roth et al. 2001). HATs mediate the acetylation of the e-amino group of the specific lysine residues in the histone N-terminal domain.

1.9.2 Sponge-Derived Proteases as Drug Targets

One of the priority approaches to discover drug candidates is the targeting of proteases, which are pertinent drug targets in cancer, cardiovascular, inflammatory, and infectious disease areas (Otto and Schirmeister 1997; Turk 2006). Proteases are enzymes that play essential functions in many signaling pathways, the development of certain types of cancer, as well as in infectious diseases such as malaria and trypanosomiasis. Approximately 32 protease inhibitors are currently in clinical use and at least 9 are in development. Examples are ritonavir, an aspartic protease inhibitor of HIV-1 in clinical use since 1996 for the AIDS treatment, and boceprevir and telaprevir approved by the FDA in 2011 for the treatment of hepatitis C virus infection (Drag and Salvesen 2010). Most of the protease inhibitors reported to date are synthetic molecules developed by structure-based design (Turk 2006). Moreover, protease inhibitors have also been found in natural sources. Miraziridine A, a pentapeptide inhibitor of cathepsins B and L, which was isolated from the marine sponge *Theonella mirabilis*, is one such example (Nakao et al. 2000). A family of aeruginosin inhibitors is active against human serine proteases and was isolated from marine sponges and cyanobacterial water blooms (Ersmark et al. 2008).

The marine sponge *Theonella swinhoei* has shown to be a source of antiprotease and anti-HIV secondary metabolites (Plaza et al. 2010). The marine sponge *Theonella aff. mirabilis* has been reported to contain the protease inhibitor miraziridine A (Nakao et al. 2000) and the papuamides A and B with anti-HIV properties (Ford et al. 1999). Miraziridine A is a secondary metabolite of particular interest due to its three structural elements, (1) (2R,3R)-aziridine-2,3-dicarboxylic acid, (2) (3S,4S)-4-amino-3-hydroxy-6-methylheptanoic acid (statine), and (3) (E)-(S)-4-amino-7-guanidino-hept-2-enoic acid (vinylogous arginine residue), which are responsible for the inhibition of three different classes of proteases, such as serine (e.g., trypsin), cysteine (e.g., cathepsins B and L), and aspartyl proteases (e.g., pepsin) (Schaschke 2004). The sponge *Theonella swinhoei* has been found to contain antifungals including cyclolithistide A, theonegramides, and theopalauamide, as well as paltolides and cytotoxic polytheonamides (Schaschke 2004).

1.9.3 Sponge-Derived Immunomodulatory Agents

Another relevant aspect in the drug discovery field is the explore for immunomodulatory agents, such as stimulators of the cells of the immune system such as T cells and NKT cells, which play significant roles in responses against microbial and tumor antigens. For example, T cells, the mediators of cellular immunity, recognize the antigens of intracellular microbes and destroy these microbes or the infected cells. T lymphocytes consist of functionally distinct populations, such as helper T lymphocytes, cytotoxic T lymphocytes, $CD4^{+}$ regulatory T cells, and gamma delta T cells (T cells). Another example of immunomodulatory agents is lectins, which are glycoproteins that participate in numerous cellular processes, such as cell communication, host defense, fertilization, and development (Rangel et al. 2011). NKT cells have been linked to microbial immunity, autoimmunity, allergy, and cancer, and, accordingly, they represent an important immunotherapeutic target with immense clinical potential (Pellicci et al. 2011). NK cell-mediated regulation of immune responses has been demonstrated to influence a great number of disease states. The substance KRN7000 (GalCer), a synthetic analogue of the natural product "agelasphin" 9b isolated originally from the sponge *Agelas mauritianus* (Natori et al. 1993), is an NKT cell stimulator and is extensively used to study the behavior of NKT cells (Park et al. 2010).

1.9.4 Sponge-Derived Natural Products as Sources of Therapeutic Agents

Since the 1960s, more than 20,000 compounds have been discovered from marine sources (Hu et al. 2011). The organisms producing these marine natural products are divided into three major biological classes: microorganisms, algae, and marine invertebrates. Amid 1985 and 2008, approximately 75 % of the compounds were isolated from marine invertebrates belonging to the phyla Porifera (sponges) and Coelenterata (coral) (Hu et al. 2011). Currently, there are four marine drugs approved by the FDA in the United States Pharmacopeia, namely, cytarabine, vidarabine, ziconotide, and Halaven (Mayer et al. 2010; Huyck et al. 2011). Cytarabine is a synthetic pyrimidine nucleoside that was developed from spongothymidine, a nucleoside firstly isolated from the Caribbean sponge *Tethya crypta*, and is used in the treatment of acute lymphocytic leukemia (Molinski et al. 2009). Vidarabine, a purine nucleoside developed from the spongouridine, originally isolated from the Caribbean sponge *Tethya crypta*, is currently obtained from *Streptomyces antibioticus* (Shen et al. 2009) and is active against the herpes simplex and encephalitis virus; it has been in clinical use for many years. The most recent marine metabolite approved by the FDA is "Halaven," a synthetic form of a chemotherapeutically active compound derived from the sponge *Halichondria okadai*, which is active in the treatment of advanced breast cancer (Huyck et al. 2011).

1.9.5 Exceptional Bioactive Secondary Metabolites from Marine Sponges

Marine sponges with an exceptionally rich chemistry have been the source of several bioactive secondary metabolites. To date, more than 5300 different products have been isolated from sponges and their associated microorganisms (Laport et al. 2009). Blunt et al. (2010) in a Natural Product Report review described 287 new compounds from marine sponges isolated in 2009. Remarkable examples of bioactive secondary metabolites isolated from marine sponges are hemiasterlin and discodermolide. Hemiasterlin is a cytotoxic tripeptide originally isolated from the marine sponge *Hemiasterella minor*, currently in Phase I clinical trials (Talpir et al. 1994). The polyketide natural product discodermolide, isolated from the marine sponge *Discodermia dissoluta*, has potent cytotoxicity to human and murine cell lines (Kingston et al.

2011). Certain marine sponges have been documented as potentially rich sources of various bioactive compounds. According to the MarinLit database, around 319 compounds have been reported from the genus *Xestospongia*, 244 compounds from the genus *Theonella*, 222 compounds from the genus Halichondria, and 118 metabolites from the genus *Aplysina*, among other sponges.

1.9.6 Defensive Enzymes from Sponge Endosymbionts

Apart from secondary metabolites, the symbionts produce defense enzymes, phospholipases as a first line of defense. The evidence for this came from the work carried out in our laboratory (Selvin 2009). The results in this study showed that another possible defense mechanism in this sessile organism can be the production of an extracellular enzyme phospholipase A2 (PLA2). The PLA2 is a ubiquitous defense enzyme found in snake and bee venoms and distributed throughout the plant and animal kingdom (Stahl et al. 1999). Recent studies envisaged that varying levels of PLA2 were found in marine invertebrates including Cnidaria (Nevalainen et al. 2004a), Porifera (Nevalainen et al. 2004b), and Echinodermata. But the PLA2 in sponge-associated bacteria has not been reported so far. The attempt to explore the possible functional role of PLA2 synthesis in sponge-associated bacteria revealed that the host sponge and the associated isolate *Streptomyces dendra* sp. nov. MSI051 yielded more or less similar phospholipase A2 activity. Thus, the enzyme may have key functional role in the ecological succession of host against predatory/fouling pressure in the habitat.

1.10 Sponge Biomaterials in Tissue Engineering and Regenerative Medicine

A powerful yet challenging approach in the tissue bioengeneering is the breakthrough understanding of siliceous spicule formation of demosponges and the hexactinellid sponges came from the discovery of axial filaments in the spicules and skeletal elements of demosponges (Cha et al. 1999), hexactinellids (Müller et al. 2009a, b) contain an enzymatically active protein which synthesizes polymeric silicate, the biosilica. This enzyme, termed silicatein, has been found to catalyze/polycondensate biosilica during the axial and radial growth of the spicules. In contrast to plant phytoliths and diatom frustules, where biosilica is deposited from a supersaturated solution onto organic templates, the siliceous spicules of sponges are formed in a hyposaturated intraorganism environment following an enzymatic mechanism by lowering the activation energy of the polycondensation reaction. The silicateins were first identified in the axial filament of the demosponge *Tethya aurantium* (Shimizu et al. 1998). They comprise a family of related protein sequences which are consisting of three isoforms, silicatein-α, silicatein-β, and silicatein-γ, in a molar ratio of 12:6:1. The silicateins belong to the papain-like cysteine protease superfamily and are the most closely associated to the cathepsin family (Müller et al. 2008). The first cathepsin in sponges was identified and cloned from the demosponge Geodia cydonium (Krasko et al. 2000). Until recently, it has been neglected that enzymes play fundamental roles during formation of these biominerals. This paradigm shift occurred after the discovery that the enzyme silicatein, which catalyzes the polycondensation of silica, and the enzyme carbonic anhydrase (CA), which catalyzes the formation of bicarbonate (HCO^{3-}/$CaCO_3$), produce solid amorphous bioglass or biocalcite. This invention suggested that in mammals, biosilica and biocalcite can act anabolically during hydroxyapatite (HA) synthesis and bone formation. Therefore, biosilica and biocalcite could be promising candidates for the fabrication of biomaterials for regenerative medicine (Wang et al. 2014). These polymers elicit cytokines that have an effect on bone mineralization (hydroxyapatite formation). In this manner, biosilica and bio-polyP cause an increased release of BMP-2, the key mediator activating the anabolic arm of the hydroxyapatite forming cells, and of RANKL. The two naturally occurring polymers that are produced by deep-sea sponges, the biogenic polyphosphate (bio-polyP) and biogenic silica (biosilica), have also been identified as promoting morphogenetic on both osteoblasts and osteoclasts. In addition, bio-polyP inhibits the progression of the pre-osteoclasts to functionally active osteoclasts.

Based on the earlier research reports, it has been anticipated that bio-polyP acts as a storage substance of energy, as a chelator for metal cations, as an inducer of apoptosis, and importantly as a stimulating agent in mineralization of bone tissue (Schroder and Müller 1999; Schröder et al. 2003). Similarly, bio-polyP also acts as a modulator of gene expression. In a recent report, Usui et al. (2010) suggested that in the osteoblast-like cell line, MC 3T3-E1, bio-polyP causes an increased gene expression of osteocalcin, osterix, bone sialoprotein, and tissue-nonspecific alkaline phosphatase, all proteins known to be crucial for bone formation (Sun et al. 2005; Sinha et al. 2010). The gene expression data in MC 3T3-E1 cells have been obtained with 1 mM polyP (Usui et al. 2010; Hacchou et al. 2007). Very recently, it was demonstrated that bio-polyP displays morphogenetic activity on bone-forming osteoblasts, SaOS-2 cells, and inhibitory activity on RAW 264.7 cells acting as osteoclasts. The osteoblast-like SaOS-2 cells form hydroxyapatite crystals, in response to exposure to bio-polyP, based on their potency to express key molecules known to control hydroxyapatite formation (Wiens et al. 2010a, b; Müller et al. 2011a, b), e.g., the bone morphogenetic protein 2 (BMP-2), osteoprotegerin (OPG), a cytokine that is expressed in osteoblasts with a noteworthy role in the maturation of osteoclasts as well as in the control of bone mineral density (Simonet et al. 1997), and the receptor activator of the NF-κB ligand (RANKL), a mediator that binds to RANK which is a receptor that mediates maturation of osteoclasts (Zhou et al. 2008). It is also important to reveal that the activity of bio-polyP can induce alkaline phosphatase, an enzyme which provides inorganic phosphate required for the synthesis of hydroxyapatite (Müller et al. 2011a, b).

Few researchers solved this interesting question, does the evolutionary oldest inorganic polymer, biosilica, share a functional relationship with the skeletal elements of the crown mammals, the calcium phosphate/hydroxyapatite (HA)-based skeletal systems? Only recently it was possible to describe the molecular level of the formation of a hard skeleton. Initial investigations were effectively performed with the siliceous sponge spicules. The key discovery was the identification of silicatein, the enzyme that initiates the biocatalytic biosilica-condensation reaction (Muller 2003; Kulaev 2004; Rao et al. 2009; Kulakovskaya et al. 2012). It initiated the resolution of the biochemical processes leading to biosilica formation. The silicateins are members of the cathepsin L and papain family of proteases. They have been discovered in the demosponge *Tethya aurantium* by the group of researchers (Müller 2003; Kulakovskaya et al. 2012) and subsequently were also identified in the demosponge *Suberites domuncula* (Kulaev et al. 2004). Based on biochemical studies, three isoforms of silicatein have been described in *T. aurantium*, silicatein-α to silicatein-γ. They have similar molecular weights (approximately 34 kDa). Among them the silicatein-α is the dominant isoform, forming the axial filament, residing in the axial canal. In *T. aurantium*, the molar ratio between silicatein-α and silicatein-β was determined to be 2:1, while in *S. domuncula*, the molar ratio amounts to 4:1. Later on the expression of the silicateins and after the first formation of silica nanoparticles, the silicatein-interacting proteins, silintaphins, are read out. Until now two silintaphins, silintaphin-1 (Rao et al. 2009) and silintaphin-2, have been described extensively. Silintaphin-1 significantly enhances the biosilica-forming activity of silicatein in vitro. A 5.3-fold increase of the biosilica-forming activity is measured at a molar ratio of 4:1 (silicatein-α/silintaphin-1) (Laitinen et al. 1997). Likewise, in *S. domuncula*, the 15-kDa protein silintaphin-2 had been identified as a second silicatein interactor. Like silintaphin-1, this protein is located in the axial filament, but particularly in the organic cylinder around the growing spicules. Silintaphin-2 is a Ca^{2+}-binding protein that complexes with four Ca^{2+} ions (Hausser and Brenner 2005).

In addition to pharmaceutical products, several other applications of sponge-derived biomaterials have also been identified. For example, the silicon skeleton of glass sponges may serve as a blueprint for the production of very efficient fiber optics (Sundar et al. 2003), and biosilica-producing enzymes from sponges have been applied in nanotechnology (Schroder et al. 2003). In fact, sponges are the only animals able to polymerize silica to produce massive skeletal elements in a single reaction at ambient temperature and pressure (Müller et al. 2009a). It has become more obvious that collagen is among the most promising sponge-derived biomaterials. Sponge collagen (e.g., from the marine sponge *Chondrosia reniformis* (Swatschek et al. 2002)) offers advantages in medical and cosmetic applications compared to mammalian connective tissue-extracted collagens because it is free of risk associated with bovine spongiform encephalopathy (BSE) (Heinemann et al. 2007). A recent study of the structural and physicochemical properties of the three-dimensional skeletal scaffold of the marine sponge *Aiolochroia crassa* showed that these fibrous scaffolds have a multilayered design and are made of chitin (Ehrlich et al. 2010). Interestingly the natural polymers like chitin are widely used in the biomedical field because of their high biocompatibility and the enriched functionalities being capable of integrating well with a variety of ligands (Rejinold et al. 2011; Ehrlich et al. 2016). The marine sponge *Theonella swinhoei*, of the order Lithistida, is typically found in deeper waters and caves of tropical oceans. *T. swinhoei* has a structurally massive and rigid morphology. Its skeleton consists of fused or interlocked spicules called desmas (Jeanteur et al. 2006).

1.11 Cultivation of Sponge Symbionts

If the bioactive compound of interest is produced by an associated microorganism, the most obvious cultivation approach would be to isolate the microorganisms from the sponge and to culture these on an appropriate growth medium. A problem encountered with this approach is that not all the associated microorganisms can be cultured, and furthermore, if they are cultured, they do not always continue to produce the bioactive compound. In order to overcome this impending problem, isolated microorganisms are grown in an attempt to induce production of bioactive compounds (Sfanos et al. 2005). Recently, Hill's group (2005) had significant success in this area. From the sponge *Acantho strongylophora*, they isolated *Micromonospora* sp., which continued to produce the bioactive compound manzamine in the absence of the host. Manzamine is now being tested in preclinical trials as a drug against malaria.

1.12 Sponge Cell Culture and Apoptosis

In vitro culture of sponges as axenic (free from microbes) dissociated sponge cells or tissue would grant a clean axenic and defined system for the production of sponge metabolites. However, to date, attempts to develop continuously proliferating cell lines from sponges were unsuccessful (Rinkevich 2005). Some successful attempts have been made using archaeocytes. Archaeocytes are totipotent sponge cells that can differentiate into other functional cell types (Simpson 1984). However, Pomponi and Willoughby (2000) were able to show sponge cell division when exposed to mitogen and phytohemagglutinin. Other options for obtaining proliferating starting materials are the use of primmorphs, gemmules, or larvae. Instead, cells from sponge embryos or larvae could be used as a source of undifferentiated cells (de Caralt et al. (2007). The cells are cultured in a basal cell-culture medium particularly formulated for primary culture of sponge cells (Pomponi 2006). Nevertheless, until now (de Caralt et al. 2007) all attempts have failed to establish a continuous cell line from sponges, because the sponge cells seem to lack the stimulus to divide regardless of the use of rich culture media. Phytohemagglutinin, a lectin that induces mitosis in mammalian cells, can be used to stimulate cell division in primary sponge cell cultures, but this resulted only in a few cell-division cycles after which the proliferation stopped wholly (Rinkevich 2005; Pomponi 2006; de Caralt et al. 2007). A promising approach to overcome this impediment could be to reduce the high apoptosis rate in sponge cells and contribute to obtaining immortalized sponge cultures (de Caralt et al. 2007). As a consequence, to develop continuous sponge cell lines, the researchers would not only use the strong capability of sponge cells to divide but also concentrate on reducing the high apoptosis activity of sponge cells. Studies on the apoptotic process in sponges are in progress. Several genes involved have been identified, and the same apoptotic molecules that have been described in mammals (members of the Bcl-2 family, members of the TNF family, caspases, transcription factors, and various proteins) have been identified in sponges (Tepsuporn et al. 2003; Wiens et al. 2003; Wiens and Muller). In sponges, a decisive first step toward obtaining transfected cells has occurred with the introduction of an immortalizing agent (human telomerase reverse transcriptase (hTERT) into *Axinella corrugata* cells (Thompson et al. 2006)). In the same study, DNA and RNA isolation revealed the presence and expression of the vector in the sponge cell suspensions. This represents a promising research line that could lead toward a continuous sponge cell culture.

1.13 Future Perspectives and Avenues

The marine environment, a virtually untapped resource to date, holds great potential as an affluent source of novel bioactive molecules, for the discovery of both novel pharmaceutical agents and nutraceuticals, which have significant overlap with pharmaceuticals as example; carotenoids can function as antioxidants or as various potent cytotoxic agents. Sponges have an enormous potential for the development of new medical drugs. Thus, although several bottlenecks remain, efforts to develop a technology for continuous cell cultures, such as the use of embryonic stem cells to form a cell line and research into the control of apoptosis, are worthwhile and might become successful in the interim. Currently, there is an upsurge interest by biotechnology companies in isolating and characterizing novel enzymes, biopolymers, and biomaterials with properties which meet the essential needs and circumvent the key barriers. Biomolecules and biomaterials from marine sources are of fastidious concern as they are likely to have novel characteristics such as increased salt tolerance, pressure tolerance, cold adaptivity, and heat tolerance and may have novel physical, chemical/stereochemical in addition to original biochemical properties. In fact, hundreds of novel enzymes without industrial applications have been identified through past projects, and enhanced focus on detailed specifications of enzymes required for new processes or for the improvement of existing ones is required. In the recent years, there has been escalating interest in the industry for isolation of novel marine microorganisms from sponges which produce fascinating molecules. Many such microorganisms produce novel chemical composition which may have applications in medicine, agriculture, and aquaculture. Perhaps the area of greatest importance will be the development of sponge-derived biomaterials which are suitable for tissue repair and regeneration. Possibly, this will necessitate a multidisciplinary approach to transform the advances in characterization and chemical modification of lead molecules and culture scaffolds to clinical applications. Identification of significant scaffolds onto which regenerative cells (stem cells, neuronal cells, osteoblasts, or chondrocytes) can be seeded and retained to generate functional three-dimensional tissue scaffolds represents a major future direction of biomaterial-based regenerative medicine. For the development of sponge-derived drugs, still major breakthroughs are necessary. This review highlighted the most promising research areas in ecology, drug discovery, natural product chemistry, and marine metagenomics and also describes advanced methodologies in the sponge biology.

Acknowledgments The author gratefully acknowledges DST-SERB, Government of India, for the financial support as under the grant no SR/FT/LS-109-2011.

References

Aizenberg J, Weaver JC, Thanawala MS, Sundar VC, Morse DE, Fratzel P (2005) Skeleton von *Euplectella* sp.: structural hierarchy from nanoscale to the macroscale. Science 309:275–278

Avise JC (2004) Molecular markers, natural history, and evolution, 2nd edn. Sinauer, Sunderland, 684 pp

Bayer K, Scheuermayer M, Fieseler L, Hentschel U (2013) Genomic mining for novel FADH2-dependent halogenases in marine sponge-associated microbial consortia. Mar Biotechnol 15(1):63–72

Blanquer A, Uriz MJ, Caujapé-Castells J (2009) Small-scale spatial genetic structure in Scopalina lophyropoda, an encrusting sponge with philopatric larval dispersal and frequent fission and fusion events. Mar Ecol Prog Ser 380:95–102

Blanquer A, Uriz MJ (2010) Population genetics at three spatial scales of a rare sponge living in fragmented habitats. BMC Evol Biol 10(1):13

Blanquer A, Uriz MJ (2011) "Living together apart": the hidden genetic diversity of sponge populations. Mol Biol Evol 28(9):2435–2438

Blunt JW, Copp BR, Munro MH, Northcote PT, Prinsep MR (2010) Marine natural products. Nat Prod Rep 27:165–237

Blunt JW, Copp BR, Keyzers RA, Munro MHG, Prinsep MR (2012) Marine natural products. Nat Prod Rep 29:144–222

Brown RR, Davis CS, Leys SP (2014) SNP discovery in a reef-forming glass sponge, *Aphrocallistes vastus*, using the Ion Torrent next generation sequencing platform. Conserv Genet Resour 6(1):49–51

Bruck WM, Sennett SH, Pomponi SA, Willenz P, McCarthy PJ (2008) Identification of the bacterial symbiont *Entotheonella* sp. in the mesohyl of the marine sponge *Discodermia* sp. ISME J 2:335–339

Brusca RC, Brusca GJ (1990) Phylum Porifera: the sponges. In: Sinauer AD (ed) Invertebrates. Sinauer Press, Sunderland

Buck MR, Karustis DG, Day NA, Honn KV, Sloane BF (1992) Degradation of extracellular-matrix proteins by human cathepsin B from normal and tumour tissues. Biochem J 282:273–278

Calderon I, Ortega N, Duran S, Becerro M, Pascual M, Turon X (2007) Finding the relevant scale: clonality and genetic structure in a marine invertebrate (*Crambe crambe*, *Porifera*). Mol Ecol 16(9):1799–1810

Cattaneo-Vietti R, Bavestrello G, Cerrano C, Sara A, Benatti U, Giovine M, Gaino E (1996) Optical fibres in an Antarctic sponge. Nature 383:397–398

Cha JN, Shimizu K, Zhou Y, Christiansen SC, Chmelka BF, Stucky GD, Morse DE (1999) Silicatein filaments and subunits from a marine sponge direct the polymerization of silica and silicones in vitro. Proc Natl Acad Sci 96(2):361–365

Chang TT, More SV, Lu IH, Hsu JC, Chen TJ, Jen YC, Lu CK, Li WS (2011) Isomalyngamide A, A-1 and their analogs suppress cancer cell migration in vitro. Eur J Med Chem 46:3810–3819

Conaco C, Bassett DS, Zhou H, Arcila ML, Degnan SM, Degnan BM, Kosik KS (2012) Functionalization of a protosynaptic gene expression network. Proc Natl Acad Sci 109(1):10612–10618

Dailianis T, Tsigenopoulos CS, Dounas C, Voultsiadou E (2011) Genetic diversity of the imperilled bath sponge *Spongia officinalis Linnaeus*, 1759 across the Mediterranean Sea: patterns of population differentiation and implications for taxonomy and conservation. Mol Ecol 20(18):3757–3772

Dayton PK (1974) Biological accommodation in the benthic community at McMurdo Sound. Antarct Ecol Monogr 44:105–128

Dayton PK (1989) Interdecadal variation in an Antarctic sponge and its predators from oceanographic climate shifts. Science 245:1484–1486

DeBiasse MB, Richards VP, Shivji MS (2010) Genetic assessment of connectivity in the common reef sponge, *Callyspongia vaginalis* (*Demospongiae*: *Haplosclerida*) reveals high population structure along the Florida reef tract. Coral Reefs 29(1):47–55

de Caralt S, Uriz MJ, Wijffels RH (2007) Cell culture from sponges: pluripotency and immortality. Trends Biotechnol 25(10):467–471

Drag M, Salvesen GS (2010) Emerging principles in protease-based drug discovery. Nat Rev Drug Discov 9(9):690–701

Duperron S, Lorion J, Samadi S, Gros O, Gaill F (2009) Symbioses between deep-sea mussels (*Mytilidae*: *Bathymodiolinae*) and chemosynthetic bacteria: diversity, function and evolution. C R Biol 332(2):298–310

Duran S, Pascual M, Estoup A, Turon X (2004a) Strong population structure in the marine sponge *Crambe crambe* (*Poecilosclerida*) as revealed by microsatellite markers. Mol Ecol 13(3):511–522

Duran S, Pascual M, Turon X (2004b) Low levels of genetic variation in mtDNA sequences over the western Mediterranean and Atlantic range of the sponge *Crambe crambe* (*Poecilosclerida*). Mar Biol 144 (1):31–35

Ehrlich H, Ilan M, Maldonado M, Muricy G, Bavestrello G, Kljajic Z, Carballo JL, Schiaparelli S, Ereskovsky A, Schupp P, Born R, Worch H, Bazhenov VV, Kurek D, Varlamov V, Vyalikh D, Kummer K, Sivkov VV, Molodtsov SL, Meissner H, Richter G, Steck E, Richter W, Hunoldt S, Kammer M, Paasch S, Krasokhin V, Patzke G, Brunner E (2010) Three-dimensional chitin-based scaffolds from Verongida sponges (Demospongiae: Porifera). Part I. Isolation and identification of chitin. Int J Biol Macromol 47:132–140

Ehrlich H, Maldonado M, Parker AR, Kulchin YN, Schilling J, Köhler B, Skrzypczak U, Simon P, Reiswig

HM, Tsurkan MV (2016) Supercontinuum generation in naturally occurring glass sponges spicules. Adv Opt Mat 4(10):1608–1613

Elliott GR, Leys SP (2007) Coordinated contractions effectively expel water from the aquiferous system of a freshwater sponge. J Exp Biol 210(21):3736–3748

Elliott GR, Leys SP (2010) Evidence for glutamate, GABA and NO in coordinating behaviour in the sponge, *Ephydatia muelleri* (*Demospongiae, Spongillidae*). J Exp Biol 213(13):2310–2321

Erpenbeck D, Breeuwer JAJ, Parra-Velandia FJ, Van Soest RWM (2006a) Speculation with spiculation?—Three independent gene fragments and biochemical characters versus morphology in demosponge higher classification. Mol Phylogenet Evol 38(2):293–305

Erpenbeck D, Hooper JNA, Wörheide G (2006b) CO1 phylogenies in diploblasts and the 'Barcoding of Life'—are we sequencing a suboptimal partition? Mol Ecol Notes 6(2):550–553

Erpenbeck D, Duran S, Rützler K, Paul V, Hooper JN, Wörheide G (2007) Towards a DNA taxonomy of Caribbean demosponges: a gene tree reconstructed from partial mitochondrial CO1 gene sequences supports previous rDNA phylogenies and provides a new perspective on the systematics of Demospongiae. J Mar Biol Assoc U K 87(06):1563–1570

Ersmark K, Del Valle JR, Hanessian S (2008) Chemistry and biology of the aeruginosin family of serine protease inhibitors. Angew Chem Int Ed 47 (7):1202–1223

Fan L, Reynolds D, Liu M, Stark M, Kjelleberg S, Webster NS, Thomas T (2012) Functional equivalence and evolutionary convergence in complex communities of microbial sponge symbionts. Proc Natl Acad Sci 109 (27):E1878–E1887

Fenical W, Jensen PR, Palladino MA, Lam KS, Lloyd GK, Potts BC (2009) Discovery and development of the anticancer agent salinosporamide A (NPI-0052). Bioorg Med Chem 17(6):2175–2180

Féral J (2002) How useful are the genetic markers in attempts to understand and manage marine biodiversity? J Exp Mar Biol Ecol 268:121–145

Fieseler L, Horn M, Wagner M, Hentschel U (2004) Discovery of the novel candidate phylum "Poribacteria" in marine sponges. Appl Environ Microbiol 70(6):3724–3732

Fieseler L, Quaiser A, Schleper C, Hentschel U (2006) Analysis of the first genome fragment from the marine sponge-associated, novel candidate *phylum Poribacteria* by environmental genomics. Environ Microbiol 8(4):612–624

Fisch KM, Gurgui C, Heycke N et al (2009) Polyketide assembly lines of uncultivated sponge symbionts from structure-based gene targeting. Nat Chem Biol 5:494–501

Ford PW, Gustafson KR, Mckee TC, Shigematsu N, Maurizi LK, Pannell LK, Williams DE, De Silva ED, Lassota P, Allen TM, Van Soest R, Andersen RJ, Boyd MR (1999) Papuamides A-D, HIV-inhibitory and cytotoxic depsipeptides from the sponges Theonella mirabilis and Theonella swinhoei collected in Papua New Guinea. J Am Chem Soc 121:5899–5909

Gehring W, Seimiya M (2010) Eye evolution and the origin of Darwin's eye prototype. Ital J Zool 77:124–136

Gerwick WH, Moore BS (2012) Lessons from the past and charting the future of marine natural products drug discovery and chemical biology. Chem Biol 19 (1):85–98

Gili J-M, Coma R (1998) Benthic suspension feeders: their paramount role in littoral marine food webs. Trends Ecol Evol 13:316–321

Guardiola M, Frotscher J, Uriz MJ (2012) Genetic structure and differentiation at a short-time scale of the introduced calcarean sponge *Paraleucilla* magna to the western Mediterranean. Hydrobiologia 687 (1):71–84

Hacchou Y, Uematsu T, Ueda O, Usui Y, Uematsu S, Takahashi M, Furusawa K (2007) Inorganic polyphosphate: a possible stimulant of bone formation. J Dent Res 86(9):893–897

Hallam SJ, Konstantinidis KT, Putnam N, Schleper C, Watanabe YI, Sugahara J, DeLong EF (2006) Genomic analysis of the uncultivated marine crenarchaeote *Cenarchaeum symbiosum*. Proc Natl Acad Sci 103 (48):18296–18301

Hardoim CCP, Costa R, Araújo FV, Hajdu E, Peixoto R, Lins U, Rosado AS, van Elsas JD (2009) Diversity of bacteria in the marine sponge *Aplysina fulva* in Brazilian coastal waters. Appl Environ Microbiol 75:3331–3343

Hausser HJ, Brenner RE (2005) Phenotypic instability of Saos-2 cells in long-term culture. Biochem Biophys Res Commun 333(1):216–222

Heinemann S, Ehrlich H, Douglas T, Heinemann C, Worch H, Schatton W, Hanke T (2007) Ultra structural studies on the collagen of the marine sponge *Chondrosia reniformis Nardo*. Biomacromolecules 8 (11):3452–3457

Hentschel J, Zündorf HJ, Hellwig FH, Schäfer-Verwimp A, Heinrichs J (2006) Taxonomic studies in Chiloscyphus Corda (Jungermanniales: Lophocoleaceae) based on nrITS sequences and morphology. Plant Syst Evol 262 (1–2):125–137

Hentschel U, Piel J, Degnam SM, Taylor MW (2012) Genomic insights into the marine sponge microbiome. Nat Rev Microbiol 10:641–654

Hildebrand M, Waggoner LE, Liu H, Sudek S, Allen S, Anderson C, Haygood M (2004) *bryA*: an unusual modular polyketide synthase gene from the uncultivated bacterial symbiont of the marine *Bryozoan Bugula neritina*. Chem Biol 11(11):1543–1552

Hill M, Hill A, Lopez N, Harriott O (2005) Sponge-specific bacterial symbionts in the Caribbean sponge, *Chondrilla nucula* (Demospongiae, Chondrosida). Mar Biol 148:1221–1230

Hirata Y, Uemura D (1986) Halichondrins-antitumor polyether macrolides from a marine sponge. Pure Appl Chem 58(5):701–710

Hochmuth T, Piel J (2009) Polyketide synthases of bacterial symbionts in sponges-evolution-based applications in natural products research. Phytochemistry 70:1841–1849

Hoffmann F, Radax R, Woebken D, Holtappels M, Lavik G, Rapp HT, Kuypers MM (2009) Complex nitrogen cycling in the sponge *Geodia barretti*. Environ Microbiol 11(9):2228–2243

Hooper JNA, van Soest RWM (2002) Systema Porifera: a guide to the classification of sponges. Kluwer Academic/Plenum Publishers, New York

Hoshino S, Saito DS, Fujita T (2008) Contrasting genetic structure of two Pacific Hymeniacidon species. Hydrobiologia 603(1):313–326

Hu GP, Yuan J, Sun L, She ZG, Wu JH, Lan XJ, Chen SP (2011) Statistical research on marine natural products based on data obtained between 1985 and 2008. Mar Drugs 9(4):514–525

Hughes TP, Baird a H, Bellwood DR, Card M, Connolly SR, Folke C et al (2003) Climate change, human impacts, and the resilience of coral reefs. Science 301:929–933

Hughes A, Inouye B, Johnson MTJ, Underwood N, Vellend M (2008) Ecological consequences of genetic diversity. Ecol Lett 11:609–623

Huyck TK, Gradishar W, Manuguid F, Kirkpatrick P (2011) Eribulin mesylate. Nat Rev Drug Discov 10 (3):173–174

Jeanteur P, Kuchino Y, Macieira-Coelho A, Rhoads RE (2006) Antifouling compounds. In: Fusetani N, Clare AS (eds) Progress in molecular and subcellular biology. Subseries marine molecular biotechnology. Springer, Berlin/Heidelberg

Jensen PR, Mincer TJ, Williams PG, Fenical W (2005) Marine actinomycete diversity and natural product discovery. Antonie Van Leeuwenhoek 87(1):43–48

Kamke J, Sczyrba A, Ivanova N, Schwientek P, Rinke C, Mavromatis K, Hentschel U (2013) Single-cell genomics reveals complex carbohydrate degradation patterns in poribacterial symbionts of marine sponges. ISME J 7:2287–2300

Kennedy J, Marchesi JR, Dobson AD (2007) Metagenomic approaches to exploit the biotechnological potential of the microbial consortia of marine sponges. Appl Microbiol Biotechnol 75(1):11–20

Kingston DGI, Qi J, Blanden AR, Bane S (2011) Design, synthesis and biological evaluation of a simplified fluorescently labeled discodermolide as a molecular probe to study the binding of discodermolide to tubulin. Bioorg Med Chem 19:5247–5254

Klisch M, Häder DP (2008) Mycosporine-like amino acids and marine toxins – the common and the different. Mar Drugs 6(2):147–163

Kosik KS (2009) Exploring the early origins of the synapse by comparative genomics. Biol Lett 5(1):108–111

Krasko A, Lorenz B, Batel R, Schröder HC, Müller IM, Müller WE (2000) Expression of silicatein and collagen genes in the marine sponge *Suberites domuncula* is controlled by silicate and myotrophin. Eur J Biochem 267(15):4878–4887

Kulaev IS, Vagabov V, Kulakovskaya T (2004) The biochemistry of inorganic polyphosphates. Wiley, New York

Kulakovskaya TV, Vagabov VM, Kulaev IS (2012) Inorganic polyphosphate in industry, agriculture and medicine: modern state and outlook. Process Biochem 47 (1):1–10

Kwon HC, Kauffman CA, Jensen PR, Fenical W (2006) Marinomycins AD, antitumor-antibiotics of a new structure class from a marine actinomycete of the recently discovered genus "Marinispora". J Am Chem Soc 128(5):1622–1632

Laitinen M, Jortikka L, Halttunen T, Böhling T, Marttinen A, Lindholm TS (1997) Soluble factors from human Saos-2 osteosarcoma cells induce ectopic bone formation and osteoblastic differentiation of cultured mesenchymal cells. J Musculoskelet Res 1(01):21–32

Laport MS, Santos OC, Muricy G (2009) Marine sponges: potential sources of new antimicrobial drugs. Curr Pharm Biotechnol 10:86–105

Laroche M, Imperatore C, Grozdanov L, Costantino V, Mangoni A, Hentschel U, Fattorusso E (2007) Cellular localisation of secondary metabolites isolated from the Caribbean sponge Plakortis simplex. Mar Biol 151:1365–1373

Li JW-H, Vederas JC (2009) Drug discovery and natural products: end of an era or an endless frontier? Science 325(5937):161–165

Li H, Su M, Hamann MT, Bowling JJ, Kim HS, Jung JH (2014a) Solution structure of a sponge-derived cystine knot peptide and its notable stability. J Nat Prod 77 (2):304–310

Li H, Bowling JJ, Su M, Hong J, Lee BJ, Hamann MT, Jung JH (2014b) Asteropsins B–D, sponge-derived knottins with potential utility as a novel scaffold for oral peptide drugs. Biochim Biophys Acta 1840 (3):977–984

Li ZY, Wang YZ, He LM, Zheng HJ (2014c) Metabolic profiles of prokaryotic and eukaryotic communities in deep-sea sponge *Lamellomorpha sp.* indicated by metagenomics. Sci Rep 4:3895

Liu M, Fan L, Zhong L, Kjelleberg S, Thomas T (2012) Metaproteogenomic analysis of a community of sponge symbionts. ISME J 6(8):1515–1525

Long X, Deng S, Mattner J, Zang Z, Zhou D, Mcnary N, Goff RD, Teyton L, Bendelac A, Savage PB (2007) Synthesis and evaluation of stimulatory properties of Sphingomonadaceae glycolipids. Nat Chem Biol 3:559–564

Lopez JV, Peterson CL, Willoughby R, Wright AE, Enright E, Zoladz S, Reed JK, Pomponi SA (2002) Characterization of genetic markers for in vitro cell line identification of the marine sponge Axinella corrugata. J Hered 93(1):27–36

Lowe CR (2000) Nanobiotechnology: the fabrication and applications of chemical and biological nanostructures. Curr Opin Struct Biol 10(4):428–434

Ludeman DA, Farrar N, Riesgo A, Paps J, Leys SP (2014) Evolutionary origins of sensation in metazoans:

functional evidence for a new sensory organ in sponges. BMC Evol Biol 14(1):3

Maldonado M, Carmona C, Velasquez Z, Puig A, Cruzado A, Lopez A, Young CM (2005) Siliceous sponges as a silicon sink: an overlooked aspect of benthopelagic coupling in the marine silicon cycle. Limnol Oceanogr 50:799–809

Mattner J, Debord KL, Ismail N, Goff RD, Cantu C 3rd, Zhou D, Saint-Mezard P, Wang V, Gao Y, Yin N, Hoebe K, Schneewind O, Walker D, Beutler B, Teyton L, Savage PB, Bendelac A (2005) Exogenous and endogenous glycolipid antigens activate NKT cells during microbial infections. Nature 434:525–529

Mayer AM, Glaser KB, Cuevas C, Jacobs RS, Kem W, Little RD, Mcintosh JM, Newman DJ, Potts BC, Shuster DE (2010) The odyssey of marine pharmaceuticals: a current pipeline perspective. Trends Pharmacol Sci 31:255–265

Moliner C, Fournier PE, Raoult D (2010) Genome analysis of microorganisms living in amoebae reveals a melting pot of evolution. FEMS Microbiol Rev 34 (3):281–294

Molinski TF, Dalisay DS, Lievens SL, Saludes JP (2009) Drug development from marine natural products. Nat Rev Drug Discov 8:69–85

Mueller UG, Wolfenbarger LL (1999) AFLP genotyping and fingerprinting. Trends Ecol Evol 14(10):389–394

Müller WEG (2003) Silicon biomineralization: biology—biochemistry—molecular biology—biotechnology (progress in molecular and subcellular biology). Springer, Berlin

Müller WEG, Wendt K, Geppert C, Wiens M, Reiber A, Schröder HC (2006) Novel photoreception system in sponges? Unique transmission properties of the stalk spicules from the hexactinellid Hyalonema sieboldi. Biosens Bioelectron 21:1149–1155

Müller WEG, Schloßmacher U, Wang XH, Boreiko A, Brandt D, Wolf SE, Tremel W, Schröder HC (2008) Poly(silicate)-metabolizing silicatein in siliceous spicules and silicasomes of demosponges comprises dual enzymatic activities (silica-polymerase and silica esterase). FEBS J 275:362–370

Müller WEG, Kasueske M, Wang XH, Schröder HC, Wang Y, Pisignano D, Wiens M (2009a) Luciferase a light source for the silica-based optical waveguides (spicules) in the demosponge *Suberites domuncula*. Cell Mol Life Sci 66:537–552. Recent advances in Sponge Genomics

Müller WE, Wang X, Cui FZ, Jochum KP, Tremel W, Bill J, Wiens M (2009b) Sponge spicules as blueprints for the biofabrication of inorganic–organic composites and biomaterials. Appl Microbiol Biotechnol 83 (3):397–413

Müller WEG, Wang XH, Schröder HC, Korzhev M, Grebenjuk VA, Markl JS, Jochum KP, Pisignano D, Wiens M (2010) A cryptochrome-based photosensory system in the siliceous sponge *Suberites domuncula* (Demospongiae). FEBS J 277:1182–1201

Müller WEG, Binder M, von Lintig J, Guo YW, Wang XH, Kaandorp JA, Wiens M, Schröder HC (2011a) Interaction of the retinoic acid signaling pathway with spicule formation in the marine sponge *Suberites domuncula* through activation of bone morphogenetic protein-1. Biochim Biophys Acta 1810:1178–1194

Müller WE, Wang X, Diehl-Seifert B, Kropf K, Schloßmacher U, Lieberwirth I, Schröder HC (2011b) Inorganic polymeric phosphate/polyphosphate as an inducer of alkaline phosphatase and a modulator of intracellular Ca < sup > 2 + </sup > level in osteoblasts (SaOS-2 cells) in vitro. Acta Biomater 7 (6):2661–2671

Müller WE, Schröder HC, Markl JS, Grebenjuk VA, Korzhev M, Steffen R, Wang X (2013) Cryptochrome in sponges a key molecule linking photoreception with phototransduction. J Histochem Cytochem 61:814–832. 0022155413502652

Nagarajan M, Maruthanayagam V, Sundararaman M (2011) A review of pharmacological and toxicological potentials of marine cyanobacterial metabolites. J Appl Toxicol 115:155–163

Nakao Y, Fujita M, Warabi K, Matsunaga S, Fusetani N (2000) Miraziridine A, a novel cysteine protease inhibitor from the marine sponge *Theonella aff. mirabilis*. J Am Chem Soc 122(42):10462–10463

Natori T, Koezuka Y, Higa T (1993) Agelasphins, novel α-galactosylceramides from the marine sponge *Agelas mauritianus*. Tetrahedron Lett 34(35):5591–5592

Nevalainen TJ, Peuravuori HJ, Quinn RJ, Llewellyn LE, Benzie JAH, Fenner PJ, Winkel KD (2004a) Phospholipase A2 in Cnidaria. Comp Biochem Physiol Part B 139:731–735

Nevalainen TJ, Quinn RJ, Hooper JNA (2004b) Phospholipase A2 in porifera. Comp Biochem Physiol Part B 137:413–420

Newman DJ, Cragg GMJ (2007) Natural products as sources of new drugs over the last 25 years. Nat Prod 70:461–477

Nichols SA, Barnes PA (2005) A molecular phylogeny and historical biogeography of the marine sponge genus Placospongia (Phylum Porifera) indicate low dispersal capabilities and widespread crypsis. J Exp Mar Biol Ecol 323(1):1–15

Nocker A, Burr M, Camper AK (2007) Genotypic microbial community profiling: a critical technical review. Microb Ecol 54:276–289

Olivera BM (2000) Conotoxin MVIIA: from marine snail venom to analgesic drug. In: Fusetani M (ed) Drugs from the sea. Karger, Basel, pp 74–85

Olivera BM, Cruz LJ, De Santos V, LeCheminant G, Griffin D, Zeikus R, Varga J (1987) Neuronal calcium channel antagonists. Discrimination between calcium channel subtypes using. omega.-conotoxin from Conus magus venom. Biochemistry 26 (8):2086–2090

Orjala J, Nagle D, Gerwick WH (1995) Malyngamide H, an ichthyotoxic amide possessing a new carbon skeleton from the Caribbean cyanobacterium Lyngbya majuscula. J Nat Prod 58:764–768

Otto HH, Schirmeister T (1997) Cysteine proteases and their inhibitors. Chem Rev 97(1):133–172

Pan X et al (2008) Ankyrin repeat proteins comprise a diverse family of bacterial type IV effectors. Science 320:1651–1654

Park JJ, Lee JH, Seo KC, Bricard G, Venkataswamy MM, Porcelli SA, Chung SK (2010) Syntheses and biological activities of KRN7000 analogues having aromatic residues in the acyl and backbone chains with varying stereochemistry. Bioorg Med Chem Lett 20(3):814–818

Park HB, Kim YJ, Park JS, Yang HO, Lee KR, Kwon HC (2011) Glionitrin B, a cancer invasion inhibitory diketopiperazine produced by microbial coculture. J Nat Prod 74:2309–2312

Paterson I, Anderson EA (2005) The renaissance of natural products as drug candidates. Science 310 (5747):451–453

Pellicci DG, Clarke AJ, Patel O, Mallevaey T, Beddoe T, Le Nours J, Rossjohn J (2011) Recognition of [beta]-linked self glycolipids mediated by natural killer T cell antigen receptors. Nat Immunol 12(9):827–833

Perović S, Krasko A, Prokic I, Müller IM, Müller WEG (1999) Origin of neuronal-like receptors in Metazoa: cloning of a metabotropic glutamate/GABA-like receptor from the marine sponge Geodia cydonium. Cell Tissue Res 296:395–404

Piel J (2002) A polyketide synthase-peptide synthetase gene cluster from an uncultured bacterial symbiont of Paederus beetles. Proc Natl Acad Sci 99 (22):14002–14007

Piel J, Hui D, Fusetani N, Matsunaga S (2004) Targeting modular polyketide synthases with iteratively acting acyltransferases from metagenomes of uncultured bacterial consortia. Environ Microbiol 6(9):921–927

Pile AJ, Patterson MR, Witman JD (1996) In situ grazing on plankton 10m by the boreal sponge *Mycale lingua*. Mar Ecol Prog Ser 141:95–102

Piel J (2004) Metabolites from symbiotic bacteria. Nat Prod Rep 21:519–538

Plaza A, Bifulco G, Masullo M, Lloyd JR, Keffer JL, Colin PL, Hooper JN, Bell LJ, Bewley CA (2010) Mutremdamide A and koshikamides C-H, peptide inhibitors of HIV-1 entry from different Theonella species. J Org Chem 75:4344–4355

Pomponi SA (2006) Biology of the porifera: cell culture. Can J Zool 84(2):167–174

Pomponi SA, Willoughby R (2000) Development of sponge cell cultures for biomedical applications. In: Austin B, Mothersill C (eds) Aquatic invertebrate cell culture. Springer, London, pp 323–336

Radwan M, Hanora A, Zan J, Mohamed NM, Abo-Elmatty DM, Abou-El-Ela SH, Hill RT (2010) Bacterial community analyses of two Red Sea sponges. Mar Biotechnol N Y N 12:350–360

Rangel TB, Rocha BA, Bezerra GA, Assreuy AM, Pires AD, Do Nascimento AS, Bezerra MJ, Do Nascimento KS, Nagano CS, Sampaio AH, Gruber K, Delatorre P, Fernandes PM, Cavada BS (2011) Crystal structure of a proinflammatory lectin from the seeds of Dioclea wilsonii Standl. Biochimie. doi:10.1016/j.biochi.2011.1009.1001

Rao NN, Gómez-García MR, Kornberg A (2009) Inorganic polyphosphate: essential for growth and survival. Annu Rev Biochem 78:605–647

Rateb ME, Ebel R (2011) Secondary metabolites of fungi from marine habitats. Nat Prod Rep 28:290–344

Reiswig HM (1971) Particle feeding in natural populations of three marine demosponges. Biol Bull 141:568–591

Rejinold NS, Chennazhi KP, Tamura H, Nair SV, Rangasamy J (2011) Multifunctional chitin nanogels for simultaneous drug delivery, bioimaging, and biosensing. ACS Appl Mater Interfaces 3:3654–3665

Rinehart K, Ryuichi S, Holt TG (1993) US Patent No. 5,256,663. US Patent and Trademark Office, Washington, DC

Rinkevich B (2005) Marine invertebrate cell cultures: new millennium trends. Mar Biotechnol 7(5):429–439

Rivera AS, Ozturk N, Fahey B, Plachetzki DC, Degnan BM, Sancar A, Oakley TH (2012) Blue-light–receptive crypto chrome is expressed in a sponge eye lacking neurons and opsin. J Exp Biol 215:1278–1286

Rosa DD, Ismael G, Lago LD, Awada A (2008) Molecular-targeted therapies: lessons from years of clinical development. Cancer Treat Rev 34(1):61–80

Roth SY, Denu JM, Allis CD (2001) Histone acetyltransferases. Annu Rev Biochem 70(1):81–120

Rua CP, Zilberberg C, Solé-Cava AM (2011) New polymorphic mitochondrial markers for sponge phylogeography. J Mar Biol Assoc U K 91 (05):1015–1022

Sakarya O, Armstrong K, Adamska M, Adamski M, Wang I-F, Tidor B, Degnan BM, Oakley TH, Kosik KS (2007) A post-synaptic scaffold at the origin of the animal kingdom. PLoS ONE 2:e506

Schaschke N (2004) Miraziridine A: natures blueprint towards protease class-spanning inhibitors. Bioorg Med Chem Lett 14:855–857

Scheuermayer M, Pimentel-Elardo S, Fieseler L, Grozdanov L, Hentschel U (2006) Microorganisms of sponges: phylogenetic diversity and biotechnological potential. In: Proksch P, Müller W (eds) Frontiers in marine biotechnology. Horizon Bioscience Norwich, Norfolk

Schippers KJ, Sipkema D, Osinga R, Smidt H, Pomponi SA, Martens DE, Wijffels RH (2012) Cultivation of sponges, sponge cells and symbionts: achievements and future prospects. Adv Mar Biol 62:273–337

Schirmer A, Gadkari R, Reeves CD, Ibrahim F, DeLong EF, Hutchinson CR (2005) Metagenomic analysis reveals diverse polyketide synthase gene clusters in microorganisms associated with the marine sponge *Discodermia dissoluta*. Appl Environ Microbiol 71 (8):4840–4849

Schmidt EW, Obraztsova AY, Davidson SK, Faulkner DJ, Haygood MG (2000) Identification of the antifungal peptide-containing symbiont of the marine sponge *Theonella swinhoei* as a novel delta-proteobacterium, '*Candidatus entotheonella palauensis*'. Mar Biol 136:969–977

Schmitt S, Tsai P, Bell J et al (2012) Assessing the complex sponge microbiota: core, variable and species-specific bacterial communities in marine sponges. ISME J 6:564–576

Schröder HC, Müller WEG (1999) Inorganic polyphosphates: biochemistry, biology, biotechnology. Springer, Berlin

Schröder HC, Kurz L, Müller WEG, Lorenz B (2000) Polyphosphate in bone. Biochem Mosc 65 (3):296–303

Schröder HC, Krasko A, Le Pennec G, Adell T, Hassanein H, Müller IM, Müller WEG (2003) Silicase, an enzyme which degrades biogenous amorphous silica: contribution to the metabolism of silica deposition in the demosponge Suberites domuncula. Prog Mol Subcell Biol 33:249–268

Selvin J (2009) Exploring the antagonistic producer Streptomyces MSI051: implications of polyketide synthase gene type II and a ubiquitous defense enzyme phospholipase A2 in the host sponge *Dendrilla nigra*. Curr Microbiol 58(5):459–463

Sfanos K, Harmody D, Dang P, Ledger A, Pomponi S, McCarthy P, Lopez J (2005) A molecular systematic survey of cultured microbial associates of deep-water marine invertebrates. Syst Appl Microbiol 28 (3):242–264

Shen W, Kim JS, Kish PE, Zhang J, Mitchell S, Gentry BG, Hilfinger J (2009) Design and synthesis of vidarabine prodrugs as antiviral agents. Bioorg Med Chem Lett 19(3):792–796

Shimizu K, Cha J, Stucky GD, Morse DE (1998) Silicatein alpha: cathepsin L-like protein in sponge biosilica. Proc Natl Acad Sci U S A 95:6234–6238

Siegl A, Kamke J, Hochmuth T, Piel J, Richter M, Liang C, Hentschel U (2011) Single-cell genomics reveals the lifestyle of Poribacteria, a candidate phylum symbiotically associated with marine sponges. ISME J 5(1):61–70

Simister R, Taylor MW, Tsai P, Fan L, Bruxner TJ, Crowe ML, Webster N (2012) Thermal stress responses in the bacterial biosphere of the Great Barrier Reef sponge, *Rhopaloeides odorabile*. Environ Microbiol 14(12):3232–3246

Simonet WS, Lacey DL, Dunstan CR, Kelley M, Chang MS, Lüthy R, Boyle WJ (1997) Osteoprotegerin: a novel secreted protein involved in the regulation of bone density. Cell 89(2):309–319

Simpson TL (1984) The cell biology of sponges. Springer, New York

Sinha KM, Yasuda H, Coombes MM, Dent SY, de Crombrugghe B (2010) Regulation of the osteoblast-specific transcription factor Osterix by NO66, a Jumonji family histone demethylase. EMBO J 29(1):68–79

Siqueira Jr JF, Fouad AF, Rocas IN (2012) Pyrosequencing as a tool for better understanding of human microbiomes. J Oral Microbiol 4:10743

Sledge GW (2005) What is targeted therapy? J Clin Oncol 23:1614–1615

SpongeBase (2010) https://octavia.vk.medizin.uni-mainz.de/login.cgi. Accessed 15 Aug 2010

Sriram V, Du W, Gervay-Hague J, Brutkiewicz RR (2005) Cell wall glycosphingolipids of Sphingomonas paucimobilis are CD1d-specific ligands for NKT cells. Eur J Immunol 35(6):1692–1701

Srivastava M, Simakov O, Chapman J, Fahey B, Gauthier MEA, Mitros T, Richards GS, Conaco C, Dacre M, Hellsten U et al (2010) The *Amphimedon queenslandica* genome and the evolution of animal complexity. Nature 466:720–726

Stahl U, Lee M, Sjodahl S, Archer D, Cellini F, Ek B (1999) Plant low molecular- weight phospholipase A2s (PLA2s) are structurally related to the animal secretory PLA2s and are present as a family of isoforms in rice (*Oryza sativa*). Plant Mol Biol 41:481–490

Staunton J, Weissman KJ (2001) Polyketide biosynthesis: a millennium review. Nat Prod Rep 18:380–416

Stevenson CS, Capper EA, Roshak AK, Marquez B, Eichman C, Jackson JR, Mattern M, Gerwick WH, Jacobs RS, Marshall LA (2002) The identification and characterization of the marine natural product scytonemin as a novel antiproliferative pharmacophore. J Pharmacol Exp Ther 303:858–866

Sun L, Blair HC, Peng Y, Zaidi N, Adebanjo OA, Wu XB, Zaidi M (2005) Calcineurin regulates bone formation by the osteoblast. Proc Natl Acad Sci U S A 102 (47):17130–17135

Sundar VC, Yablon AD, Grazul JL, Ilan M, Aizenberg J (2003) Fibre-optical features of a glass sponge. Nature 424(6951):899–900

Swatschek D, Schatton W, Kellermann J, Müller WE, Kreuter J (2002) Marine sponge collagen: isolation, characterization and effects on the skin parameters surface-pH, moisture and sebum. Eur J Pharm Biopharm 53(1):107–113

Swinney DC, Anthony J (2011) How were new medicines discovered? Nat Rev Drug Discov 10(7):507–519

Talpir R, Benayahu Y, Kashman Y, Pannell L, Schleyer M (1994) Hemiasterlin and geodiamolide TA; two new cytotoxic peptides from the marine sponge *Hemiasterella Minor* (Kirkpatrick). Tetrahedron Lett 35:4453–4456

Taylor MW, Thacker RW, Hentschel U (2007) Genetics. Evolutionary insights from sponges. Science 316:1854–1855

Tepsuporn S, Kaltenbach JC, Kuhns WJ, Burger MM, Fernandez-Busquets X (2003) Apoptosis in *Microciona prolifera* allografts. Biol Bull 205(2):199–201

Thomas TR, Kavlekar DP, Lokabharathi PA (2010a) Marine drugs from sponge-microbe association – a review. Mar Drugs 8:1417–1468

Thomas T, Rusch D, DeMaere MZ, Yung PY, Lewis M, Halpern A, Kjelleberg S (2010b) Functional genomic signatures of sponge bacteria reveal unique and shared features of symbiosis. ISME J 4 (12):1557–1567

Thompson JH et al (2006) Transfection of marine sponge cells to produce a cell line. In: Custodio MR et al (eds) Book of abstracts 7th international sponge symposium. Biodiversity, innovation, sustainability. Armacao de Buzios. Springer, New York

Thorpe JP, Solé-Cava AM (1994) The use of allozyme electrophoresis in invertebrate systematics. Zool Scr 23(1):3–18

Trindade-Silva AE, Rua C, Silva GG, Dutilh BE, Moreira APB, Edwards RA, Thompson FL (2012) Taxonomic and functional microbial signatures of the endemic marine sponge *Arenosclera brasiliensis*. PLoS ONE 7(7):e39905

Turk B (2006) Targeting proteases: successes, failures and future prospects. Nat Rev Drug Discov 5 (9):785–799

Unson MD, Holland ND, Faulkner DJ (1994) A brominated secondary metabolite synthesized by the cyanobacterial symbiont of a marine sponge and accumulation of the crystalline metabolite in the sponge tissue. Mar Biol 119:1–11

Uriz MJ, Turon X (2012) Sponge ecology in the molecular era. Adv Mar Biol 61:345–410

Usui Y, Uematsu T, Uchihashi T, Takahashi M, Ishizuka M, Doto R, Furusawa K (2010) Inorganic polyphosphate induces osteoblastic differentiation. J Dent Res 89(5):504–509

van Oppen MJ, Catmull J, McDonald BJ, Hislop NR, Hagerman PJ, Miller DJ (2002) The mitochondrial genome of Acropora tenuis (Cnidaria; Scleractinia) contains a large group I intron and a candidate control region. J Mol Evol 55(1):1–13

Vargas S, Schuster A, Sacher K, Büttner G, Schätzle S, Läuchli B, Wörheide G (2012) Barcoding sponges: an overview based on comprehensive sampling. PLoS ONE 7(7):e39345

Venter JC, Remington K, Heidelberg JF, Halpern AL, Rusch D, Eisen JA, Smith HO (2004) Environmental genome shotgun sequencing of the Sargasso Sea. Science 304(5667):66–74

Verweij J (2009) Soft tissue sarcoma trials: one size no longer fits all. J Clin Oncol 27(19):3085–3087

Vogel S (1977) Current-induced flow through living sponges in nature. Proc Natl Acad Sci U S A 74:2069–2071

Wang XH, Schröder HC, Wang K, Kaandorp JA, Müller WEG (2012) Genetic, biological and structural hierarchies during sponge spicule formation: from soft sol-gels to solid 3D silica composite structures. Soft Matter 5:3657–3662

Wang X, Schröder HC, Feng Q, Draenert F, Müller WE (2013) The deep-Sea natural products, biogenic polyphosphate (Bio-PolyP) and biogenic silica (bio-silica), as biomimetic scaffolds for bone tissue engineering: fabrication of a morphogenetically-active polymer. Mar Drugs 11(3):718–746

Wang X, Schröder HC, Müller WE (2014) Enzyme-based biosilica and biocalcite: biomaterials for the future in regenerative medicine. Trends Biotechnol 32 (9):441–447

Webster NS, Negri AP, Munro MMHG, Battershill CN (2004) Diverse microbial communities inhabit Antarctic sponges. Environ Microbiol 6:288–300

Webster NS, Taylor MW, Behnam F, Lücker S, Rattei T, Whalan S, Horn M, Wagner M (2010) Deep sequencing reveals exceptional diversity and modes of transmission for bacterial sponge symbionts. Environ Microbiol 12:2070–2082

Webster NS, Taylor MW (2012) Marine sponges and their microbial symbionts: love and other relationships. Environ Microbiol 14(2):335–346

Wichels A, Würtz S, Döpke H, Schütt C, Gerdts G (2006) Bacterial diversity in the breadcrumb sponge *Halichondria panicea* (Pallas). FEMS Microbiol Ecol 56:102–118

Wiens M, Batel R, Korzhev M, Müller WE (2003) Retinoid X receptor and retinoic acid response in the marine sponge Suberites domuncula. J Exp Biol 206 (18):3261–3271

Wiens M, Wang X, Schloßmacher U, Lieberwirth I, Glasser G, Ushijima H, Müller WE (2010a) Osteogenic potential of biosilica on human osteoblast-like (SaOS-2) cells. Calcif Tissue Int 87(6):513–524

Wiens M, Wang X, Schröder HC, Kolb U, Schloßmacher U, Ushijima H, Müller WE (2010b) The role of biosilica in the osteoprotegerin/RANKL ratio in human osteoblast-like cells. Biomaterials 31(30):7716–7725

Wilkinson CR (1983) Net primary productivity in coral reef sponges. Science 219:410–412

Wilkinson CR, Garrone R, Vacelet J (1984) Marine sponges discriminate between food bacteria and bacterial symbionts: electron microscope radioautography and in situ evidence. Proc R Soc Lond B 220:519–528

Wörheide G, Nichols SA, Goldberg J (2004) Intragenomic variation of the rDNA internal transcribed spacers in sponges (Phylum Porifera): implications for phylogenetic studies. Mol Phylogenet Evol 33(3):816–830

Wörheide G (2006) Low variation in partial cytochrome oxidase subunit I (COI) mitochondrial sequences in the coralline demo sponge *Astrosclera willeyana* across the Indo-Pacific. Mar Biol 148(5):907–912

Yabuuchi E, Yano I, Oyaizu H, Hashimoto Y, Ezaki T, Yamamoto H (1990) Proposals of Sphingomonas paucimobilis gen. nov. and comb. nov., Sphingomonas parapaucimobilis sp. nov., Sphingomonas yanoikuyae sp. nov., Sphingomonas adhaesiva sp. nov., Sphingomonas capsulata comb. nov., and two genospecies of the genus Sphingomonas. Microbiol Immunol 34:99–119

Yu Y, DesJardins C, Saxton P, Lai G, Schuck E, Wong YN (2013) Characterization of the pharmacokinetics of a liposomal formulation of eribulin mesylate (E7389) in mice. Int J Pharm 443(1):9–16

Zhang JY, Tao LY, Liang YJ, Chen LM, Mi YJ, Zheng LS, Wang F, She ZG, Lin YC, To KKW, Fu LW (2010) Anthracenedione derivatives as anticancer agents isolated from secondary metabolites of the mangrove endophytic fungi. Mar Drugs 8:1469–1481

Zhou Z, Han JY, Xi CX, Xie JX, Feng X, Wang CY, Xiong WC (2008) HMGB1 regulates RANKL-induced osteoclastogenesis in a manner dependent on RAGE. J Bone Miner Res 23(7):1084–1096

2 Global Constraints, Prospects, and Perspectives of Marine Sponge Research

Baboucarr Lowe, Jayachandran Venkatesan, Hermann Ehrlich, and Se-Kwon Kim

Abstract

Marine sponges play important roles in maintaining a stable marine microenvironment by factor of their symbiosis with other organisms. These ecological roles are indispensable. They are source of numerous bioactive compounds of therapeutics and biotechnology importance, revolutionizing research in these sectors at great depth. In this chapter, we highlighted the perspectives in research limitations, constraints, and global response towards marine sponge research and conservation. We examine the challenges faced by scientists in their quest to explore the ecological roles played by these ubiquitous benthos metazoans in marine ecosystems.

Keywords

Microbes • Climate change • Conservation and sponge disease

2.1 Introduction

Coral reefs are one of the most important and flourishing ecosystems on the earth surface. They are composed of high density of marine lives each with unique ecological significances and attributes. In the coral reefs, you will find different species of marine sponges, which are the most ancient metazoans with ubiquitous benthic distribution. They are distributed along all latitudes from intertidal regions to deep sea (Pallela et al. 2011). They have active metabolites which are toxic and help the sponge to live the ecologically competitive natural marine environment. Sponges are much primitive aquatic

B. Lowe
Department of Marine Bio Convergence Science, Pukyong National University, 365, Sinseon-ro, Nam-gu, Busan 608-739, South Korea
e-mail: baboucarr18@yahoo.com

J. Venkatesan • S.-K. Kim (✉)
Department of Marine Bio Convergence Science, Pukyong National University, 365, Sinseon-ro, Nam-gu, Busan 608-739, South Korea

Marine Bioprocess Research Center, Pukyong National University, 365, Sinseon-ro, Nam-gu, Busan 608-739, South Korea
e-mail: venkatjchem@gmail.com; sknkim@pknu.ac.kr

H. Ehrlich
Institute of Experimental Physics, TU Bergakademie Freiberg, Leipziger 23, Freiberg 09599, Germany
e-mail: Herman.Ehrlich@physik.tu-freiberg.de

R. Pallela, H. Ehrlich (eds.), *Marine Sponges: Chemicobiological and Biomedical Applications*,
DOI 10.1007/978-81-322-2794-6_2

animals that lack organs but have specialized cell and collagenous matrix. Their bodies are organized around a simple or complex water system, making them highly effective filter feeders (Campbell et al. 2009). Sponges have a wide array of colors, textures, sizes, and shape. They can be brightly colored (with a hue of red, orange, green, or yellow) with highly differentiated mesophyl containing spicules of different shapes and specialized cells of diverse function. Excluding the ancient class of sponges (Archeocyatha, fossils from cambrian period), the three types of marine sponges are classified based on the type of spicules they possess: bony (Calcarea), glass (Haxactinellida), and sponging (Demospongiae). Marine sponges, especially the Demospongiae, have attracted significant attention from various scientific disciplines because of their possession of chemical molecules and protein such as collagen. They produce various novel chemical compounds and are of great interest to chemists. Sponges can be considered as microbial fermenters that create new avenues in marine microbiology and biotechnology, by harboring symbionts including cyanobacteria, red algae, green algae, diatoms and dinoflagellates, fungi, worms, clams, bacteria, and nematodes (Webster 2007; Pallela et al. 2011). In this chapter, we shall discuss the major research constraints and challenges faced by marine biologist in their studies to understand the ecology, biochemical features, and functional roles played by sponges in the marine ecosystem.

2.2 Current Research Limitations

2.2.1 Sponge Microbial Symbioses

Marine sponges have an enormous diversity of secondary metabolites believed to be produced from the symbiotic relationships it builds with other marine microbes such as zooxanthellate and cyanobacteria. The resident diversity of these microbes in sponge bodies is of great importance to pharmacologist and biotechnologist. Microbial diversity and distribution in sponges' bodies have generally been less extensively studied to understand their roles in metabolism. Here we shall discuss their importance in studying marine sponge diversity and ecology. Thus, it is viewed as one of the challenges facing research on marine sponges. Marine sponges have wide distributions, and their evolutionary emergence period extends to as far back as the Precambrian period, some 600 million years ago. They have a rich biomass and can be considered the most important species in deep and shallow water communities. The variety and diversity of microbial symbionts at great depth have been linked to their evolutionary and ecological success as benthos filter feeders. There are less ample evidences confirming the contribution of symbiotic microbes to sponge survival. Cyanobacteria are known to translocate synthase to host sponge, and a decline in sponge health as a result of loss of cyanobacteria symbiosis has been experimentally confirmed. Meanwhile, *Vibrio* sp. associated with the sponge *Dysidea* sp. *were* shown to synthesize cytotoxic and antibacterial tetrabromodiphenyl ether. The diketopiperazines associated with the sponge *Tedania ignis* were found to be produced by *Micrococcus* sp., and the antifungal peptide theopalauamide, isolated from the marine sponge *Theonella swinhoei*, was shown to be contained a novel δ-proteobacterial symbiont (Webster et al. 2001). However, microbes with specific phenotypes provide us information in determining the state of sponge health. This has been observed in microbial populations vertical transfer from parent sponge to larval offspring among ammonium-oxidizing archaea, nitrite-oxidizing phylum *Nitrospira*, sulfate-reducing bacteria, and anaerobic phototrophs communities (Webster and Blackall 2009). Studies based on fluorescence in situ hybridization or 16S rRNA sequencing revealed that sponges are host to a dense microbial population of phylogenetic complexity much different from marine planktons. However, the huge number of great variety in bacteria sheltered in sponge tissue; microbial sponge communities' genomes are not fully sequenced (Schirmer et al. 2005). Many of the bacteria cannot be cultured; thus, biochemical characterization and the symbiotic-ecological relationship of these microbes are not well known. There is a need for the development of alternative strategies to culture them. For now, genomic approaches aim at isolating biosynthetic genes and expressing them in surrogate hosts are such options. Usually, this process is characterized by the construction and screening of metagenomic libraries, in which large species of DNA isolated from mixed population without prior cultivation are cloned and screened for target genes or bioactivities (Schirmer et al. 2005; Webster et al. 2001). There is lack of sufficient experimental data to fully understand microbial activities in sponges. It limits evaluation efforts to make reliable assessment on sponge response to stress and diseases being reported in coral reefs. Bacterial-sponge symbiosis is known to indicate a degree of toxin release as a result of metabolism. Therefore we need to know what is the host response mechanism and its regeneration approaches if there are environmental or ecological changes in its habitat. Such changes are quite phenomenal nowadays as a result of human activities in the coral ecosystems. It is worth noting that the putative benefits of symbionts to their host also means their part of the nutritional chain of the host, either by intracellular digestion or by translocation of metabolites (Friedrich et al. 1999). All these metabolic events can have an overall impact on the sponge's ability to respond to environmental changes in its surrounding.

2.2.2 Sponge Disease

The emergence of a disease causes consequential effects on population and functioning of marine ecosystems through the reduction of important habitat players like habitat-forming foundation species. In the 1930s, pandemic wasting disease of the eelgrass resulted to pervasive losses in many areas about the Atlantic coast across the United States, Europe, and Canada. This was less than a percentage of their normal abundance. Also, the pathogen *Perkinsus marinus* has brought about 24–57 % yearly mortality in the Chesapeake Bay, inspiring the loss of huge commercial value to the regional oyster reef habitat. This is much closely related to the same level of destruction caused to keystone herbivore *Diadema antillarum* in the Caribbean in the 1980s. During those periods, white band disease caused a drastic decline in *Acropora palmate* and *A. cervicornis* which are prolific Caribbean corals, resulting immeasurable damage to the reef structure in the last three millennia (Bruno et al. 2007). Recently, there has been worldwide increase of diseases affecting marine organisms, including sponges, seagrasses, sea urchins, shellfish, fish, marine mammals, and corals (Jones et al. 2004). As evidenced in Fig. 2.1, the impact of disease outbreak in the marine ecosystem could be disastrous to the overall biodiversity in the marine waters. Ecological interrelationships between species is highly interdependent in nature, therefore the rate of transmission would be quite enormous and henceforth resulting in a system of paired imbalance of nutrient supply in the food chains of these organisms.

Long-term impact of sponge disease epidemics will continue to affect the long-lived, slow-growing communities of sponges, attributing largely to commercial sponge harvesting populations that are prone to numerous diseases. The severe epidemic of 1938 caused the disapperance of about 70–95 % in sponge population. In 1987, commercial harvesting of sponges in the Mediterranean averaged 100 tonnes per year. But in the Ligurian Sea, in the year 1987, more than 60 % of mercantile sponges were diseased. However, limited microbiological research has been conducted to identify aetiological agents and/or environmental factors, causing high sponge mortalities (Webster 2007). Meanwhile, outbreaks of disease in coral reefs have been attributed to common phenomena such as bleaching, thermal stress (Fig. 2.2 represents correlations of environment stress response to disease emergence and host resistance), microbial symbioses, ocean acidification, sedimentation, toxic chemicals, nutrient imbalance, ultraviolet radiation, and biotic factors (e.g., predation, overgrowth of algae, infectious diseases).

Significant coral diseases have been reported and well documented in the Caribbean and other regions as a result of preliminary survey in

Fig. 2.1 Underwater photograph of representative *Xestospongia muta* individuals: healthy individual and (**a**) individuals at advanced stage of disease (**b**, **c**, **d**) underwater photography by Hilde Angermeier (Figure adopted from Angermeier et al. 2011)

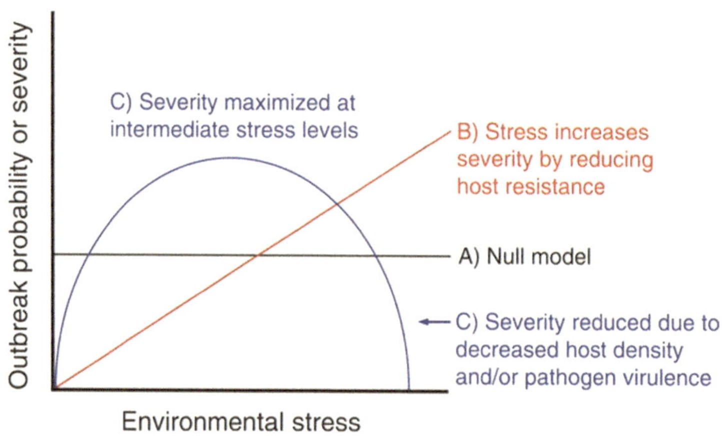

Fig. 2.2 Conceptual model of potential effects of environmental stress (magnitude or frequency) on the probability or severity (e.g., prevalence or impacts on host populations) of disease outbreaks. The model includes three possible scenarios: (**a**) the null model of no effect, (**b**) a positive, linear effect of stress such as when host density is unrelated to incidence and when the pathogen is not negatively affected by the stress, and (**c**) a parabolic stress effect (Bruno et al. 2007)

Australia, Philippines, Palau, and East Africa (Harvell et al. 1999; 2007). The barrel sponge *Xestospongia muta* (Demospongiae, Haplosclerida) which is a common member of the Caribbean coral reef communities has shown a decrease in population. *X. muta* is among the group of high microbial abundance sponges (HMA), it harbors a dense community of microbes with distinguished phylogenetic difference within its mesophyll. The upper layers of the mesophyll are surrounded by bacteria of the *synechococcus/prochlorococcus* clade of cyanobacteria symbiont, which gives it a reddish-brown appearance. Reproductive elements by vertical transmission of symbionts are evident. Report of widespread bleaching of *X. muta* has been documented in Puerto Rico, Belize, Florida Keys, Curacao, as well as Cuba and the reef of Cozumel (Mexico). Two types of bleaching has been described by investigators including cyclic bleaching where sponges can recover from with in time; it is less fatal than sponge orange band, resulting death of many sponges. Sponge orange band appears as a lesion and spreads over the sponge's entire body, leaving the skeletal framework compromised by sorts of orange transition band, hence its name.

Bleaching typically begins as isolated patches making way deeper in to the tissue spreading to the entire sponge body. This transition can be accompanied by an orange band. Loss of Color, massive tissue destruction, and erosion usually lead to the collapse of the entire sponge (Angermeier et al. 2011). Fig. 2.3 highlights the occurrences of white syndrome, bleaching intensities, as well as weekly sea surface temperature abnormalities. Contagious sponge disease can have catastrophic effects on natural sponge populations and mariculture. In 1938–1939 and 1947–1948, a great deal of the commercial sponges in the Gulf of Mexico were killed by such disease. It was observed that survival in low-density populations was much better than in high-density cultivation sites. Hence, such contagious diseases could be extremely harmful when introduced in an in vitro culture. Osinga et al. (1998) cultured four tropical marine demosponges in a close system, once observed that all specimens of one species declined within 1 week, while other species remained unaffected. They reported that a species-specific disease is the cause of this mortality. Meanwhile it is also assumed that high temperatures increase the susceptibility of sponges to pathogens. The

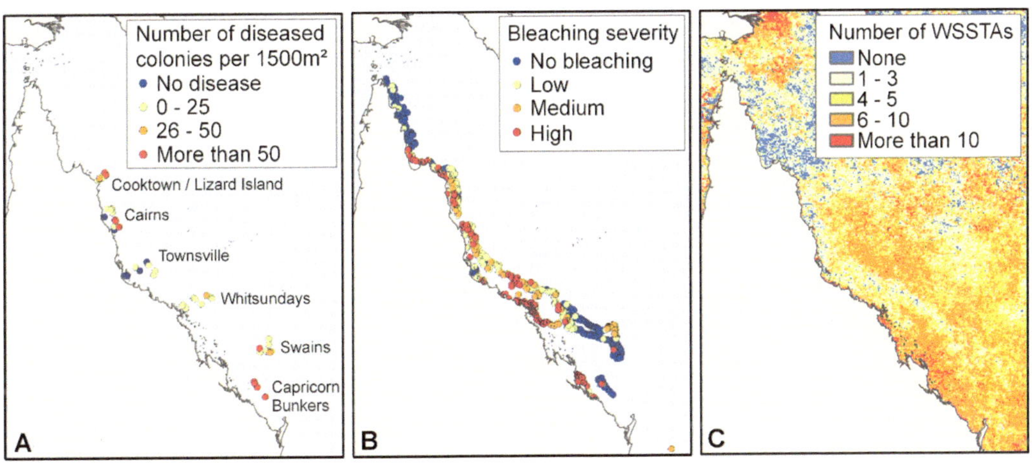

Fig. 2.3 (**a**) Frequency of white syndrome cases from March 2002 to March 2003; (**b**) bleaching intensity for scleractinian coral in March 2002 (Modified from Berkelmans et al. 2004); and (**c**) WSSTAs in 2002. (Adopted from Bruno et al. 2007)

1986–1990 outbreak of disease in the Mediterranean mainly affected populations in the warmest regions (Osinga et al. 1999).

2.2.3 Climate Change on Marine Ecosystem

Adaptive response to large-scale perturbations such as climate change affects all biological levels. They initially take place at the individual level but are integrated and translated to upper levels of the ecosystem. Environmental parameters, organismal response, and species interactions constitute the framework of the ecosystem's dynamic equilibrium. Any change in these parameters would lead to functional consequences, especially when they involve structuring, key, and/or engineer species. However, these types of impacts are poorly documented in the Mediterranean. Since the Mediterranean is markedly oligotrophic, benthic littoral ecosystems are under strong nutritional forces. For some benthic suspension feeders, seasonality in food uptake is characterized by summer dormancy due to low food availability. Late summer is also the time when most mass-mortality events have been reported, mainly affecting species experiencing summer energy shortage, which includes many structuring species (e.g., anthozoan and sponges). The combination of thermal stress and food shortage results in mortality events likely to disrupt benthic-pelagic coupling (Lejeusne et al. 2010). Coral reefs account for one sixth of the world's coastline and support hundreds of thousands of animals and plant species. Fifty-eight percent of the world's reefs are reported to be threatened by human activities. Terrestrial agriculture, deforestation, and infrastructural developments are introducing large quantities of sediments, nutrients, and other pollutants in to coastal waters, causing widespread eutrophication and degradation of biologically productive habitats. Coral reefs are often fished intensively; in the regions of the Indian and Pacific Oceans, fishing with dynamite and poisons have devastated the reef habitats. Coral reefs are also susceptible to climate change. About 25 % of the world's coral reefs have already been destroyed or severely degraded through problems associated with climate change (Doney et al. 2012). There is more than enough available information about climate change impacts on our global polar, terrestrial and tropical marine environments.

Global greenhouse emissions have risen from 0.5% to 1% over the past few decades (Karl and Trenberth 2003). Between the years of 1910, 1945, and 1976, the earth has warmed an unprecedented level; an approximated 0.6 °C over the past century. The rise in global temperatures increased about two times higher than the latter periods. Such a record increase is greater than any other time period within the last millennium. Individual organisms, populations, and ecological communities do not adapt to these changes of approximated global averages (Walther et al. 2002). For example, coral bleaching occurs when photosymbiotic coral loss or expel major portions of their dinoflagellate (Zooxanthellae) flora, when the concentration of photosynthetic pigments in the zooxanthellate decline drastically, or when there is some combination of these events. As such times, the coral host becomes pale or bleached due to the low concentration of plant pigments and the increased visibility of the coral's white calcareous skeleton is witnessed. Bleaching is not limited to reef building and scleractinian corals but also occurs in a variety of other zooxanthellae species including hydrocorals, alcyonarians, sea anemones, soft corals and bivalve molluscs. Sponges that host photosynthetic cyanobacteria are also bleached. Bleaching is a stress reaction that can be induced by many conditions, such as high or low water temperatures, high fluxes of visible and ultraviolet radiation, prolonged aerial exposure, freshwater dilution, high sedimentation, and various pollutants. High temperature and high light intensity increase the production of active oxygen species. Several sites affected in the early 1980s in the Great Barrier Reef region of Australia and at five other sites were again disturbed within 1 year of the moderate ENSO event in 1987. During the 1986–1988 bleaching complex, there was disturbance at 12 new sites, including reefs throughout the red sea and nearly the entire extended Caribbean region, including the flower Garden Banks (Gulf of Mexico) and Bermuda (1988). Meanwhile, if sea warming is thought to be an important cause of the recent disturbance, then it is necessary to determine if there has been a significant increase in the frequency of mortality events (Glynn 1991). Elevated sea temperatures as small as 1 °C above long-term summer averages lead to bleaching (loss of coral algal symbionts), and global SST (sea surface temperature) has risen by an average of 0.1–0.2 °C since 1976. A more acute problem for coral reefs is the increase in extreme temperature events. A comparative analysis of the mean global temperature and mean carbon dioxide was conducted after 1958 in Mauna Loa. El Niño events have been increasing in frequency and severity since records began in the early 1900s, and researchers expect this trend to continue over the coming decades. A particular strong El Niño in 1997–1998 caused bleaching in every ocean (up to 95 % of corals bleached in the Indian Ocean), ultimately resulting in 16 % of corals rendered extinct globally. Recent evidence of genetic variation among the obligate algal symbionts measured in terms of temperature thresholds suggest that some evolutionary response to higher water temperatures may be possible. Changes in genotype frequencies toward increased frequency of higher-temperature-tolerant symbionts appear to have occurred within some coral populations between the mass bleaching events of 1997–1998 and 2000–2001. However, other studies have indicated that many entire reefs are already at their thermal tolerance limits. Coupled with poor dispersal of symbionts between reefs, this has led several researchers to conclude that local evolutionary responses are unlikely to mitigate the negative impacts of future temperature rises (Parmesan 2006). Species distribution and changes in climatic parameters at a global level, lack of a comparative literature to trace this trend of biogeographic drift in species distribution, as well as lack of a universal consensus on impact assessment mechanisms of climate change continue to widen client gaps in ecological studies (Walther et al. 2005).

2.3 Global Response and Conservation Efforts

It was in the 1960s and early 1970s that the international community realized the large-scale ecological destruction emerging from the influence of industrial capitalism. During that time, clusters of international institutions and conventions were set up, aimed at conserving ecological systems and species. In the late 1960s, environmental discourses began to emerge, explaining human-environmental relationships encapsulated in modernism. In 1981, the World Conservation Strategy (WCS) was established by the International Union for the Conservation of Nature (IUCN), stressing the need for environmental and ecological conservation, protection, and management. At the Rio Earth Summit under the auspices of the United Nations Conference on Environment and Development (UNCED) held in Rio de Janeiro in 1992, the concept of sustainable development emerged, prioritizing the consideration of all three dimensions of development, namely, environmental, social, and economic imperatives. The Convention on Biological Diversity (CBN; UN 1992) was also an outcome of the Rio Summit and has a scientific mandate to conserve biodiversity and promote the sustainable use of natural resources (Scott 2013). In 2004, UN Resolution 59/25 was adopted among a host of other resolutions in response to the urgent need to protect vulnerable marine ecosystems (VMEs) from destructive fishing practices, including bottom trawl fishing, in areas beyond national jurisdiction. But reports from the UN Secretary General in 2006 do indicate that the implementation progress of 2004 resolution was less, for little actions had been taken to protect deep-sea ecosystem (UN Resolution 59/25). Thus, UN Resolution 61/105 and UN Resolution 64/72 were adopted in the latter years. Despite the adoption, sets of International Guidelines for the Management of Deep-Sea Fisheries in the

High Seas were negotiated under the fringes of the United Nations Food and Agriculture Organization (UN FAO) to, inter alia, further define and agree to criteria for the conduct of impact assessment of high sea bottom fisheries, identify VMEs, and then assess whether deep-sea fisheries would have significant adverse impacts on VMEs. Subsequently the following year in 2009, UN Resolution 64/72 was adopted as a result of less progress in implementation of the previous years' adopted resolutions (Deep-Sea Conservation Coalition). Despite all these important efforts, the implementation of gains registered is far less than the global demand and need to protect marine ecosystems (Rogers and Gianni 2010; Gianni 2011).

Until now marine biologists, conservation biologists, international conservation organizations, and marine ecology enthusiasts are grately concerned that human activities are causing unprecedented levels of damage to deep-sea corals and sponge ecosystems. This can be widely observed in many parts of the world's marine reserves and biodiversity hotspots. This have sparked global response to help in the conservation of the fragile and dense microenvironment of deep seas and protect species from extinction. Deep-sea coral ecosystems take a great period of time to build and recover from damage. About 75 % of the world coral is endangered, scientist have begun to understand the diversity and state of vulnerability of marine coral reefs. There are growing concerns about deep-sea oil and gas exploration, mining, and global warming, as we have realized these are great forms of threat to coral and sponge communities. Today, bottom trawlers that drag large and heavy weighed nets across seafloors at depths greater than 1000 meters are worrisome. Scientific findings have shown that the sophisticated use of trawlers move away fish from shallow waters and this practice is more devastating to sponge communities as well. Similar distructions of deep-sea coral ecosystems from bottom-tending gears has been reported in most regions in the USA and in other areas of the world where they were studied. Also, bottom trawling is widespread and considered one of the major threats to deep-sea corals in most part of the USA. The ratification of international treaties and their subsequent domestication into laws in individual countries toward the use of deep-sea trawlers has been reechoed by scientist and conservationist; in the meantime, countries like Australia, New Zealand, Canada, and Norway have taken steps toward protecting some coral and sponge ecosystems around their jurisdiction. To date we continue to see influx of long-transported materials entering the open ocean system, and concern over the effect of organo-chlorine compounds. The effects of these contaminants can be highly devastating and could lead to the extinction of fragile marine breeding benthos organisms like marine sponges, if not checked. The comparative consequences are shown in Fig. 2.4. It will be quite important to note that despite the numerous reviews that have been conducted on biodiversity and conservation approaches aim at giving relevant strategic information to spearhead conservation efforts where necessary, there is still no concise synthesis for marine biodiversity in relation to conservation (Gray 1997; Avise 1998). Furthermore, a study carried out by the National research council concluded that bottom trawling and dredging causes reduction in habitat complexity by removing the damage of physical structure of the sea floor and change in species composition. Hence, the National Oceanic and Atmospheric Administration (NOAA) strategic plan for deep-sea coral and sponge ecosystem aimed at boosting research, management, and international cooperation activities on deep-sea coral and sponge ecosystem from fiscal year 2010 to 2019. Their research and exploration activities are designed to locate and characterize deep-sea coral and sponge ecosystem, bioecological biodiversity, and impact of human activities such as fishing. They will also study past ocean waters condition to reliably and effectively predict impacts of climate change using deep-sea corals. Still, there is an immense need to widen our understanding about the bioecological and development patterns of corals and sponges. Generally, there is little against ample knowledge about the basic biology and/or life history of

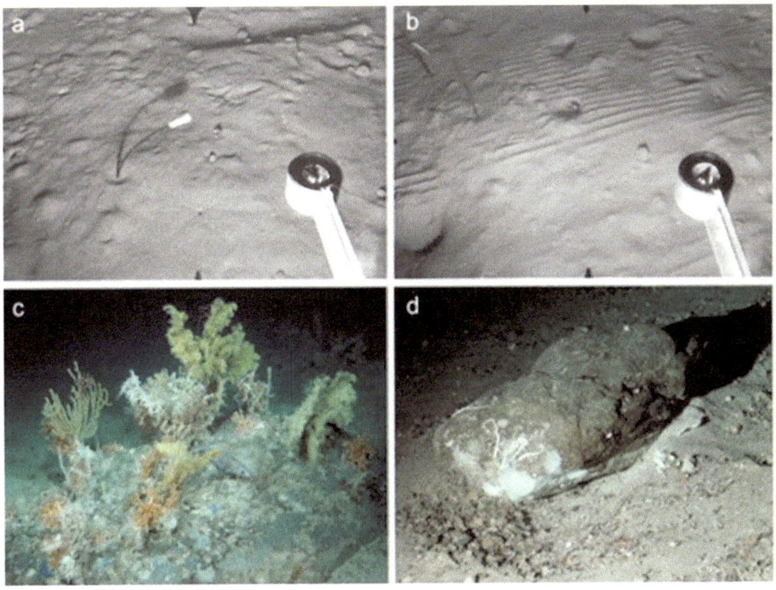

Fig. 2.4 Images showing the effects of bottom trawling on benthic habitats. (**a**) Untrawled soft sediment with intact glass sponge (*Hyalonema* sp.), 1295 m. (**b**) Trawled soft sediment with broken glass sponge stalk, 1298 m. (**c**) Untrawled dropstone with diverse coral and sponge epifauna, 970 m Porcupine Bank (**d**) Trawled dropstone, 590 m (Modified Ref. (Davies et al. 2007))

this organism and other coral reef dwellers. For example, we do not fully understand the dispersal mechanism, age and growth, niche and/or nutritional formulas, and recruitment strategies of these organisms. This type of information is necessary to evaluate the influence of stressors and predict long-term global survival chains in marine sponges. Without these information, it will be difficult to determine resilience levels and assess the ability of deep-sea coral and sponge species to recover from damages. Our understanding of these ecological and environmental factors will therefore enrich our conservation approaches and make management strategies more efficient (Oceanic et al. 2010). Global 200 model findings aimed at conserving the earth's most valuable ecoregion from terrestrial to marine ecosystem globally have shown that upwelling areas are heavily overfished, enclosed seas degraded, and coral reefs severely affected by habitat destruction and overfishing around the world (Olson and Dinerstein 1998). It is prudent to note that national initiatives represent an initial step for deep-sea conservation, although much of deep sea lies beyond national jurisdictions; unless we work together quickly to establish global marine-protected areas, we might realize the Victorian vision of a deep sea that is devoid of larger life—the kingdom of the worms—once more (Roberts et al. 2002).

2.4 Recommendations and Conclusions

In 2004, a petition signed by a team of 1,452 scientists and/or conservationist from 69 countries was presented at the annual meeting of the American Association for the Advancement of Science (AAAS). Reiterating calls for the urgent need to protect our coral and marine sponge ecosystems from adverse destruction chiefly emanating from the action of deep-sea trawling. They further urged the United Nations and other appropriate institutions to establish a moratorium on bottom trawling in the high sea. It also highlighted the discovery of deep-sea corals by scientists in Japan, Tasmania, New Zealand, Alaska, California, Nova Scotia, Maine, North Carolina, Florida, Colombia, Brazil, Norway, Sweden, the UK, Ireland, and Mauritania. Therefore, it is strongly recommend that more rigorous research is needed to enable us clearly to understand the life features, eco-physiochemical composition, species characteristics, and symbiotic interrelationships in the coral reefs to ensure comprehensive analyses in risk and impact assessments of global environmental factors and humankind influences that continue to threaten the myriad of species in coral reefs. Meanwhile, it is recommend that scientists explore more into the disease causation factors, transmission routes, regenerative conservation techniques and approaches to solve the problem of coral diseases. Marine sponges are important contributors to a nutrient delivery chain in the marine ecosystem. Over the past two decades, there have been great revelations from numerous research about the potentials of marine sponge metabolites' applications in biotechnology and pharmacology. These challenges aforementioned will continue to linger in our minds as we work to explore alternatives to solve these challenges.

Acknowledgments This research was supported by a grant from the Marine Bioprocess Research Center of the Marine Biotechnology Program, funded by the Ministry of Oceans and Fisheries, Republic of Korea.

References

Angermeier H, Kamke J, Abdelmohsen UR, Krohne G, Pawlik JR, Lindquist NL, Hentschel U (2011) The pathology of sponge orange band disease affecting the Caribbean barrel sponge Xestospongia muta. FEMS Microbiol Ecol 75(2):218–230

Avise J (1998) Conservation genetics in the marine realm. J Hered 89(5):377–382

Berkelmans R, De'ath G, Kininmonth S, Skirving WJ (2004) A comparison of the 1998 and 2002 coral bleaching events on the Great Barrier Reef: spatial correlation, patterns, and predictions. Coral Reefs 23 (1):74–83

Bruno JF, Selig ER, Casey KS, Page CA, Willis BL, Harvell CD, Sweatman H, Melendy AM (2007) Thermal stress and coral cover as drivers of coral disease outbreaks. PLoS Biol 5(6):e124

Campbell J, Simms J (2009) Status report on coral and sponge conservation in Canada. Fisheries and Oceans Canada, vii + 87 p

Davies AJ, Roberts JM, Hall-Spencer J (2007) Preserving deep-sea natural heritage: emerging issues in offshore conservation and management. Biol Conserv 138 (3):299–312

Doney SC, Ruckelshaus M, Duffy JE, Barry JP, Chan F, English CA, Galindo HM, Grebmeier JM, Hollowed AB, Knowlton N (2012) Climate change impacts on marine ecosystems. Ann Rev Mar Sci 4:11–37

Friedrich A, Merkert H, Fendert T, Hacker J, Proksch P, Hentschel U (1999) Microbial diversity in the marine sponge Aplysina cavernicola (formerly Verongia cavernicola) analyzed by fluorescence in situ hybridization (FISH). Mar Biol 134(3):461–470

Gianni M, Currie, DEJ, Fuller S, Speer L, Ardron J, Weeber B, Gibson M, Roberts G, Sack K, Owen S, Kavanagh A (2011) Unfinished business: a review of the implementation of the provisions of UNGA resolutions 61/105 and 64/72 related to the management of bottom fisheries in areas beyond national jurisdiction. Deep Sea Conserv Coalition.

Glynn PW (1991) Coral reef bleaching in the 1980s and possible connections with global warming. Trends Ecol Evol 6(6):175–179

Gray JS (1997) Marine biodiversity: patterns, threats and conservation needs. Biodivers Conserv 6(1):153–175

Harvell C, Kim K, Burkholder J, Colwell R, Epstein PR, Grimes D, Hofmann E, Lipp E, Osterhaus A, Overstreet RM (1999) Emerging marine diseases – climate links and anthropogenic factors. Science 285 (5433):1505–1510

Harvell D, Jordán-Dahlgren E, Merkel S, Rosenberg E, Raymundo L, Smith G, Weil E, Willis B (2007) Coral disease, environmental drivers, and the balance between coral and microbial associates. Oceanography 20:172–195

Jones RJ, Bowyer J, Hoegh-Guldberg O, Blackall LL (2004) Dynamics of a temperature-related coral disease outbreak. Mar Ecol Prog Ser 281:63–77

Karl TR, Trenberth KE (2003) Modern global climate change. Science 302(5651):1719–1723

Lejeusne C, Chevaldonné P, Pergent-Martini C, Boudouresque CF, Pérez T (2010) Climate change effects on a miniature ocean: the highly diverse, highly impacted Mediterranean Sea. Trends Ecol Evol 25(4):250–260

Oceanic N, Administration A, Program CRC (2010) NOAA strategic plan for deep-sea coral and sponge ecosystems: research, management, and international cooperation. Silver Spring, MD, NOAA Coral Reef Conservation Program. NOAA Technical Memorandum CRCP 11.67 pp

Olson DM, Dinerstein E (1998) The Global 200: a representation approach to conserving the Earth's most biologically valuable ecoregions. Conserv Biol 12 (3):502–515

Osinga R, Planas Muela E, Tramper J, Wijffels R (1998) In vitro cultivation of four marine sponge species. Determination of the nutritional demands. Actes Colloq Ifremer 21:121–127

Osinga R, Tramper J, Wijffels RH (1999) Cultivation of marine sponges. Mar Biotechnol 1(6):509–532

Pallela R, Koigoora S, Gunda VG, Sunkara MS, Janapala VR (2011) Comparative morphometry, biochemical and elemental composition of three marine sponges (Petrosiidae) from Gulf of Mannar, India. Chem Speciat Bioavailab 23(1):16–23

Parmesan C (2006) Ecological and evolutionary responses to recent climate change. Annu Rev Ecol Evol Syst 37:637–669

Roberts CM, McClean CJ, Veron JEN, Hawkins JP (2002) Marine biodiversity hotspots and conservation priorities for tropical reefs. Science 295(5558): 1280–1284

Rogers AD, Gianni M (2010) Implementation of UNGA resolutions 61/105 and 64/72 in the management of deep-sea fisheries on the high seas. International Programme on the State of the Ocean, London, 97pp

Schirmer A, Gadkari R, Reeves CD, Ibrahim F, DeLong EF, Hutchinson CR (2005) Metagenomic analysis reveals diverse polyketide synthase gene clusters in microorganisms associated with the marine sponge Discodermia dissoluta. Appl Environ Microbiol 71 (8):4840–4849

Scott D (2013) Science, transformation and society: a contextual analysis of South Africa's SANCOR-managed marine and coastal research programmes. Afr J Mar Sci 35(3):361–383

Walther G-R, Post E, Convey P, Menzel A, Parmesan C, Beebee TJ, Fromentin J-M, Hoegh-Guldberg O, Bairlein F (2002) Ecological responses to recent climate change. Nature 416(6879):389–395

Walther G-R, Berger S, Sykes MT (2005) An ecological 'footprint' of climate change. Proc R Soc Lond B Biol Sci 272(1571):1427–1432

Webster NS (2007) Sponge disease: a global threat? Environ Microbiol 9(6):1363–1375

Webster NS, Blackall LL (2009) What do we really know about sponge-microbial symbioses. ISME J 3 (1):1–3

Webster NS, Wilson KJ, Blackall LL, Hill RT (2001) Phylogenetic diversity of bacteria associated with the marine sponge Rhopaloeides odorabile. Appl Environ Microbiol 67(1):434–444

Chemical Ecology of Marine Sponges

3

Narsinh L. Thakur and Anshika Singh

Abstract

Sponges successfully inhabit diverse habitats ranging from hard- to soft-bottom communities, tropical to polar latitudes, intertidal to deep-sea environments and fresh- to saltwaters, which are shared by other organisms. It is important to study sponges due to their high abundance, longevity, plasticity and ability to produce bioactive secondary metabolites. They use their infochemicals and allelochemicals in multiple ways in order to interact with other organisms to maintain their space, deter predator and prevent epibiont growth on their surfaces. These compounds are highly bioactive and have been explored for their possible therapeutic applications. As compared to the biomedical applications of these compounds, their possible ecological roles in competition for space, defence against predator and prevention of epibiosis have received little attention. Knowledge of the ecological roles of sponge metabolites will contribute significantly to plan effective and sustainable wild harvests to obtain novel compounds. This review highlights the importance of sponge chemical ecology in marine bioprospecting.

Keywords

Chemical ecology • Marine sponge • Competition for space • Infochemicals • Allelochemicals

3.1 What Is Chemical Ecology?

No living species live in isolation. The environment in which they thrive affects their distribution, diversity, biomass and behaviour. Living organisms interact among themselves and with their environment using the language of chemistry. They respond to other species and abiotic and biotic factors of the habitat by producing chemical metabolites (Hay 1996). The study of chemically mediated interactions of an organism with other organisms (sharing the same habitat) and its environment is defined as chemical ecology. There are two different categories of chemicals

N.L. Thakur (✉) • A. Singh
CSIR-National Institute of Oceanography, Dona Paula, Goa 403 004, India
e-mail: thakurn@nio.org

R. Pallela, H. Ehrlich (eds.), *Marine Sponges: Chemicobiological and Biomedical Applications*,
DOI 10.1007/978-81-322-2794-6_3

known as primary metabolites and secondary metabolites. As their names suggest, primary metabolites are directly involved in the prime functions such as growth, development and reproduction, whereas the secondary metabolites act as infochemicals (messenger molecules) and allelochemicals (toxic compounds). Living organisms are under constant ecological pressure due to their competitors, predators and pathogens, which in turn affect their primary processes (growth and reproduction). They produce infochemicals for communication with other organisms and allelochemicals for defence purpose. The infochemicals act as signal molecules and help in successful larval settlement, alarming the organisms against possible dangers and communicating with their symbionts. Allelochemicals, on the other hand, fight against competitors, predators, fouling organisms and pathogens. These chemicals reduce the ecological stress over the organism. The researchers have mostly focussed on utilization of these chemicals in the field of marine bioprospecting. Fortunately, in recent past, various collaborative efforts of both chemists and biologists have succeeded in diverting the focus to understand the natural role of these chemicals (Pawlik 1992, 1993; Fenical 1993; McClintock et al. 1994; Hay 1996, 2014). To study the ecology of the organisms and harness the information, it becomes important to decode their language which is in the form of chemical molecules. The knowledge of chemical defence of the organisms is very useful for obtaining novel bioactive compounds of therapeutic value. However, the production of secondary metabolites in the organisms does not occur continuously and is subjected to spatio-temporal variation (Sacristán-Soriano et al. 2012). Various reports have postulated two prominent hypotheses to explain variability in the production of secondary metabolites: (1) growth-differentiation balance hypothesis (GDBH) and (2) optimal defence theory (ODT). The GDBH assumes that an organism maintains a balance between resources invested in primary and secondary functions, whereas ODT postulates that allelochemical production is primarily restricted to areas that are under threat or areas of higher importance (e.g. reproductive organs) (Eder et al. 1998; Schupp et al. 1999). The ODT is applicable to various marine invertebrates including sponges (Thoms and Schupp 2007). According to ODT, the production of allelochemical is a costly process. The total energy reservoirs of the organisms are fixed/limited, and the production of any secondary metabolites take place at the expense of other primary functions such as growth and reproduction (Leong and Pawlik 2010).

3.2 Marine Sponges

Marine sponges are important components of benthic community due to their longevity, abundance and remarkable survivorship. The sponges are noteworthy because they lack specialized organs and behaviours and yet are ubiquitous throughout temperate, tropical and polar habitats. Most of the sponges have siliceous spicules which provide a cutting edge over other calcareous organisms, subjected to problems of ocean acidification and global warming (Bell et al. 2013). Marine sponges play various functional roles in the ecosystem (Bell 2008). Some of these roles are enumerated here:

1. They act as link between benthic and pelagic communities due to their efficient filtration rate.
2. They provide shelter to many invertebrates due to high porosity of body structure.
3. Sponges play significant role in carbon and nitrogen cycle due to presence of their microbial symbionts. Siliceous sponges have been postulated to contribute significantly to global silicon cycling (Maldonado et al. 2005).
4. Sponges are well-known producers of various bioactive compounds. A large number of compounds with lethal or growth inhibitory properties have been isolated from sponges (Blunt et al. 2009). They produce defensive compounds to protect their occupied space and deter predators (Paul et al. 2006; Faulkner

Table 3.1 List of selective compounds from different sponge species with their ecological roles and bioactivity result

S. no.	Compound with structure	Source	Ecological functions	Bioactivity	References
1.	Avarol 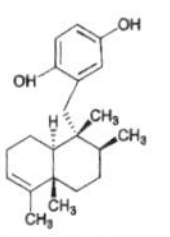	*Dysidea avara*	Useful in competition for space, inhibition of settlement of *B. amphitrite* cyprids	Anticancer and antifouling	De Caralt et al. (2013)
2.	Norselic acid A 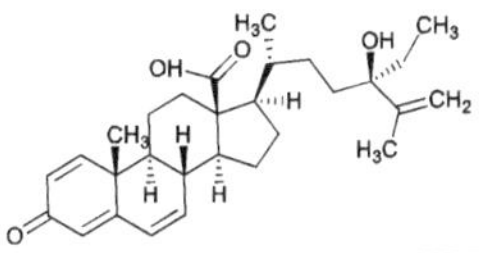	*Crella* sp.	Predator deterrence	Antimicrobial	Ma et al. (2009)
3.	1-O-Hexadecyl-sn-glycero-3-phosphocholine (R=H) and 1-O-octadecyl-sn-glycero-3-phosphocholine (R=CH_2CH_3) 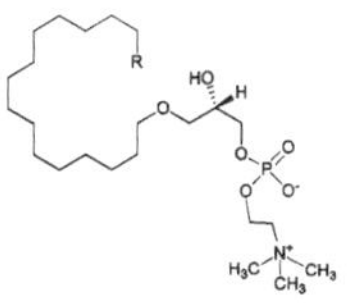	*Suberites domuncula*	Prevention of epibiont growth and controlling growth of surface bacteria	Platelet-activating factor	Müller et al. (2004)

2000). Table 3.1 lists out few compounds from sponges whose ecological roles have been studied along with their bioactivity.

3.3 Competition for Space

Living space is one of the basic requirements of any organism as it is essential for successful recruitment and post-settlement growth of its larval stage. The suitability of space is determined in terms of its exposure to water currents carrying food and adequate irradiance in case of phototrophic sessile organisms (Connell et al. 2004; Chadwick and Morrow 2011; Birrell et al. 2008; Foster et al. 2008). In space-limited environment, such as coral reefs and rocky intertidal areas, diverse sessile marine organisms struggle for their occupied space. Competition for substratum among sessile marine invertebrates is a major process, which affects diversity, abundance and zonation in an ecosystem. Marine invertebrates acquire living space by employing various strategies such as (1) growth interactions, (2) aggressive behaviour, (3) feeding interactions, and (4) allelopathy (secretion of toxic chemicals). The allelochemicals take part in both intra- and interspecies spatial competition. The allelochemicals can be water soluble or hydrophobic depending on the requirement of the organisms. Several sessile marine organisms such as anemones, soft corals (Coelenterata, Alcyonacea), tunicates, sponges, etc. have been reported to utilize chemical defence to protect their space from other aggressive organisms (Alino et al. 1992; Aerts 2000). Among all the sessile invertebrates, allelopathy or chemical defence is the most important in case of soft-bodied sessile invertebrates especially sponges, which lack any physical defences (Lindquist 2002). Sponges are considered as one of the longest living animals (McMurray et al. 2008); thus, these are persistent competitor of other sessile invertebrates. Sponges have high regenerative and morphological plasticity which allow them to grow along with other fast-growing sessile organisms (Garrabou and Zabala 2001). Remarkable allelopathic behaviour along with longevity and plasticity makes sponges an ideal model to study competitive interactions in context of their ecological significance.

Most of the studies on chemical-mediated spatial competition have been carried out on sponges and corals. Sponges compete with corals by (1) undermining the skeletal integrity of corals using notorious compounds and (2) reducing their photosynthetic efficiency, respiration and growth rate (Sullivan et al. 1983; Pawlik et al. 2007). For instance, the bio-eroding sponge *Siphonodictyon* sp. produced a chemical 'siphonodictidine' in its mucus that prevented growth and reduced respiration rates of adjacent coral polyps (Sullivan et al. 1983). In another example, the Caribbean sponge *Plakortis halichondroides* inhibited the metabolic processes in 14 different species of corals, both in contact and at distant by producing waterborne allelochemical (Porter and Targett 1988). These allelochemical-mediated interactions are highly specific. For example, in a study on Indonesian sponges, the sponge's allelochemicals affected only specific set of neighbours (de Voogd et al. 2004). The compound 'halichonacyclamine A' secreted by the sponge *Haliclona* sp. resulted in cytotoxicity in coral *Acropora nobilis* creating a zone of clearance around the sponge (Garson et al. 1999). Later, Russell et al. (2003) postulated the production of this cytotoxic compound at larval stage of the sponge *Haliclona* sp. Another pure compound 'clionapyrrolidine A' isolated from the encrusting sponge *Cliona tenuis* killed live coral tissue in 1–4 days (Chaves-Fonnegra et al. 2008). Later, this compound was found to affect corals through deep-tissue contact rather than superficial external contact (López-Victoria et al. 2006). Under stressful conditions, sponges adapted well and outcompeted the corals (Aerts and Van Soest 1997; Aerts 1998). Moreover, several Caribbean sponges affected the photosynthesis in microalgal symbionts, resulting in partial or complete bleaching in the massive coral *Diploria labyrinthiformis* (Pawlik et al. 2007). In this study, the corals showed reduction in both, their baseline fluorescence (correlated with the production of chlorophyll a), and potential quantum yield (measure of photosynthetic efficiency) when subjected to the crude extract of sponge *Agelas clathrodes*. The sponge impaired the photosynthesis in microalgal symbionts (zooxanthellae) that resulted in partial or sometimes complete bleaching. The compound belonging to class of pyrrole–imidazole alkaloids was found to be responsible for inefficient photosynthetic ability in coral symbionts (zooxanthellae) (Pawlik et al. 2007). Besides hard corals, soft corals (Zoanthids) and sea anemones are found to interact with sponges using their aggressive organs such as sweeper tentacles (Langmead and Chadwick-Furman 1999) or releasing of allelochemicals (Aceret et al. 1995; Maida et al. 1995). Several species of Zoanthids such as *Palythoa* and *Zoanthus* are reported to be abundant in nutrient-rich regions such as coastal areas near Brazil (Costa Jr et al. 2008) and Venezuela (Bastidas and Bone 1996). Thus, they get selective advantage over other sessile organisms such as sponges (Costa et al. 2008). In intertidal regions of the Indian west coast, the sponge *Cinachyrella cf. cavernosa* withstands in the competition for space by soft coral *Zoanthus sansibaricus* (Fig. 3.1). This standoff competition (with no visible outcome) for space might be possibly mediated via chemical interactions among these species (Singh and Thakur, unpublished data). Spatial completion via

Fig. 3.1 Competition for space by sponge *Cinachyrella cf. cavernosa* against cnidarian *Zoanthus sansibaricus*

chemicals occurred among a diverse group of encrusting bryozoans and sponges in coral reefs (Jackson and Buss 1975; Nandakumar et al. 1993). Another study on nine sympatric sponges displayed species-specific chemical responses against bryozoan species (Jackson and Buss 1975). However, the experiments failed to estimate the natural concentrations due to the use of whole organism extracts. Moreover, there was differential production and release of toxic chemicals as per intensity of competitive interactions (Jackson and Buss 1975). Allelochemical-mediated interactions among sponges and bryozoans have been reported by several authors (Buss 1976; Coll et al. 1983). Besides acquisition of living space, availability of food has been demonstrated as important parameter determining the outcome of competitive interaction among bryozoans and sponges (Jackson and Buss 1975). Some studies have reported spatial competition among sponges. For instance, the Mediterranean sponge *Crambe crambe* prevented growth of any other sponges in its surroundings (Turon et al. 1996). Similarly, the coral reef sponge *Dysidea* sp. in Guam produced allelochemical '7-deacetoxyolepupuane' causing necrosis in other species of sponges (Thacker et al. 1998).

3.4 Defence Against Predators

The 'life-dinner principle' suggests that there is significant amount of pressure on the organisms to defend themselves against predators (Dawkins and Krebs 1979). This is because if the predator loses the prey, the loss is only 'one-time meal', but if a particular organism is preyed, the loss is huge in terms of their total survivorship and chances of propagation in future. Predation cannot only influence the population dynamics of prey but also that of predators to some extent. The soft-bodied, sessile sponges lack any physical defences and are physically vulnerable to predation in their highly diverse environment (Fig. 3.2). The major predators of sponges include a few fish species (Wulff 1995; Dunlap and Pawlik 1996), hawksbill turtles (Bjorndal and Jackson 2003), molluscs (e.g. nudibranch) which usually feed on few sponge species (Pawlik et al. 1988; Paul 1992), echinoderms (Birenheide et al. 1993; Waddell and Pawlik 2000) and sea urchins (Ayling 1981). To tackle these predators, sponges employ various strategies such as:

Fig. 3.2 The sponges *Cinachyrella cf. cavernosa* under constant stress due to predatory fish *Istigobius ornatus* in highly diverse intertidal environment

1. Physical and structural defences like the presence of spicules and spongin fibres that contribute to increased toughness and reduced nutritional quality of tissue
2. Chemical defences

The biosynthesis or dietary sequestration of toxic, noxious or distasteful metabolites by organisms against predators is more common in marine environments (Paul 1992; Hay 1996; McClintock and Baker 2010), and marine sponges are no exception to this (Pawlik et al. 1995; Assmann et al. 2000; Wilson et al. 1999). Earlier reports suggest that 69 % of tropical Atlantic sponges, 100 % of Mediterranean sponges and 78 % of Antarctic sponges produce chemicals to deter some of their predators (Peters et al. 2009). We reviewed here some of the examples of sponges from each of these regions.

Antarctic sponges have been reported to use wide range of secondary metabolites to deter their predators. The major spongivores of

Antarctic regions are starfish and nudibranch (McClintock et al. 2005). The role of secondary metabolites in deterring the predators was first studied in Antarctic sponges from McMurdo Sound, Ross Sea (Dayton et al. 1974). This feeding deterrence was observed indirectly by studying the chemotactile retraction of sensory tube feet of the common predator sea star *Perknaster fuscus* (McClintock and Baker 1997; McClintock et al. 1994, 2000; Amsler et al. 2001). This behavioural assay which was based on response of the chemosensory tube feet of spongivorous sea stars was applicable to only few omnivorous Antarctic sea stars (e.g. *Odontaster validus*) (McClintock and Baker 1997; Núñez-Pons and Avila 2014). However, the most common spongivore *P. fuscus* displayed erratic response of sensory tube feet when subjected to food pellet assays. For instance, *P. fuscus* did not cause tube foot retractions when assayed for the rapidly growing, space-dominating sponge *Mycale acerata*; however, they displayed acceptance for coloured sponge *Kirkpatrickia variolosa* which was avoided/unpalatable by it under natural circumstances (Amsler et al. 2000, 2001). It was observed that the unpalatable food or sponge extract can cause tube foot to remain retracted for up to 60 s or more, whereas palatable foods are immediately accepted; thus, this behavioural bioassay was used successfully to determine sea star feeding preferences. Recently, a reliable and direct method has been adopted to observe the deterrence against the common omnivorous predator sea star *Odontaster validus* in 27 common Antarctic sponges at Palmer Station, Anvers Island (Peters et al. 2009). The feeding assay was conducted by positioning a fresh sponge tissue or sponge extract embedded in food pellet onto the tube feet ambulacral feeding groove equidistant between the mouth and the tip of the arm. By tracing the movement of tissue or food pellet to the mouth or its discharge via ambulacral groove, it was possible to evaluate the deterrent properties of the sponge.

The chemically mediated predator deterrence involved high energetic cost, and there is trade-off among the life history processes such as growth, reproduction and defence. The optimal defence theory (ODT) provides a framework to test the variations in defences within organisms. The feeding assays were conducted on the outer and inner tissues of sponges to find if there exists any difference in feeding preference due to the different concentration of chemical deterrents (Peters et al. 2009). It was found that the outer tissue of 78 % of the experimental sponges ($n = 27$) were deterrent to sea stars. However, both the inner and outer tissue of 62 % of sponges were deterred by sea stars. Thus, these observations do not support ODT completely.

Many secondary metabolites having deterrence properties have been reported from Antarctica sponges. The lipophilic extracts from the Antarctic sponge *Dendrilla membranosa* displayed feeding deterrence in the amphipod *Gondogeneia antarctica* at natural concentrations (Amsler et al. 2000). In contrast, norselic acids (e.g. norselic acid A) extracted from Antarctica sponge *Crella* sp. did not show feeding deterrence in sympatric amphipod and the sea star *O. validus* at tissue-level concentration (Ma et al. 2009). However, when the feeding assays were conducted at 3–10 times of the tissue-level concentration, there was 18 % reduction in spongivory by amphipods and slight deterrence in sea star, respectively. This was attributed to the inefficiency of traditional protocols of chemical isolation, and the estimation of natural concentration might not be a true representative of actual concentrations of the natural product in this sponge (Ma et al. 2009). Besides chemical clues, visual clues such as colour of the prey can also determine the feeding behaviour. One of the major predators of Antarctic sponges is sea stars which lack visual organ. It was hypothesized that the present-day Antarctic sponges still retain visual pigments (e.g. discorhabdins and variolins) which was genetically transferred by the ancestral lineages that lived in ancient warmer Antarctic with dominance of visual predators such as fish and turtles (McClintock and Baker 1998). Along with these visual pigments, sponges seemed to inherit antipredatory chemicals which have further evolved according the existing predation pressure. However, this hypothesis that the sponges retained their inherited colour depending on the selective pressure due to the presence or absence of visual predators did not seem to follow a general trend. The example of Antarctic sponge of genera *Haliclona* and *Calyx* that lack colour supports the hypothesis as their temperate and tropical counterparts are brightly coloured, might be, due to the presence of visual predators. Similarly, species of Antarctic genus *Leucetta* are colourless, while their tropical counterparts are coloured. In contrast, the Antarctic sponge *Latrunculia apicalis* is dark green in colour and reported to cause behavioural response (tube feet retraction) in *P. fuscus* (McClintock et al. 1994, 2000). Two bioactive compounds discorhadbin alkaloids (discorhabdin C and G) have been isolated from *L. apicalis* (Yang et al. 1995). However, these compounds were not studied for their ecological role in antipredatory activity. Various other species of Antarctic sponges such as *Suberites* sp., *Latrunculia apicalis*, *Leucetta leptorhapsis*, *Dendrilla membranosa* and *Kirkpatrickia variolosa* have been documented to produce various different types of secondary metabolites with highly variable nature (Perry et al. 1994; Trimurtulu et al. 1994; Shin et al. 1995; Yang et al. 1995; Jayatilake et al. 1997; Moon et al. 2000; Ankisetty and Slattery 2008). However, the ecological role of these compounds in predator deterrence, antifouling and defending the occupied space has not been investigated yet. The understanding of the ecological role might assist in accounting the variability in type and concentration of the secondary metabolites. In general, tropical waters experience higher levels of predation than Antarctic and temperate regions, thereby creating greater potential for the evolution of chemical defences. According to latitudinal gradient hypothesis, there exists an inverse relationship between latitude and ichthyotoxicity in sponges. However, a recent report contradicted latitudinal gradient hypothesis in case of marine sponges (Becerro et al. 2003). There have been extensive studies on the role of chemicals in feeding deterrence in tropical sponges. The feeding experiments on 17 common Red Sea sponge species and 17 common Caribbean sponge species were carried out. In this investigation, it was found that 41 % of Red sea sponges deterred the fish *T. klunzingeri*, and 65 % were deterrent to the sea urchin *Diadema setosum*. On the other hand, 17 Caribbean sponges displayed variable level deterrence to the predators. However, they showed a general trend in deterrence to fish predators which was not influenced by geographic origin (Burns et al. 2003). Reports from the feeding assay conducted on Caribbean sponges suggested that 69 % of tested sponge species were able to deter the common generalist reef fish *Thalassoma bifasciatum* (Pawlik et al. 1995). This demonstrates that the Caribbean sponges mostly rely on the chemically mediated defence to avoid their predators. Similarly, researchers have demonstrated that the crude extracts from several Caribbean sponge species when tested against the unnatural predators of these sponges (hermit crab *Paguristes punticeps* and non-spongivorous fish *Thalassoma bifasciatum*) successfully deterred predation (Pawlik et al. 1995; Waddell and Pawlik 2000). Some sponge-feeding nudibranchs have evolved a mechanism to accumulate the defensive compounds through de novo synthesis.

The chemical defences sometimes act as a clue for the predator. This was illustrated in an experiment conducted to study prey–predator relationship in octocorals and its natural fish predator *Chaetodon melannotus* (Alino et al. 1992). The octocorals produced highly toxic compound that acted as clue for its predator *Chaetodon melannotus* to locate the prey. However, in feeding experiments, this toxic compound might be able to deter by generalist predators like *T. bifasciatum*, thus providing misleading ecological information about chemical defence in octocorals. These findings suggested that feeding assay test should be performed on natural predators of the test organisms, because the chemical defences are highly specific and they are effective mostly against the natural predators (Pennings et al. 1994). There are reports on the differential distribution of secondary metabolites in different tissue layer of the sponges. The Micronesian sponge *Oceanapia* sp. which produced two major deterrent compounds kuanoniamine C and D demonstrated highest concentration on outer parts (area exposed to predators). This example supports the ODT that the tissue or body parts which are under threat or have significant role in reproduction are protected well by investigating more energy in terms of secondary metabolites. In contrast, both the outer and the inner tissue of the common Caribbean sponges *Xestospongia muta* and *Chondrilla nucula* were found to deter predators showing no evidence in support of ODT (Chanas et al. 1997; Swearingen and Pawlik 1998).

3.5 Epibiotic Defence

Sponges are filter-feeding organisms. The outer and inner body surfaces are continuously exposed to various microorganisms and larvae of other invertebrates. These epibionts have tendency to populate on the sponge surface. The bacteria and diatoms are the first organisms that attach to the surface in the process of microfilm formation. Later, this microbial film acts as conditioned layer for the settlement of macrofouling organisms, such as macroalgae, invertebrates and propagules of other organisms. For efficient filter feeding, sponges need to keep their pores free from clogging. The fouling by epibionts can negatively affect the feeding efficiency of the host by blocking the canal system. Moreover, it may dislodge the sponges from substratum by increasing hydrodynamic drag (Lesser et al. 1992). Additionally, epibionts compete with their host for resources such as food and irradiation. Epibionts can also cause growth inhibition, tissue necrosis and death of host organisms (Wahl and Mark 1999). Fortunately, sponges are able to prevent biofouling by producing bioactive metabolites produced either by itself or by symbiotic microorganisms (Kelly et al. 2003; Lee et al. 2006). This aids in maintaining the optimal feeding rate and preventing other harmful effects of epibiosis.

There are studies on sponge crude extracts that have been successful in preventing the larval settlement and growth (Thakur and Anil 2000; Hellio et al. 2005; Lee et al. 2006). The crude extract from the sponges *Haliclona cymaeformis*, *Haliclona* sp. and *Callyspongia* sp. was found to suppress the density of macrofoulers and diatoms on the surface of the sponges (Dobretsov et al. 2005). However, the bacterial densities on the surfaces of *H. cymaeformis* and *Callyspongia* sp. were found to be unaffected. Molecular identification of the bacterial communities revealed that there was drastic decrease in diversity of bacterial community. Moreover, the sponge extracts were also found to reduce the diversity and species richness in case of diatoms which imply that sponge metabolites can not only prevent the settlement of propagules of invertebrates but also control the diversity of micro- and macro-fouling communities to combat the harmful effects of biofouling. Similarly, the crude extract of seven dominant sponges in Hong Kong waters showed drastic reduction in settlement and recruitment of bacteria and diatoms under both laboratory and field experiments (Dobretsov et al. 2005). The sponge chemicals were found to inhibit fouling of nearby nonliving surfaces showing the chemical defence as the main tool to combat epibiosis (Dobretsov et al. 2005). Several novel compounds from marine sponges and their synthetic analogs have been reported in various studies as natural antifouling agents. Different species of genus *Axinyssa* contained sesquiterpene carbonimide dichlorides and a guaiane-type sesquiterpene peroxide that have shown to inhibit recruitment of larvae of the barnacle *Balanus amphitrite*, thus promising to be the potent antifouling compounds (Fusetani 2004; Nogata and Kitano 2006). Similarly, a sesquiterpenoid hydroquinone 'avarol' (isolated from sponge *Dysidea avara*) and its synthetic analogs displayed high antifouling activity against cyprids of *B. amphitrite*. Out of all synthetic analogs, 30-p-chlorophenylamino avarone and 40-propylthioavarone showed highest therapeutic ratios, i.e. maximum inhibition at lowest concentration (Tsoukatou et al. 2007). In addition to this, all 14 terpenoids isolated from Mediterranean sponges also showed antifouling activity against barnacle larvae. Out of these 14 terpenoids, hydroquinone A acetate and dihydrospongin II displayed maximum toxicity (Hellio et al. 2005). Another study on the marine sponge *Luffariella variabilis* demonstrated that manoalide and its analogs sesterterpenoids isolated from this sponge inhibited bacterial quorum sensing, thus suggesting their potential application as QS inhibitors to control QS activities during microfilm formation (Skindersoe et al. 2008). The compound 'A-norsteroids' isolated from the marine sponge *Acanthella cavernosa* showed moderate anti-barnacle activity (Qiu et al. 2008). The bromotyrosine derivatives derived from this sponge were found to be effective antifouling agent (Fusetani 2004). From the class of bastadins, 'bastadin-9' exhibited potent AF at minimum inhibitory concentration (Ortlepp et al. 2007). Sponge originated compounds Oroidin (a pyrroloimidazole alkaloid) and its analogs were found to be highly effective against general group of fouling bacteria such as *Pseudomonas aeruginosa* (Kelly et al. 2003). The compounds from *Haliclona* sp., haliclonacyclamine A and halaminol A did not affect the ascidian larval settlement but prevented further metamorphosis of settled larva, leading to their mortality. However, these compounds inhibited both settlement and metamorphosis when tested against the larvae of sponges, polychaete, gastropod and bryozoan (Roper et al. 2009). Polybrominated diphenyl ethers isolated from marine sponge *Dysidea* sp., and their synthetic analogs were found to be highly toxic against barnacle larvae (Ortlepp et al. 2008). Also, barettin and 8,9-dihydrobarettin, brominated DKPs, isolated from marine sponge *Geodia baretti* displayed toxicity against barnacle larvae by reducing their recruitment rate by 81–89 % (Sjögren et al. 2004).

The sponges have been reported to have direct (by producing antifouling compounds) and indirect (by promoting associated bacteria for producing antifouling) epibiotic chemical defence (Thakur and Anil 2000; Thakur et al. 2003). The crude extract from the sponge *Ircinia ramosa* and its associated bacteria showed antibacterial activity (Thakur and Anil 2000). In the same study, it was shown that there was seasonal change in the polarity of active metabolite. In colder months, the activity was located in polar fraction, whereas the non-polar extract was found to be active in the warmer months. In warmer season, the active non-polar fraction aids in preventing extreme epibiosis (due to increased microbial diversity) faced by the sponges. This is because non-polar active metabolites are able to persist on sponge surface for a longer time due to their hydrophobic nature and slower dissolution rates. In subsequent studies on the sponge *Suberites domuncula*, it was found that recombinant perforin-like protein was responsible for high antibacterial activity in the sponges which lead to the hypothesis that both direct (by sponges' metabolites) and indirect (by associated bacterial community) chemical-mediated defences are utilized in sponges against epibiosis (Thakur et al. 2003). Further investigation was done to find out role of sponge-associated bacteria in sponge's epibiotic defence in relation to temporal changes in environmental factors (Thakur et al. 2004). The researchers identified the culturable bacteria associated with the sponge by 16S RNA sequencing and evaluated the seasonal variation in their abundance and antifouling properties. It was demonstrated that about 60 % sponge-associated bacteria with antifouling properties belonged to the class of Gram-positive (Thakur et al. 2004) which are known for sporulation under adverse conditions and antibiotic production (Marahier et al. 1993). In the same study, it was found that the metabolites of both host sponge and associated bacteria can play multiple roles such as defence against epibionts and regulating the internal bacterial community via intra and interspecific bacterial inhibition. The studies on the demosponge *Suberites domuncula* demonstrated that this sponge produced lyso-PAF (platelet-activating factor) compounds, 1-O-hexadecyl-sn-glycero-3-phosphocholine and 1-O-octadecyl-sn-glycero-3-phosphocholine with antibacterial activity when subjected to external stress molecule endotoxin lipopolysaccharide (LPS) (Müller et al. 2004). The results suggested the production of these bioactive lipid derivatives triggered by exposure to endotoxin (LPS), which is usually present in outer layer of Gram-negative bacteria.

This emphasizes on the adaptive role of these molecules in epibiotic defence against the fouling organisms. Moreover, *S. domuncula* was found to produce hydroxylated aromatic compounds protocatechuate which acts as a carbon source for its surface-associated bacteria, thereby illustrating a symbiotic relationship between sponge and its associated bacteria (Müller et al. 2004).

By using these techniques, sponges prevent the settlement of other organisms on their surfaces. The importance of sponge-associated-microbial communities have also been depicted in preventing the larval settlement on the surface of the sponge (Hentschel et al. 2003; Thoms et al. 2003; Lee et al. 2006; Taylor et al. 2004). The study on the sponges *Aplysina aerophoba* and *Theonella swinhoei* illustrated that the associated microbial community of these sponges is unique and different from the microbial community in the ambient water (Hentschel et al. 2002).

Chemical defences can be very specific to particular fouling organisms or may have a broad spectrum of bioactivity to fouling organisms, with additional role as deterrent to predators, and cytotoxicity against competitors and pathogens. To defend against fouling organisms, marine sponge uses various strategies such as (1) the use of secondary metabolites, produced by itself or the symbionts, (2) innate immune system (Thakur et al. 2005) and (3) by cell shedding or removal of outer tissue (Barthel and Wolfrath 1989). The innate immune systems of sponges have been also believed to play a role in the prevention of microbial invasion (Schröder et al. 2003). The presence of a tachylectin-related protein (*Suberites* lectin) in the demosponge *Suberites domuncula* was demonstrated which displayed high antibacterial activity (Schröder et al. 2003). It was found to be similar to horseshoe crab lectins and was hypothesized to be a key molecule of sponges' innate immune system to defend against bacterial epibiosis. Additionally, the molecular basis of innate defence in the sponge *Suberites domuncula* was studied in which the sponge displayed increased expression of adaptor gene (AdaPTin-1) and production of lysozyme when exposed to peptidoglycan (PPG) (cell wall component of Gram-positive bacteria) (Thakur et al. 2005). The sponge and its associates work in harmony by means of various signalling molecules and wide range of chemical molecules. Recently, various chemical molecules and related genes which are responsible for (1) effective communication between the sponge host and its associated microbial community and (2) response to the external stimuli have been reviewed in detail (Wang et al. 2013).

3.6 Defence Against Pathogens

Marine sponges are efficient filter-feeding organisms (Stabili et al. 2006). They obtain their food by pumping large amount of sea water and retaining the nutrients from filtered water. They are able to filter 72,000 times of their volume and retain 58–99 % particle (Pile et al. 1997) and about 96.1 % of bacteria from the filtered water (Reiswig 1971). The sea water has a rich source of both harmful and beneficial microbes. Reports have suggested that sometimes 40 % of sponge biomass is due to associated bacteria (Vacelet and Donadey 1977; Friedrich et al. 2001). Sponges provide a shelter for various microorganisms such as *Cyanobacteria*, diverse heterotrophic bacteria, unicellular algae and zoochlorellae. The porous body of the sponge is prone to colonization by the microbes, some of which may be pathogenic and some can cause tissue necrosis. The microbes can affect the sponges both (1) directly by the pathogenic and/or parasitic behaviour and (2) indirectly by causing microbial films (as discussed in previous section). Most of diseases in sponges are due to bacterial or fungal infections (Webster 2007; Webster and Blackall 2009). Besides bacterial and fungal infection, there are reports on parasitic behaviour of diatoms associated with several Antarctic species (Bavestrello et al. 2008; Cerrano et al. 2000). In most of the investigations, the causative organisms have not been reported except a few cases (Webster et al. 2002; Mukherjee et al. 2009; Cervino et al. 2006). The identification of primary causative agents of diseases in sponges is difficult due to the presence of large number of associates/symbionts. Reports have suggested that there is change in symbiotic bacterial community in diseased sponges (Bourne 2005; Cervino et al. 2006; Webster et al. 2008; Angermeier et al. 2011). Marine sponges have been hypothesized to act as disease reservoir for opportunistic bacteria and coral pathogens (Negandhi et al. 2010).

Sponges develop resistance against these microbes by producing unique and diverse secondary metabolites. The antimicrobial compounds in sponges are highly specific in effect, thus allowing the sponge host to select beneficial microbes and prevent infections from harmful ones. As described earlier, the associated microbial community in sponge can efficiently participate in host defence (Thakur et al. 2004). Several antibiotics have been isolated from marine sponges which are highly active against human pathogens or several other marine bacteria. Discodermin A and its variants (B-D, F-H) which were isolated from the marine sponge *Discodermia kiiensis* (Fusetani and Matsunaga 1993) exhibited high antibacterial, antifungal and selective cytotoxic properties in vitro. The compound theonellamide F, isolated from *Theonella* sponge displayed high antifungal and cytotoxic activity in laboratory (Matsunaga et al. 1989). Halicylindramides (A–C) from the marine sponge *Halichondria cylindrata* showed antifungal activity against *Mortierella ramanniana* and high cytotoxicity in P388 murine leukaemia cells in vitro (Li et al. 1995). The sponge *Aciculites orientalis* produced three cyclic peptides, aciculitins AC (1–3) that inhibited the growth of fungus *Candida albicans* and were cytotoxic towards the human colorectal cancer cell line HCT-116 (Bewley et al. 1996). Similarly, another compound named cyclolithistide A, isolated from a marine sponge *Theonella swinhoei*, displayed potent antifungal activity *against C. albicans* but not against *Escherichia coli* or *Bacillus subtilis* (Clark et al. 1998). The Australian sponges *Phoriospongia* sp. and *Callyspongia bilamellata* contained two depsipeptides, phoriospongin A and B that were found effective against the livestock parasite nematode *Haemonchus contortus* (Capon et al. 2002). These examples suggest that these antibiotics and cytotoxic compounds produced by sponges might be useful in fighting against microbial infections and parasites in their natural environment. However, they have not been tested against the naturally occurring bacteria in order to elucidate their ecological role. Few studies have been undertaken to study the ecological role of these compounds by testing the antimicrobial activity against ecologically relevant bacteria. Results have shown that sponges use chemical defence to combat the microbial infection and maintain their well-being (Becerro et al. 1994; Kelman et al. 2001). Additionally, these metabolites aid in regulating symbiotic bacterial populations inside the sponge body (Bergquist 1978). Crude organic extract of 11 species of the Red Sea reef sponges showed antibacterial activity when tested against the bacteria found in their natural habitat (Kelman et al. 2001). Among these, the sponge *Amphimedon viridis* yielded highly bioactive pyridinium alkaloids that showed selective activity against specific group of bacteria. These compounds were highly active against some of the pathogenic bacteria, whereas they did exhibit activity against beneficial bacteria associated with the sponge. This shows that sponges' metabolites are specific in their nature and they are produced according to the requirement of the organism.

3.7 Conclusions

This review highlights the remarkable ability of sponges to defend the occupied space from other sessile invertebrates, deter predators, maintain optimal feeding rate by preventing settlement of epibionts and fight against microbial infection by using chemical weapons. Many researchers have demonstrated the use of secondary metabolites in the medical field to treat various deadly diseases, but few have studied the natural roles of these compounds to defend sponges against predators, microbes, aggressive neighbours and fouling organisms. These compounds hinder the growth of the competitors, fouling organisms and harmful microbes and cause deterrence in predator by impairing their cells by cytotoxic or antibiotic activities, thereby establishing themselves as potential drug candidates. However, the outcome of these ecological interactions in many of these sessile invertebrates including marine sponges depends on environmental factors (e.g. levels of nutrients, temperature, irradiance, etc.) that alter their life history processes such as growth and reproduction. Production and secretion of secondary metabolites by the organisms depends on their necessity. In such a scenario, understanding the ecological roles of sponge secondary metabolites will aid in discovery of many novel compounds and their sustainable utilization for mankind. Future studies should focus more on identifying the novel compounds and understanding their ecological importance thereby linking chemical ecology with bioprospecting

Acknowledgements We are grateful to the Director, CSIR-National Institute of Oceanography, for his support and encouragement. We are thankful to Dr. P.A. Thomas of Central Marine Fisheries Research Institute, Trivandrum, India, for identification of sponges described in the photographs. Our special thanks to Dr. John N.A. Hooper and Dr. Kathryn Hall from Queensland Museum, Australia for their valuable inputs in identification of Cinachyrella sp. The author A. Singh greatly acknowledges CSIR, India, for the award of senior research fellowships. This work was supported by CSIR funded project 'Ocean Finder' (PSC0105).

References

Aceret T, Sammarco P, Coll J (1995) Effects of diterpenes derived from the soft coral *Sinularia flexibilis* on the eggs, sperm and embryos of the scleractinian corals *Montipora digitata* and *Acropora tenuis*. Mar Biol 122(2):317–323

Aerts L (1998) Sponge/coral interactions in Caribbean reefs: analysis of overgrowth patterns in relation to species identity and cover. Mar Ecol Prog Ser 175: 241–249

Aerts LA (2000) Dynamics behind standoff interactions in three reef sponge species and the coral *Montastraea cavernosa*. Mar Ecol 21(3-4):191–204

Aerts L, Van Soest R (1997) Quantification of sponge/coral interactions in a physically stressed reef community, NE Colombia. Mar Ecol Prog Ser Oldendorf 148(1):125–134

Alino PM, Sammarco PW, Coll J (1992) Competitive strategies in soft corals (Coelenterata, Octocorallia). IV. Environmentally induced reversals in competitive superiority. Mar Ecol Prog Ser Oldendorf 81(2): 129–145

Amsler CD, Moeller CB, McClintock JB, Iken KB, Baker BJ (2000) Chemical defenses against diatom fouling in Antarctic marine sponges. Biofouling 16(1):29–45

Amsler CD, McClintock JB, Baker BJ (2001) Secondary metabolites as mediators of trophic interactions among Antarctic marine organisms. Am Zool 41(1):17–26

Angermeier H, Kamke J, Abdelmohsen UR, Krohne G, Pawlik JR, Lindquist NL, Hentschel U (2011) The pathology of sponge orange band disease affecting the Caribbean barrel sponge *Xestospongia muta*. FEMS Microbiol Ecol 75(2):218–230

Ankisetty S, Slattery M (2008) Site specific comparison of secondary metabolites of *Stylissa massa* by Hplc chemical fingerprinting. Planta Med 74(03):P-123. doi:10.1055/s-2008-1075319

Assmann M, Lichte E, Pawlik JR, Köck M (2000) Chemical defenses of the Caribbean sponges *Agelas wiedenmayeri* and *Agelas conifera*. Mar Ecol Prog Ser 207: 255–262

Ayling A (1981) The role of biological disturbance in temperate subtidal encrusting communities. Ecology 62:830–847

Barthel D, Wolfrath B (1989) Tissue sloughing in the sponge *Halichondria panicea*: a fouling organism prevents being fouled. Oecologia 78(3):357–360

Bastidas C, Bone D (1996) Competitive strategies between *Palythoa caribaeorum* and *Zoanthus sociatus* (Cnidaria: Anthozoa) at a reef flat environment in Venezuela. Bull Mar Sci 59(3):543–555

Bavestrello G, Cerrano C, Di Camillo C, Puce S, Romagnoli T, Tazioli S, Totti C (2008) The ecology of protists epibiontic on marine hydroids. J Mar Biol Assoc U K 88(08):1611–1617

Becerro MA, Lopez NI, Turon X, Uriz MJ (1994) Antimicrobial activity and surface bacterial film in marine sponges. J Exp Mar Biol Ecol 179(2):195–205

Becerro MA, Thacker RW, Turon X, Uriz MJ, Paul VJ (2003) Biogeography of sponge chemical ecology: comparisons of tropical and temperate defenses. Oecologia 135(1):91–101

Bell JJ (2008) The functional roles of marine sponges. Estuar Coast Shelf Sci 79(3):341–353

Bell JJ, Davy SK, Jones T, Taylor MW, Webster NS (2013) Could some coral reefs become sponge reefs as our climate changes? Glob Chang Biol 19(9): 2613–2624

Bergquist PR (1978) Sponges. University of California Press, Berkeley/Los Angeles, 268 pp

Bewley CA, He H, Williams DH, Faulkner DJ (1996) Aciculitins AC: cytotoxic and antifungal cyclic peptides from the lithistid sponge *Aciculites orientalis*. J Am Chem Soc 118(18):4314–4321

Birenheide R, Amemiya S, Motokawa T (1993) Penetration and storage of sponge spicules in tissues and coelom of spongivorous echinoids. Mar Biol 115(4):677–683

Birrell CL, McCook LJ, Willis BL, Diaz-Pulido GA (2008) Effects of benthic algae on the replenishment of corals and the implications for the resilience of coral reefs. Oceanogr Mar Biol Annu Rev 46:25–63

Bjorndal KA, Jackson JB (2003) IU roles of sea turtles in marine ecosystems: reconstructing the past. Biol Sea Turtles 2:261

Blunt JW, Copp BR, Hu W-P, Munro M, Northcote PT, Prinsep MR (2009) Marine natural products. Nat Prod Rep 26(2):170–244

Bourne D (2005) Microbiological assessment of a disease outbreak on corals from Magnetic Island (Great Barrier Reef, Australia). Coral Reefs 24(2):304–312

Burns E, Ifrach I, Carmeli S, Pawlik J, Ilan M (2003) Comparison of anti-predatory defenses of Red Sea and Caribbean sponges. I. Chemical defense. Mar Ecol Prog Ser 252:105–114

Buss LW (1976) Better living through chemistry: the relationship between allelochemical interactions and competitive networks. In: Harrison FW, Cowden RR (eds) Aspects of sponge biology. Academic Press, New York, pp 315–327

Capon RJ, Ford J, Lacey E, Gill JH, Heiland K, Friedel T (2002) Phoriospongin a and B: two new nematocidal depsipeptides from the Australian marine sponges *Phoriospongia* sp. and *Callyspongia b ilamellata.* J Nat Prod 65(3):358–363

Cerrano C, Bavestrello G, Bianchi C, Cattaneo-Vietti R, Bava S, Morganti C, Morri C, Picco P, Sara G, Schiaparelli S (2000) A catastrophic mass-mortality episode of gorgonians and other organisms in the Ligurian Sea (North-western Mediterranean), summer 1999. Ecol Lett 3(4):284–293

Cervino JM, Winiarski-Cervino K, Polson SW, Goreau T, Smith GW (2006) Identification of bacteria associated with a disease affecting the marine sponge *Ianthella basta* in New Britain, Papua New Guinea. Mar Ecol Prog Ser 324:139–150

Chadwick NE, Morrow KM (2011) Competition among sessile organisms on coral reefs. In: Dubinsky Z, Stambler N (eds) Coral Reefs: an ecosystem in transition. Springer, Doedrecht, pp 347–371

Chanas B, Pawlik JR, Lindel T, Fenical W (1997) Chemical defense of the Caribbean sponge *Agelas clathrodes* (Schmidt). J Exp Mar Biol Ecol 208(1):185–196

Chaves-Fonnegra A, Castellanos L, Zea S, Duque C, Rodríguez J, Jiménez C (2008) Clionapyrrolidine a—a metabolite from the encrusting and excavating sponge *Cliona tenuis* that kills coral tissue upon contact. J Chem Ecol 34(12):1565–1574

Clark DP, Carroll J, Naylor S, Crews P (1998) An antifungal cyclodepsipeptide, cyclolithistide A, from the sponge *Theonella swinhoei.* J Org Chem 63(24): 8757–8764

Coll J, Tapiolas D, Bowden B, Webb L, Marsh H (1983) Transformation of soft coral (Coelenterata: Octocorallia) terpenes by *Ovula ovum* (Mollusca: Prosobranchia). Mar Biol 74(1):35–40

Connell JH, Hughes TP, Wallace CC, Tanner JE, Harms KE, Kerr AM (2004) A long-term study of competition and diversity of corals. Ecol Monogr 74(2): 179–210

Costa OS Jr, Nimmo M, Attrill MJ (2008) Coastal nutrification in Brazil: a review of the role of nutrient excess on coral reef demise. J S Am Earth Sci 25(2):257–270

Dawkins R, Krebs JR (1979) Arms races between and within species. Proc R Soc London, Ser B 205 (1161):489–511

Dayton PK, Robilliard GA, Paine RT, Dayton LB (1974) Biological accommodation in the benthic community at McMurdo Sound. Antarct Ecol Monogr 44:105–128

De Caralt S, Bry D, Bontemps N, Turon X, Uriz MJ, Banaigs B (2013) Sources of secondary metabolite variation in *Dysidea avara* (porifera: demospongiae): the importance of having good neighbors. Mar Drugs 11(2):489–503

de Voogd NJ, Becking LE, Hoeksema BW, van Soest R (2004) Sponge interactions with spatial competitors in the Spermonde Archipelago. Bolletino di Museo e Istituto di Biologia dell'Universita di Genova 68: 253–261

Dobretsov S, Dahms H-U, Tsoi MY, Qian P-Y (2005) Chemical control of epibiosis by Hong Kong sponges: the effect of sponge extracts on micro-and macrofouling communities. Mar Ecol Prog Ser 297:119–129

Dunlap M, Pawlik J (1996) Video-monitored predation by Caribbean reef fishes on an array of mangrove and reef sponges. Mar Biol 126(1):117–123

Eder C, Schupp P, Proksch P, Wray V, Steube K, Müller CE, Frobenius W, Herderich M, van Soest RW (1998) Bioactive pyridoacridine alkaloids from the Micronesian sponge *Oceanapia* sp. J Nat Prod 61 (2):301–305

Faulkner DJ (2000) Marine natural products. Nat Prod Rep 17(1):7–55

Fenical W (1993) Chemical studies of marine bacteria: developing a new resource. Chem Rev 93(5): 1673–1683

Foster NL, Box SJ, Mumby PJ (2008) Competitive effects of macroalgae on the fecundity of the reef-building coral *Montastraea annularis.* Mar Ecol Prog Ser 367: 143–152

Friedrich AB, Fischer I, Proksch P, Hacker J, Hentschel U (2001) Temporal variation of the microbial community associated with the Mediterranean sponge *Aplysina aerophoba.* FEMS Microbiol Ecol 38(2-3):105–115

Fusetani N (2004) Biofouling and antifouling. Nat Prod Rep 21(1):94–104

Fusetani N, Matsunaga S (1993) Bioactive sponge peptides. Chem Rev 93(5):1793–1806

Garrabou J, Zabala M (2001) Growth dynamics in four Mediterranean demosponges. Estuar Coast Shelf Sci 52(3):293–303

Garson M, Clark R, Webb R, Field K, Charan R (1999) Ecological role of cytotoxic alkaloids: *Haliclona* n. sp., an unusual sponge/dinoflagellate association. In: Hooper JNA (ed) 5th international sponge conference, Qld Museum, pp 205–213

Hay ME (1996) Marine chemical ecology: what's known and what's next? J Exp Mar Biol Ecol 200(1):103–134

Hay M (2014) Challenges and opportunities in marine chemical ecology. J Chem Ecol 40(3):216–217. doi:10.1007/s10886-014-0393-5

Hellio C, Tsoukatou M, Maréchal J-P, Aldred N, Beaupoil C, Clare AS, Vagias C, Roussis V (2005) Inhibitory effects of Mediterranean sponge extracts and metabolites on larval settlement of the barnacle *Balanus amphitrite*. Mar Biotechnol 7(4):297–305

Hentschel U, Hopke J, Horn M, Friedrich AB, Wagner M, Hacker J, Moore BS (2002) Molecular evidence for a uniform microbial community in sponges from different oceans. Appl Environ Microbiol 68(9):4431–4440

Hentschel U, Fieseler L, Wehrl M, Gernert C, Steinert M, Hacker J, Horn M (2003) Microbial diversity of marine sponges. In: Sponges (Porifera). Springer, Berlin, pp 59–88

Jackson J, Buss L (1975) Allelopathy and spatial competition among coral reef invertebrates. Proc Natl Acad Sci 72(12):5160–5163

Jayatilake GS, Baker BJ, McClintock JB (1997) Rhapsamine, a cytotoxin from the Antarctic sponge *Leucetta leptorhapsis*. Tetrahedron Lett 38(43): 7507–7510

Kelly SR, Jensen PR, Henkel TP, Fenical W, Pawlik JR (2003) Effects of Caribbean sponge extracts on bacterial attachment. Aquat Microb Ecol 31(2):175–182

Kelman D, Kashman Y, Rosenberg E, Ilan M, Ifrach I, Loya Y (2001) Antimicrobial activity of the reef sponge *Amphimedon viridis* from the Red Sea: evidence for selective toxicity. Aquat Microb Ecol 24(1):9–16

Langmead O, Chadwick-Furman N (1999) Marginal tentacles of the corallimorpharian *Rhodactis rhodostoma*. 1. Role in competition for space. Mar Biol 134(3):479–489

Lee OO, Lau SC, Qian P-Y (2006) Defense against epibiosis in the sponge Mycale adhaerens: modulating the bacterial community associated with its surface. Aquat Microb Ecol 43(1):55–65

Leong W, Pawlik JR (2010) Evidence of a resource tradeoff between growth and chemical defenses among Caribbean coral reef sponges. Mar Ecol Prog Ser 406:71–78

Lesser MP, Shumway SE, Cucci T, Smith J (1992) Impact of fouling organisms on mussel rope culture: interspecific competition for food among suspension-feeding invertebrates. J Exp Mar Biol Ecol 165(1): 91–102

Li H-y, Matsunaga S, Fusetani N (1995) Halicylindramides A–C, antifungal and cytotoxic depsipeptides from the marine sponge *Halichondria cylindrata*. J Med Chem 38(2):338–343

Lindquist N (2002) Chemical defense of early life stages of benthic marine invertebrates. J Chem Ecol 28(10): 1987–2000

López-Victoria M, Zea S, Weil E (2006) Competition for space between encrusting excavating Caribbean sponges and other coral reef organisms. Mar Ecol Prog Ser 312:113–121

Ma WS, Mutka T, Vesley B, Amsler MO, McClintock JB, Amsler CD, Perman JA, Singh MP, Maiese WM, Zaworotko MJ (2009) Norselic acids A– E, highly oxidized anti-infective steroids that deter mesograzer predation, from the Antarctic sponge *crella* sp. J Nat Prod 72(10):1842–1846

Maida M, Sammarco PW, Coll JC (1995) Preliminary evidence for directional allelopathic effects of the soft coral *Sinularia flexibilis* (Alcyonacea: Octocorallia) on scleractinian coral recruitment. Bull Mar Sci 56(1):303–311

Maldonado M, Carmona MC, Velásquez Z, Puig A, Cruzado A, López A, Young CM (2005) Siliceous sponges as a silicon sink: an overlooked aspect of benthopelagic coupling in the marine silicon cycle. Limnol Oceanogr 50(3):799–809

Marahier MA, Nakano MM, Zuber P (1993) Regulation of peptide antibiotic production in Bacillus. Mol Microbiol 7(5):631–636

Matsunaga S, Fusetani N, Hashimoto K, Walchli M (1989) Theonellamide F. A novel antifungal bicyclic peptide from a marine sponge *Theonella* sp. J Am Chem Soc 111(7):2582–2588

McClintock JB, Baker BJ (1997) A review of the chemical ecology of Antarctic marine invertebrates. Am Zool 37(4):329–342

McClintock JB, Baker BJ (1998) Chemical ecology in Antarctic seas: chemical interactions can lead to unusual arrangements between species. Am Sci 86 (3):254–263

McClintock JB, Baker BJ (2010) Marine chemical ecology. CRC press, Boca Raton

McClintock JB, Baker BJ, Slattery M, Hamann M, Kopitzke R, Heine J (1994) Chemotactic tube-foot responses of a spongivorous sea star *Perknaster fuscus* to organic extracts from antarctic sponges. J Chem Ecol 20(4):859–870

McClintock JB, Baker BJ, Amsler CD, Barlow TL (2000) Chemotactic tube-foot responses of the spongivorous sea star *Perknaster fuscus* to organic extracts of sponges from McMurdo Sound, Antarctica. Antarct Sci 12(01):41–46

McClintock JB, Amsler CD, Baker BJ, Van Soest RW (2005) Ecology of Antarctic marine sponges: an overview. Integr Comp Biol 45(2):359–368

McMurray S, Blum J, Pawlik J (2008) Redwood of the reef: growth and age of the giant barrel sponge *Xestospongia muta* in the Florida keys. Mar Biol 155(2):159–171

Moon B, Park YC, McClintock JB, Baker BJ (2000) Structure and bioactivity of erebusinone, a pigment from the Antarctic sponge *Isodictya erinacea*. Tetrahedron 56(46):9057–9062

Mukherjee J, Webster N, Llewellyn LE (2009) Purification and characterization of a collagenolytic enzyme from a pathogen of the Great Barrier Reef sponge, *Rhopaloeides odorabile*. PLoS One 4(9):e7177

Müller W, Klemt M, Thakur N, Schröder H, Aiello A, D'esposito M, Menna M, Fattorusso E (2004) Molecular/chemical ecology in sponges: evidence for an adaptive antibacterial response in *Suberites domuncula*. Mar Biol 144(1):19–29

Nandakumar K, Tanaka M, Kikuchi T (1993) Interspecific competition among fouling organisms in Tomioka Bay, Japan. Mar Ecol Prog Ser 94:43–43

Negandhi K, Blackwelder PL, Ereskovsky AV, Lopez JV (2010) Florida reef sponges harbor coral disease-associated microbes. Symbiosis 51(1):117–129

Nogata Y, Kitano Y (2006) Isocyano compounds as non-toxic antifoulants. In: Antifouling compounds. Springer, Berlin, pp 87–104

Núñez-Pons L, Avila C (2014) Deterrent activities in the crude lipophilic fractions of Antarctic benthic organisms: chemical defences against keystone predators. Polar Res 33:21624

Ortlepp S, Sjögren M, Dahlström M, Weber H, Ebel R, Edrada R, Thoms C, Schupp P, Bohlin L, Proksch P (2007) Antifouling activity of bromotyrosine-derived sponge metabolites and synthetic analogues. Mar Biotechnol 9(6):776–785

Ortlepp S, Pedpradap S, Dobretsov S, Proksch P (2008) Antifouling activity of sponge-derived polybrominated diphenyl ethers and synthetic analogues. Biofouling 24(3):201–208

Paul VJ (1992) Chemical defenses of benthic marine invertebrates. In: Ecological roles of marine natural products. Comstock Publishing Associates, Ithaca, pp 164–188

Paul VJ, Puglisi MP, Ritson-Williams R (2006) Marine chemical ecology. Nat Prod Rep 23(2): 153–180

Pawlik JR (1992) Chemical ecology of the settlement of benthic marine invertebrates. Oceanogr Mar Biol Annu Rev 30:273–335

Pawlik JR (1993) Marine invertebrate chemical defenses. Chem Rev 93(5):1911–1922

Pawlik J, Chanas B, Toonen R, Fenical W (1995) Defenses of Caribbean sponges against predatory reef fish. I. Chemical deterrency. Mar Ecol Prog Ser Oldendorf 127(1):183–194

Pawlik JR, Kernan MR, Molinski TF, Harper MK, Faulkner DJ (1988) Defensive chemicals of the Spanish dancer nudibranch *Hexabranchus sanguineus* and its egg ribbons: macrolides derived from a sponge diet. J Exp Mar Biol Ecol 119(2):99–109

Pawlik JR, Steindler L, Henkel TP, Beer S, Ilan M (2007) Chemical warfare on coral reefs: sponge metabolites differentially affect coral symbiosis in situ. Limnol Oceanogr 52(2):907–911

Pennings SC, Pablo SR, Paul VJ, Emmett Duffy J (1994) Effects of sponge secondary metabolites in different diets on feeding by three groups of consumers. J Exp Mar Biol Ecol 180(1):137–149

Perry NB, Ettouati L, Litaudon M, Blunt JW, Munro MH, Parkin S, Hope H (1994) Alkaloids from the antarctic sponge *Kirkpatrickia varialosa*.: Part 1: Variolin b, a new antitumour and antiviral compound. Tetrahedron 50 (13):3987–3992

Peters KJ, Amsler CD, McClintock JB, van Soest RW, Baker BJ (2009) Palatability and chemical defenses of sponges from the western Antarctic Peninsula. Mar Ecol Prog Ser 385:77–85

Pile A, Patterson M, Witman J (1997) Finding Reiswig's missing carbon: quantification of sponge feeding using dual-beam flow cytometry. In: Proceedings of the 8th international coral reef symposium, 1997, pp 1403–1410

Porter JW, Targett NM (1988) Allelochemical interactions between sponges and corals. Biol Bull 175(2):230–239

Qiu Y, Deng ZW, Xu M, Li Q, Lin WH (2008) New A-nor steroids and their antifouling activity from the Chinese marine sponge *Acanthella cavernosa*. Steroids 73(14): 1500–1504

Reiswig HM (1971) Particle feeding in natural populations of three marine demosponges. Biol Bull 141 (3):568–591

Roper K, Beamish H, Garson M, Skilleter G, Degnan B (2009) Convergent antifouling activities of structurally distinct bioactive compounds synthesized within two sympatric *Haliclona* demosponges. Mar Biotechnol 11(2):188–198

Russell B, Degnan B, Garson M, Skilleter G (2003) Distribution of a nematocyst-bearing sponge in relation to potential coral donors. Coral Reefs 22(1):11–16

Sacristán-Soriano O, Banaigs B, Becerro MA (2012) Temporal trends in the secondary metabolite production of the sponge *Aplysina aerophoba*. Mar Drugs 10 (4):677–693

Schröder H, Ushijima H, Krasko A, Gamulin V, Thakur N, Diehl-Seifert B, Muller I, Muller W (2003) Emergence and disappearance of an immune molecule, an antimicrobial lectin, in basal Metazoa-A tachylectin-related protein in the sponge *Suberites domuncula*. J Biol Chem 278(35):32810–32817

Schupp P, Eder C, Paul V, Proksch P (1999) Distribution of secondary metabolites in the sponge *Oceanapia* sp. and its ecological implications. Mar Biol 135(4):573–580

Shin J, Seo Y, Rho J-R, Baek E, Kwon B-M, Jeong T-S, Bok S-H (1995) Suberitenones A and B: sesterterpenoids of an unprecedented skeletal class from the Antarctic sponge *Suberites* sp. J Org Chem 60 (23):7582–7588

Sjögren M, Göransson U, Johnson A-L, Dahlström M, Andersson R, Bergman J, Jonsson PR, Bohlin L (2004) Antifouling activity of brominated cyclopeptides from the marine sponge *Geodia barretti*. J Nat Prod 67(3):368–372

Skindersoe ME, Alhede M, Phipps R, Yang L, Jensen PO, Rasmussen TB, Bjarnsholt T, Tolker-Nielsen T, Høiby N, Givskov M (2008) Effects of antibiotics on quorum sensing in *Pseudomonas aeruginosa*. Antimicrob Agents Chemother 52(10):3648–3663

Stabili L, Licciano M, Giangrande A, Longo C, Mercurio M, Marzano CN, Corriero G (2006) Filtering activity of *Spongia officinalis* var. *adriatica* (Schmidt) (Porifera, Demospongiae) on bacterioplankton: Implications for bioremediation of polluted seawater. Water Res 40(16):3083–3090

Sullivan B, Faulkner DJ, Webb L (1983) Siphonodictidine, a metabolite of the burrowing sponge *Siphonodictyon* sp. that inhibits coral growth. Science 221(4616): 1175–1176

Swearingen D III, Pawlik J (1998) Variability in the chemical defense of the sponge *Chondrilla nucula* against predatory reef fishes. Mar Biol 131(4): 619–627

Taylor MW, Schupp PJ, Dahllöf I, Kjelleberg S, Steinberg PD (2004) Host specificity in marine sponge-associated bacteria, and potential implications for marine microbial diversity. Environ Microbiol 6(2): 121–130

Thacker RW, Becerro MA, Lumbang WA, Paul VJ (1998) Allelopathic interactions between sponges on a tropical reef. Ecology 79(5):1740–1750

Thakur NL, Anil A (2000) Antibacterial activity of the sponge *Ircinia ramosa*: importance of its surface-associated bacteria. J Chem Ecol 26(1):57–71

Thakur NL, Hentschel U, Krasko A, Pabel CT, Anil AC, Müller WE (2003) Antibacterial activity of the sponge *Suberites domuncula* and its primmorphs: potential basis for epibacterial chemical defense. Aquat Microb Ecol 31(1):77–83

Thakur NL, Anil AC, Müller WE (2004) Culturable epibacteria of the marine sponge *Ircinia fusca*: temporal variations and their possible role in the epibacterial defense of the host. Aquat Microb Ecol 37(3):295–304

Thakur N, Perović-Ottstadt S, Batel R, Korzhev M, Diehl-Seifert B, Müller I, Müller W (2005) Innate immune defense of the sponge *Suberites domuncula* against gram-positive bacteria: induction of lysozyme and AdaPTin. Mar Biol 146(2):271–282

Thoms C, Schupp PJ (2007) Chemical defense strategies in sponges: a review. Porifera Res Biodivers Innov Sustain 28:627–637

Thoms C, Horn M, Wagner M, Hentschel U, Proksch P (2003) Monitoring microbial diversity and natural product profiles of the sponge *Aplysina cavernicola* following transplantation. Mar Biol 142(4):685–692

Trimurtulu G, Faulkner DJ, Perry NB, Ettouati L, Litaudon M, Blunt JW, Munro MH, Jameson GB (1994) Alkaloids from the antarctic sponge *Kirkpatrickia varialosa*. Part 2: variolin A and *N* (3′)-methyl tetrahydrovariolin B. Tetrahedron 50(13): 3993–4000

Tsoukatou M, Maréchal JP, Hellio C, Novaković I, Tufegdzic S, Sladić D, Gašić MJ, Clare AS, Vagias C, Roussis V (2007) Evaluation of the activity of the sponge metabolites avarol and avarone and their synthetic derivatives against fouling micro-and macroorganisms. Molecules 12(5):1022–1034

Turon X, Becerro MA, Uriz MJ (1996) Seasonal patterns of toxicity in benthic invertebrates: the encrusting sponge *Crambe crambe* (Poecilosclerida). Oikos 75: 33–40

Vacelet J, Donadey C (1977) Electron microscope study of the association between some sponges and bacteria. J Exp Mar Biol Ecol 30(3):301–314

Waddell B, Pawlik JR (2000) Defenses of Caribbean sponges against invertebrate predators. 11. Assays with sea stars. Mar Ecol Prog Ser 195:133–144

Wahl M, Mark O (1999) The predominantly facultative nature of epibiosis: experimental and observational evidence. Mar Ecol Prog Ser 187(1):59–66

Wang X, Brandt D, Thakur NL, Wiens M, Batel R, Schröder HC, Müller WE (2013) Molecular cross-talk between sponge host and associated microbes. Phytochem Rev 12(3):369–390

Webster NS (2007) Sponge disease: a global threat? Environ Microbiol 9(6):1363–1375

Webster NS, Blackall LL (2009) What do we really know about sponge-microbial symbioses. ISME J 3(1):1–3

Webster NS, Negri AP, Webb RI, Hill RT (2002) A spongin-boring α-proteobacterium is the etiological agent of disease in the Great Barrier Reef sponge *Rhopaloeides odorabile*. Mar Ecol Prog Ser 232:305–309

Webster NS, Xavier JR, Freckelton M, Motti CA, Cobb R (2008) Shifts in microbial and chemical patterns within the marine sponge *Aplysina aerophoba* during a disease outbreak. Environ Microbiol 10(12):3366–3376

Wilson DM, Puyana M, Fenical W, Pawlik JR (1999) Chemical defense of the Caribbean reef sponge *Axinella corrugata* against predatory fishes. J Chem Ecol 25(12):2811–2823

Wulff L (1995) Sponge-feeding by the Caribbean starfish *Oreaster reticulatus*. Mar Biol 123(2):313–325

Yang A, Baker BJ, Grimwade J, Leonard A, McClintock JB (1995) Discorhabdin alkaloids from the Antarctic sponge *Latrunculia apicalis*. J Nat Prod 58(10): 1596–1599

Bioeroding Sponges in Aquaculture Systems

4

P. Sunil Kumar

Abstract

A total of nine species of boring sponges were found to infest the molluscs in marine culture systems, and various species as per their numerical abundance in total shells examined are *Cliona vastifica*, *Cliona lobata*, *Cliona margaritifera*, *Cliona celata*, *Cliona carpenteri*, *Thoosa hancocki*, *Thoosa armata*, *Aka minuta* and *Alectona millari*. *Cliona vastifica*, a euryhaline species, is distributed in the estuaries of the west coast of India. Of the above nine species, *C. margaritifera* and *C. lobata*, two dreadful pests, elsewhere, have migrated to Vizhinjam area of Thiruvananthapuram, Kerala (culture rafts) around 1980. Since then, these species have migrated to wild molluscan stocks along the southwest coast and thence to Gulf of Mannar, causing considerable hike in the infestation pattern, species composition, etc. The contact of sponge canal system with water is ensured by portions of the sponge, called papillae, protruding from its substratum surface. This growing form known as α-stage is, in some species, substituted by a complete removal of the substratum such that the sponge becomes a free-living organism (γ-stage). In other cases, the epilithic portion continues to develop until the papillae are connected by a more or less thick crust of sponge (β-stage). In this chapter, distribution and migration patterns of boring sponges has been presented by specifying the eroding indication by these organisms.

Keywords

Bioeroding • Marine sponges • *Cliona* spp. • Culture rafts

4.1 Introduction

Molluscan aquaculture is gaining momentum in coastal areas in India. About 12 major taxa of marine algae and invertebrates have been included under the category of boring organisms.

P.S. Kumar
Kerala University of Fisheries and ocean Studies, Panangad, Kochi, Kerala 682506, India
e-mail: sukkuedavetty2012@gmail.com; sukkukumarp06@rediffmail.com

R. Pallela, H. Ehrlich (eds.), *Marine Sponges: Chemicobiological and Biomedical Applications*, DOI 10.1007/978-81-322-2794-6_4

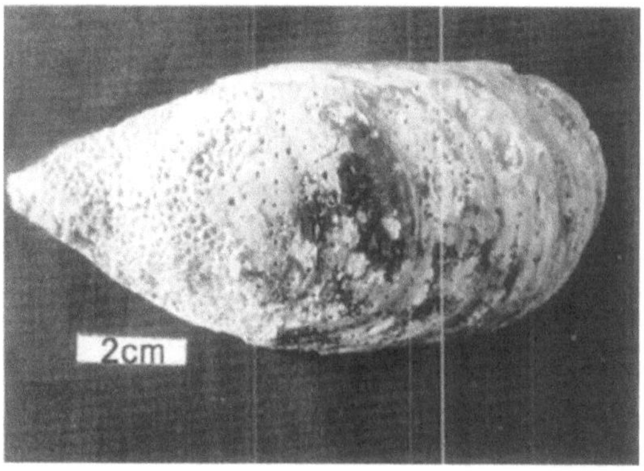

Fig. 4.1 Perna indica shell (*left valve*) infested by boring sponge *Cliona lobata* (scale – 2 cm) (Adopted from Sunil Kumar 2002)

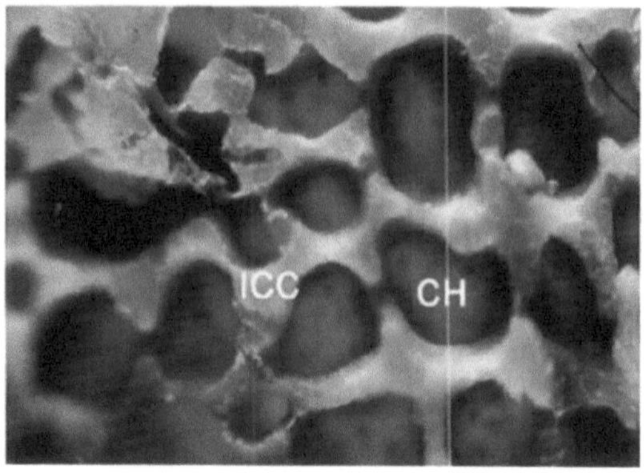

Fig. 4.2 HS through ULC viewed downwards in the direction of *arrow* 3 (in Fig. 4.2) showing magnified view of chambers and inter-chamberal canals (CH and ICC) (Adopted from Sunil Kumar 2002)

Among these clionid sponges, polychaetes and boring molluscs are the most common among boring organisms. Knowledge of the boring and fouling organisms is inevitable from molluscan culture point of view. The borers cause external damage as well as affect growth and nutritional quality of cultured molluscs. Sponges infesting cultured molluscan species are identified by the method described by Old (1941).

Boring sponges were discovered in French oyster beds in the 1800s. They were later reported from edible oyster culture systems in the Indian seas (Thomas 1979, 1983a) and mussel culture farms in Japan and India (Alagarswami and Chellam 1976; Thomas 1983a). Thomas (1972, 1979) reported 20 species of boring sponges from the Gulf of Mannar and Palk Bay (Thomas 1972, 1979).

The pearl oyster *Pinctada fucata* culture rafts at Tuticorin were infested with two conventional species, *Cliona vastifica* (Fig. 4.1) and *Cliona celata* (Fig. 4.2). However, later by 1982, *Cliona celata* was found to be the dominant species. In Vizhinjam area, two conventional species *Cliona vastifica* and *C. celata* were identified to be the dominant species. Moreover, *Cliona margaritifera* and *Cliona lobata* first appeared in raft-cultured pearl oysters in 1980, where severe competition for space occurs between these new invaders *Cliona margaritifera* and *Cliona lobata,* and the already existing conventional species *Cliona celata, C. vastifica* and *C. carpenteri*.

Boring sponges are ancient biological players in various geologic phenomena such as the destruction of coastal limestone which leaves a signature that is an important tool for pale environmental reconstruction. These signatures can be used by measuring the size of bore holes. The external surface of shell is bored by the chipping activity of boring sponges (Fig. 4.1). The chamber canal pattern formed inside the calcareous shell is shown in Fig. 4.2.

4.2 Distribution of Boring Sponges

Many species of sponges are known to bore into submerged calcareous objects like coral rocks, molluscan shells, calcareous algae, etc. A detailed survey made by Thomas (1972, 1975) revealed the presence of 32 species of boring sponges in Indian waters, and it was concluded that this is an area which harbours the maximum number of boring sponges in the world. Besides, one species (*Cliona vastifica* Hancock) which is rather common in the marine environment is unique in its distribution since it has succeeded in colonising the estuarine areas, posing a serious threat to the gregarious molluscs found in the estuaries (Thomas 1972, 1975). Thomas (1983a) described 481 species of marine sponges availble in the Indian marine arena covering wide zoogeographical areas such as the Atlantic Ocean, Mediterranean Sea, Red Sea, Australian region, Pacific Ocean, the Antarctic and Arctic Oceans.

Sponges constitute a major group among 12 different taxa of marine animals and plants which can cause considerable damage to the calcium carbonate-secreting animals such as molluscs, corals, barnacles, etc. The biological, chemical and geological changes that these organisms would bring about in the marine and estuarine environments are by no way insignificant as they cause bioerosion and influence calcium balance in the sea and control the structure of calcium carbonate-secreting animals.

A systematic account on the coral boring sponges infecting the fringing reefs of the Gulf of Mannar and Palk Bay revealed the presence of 20 species belonging to 3 orders, 4 families and 9 genera (Thomas 1972). Of the various genera, species belonging to the genus *Cliona* were found widespread in the reef system of the area. This was followed by a detailed study on the boring sponges infecting the economically important molluscs of the Indian seas, and 32 species belonging to 3 orders, 4 families and 13 genera were recorded (Thomas 1979). This number is very high (32) when compared to any other part of world oceans, and this indicates that the Indian seas are worst affected by boring sponges. Abundances and biodiversities of these bioeroding sponge species are influenced by various factors like water depth, sediment quality, and by the availability of suitable substrate for attachment (Schönberg 2015). This could be the significant reason for spreading these species from place to place through a diverse mechanisms.

4.3 Species Composition and Migration Pattern

Of the six species of boring sponges, two species, viz., *Cliona margaritifera* Dendy and *C. lobata* Hancock, require special mention in the context as the former, after its first appearance in Sri Lankan pearl banks in an epidemic level in 1902, disappeared totally from the beds. The reappearance of this highly dangerous species on raft-cultured pearl oysters at Vizhinjam around 1980, hence, is very interesting as it forms a major invasion on pearl oysters of Indian seas after a lapse of about 80 years. Since 1980, the incidence of *C. margaritifera* in various natural molluscan beds along the southwest coast of India has generally been on the increase and by 1982 a sizeable fraction of the boring sponge population was constituted by this species. The other species, *C. lobata*, which is a widespread oyster pest in the Atlantic Ocean, was first recorded from the Gulf of Mannar in 1937 by Burton. But all subsequent surveys failed to record the same from the Indian seas. As in the case of *C. margaritifera*, the impact of this species was also felt among all gregarious molluscs of the southwest coast of India (Thomas 1983a). The incidence of boring sponges (infection/100 shells) was very high (up to 75 %) on raft-cultured pearl oysters both at Tuticorin and at Vizhinjam against a meagre rate of 3–7 % noted in natural beds (Appukuttan 1987). But the incidence was found high during the subsequent season, i.e. 47 % in 1980 and 60 % in 1981 (Thomas et al.1993) among raft culture pearl oysters. When the above two new invaders, *Cliona lobata* and *C. margaritifera*, spread to natural beds after 1980, there was an abrupt increase in the rate of incidence in various natural beds initially, but the percentage came down gradually to an equilibrium level within a few years due to the slackening in the activity of less competent conventional boring species of various sea beds.

4.4 Conclusion

The higher incidence of boring species noted on culture rafts for a prolonged period, and on the contrary, it is an indication that the ecological equilibrium in nature is no longer in operation in the man-made system—the culture raft. Therefore, any management system which gives more importance to ecological aspects would help in cutting down the higher incidence of boring sponges that can be seen on artificial systems to a lower level, as noticed in the natural beds (Thomas 1983b).

References

Alagarswami K, Chellam A (1976) On fouling and boring organisms and mortality of pearl oysters in the farm at Veppalodai, Gulf of Mannar. Indian J Fish 23 (1&2):10–22

Appukuttan K (1987) Pearl oyster culture in Vizhinjam Bay. CMFRI Bull Pearl Cult 39:54–61

Old MC (1941) The taxonomy and distribution of the boring sponges (Clionidae) along the Atlantic coast of North America, vol 44. Chesapeake Biological Laboratory, Solomons Island, pp 1–30

Schönberg CH (2015) Monitoring bioeroding sponges: using rubble, quadrat, or intercept surveys? Biol Bull 228(2):137–155

Sunil Kumar P (2002). Bioeroding sponge infestation on the mussel Perna India Kuriakose & Nair 1976, from the southwest coast of India. PhD thesis, CMFRI library eprints, pp 268

Thomas PA (1972) Boring sponges of the reefs of Gulf of Mannar and Palk Bay. Symp Corals Coral Reefs Mar Biol Ass India: 333–362

Thomas PA (1975) Boring sponges of Zuari and Mandovi estuaries. Bull Dep Mar Sci CUSAT 7(1):117–126

Thomas PA (1979) Boring sponges destructive to economically important molluscan beds and coral reefs in Indian seas. Indian J Fish 26(1&2):163–200

Thomas PA (1983a) Distribution and affinities of the sponge fauna of the Indian region. J Mar Biol Assoc India 25(1&2):7–16

Thomas PA (1983b) Some pathological aspects akin to sponge boring in molluscan shells. Proc Symp Coast Aquac 2:671–676

Thomas PA, Ramadoss K, Vincent S (1993) Invasion of Cliona margaritifera Dendy and C. lobata Hancock on the molluscan beds along the Indian coast. J Mar Biol Assoc India 35(1&2):145–156

Marine Sponge-Associated Actinobacteria and Their Biological Properties 5

Panchanathan Manivasagan and Se-Kwon Kim

Abstract

Marine sponge-associated actinobacteria have become an important resource of bioactive compounds. Recent findings from culture-dependent and culture-independent methods have demonstrated that indigenous marine actinobacteria exist in the oceans. There is tremendous biodiversity and novelty among the marine actinobacteria present in marine sponges. Progress has been made to isolate new actinobacteria from samples collected at different marine sponges and habitats. These marine actinobacteria produce different types of novel secondary metabolites. Many of these metabolites possess biological activities and have the potential to be developed as therapeutic agents. Marine actinobacteria are a prolific but underexploited resource for the discovery of new secondary metabolites.

Keywords

Marine actinobacteria • Bioactive natural products • Marine sponges • Biological activities • Secondary metabolites

5.1 Introduction

The marine environment, particularly with sponges, is a rich resource of novel bioactive metabolites—287 novel metabolites were isolated from marine sponges in 2008 (Blunt et al. 2010). The availability of biomass is a limiting factor for isolating marine natural products. The widespread isolation of typical microbial metabolites from sponges leads to the hypothesis that these metabolites are in fact the products of microbial metabolism (Fortman and Sherman 2005). The isolation of secondary metabolite-producing bacteria from sponges and of microbial secondary metabolism gene clusters from the metagenome of sponges has led to the general understanding that these metabolites are,

P. Manivasagan • S-K. Kim (✉)
Specialized Graduate School Science & Technology Convergence, Department of Marine-Bio, Convergence Science and Marine Bioprocess Research Center, Pukyong National University, Busan 608-739, South Korea
e-mail: manimaribtech@gmail.com; sknkim@pknu.ac.kr

R. Pallela, H. Ehrlich (eds.), *Marine Sponges: Chemicobiological and Biomedical Applications*,
DOI 10.1007/978-81-322-2794-6_5

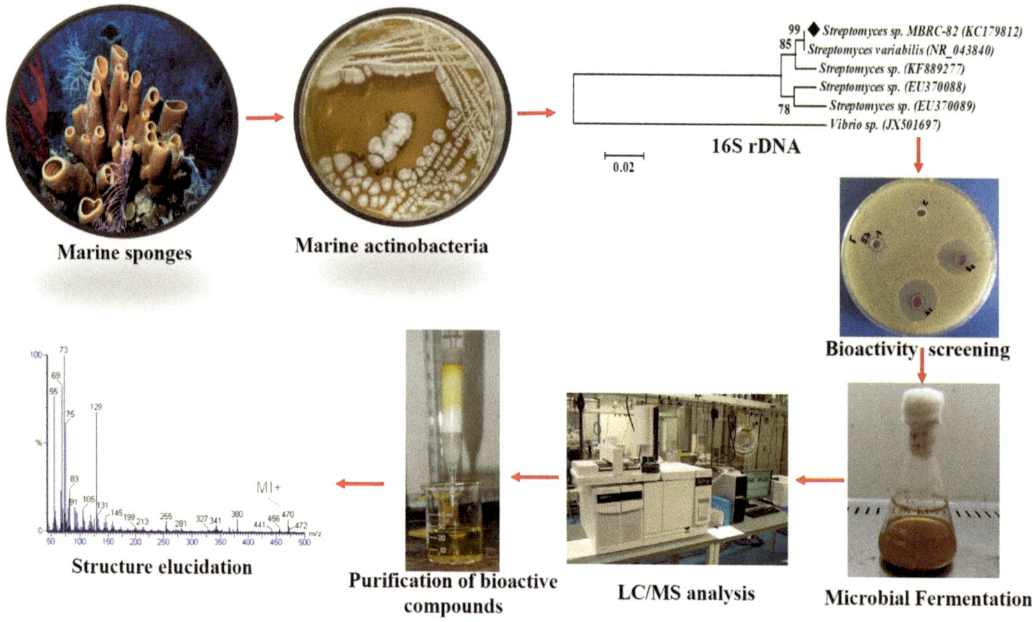

Fig. 5.1 Flow chart depicting the main techniques of isolation, culture, and biological properties of marine sponge-associated actinobacteria

in many cases, the products of microbial symbionts and are not derived from the microbial diet of sponges (Kennedy et al. 2009). Thus, marine organism-associated microbes have been attracting increasing interest as potential sources of marine natural products in order to solve the supply shortage. A number of reports have been published on the isolation of actinobacteria from marine organisms (Li 2009). Screening bioactive substances from these marine-derived actinobacteria has yielded several novel bioactive metabolites (Lin et al. 2010; Oh et al. 2011; Piel et al. 2005).

Sponge-associated actinobacteria are of particular interest in producing antibiotics and other therapeutically significant compounds (Takahashi and Omura 2003). Several antibiotics have been isolated from marine actinobacteria (Woo et al. 2002; Maskey et al. 2003, 2004; Sujatha et al. 2005; Li et al. 2005; Lombó et al. 2006). Figure 5.1 depicts a brief process of extracting, purifying, and characterizing the bioactive compounds from marine sponge-associated actinobacteria.

5.2 Marine Sponges

Marine sponges (Phylum Porifera) are multicellular invertebrate sessile filter feeders that provide unique and favorable environmental conditions for microbial colonization and often harbor abundant and diverse microbes. Microbial communities associated with marine sponges are very complex, contributing up to 40 % of the sponge biomass (Friedrich et al. 2001; Taylor et al. 2007). Marine sponge-associated bacterial communities include the following taxa: *Acidobacteria*, *Actinobacteria*, *Bacteroidetes*, *Chlamydiae*, *Chloroflexi*, *Cyanobacteria*, *Deinococcus-Thermus*, *Firmicutes*, *Gemmatimonadetes*, *Nitrospira*, *Planctomycetes*, *Proteobacteria*, *Spirochaetes*, and *Verrucomicrobia* (Hardoim et al. 2009; Kamke et al. 2010). Among the bacterial associates, members of *Actinobacteria* are often sponge-specific (Webster et al. 2001; Selvin et al. 2009) and have been identified as dominant producers of biologically active compounds (Lang et al. 2004; Thomas et al. 2010). There is evidence that the presence of biosynthesis genes encoding polyketide synthases (PKSs) and nonribosomal peptide synthetases (NRPSs) in marine sponge-associated actinomycetes is a useful indicator for the selection of strains to isolate new natural products (Schneemann et al. 2010).

5.3 Marine Sponge-Associated Actinobacteria

The class actinobacteria consists of a diverse range of Gram-positive bacteria with high G + C DNA content. Actinobacteria were originally isolated from soils and are of great interest as resources of lead bioactive compounds for the biomedical industry, since over two-thirds of naturally occurring antibiotics are produced by the order *Actinomycetales* in the *Actinobacteria* (Okami and Hotta 1988). Natural products are the major source of novel drugs. This is demonstrated by the fact that approximately 60 % of those compounds commercially available or in the late stages of clinical trials for the treatment of infectious diseases or cancer are being derived from natural products (Cragg et al. 1997). Because of the good track record of actinomycetes in this regard, much effort is focused on the isolation of new actinomycetes for drug screening programs.

Members of the phylum *Actinobacteria* and specifically the order *Actinomycetales* have been identified as abundant members of sponge-associated microbial communities (Hentschel et al. 2002; Kim et al. 2005). Their existence in the marine environment has been further shown in marine sediments as well as in the deepest ocean trenches (Bredholdt et al. 2007; Fenical and Jensen 2006; Mincer et al. 2002; Pathom-Aree et al. 2006; Maldonado et al. 2005). Actinomycetes are of considerable interest due to their ability to produce novel chemical entities with diverse pharmacological activities. Marine actinobacteria in particular have yielded numerous new secondary metabolites (Lam 2006). Novel actinomycete taxa of marine origin have also been recovered as best exemplified by *Salinispora*, the first marine obligate actinomycete isolated from the marine sediments (Maldonado et al. 2005) as well as from a sponge (Kim et al. 2006).

5.4 Distributions of Marine Actinobacteria

New actinomycete groups have been found in the Great Barrier Reef sponges *Rhopaloeides odorabile*, *Pseudoceratina clavata*, and *Candidaspongia flabellate* and the Mediterranean sponges *Aplysina aerophoba* and *Theonella swinhoei* (Kim et al. 2005; Hentschel et al. 2002; Webster et al. 2001). Unusual actinomycetes, belonging to *Micrococceae*, *Dermatophilaceae*, and *Gordoniaceae*, have been isolated from sponges (Hill 2004). Jensen et al. reported the isolation of actinomycetes from algal and sponge samples and observed different rates of recovery of actinomycetes, which might have been caused by different methodologies used to process the samples (Jensen et al. 2005). Novel bioactive metabolites have been obtained from actinomycetes isolated from sponges (Hill 2004).

Actinobacteria isolated from the samples collected at the marine environments mentioned above, such as the deep sea floor, marine invertebrates, and marine snow, all represent unique ecosystems that cannot be found anywhere else in the world. The isolation of marine actinobacteria that evolved from, and adapted to, these unique ecosystems is a prolific source for the discovery of new secondary metabolites. For example, marine invertebrates such as corals and sponges are hosts to a multitude of microorganisms. These animals can be considered as miniature ecosystems in which different species of microorganisms compete with each other for resources. Therefore, secondary metabolites originating from symbiotic or commensal microorganisms of marine invertebrates might not necessarily have a function in the chemical ecology of the host animal itself, but they could instead be chemical weapons or signaling agents employed in the fight for growth and survival that is going on among these microorganisms (Lam 2006).

5.5 Diversity of Culturable Actinobacteria

Actinobacteria are widely distributed in marine sponges. At the time of writing, over 30 sponge genera had been reported to be hosts of actinomycetes, with ten genera having each been collected in different sea areas (Kamke et al. 2010; Radwan et al. 2010; Zhu et al. 2008; Selvin et al. 2009). Among the nearly 10,000 sponge-associated microbial sequences submitted to public databases, about one-sixth belong to actinobacteria (Webster and Taylor 2012), indicating that this is an important group among sponge-associated microorganisms. Actinomycetes abundance in marine sponges is variable but can make up over 20 % of the total microorganisms in some marine sponges (Montalvo et al. 2005). The study of marine-sponge-associated actinobacterial diversity involves both culture-dependent and culture-independent methods. In the past decade, a large number of marine sponge-associated actinomycetes have been identified using culture methods, spanning 26 genera (Abdelmohsen et al. 2010; Jiang et al. 2008). The use of culture-independent methods has enabled the detection of an additional five genera of actinomycetes in marine sponges, as well as many unculturable novel actinobacterial taxa (Gerçe et al. 2009; Xin et al. 2008). Although both of the abovementioned methods have defects and bias, the culture-dependent method is still popular even in the "omics" age (Giovannoni and Stingl 2007). This is partly because the isolates yielded from this method provide very useful phenotypic and genotypic information (Galkiewicz et al. 2011), such as physiological traits and biosynthetic potential, for further ecological investigation and bioprospecting.

Xi et al. reported the diversity and secondary metabolite potential of culturable actinomycetes associated with eight different marine sponges collected from the South China Sea and the Yellow sea. A total of 327 strains were isolated, and 108 representative isolates were selected for phylogenetic analysis. Ten families and 13 genera of *Actinomycetales* were detected, among which five genera represent first records isolated from marine sponges. Oligotrophic medium M5 (water agar) proved to be efficient for selective isolation, and *Micromonospora–Streptomyces* was proposed as the major distribution group of sponge-associated actinomycetes from the China Seas. Ten isolates are likely to represent novel species. Sponge *Hymeniacidon perleve* was found to contain the highest genus diversity (seven genera) of actinomycetes. Housekeeping gene phylogenetic analyses of the isolates indicated one ubiquitous *Micromonospora* species, one unique *Streptomyces* species, and one unique *Verrucosispora* phylogroup. Of the isolates, 27.5 % displayed antimicrobial activity and 91 % contained polyketide synthase and/or nonribosomal peptide synthetase genes, indicating that these isolates had a high potential to produce secondary metabolites. The isolates from sponge *Axinella* sp. contained the highest presence of both antimicrobial activity and NRPS genes, while those from isolation medium DNBA showed the highest presence of antimicrobial activity and PKS I genes (Xi et al. 2012). Jiang et al. reported the diversity of actinobacteria isolated from the marine sponge *Iotrochota* sp. collected in the South China Sea belong to three actinobacteria genera, and one isolate may be a new species. *Streptomyces* appears to be the dominant genus among symbionts and adherents present in *Iotrochota* sp. in the South China Sea. Prescreening for PKS and NRPS revealed extensive metabolic potential in this group of diverse actinobacteria (Jiang et al. 2008).

Zhang et al. reported that a total of 106 actinobacteria associated with the marine sponge *Hymeniacidon perleve* collected from the Yellow Sea, China, were isolated using eight different media. The number of species and genera of actinobacteria recovered from the different media varied significantly, underlining the importance of optimizing the isolation conditions. The phylogenetic diversity of the actinobacteria isolates was assessed using 16S rRNA gene amplification – restriction fragment length polymorphism (RFLP) analysis of the 106 strains with different morphologies. The RFLP fingerprinting of selected strains by HhaI digestion of the 16S rRNA genes resulted in eleven different patterns. The HhaI-RFLP analysis gave good resolution for the identification of the actinobacteria isolates at the genus level. A phylogenetic analysis using 16S rRNA gene sequences revealed that the isolates belonged to seven genera of culturable actinobacteria including *Actinoalloteichus*, *Micromonospora*, *Nocardia*, *Nocardiopsis*, *Pseudonocardia*, *Rhodococcus*, and *Streptomyces*. The dominant genus was *Streptomyces*, which represented 74 % of the isolates. Three of the strains identified are candidates for new species (Zhang et al. 2006).

5.6 Bioactive Natural Products

Actinobacterial natural products continue to provide both the raw materials and the design inspiration for the majority of pharmaceutical lead discovery and drug development (Newman and Cragg 2007). However, after decades of fruitful bioprospecting and drug discovery, the traditional sources of natural products (e.g., plants and terrestrial actinomycetes) are realizing the "law of diminished returns" (Fischbach and Walsh 2009). In the face of declining antibiotic and anticancer drug discovery rates, the marine environment has emerged as an important source of bioactive natural products. There are, for example, several exciting marine-derived molecules currently on the pharmaceutical market and dozens more progressing through the development pipeline (Mayer et al. 2010). In many cases the source organisms are as diverse as the molecular structures, yet bacteria living in close association with the larger "host" organism are often found to produce the metabolites of interest (Simmons et al. 2008). Marine sponges (Porifera) in particular harbor extremely rich and diverse populations of microorganisms and have yielded many bioactive natural products (Simmons et al. 2008; Blunt et al. 2010).

5.6.1 Antimicrobial Activity

Marine microorganisms, particularly marine actinobacteria, have attracted considerable attention as one of the most important resources for new biologically active metabolites (Fenical and Jensen 2006). For example, new compounds have been isolated from actinobacteria of sponge origin (Lee et al. 2005). Gandhimathi et al. isolated *Streptomyces* spp. from the Bay of Bengal region of the Indian peninsular coastal by scuba diving at 10–15 m depth. The endosymbiotic marine actinomycetes exhibited potent antimicrobial activity against the growth of human pathogens. Particularly, the strains CPI 3, CPI 9, CPI 12, and CPI 13 showed the highest antimicrobial activity (Gandhimathi et al. 2008). Recently, actinobacteria associated with marine sponges have been reported as richest source of potential antagonists (Faulkner 2001; Piel et al. 2004).

Marine *Streptomyces* sp. DA11 isolated from South China, found to be associated with sponge *Craniella australiensis*, produced the enzyme chitinase and showed antifungal activities against *Aspergillus niger* and *Candida albicans* (Han et al. 2009).

5.6.2 Anti-inflammatory Activity

The major goal of research is to discover novel anti-infective agents such as those against the parasites *Leishmania major* and *Trypanosoma brucei* that cause leishmaniasis and African sleeping sickness, respectively. These parasites currently affect around 12 million people living in tropical and subtropical areas (Natera et al. 2007). The alarming death rate caused by these parasites and the emergence of antibiotic resistance underline the need for new and effective drugs. Many research programs focus on the discovery of anti-infective agents from marine sponges and their associated microorganisms (Pimentel-Elardo et al. 2009). Pimentel-Elardo et al. (2010) isolated *Streptomyces* sp. from Mediterranean sponges, studied their secondary metabolite production, and screened for anti-infective activities. Bioassay-guided isolation and purification yielded three previously known compounds, namely, cyclic depsipeptide valinomycin, indolocarbazole alkaloid staurosporine, and butenolide. These compounds exhibited novel antiparasitic activities specifically against *Leishmania major* (valinomycin IC_{50} <0.11 μM; staurosporine IC_{50} 5.30 μM) and *Trypanosoma brucei brucei* (valinomycin IC_{50} 0.0032 μM; staurosporine IC_{50} 0.022 μM; butenolide IC_{50} 31.77 μM) (Pimentel-Elardo et al. 2010).

Abdelmohsen et al. (2010) isolated 90 actinomycetes from 11 different species of marine sponges. Testing for anti-infective activities was performed against clinically relevant, Gram-positive (*Enterococcus faecalis*, *Staphylococcus aureus*) and Gram-negative (*Escherichia coli*, *Pseudomonas aeruginosa*) bacteria, fungi (*Candida albicans*), and human parasites (*Leishmania major*, *Trypanosoma brucei*). It showed a high diversity of actinomycetes associated with marine sponges as well as highlights their potential to produce anti-infective agents (Abdelmohsen et al. 2010).

5.6.3 Antitumor Activity

Zheng et al. reported the detection of antitumor and antimicrobial activities in marine organism-associated actinomycetes isolated from the Taiwan Strait, China. Antitumor activity was studied by the MTT assay, and DNA target activity was studied by the biochemical induction assay, while antimicrobial activity was determined by observing bacterial and fungal growth inhibition. 20.6 % of marine actinomycete cultures displayed cytotoxic activity on P388 cells at dilutions at and below 1:320 and 18.6 % on KB cells. 2.96 % of marine actinomycete cultures displayed inducing activity. Among all marine actinomycetes isolated, the genus *Micromonospora* has the highest positive rate of inducing activity. However, most antimicrobial activity was found in the genus *Streptomyces*. These results indicate that marine organism-associated actinomycetes could be a promising source for antitumor and antimicrobial bioactive agents (Zheng et al. 2000).

5.6.4 Anticancer Activity

Cancer is one of the most important causes of mortality in the modern world, with more than ten million new cases reported every year (Conde et al. 2011). It is well established that cancer is a multifactorial disease caused by a complex mixture of genetic and environmental factors (Balmain et al. 2003; Hanahan and Weinberg 2000; Ponder 2001), where considerable advances have led to a more comprehensive understanding of cancer at the genetic, molecular, and cellular levels, providing with new targets and strategies for therapy (Praetorius and Mandal 2007). Nevertheless, these advances have yet to be effectively translated into functioning diagnostics and therapy. For example, effectiveness of many anticancer drugs is limited due to their inability to reach the target site in sufficient concentrations and efficiently exert the pharmacological effect without causing irreversible unwanted injury to the healthy tissues and cells (Ferrari 2005; Peer et al. 2007). Bendigoles D–F (**1**) (Fig. 5.2) are bioactive sterols isolated from the new marine sponge-associated actinobacterium, *Actinomadura* sp. SBMs009. Isolation of these compounds was guided by a novel high-content screen for NF-κB and glucocorticoid receptor (GR) activity, and cytotoxicity assays. Interestingly, D displayed cytotoxicity against the L929 (mouse fibroblast) cell line with an IC_{50} approximated to 30 μM and was the most active inhibitor of GR-translocation,

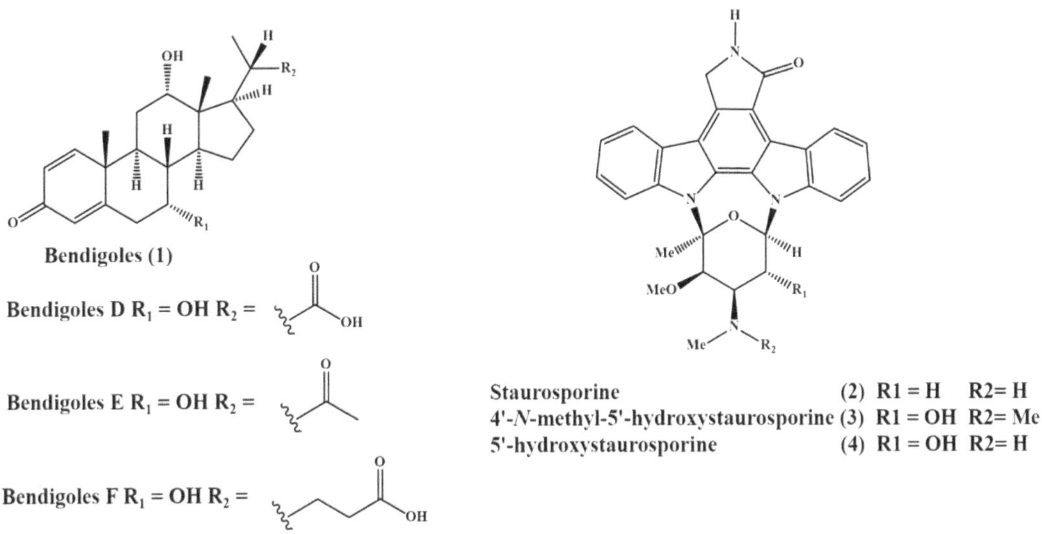

Fig. 5.2 Chemical structure of bendigoles D–F, staurosporine, 4′-*N*-methyl-5′-hydroxystaurosporine, and 5′-hydroxystaurosporine

while D–F was the most effective inhibitor of NF-κB nuclear translocation with an IC_{50} of 71 μM (Simmons et al. 2011).

Two new indolocarbazole alkaloids, 4′-*N*-methyl-5′-hydroxystaurosporine (**3**) and 5′-hydroxy staurosporine (**4**), were isolated together with the known staurosporine (**2**) (Fig. 5.2) from the culture broth of a marine *Micromonospora* sp. L-31-CLCO-002 obtained from a homogenate of the sponge *Clathrina coriacea* collected from the coast of Fuerteventura Island in the Canary Islands archipelago (Hernandez et al. 2000).

Two new anthracyclines, tetracenoquinocin (**5**) and 5-iminoaranciamycin (**6**), together with the known compounds aranciamycin (**7**) (Fig. 5.3) and antibiotic SM 173B were isolated from the culture of *Streptomyces* sp. -Sp080513GE-26 associated with a marine sponge, *Haliclona* sp. These compounds were evaluated for cytotoxicity against two cancer cell lines. Cytotoxic activities of these compounds against human cervical carcinoma HeLa cells and human acute myelogenous leukemia LH-60 cells were examined. Aranciamycin showed cytotoxicity with IC_{50} values of 2.7 and 4.1 μM against HeLa and HL-60 cells, respectively, while tetracenoquinocin exhibited weaker cytotoxicities with IC_{50} values of 120 and 210 μM, respectively, and 5-iminoaranciamycin was inactive to these cancer cells (IC_{50} >200 μM). On comparing the cytotoxic activity of these compounds, it was found that the ketone functional group at C-5 is essential for the cytotoxicity against the cancer cells (Motohashi et al. 2010).

Apart from the anticancer compounds produced by the marine actinomycetes depicted above, there are several additional compounds with antitumor activity like the topoisomerase I inhibitors cyclopropane and 14-methylhexadecanoic fatty acids produced by *Streptomyces* sp. strain KM86-913, isolated from a marine sponge collected from the seashore of Keomun Island, Korea (Lee et al. 1998). In other cases, compounds identified are yet to be uncharacterized as is the case of light-activated cytotoxic compounds produced by different microorganisms, including actinomycetes isolated from marine sponges collected from various places along the coast of Peninsular Malaysia (Kamal et al. 2009). IB-96212, a 26-membered macrolide that contains a spiroketal lactone structure, is produced by the actinomycete, *Micromonospora* sp. L-25-ES25-008, isolated from a sponge, collected from the Indian Ocean near the coast of Mozambique

Fig. 5.3 Chemical structure of tetracenoquinocin, 5-iminoaranciamycin, and aranciamycin

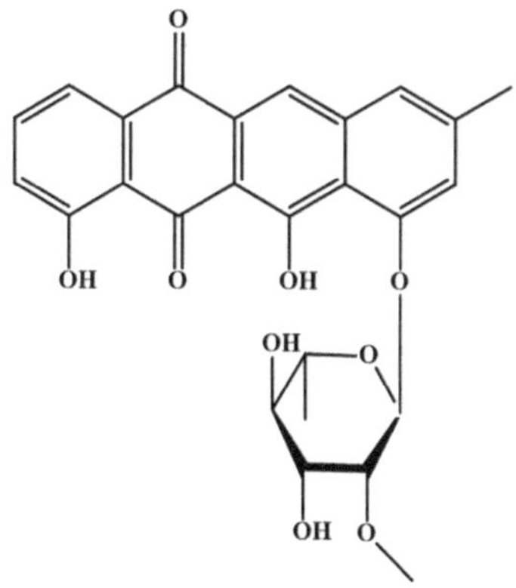

Tetracenoquinocin (5)

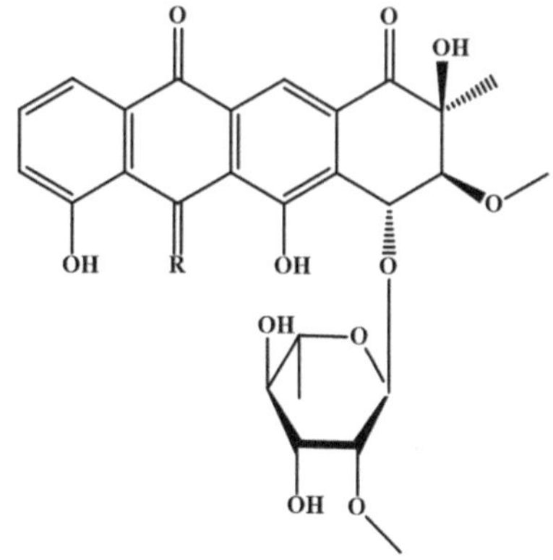

5-Iminoaranciamycin (6) R = NH

Aranciamycin (7) R = O

(Canedo et al. 2000). This compound showed cytotoxic activity against mouse leukemia P388 and human non-small cell lung cancer A-549, colon adenocarcinoma HT-29, and melanoma MEL-28 cell lines. The activity against P388 cell line was four orders of magnitude higher than the activity against A-549, HT-29, and MEL-28 cell lines (Fernandez-Chimeno et al. 2000).

5.7 Conclusion

Sponge-actinobacterial associations are found to be very specific in the production of particular bioactive compounds. However, the mutual mechanism between host and the actinobacterial associate, in compound production, is not well understood. The easiest and best way for commercial production of these compounds is by culturing the host and/or the associated actinobacteria under controlled conditions. But the ability of the symbiont to produce the compound consistently for several generations in culture media has to be tested and standardized. Moreover, there is a need for quantifying the role of sponge ecology in orchestrating the production of specific compounds. Metagenomic approaches are also being increasingly used for targeting putative genes encoding potential metabolites in uncultured microbial biota. These approaches would help in delineating the contribution of either the host or actinobacterial associate or both partners in the production of metabolites. A few compounds have been found to be produced both in terrestrial and marine ecosystems by different groups of host-symbiont association. This suggests the possibility of horizontal gene transfer through evolution. Discovery of potent actinobacterial associates producing therapeutic compounds has opened up a new era in marine pharmacology. Understanding the optimum ecological conditions which drives the sustainable production of bioactive compounds from sponges and their actinobacterial associates would help in formulating various production strategies. Adopting different cultivation strategies and metagenomic approaches would be the need of the hour in discovering new genes, enzymes, and natural products and in enhancing the commercial production of marine drugs.

Acknowledgments This research was supported by a grant from Marine Bioprocess Research Center of the Marine Biotechnology Program funded by the Ministry of Oceans and Fisheries, Republic of Korea.

References

Abdelmohsen UR, Pimentel-Elardo SM, Hanora A, Radwan M, Abou-El-Ela SH, Ahmed S, Hentschel U (2010) Isolation, phylogenetic analysis and anti-infective activity screening of marine sponge-associated actinomycetes. Mar Drugs 8(3):399–412

Balmain A, Gray J, Ponder B (2003) The genetics and genomics of cancer. Nat Genet 33:238–244

Blunt JW, Copp BR, Munro MH, Northcote PT, Prinsep MR (2010) Marine natural products. Nat Prod Rep 27 (2):165–237

Bredholdt H, Galatenko OA, Engelhardt K, Fjærvik E, Terekhova LP, Zotchev SB (2007) Rare actinomycete bacteria from the shallow water sediments of the Trondheim fjord, Norway: isolation, diversity and biological activity. Environ Microbiol 9 (11):2756–2764

Canedo LM, Fernández-Puentes JL, Baz JP (2000) IB-96212, a novel cytotoxic macrolide produced by a marine *Micromonospora*. II. Physico-chemical properties and structure determination. J Antibiot 53 (5):479

Conde J, Doria G, Baptista P (2011) Noble metal nanoparticles applications in cancer. J Drug Deliv 2012:1–12

Cragg GM, Newman DJ, Snader KM (1997) Natural products in drug discovery and development. J Nat Prod 60(1):52–60

Faulkner DJ (2001) Marine natural products. Nat Prod Rep 18(1):1–49

Fenical W, Jensen PR (2006) Developing a new resource for drug discovery: marine actinomycete bacteria. Nat Chem Biol 2(12):666–673

Fernandez-Chimeno RI, Canedo L, Espliego F, Grávalos D, De La Calle F, Fernández-Puentes JL, Romero F (2000) IB-96212, a novel cytotoxic macrolide produced by a marine *Micromonospora*. I. Taxonomy, fermentation, isolation and biological activities. J Antibiot 53(5):474–478

Ferrari M (2005) Cancer nanotechnology: opportunities and challenges. Nat Rev Cancer 5(3):161–171

Fischbach MA, Walsh CT (2009) Antibiotics for emerging pathogens. Science 325(5944):1089–1093

Fortman J, Sherman DH (2005) Utilizing the power of microbial genetics to bridge the gap between the promise and the application of marine natural products. Chembiochem 6(6):960–978

Friedrich AB, Fischer I, Proksch P, Hacker J, Hentschel U (2001) Temporal variation of the microbial community associated with the Mediterranean sponge *Aplysina aerophoba*. FEMS Microbiol Ecol 38(2-3):105–115

Galkiewicz JP, Pratte ZA, Gray MA, Kellogg CA (2011) Characterization of culturable bacteria isolated from the cold-water coral *Lophelia pertusa*. FEMS Microbiol Ecol 77(2):333–346

Gandhimathi R, Arunkumar M, Selvin J, Thangavelu T, Sivaramakrishnan S, Kiran GS, Shanmughapriya S, Natarajaseenivasan K (2008) Antimicrobial potential of sponge associated marine actinomycetes. J Med Mycol 18(1):16–22

Gerçe B, Schwartz T, Voigt M, Rühle S, Kirchen S, Putz A, Proksch P, Obst U, Syldatk C, Hausmann R (2009) Morphological, bacterial, and secondary metabolite changes of *Aplysina aerophoba* upon long-term maintenance under artificial conditions. Microb Ecol 58(4):865–878

Giovannoni S, Stingl U (2007) The importance of culturing bacterioplankton in the 'omics' age. Nat Rev Microbiol 5(10):820–826

Han Y, Yang B, Zhang F, Miao X, Li Z (2009) Characterization of antifungal chitinase from marine *Streptomyces* sp. DA11 associated with south China Sea sponge *Craniella australiensis*. Marine Biotechnol 11(1):132–140

Hanahan D, Weinberg RA (2000) The hallmarks of cancer. Cell 100(1):57–70

Hardoim C, Costa R, Araujo F, Hajdu E, Peixoto R, Lins U, Rosado A, Van Elsas J (2009) Diversity of bacteria in the marine sponge *Aplysina fulva* in Brazilian coastal waters. Appl Environ Microbiol 75 (10):3331–3343

Hentschel U, Hopke J, Horn M, Friedrich AB, Wagner M, Hacker J, Moore BS (2002) Molecular evidence for a uniform microbial community in sponges from different oceans. Appl Environ Microbiol 68(9):4431–4440

Hernandez L, Blanco J, Baz JP, Puentes J, Millán FR, Vázquez FE, Fernández-Chimeno RI, Grávalos DG (2000) 4′-N-methyl-5′-hydroxystaurosporine and 5-′-hydroxystaurosporine, new indolocarbazole alkaloids from a marine *Micromonospora* sp. strain. J Antibiot 53(9):895–902

Hill RT (2004) Microbes from marine sponges: a treasure trove of biodiversity for natural products discovery. In: Bull AT (ed) Microbial diversity and bioprospecting. ASM Press, Washington, DC, pp 177–190

Jensen PR, Gontang E, Mafnas C, Mincer TJ, Fenical W (2005) Culturable marine actinomycete diversity from tropical Pacific Ocean sediments. Environ Microbiol 7 (7):1039–1048

Jiang S, Li X, Zhang L, Sun W, Dai S, Xie L, Liu Y, Lee KJ (2008) Culturable actinobacteria isolated from marine sponge *Iotrochota* sp. Mar Biol 153 (5):945–952

Kamal N, Sabaratnam V, Abdullah N, Ho AS, Teo SH, Lee HB (2009) Light-activated cytotoxic compounds from Malaysian microorganisms for photodynamic therapy of cancer. Antonie Van Leeuwenhoek 95 (2):179–188

Kamke J, Taylor MW, Schmitt S (2010) Activity profiles for marine sponge-associated bacteria obtained by 16S rRNA vs 16S rRNA gene comparisons. ISME J 4 (4):498–508

Kennedy J, Baker P, Piper C, Cotter PD, Walsh M, Mooij MJ, Bourke MB, Rea MC, O'Connor PM, Ross RP (2009) Isolation and analysis of bacteria with antimicrobial activities from the marine sponge *Haliclona simulans* collected from Irish waters. Marine Biotechnol 11(3):384–396

Kim TK, Garson MJ, Fuerst JA (2005) Marine actinomycetes related to the '*Salinospora*' group from the Great Barrier Reef sponge *Pseudoceratina clavata*. Environ Microbiol 7(4):509–518

Kim TK, Hewavitharana AK, Shaw PN, Fuerst JA (2006) Discovery of a new source of rifamycin antibiotics in marine sponge actinobacteria by phylogenetic prediction. Appl Environ Microbiol 72(3):2118–2125

Lam KS (2006) Discovery of novel metabolites from marine actinomycetes. Curr Opin Microbiol 9 (3):245–251

Lang S, Beil W, Tokuda H, Wicke C, Lurtz V (2004) Improved production of bioactive glucosylmannosyl-glycerolipid by sponge-associated *Microbacterium* species. Marine Biotechnol 6(2):152–156

Lee HK, Lee D-S, Lim J, Kim JS, Im KS, Jung JH (1998) Topoisomerase i inhibitors from the *Streptomyces* sp. strain KM86-9B isolated from a marine sponge. Arch Pharm Res 21(6):729–733

Lee H-S, Shin HJ, Jang KH, Kim TS, Oh K-B, Shin J (2005) Cyclic peptides of the nocardamine class from a marine-derived bacterium of the genus *Streptomyces*. J Nat Prod 68(4):623–625

Li Z (2009) Advances in marine microbial symbionts in the China Sea and related pharmaceutical metabolites. Mar Drugs 7(2):113–129

Li F, Maskey RP, Qin S, Sattler I, Fiebig HH, Maier A, Zeeck A, Laatsch H (2005) Chinikomycins A and B: isolation, structure elucidation, and biological activity of novel antibiotics from a marine *Streptomyces* sp. Isolate M045. J Nat Prod 68(3):349–353

Lin Z, Antemano RR, Hughen RW, Tianero MDB, Peraud O, Haygood MG, Concepcion GP, Olivera BM, Light A, Schmidt EW (2010) Pulicatins A– E, neuroactive thiazoline metabolites from cone snail-associated bacteria. J Nat Prod 73(11):1922–1926

Lombó F, Velasco A, Castro A, De la Calle F, Braña AF, Sánchez-Puelles JM, Méndez C, Salas JA (2006) Deciphering the biosynthesis pathway of the antitumor thiocoraline from a marine actinomycete and its expression in two *Streptomyces* species. Chembiochem 7(2):366–376

Maldonado LA, Fenical W, Jensen PR, Kauffman CA, Mincer TJ, Ward AC, Bull AT, Goodfellow M (2005) *Salinispora arenicola* gen. nov., sp. nov. and *Salinispora tropica* sp. nov., obligate marine actinomycetes belonging to the family *Micromonosporaceae*. Int J Syst Evol Microbiol 55(5):1759–1766

Maskey RP, Helmke E, Laatsch H (2003) Himalomycin A and B: isolation and structure elucidation of new fridamycin type antibiotics from a marine *Streptomyces* isolate. J Antibiot 56(11):942–949

Maskey RP, Helmke E, Kayser O, Fiebig HH, Maier A, Busche A, Laatsch H (2004) Anti-cancer and antibacterial trioxacarcins with high anti-malaria activity from a marine Streptomycete and their absolute stereochemistry. J Antibiot 57(12):771–779

Mayer A, Glaser KB, Cuevas C, Jacobs RS, Kem W, Little RD, McIntosh JM, Newman DJ, Potts BC, Shuster DE (2010) The odyssey of marine pharmaceuticals: a current pipeline perspective. Trends Pharmacol Sci 31(6):255–265

Mincer TJ, Jensen PR, Kauffman CA, Fenical W (2002) Widespread and persistent populations of a major new marine actinomycete taxon in ocean sediments. Appl Environ Microbiol 68(10):5005–5011

Montalvo NF, Mohamed NM, Enticknap JJ, Hill RT (2005) Novel actinobacteria from marine sponges. Antonie Van Leeuwenhoek 87(1):29–36

Motohashi K, Takagi M, Shin-ya K (2010) Tetracenoquinocin and 5-iminoaranciamycin from a sponge-derived *Streptomyces* sp. Sp080513GE-26. J Nat Prod 73(4):755–758

Natera S, Machuca C, Padrón-Nieves M, Romero A, Díaz E, Ponte-Sucre A (2007) *Leishmania* spp.: proficiency of drug-resistant parasites. Int J Antimicrob Agents 29(6):637–642

Newman DJ, Cragg GM (2007) Natural products as sources of new drugs over the last 25 years. J Nat Prod 70(3):461–477

Oh D-C, Poulsen M, Currie CR, Clardy J (2011) Sceliphrolactam, a polyene macrocyclic lactam from a wasp-associated *Streptomyces* sp. Org Lett 13 (4):752–755

Okami Y, Hotta K (1988) Search and discovery of new antibiotics. In: Good fellow M, Williams ST, Mordarski M (eds) Actinomycetes in biotechnology. Academic, San Diego, pp 33–67

Pathom-Aree W, Stach JE, Ward AC, Horikoshi K, Bull AT, Goodfellow M (2006) Diversity of actinomycetes isolated from challenger deep sediment (10,898 m) from the Mariana Trench. Extremophiles 10 (3):181–189

Peer D, Karp JM, Hong S, Farokhzad OC, Margalit R, Langer R (2007) Nanocarriers as an emerging platform for cancer therapy. Nat Nanotechnol 2 (12):751–760

Piel J, Hui D, Wen G, Butzke D, Platzer M, Fusetani N, Matsunaga S (2004) Antitumor polyketide biosynthesis by an uncultivated bacterial symbiont of the marine sponge *Theonella swinhoei*. Proc Natl Acad Sci U S A 101(46):16222–16227

Piel J, Butzke D, Fusetani N, Hui D, Platzer M, Wen G, Matsunaga S (2005) Exploring the chemistry of uncultivated bacterial symbionts: antitumor polyketides of the Pederin family. J Nat Prod 68(3):472–479

Pimentel-Elardo SM, Scheuermayer M, Kozytska S, Hentschel U (2009) *Streptomyces axinellae* sp. nov., isolated from the Mediterranean sponge *Axinella polypoides* (Porifera). Int J Syst Evol Microbiol 59 (6):1433–1437

Pimentel-Elardo SM, Kozytska S, Bugni TS, Ireland CM, Moll H, Hentschel U (2010) Anti-parasitic compounds from *Streptomyces* sp. strains isolated from Mediterranean sponges. Mar Drugs 8(2):373–380

Ponder BA (2001) Cancer genetics. Nature 411 (6835):336–341

Praetorius NP, Mandal TK (2007) Engineered nanoparticles in cancer therapy. Recent Pat Drug Deliv Formul 1(1):37–51

Radwan M, Hanora A, Zan J, Mohamed NM, Abo-Elmatty DM, Abou-El-Ela SH, Hill RT (2010)

Bacterial community analyses of two Red Sea sponges. Marine Biotechnol 12(3):350–360

Schneemann I, Nagel K, Kajahn I, Labes A, Wiese J, Imhoff JF (2010) Comprehensive investigation of marine actinobacteria associated with the sponge *Halichondria panicea*. Appl Environ Microbiol 76 (11):3702–3714

Selvin J, Gandhimathi R, Kiran GS, Priya SS, Ravji TR, Hema T (2009) Culturable heterotrophic bacteria from the marine sponge *Dendrilla nigra*: isolation and phylogenetic diversity of actinobacteria. Helgol Mar Res 63(3):239–247

Simmons TL, Coates RC, Clark BR, Engene N, Gonzalez D, Esquenazi E, Dorrestein PC, Gerwick WH (2008) Biosynthetic origin of natural products isolated from marine microorganism–invertebrate assemblages. Proc Natl Acad Sci 105(12):4587–4594

Simmons L, Kaufmann K, Garcia R, Schwär G, Huch V, Müller R (2011) Bendigoles D–F, bioactive sterols from the marine sponge-derived *Actinomadura* sp. SBMs009. Bioorg Med Chem 19(22):6570–6575

Sujatha P, Bapi Raju K, Ramana T (2005) Studies on a new marine streptomycete BT-408 producing polyketide antibiotic SBR-22 effective against methicillin resistant *Staphylococcus aureus*. Microbiol Res 160(2):119–126

Takahashi Y, Omura S (2003) Isolation of new actinomycete strains for the screening of new bioactive compounds. J Gen Appl Microbiol 49(3):141–154

Taylor MW, Radax R, Steger D, Wagner M (2007) Sponge-associated microorganisms: evolution, ecology, and biotechnological potential. Microbiol Mol Biol Rev 71(2):295–347

Thomas TRA, Kavlekar DP, LokaBharathi PA (2010) Marine drugs from sponge-microbe association—a review. Mar Drugs 8(4):1417–1468

Webster NS, Taylor MW (2012) Marine sponges and their microbial symbionts: love and other relationships. Environ Microbiol 14(2):335–346

Webster NS, Wilson KJ, Blackall LL, Hill RT (2001) Phylogenetic diversity of bacteria associated with the marine sponge *Rhopaloeides odorabile*. Appl Environ Microbiol 67(1):434–444

Woo J-H, Kitamura E, Myouga H, Kamei Y (2002) An antifungal protein from the marine bacterium *streptomyces* sp. Strain AP77 is specific for Pythium porphyrae, a causative agent of red rot disease in *Porphyra* spp. Appl Environ Microbiol 68 (6):2666–2675

Xi L, Ruan J, Huang Y (2012) Diversity and biosynthetic potential of culturable actinomycetes associated with marine sponges in the China seas. Int J Mol Sci 13 (5):5917–5932

Xin Y, Huang J, Deng M, Zhang W (2008) Culture-independent nested PCR method reveals high diversity of actinobacteria associated with the marine sponges *Hymeniacidon perleve* and *Sponge* sp. Antonie Van Leeuwenhoek 94(4):533–542

Zhang H, Lee YK, Zhang W, Lee HK (2006) Culturable actinobacteria from the marine sponge *Hymeniacidon perleve*: isolation and phylogenetic diversity by 16S rRNA gene-RFLP analysis. Antonie Van Leeuwenhoek 90(2):159–169

Zheng Z, Zeng W, Huang Y, Yang Z, Li J, Cai H, Su W (2000) Detection of antitumor and antimicrobial activities in marine organism associated actinomycetes isolated from the Taiwan Strait, China. FEMS Microbiol Lett 188(1):87–91

Zhu P, Li Q, Wang G (2008) Unique microbial signatures of the alien Hawaiian marine sponge *Suberites zeteki*. Microb Ecol 55(3):406–414

Novel Insights on the Symbiotic Interactions of Marine Sponge-Associated Microorganisms: Marine Microbial Biotechnology Perspective

6

P.V. Bramhachari, Satish Mutyala, Ira Bhatnagar, and Ramjee Pallela

Abstract

Marine sponges are the most dominant group responsible for discovering a large number of natural products that have been used as template to develop therapeutic drugs. These natural products have a wide range of therapeutic properties, including antimicrobial, antioxidant, antihypertensive, anticoagulant, anticancer, anti-inflammatory, wound healing, and immune modulators. Marine sponges and their symbionts are the most primitive of the multicellular organisms, produce a plethora of secondary metabolites, and accumulate large populations of microbes within the mesohyl. These microbes are believed to exist as both intracellular and extracellular symbionts and have been proposed to be involved in matrix stabilization, waste processing, and producing secondary metabolites for defense. Many of bioactive compounds are secreted by sponge-associated microbes, such as bacteria, fungi, blue-green algae, and actinomycetes. It includes antibiotics, peptides and non-ribosomal peptides (NRPs), polyketides, enzymes, quinones and quinolone derivatives, alkaloids, and pigments. In the present review, we highlighted the new developments in the field of marine sponge metabolite research and symbiotic and functional interactions between associated microbes and host sponges and its potentials in microbial biotechnology approaches.

Keywords

Secondary metabolites • Bioactive compounds • Sponge-associated microbes • Functional interactions • Microbial biotechnology

P.V. Bramhachari (✉) • S. Mutyala
Department of Biotechnology, Krishna University, Machilipatnam 521 001, Andra Pradesh, India
e-mail: veerabramha@gmail.com

I. Bhatnagar
Center for Cellular and Molecular Biology (CCMB), Hyderabad 500007, Telangana, India

R. Pallela
IKP Knowledge Park, Genome Valley, Turkapally, Hyderabad 500078, Telangana, India
e-mail: rpallela@gmail.com

R. Pallela, H. Ehrlich (eds.), *Marine Sponges: Chemicobiological and Biomedical Applications*, DOI 10.1007/978-81-322-2794-6_6

6.1 Introduction

Associations between sponges and bacteria have existed for 600 million years making them one of the most ancient of all the symbioses (Wilkinson 1984). Most sponges host diverse and abundant communities of microorganisms which contribute to host health, ecology, and evolution (Hentschel et al. 2006; Taylor et al. 2007; Webster and Taylor 2012). Bacteria, fungi, and microalgae are the common microbial associates of a sponge; they may live like symbionts or pathogens or parasite (Bavestrello et al. 2000; Hummel et al. 1988; Webster et al. 2002) (Fig. 6.1). Among the plethora of microbes in the three domains of life that may inhabit these animals, bacteria conspicuously emerge as the most dominant and diverse (Taylor et al. 2007) (Tables 6.1, 6.2, and 6.3).

Many microbes are symbiotically harbored by sponges, since they engross themselves in metabolism of nutrients, cycling of elements (Thomas et al. 2010; Webster and Taylor 2012), production of bioactive metabolites for chemical defense (Piel et al. 2004a, b; Hochmuth and Piel 2009), and maintaining of the structure of sponges (Wilkinson 1978). Diverse sponge microbiota may contribute to the host's metabolism with photosynthesis (Usher et al. 2007), major nitrogen cycle events (Bayer et al. 2008; Hoffmann et al. 2009; Wilkinson and Fay 1979), sulfate reduction (Hoffmann et al. 2005), and carbon fixation (Wilkinson 1983; Thacker 2005), and in return, the host provides an enriched ecological niche for its microbial partners (Taylor et al. 2007; Hentschel et al. 2012). Hitherto 32 phyla of bacteria and candidate phyla were identified in sponges (Taylor et al. 2007; Schmitt et al. 2012; Webster and Taylor 2012). Nevertheless a stable association or symbiotic association is observed between sponges and bacterial phyla of Poribacteria, Acidobacteria, Proteobacteria, Actinobacteria,

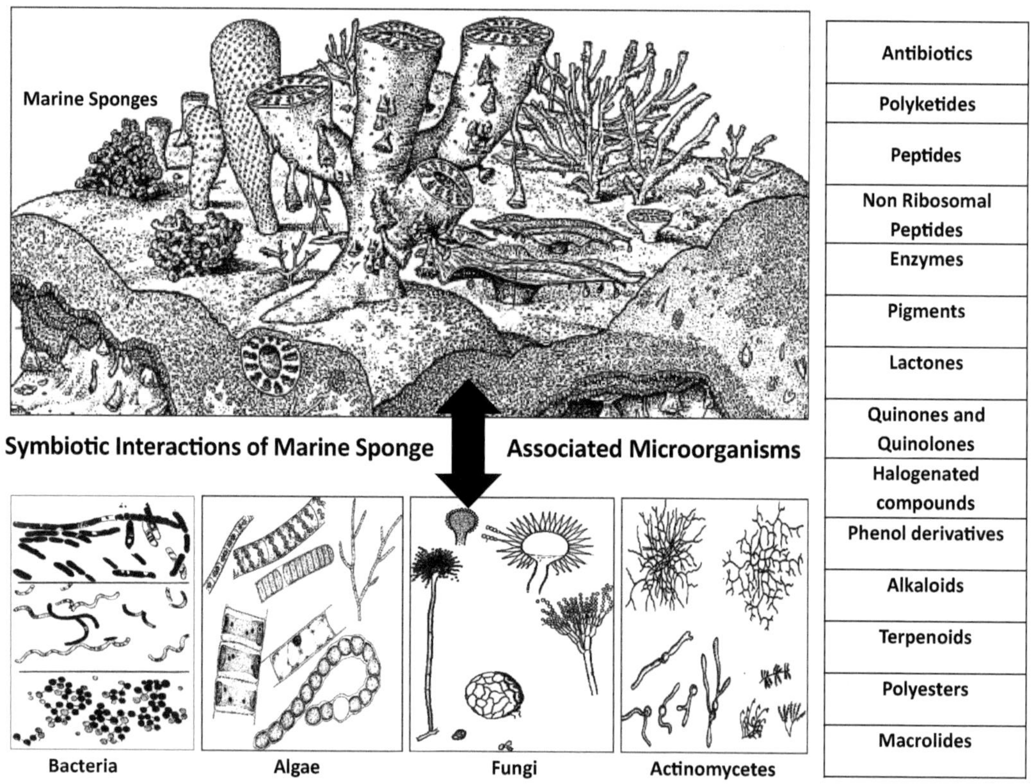

Fig. 6.1 Symbiotic association of marine sponge associated microorganisms

Nitrospira, Chloroflexi, Gemmatimonadetes, and Cyanobacteria (Taylor et al. 2007; Schmitt et al. 2012). Research over the past decade has revealed that many sponges use both vertical (Schmitt et al. 2007) and horizontal (Taylor et al. 2007; Schmitt et al. 2008; Webster et al. 2010) transmission strategies to maintain their complex and diverse microbial communities.

Marine sponges haven microorganisms on their surfaces, in their canal systems, in their intercellular spaces and contribute up to 40 % of the total cellular content of a sponge. Marine invertebrates, particularly sponges, represent a vital source for potentially active and biologically functional natural products (Osinga et al. 2001). Wide varieties of metabolites obtained from such sponges have exhibited pharmaceutical products such as novel enzyme inhibitors, cell division inhibitors, antiviral, antifungal, antimicrobial, anti-inflammatory, antitumor, cytotoxic, or cardiovascular properties (Boopathy and Kathiresan 2010). Several studies have reported the discovery of new bioactive compounds from marine organisms, focusing primarily on chemistry of secondary metabolites, which include more than 15,000 structurally diverse bioactive compounds isolated during the last 30 years (Salomon et al. 2004). In spite of the wide industrial applications on marine sponge-associated microbial metabolites, the ecological and symbiotic associations occurring between the microorganisms and the marine substrates have been greatly neglected (Kurtboke 2005). Therefore, the investigations on the secondary metabolites of marine-derived microorganisms have geared up rapidly (Petchi et al. 2013). These symbiotic bacteria may also facilitate the sponges in digestion, in waste removal, and in nutritional process either by intracellular digestion or by translocation of metabolites including photosynthesis, nitrification, and nitrogen fixation. They also stabilize the sponge skeleton and participate in the host's chemical defense. Marine sponges are usually sessile; they need defense system to overcome its predators in its surrounding environment. Generally they produce chemicals to defend its enemies and competitors. The physiological behavior of accumulating microorganisms and metabolite production has drawn special attention to sponges as a source of novel marine metabolites and as systems for studying microbial host interactions. More and more novel compounds are explored from marine sponges every year than any other taxon, and its pharmaceutical applications are well demonstrated (Blunt et al. 2006; Munro et al. 1999). Some of the research findings evidenced that these compounds are produced by associated microbes rather than sponges (Bewley and Faulkner 1998; Piel et al. 2004a, b; Schmidt et al. 2000). These investigations created specific interest in the sponge–microbe association in the production of pharmacologically active compounds. Interestingly the bacteria are determined to be the major source of natural products that provide rich opportunities for both biochemical investigations. Sponges are one of the richest sources of biologically active secondary metabolites with vast chemical diversity. Although the majority of the known bacterial metabolites derived from free-living organisms, increasing evidence supports the widespread existence of chemically prolific bacteria living in symbioses.

The assessment of sponge–microbe specificity and the diversity of sponge-associated bacterial communities have recently attained more attention to better understand the global marine biodiversity (Taylor et al. 2004; Steinberg and Kjelleberg 2010). The diversity of microorganisms living inside sponges covers most bacterial phyla, and a current paradigm is that certain bacterial clades are exclusively found in marine sponges. The presence of a very diverse microbial community inside sponges aroused suspicion that many of the cytotoxic compounds discovered in sponges were actually produced by their associated microorganisms, and for a number of compounds previously ascribed to sponges, this has indeed been established in numerous sponge species. Sponges show specificity towards the kind of bacteria with symbiotic association, even if there is no possibility of finding those bacteria in the surrounding water of that particular sponge (Hentschel et al. 2002; Mehbub et al. 2014; Taylor et al. 2007). Recently, the concept of sponge-specific microbes was meticulously studied by performing molecular phylogenetic analyses of all existing 16S and 18S rRNA gene sequence databases that originated from

Table 6.1 Bioactive metabolites produced from Marine sponge and Bacteria association

Sponge name	Microorganism	Bioactive metabolites	Effect	References
Dysidea species	*Vibrio* sp.	Tetrabromodiphenyl ethers	Antibacterial and cytotoxic activity	Elyakov et al. (1991)
Hyatella sp.	*Vibrio* sp.	Andrimid, a peptide antibiotic	Antibiotic against *Bacillus*	Oclarit et al. (1994)
Hyrtios altum	*Vibrio* sp.	Trisindoline	Antibiotic against *Bacillus subtilis, Escherichia coli*, and *Staphylococcus aureus*	Kobayashi et al. (1994) and Braekman and Daloze (2004)
Stelletta tenuis	*Alcaligenes faecalis* strain A72	L,L-Diketopiperazine	Moderate antimicrobial activity	Li (2009)
Isodictya setifera	*Pseudomonas aeruginosa*	Cyclo-(L-proline-L-methionine	Against *Micrococcus luteus, Staphylococcus aureus*, and *Bacillus subtilis*	Jayatilake et al. (1996)
Leucetta microraphis	*Cyanobacteria sp.*	Leucamide A	Inhibit the growth of cell lines of tumors	König et al. (2005)
Halichondria okadai	*Alteromonas* sp.	Tetracyclic alkaloid alteramide A	Cytotoxic effect on various kinds of cancer cell lines	Bhalla et al. (2008)
Homophymia sp.	*Pseudomonas* sp.	Five different active quinolones	2-Undecyl-4-quinolone is active on malarial parasite *Plasmodium falciparum* and HIV-1. 2-Nonyl-4-hydroxyquinoline *N*-oxide is active against *Staphylococcus aureus*	Bultel-ponce et al. (1999)
Halichondria okadai and *Halichondria melanodocia*	*Pseudomonas* and *Alteromonas*	Okadaic acid	Protein phosphatase inhibitor	Wang (2006)
Halichondria okadai	*Alteromonas* sp.	Macrolactam and amide esters	Antimicrobial and cytotoxic activity	Kelecom (2002), Shigemori et al. (1992), and Bhalla et al. (2008)
Halichondria panacea	*Microbacterium* sp.	2(1-*O*-acyl-3-[*R*-glucopyranosyl-(1-3)-(6-*O*-acyl-*R*-mannopyranosyl)]glycerol) (a glycoglycerolipid)	Antitumor agent	Wicke et al. (2000)
Acanthella acuta	*Bacillus pumilus* AAS3	Diglucosyl–glycerol GG11	Strongly inhibiting tumor cell lines HM02 and Hep G2 growth	Ramm et al. (2004)
Xestospongia sp.	*Micrococcus luteus* R-1588-10	Acyl-1-(acyl-6′-mannobiosyl)-3-glycerol (lutoside) and 2,4,4′-trichloro-2′-hydroxydiphenylether (triclosan)	Antimicrobial agents	Bultel-Poncé et al. (1998)
Aplysina aerophoba	*Bacillus subtilis* A184, A190, and A202	Fengycins, iturins, and surfactins	Effect on multidrug-resistant pathogens	Pabel et al. (2003)

Aplysina aerophoba	*Bacillus pumilus* A586	Surfactin-like compound (plumilacidin containing β-hydroxy fatty acid)	Act on *Staphylococcus aureus*	Pabel et al. (2003)
Halichondria Okadai	*Pseudomonas* sp. KK10206C	Okadaxanthine and C50 carotenoid	Oxygen quenchers	Thakur and Müller (2005), Miki et al. (1994), and Kelecom (2002)
Xestospongia testudinaria	*Marinobacter litoralis*	Urease	Metabolize nitrogenous waste urea	Su et al. (2013)
Dysidea avara	*Bacillus atrophaeus* c89	Bacillamide C and neobacillamide A	Non-ribosomal peptide synthetases	Liangjie et al. (2011) and Liu et al. (2012)
Hyatella sp.	*Vibrio* sp.	Andrimid (polyketide peptide)	Anti-*bacillus* properties	Oclarit et al. (1994)
Tedania ignis	*Micrococcus sp.*	DKPs	Antitumor, antibiotic, and antiviral activities	Stierle et al. (1988) and Rhee (2004)
Isodictya setifera	*Pseudomonas aeruginosa*	Alkaloids of phenazine and DKPs	Active against *Bacillus cereus*, *M. luteus*, and *S. aureus*	Thornton (1995), Liu et al. (2007), and Giddens and Bean (2007)
Suberea creba	*Pseudomonas sp.*	2-n-Heptylquinol-4-one	Active against *S. aureus* (antibiotic)	Debitus et al. (1998)
Acanthostrongylophora sp.	*Micromonospora*	Manzamine A alkaloid	Anti-infective, antitumor, and antimalarial	Sakai et al. (1986), Ang et al. (2000), and Dunlap et al. (2007);
Mycale plumose	Novel strain of *Actinobacterium*, *Saccharopolyspora* sp.	Undecylprodigiosin and metacycloprodigiosin	Exhibit cytotoxic effect on cell lines of cancer such as SPCA4, P388, BEL-7402, HL60, and A-549	Liu et al. (2005)
Craniella australiensis	*Streptomyces* sp. DA11	Chitinase	Antifungal	Han et al. (2009)
Suberea clavata	*Salinospora*	Rifamycins (rifamycin SV and rifamycin B) and polyketides	Antibiotic	Kim et al. (2005, 2006)
Halichondria okadai	*Rubritalea squalenifasciens* HOact23T	Diapolycopenedioic acid xylosyl esters A, B, and C (acyl glycol–carotenoic acids)	Act as antioxidant	Kasai et al. (2007), Blunt et al. (2009), and Shindo et al. (2008)
Tethya seychellensis	*Sphingomonas phyllosphaerae KODA19-6*	Zeaxanthin	Wide range of pharmacological applications	Beatty et al. (1999), Landrum and Bone (2001), Loane et al. (2008), Mares-perlman et al. (2002), and Sajilata et al. (2008)
Dysidea avara	*Bacillus atrophaeus*	Bacillamide C and a new compound of neobacillamide A	Antimicrobial compound	Liu et al. (2012)
Aplysina aerophoba	*Bacillus subtilis* strains A184, A190, and A202 and *Bacillus pumilus* A586	*Bacillus subtilis* A184 synthesized surfactins, iturins, and fengycins, while strain A190 synthesized surfactin and strain A202 synthesized iturin and *Bacillus pumilus* A586	A184 was effective against the multidrug-resistant strains of *Staphylococcus aureus* and *Staphylococcus epidermidis*. *Bacillus pumilus* A586 was active against *Staphylococcus aureus*	Pabel et al. (2003)

(continued)

Table 6.1 (continued)

Sponge name	Microorganism	Bioactive metabolites	Effect	References
Dysidea avara	*Bacillus atrophaeus*	Bacillamide C and neobacillamide A	Antibiotics	Liu et al. (2012)
Dysidea	*Vibrio* sp.	Tetrabromodiphenyl ethers	Antibacterial and cytotoxic activity	Elyakov et al. (1991)
Stelletta tenuis	*Alcaligenes faecalis* strain A72	L,L-Diketopiperazine	Antimicrobial activity	Li (2009)
Theonella swinhoei and *Entotheonella palauensis*	δ-Proteobacterium	Theonegramide and theonellamide F	Theonegramide is an antifungal compound	Bewley and Faulkner (1994) and Piel (2004a, b)
Aplysina aerophoba and *Aplysina cavernicola*	*Pseudoalteromonas*	Brominated alkaloids	Cytotoxic activities, repellent properties against predators and antimicrobial activities	Zheng et al. (2000) and Proksch et al. (2002)
K. variolosa	*Pseudomonas aeruginosa*	Diketopiperazines and phenazine alkaloid	Antibiotics	McClintock et al. (2005)
Dysidea avara	α-Proteobacteria related bacteria	2-Methylthio-1,4-naphthoquinone	Antibacterial and antiangiogenic activity	Thakur and Müller (2005)
G. cynodium and *D. avara*	*Vibrio*, *Pseudoalteromonas* and *Photobacterium*	Cyclicdipeptide	Quorum-sensing bioreporters and quorum quenching	Abbamondi et al. (2014)
Antarctic sponge	*Arthrobacter sp.*	Volatile organic compounds	Antibacterial activity	Orlandini et al. (2014)
Callyspongia diffusa	*Shewanella algae*	Antimicrobial compound	Bioactivity against the growth of bacterial and fungal pathogens	Rachanamol et al. (2014)
Fasciospongia cavernosa	*Bacillus subtilis*	Anticholinesterase	Anticholinesterase inhibitor compound	Pandey et al. (2014)
Spongia officinalis	*Streptomyces sp. MAPS15*	Pyrrolidone antimicrobial agent	Antagonistic activity against various bacterial and fungal pathogens	Sathiyanarayanan et al. (2014)
Mediterranean sponge Dysidea avara and Red Sea sponge, Spheciospongia vagabunda	*Actinokineospora sp. EG49 and Nocardiopsis sp. RV163*	N-(2-Hydroxyphenyl)-acetamide (11), 1,6-dihydroxyphenazine (12) and 5a,6,11a,12-tetrahydro-5a,11a-dimethyl[1,4]benzoxazino [3,2-b] [1,4]benzoxazine (13a) angucycline, diketopiperazine and β-carboline derivatives	Antagonistic activity against *Bacillus* sp. P25, *Trypanosoma brucei*	Dashti et al. (2014)

Table 6.2 Bioactive metabolites produced from Marine sponge and Fungi association

Sponge name	Fungi	Bioactive metabolites	Effect	References
Halichondria Japonica	*Phoma* sp. Q60596	Antifungal antibiotic (YM-202204)	Against *Aspergillus fumigates, Candida albicans*, and *Cryptococcus neoformans*	Nagai et al. (2002)
Jaspis aff. Johnstoni	Fungal class hyphomycetes	Coriolin B, dihydrocoriolin C, and chloriolines A, B, and C	Inhibitory effect on human breast cell lines and CNS cell lines	Bernan et al. (1997), Zhang et al. (2009), Biabani and Laatsch (1998), and Cheng et al. (1994)
Hyrtios sponge	*Aspergillus niger*	Asperazine (diketopiperazine alkaloid family)	Antileukemic	Varoglu et al. (1997)
Mycale plumose	*Penicillium aurantiogriseum*	Quinazoline alkaloids (aurantiomides B and C)	Moderate cytotoxic effect	Xin et al. (2007)
Chondrosia reniformis	*Penicillium rugulosum*	Polyketides (prugosenes A1–A3, B2, C1, and C2)	Templates for anti-infective compounds	Sufrin et al (2009), Lang et al. (2007), and Blunt et al. (2009)
Zyzzya sp.	*Penicillium brocae*	Polyketides (brocaenols A–C)	Cytotoxic polyketides	Bugni et al. (2003) and Ebel (2006)
Haliclona valliculata	*Emericella variecolor*	Evariquinone and anthraquinone	Antiproliferative agents on ATCC CCL17, human cervix carcinoma (KB), and NCI 503473, non-small cell lung cancer (NCI-H460) cells	Bringmann et al. (2003a, b)
Niphates olemda	*Curvularia lunata*	Two anthraquinones (lunatin and cytoskyrin A)	Active against *Bacillus subtilis, Staphylococcus aureus*, and *Escherichia coli*	Bhadury et al. (2006) and Jadulco et al. (2002)
Ircinia fasciculata	*Penicillium chrysogenum*	Sorbicillactone A	Cytostatic effect on murine leukemic lymphoblasts and also saves human T cells from HIV-1 cytopathic effects	Bringmann et al. (2003a, b) and Bringmann et al. (2007)
Xestospongia exigua	*Penicillium* cf. *montanense*	Xestodecalactones A, B, and C	Xestodecalactone B is active against *Candida albicans*	Thakur et al. (2003)
Axinella sp.	*Myrothecium* sp. JS9	Roridin A and D	Antimicrobial, cytotoxic, phytotoxic, antimalarial, and antileukemic properties	Xie et al. (2008)
Halichondria okadai	*Trichoderma harzianum* OUPSN115	Trichodenone A, B, and C	Leukemia cell lines	Thakur et al. (2003), Amagata et al. (1998a, b), and Usami et al. (2000)
Callyspongia aerizusa	*Cladosporium herbarum*	Acetyl Sumiki's acid	Active against *Staphylococcus aureus* and *Bacillus subtilis*	Jadulco et al. (2001)

(continued)

Table 6.2 (continued)

Sponge name	Fungi	Bioactive metabolites	Effect	References
Petrosia sp.	*Aspergillus versicolor*	Polyketides (decumbenones A, B, and versiol) and lipopeptide fellutamide C	Decubenone A acts a potent inhibitor of melanin synthesis	Lee et al. (2007) (2010)
Aplysina aerophoba	*Microsphaeropsis*	10-Hydroxy-18-methoxylbetaenone	Inhibitor of protein kinase C	Brauers et al. (2000)
Callyspongia cf. *C. flammea*	*Stachylidium sp.*	Derivatives of phthalimidine such as marilines	Potent inhibitors of elastase of human leukocytes	Almeida et al. (2012)
Melophlus sp.	*Penicillium* sp. FF001	Citrinin	Antibiotic activity on multidrug-resistant strains	Subramani et al. (2013)
Axinella sp.	*Acremonium* sp. 021172C	Octapeptides (RHM1, RHM2, RHM3, and RHM4)	RHM1 showing antibacterial activity as well as cytotoxic effect on cancer cell lines	Boot et al. (2006) (2007)
Pseudoceratina purpurea	*Metarhizium* sp. 001103	Desmethyl B; destruxins A, B2, and E; and chlorohydrins	Inhibition of cell lines of human tumor	Boot et al. (2007)
Halichondria japonica	*Gymnascella dankaliensis* OUPS-N134	Dankastatins A and B and gymnastatins A, B, C, F, G, Q, and R	Cytotoxic and growth-inhibiting agent on P388 stem cell lines of leukemia	Bugni and Ireland (2004), Numata et al. (1997), Amagata et al. (1998a, b, 1999, 2006, 2008), and Mayer (1999)
Ectyoplasia ferox	*Phoma*	Epoxyphomalin A and B	Antagonistic effect on different tumor cell lines	Mohamed et al. (2009)
Ectyoplasia ferox	*Spicellum roseum* 193H15	8-Deoxytrichothecin and trichodermol	Inhibiting LacCer synthase	Kralj et al. (2007) and Chatterjee and Kolmakova (2004)
Axinella sp.	*Penicillium citrinum*	Isocyclocitrinols A and 22-acetylisocyclocitrinol A	Antibacterial activity against *Staphylococcus epidermidis* and *Enterococcus durans*	Amagata et al. (2003)
Axinella verrucosa	*Penicillium* sp.	Communesins B, C, and D, oxaline, griseofulvin and dechlorogriseofulvin	Communesins and oxaline exhibited an antiproliferative activity. Griseofulvin is an antifungal agent	Koizumi et al. (2004), Kolachana and Smith (1994), and Jadulco et al. (2004)
Xestospongia exigua	*Aspergillus versicolor*	Aspergillitine	Antibacterial activity against *Bacillus subtilis*	Bhadury et al. (2006) and Lin et al. (2002)
Ectyoplasia ferox	*Coniothyrium* sp. 193477	(3*S*)-(3′,5′-Dihydroxyphenyl) butan-2-one and 2-(1′(*E*)-propenyl)-octa-4(*E*),6(*Z*)-diene-1,2-diol, (3*R*)-6-methoxymellein, (3*R*)-6-methoxy-7-chloromellein and cryptosporiopsinol	Cryptosporiopsinol exhibited noteworthy antimicrobial activity	Höller et al. (1999a, b)

(continued)

Table 6.2 (continued)

Sponge name	Fungi	Bioactive metabolites	Effect	References
Myxilla incrustans	*Microsphaeropsis* sp. H5-50	Microsphaeropsisin and (*R*)-mellein, (3*R*,4*S*)-hydroxymellein, (3*R*,4*R*)-hydroxymellein and 4,8-dihydroxy-3,4-dihydro-2H-naphthalen-1-one	Microsphaeropsin and its derivative exhibited antifungal activity	Thakur et al. (2003)
Callyspongia vaginalis	*Ulocladium botrytis* 193A4	1-Hydroxy-6-methyl-8-(hydroxyl methyl) xanthone	Antifungal agent	Höller et al. (1999a, b) and König et al. (2005)
Axinella verrucosa	*Penicillium* sp.	Communesin B, C, and D, oxaline, griseofulvin, and dechlorogriseofulvin	Antifungal and antiproliferative activity	Jadulco et al. (2004)
Petrosia ficiformis	*Penicillium brevicompactum*	Petrosifungins A and B, brevianamide A, mycophenolic acid, and asperphenamate	Mycophenolic acid is immunosuppressive agent	Bringmann et al. (2004)
Petrosia ficiformis	*Aspergillus insuetus*	Terretonins E and F	Inhibitors of mammalian mitochondrial respiratory chain	Lopez Gresa et al. (2009)
Axinella damicornis	*Aspergillus niger*	Bicoumanigrin A, aspernigrin B	Bicoumanigrin A exhibited cytotoxicity against human cancer cell lines, aspernigrin B showed neuroprotective effect	Hiort et al. (2004)
Marine sponge	*Aspergillus versicolor Hmp-F48*	Cytotoxic polyphenols, dibenzo[1,4]dioxin 1, prenylated diphenyl ethers, 2 and 3, together with six known compounds, 4–9	Cytotoxicity	Wang et al. (1998)

sponges. Overall, 27 % of all the sponge-derived sequences fall into monophyletic, sponge-specific sequence clusters within the bacteria, archaea, and fungi (Simister et al. 2012). It is well established that particular sponge species can host stable microbial populations that are different to the communities in other species (Schmitt et al. 2012), and a recent molecular evidence also supports the potential for host–symbiont coevolution (Fan et al. 2012). Additional examples include a metagenomic sequence analysis that recently highlighted how core symbiont functions can be provided in different sponge species by functionally equivalent microbes and analogous enzymes or biosynthetic pathways (Fan et al. 2012).

This chapter compiles the most relevant studies performed in order to comply with the development of peptides and depsipeptides derived from marine animals as anticancer drugs. With the latest increase in peptide research, the purpose of this review is to facilitate discussion on this issue since marine peptides are one of the recent perspectives in the development of new compounds for further drugs and therapeutic use in the treatment of cancer. Bioactive peptides and depsipeptides, most currently studied from animal marine species with anticancer potential and which reached clinical trials, have therefore been examined. Continued investigations of sponge-derived metabolites and their biotechnological implications should guarantee vital

Table 6.3 Bioactive metabolites produced from Marine sponge and Algae association

Sponge name	Algae	Bioactive metabolites	Effect	References
Lamellodysidea herbacea	*Oscillatoria spongeliae*	Polybrominated biphenyl ether (2-(2′,4′-dibromophenyl)-4,6-dibromophenol	Antibacterial activity on *Staphylococcus aureus*, *Escherichia coli*, *Bacillus subtilis*, etc.	Hentschel et al. (2001)
Lamellodysidea herbacea	*Oscillatoria spongeliae*	Dihydrodysamide C, didechlorodihydrodysamide C, and piperazines (DKPs)	Receptor probes as well as therapeutic agents	Haygood et al. (1999), Flowers et al. (1998), Unson and Faulkner (1993), and Besada et al. (2005)
Ptilocaulis trachys	*Lyngbya majuscula*	Majusculamide C	Antifungal activity against pathogens of commercially important plants	Williams et al. (1993) and Dunlap et al. (2007)
Halichondria okadai and *Halichondria melanodocia*	Dinoflagellate *Prorocentrum lima*	Okadaic acid	Protein phosphatase inhibitor	Kobayashi and Ishibashi (1993) and Wang (2006)

interest in sponge-associated symbiotic microbial interactions. Recent data analysed by Mehbub et al. (2014) addressed significant future trends and opportunities in the search for novel leads either from marine sponges or their symbionts.

6.2 Bioactive Compounds from Marine Sponge-Associated Microbes

6.2.1 Antibiotics

Many bacteria developed resistance to antibiotics, especially strains belonging to *Staphylococcus aureus*, and few other bacterial strains developed resistances to existing antibiotics (Rice 2006). Obviously, there is a need for the rapid development of novel antibiotics to these resistant strains. In addition to that, marine sponges were considered for years as a rich source of natural products and metabolites for antibiotics with strong inhibitory activity against bacteria, fungi, and microbes. Research showed that many bioactive compounds from various sponge species can be useful for the development of novel antibiotics and antimicrobial drugs. Symbiotic bacteria associated with sponges produce potential antimicrobial compounds and bioactive compounds. Many bioactive compounds extracted from the marine sponges *Dendrilla nigra* (Ivanova et al. 1999), *Axinella donnani*, and *Clathria gorgonoides* (Aishwarya et al. 2013) have demonstrated significant antimicrobial activities against various Gram-positive pathogenic bacteria. In a similar report, Venkateswarlu and Biabani (1995) also exhibited antibacterial activity from dichloromethane-methanol (1:1) extract of sponge *Phycopsis* sp. collected from Tuticorin coast of India. Several bacterial metabolites have been isolated from the marine sponges: cryptophycin I and chondramide were isolated from *Dysidea* spp. and *Jaspis* spp., respectively (Sabdono and Radjasa 2008). In addition to therapeutic applications of bioactive compounds from marine sponges against various bacteria, there are many reports which reveal that antitumor drugs such as Kendarimide A, isolated from sponge *Haliclona* sp., have been shown to reverse multidrug resistance in tumor cells (Roser et al. 2005). Several bacterial species *Bacillus megaterium, Pseudoalteromonas aurantia*, *Pseudoalteromonas piscicida*, and *Pseudoalteromonas rubra* isolated from the sponges *Euplexaura curvata* and *Hymeniacidon perleve* were shown to produce bioactive compounds with wide range of antimicrobial spectrum, respectively (Zheng et al. 2005a, b). Similarly, *Bacillus atrophaeus*, isolated from the marine sponge *Dysidea avara* in the South China Sea, produced two different antibiotics, viz., Bacillamide C and Neobacillamide A (Liu et al. 2012). However, *Dysidea* sponge (Eastern Samoa), harboring *Vibrio* sp., was shown to produce tetrabromodiphenyl ethers which have antibacterial and cytotoxic activities (Elyakov et al. 1991). Many antibiotics including cyclic peptides, cyclic lipopeptides, and novel thiopeptides have been reported from marine *Bacillus* spp. (Nagai et al. 2003; Suzumura et al. 2003). It is exciting to note that a novel antibiotic trisindole, a derivative of trisindoline, was characterized from a marine Okinawan sponge *Hyrtios altum* associated with *Vibrio* sp. and was shown to exhibit potential antibiotic activity against *Escherichia coli*, *Bacillus subtilis*, and *Staphylococcus aureus* (Kobayashi et al. 1994; Braekman and Daloze 2004). Strikingly two novel antibiotics, YM-266183 and YM-266184, were discovered in the culture broth of *Bacillus cereus* QN03323, which was symbiotically associated with the marine sponge *Halichondria japonica* (Laatsch 2006).

Micrococcus luteus R-1588-10, associated with marine sponge *Xestospongia* sp. of New Caledonia, Southwest Pacific, synthesized two active antimicrobial agents such as acyl-1-(acyl-6′-mannobiosyl)-3-glycerol (lutoside) and 2,4,4′-trichloro-2′-hydroxydiphenylether (triclosan) (Bultel-Poncé et al. 1998). Extracts of *Pseudomonas* species harbored on marine sponge *Suberites domuncula* exhibited cytotoxic, hemolytic, antiangiogenic, and antimicrobial activity against multidrug-resistant strains, such as *Staphylococcus epidermidis* and *Staphylococcus aureus* (Thakur et al. 2003, 2005). Interestingly, two metabolites known as Roridin A and D isolated from *Myrothecium* sp. JS9 were associated with *Axinella* sp. of South China Sea and exhibited antimicrobial, cytotoxic, phytotoxic, antimalarial, and antileukemic properties (Xie et al. 2008).

The sponge *Dendrilla nigra* is a rich source of cultivable marine actinomycetes. Investigations on a sponge specimen collected from the Vizhinjam coast (west coast of India) revealed that *Micromonospora–Saccharomonospora–Streptomyces* group was the major cultivable actinobacteria found in the sponge (Selvin et al. 2009). The species *Streptomyces dendra* sp. nov. MSI051 isolated from the sponge *Dendrilla nigra* exhibited a broad spectrum of antibacterial activity. The host sponge, as well as the associated bacterial symbiont MSI051, contained high levels of PLA2 (phospholipase A2) which acts as an antibacterial protein in the defense system of higher animals (Selvin 2009). Another strain, *Streptomyces* sp. BLT7 isolated from *Dendrilla nigra* obtained from Kanyakumari (south east coast of India) also showed potential antibacterial activity in their extracellular products (Selvin et al. 2004). In a recent report, an interesting bioactive compound was also characterized from a marine sponge *Dendrilla nigra* associated with *Nocardiopsis dassonvillei* MAD08 and was shown to exhibit a potential antibacterial and anticandidal activity against the multidrug-resistant pathogenic microbial strains (Selvin et al. 2009).

A fungal species *Trichoderma harzianum* OUPSN115 associated with *Halichondria okadai* that exists in Japanese water is responsible for the production of novel bioactive cytotoxic compounds (trichodenone A, B, and C); these compounds are active against leukemia cell lines (P388) (Thakur et al. 2003; Amagata et al. 1998a, b; Usami et al. 2000). A fungal species *Cladosporium herbarum* isolated from *Callyspongia aerizusa* from Indonesia synthesizes an antimicrobial compound, acetyl Sumiki's acid. This compound is active against *Staphylococcus aureus* and *Bacillus subtilis* (Jadulco et al. 2001). A fungal strain *Phoma* sp. Q60596, obtained from the symbiotic association of marine sponge *Halichondria Japonica*, resulted a novel antifungal antibiotic, YM-202204, and exhibited potent activity against *Candida albicans*, *Cryptococcus neoformans*, and *Aspergillus fumigatus* (Nagai et al. 2002). Species belongs to genus *Salinospora* associated with *Suberea clavata* (the Great Barrier Reef sponge) and synthesizes rifamycins (rifamycin SV and rifamycin B) and polyketides (Kim et al. 2006). A *Penicillium* sp. FF001 associated with Fijian marine sponge *Melophlus* sp. synthesizes a crystalline compound; its structural formula is $C_{13}H_{14}O_5$. Database search (AntiBase) related to natural products and spectroscopy analysis revealed that this compound closely matched with citrinin. Apparently these compounds have broad spectrum of antibiotic activity on multidrug-resistant strains of pathogens, such as *Staphylococcus aureus,* which is resistant to methicillin, rifampin, and vancomycin resistant *Cryptococcus neoformans* (yeast) and *E. faecium* (Subramani et al. 2013). The sponge *Callyspongia vaginalis*, isolated from the Caribbean Sea, yielded a new tyrosine kinase inhibitor and the antimicrobial compound ulocladol together with the antifungal agent 1-hydroxy-6-methyl-8-(hydroxyl methyl) xanthone. These compounds have been extracted from sponge-associated fungi *Ulocladium botrytis* 193A4 (Höller et al. 1999a, b; König et al. 2005). The ethyl acetate extracts of *Penicillium* sp., derived from the Mediterranean sponge *Axinella verrucosa*, yielded a group of bioactive compounds, viz., communesin B, C, and D, oxaline, griseofulvin, and dechlorogriseofulvin which sequentially exhibited antifungal activity and antiproliferative activity (Jadulco et al. 2004). Aspergillitine was also isolated from the sponge *Xestospongia exigua* in symbiotic association with the fungus *Aspergillus versicolor* and exhibited antibacterial activity against *Bacillus subtilis* (Jadulco et al. 2002; Bhadury et al. 2006). A fungal strain *Coniothyrium* sp. 193477, isolated from the sponge *Ectyoplasia ferox* from the Caribbean Islands of Dominica, yielded novel antimicrobial compounds such as (3*S*)-(3′,5-′-dihydroxyphenyl)butan-2-one and 2-(1′(*E*)-propenyl)-octa-4(*E*),6(*Z*)-diene-1,2-diol. Among these, cryptosporiopsinol demonstrated significant antimicrobial activity (Höller et al. 1999a, b). Additionally the same research group also discovered a novel antifungal metabolite known as microsphaeropsisin obtained from the fungal strain *Microsphaeropsis* sp. H5-50 that was symbiotically associated with the marine sponge *Myxilla incrustans*. A fungal species, *Stachylidium*, associated with marine sponge *Callyspongia* cf. *C. flammea* produces derivatives of phthalimidine such as marilines A (1) (1a), A (2) (1b), B (2), and C (3). They are potent inhibitors of elastase of human leukocytes (Almeida et al. 2012). Dankastatins A and B and gymnastatins A, B, C, F, G, Q, and R purified from *Gymnascella dankaliensis* OUPS-N134, which associates with *Halichondria japonica*, act as cytotoxic and growth-inhibiting agent on P388 stem cell lines of leukemia. Human and breast cell lines are also equally affected by gymnastatin Q (Bugni and Ireland 2004; Numata et al. 1997; Amagata et al. 1998a, b, 1999, 2006, 2008; Mayer 1999).

The *Oscillatoria spongeliae* (cyanobacterium), which is an endosymbiont of *Dysidea herbacea* (marine sponge), was shown to produce a polybrominated biphenyl ether antibiotic (Newman and Hill 2006; Hentschel et al. 2001). A specimen of cyanobacteria *Oscillatoria spongeliae* that was symbiotically associated with the marine sponge *Lamellodysidea herbacea* yielded a polybrominated biphenyl ether such as 2-(2′,4′-dibromophenyl)-4,6-dibromophenol. However, *Oscillatoria spongeliae was also* observed as an endosymbiont in different sponge mesohyl. Polybrominated biphenyl ethers from *Lamellodysidea herbacea* were shown with antibacterial activity against both Gram-negative and Gram-positive bacteria and unicellular marine cyanobacteria. The compound 2-(2′,4′-dibromophenyl)-4,6-dibromophenol also showed antibacterial activity against *Staphylococcus aureus*, *Escherichia coli*, *Bacillus subtilis*, etc. Andrimid, a peptide antibiotic, act against bacillus, obtained from *Vibrio* species associated with *Hyatella* sp. (Oclarit et al. 1994).

6.2.2 Quinone and Quinolone Derivatives

Muller and group reported that α-Proteobacteria associated with *Dysidea avara* was found to secrete a quinolone compound 2-methylthio-1,4-naphthoquinone(Müller et al. 2004a, b). This compound distinctively possessed both antibacterial and antiangiogenic activity (Thakur and Müller 2005). Additional example includes a *Pseudomonas* species associated with marine sponge *Homophymia* sp. which produced five different active quinolones. Among these 2-undecyl-4-quinolone is active against malarial parasite *Plasmodium falciparum* and HIV-1. And the second molecule 2-nonyl-4-hydroxyquinoline *N*-oxide exhibited antibacterial activity against *Staphylococcus aureus* (Bultel-Ponce et al. 1999). Another interesting antibiotic quinine, 2-n-heptylquinol-4-one, was purified from *Pseudomonas* sp. associated with marine sponge *Suberea creba* which exhibited antibacterial activity against *S. aureus* (Debitus et al. 1998). A fungal strain *Emericella variecolor* obtained from *Haliclona valliculata* synthesized two novel compounds such as evariquinone and anthraquinone, and these compounds exhibited strong antiproliferative activity against (KB) human cervix carcinoma ATCC CCL17 and NCI 503473 and non-small cell lung cancer (NCI-H460) cells (Bringmann et al. 2003a, b). In another study a fungal species *Curvularia lunata* symbiotically associated with marine sponge *Niphates olemda* of Indonesian marine water synthesized two anthraquinones (lunatin and cytoskyrin A), which are active against *Bacillus subtilis*, *Staphylococcus aureus*, and *Escherichia coli* (Bhadury et al. 2006; Jadulco et al. 2002).

6.2.3 Peptides and Non-ribosomal Peptides

Marine sponges are excellent resource for unique peptides and most of them are bioactive. Some of them are cyclic and a few are linear, consisting of unique amino acids, which are very rare in terrestrial organisms, and most of them are novel peptide molecules (Aneiros and Garateix 2004). Many bioactive peptides and depsipeptides with anticancer potential have been extracted from various marine invertebrates like tunicates, sponges, soft corals, sea hares, nudibranchs, bryozoans, sea slugs, and other marine organisms (Haefner 2003; Naqash and Nazeer 2010). An extensive group of peptides and depsipeptides are extracted from marine organisms (Holzinger and Meindl 1997). Biologically all the active peptides obtained from marine invertebrate species are considered to have diverse activities, including opioid agonistic, mineral binding, immunomodulatory, antimicrobial, antioxidant, antithrombotic, hypocholesterolemic, and antihypertensive actions (Kim and Wijesekara 2010). By modulating and improving physiological functions, bioactive peptides may provide new therapeutic applications for the prevention and/or treatment of chronic diseases.

Alcaligenes faecalis strain A72 which was reported to be symbiotically associated with *Stelletta tenuis* (South China Sea sponge) produced L,L-diketopiperazine which evidenced antimicrobial activity (Li 2009). Similarly *Pseudomonas aeruginosa*, isolated from the Antarctic marine symbiont sponge *Isodictya setifera*, synthesized a cyclo-(L-proline-L-methionine), which was active against various microbial pathogens *Micrococcus luteus*, *Staphylococcus aureus*, and *Bacillus subtilis* (Jayatilake et al. 1996). Similarly *Bacillus atrophaeus* strain c89, obtained from a marine sponge *Dysidea avara*, synthesized two different peptide derivatives, such as bacillamide C and neobacillamide A (Liu et al. 2012) by using non-ribosomal peptide synthetases (NRPs) (Liangjie et al. 2011). Interestingly a *Micrococcus* sp. isolated from the marine sponge *Tedania ignis* was reported to synthesize three different diketopiperazines, cyclo-(L-Pro-L-Val), cyclo-(L-Pro-L-Ala), and cyclo (L-Pro-L-Leu) (Stierle et al. 1988). This work clearly depicted that these dipeptides possessed antitumor, antibiotic, and antiviral activities (Rhee 2004; Scopel et al. 2013). In a recent investigation by Fan et al. (2012), the genomes of bacterial sponge symbionts carry a large number of proteins that are similar to eukaryotic proteins controlling phagocytosis and cytoskeletal formation. An antifungal glycopeptide known as theonegramide was previously isolated from *Theonella swinhoei*, collected from the Philippines (Bewley and Faulkner 1994). Interestingly, filamentous δ-proteobacterium was detected in their symbiotic partners *Entotheonella palauensis and Theonella swinhoei* and yielded two interesting glycopeptide metabolites theonegramide and theonellamide F (Piel 2004a, b). A *Pseudoalteromonas* sp. associated with the sponge *Halisarca ectofibrosa* has been isolated for its ability to inhibit S. *aureus*, *Micrococcus luteus*, *Bacillus subtilis*, *Escherichia coli*, and *V. anguillarum* (Rungprom et al. 2008). A filamentous bacterial symbiont associated with marine lithistid sponge *Theonella swinhoei* produced a cyclic glycopeptide molecule theopalauamide and displayed numerous biological activities (Schmidt et al. 1998; Bewley and Faulkner 1998). In 1996, Bewley and Faulkner extracted and identified two new antifungal molecules, microsclerodermin A and B, from a lithistid sponge *Microscleroderma* sp. (Bewley et al. 1996).

A fungal strain *Acremonium* sp. 021172C which is associated with marine sponge *Axinella* sp. isolated from Papua New Guinea resulted in four novel octapeptides (RHM1, RHM2, RHM3, and RHM4). Interestingly these RHM peptides along with few other efrapeptins exhibited potent antibacterial activity and cytotoxic effect on cancer cell lines (Boot et al. 2007). In another report, the same research group explored that a fungal strain, *Metarhizium* sp. 001103, that was symbiotically associated with Fijian marine sponge *Pseudoceratina purpurea* synthesized a N-methylated cyclic destruxin family depsipeptides. The following metabolites desmethyl B; destruxins A, B2, and E; and chlorohydrins exhibited specific inhibition of cell lines of human tumor. E chlorohydrins inhibited growth of tumor cells and also showed cytotoxic effect on cell lines of murine c38 (Boot et al. 2007). A strain of *Penicillium brevicompactum* derived from the specimen of *Petrosia ficiformis* provided two new cyclopentadepsipeptides, petrosifungins A and B, along with the known fungal metabolites brevianamide A, mycophenolic acid (a well-known immunosuppressive agent), and asperphenamate (Bringmann et al. 2004). Interestingly a strain of *Aspergillus insuetus* obtained from the surface of *Petrosia ficiformis* also yielded two new compounds, terretonins E and F. They are potent inhibitors of mammalian mitochondrial respiratory chain (Lopez Gresa et al. 2009). Kralj and colleagues reported that a fungus *Spicellum roseum* 193H15, derived from *Ectyoplasia ferox*, was found to produce trichothecenes such as trichodermol and 8-deoxytrichothecin (Kralj et al. 2007). These molecules considerably inhibited the activity of LacCer synthase (role in oncogene expression and cell proliferation) in neuroblastoma cells. The fungus also yielded two cyclohexadepsipeptides, spicellamides A and B (Kralj et al. 2007).

Shallow water marine sponge *Lamellodysidea herbacea* of Indo-Pacific region was reported to be symbiotically associated with non-heterocystous cyanobacteria *Oscillatoria spongeliae* (Haygood et al. 1999). These cyanobacteria induces host to produce wide range of secondary metabolites (Arillo et al. 1993). Structural studies revealed that symbiotic association of *Lamellodysidea herbacea* and *Oscillatoria spongeliae* of Great Barrier Reef, Australia, was shown to produce chlorinated dihydrodysamide C, didechlorodihydrodysamide C, and diketopiperazines (DKPs) (Unson and Faulkner 1993; Flowers et al. 1998). Since DKPs are general motifs in different biologically active products, they are used in making of receptor probes as well as therapeutic agents (Besada et al. 2005). A species of cyanobacteria associated with *Leucetta microraphis* of the Great Barrier Reef of Australia can synthesize leucamide A (non-ribosomal cyclic peptide), which was known to be a potent inhibitor of the growth of tumor cell lines with a mutation in p53 gene (König et al. 2005). A cyclic depsipeptide, majusculamide C, isolated from the metabolites of the sponge *Ptilocaulis trachys* symbiotically associated with toxic blue-green alga *Lyngbya majuscula* exhibited antifungal activity against pathogens of commercially important plants derived (Williams et al. 1993; Dunlap et al. 2007).

6.2.4 Enzymes

An antifungal chitinase extracted from *Streptomyces* Sp. DA11 growing as symbiont on marine sponge *Craniella australiensis* exhibited unique physical properties with high pH and salinity tolerance (Han et al. 2009). A species of fungal genus *Microsphaeropsis* associated with Mediterranean sponge *Aplysina aerophoba* synthesized a protein kinase C inhibitor (10-hydroxy-18-methoxylbetaenone) (Brauers et al. 2000) and is potentially used in cancer therapies (Mackay and Twelves 2003).

In the earlier days, many researchers felt that *Marinobacter litoralis* associated with *Xestospongia testudinaria* produced urease to metabolize urea. Later, they found that few of the *ureC* genes expressed in *Xestospongia testudinaria* are very similar to α-subunit of urease from the members of Proteobacteria, Cyanobacteria, Actinobacteria, and *Magnetococcus*. This is the first example for symbiotic production and utilization of metabolite by sponges and its associates (Su et al. 2013).

Marine sponges *Halichondria okadai* and *Halichondria melanodocia* provide good examples for the importance of microalgal association in the production of natural compounds recovered from these invertebrates. Both species of *Halichondria* contain the protein phosphatase inhibitor okadaic acid (Wang 2006). It was first isolated from the sponge *Halichondria okadai*, but, later, it was found out that a dinoflagellate *Prorocentrum lima* produced the inhibitor okadaic acid (Kobayashi and Ishibashi 1993). Apparently *Halichondria okadai* and *Halichondria melanodocia* was shown to act as reservoirs for associate symbiotic bacteria *Pseudomonas* and *Alteromonas* and produces okadaic acid (Wang 2006), which is a well-known example for protein phosphatase inhibitor.

6.2.5 Polyketides

Polyketides are a large family of natural products found in microbes and include many clinically vital drugs such as tetracycline, daunorubicin, erythromycin, rapamycin, and lovastatin. They are biosynthesized from acyl CoA precursors by polyketide synthases (PKSs). Firstly, much of the contemporary research on polyketide biosynthesis is determined by unparalleled biological activities and colossal commercial value of these natural products, which remain as successful candidate molecules for novel drug discovery; secondly, the extraordinary structure, mechanism, and catalytic reactivity of PKSs provide an exceptional opportunity to explore the molecular mechanisms of enzyme catalysis, molecular recognition, and protein–protein interaction; and thirdly, the significant versatility and amenability of PKSs allow the exploration of novel compounds, by using combinatorial methods in biosynthesis. Polyketides and non-ribosomal peptides are often evoked as examples of metabolites found in sponges with a likely (bacterial) symbiont origin. These molecules, synthesized by large multifunctional enzymes called polyketide synthases (PKS) and non-ribosomal peptide synthetases (NRPS), encompass substance classes that are typical for microorganisms (Piel 2004a, b). Many marine microbial metabolites possess both potent bioactivity and interesting polyketide and/or peptide structures and are therefore attractive targets for molecular genetic studies.

The marine sponge *Theonella swinhoei* from Palau contains a cytotoxic polyketide, swinholide A and the bicyclic glycopeptide antifungal compound theopalauamide and symbiotically associated bacteria within this sponge include unicellular cyanobacteria, unicellular bacteria and filamentous bacteria (Bewley and Faulkner 1998). *Penicillium rugulosum* associated with marine sponge *Chondrosia reniformis* was shown to produce seven polyketides (prugosenes A1–A3, B2, C1, and C2) having potential to act as template for anti-infective compounds (Lang et al. 2007; Sufrin et al. 2009; Blunt et al. 2009). *Penicillium brocae* isolated from the marine sponge *Zyzzya* sp. of Fiji was reported to synthesize three polyketides (Brocaenols A–C), having a functional property of cytotoxicity when tested on HCT-116 (Bugni et al. 2003; Ebel 2006). In another research study, a fungal species *Aspergillus versicolor* was shown to be symbiotically associated with marine sponge *Petrosia* sp. which synthesized three different polyketides (decumbenones A and B and versiol) and a cytotoxic lipopeptide fellutamide C, respectively. Interestingly the molecule Decubenone A acts as a potent inhibitor of melanin synthesis (Lee et al. 2007, 2010). Few other studies have reported on the isolation of fungi from three Hawaiian sponges (Li and Wang 2009), Mediterranean sponges *Psammocinia* sp. (Paz et al. 2010) and *Tethya aurantium* (Wiese et al. 2011), and China Sea sponges (Zhou et al. 2011) and on the secondary metabolite genes in the obtained isolates and found 15 PKS genes and 4 NRPS genes (Zhou et al. 2011). However, despite the numerous associations documented by marine sponges, two potent cytotoxic compounds, epoxyphomalin A and B, were discovered from *Phoma* sp. and showed activity against various human tumor cell lines (Mohamed et al. 2009).

An interesting group of microorganisms are the *Actinobacteria* because of the fact that they are known to possess a wide diversity of secondary metabolite genes, particularly NRPS and PKS (Goodfellow and Fiedler 2010; Schneemann et al. 2010). A *Rhodococcus* sp. has been shown to harbor 24 NRPS and 7 PKS genes (McLeod et al. 2006), while the genome of the industrial-scale erythromycin producer *Saccharopolyspora erythraea* contains at least 25 secondary metabolite genes for the production of terpenes, polyketides, and non-ribosomal peptides (Oliynyk et al. 2007). The genome of *Salinispora tropica*, the source of the proteasome inhibitor salinosporamide A (Feling et al. 2003), harbors 17 PKS, NRPS, or hybrid gene clusters (Udwary et al. 2007). *Actinobacteria*, especially *Streptomyces* spp. have been shown to host a wide range of secondary metabolite genes. Thirty such gene clusters have been identified in the genome of the avermectin producer *Streptomyces avermitilis* (Ikeda et al. 2003), and one of the most medically important bacterium, *Streptomyces coelicolor*, is known to harbor at least 20 such genes (Bentley et al. 2002). Cell separation experiments inferred a cyanobacterial sponge associate to be the actual producer of the peptide (Flowers et al. 1998), and closely related compounds, e.g., pseudosysidenin (Jiménez and Ribes 2007), were also found in free-living Cyanobacteria. It could be envisaged, therefore, that by comparison of PKS genes of a Beetle's *Pseudomonas* symbiont (Piel 2002), which produces pederin, a compound related to the *Theonella swinhoei's* derived onnamide A (Piel et al. 2004a, b).

Species belonging to *Vibrio* and *Pseudovibrio* associated with marine sponge *Irciniidae* family have genes for polyketides synthase (Esteves et al. 2013). Andrimid, a polyketide peptide compound derived from *Vibrio* sp. associated with marine sponge *Hyatella* sp., shows anti-*bacillus* properties (Oclarit et al. 1994). The discovery of *onn* genes encoding the biosynthesis of onnamide A in the microbial metagenome of the sponge *Theonella swinhoei* was made by Piel et al. (2004a, b). This polyketide exhibited extremely potent antitumor activities. This provides the first experimental proof for bacterial origin of marine sponge-derived natural compounds (Grozdanov and Hentschel 2007).

6.2.6 Alkaloids

Alkaloids are a class of secondary metabolites and exhibit interesting structural complexity and progressively more studied biological activities; thus, these compounds are attracting the attention of upcoming researchers from numerous disciplines worldwide (Forte et al. 2009). The alkaloids are relatively simple and are thus suitable candidate molecules for optimization using established medicinal chemistry strategies. Because of its relatively low molecular mass and simple structure, alkaloids offer several possibilities for chemical optimization through addition of side chains or functional groups. Marine alkaloids from *Agelas* sponges and their synthetic analogues have been extensively studied as inhibitors of bacterial biofilm formation and antibacterial, antifungal (Rogers et al. 2010), and antiprotozoal agents (Scala et al. 2010).

Hyphomycetes associated with *Jaspis aff. Johnstoni* (Indo-Pacific sponge) can synthesize tricyclic sesquiterpenes coriolin B, dihydrocoriolin C, and chloriolines A, B, and C; among these, coriolin B showed efficient inhibitory effect on human breast cell lines and CNS cell lines (Bernan et al. 1997; Zhang et al. 2009; Biabani and Laatsch 1998; Cheng et al. 1994). Asperazine (diketopiperazine alkaloid family), an antileukemic compound, was extracted from *Aspergillus niger* which are harbored on *Hyrtios* sponge (Varoglu et al. 1997). Quinazoline alkaloids (aurantiomides B and C) purified from *Penicillium aurantiogriseum* associated with *Mycale plumose* sponge (China) exhibit moderate cytotoxic effect (Xin et al. 2007). From a static culture of the fungal strain *Aspergillus niger* isolated from the Mediterranean sponge *Axinella damicornis*, eight secondary metabolites belonging to four entirely different structural classes were obtained. Among these, the new compound 3,3′-bicoumarin (bicoumanigrin A) showed moderate cytotoxicity against human cancer cell lines in vitro. Another compound, aspernigrin B, displayed a strong neuroprotective effect by significantly reducing the increase of intracellular calcium concentration in rat cortical neurons stimulated with glutamic acid or quisqualic acid (Hiort et al. 2004).

In a recent report, *Pseudoalteromonas* was shown to have isolated from marine sponge species *Aplysina aerophoba*, and *Aplysina cavernicola* produced brominated alkaloids with cytotoxic activities, repellent properties against predators, and antimicrobial activities (Zheng et al. 2000; Proksch et al. 2002). A tetracyclic alkaloid alteramide A, purified from *Alteromonas* sp. which is symbiotically associated with *Halichondria Okadai*, showed cytotoxic effect on various cancer cell lines (Bhalla et al. 2008). In a similar report, *Pseudomonas aeruginosa* which was shown to have associated with Antarctic marine sponge *Isodictya setifera* produced different types of diketopiperazine (DKPs) and phenazine alkaloids (Thornton 1995); among these phenazine alkaloids exhibited antibacterial activity against *Bacillus cereus, M. luteus*, and *S. aureus* (Liu Giddens and Bean 2007). Another striking example is *Pseudomonas aeruginosa* which was shown to colonize *K. variolosa and* produced diketopiperazines and phenazine alkaloid antibiotics (McClintock et al. 2005). An actinobacterium, *Micromonospora*, isolated from the marine sponge *Acanthostrongylophora* sp. of Indonesia was shown to synthesize an anti-infective manzamine A alkaloid (Dunlap et al. 2007). Manzamine A possesses antitumor (Sakai et al. 1986) as well as antimalarial activity (Ang et al. 2000).

6.2.7 Lactones

A bicyclic lactone derivative known as sorbicillactone A was extracted from *Penicillium chrysogenum* associated with Mediterranean sponge *Ircinia fasciculata* and exhibited selective cytostatic effect on murine leukemic lymphoblasts and also protected human T cells from HIV-1 cytopathic effects (Bringmann et al. 2003a, b, 2007). However, in a similar study, a fungal strain *Penicillium* cf. *montanense* isolated from marine sponge *Xestospongia exigua* of Bali Sea, Indonesia, synthesized three decalactone novel metabolites (xestodecalactones A, B, and C). Out of these, only xestodecalactone B showed activity against *Candida albicans* (Thakur et al. 2003). In another study *Alteromonas* sp. associated with marine sponge, *Halichondria Okadai*, produced macrolactam and amide esters which are responsible for antimicrobial and cytotoxic activity (Kelecom 2002; Shigemori et al. 1992; Bhalla et al. 2008).

6.2.8 Steroids

Structurally unique steroids, isocyclocitrinols A and 22- cetylisocyclocitrinol A, were isolated from the extract of a saltwater culture of sponge-derived fungus *Penicillium citrinum*, separated from the sponge *Axinella* sp., collected in Papua New Guinea. Both the steroid compounds exhibited weak antibacterial activity against *Staphylococcus epidermidis* and *Enterococcus durans* (Amagata et al. 2003).

6.2.9 Lipids

Microbacterium sp. obtained from the marine sponge *Halichondria panacea* in Adriatic coast, Croatia, which synthesized a glycoglycerolipid derivative molecule 2(1-*O*-acyl-3-(*R*-glucopyranosyl-(1-3)-(6-*O*-acyl-*R*-mannopyranosyl)) glycerol), was shown to be an antitumor agent (Wicke et al. 2000). An interesting lipid derivative molecule diglucosyl–glycerolipid, GGL11 extracted from *Bacillus pumilus* AAS3, which was shown to be associated with Mediterranean sponge *Acanthella acuta*, strongly inhibited the growth of tumor cell lines HM02 and Hep G2 (Ramm et al. 2004). Strains of *Bacillus subtilis* A184,

A190, and A202, associated with *Aplysina aerophoba* of Mediterranean coast of France, synthesized different antimicrobial compounds. A184 strain synthesizes fengycins, iturins, and surfactins; these compounds have integrated effect on multidrug-resistant pathogens such as *Staphylococcus aureus* and *Staphylococcus epidermidis* (Pabel et al. 2003). Another bacterial strain *Bacillus pumilus* A586 isolated from *Aplysina aerophoba* synthesizes surfactin-like compound (plumilacidin containing β-hydroxy fatty acid), which have high potency to act on *Staphylococcus aureus* (Pabel et al. 2003).

6.2.10 Pigments

Pseudomonas sp. KK10206C associated with *Halichondria Okadai* produced two potential oxygen quenchers such as okadaxanthine and C50 carotenoid (Thakur and Müller 2005; Miki et al. 1994; Kelecom 2002). Gram-negative bacteria, *Rubritalea squalenifasciens* HOact23T, isolated from *Halichondria okadai* is a good source for diapolycopenedioic acid xylosyl esters A, B, and C (acyl glycol-carotenoic acids), which act as antioxidant (Kasai et al. 2007; Blunt et al. 2009; Shindo et al. 2008). Interestingly novel strain of actinobacterium, *Saccharopolyspora* sp., isolated from Qingdao coast sponge *Mycale plumose* which produced two pyrrole pigments such as undecylprodigiosin and metacycloprodigiosin, was shown to exhibit cytotoxic effect on cell lines of cancer such as SPCA4, P388, BEL-7402, HL60, and A-549 (Liu et al. 2005). Two bacterial strains *Sphingomonas phyllosphaerae* FA2 (T) and *Sphingomonas natatoria* DSM 3183 (T) associated with marine sponge *T. seychellensis* synthesized a pigment zeaxanthin having potential pharmacological applications, macular degeneration (Beatty et al. 1999; Sajilata et al. 2008) and cancer (Nishino et al. 2002), and additives for fish and poultry feed.

6.3 Conclusions

These sponge-derived compounds include a wide variety of different chemical classes such as alkaloids, polyketides, and terpenoids among others. The occurrence of structural similarities between some of these compounds from sponges and those from the sponge microbiota has led to the hypothesis that at least some of these bioactive compounds may in fact be of microbial origin (Wang 2006). Thus, it is clear that the sponge–microbe association makes sponges an ideal source for biologically active microorganisms producing potentially novel chemicals.

Recent advances in our understanding of sponge–microorganism symbiotic associations were meticulously highlighted in this chapter. It is startling to study the fundamental aspects of sponge symbiont biology, particularly in the areas of symbiont metabolism and evolution. Conversely, the escalating research interest in this topic promises a bright future for the marine biotechnology. Detailed studies of symbiont transmission, sponge-specific microorganisms, sponge–microbe associations, and improved host phylogenies will greatly facilitate our understanding of the evolution of these systems. Enhanced understanding of the ecological and evolutionary implications of sponge-bacterial symbioses gained over the past decade has prompted considerable new research in this field. The function and physiology of sponge-associated microbes are increasingly important research topics, reflecting our current paucity of knowledge about many of the microbial associates of sponges. Chemical synthesis of sponge-derived compounds or their simpler derivatives offers the most reliable option for sustainable, long-term drug supply if the said compounds can be produced cost effectively. Emerging technologies such as metagenomics and high-throughput microbial cultivation approaches put forward exciting potential for accessing those compounds which are produced by microorganisms. Despite of the various efforts made using molecular phylogenetics and metagenomic analyses, a solid framework was developed by many researchers for the application of ecophysiological analyses of uncultivated microbes Mohamed (2007).

Cracking the codes of communication between sponge and symbiont remains the key to unlocking the surreptitious of symbiosis function and will greatly benefit all areas of research from biodiversity and biotechnology. The metabolic options offered by symbiotic associations provide exciting potential for drug development and highlight the need for new discovery strategies applicable to these complexes systems. It is increasingly being recognized that biosynthetic pathways leading to the synthesis of specialized metabolites may play key roles in the biology of symbiosis. However, additional research is required to reveal the cellular and molecular interactions of sponge–microbe symbiotic relationships. Moreover, future combinations of metagenomic data warrants further attention on the molecular and biochemical bases of sponge–microbe symbiotic interactions. Even though many aspects of sponge–microbe associations are fascinating and important from a basic research point of view, we concede that it is largely biotechnological interest that will sustain this field into the future. The biotechnological potential of sponge–microbe associations has been widely customary, yet the transition from laboratory discovery to large-scale commercial production remains exceptionally intricate. Looking to the future, it is apparent that even greater integration among biotechnologists, microbiologists, chemists, geneticists, zoologists, and aquaculture experts will be crucial in commercialization of the bioactive metabolites derived from sponges and their microbial partners.

Among marine invertebrates, marine sponges from phylum Porifera is the most dominant group responsible for discovering a large number of natural products that have been used as template to develop therapeutic drugs. These natural products have a wide range of therapeutic properties, including antimicrobial, antioxidant, antihypertensive, anticoagulant, anticancer, anti-inflammatory, wound healing and immune modulator, and other medicinal effects. Therefore, marine sponges are considered a rich source of chemical diversity and health benefits for developing drug candidates, cosmetics, nutritional supplements, and molecular probes that can be supported to increase the healthy life span of humans.

The metagenomic libraries from sponge-associated microbial consortia serve as “genomic store houses” that are probed for metabolic and/or other functional genes of interest. These efforts have revealed microbial nitrification as a major pathway in sponges, and we are currently investigating other key pathways of the nitrogen and carbon cycle. These results will provide information about symbiont diversity and metabolic fluxes in the sponge–microbe association which are key themes in symbiosis research.

In summary, the application of high-throughput sequencing has contributed significantly to our understanding of sponge–microbe associations. Even though the main groups of microbes present in marine sponges have not changed, the approach that has to date been employed has helped to describe the rare microbial biosphere in marine sponges. It also demonstrated the presence of sponges with an extraordinarily high diversity of associated microorganisms and sponges which do not appear to have such highly diverse symbiont communities. Especially interesting, and novel, is the observation of a large amount of sponge species-specific microbial organisms which is in stark contrast to a previously widely held theory of global, sponge-specific community. Thus, the main contribution the application of high-throughput sequencing studies has had to date is to increase our understanding of sponge microbiology through the provision of a more comprehensive view of the community structures within various sponges and to provide initial insights into the subsets of these populations which may be playing metabolic roles within the sponge ecosystem. In addition it has enabled us to compare many sponge samples in a more exhaustive way than before, thereby making it possible to infer more robust conclusions with respect to sponge–microbe associations. In the future the use of metatranscriptomics holds much promise in providing further insights into sponge symbiont function.

Acknowledgments The authors are thankful to (DST-SERB), for providing financial assistance under the GRANT NO: SR/FT/LS-129/2011, Govt. of India, New Delhi.

References

Abbamondi GR, De Rosa S, Iodice C, Tommonaro G (2014) Cyclic dipeptides produced by marine spongeassociated bacteria as quorum sensing signals. Nat Prod Commun 9(2):229–232

Aishwarya MS, Lipton AP, Sarika AR (2013) Phylogenetic appraisal of the drug bearing marine sponge *Callyspongia subarmigera* (Ridley, 1884) from South India. Indian J Geol Mar Sci 42:139–145

Almeida C, Hemberger Y, Schmitt SM, Bouhired S, Natesan L, Kehraus S, König GM (2012) Marilines A–C: novel phthalimidines from the sponge-derived fungus *Stachylidium* sp. Chemistry-A Eur J 18 (28):8827–8834

Amagata T, Doi M, Ohta T, Minoura K, Numata A (1998a) Absolute stereostructures of novel cytotoxic metabolites, gymnastatins A–E, from a *Gymnascella* species separated from a *Halichondria* sponge. J Chem Soc Perkin Trans 1(1):3585–3599

Amagata T, Usami Y, Minoura K, Ito T, Numata A (1998b) Cytotoxic substances produced by a fungal strain from a sponge: physico-chemical properties and structures. J Antibiot 51:33–40

Amagata T, Doi M, Tohgo M, Minoura K, Numata A (1999) Dankasterone, a new class of cytotoxic steroid produced by a *Gymnascella* species from a marine sponge. Chem Commun 14:1321–1322

Amagata T, Amagata A, Tenney K, Valeriote FA, Lobkovsky E, Clardy J, Crews P (2003) Unusual C25 steroids produced by a sponge-derived *Penicillium citrinum*. Org Lett 5:4393–4396

Amagata T, Minoura K, Numata A (2006) Gymnastatins F-H, cytostatic metabolites from the sponge-derived fungus *Gymnascella dankaliensis*. J Nat Prod 69:1384–1388

Amagata T, Tanaka M, Yamada T, Minoura K, Numata A (2008) Gymnastatins and Dankastatins, growth inhibitory metabolites of a *Gymnascella* species from a *Halichondria* sponge. J Nat Prod 71:340–345

Aneiros A, Garateix A (2004) Bioactive peptides from marine sources: pharmacological properties and isolation procedures. J Chromatogr B 803(1):41–53

Ang KKH, Holmes MJ, Higa T, Hamann MT, Kara UAK (2000) *In vivo* antimalarial activity of the beta-carboline alkaloid manzamine A. Antimicrob Agents Chemother 44:1645–1649

Arillo A, Bavestrello G, Burlando B, Sara M (1993) Metabolic integration between symbiotic cyanobacteria and sponges: a possible mechanism. Mar Biol 117:159–162

Bavestrello G, Arillo A, Calcinai B, Cattaneo-Vietti R, Cerrano C, Gaino E, Sara M (2000) Parasitic diatoms inside Antarctic sponges. Biol Bull 198(1):29–33

Bayer K, Schmitt S, Hentschel U (2008) Physiology, phylogeny and *in situ* evidence for bacterial and archaeal nitrifiers in the marine sponge *Aplysina aerophoba*. Environ Microbiol 10:2942–2955

Beatty S, Boulton M, Henson D, Koh HH, Murray IJ (1999) Macular pigment and age-related macular degeneration. Br J Ophthalmol 83:867–877

Bentley SD, Chater KF, Cerdeno-Tarraga AM, Challis GL, Thomson NR, James KD, Harris DE, Quail MA, Kieser H, Harper D, Bateman A, Brown S, Chandra G, Chen CW, Collins M, Cronin A, Fraser A, Goble A, Hidalgo J, Hornsby T, Howarth S, Huang CH, Kieser T, Larke L, Murphy L, Oliver K, O'Neil S, Rabbinowitsch E, Rajandream MA, Rutherford K, Rutter S, Seeger K, Saunders D, Sharp S, Squares R, Squares S, Taylor K, Warren T, Wietzorrek A, Woodward J, Barrell BG, Parkhill J, Hopwood DA (2002) Complete genome sequence of the model actinomycete *Streptomyces coelicolor* A3 (2). Nature 417:141–147

Bernan VS, Greenstein M, Maiese WM (1997) Marine microorganisms as a source of new natural products. Adv Appl Microbiol 43:57–90

Besada P, Mamedova L, Thomas CJ, Costanzi S, Jacobson KA (2005) Design and synthesis of new bicyclic diketopiperazines as scaffolds for receptor probes of structurally diverse functionality. Org Biomol Chem 3:2016–2025

Bewley CA, Faulkner DJ (1994) Theonegramide, an antifungal glycopeptide from the Philippine lithistid sponge *Theonella swinhoei*. J Org Chem 59 (17):4849–4852

Bewley CA, Faulkner DJ (1998) Lithistid sponges: star performers or hosts to the stars. Angew Chem Int Ed 37:2162–2178

Bewley CA, Faulkner DJ (1994) Theonegramide, an antifungal glycopeptide from the Philippine lithistid sponge *Theonella swinhoei*. J Org Chem 59 (17):4849–4852

Bhadury P, Mohammad BT, Wright PC (2006) The current status of natural products from marine fungi and their potential as anti-infective agents. J Ind Microbiol Biotechnol 33:325–337

Bhalla TC, Sharma M, Sharma NN (2008) Microbial production of flavours and fragrances; fats and oils; dyes; bioplastics (PHAS); polysaccharides; pharmacologically active substances from marine microbes; anticancer agents and microbial transformation. In: Satyanarayana T, Chand S (eds) Applied microbiology, vol 7. National Science Digital Library NISCAIR, New Delhi, pp 1–34

Biabani MAF, Laatsch H (1998) Advances in chemical studies on low-molecular weight metabolites of marine fungi. J Prakt Chem/Chemiker Ztg 340 (7):589–607

Blunt JW, Copp BR, Munro MH, Northcote PT, Prinsep MR (2006) Marine natural products. Nat Prod Rep 23:26–78

Blunt JW, Copp BR, Hu W, Munro MHG, Northcote PT, Prinsep MR (2009) Marine natural products. Nat Prod Rep 26:170–244

Boopathy NS, Kathiresan K (2010) Anticancer drugs from marine flora: an overview. J Oncol 2010:1–18, 214186

Boot CM, Tenney K, Valeriote FA, Crews P (2006) Highly N-methylated linear peptides produced by an atypical sponge-derived *Acremonium sp.* J Nat Prod 69:83–92

Boot CM, Amagata T, Tenney K, Compton JE, Pietraszkiewicz H, Valeriote FA, Crews P (2007) Four classes of structurally unusual peptides from two marine-derived fungi: structures and bioactivities. Tetrahedon 63:9903–9914

Braekman J, Daloze D (2004) Chemical and biological aspects of sponge secondary metabolites. Phytochem Rev 3:275–283

Brauers G, Edrada RA, Ebel R, Proksch P, Wray V, Berg A, Grafe U, Schachtele C, Totzke F, Finkenzeller G, Marme D, Kraus J, Munchbach M, Michel M, Bringmann G, Schaumann K (2000) Anthraquinones and betaenone derivatives from the sponge-associated fungus *Microsphaeropsis* species: novel inhibitors of protein kinases. J Nat Prod 63:739–745

Bringmann G, Lang G, Muhlbacher J, Schaumann K, Steffens S, Rytik PG, Hentschel U, Morschhauser J, Müller WEG, Sorbicillactone A (2003a) A structurally unprecedented bioactive novel-type alkaloid from a sponge-derived fungus. Prog Mol Subcell Biol 37:231–253

Bringmann G, Lang G, Steffens S, Gunther E, Schaumann K (2003b) Evariquinone, isoemericellin, and stromemycin from a sponge derived strain of the fungus *Emericella variecolor*. Phytochemistry 63:437–443

Bringmann G, Lang G, Steffens S, Schaumann K (2004) Petrosifungins A and B, novel cyclodepsipeptides from a sponge-derived strain of Penicillium brevicompactum. J Nat Prod 67:311–315

Bringmann G, Gulder TAM, Lang G, Schmitt S, Stöhr R, Wiese J, Nagel K, Imhoff JF (2007) Large–scale biotechnological production of the antileukemic marine natural product sorbicillactone A. Mar Drugs 5:23–30

Bugni TS, Ireland CM (2004) Marine derived fungi: a chemically and biologically diverse group of microorganisms. Nat Prod Rep 21:143–163

Bugni TS, Bernan VS, Greenstein M, Janso JE, Maiese WM, Mayne CL, Ireland CM (2003) Brocaenols A-C: novel polyketides from a marine-derived *Penicillium brocae*. J Org Chem 68:2014–2017

Bultel-Poncé V, Debitus C, Berge J, Cerceau C, Guyot M (1998) Metabolites from the sponge associated bacterium *Micrococcus luteus*. J Mar Biotechnol 6:233–236

Bultel-Ponce V, Berge JP, Debitus C, Nicolas JL, Guyot M (1999) Metabolites from the sponge-associated bacterium Pseudomonas species. Mar Biotechnol 1:384–390

Chatterjee S, Kolmakova A (2004) Lactosylceramide synthase: from molecular biochemistry to biological function. Lipids (sphingolipid metabolizing enzymes 2004). Research Signpost, Trivandrum, pp 33–41

Cheng XC, Varoglu M, Abrell L, Crews P, Lobkovsky, Clardy J (1994) Chloriolins AC, chlorinated sesquiterpenes produced by fungal cultures separated from a *Jaspis* marine sponge. J Org Chem 59 (21):6344–6348

Dashti Y, Grkovic T, Abdelmohsen UR, Hentschel U, Quinn RJ (2014) Production of induced secondary metabolites by a co-culture of sponge-associated Actinomycetes, Actinokineospora sp. EG49 and Nocardiopsis sp. RV163. Mar Drugs 12(5):3046–3059

Debitus C, Guella G, Mancini I, Waikedre J, Guemas JP, Nicolas JL, Pietra F (1998) Quinolones from a bacterium and tyrosine metabolites from its host sponge, *Suberea creba* from the Coral Sea. J Mar Biotechnol 6:136–141

Dunlap WC, Battershill CN, Liptrot CH, Cobb RE, Bourne DG, Jaspars M, Long PF, Newman DJ (2007) Biomedicinals from the phytosymbionts of marine invertebrates: a molecular approach. Methods 42:358–376

Ebel R (2006) Secondary metabolites from marine-derived fungi. In: Proksch P, Müller WEG (eds) Frontiers in marine biotechnology. Horizon Scientific Press, Norwich, pp 73–143

Elyakov GB, Kuznetsova T, Mikhailov VV, Maltsev II, Voinov VG, Fedoreyev SA (1991) Brominated diphenyl ethers from a marine bacterium associated with the sponge *Dysidea* sp. Cell Mol Life Sci 47:632–633

Esteves AI, Hardoim CC, Xavier JR, Gonçalves J, Costa R (2013) Molecular richness and biotechnological potential of bacteria cultured from Irciniidae sponges in the north-east Atlantic. FEMS Microbiol Ecol 85 (3):519–536

Fan L, Reynolds D, Liu M et al (2012) Functional equivalence and evolutionary convergence in complex communities of microbial sponge symbionts. PNAS USA 109(27):E1878–E1887

Feling RH, Buchanan GO, Mincer TJ, Kauffman CA, Jensen PR, Fenical W (2003) Salinosporamide A: a highly cytotoxic proteasome inhibitor from a novel microbial source, a marine bacterium of the new genus *Salinospora*. Angew Chem Int Ed Engl 42:355–357

Flowers AE, Garson MJ, Webb RI, Dumdei EJ, Charan RD (1998) Cellular origin of chlorinated diketopiperazines in the dictyoceratid sponge *Dysidea herbacea* (Keller). Cell Tissue Res 292:597–607

Forte B, Malgesini B, Piutti C, Quartieri F, Scolaro A, Papeo G (2009) A submarine journey: the pyrrole-imidazole alkaloids. Mar Drugs 7(4):705–753

Giddens SR, Bean DC (2007) Investigations into the *in vitro* antimicrobial activity and mode of action of the phenazine antibiotic D-alanylgriseoluteic acid. Int J Antimicrob Agents 29:93–97

Goodfellow M, Fiedler HP (2010) A guide to successful bioprospecting: informed by actinobacterial systematics. Antonie Van Leeuwenhoek 98:119–142

Grozdanov L, Hentschel U (2007) An environmental genomics perspective on the diversity and function of marine sponge-associated microbiota. Curr Opin Microbiol 10:215–220

Haefner B (2003) Drugs from the deep: marine natural products as drug candidates. Drug Discov Today 8:536–544

Han Y, Yang B, Zhang F, Miao X, Li Z (2009) Characterization of antifungal chitinase from marine *Streptomyces* sp. DA11 associated with south China sea sponge *Craniella australiensis*. Mar Biotechnol 11:132–140

Haygood MG, Schmidt EW, Davidson SK, Faulkner DJ (1999) Microbial symbionts of marine invertebrates: opportunities for microbial biotechnology. J Mol Microbiol Biotechnol 1:33–43

Hentschel U, Schmid M, Wagner M, Fieseler L, Gernert C, Hacker J (2001) Isolation and phylogenetic analysis of bacteria with antimicrobial activities from the Mediterranean sponges *Aplysina aerophoba* and *Aplysina cavernicola*. FEMS Microbiol Ecol 35 (3):305–312

Hentschel U, Hopke J, Horn M, Friedrich AB, Wagner M, Hacker J, Moore BS (2002) Molecular evidence for a uniform microbial community in sponges from different oceans. Appl Environ Microbiol 68(9):4431–4440

Hentschel U, Usher KM, Taylor MW (2006) Marine sponges as microbial fermenters. FEMS Microbiol Ecol 55(2):167–177

Hentschel U, Piel J, Degnam SM, Taylor MW (2012) Genomic insights into the marine sponge microbiome. Nat Rev Microbiol 10:641–654

Hiort J, Maksimenka K, Reichert M, Perovic-Ottstadt S, Lin WH, Wray V, Steube K, Schaumann K, Weber H, Proksch P, Ebel R, Müller WEG, Bringmann G (2004) New natural products from the sponge-derived fungus *Aspergillus niger*. J Nat Prod 67:1532–1543

Hochmuth T, Piel J (2009) Polyketide synthases of bacterial symbionts in sponges-evolution-based applications in natural products research. Phytochemistry 70:1841–1849

Hoffmann F, Larsen O, Thiel V, Rapp HT, Pape T et al (2005) An anaerobic world in sponges. Geomicrobiol J 22:1–10

Hoffmann F, Radax R, Woebken D, Holtappels M, Lavik G et al (2009) Complex nitrogen cycling in the sponge *Geodia barretti*. Environ Microbiol 11:2228–2243

Höller U, König GM, Wright AD (1999a) A new tyrosine kinase inhibitor from a marine isolate of *Ulocladium botrytis* and new metabolites from the marine fungi *Asteromyces cruciatus* and *Varicosporina ramulosa*. Eur J Org Chem 11:2949–2955

Höller U, König GM, Wright AD (1999b) Three new metabolites from marine derived fungi of the genera *Coniothyrium* and *Microsphaeropsis*. J Nat Prod 62:114–118

Holzinger A, Meindl U (1997) Alga Micrasterias. Cell Motil Cytoskeleton 38:365–372

Hummel H, Sepers ABJ, De Wolf L, Melissen FW (1988) Bacterial growth on the marine sponge *Halichondria panicea* induced by reduced water flow rate. Mar Ecol Prog Ser 42:195–198

Ikeda H, Ishikawa J, Hanamoto A, Shinose M, Kikuchi H, Shiba T, Sakaki Y, Hattori M, Ōmura S (2003) Complete genome sequence and comparative analysis of the industrial microorganism *Streptomyces avermitilis*. Nat Biotechnol 21:526–531

Ivanova EP, Vysotskii MV, Svetashev VI, Nedashkovskayal OI, Gorshkoval NM, Mikhailovl VV, Yumoto N, Shigeri Y et al (1999) Characterization of *Bacillus* strains of marine origin. Int Microbiol 2:267–271

Jadulco R, Proksch P, Wray V, Sudarsono, Berg A, Grafe U (2001) New macrolides and furan carboxylic acid derivative from the sponge derived fungus *Cladosporium herbarum*. J Nat Prod 64:527–530

Jadulco R, Brauers G, Edrada RU, Ebel R, Wray V, Sudarsono, Proksch P (2002) New metabolites from sponge derived fungi *Curvularia lunata* and *Cladosporium herbarum*. J Nat Prod 65:730–733

Jadulco R, Edrada RA, Ebel R, Berg A, Schaumann K, Wray V, Steube K, Proksch P (2004) New communesin derivatives from the fungus *Penicillium* sp. derived from the Mediterranean sponge *Axinella verrucosa*. J Nat Prod 67:78–81

Jayatilake GS, Thornton MP, Leonard AC, Grimwade JE, Baker BJ (1996) Metabolites from an Antarctic sponge associated bacterium, *Pseudomonas aeruginosa*. J Nat Prod 59:293–296

Jimenez E, Ribes M (2007) Sponges as a source of dissolved inorganic nitrogen: nitrification mediated by temperate sponges. Limnol Oceanogr 52:948–958

Kurtboke DI (2005) Actinophages as indicators of actinomycete taxa in marine environments. Anton Leeuw Int J Gen Mol Microbiol 87(1):19–28

Liangjie Jin, Wen Ma, Chongsheng Peng, Ying Yin, Bailiang Xu, Fengli Zhang, Yuewei Guo, Zhiyong Li (2011) Bacillamide C production by the optimized cultivation of the *Bacillus atrophaeus* strain C89 associated with the South China Sea sponge *Dysidea avara*. Process Biochem 46(5):1153–1159

Kasai H, Katsuta A, Sekiguchi H, Matsuda S, Adachi K, Shindo K, Yoon J, Yokota A, Shizuri Y (2007) *Rubritalea squalenifaciens* sp. nov., a squalene-producing marine bacterium belonging to subdivision 1 of the phylum '*Verrucomicrobia*'. Int J Syst Evol Microbiol 57:1630–1634

Kelecom A (2002) Secondary metabolites from marine microorganisms. An Acad Bras Cienc 74:151–170

Kim TK, Garson MJ, Fuerst JA (2005) Marine actinomycetes related to the '*Salinospora*' group from the Great Barrier Reef sponge *Pseudoceratina clavata*. Environ Microbiol 7:509–518

Kim TK, Hewavitharana AK, Shaw PN, Fuerst JA (2006) Discovery of a new source of rifamycin antibiotics in marine sponge actinobacteria by phylogenetic prediction. Appl Environ Microbiol 72:2118–2125

Kim SK, Wijesekara I (2010) Development and biological activities of marine-derived bioactive peptides: a review. J Funct Foods 2(1):1–9

Kobayashi J, Ishibashi M (1993) Bioactive metabolites of symbiotic marine microorganism. Chem Rev 93:1753–1769

Kobayashi M, Aoki S, Gato K, Matsunami K, Kurosu M, Kitagawa I (1994) Marine natural products. XXXIV. Trisindoline, a new antibiotic indole trimer, produced by a bacterium of *Vibrio* sp. separated from the marine sponge *Hyrtios altum*. Chem Pharm Bull 42:2449–2451

Koizumi Y, Arai M, Tomoda H, Omura S (2004) Oxaline, a fungal alkaloid, arrests the cell cycle in M phase by inhibition of tubulin polymerization. Biochim Biophys Acta 1693:47–55

Kolachana P, Smith MT (1994) Induction of kinetochore-positive micronuclei in human lymphocytes by the anti-fungal drug griseofulvin. Mutat Res 322:151–159

König GM, Kehraus S, Seibert SF, Abdel-Lateff A, Müller D (2005) Natural products from marine organisms and their associated microbes. Chembiochem 7:229–238

Kralj A, Gurgui M, König GM, van Echten-Deckert G (2007) Trichothecenes induce accumulation of glucosylceramide in neural cells by interfering with lactosylceramide synthase activity. Toxicol Appl Pharmacol 225:113–122

Laatsch H (2006) Marine bacterial metabolite. In: Proksch P, Müller WEG (eds) Frontiers in marine biotechnology. Horizon Bioscience, Norfolk, pp 225–288

Landrum JT, Bone RA (2001) Arch Biochem Biophys 385:28–40

Lang G, Wiese J, Schmaljohann R, Imhoff JF (2007) New pentaenes from the sponge-derived marine fungus *Penicillium rugulosum*: structure determination and biosynthetic studies. Tetrahedron 63:11844–11849

Lee YM, Mansoor TA, Hong J, Lee C-O, Bae KS, Jung JH (2007) Polyketides from a sponge-derived fungus, *Aspergillus versicolor*. Nat Prod Sci 13:90–96

Lee YM, Dang HT, Hong J, Lee C-O, Bae KS, Kim DK, Jung JH (2010) A cytotoxic lipopeptide from the sponge-derived fungus *Aspergillus versicolor*. Bull Kor Chem Soc 31:205–208

Li Z (2009) Advances in marine microbial symbionts in the China Sea and related pharmaceutical metabolites. Mar Drugs 7(2):113–129

Li Q, Wang G (2009) Diversity of fungal isolates from three Hawaiian marine sponges. Microbiol Res 164 (2):233–241

Lin W, Brauers G, Ebel R, Wray V, Sudarsono, Berg A, Proksch P (2002) Novel chromone derivatives from the fungus *Aspergillus versicolor* isolated from the marine sponge *Xestospongia exigua*. J Nat Prod 66:57–61

Liu R, Cui C, Duan L, Gu Q, Zhu W (2005) Potent *in vitro* anticancer activity of metacycloprodigiosin and undecylprodigiosin from a sponge-derived actinomycete *Saccharopolyspora* sp. *nov*. Arch Pharm Res 28:1341–1344

Liu H, He Y, Jiang H, Peng H, Huang X, Zhang X, Thomashow LS, Xu Y (2007) Characterization of a phenazine-producing strain *Pseudomonas chlororaphis* GP72 with broad-spectrum antifungal activity from green pepper rhizosphere. Microbiology 54:302–306

Liu F, Sun W, Su F, Zhou K, Li Z (2012) Draft genome sequence of the sponge-associated strain *Bacillus atrophaeus* C89, a potential producer of marine drugs. J Bacteriol 194(16):4454

Loane E, Nolan JM, Donovan O, Bhosale P, Bernstein PS, Beatty S (2008) Surv Ophthalmol 53:68–81

Lopez-Gresa MP, Cabedo N, González-Mas MC, Ciavatta MA, Avila C, Primo J (2009) Terretonins E and F, inhibitors of the mitochondrial respiratory chain from the marine-derived fungus *Aspergillus insuetus*. J Nat Prod 72:1348–1351

Mackay HJ, Twelves CJ (2003) Protein kinase C: a target for anticancer drugs? Endocr Relat Cancer 10:389–396

Mares-Perlman JA, Millen AE, Ficek TL, Hankinson SE (2002) J Nutr 132:518S–524S

Mayer AMS (1999) Marine pharmacology in 1998: antitumor and cytotoxic compounds. Pharmacologist 41:159–164

McClintock JB, Charles DA, Bill JB, Rob WMV (2005) Ecology of Antarctic marine sponges: an overview. Integr Comp Biol 45:359–368

McLeod MP, Warren RL, Hsiao WWL, Araki N, Myhre M, Fernandes C, Miyazawa D, Wong W, Lillquist AL, Wang D, Dosanjh M, Hara H, Petrescu A, Morin RD, Yang G, Stott JM, Schein JE, Shin H, Smailus D, Siddiqui AS, Marra MA, Jones SJM, Holt R, Brinkman FSL, Miyauchi K, Fukuda M, Davies JE, Mohn WW, Eltis LD (2006) The complete genome of *Rhodococcus* sp. RHA1 provides insights into a catabolic powerhouse. Proc Natl Acad Sci USA 103:15582–15587

Mehbub MF, Lei J, Franco C, Zhang W (2014) Marine Sponge derived natural products between 2001 and 2010: trends and opportunities for discovery of bioactives. Mar Drugs 12(8):4539–4577

Miki W, Otaki N, Yokoyama A, Izumida H, Shimidzu N (1994) Okadaxanthin, a novel C50-carotenoid from a bacterium *Pseudomonas* sp. KK10206C associated with a marine sponge *Halichondria okadai*. Experientia 50:684–686

Mohamed NM (2007) Ecophysiology of microbial communities associated with Marine Sponges Ircinia Strobilina and Mycale Laxissima. Ph.D. thesis, University of Maryland, 251 p

Mohamed IE, Gross H, Pontius A, Kehraus S, Krick A, Kelter G, Maier A, Fiebig H, König GM (2009) Epoxyphomalin A and B, prenylated polyketides with potent cytotoxicity from the marine-derived fungus *Phoma* sp. Org Lett 11:5014–5017

Müller WEG, Grebenjuk VA, Thakur NL, Thakur AN, Batel R, Krasko A, Breter HJ (2004a) Oxygen-controlled bacterial growth in the sponge *Suberites domuncula*: toward a molecular understanding of the symbiotic relationships between sponge and bacteria. Appl Environ Microbiol 70(4):2332–2341

Müller WEG, Thakur NL, Ushijima H, Thakur AN, Krasko A, Pennec G, Indap MM, Perovic-Ottstadt S, Schröder HC, Lang G, Bringmann G (2004b) Matrix-mediated canal formation in primmorphs from the sponge *Suberites domuncula* involves the expression of a CD36 receptor-ligand system. J Cell Sci 117:2579–2590

Munro MH, Blunt JW, Dumdei EJ, Hickford SJ, Lill RE, Li S, Battershill CN, Duckworth AR (1999) The discovery and development of marine compounds with pharmaceutical potential. J Biotechnol 70 (1):15–25

Nagai K, Kamigiri K, Matsumoto H, Kawano Y, Yamaoka M, Shimoi H, Watanabe M, Suzuki K (2002) YM-202204, a new antifungal antibiotic produced by marine fungus *Phoma sp*. J Antibiot 55:1036–1041

Nagai K, Kamigiri K, Arao N, Suzumura K, Kawano Y, Yamaoka M, Zhang H, Watanabe M, Suzuki K (2003) YM-266183 and YM-266184, novel thiopeptide antibiotics produced by *Bacillus cereus* isolated from a marine sponge. I. Taxonomy, fermentation, isolation, physico-chemical properties and biological properties. J Antibiot 56:123–128

Naqash SY, Nazeer RA (2010) Antioxidant activity of hydrolysates and peptide fractions of *Nemipterus japonicus* and *Exocoetus volitans* muscle. J Aquat Food Prod Technol 19(3–4):180–192

Newman DJ, Hill RT (2006) New drugs from marine microbes: the tide is turning. J Ind Microbiol Biotechnol 33(7):539–544

Nishino H, Murakoshi M, Ii T, Takemura M, Kuchide M, Kanazawa M, Mou XY, Wada S, Webster NS, Negri AP, Webb RI, Hill RT (2002) A spongin-boring alphaproteobacterium is the etiological agent of disease in the Great Barrier Reef sponge *Rhopaloeides odorabile*. Mar Ecol Prog Ser 232:305–309

Numata A, Amagata T, Minoura K, lto T (1997) Gymnastatins, novel cytotoxic metabolites produced by a fungal strain from a sponge. Tetrahedon Lett 38:5675–5678

Oclarit JM, Okada H, Ohta S, Kaminura K, Yamaoka Y, Iizuka T, Miyashiro S, Ikegami S (1994) Anti-*Bacillus* substance in the marine sponge, *Hyatella* species, produced by an associated *Vibrio* species bacterium. Microbios 78:7–16

Oliynyk M, Samborskyy M, Lester JB, Mironenko T, Scott N, Dickens S, Haydock SF, Leadlay PF (2007) Complete genome sequence of the erythromycin-producing bacterium *Saccharopolyspora erythraea* NRRL23338. Nat Biotechnol 25:447–453

Orlandini V, Maida I, Fondi M, Perrin E, Papaleo MC, Bosi E, Fani R (2014) Genomic analysis of three sponge-associated Arthrobacter Antarctic strains, inhibiting the growth of *Burkholderia cepacia* complex bacteria by synthesizing volatile organic compounds. Microbiol Res 169(7):593–601

Osinga R, Armstrong E, Burgess JG, Hoffmann F, Reitner J, Schumann-Kindel G (2001) Sponge microbe associations and their importance for sponge bioprocess engineering. Hydrobiologia 461:55–62

Pabel CT, Vater J, Wilde C, Franke P, Hofemeister J, Adler B, Bringmann G, Hacker J, Hentschel U (2003) Antimicrobial activities and matrix-assisted laser desorption/ionization mass spectrometry of *Bacillus* isolates from the marine sponge *Aplysina aerophoba*. J Mar Biotechnol 5:424–434

Pandey S, Sree A, Sethi DP, Kumar CG, Kakollu S, Chowdhury L, Dash SS (2014) A marine sponge associated strain of *Bacillus subtilis* and other marine bacteria can produce anticholinesterase compounds. Microb Cell Factories 13(1):24

Paz Z, Komon-Zelazowska M, Druzhinina IS, Aveskamp MM, Shnaiderman A, Aluma Y, Carmeli S, Ilan M, Yarden O (2010) Diversity and potential antifungal properties of fungi associated with a Mediterranean sponge. Fungal Divers 42:17–26

Petchi RR, Vijaya C, Parasuraman S (2013) Antiarthritic activity of ethanolic extract of *Tridax procumbens* (Linn.) in Sprague Dawley rats. Pharmacognosy Res 5:113–117

Piel J (2002) A polyketide synthase-peptide synthetase gene cluster from an uncultured bacterial symbiont of *Paederus* beetles. Proc Natl Acad Sci U S A 99:14002–14007

Piel J (2004) Metabolites from symbiotic bacteria. Nat Prod Rep 21:519–538

Piel J, Hui D, Fusetani N, Matsunaga S (2004a) Targeting modular polyketide synthases with iteratively acting acyltransferases from metagenomes of uncultured bacterial consortia. Environ Microbiol 6:921–927

Piel J, Hui D, Wen G, Butzke D, Platzer M, Fusetani N, Matsunaga S (2004b) Antitumor polyketide biosynthesis by an uncultivated bacterial symbiont of the marine sponge *Theonella swinhoei*. PNAS USA 101:16222–16227

Proksch P, Edrada RA, Ebel R (2002) Drugs from the seas: current status and microbiological implications. Appl Microbiol Biotechnol 59:125–134

Ramm W, Schatton W, Wagner-Dobler I, Wray V, Nimtz M, Tokuda H, Enjyo F, Nishino H, Beil W, Heckmann R, Lurtz V, Lang S (2004) Diglucosyl-glycerolipids from the marine sponge-associated *Bacillus pumilus* strain AAS3: their production, enzymatic modification and properties. Appl Microbiol Biotechnol 64:497–504

Rachanamol RS, Lipton AP, Thankamani V, Sarika AR, Selvin J (2014) Molecular characterization and bioactivity profile of the tropical spongeassociated bacterium *Shewanella* algae VCDB. Helgol Mar Res 68(2):263–269

Rhee KH (2004) Cyclic dipeptides exhibit synergistic, broad spectrum antimicrobial effects and have anti-mutagenic properties. Int J Antimicrob Agents 24:423–427

Rice LB (2006) Antimicrobial resistance in gram-positive bacteria. Am J Infect Control 34(5):S11–S19

Rogers SA, Huigens RW, Cavanagh J, Melander C (2010) Synergistic effects between conventional antibiotics and 2-aminoimidazole-derived anti-biofilm agents. Antimicrob Agents Chemother 54:2112–2118

Roser DJ, Ashbolt N, Ho G, Mathew K, Nair J, Ryken-Rapp D, Toze S (2005) Hydrogen sulphide production tests and the detection of groundwater faecal contamination by septic seepage. Water Sci Technol 51:291–300

Rungprom W, Siwu ER, Lambert LK, Dechsakulwatana C, Barden MC, Kokpol U, Garson MJ (2008) Cyclic tetrapeptides from marine bacteria associated with the seaweed *Diginea* sp. and the sponge *Halisarca ectofibrosa*. Tetrahedron 64(14):3147–3152

Sabdono A, Radjasa OR (2008) Microbial symbionts in marine sponges: marine natural product factory. J Coast Dev 11:57–62

Sajilata MG, Singhal RS, Kamat MY (2008) The carotenoid pigment zeaxanthin—a review. Compr Rev Food Sci Food Saf 7:29–49

Sakai R, Higa T, Jefford CW, Bernardinelli G (1986) Manzamine A, a novel antitumor alkaloid from a sponge. J Am Chem Soc 108:6404–6405

Salomon CE, Magarvey NA, Sherman DH (2004) Merging the potential of microbial genetics with biological and chemical diversity: an even brighter future for marine natural product drug discovery. Nat Prod Rep 21(1):105–121

Sathiyanarayanan G, Gandhimathi R, Sabarathnam B, Kiran GS, Selvin J (2014) Optimization and production of pyrrolidone antimicrobial agent from marine sponge-associated *Streptomyces* sp. MAPS15. Bioprocess Biosyst Eng 37(3):561–573

Scala F, Fattorusso E, Menna M, Taglialatela-Scafati O, Tierney M, Kaiser M, Tasdemir D (2010) Bromopyrrole alkaloids as lead compounds against protozoan parasites. Mar Drugs 8(7):2162–2174

Schmidt EW, Bewley CA, Faulkner DJ (1998) Theopalauamide, a bicyclic glycopeptide from filamentous bacterial symbionts of the lithistid sponge *Theonella swinhoei* from Palau and Mozambique. J Org Chem 63:1254–1258

Schmidt EW, Obraztsova AY, Davidson SK, Faulkner DJ, Haygood MG (2000) Identification of the antifungal peptide-containing symbiont of the marine sponge *Theonella swinhoei* as a novel δ-proteobacterium, "*Candidatus entotheonella palauensis*". Mar Biol 136(6):969–977

Schmitt S, Weisz JB, Lindquist N, Hentschel U (2007) Vertical transmission of a phylogenetically complex microbial consortium in the viviparous sponge *Ircinia felix*. Appl Environ Microbiol 73(7):2067–2078

Schmitt S, Angermeier H, Schiller R, Lindquist N, Hentschel U (2008) Molecular microbial diversity survey of sponge reproductive stages and mechanistic insights into vertical transmission of microbial symbionts. Appl Environ Microbiol 74 (24):7694–7708

Schmitt S, Tsai P, Bell J, Fromont J, Ilan M, Lindquist N (2012) Assessing the complex sponge microbiota: core, variable and species specific bacterial communities in marine sponges. ISME J 6:564–576

Schneemann I, Nagel K, Kajahn I, Labes A, Wiese J, Imhoff JF (2010) Comprehensive investigation of marine actinobacteria associated with the sponge *Halichondria panicea*. Appl Environ Microbiol 76:3702–3714

Scopel M, Abraham WR, Henriques AT, Macedo AJ (2013) Dipeptide *cis-cyclo* (Leucyl-Tyrosyl) produced by sponge associated *Penicillium* sp. F37 inhibits biofilm formation of the pathogenic *Staphylococcus epidermidis*. J Bioorg Med Chem Lett 23(3):624–626

Selvin J (2009) Exploring the antagonistic producer *Streptomyces* MSI051: implications of polyketide synthase gene type II and a ubiquitous defense enzyme phospholipase A2 in the host sponge *Dendrilla nigra*. Curr Microbiol 58:459–463

Selvin J, Joseph S, Asha KRT, Manjusha WA, Sangeetha VS, Jayaseema DM, Antony MC, Vinitha AJD (2004) Antibacterial potential of antagonistic *Streptomyces* sp. isolated from marine sponge *Dendrilla nigra*. FEMS Microbiol Ecol 50:117–122

Selvin J, Shanmughapriya S, Gandhimathi R, Kiran GS, Ravji TR, Natarajaseenivasan K, Hema TA (2009) Optimization and production of novel antimicrobial agents from sponge associated marine actinomycetes *Nocardiopsis dassonvillei* MAD08. Appl Microbiol Biotechnol 83:435–445

Shigemori H, Bae MA, Yazawa K, Sasaki T, Kobayashi J (1992) Alteramide A, a new tetracyclic alkaloid from a bacterium *Alteromonas* sp. associated with the marine sponge *Halichondria okadai*. J Org Chem 57:4317–4320

Shindo K, Asagi E, Sano A, Hotta E, Minemura N, Mikami K, Tamesada E, Misawa N, Maoka T (2008) Diapolycopenedioic acid xylosyl esters A, B, and C, novel antioxidative glyco-C30-carotenoic acids produced by a new marine bacterium *Rubritalea squalenifaciens*. J Antibiot 61:185–191

Simister RL, Deines P, Botté ES, Webster NS, Taylor MW (2012) Sponge-specific clusters revisited: a comprehensive phylogeny of sponge-associated microorganisms. Environ Microbiol 14:517–524

Steinberg PD, Kjelleberg S (2010) Functional genomic signatures of sponge bacteria reveal unique and shared features of symbiosis. ISME J 4:1557–1567

Stierle AC, Cardellina JH, Singleton FL (1988) A marine Micrococcus produces metabolites ascribed to the sponge *Tedania ignis*. Experientia 44:1021–1022

Su J, Jin L, Jiang Q, Sun W, Zhang F, Li Z (2013) Phylogenetically diverse *ureC* genes and their expression suggest the urea utilization by bacterial symbionts in marine sponge *Xestospongia testudinaria*. PLoS One 8(5):e64848

Subramani R, Kumar R, Prasad P, Aalbersberg W (2013) Cytotoxic and antibacterial substances against multidrug resistant pathogens from marine sponge symbiont: citrinin, a secondary metabolite of *Penicillium* sp. Asian Pac J Trop Biomed 3(4):291–296

Sufrin JR, Finckbeiner S, Oliver CM (2009) Marine-derived metabolites of *S*-adenosylmethionine as templates for new anti-infectives. Mar Drugs 7:401–434

Suzumura K, Yokoi T, Funatsu M, Nagai K, Tanaka K, Zhang H, Suzuki K (2003) YM-266183 and YM-266184, novel thiopeptide antibiotics produced by *Bacillus cereus* isolated from a marine sponge. II. Structure elucidation. J Antibiot 56:129–134

Taylor MW, Schupp PJ, Dahllof I, Kjelleberg S, Steinberg PD (2004) Host specificity in marine sponge-associated bacteria, and potential implications for marine microbial diversity. Environ Microbiol 6 (2):121–130

Taylor MW, Radax R, Steger D, Wagner M (2007) Sponge-associated microorganisms: evolution, ecology and biotechnological potential. Microbiol Mol Biol Rev 71:295–347

Thacker RW (2005) Impacts of shading on sponge-cyanobacteria symbioses: a comparison between host-specific and generalist associations. Integr Comp Biol 45:369–376

Thakur NL, Müller WEG (2005) Sponge-bacteria association: a useful model to explore symbiosis in marine invertebrates. Symbiosis 39:109–116

Thakur NL, Hentschel U, Krasko A, Pabel CT, ANR AC, Müller WEG (2003) Antibacterial activity of the sponge *Suberites domuncula* and its primmorphs: potential basis for epibacterial chemical defense. Aquat Microb Ecol 31:77–83

Thakur AN, Thakur NL, Indap MM, Pandit RA, Datar VV, Müller WEG (2005) Antiangiogenic, antimicrobial and cytotoxic potential of sponge-associated bacteria. Mar Biotechnol 7:245–252

Thomas T, Rusch D, DeMaere MZ, Yung PY, Lewis M, Halpern A, Heidelberg KB, Egan S (2010) Functional genomic signatures of sponge bacteria reveal unique and shared features of symbiosis. ISME J 4 (12):1557–1567

Thornton MP (1995) MS thesis. Florida Institute of Technology, Melbourne, 51

Udwary DW, Zeigler L, Asolkar RN, Singan V, Lapidus A, Fenical W, Jensen PR, Moore BS (2007) Genome sequencing reveals complex secondary metabolome in the marine actinomycete *Salinispora tropica*. Proc Natl Acad Sci U S A 104:10376–10381

Unson MD, Faulkner DJ (1993) Cyanobacterial symbiont biosynthesis of chlorinated metabolites from *Dysidea herbacea* (Porifera). Cell Mol Life Sci 49:349–353

Usami Y, Ikura T, Amagata T, Numata A (2000) First total syntheses and configurational assignments of cytotoxic trichodenones A–C. Tetrahedron Asymmetry 11:3711–3725

Usher KM, Bergman B, Raven JA (2007) Exploring Cyanobacterial Mutualisms. Annu Rev Ecol Evol Syst 38:255–273

Varoglu M, Corbett TH, Valeriote FA, Crews P (1997) Asperazine, a selective cytotoxic alkaloid from a sponge-derived culture of *Aspergillus niger*. J Org Chem 62:7078–7079

Venkateswarlu Y, Biabani MAF (1995) Phycopsisenone A new phenolic secondary metabolic from the sponge *Physopsis* spp. J Nat Prod 58:269–270

Wang G (2006) Diversity and biotechnological potential of the sponge-associated microbial consortia. J Ind Microbiol Biotechnol 33:545–551

Wang G, Abrell LM, Avelar A, Borgeson BM, Crews P (1998) New hirsutane based sesquiterpenes from salt water cultures of a marine sponge derived fungus and the terrestrial fungus *Coriolus consors*. Tetrahedron 54:7335–7342

Webster NS, Negri AP, Webb RI, Hill RT (2002) A spongin-boring alpha-proteobacterium is the etiological agent of disease in the great barrier reef sponge Rhopaloeides odorabile. Mar Ecol Prog Ser 232:305–309

Webster NS, Taylor MW, Behnam F, Lücker S, Rattei T, Whalan S, Horn M, Wagner M (2010) Deep sequencing reveals exceptional diversity and modes of transmission for bacterial sponge symbionts. Environ Microbiol 12(8):2070–2082

Webster NS, Taylor MW (2012) Marine sponges and their microbial symbionts: love and other relationships. Environ Microbiol 14:335–346

Wicke C, Hners M, Wray V, Nimtz M, Bilitewski U, Lang S (2000) Production and structure elucidation of glycoglycerolipids from a marine sponge associated *Microbacterium* species. J Nat Prod 63:621–626

Wiese J, Ohlendorf B, Blümel M, Schmaljohann R, Imhoff JF (2011) Phylogenetic Identification of Fungi Isolated from the Marine Sponge *Tethya aurantium* and Identification of Their Secondary Metabolites. Mar Drugs 9:561–585

Wilkinson CR (1978) Microbial associations in sponges II. Numerical analysis of sponge and water bacterial populations. Mar Biol 49:169–176

Wilkinson CR (1983) Net primary productivity in coral reef sponges. Science 219:410–412

Wilkinson CR (1984) Immunological evidence for the Precambrian origin of bacterial symbioses in marine sponges. Proc R Soc Lond B 220(1221):509–517

Wilkinson CR, Fay P (1979) Nitrogen fixation in coral reef sponges with symbiotic cyanobacteria. Nature 279:527–529

Williams DE, Burgoyne DL, Rettig SJ, Andersen RJ, Fathi-Afshar ZR, Allen TM (1993) The isolation of Majusculamide C from the sponge *Ptilocaulis trachys* collected in Enewetak and determination of the absolute configuration of the 2-methyl-3-aminopentanoic acid residue. J Nat Prod 56:545–551

Xie LW, Jiang SM, Zhu HHL, Sun W, Ouyang YC, Dai SK, Li X (2008) Potential inhibitors against *Sclerotinia sclerotiorum*, produced by the fungus *Myrothecium* sp. associated with the marine sponge *Axinella* sp. Eur J Plant Pathol 122:571–578

Xin ZH, Fang Y, Du L, Zhu T, Duan L, Chen J, Gu Q, Zhu W (2007) Aurantiomides A–C, quinazoline alkaloids from the sponge-derived fungus *Penicillium aurantiogriseum* SP0–19. J Nat Prod 70:853–855

Zhang Y, Mu J, Feng Y, Kang Y, Zhang J, Gu P, Wang Y, Ma L, Zhu Y (2009) Broad-spectrum antimicrobial epiphytic and endophytic fungi from marine organisms: isolation, bioassay and taxonomy. Mar Drugs 7(2):97–112

Zheng Z, Zeng W, Huang Y, Yang Z, Li J, Cai H, Su W (2000) Detection of antitumor and antimicrobial activities in marine organism associated actinomycetes isolated from the Taiwan Strait, China. FEMS Microbiol Lett 188:87–91

Zheng L, Chen H, Han X, Lin W, Yan X (2005a) Antimicrobial screening and active compound isolation from marine bacterium NJ6-3-1 associated with the sponge *Hymeniacidon perleve*. World J Microbiol Biotechnol 21:201–206

Zheng L, Yan X, Xu J, Chen H, Lin W (2005b) *Hymeniacidon perleve*, associated bioactive *Pseudomonas* spp. NJ-6-3-1. Prikl Biokhim Mikrobiol 41:35–39

Zhou K, Zhang X, Zhang F, Li Z (2011) Phylogenetically diverse cultivable fungal community and polyketide synthase (PKS), non-ribosomal peptide synthase (NRPS) genes associated with the South China Sea sponges. Microb Ecol 62:644–654

7 Remarks on the Chemo Biological Applications of Marine Sponges

P. Sunil Kumar

Abstract

Since many years, marine sponges have been ranked at the top with respect to the discovery of bioactive compounds of potential pharmaceutical applications. The diversity in chemical structures of sponge-derived metabolites is related to an equally diverse pattern of activities ranging from antifouling to anti-HIV properties. These discoveries have attracted the attention of scientists to develop feasible strategies for obtaining sponge-derived metabolites on commercial scale. Sponges have the potential to provide future drugs against important diseases, such as cancer, a range of viral diseases, malaria and inflammations. The molecular mode of action of most metabolites is still unclear. However, the way by which they interfere with the pathogenesis of a wide range of diseases has been reported by many workers. This chapter majorly focuses the important chemo-active molecules from marine sponges and their applications in various disease therapies.

Keywords

Sponge • Biological properties • Secondary metabolites • Bioactive compounds

7.1 Introduction

Marine organisms are a rich source of natural products differing from those of terrestrial organisms in both chemical structures and peculiarities of biological actions including antibacterial, antiviral, antifungal and antiparasitic substances. So far about 8200 new marine natural products of low molecular weight have been isolated from marine organisms. The compounds isolated from these organisms have unique chemical structures, unusually high biological activity and participation in intra- and interspecific relationships in under water communities that focus attention on bioactive substances.

More than 5300 different products from marine sponges and their associated microorganisms have now been described, and more than 200 new metabolites from these organisms

P.S. Kumar (✉)
Kerala University of Fisheries and Ocean Studies, Panangad, Kochi, Kerala 682506, India
e-mail: sukkuedavetty2012@gmail.com

R. Pallela, H. Ehrlich (eds.), *Marine Sponges: Chemicobiological and Biomedical Applications*,
DOI 10.1007/978-81-322-2794-6_7

are reported each year (Faulkner 1992). Marine sponges are a gold mine with respect to the diversity of their secondary metabolites and can provide potential drugs against a variety of diseases (Jain and Tiwari 2007). Some substances have the potential for clinically effective treatments. These include ara-A (vidarabine), an antiviral drug used against the herpes simplex encephalitis virus; manzamine, with activity against malaria, tuberculosis, and HIV; lasonolides, with antifungal activity; psammaplin A, with antibacterial activity; spongistatins, with anticancer effects; mycalamides A and B, which inhibit protein synthesis causing apoptosis; pateamines, which harbour immunosuppressive and apoptotic properties; peloruside, with potential antibiotic activity; manolide and luffariellolide with anti-inflammatory effects; chondropsin class, which inhibited the growth of tumours; and halichondrin and geodiamolides A, B and H with antiproliferative activity. The natural products spongouridine and spongothymidine from *Cryptotheca* served as a template for the synthesis of the antiviral drug ara-A (vidarabine) that led to the development of acyclovir, which has effects against herpes virus and is used as medicinal drug (azidothymidine) for AIDS.

Scientists and pharmaceutical manufacturers have moved on to synthetic organic molecules over natural products. Sponges contain many peculiar chemical compounds that are not seen in other animals, but their biotechnological potentials have not yet been fully assessed or documented.

7.2 Properties of Sponges

Sponges occupy a unique position in the evolutionary ladder, and hence they exhibit unique physical and physiological properties.

7.2.1 Cell Aggregation and Regeneration

Cell aggregation is a unique property of sponges. Squeezed-out individual cells of sponge when kept in fresh sea water may aggregate and form a new individual with all features of the original sponge. This property of sponges is species specific. This crucial and exciting property of sponge will one day provide some clue to fight the dreadful diseases like AIDS. The regeneration capacity of sponge is well known for the last several years. The information gathered in this line can be utilized in culturing sponge with drug potential.

7.2.2 Sponge: A Group with Only Functional Systems

In sponges, true germ layers are absent. Histological systems are absent and are represented by functional systems like myocytes, the cells for local contraction, sclerocytes for secreting spicules, amoebocytes for budding and choanocytes for water circulation. Archaeocytes, a specific type of cells, are totipotent in nature *i.e.*, they are capable of undergoing any change to suit the structure or function of any other type of cells. These properties of sponges enable them to serve as an efficient tool in cytological studies.

7.2.3 Microsymbiont Involvement in Drug Production

The association of sponge with higher invertebrates is well documented, but their association with uni- and multicellular algae and bacteria is poorly studied. In sponges about 3–40 % of their total volume is occupied by bacteria. It is studied that sponge phagocytoses the bacteria when scarcity of food occurs. But it is difficult to culture the bacteria that harbour sponge. Zoochlorella is a common symbiotic bacterium seen inside sponge. They release photosynthetic products and other metabolites on receiving signals from the host sponge.

7.2.4 Episymbiont Isolation and Enzyme Production

Sponges are a good source of epibiotic bacteria and fungi. One hundred three aerobic heterophilic bacteria and 21 fungi were isolated from sponges. Protease endoglucanase, amylase and asparaginase occur in association with sponges and are capable of producing enzymes with pharmaceutical applications.

7.3 Classification of Bioactive Compounds from Marine Sponges

Bioactivity in marine organisms is now a field of great interest to scientists all over the world, and this has emanated with the discovery of arabinose nucleosides from marine sponge *Tethya crypta* (de Laubenfels). A group of natural products including steroids, glycosides, quinoid compounds, alkaloids, peptides, etc. were isolated from sponges. Some of these compounds possess antimicrobial, antitumour and antifungal properties, and even some act as immunostimulants and inhibitors of enzymes (Belarbi et al. 2003). Many sponges have evolved biochemical and physiological mechanisms that include the production of bioactive compounds for purposes such as reproduction, communication, protection against predation, infection and competition.

7.3.1 Halichondrin Compounds

Halichondrin B and halichondrin B12 are the common halichondrin compounds so far isolated from marine sponge *Halichondria* (Fig. 7.1). Halichondrin B12 is more active over its counterparts like halichondrin B. Halichondrin B12 has reached preclinical stage as anticancer compound. Okadaic acid, another compound, was isolated from *H. okadai*. It exhibits marked in vitro cytotoxicity. Glycookadaic acid obtained exhibits anticachexia activity. Discoderm A, the first antimicrobial peptide isolated from the sponge *Discodermia dissoluta*, inhibits the growth of *Bacillus subtilis*, *S. aureus*, *E. coli* and *P. aeruginosa*.

7.3.2 Discodermolides

Discodermolides (Fig. 7.2), a powerful immunosuppressive agent, might have a future role in suppressing rejection after transplant surgery. Discodermolide was isolated from rare deep-water sponges, which are found only in the Bahamas. Discodermolide has immunosuppressive effects. However, trials using this substance were discontinued due to lack of efficacy and toxicity problems. The potential remains for its use in combination drug therapy. Five novel cyclic depsipeptides were isolated from *Theonella* spp.

7.3.3 Other Compounds

Sponge products ara-A (vidarabine), the antiviral drug, manzamine A (activity against malaria, tuberculosis, HIV and others), lasonolides

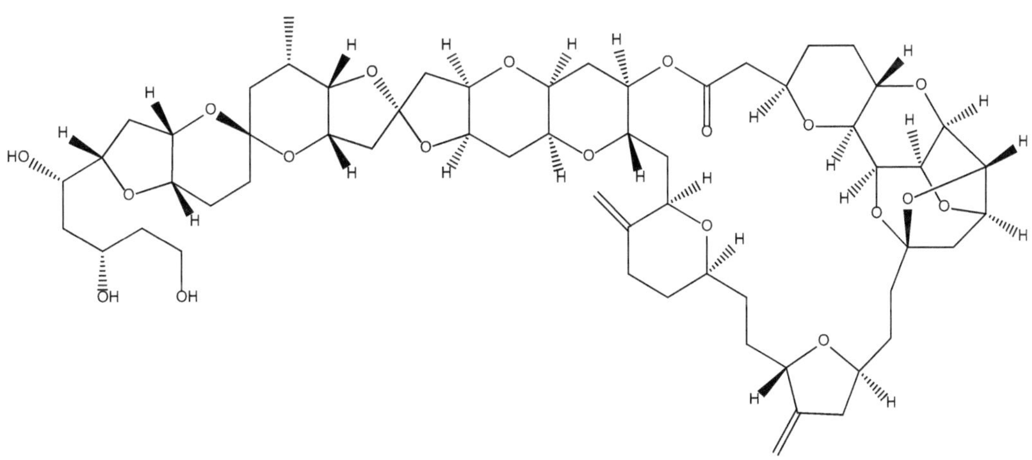

Fig. 7.1 Halichondrin B (Source: (Hirata and Uemura 1986))

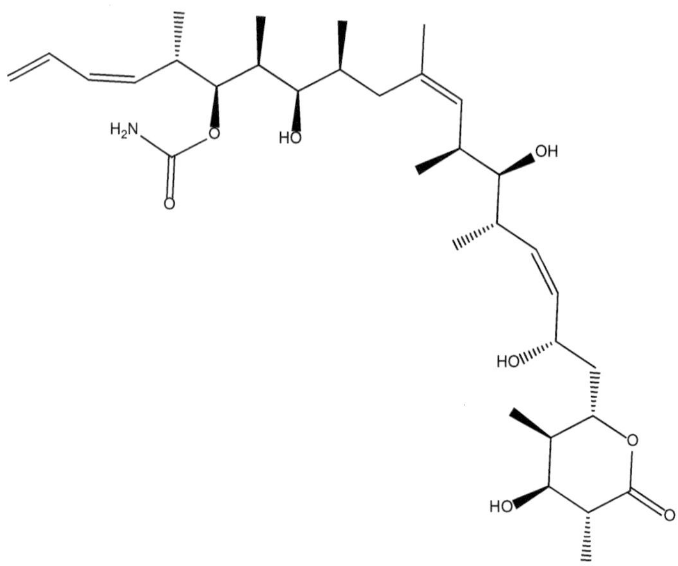

[(3Z,5S,6S,7S,8R,9S,11Z,13S,14S,15S,16Z,18S)-8,14,18-Trihydroxy-19-[(2S,3R,4S,5R)-4-hydroxy-3,5-dimethyl-6-oxooxan-2-yl]-5,7,9,11,13,15-hexamethylnonadeca-1,3,11,16-tetraen-6-yl] carbamate

Fig. 7.2 Discodermolide (Source: (Gunasekara et al. 1991))

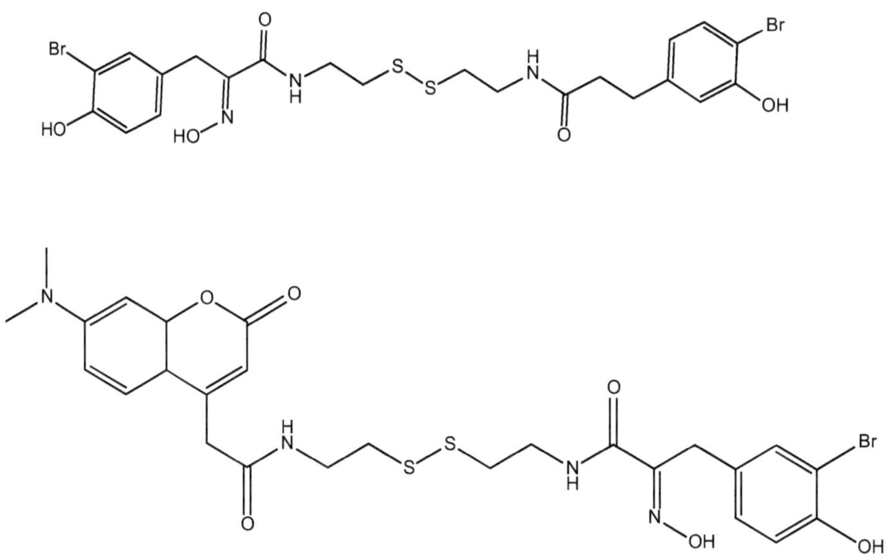

Fig. 7.3 Psammaplin A (Source: (Shim et al. 2004))

(antifungal activity) and psammaplin A (antibacterial activity) are considered as promising leads (Fig. 7.3).

Jaspamide, another compound, was isolated from *Jaspis* spp. (Fig. 7.4). This compound exhibited potential insecticidal activity and inhibits the growth of the fungus *Candida albicans*. *Geodia* sp. is a good source of geodiamolides A and B. Clionamide extracted from *Cliona celata* yields clionamide which

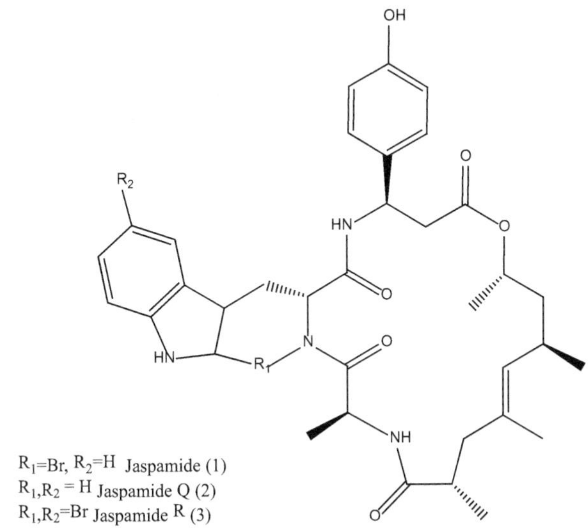

Fig. 7.4 Jaspamide (Source: (Ebada et al. 2009))

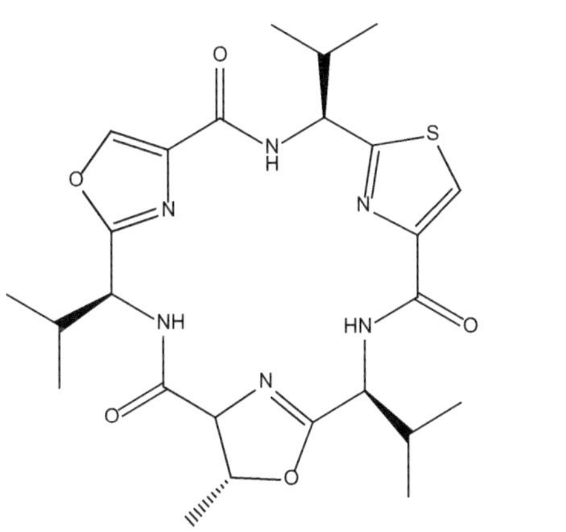

Fig. 7.5 Bistratamide D (Source: (Total synthesis of bistramide 1994))

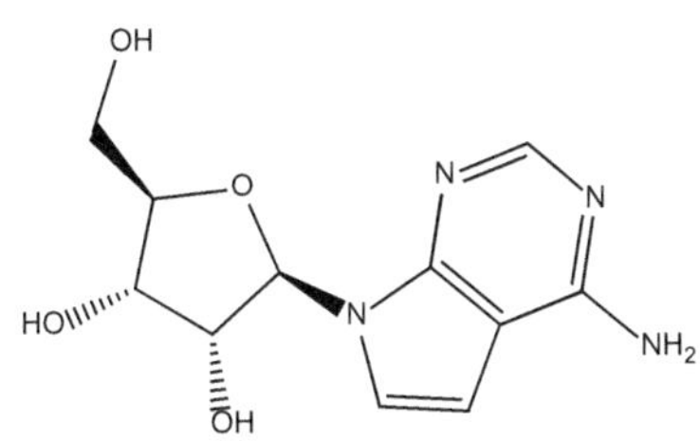

Fig. 7.6 Tubercidin (Source: (Jack et al. 1989))

exhibits antimicrobial activity against *Staphylococcus aureus*. Sesquiterpenes have been isolated from Chondrosida, *Halichondria* spp. and *Strongylophora* spp. Tetracyclic terpenes have been reported from *Siphonochalina siphonella*, *Axinella* spp. and *Raspaciona aculeata*.

Later, ara-A and spongouridine were discovered as natural metabolites from the Mediterranean gorgonian *Eunicella cavolini* (Dunlap et al. 2007). Cytarabine from *Cryptotheca crypta* is routinely used in the treatment of leukaemia and lymphoma. Its fluorinated derivative, gemcitabine, has been approved for use in patients with pancreatic, breast, bladder and non-small cell lung cancer. Isocyano terpenes kalihinol A and kalihinol F extracted from the sponge *Acanthella cavernosa* inhibit the growth of the bacteria *Bacillus subtilis*, *Staphylococcus aureus* and *Candida albicans*. Manoalide from the Palauan sponge *Luffariella variabilis* has antibiotic effects against *Streptomyces pyogenes* and *Staphylococcus aureus* (Ebada et al. 2010) (Figs. 7.5 and 7.6).

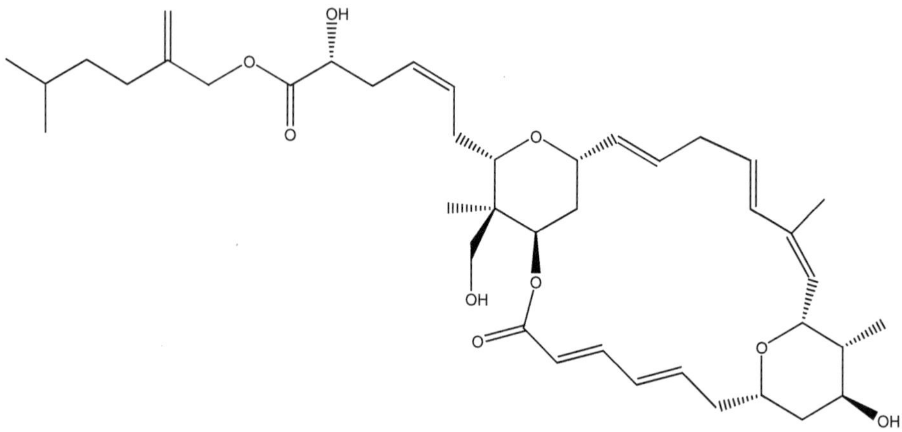

Fig. 7.7 Lasonolide A (Source: (Zhang et al. 2012))

The growth of *Bacillus subtilis* and *Staphylococcus aureus* is inhibited by 3-formamido-1(10)-cadinene from the Paulan sponge *Axinyssa aplysinoides*. The isocyano terpenes kalihine A and B from *Acanthella klethra* have antifungal effects against *Mortierella ramannianus* and *Penicillium chrysogenum* (Paul et al. 2007) (Fig. 7.7).

Mycalamides A and B, which inhibit protein synthesis causing apoptosis; pateamines, which harbour immunosuppressive and apoptotic properties; peloruside, with potential antibiotic activity; manolide (Dunlap et al. 2007) and luffariellolide (Ebada et al. 2010), with anti-inflammatory effects; chondropsin class, which inhibited the growth of tumours (Dunlap et al. 2007); and halichondrin (Molinski et al. 2009) and geodiamolides A, B and H (Mayer and Gustafson 2008), with antiproliferative activity have promising applications.

7.4 Conclusion

Chemistry of bioactive compounds is relatively a new branch of modern pharmacology. The antiviral and antitumour activity of these compounds necessitates bulk production of these species by cell culture (Donia and Hamann 2003). Further research in this area focuses on isolating and identifying organisms that are new to science for the production of useful new compounds, determining their chemical structures, the biochemical pathways by which they are produced and the environmental and physiological mechanisms that trigger their production. The isolation of a pure compound from crude extract of sponges is a laborious process. This branch of pharmacology involves isolation of lure compounds and in vitro screening to ascertain their bioactive potential which requires coordinated research work among different groups consisting of chemists and biologists. It is difficult to cultivate sponges and their microbial fauna. Valuable compounds must be extracted and purified from samples. The samples are collected by hand using scuba diving or with the aid of submersibles equipped with robotic arms. The awkwardness of both these methods complicates their use in the modern pharmaceutical industry (Molinski et al. 2009). Much research in the field of natural products has focused on sponges (Table 7.1).

Table 7.1 Compounds of bioactive potential from marine sponges

Compound	Source	Property
Diterpenes	*Spongia officinalis*	Antifungal
Adociaquinone B	*Xestospongia* sp.	Antitumour
Bistratamide D	*Lissoclinum bistratum*	Antitumour
Makaluvamine N	*Zyzzya fuliginosa*	Anti-catalytic
Tubercidin	*Didemnum voettzkowi*	Adenosine kinase inhibitors
Euryspongiols	*Euryspongia* sp.	Antihistamine
Daytronic acid	*Dactylospongia elegans*	Terpenoid
Didemnum B	*Tridemnum* sp.	Antitumour
Arabinosides	*Tethya crypta*	Anticancer and antiviral
Bryostatins	*Bugula neritina*	Antitumour lactones
Echinosulphonic acid	*Echinodactylum* sp.	Bromondole sulphonic acids
Euryspongiols	*Euryspongia* sp.	Antihistaminic
Discodermolide	*Discodermia dissoluta*	Immunosuppressive agent
Pseudopterosin E	*Pseudopterogorgia elisabethae*	Anti-inflammatory agent
Dolastatin	*Dolabella auricularia*	Anticancer
Halichondrin B	*Lissodendoryx* spp.	Anticancer
Makaluvamine	*Zyzzia* sp.	Neurotoxin
Aerothionine	*Verongia aerophoba*	Antibiotic
Onnamide A	*Theonella* spp.	Antitumour

References

Lowe JT, Wrona IE, Panek JS (2007) Total synthesis of bistramide A. Org Lett 9(2):327–330

Belarbi EH, Gómez AC, Chisti Y, Garcí F, Grima EM (2003) Producing drugs from marine sponges. Biotechnol Adv 21(7):585–598

Donia M, Hamann MT (2003) Marine natural products and their potential applications as anti-infective agents. Lancet Infect Dis 3(6):338–348

Dunlap WC, Battershill CN, Liptrot CH, Cobb RE, Bourne DG, Jaspars M, Long PF, Newman DJ (2007) Biomedicinals from the phytosymbionts of marine invertebrates: a molecular approach. Methods 42(4):358–376

Ebada SS, Wray V, de Voogd NJ, Deng Z, Lin W, Proksch P (2009) Two new Jaspamide derivatives from the marine sponge *Jaspis splendens*. Mar Drugs 7(3):434–444

Ebada SS, Lin W, Proksch P (2010) Bioactive sesterterpenes and triterpenes from marine sponges: occurrence and pharmacological significance. Mar Drugs 8(2):313–346

Faulkner D (1992) Marine natural products. Nat Prod Rep 9(4):323–364

Gunasekara SP, Gunasekera M, Longley RE, Schutte GK (1991) Discodermolide: a new bioactive polyhydroxylated lactone from the marine sponge Discodermia dissoluta. J Org Chem 6(3):1346–1346 [Erratum document cited in CA113(9):75187b]

Hirata Y, Uemura D (1986) Halichondrins- antitumour polyethermacrolides from a marine sponge. Pure Appl Chem 58(5):701–710

Jack DA, Roger JB, Stewart G. Howard BC, Steven BL, Steven SM, Donald FS, Roland KR (1989) Synthesis of tubercidin, 6-chlorotubercidin and related nucleosides. Nucleosides Nucleotides 8(7):1201–1216

Jain R, Tiwari A (2007) Sponges: an invertebrate of bioactive potential. Cur Sci Assoc Bangalore, India 93(4): 444

Zhang YW, Ghosh AK, Pommier Y (2012) Lasonolide A, a potent and reversible inducer of chromosome condensation. Cell Cycle 11(23):4424–4435

Mayer AM, Gustafson KR (2008) Marine pharmacology in 2005–2006: antitumour and cytotoxic compounds. Eur J Cancer 44(16):2357–2387

Molinski TF, Dalisay DS, Lievens SL, Saludes JP (2009) Drug development from marine natural products. Nat Rev Drug Discov 8(1):69–85

Paul VJ, Arthur KE, Ritson-Williams R, Ross C, Sharp K (2007) Chemical defenses: from compounds to communities. Biol Bull 213(3):226–251

Shim JS, Lee HS, Shin J, Kwon HJ (2004) Psammaplin A, a marine natual product inhibits aminopeptidase and suppresses angiogenesis in vitro. Cancer Lett 203(2): 163–169

8 Sponges as Biomonitors of Metal Toxicity in the Aquatic Systems

Koigoora Srikanth and Janapala Venkateswara Rao

Abstract

Aquatic ecosystems around the world are subjected to unrelenting stress caused by urban sprawl, discharge of effluents from domestic and other ecological impacts such as infrastructure, land reclamation for port and industrial development and habitat destruction. Sponges are efficient to accumulate the metals in their tissues if the concentration of these metals is low in the ambient environment. Difference in the accumulation of metals was observed in different species of sponges collected from the studied areas. Their strong capacity to accumulate metals and the diversity of sponges in the different coastal regions of the world make them ideal biomonitors of metal contamination of the confined region.

Keywords

Bioindicators • Metal accumulation • Sponges • Biomonitoring

8.1 Introduction

These days there is a lot of awareness on the complexity of aquatic environment and the delicate balance which is seen maintained with the existing ecosystem. Measures should be undertaken to understand the complex interactions of the nature so as to prevent or minimise its future degradation. Monitoring of contaminants in the aquatic ecosystem has been paid much attention due to the high risk involved upon exposure to different organisms confined to that area/coast. The existing contaminants get transformed into more toxic form upon exposure to suitable environmental conditions such as pH, salinity and temperature and are bioaccumulated in the organisms, magnified in the food chain and causing deleterious effects on human health. Every year a new chemical is introduced in the market and is finding its way into the aquatic environment (Srikanth and Rao 2014). There

K. Srikanth (✉)
CESAM, Department of Chemistry, University of Averio, Averio 3810193, Portugal

Toxicology Unit, Biology Division, CSIR-Indian Institute of Chemical Technology, Hyderabad 500607, India
e-mail: koigooras@ua.pt

J.V. Rao
Toxicology Unit, Biology Division, CSIR-Indian Institute of Chemical Technology, Hyderabad 500607, India

R. Pallela, H. Ehrlich (eds.), *Marine Sponges: Chemicobiological and Biomedical Applications*, DOI 10.1007/978-81-322-2794-6_8

are an increased number of contaminants entering into the aquatic environment which usually include heavy metals, hydrocarbons both aliphatic and aromatic, radionuclides, polyaromatic hydrocarbons, etc. Among all these contaminants, the pollution by heavy metals is considered potentially hazardous to all the aquatic biota and humans (Cachada et al. 2012). The contamination of the aquatic environment by heavy metals has become a global phenomenon due to their persistence and accumulation of the inhabiting organisms (Gochfeld 2003). The heavy metals belong to the priority of contaminants according to the European Water Framework Directive 2000/60/EC (Arsene et al. 2009; Srikanth and Rao 2014). Heavy metals are the results of both natural and anthropogenic process. Heavy metal pollution usually results from atmospheric deposition, geological weathering and disposal from agricultural, municipal, industrial or residential products (Demirak et al. 2006). The heavy metals entering the aquatic environment can be absorbed from the water column onto the surface of the fine particles of sediment and are moving along with it. Heavy metals are precipitated in different biogeochemical mechanisms and are effecting the aquatic ecosystem through bioaccumulation and biomagnification process (Onen et al. 2011). The current trend is the use of various indicator organisms in environmental research studies for biomonitoring. The organism selected for monitoring of heavy metals in coastal water facilitates spatial and temporal comparison (Rainbow and Blackmore 2001; Onen et al. 2011). The use of indicator organisms to assess levels of contaminants particularly heavy metals in aquatic ecosystem is very common these days. There are a number of biomonitors in aquatic environments for which the kinetic and accumulation processes are well known. Those aquatic organisms which take up particulate and dissolved contaminants may be used as the indicators of the bioavailability, over time, of a specific pollutant (Desideri et al. 2010). Recently, much attention is paid on the risk involved to human health while considering the environment itself; hence, biomonitors are used to evaluate the impact of different contaminants on the human health. There are feasible problems linked with biomonitoring research that can assist in mitigating the misunderstanding and mistakes before the investigation is being conducted. Invertebrates which are sessile and benthic are said to be the ideal tools for evaluating the local pollution monitoring since they cannot avoid from the waterborne contaminants released in that particular region (Cebrian et al. 2007). Among them sponges are the Preffered biomonitors because of their most ideal features which satisfy the ideal biomonitors. The main aim of this chapter is to present the use of sponges as ideal biomonitors for monitoring contamination in different coastal regions of the world. This study provides how appropriate sponges are for knowing the contamination status of a particular environment.

8.2 Biomonitoring

The use of various organisms and biomaterials to test its feasibility for monitoring various contaminants under different exposure conditions downstream from the discharge is defined as a biomonitor. This is typically performed to obtain information on the environmental characteristics of the inhabiting environment. This is usually done to obtain relevant information in the biomonitoring of chemical analysis of the environmental matrix such as water, sediment and tissues which is frequently performed to reveal the contamination status of the environment; however, it does not provide any relevant information relating to the toxicity of such pollutants on the organisms. The organisms used for biomonitoring should follow few of the criteria of which the prime most important thing is their permanent and widespread distribution of the organism even in the remote areas. Biomonitoring is a scientific approach for monitoring the environment along with the human exposure to both natural and synthetic chemicals, which is relied on the individual organism's different tissues and their fluids. This is seen evident by the increase in the levels of the marker chemicals which usually include enzymes or some other breakdown product of the chemicals or some other changes occurring in the organism reflecting their exposure. The analysis of these chemicals provides the evidence of the contaminants that are bioconcentrated or bioaccumulated in the tissues of the organisms and the corresponding effects induced. The obtained results from the measurements of the above-mentioned chemicals are due to the cumulative effects of both natural and man-made chemicals that have entered and got accumulated in the organisms and have induced the corresponding changes. The cosmopolitan distribution of the selected organism and their inhabiting space for biomonitoring provide the data on the potential effects and actual integrated toxicities of contaminants, indicating the corresponding deleterious effects on the environment.

The available literature states that most of the publications or review articles on bioindicators are about the metal contamination, and the most extensively used organisms for these studies are invertebrates, fish and mammal species (Burger); similarly in the case of aquatic pollution monitoring, the most commonly used biomonitors are insect, molluscs, fish, plankton, plant and birds. Each of the species used has its own advantages for biomonitoring of metal pollution in aquatic environment when compared to others, but the use of sponges is very limited or scanty hence in the current chapter the task of using marine sponges as biomonitors because of its unique features which satisfy most of the characters required for biomonitoring studies. During the recent past some of the researchers have shown interest on using the marine sponges for biomonitoring studies. Sponges do show some of the unique characters which make them convenient tools for characterising the state of the aquatic ecosystem (Rao et al. 2006, 2007, 2009; Pan et al. 2011; de Mestre et al. 2012; Srikanth and Rao 2014). The environmental crises in temperate and tropical ecosystem are usually evaluated by using sponge communities (Perez et al. 2005); the body of sponge is seen surrounded by a number of inhalant pores through which they are seen actively pumping the water through their porous body, and during this process they are seen trapping the food material which is seen present in both suspended and dissolved phases; hence, these are considered as "biological particle traps". In this manner, these organisms can accumulate and concentrate the contaminants from both suspension and dissolved phases. High concentration of different contaminants has been reported in different species of sponges: metals (Patel et al. 1985; Verdenal et al. 1990; Richelle-Maurer et al. 2003; Hansen et al. 1995; Perez et al. 2005; Cebrian et al. 2006; Rao et al. 2006, 2007, 2009; Pan et al. 2011; de Mestre et al. 2012; Genta-Jouve et al. 2012), hydrocarbons (Zahn et al. 1981) and organochlorinated compounds (Verdenal et al. 1990; Arnoux et al. 1992; Perez et al. 2003). Biomonitoring studies performed indicate that accumulation is a function of the quantity present in the environment and that bioconcentration factors may be very high (Richelle-Maurer et al. 1993; Hansen et al. 1995; Cebrian et al. 2003; Perez et al. 2003). Biomonitoring studies have used a number of aquatic organisms including mussels. "Mussel Watch Programme" is seen practised in the Columbia River Basin, but a similar type of proposal "Sponge Watch Programme" was proposed by Patel et al. (1985) and Hansen et al. (1995); surprisingly, these organisms have been used little in the estimation of contamination levels.

8.3 Marine Animals as Biomonitors

Algae are recognised as one of the important primary producers in the aquatic ecosystems which have an important role in cycling of dissolved oxygen in the water through the process of photosynthesis. Algae mostly satisfy the basic requirements of biomonitor: sedentary, easy to collect, cosmopolitan distribution and the concentration of metals up to a satisfactory degree (Conti et al. 2002). Recently, the focus has turned on the use of algae for biomonitoring of metal contaminants in the aquatic ecosystem (Nouri et al. 2009; Girgin et al. 2010). Although biomonitoring of algae has provided significant results, limited usage of such species is expected in the actual investigations due to some of their biological characters. The complexity of algae makes the analysis more difficult, and still more information about the physiology and further studies are necessary to clarify the accumulation patterns. Moreover, the high tolerance, ease of sampling and sedentary nature of the macrophytes make them ideal for monitoring aquatic contamination. The changes in the photosynthesis, respiration and growth all are the indicators of plants exposed to model pollutants (Kumar et al. 2006). The biomonitoring ability of macrophytes is weakened by the existence of different species in different environmental conditions. However, zooplankton play an important role in the biogeochemical cycling of metals in the aquatic system (Stewart et al. 2005); for this reason in several studies, zooplankton organisms have been specifically used for biomonitoring purpose of the bioavailability of elements in the aquatic systems (Kahle and Zauke 2003). These include protozoa, crustacean, amphipod, copepod, etc., of the aquatic ecosystem; these species are seen accumulating and metabolising a reasonable amount of contaminants from the surrounding ambient environment and offer as the most feasible organism for biomonitoring. Different zooplankton species are seen widely employed in the biomonitoring process in spite of their limitations because of their validity as biomonitors is very low.

Protozoa constitute an important component of the food chain among the aquatic ecosystems; it does reflect the characters of function and structure of the entire aquatic ecosystem. Many protozoan species are very sensitive to pollutants and seen responding towards the variation in the environment (Xu et al. 2002). Protozoans are easy to handle, convenient for sampling and suitable for performing the various bioassays; these organisms serve as the perfect bioindicators because of their diversity, configuration and distribution characters; they are the ideal choice for biomonitoring. Because of their lower level of organisation, biomonitoring studies have been limited using protozoans. However, crustaceans are seen typically living in the strandlines of shores, which are seen absorbing the metals from both solution and food (Weeks and Rainbow 1993); these crustaceans are usually consuming their food from the plant material which is usually macrophytes. These macrophytes receive the metals from the surrounding ambient environment, and the most predominant organisms of this particular area are crustaceans (Rainbow et al. 2006).

The representative organisms for biomonitoring among crustaceans are *Daphnia magna* and *Artemia salina* which are most sensitive to the environmental pollutants such as metals, pesticides and other contaminants. The most predominant features observed in the exposed organism usually include morphology, behaviour, growth and fertilisation (Adema 1978). This species has also its own limitations such as generation time and weakness; the lack of development markers would prevent this from using them as the possible biomonitoring organisms to some extent. Moreover, the use of aquatic organisms especially macrobenthic invertebrates as the biomonitoring organisms has gained much significance. Amphipods and crustaceans are among the organisms widely studied in this respect (Fialkowski et al. 2003), and the accumulation of metals by these organisms is well known. Amphipods are important components of freshwater, estuarine and eulittoral ecosystem and are widespread and are thus interesting candidates for biomonitoring studies (Rinderhagen et al. 2000). These organisms show a net accumulation which is one of the prerequisites for using them as biomonitors (Clason et al. 2004); however, when compared to bivalves, their usage as biomonitors is quite less. Different species of amphipods from different locations have shown exponential relationships between different metals and their body lengths, while in some other species, no length dependency was required. There is seen difference in between the juveniles and adults in response to the toxic metals, but these are not prominent as that of amphipods for biomonitoring purpose.

Biomonitors provide integrated measures of the ecotoxicologically significant fraction of ambient metal in those waters (Rainbow 1995); these offer the most direct measure of metal pollution as concerned with the existing environment. Mussels do have a number of characters which make them one of the ideal biomonitoring organisms: they have cosmopolitan distribution, are sedentary and stable for different environmental contaminants, have high bioaccumulation factors of pollutants, can live long, have slow metabolising enzyme activities and have the ability to survive in both laboratory and field conditions. Different bivalve molluscs such asclams, oyster and mussels are extensively used in the monitoring of the environmental concentration of metals. In biomonitoring bioaccumulation is the key factor using this species. Biomonitoring of Cu, Zn, Fe, Cd, Pb, Mn and Ni is monitored using *M. trossulus* from the Baltic Sea (Rainbow et al. 2000). Most of the bivalve molluscs live along shoreline which limits their use as the biomonitoring organism. However, similar to bivalve molluscs, gastropods are often employed as biomonitoring organisms to metals, organotin compounds and also endocrine-disrupting chemicals because they are available round the year and easy to collect (Bayen et al. 2004). The monitoring of contaminants using gastropods is preferred because these take up the contaminants from all the environmental compartments such as the medium in which they are surviving; through ingestion of food materials, it is able to concentrate the contaminants in their body. The two most common gastropods of the genera *Monodonta* (snails) and *Patelloida* (limpets) are among the most commonly used organisms for biomonitoring (Cubadda et al. 2001). Most of the studies indicated that in the Mediterranean Basin, the gastropods are seen extensively used for biomonitoring of contaminants. The accumulation of various contaminants varies among different species of gastropods which offer them as the most feasible organisms for monitoring pollution (Liang et al. 2004). The different species of gastropods based on their age are being used for monitoring the contaminants which include *M. turbinate*, *M. mutabilis*, *P. caerulea* and *P. lusitanica* (Cubadda et al. 2001). Heavy metals (Cd, Cu, Zn, Fe, Pb, Ni and Co) using *P. caerulea* from both polluted and nonpolluted waters from the Iskenderun of Gulf of Turkey are monitored (Yuzereroglu et al. 2010). Similar kind of studies using three species of gastropods *R. rapiformis*, *C. virgineus* and *H. pugilinus* is seen used for monitoring the metal contamination in the coastal regions of Pondicherry, India (Kaviarasan et al. 2012). Alteration in the morphology of the gastropods exposed to organotin compounds is now the matter of considerable research; the size of the penis and the vas deferens index exposed to TBT are monitored in the species of *T. clavigera*. All the above features mentioned make them one of the useful biomonitoring organisms for monitoring pollution in the aquatic ecosystem (Horiguchi et al. 1997).

Fish are seen extensively used in the biomonitoring studies because of its large size, long life cycle, easy to rare, etc. However, the fish species are at the topmost position in the aquatic food chain and may directly affect the humans as it is the major food source. The change in the behavioural response when exposed to different contaminants is also considered as one of the important factors in biomonitoring studies (Rao et al. 2006); the other parameters which are seen used for the biomonitoring purpose include growth, reproduction, metabolism and fecundity. A number of fish species are seen involved in the biomonitoring programme which usually include zebra fish, medaka, rainbow trout, *Catla catla* and rohu (Svecevicius and Kazlauskiene 2011; Momoshima et al. 2007; Garg et al. 2009; Malik et al. 2010). The long duration, its mobility and high experimental cost limit the use of the fish species in biomonitoring studies.

8.4 Sponges as Biomonitors

Sponges are primitive poriferans still existing on the planet earth. All the sponges are aquatic and usually found in both fresh and marine waters. These sponges are adherent to the substratum filtering large quantity of waters through their inherent pores. They take up their food through this filtering mechanism during this process, and they take up microscopic food particles and also the dissolved organic matter present in the surrounding environment (Srikanth and Rao 2014). Sponges are primitive multicellular animals which do have basic level of cellular organisation. Sponges constitute a number of cells and the outer layer of sponge is covered by specialised cells called pinacocytes. Sponges internally constitute canals and chambers through which water is circulated continuously. The chambers constitute flagellated cells called choanocytes which are also involved in the circulation of water and food capture. Sponges constitute a collagenous matrix called mesohyl which is filled in the spaces between the canals and chambers, and this matrix has a number of cells supporting the spicules and fibres. Sponges are seen growing in different sizes and shape, and these are based on the spicules and internal minerals secreted by the specialised cells. Skeletons of sponges usually constitute siliceous or calcareous spicules and organic collagenous fibres called spongin. Based on these materials the sponges are usually soft, hard, fragile, crispy or compressible. Sponges are occurring in different shapes: ramose, flat, branched, bulbous and cushion type. The entire body of sponges contain numerous openings called the incurrent pores and larger pores called excurrent pores. Sponges exist in different shapes among different species and genera which is based on the number of environmental factors such as hydrodynamics, light and turbidity.

Sponges are among the major benthic and sedentary groups with prominent role in many coral reef communities around the world (Ilan et al. 2004); these sponges act as biogenic habitats that support abundant and highly diverse epifaunal and infaunal microbial communities which make up significant biomass of their host. Despite the huge potential of sponges as biomonitors, their study is still limited till date. Sponges do have some of the unique characters of a biomonitor like its cosmopolitan distribution, typical multicellular animal, filter feeding, sedentary nature, long life span, tolerance to both physical and chemical fluctuations and ability to remain stable for longer period. Sponges filter large quantities of water per hour (Vogel 1977), and during this process, they accumulate a number of contaminants in their tissues which makes them a potential biomonitor, and also because of their adaptability to a variety of ecological niches both freshwater and marine ecosystems, it makes them ideal for biomonitoring (Barnes 2009). Sedentary nature of sponges helps to monitor the quality of that particular environment where they are inhabiting when compared to a mobile organism, and because of their sessile nature, they are readily transplanted, and many will grow successfully from fragments (Roberts et al. 2006).

The aquatic ecosystem is continuously flooded with different contaminants such as metals and PAH, originated from the anthropogenic activities which are usually released from the urban communities and industries which are released via the riverine inputs. The effects and the fate of these contaminants on the existing aquatic organisms have to be extensively studied by the aquatic toxicologist (Rao et al. 2006, 2007, 2009). To monitor the level of contaminants, a large array of biomonitoring methods are followed that allow time and biological integrated measures of the level of contaminants and their potential effects on the existing organisms (Rainbow and Phillips 1993). Usually the most preferred biomonitors for evaluating the environmental crisis need the following characters: cosmopolitan distribution, high tissue content, able to sustain under varying environmental conditions, easy identification, long life span, sedentary nature, etc. The survival or the absence of a monitoring organism is first used as the clue for monitoring long-term effects of contamination on communities and population. The monitoring of biochemical changes at the infra-individual scale can also be used as a biomarker of exposure effects on a short-term basis (Lyons et al. 2010). Those organisms which are seen accumulating the contaminants and reflecting the same in their tissues are used to monitor the bioavailable fraction of contaminants in the ambient environment (Zhou et al. 2008). The evaluation of the contamination state of the aquatic ecosystem is dependent on several species which exactly reflect the major sources of pollutants (waterborne, particular, sediments). As said above a number of organism are being used for monitoring the contaminants such as protozoans, crustacean, amphipod and copepod; bivalve molluscs, gastropods and fish have been commonly used as biomonitors of metal contamination in the aquatic ecosystem (Roberts et al. 2006). Considering the above-renowned characteristics of appropriate biomonitors, marine sponges of the phylum Porifera have been recommended as undoubtedly suitable candidates for more detailed investigations. Sponges are sessile invertebrates found in both freshwater and marine water systems and are seen available along the shore and in deep oceans; they are widely distributed in the sublittoral area, and they can dominate in both diversity and biomass in temperate rocky bottoms to polar continental shelf. These are seen filtering large volumes of water through their inhalant pores and canal system, which is seen favouring metal accumulation from the dissolved and suspended phases (Vogel 1977; Rao et al. 2006, 2007, 2009; Pallela et al. 2011; Srikanth and Rao 2014). Earlier studies already indicated them as good biomonitors (Patel et al. 1985; Perez et al. 2005; Cebrian et al. 2007; Rao et al. 2006, 2007, 2009; Pan et al. 2011; de Mestre et al. 2012; Genta-Jouve

Table 8.1 Marine sponge species studied as environmental biomonitors and bioindicators

Sponge species	Contaminant	References
T. lyncurium	Benzopyrene	Zahn et al. (1981)
S. and *P. foetida*	Cd, Cr, Sn, Co, Zn, Ti, Cu, Mn	Patel et al. (1985)
T. charcoti	Cd, Zn	Capon et al. (1993)
H. panacea Pallas	Cd	Olesen and Weeks (1994)
H. panacea Pallas	Cu, Zn, Cd, Cr	Hansen et al. (1995)
T. charcoti	Cd	Bargagli et al. (1996)
L. fusifera and *L. abietina*	Pb, Cu and Zn	Efremova et al. (2002)
Heteromyenia sp., *E. fragilis*	Ethylbenzene, nonylphenol, bisphenol A	Hill et al. (2002)
C. crambe	Cu, Pb	Cebrian et al. (2003)
S. officinalis	Polychlorobiphenyls	Perez et al. (2003)
H. bowerbanki, D. camera, H. panicea	Cd, Hg, Pb, Cu, Zn, As	Philip et al. (2003)
Several Mediterranean sponges	Heavy metals	Perez et al. (2004)
S. officinalis	Fe, Pb, Cr	Perez et al. (2005)
S. officinalis	Cu, Zn, Ag	Berthet et al. (2005)
P. testudinaria	Al, Fe, Mn, As, Ni, Co, Cu, Se	Rao et al. (2006)
H. balfourensis, M. acerata, S. antarcticus	Cu, Zn, Cd, Pb, Hg, As	Negri et al. (2006)
C. reniformis	Cu	Cebrian et al. (2006)
C. crambe, C. reniformis, P. tenacior, D. avara	Pb, Cu	Cebrian et al. (2007)
S. fibulata	Fe, Al, Ni, Mn, Cu, Cr, Co, Ba, Zn, V, Pb, Cd	Rao et al. (2007)
H. tenuiramosa	As, Cd, Co, Cu, Fe, Mn, Ni	Rao et al. (2009)
C. clathrus	Cd	Ledda et al. (2011)
H. erectus, Hyrtios sp., *S. carteri, Chalinula* sp., *X. testudinaria, P. papyracea, Amphimedon* sp., *S. arabica, S. incontans*	Cd, Zn, Ag, Cu, Pb, As, Hg	Pan et al. (2011)
Mycale sp. *S.* cf. *diversicolor*	Cd, Zn, Cu, Pb, Hg, Se	de Mestre et al. (2012)
C. reniformis, A. oroides, I. variabilis, A. acuta, C. damicornis, C. verrucosa	^{110m}Ag, ^{241}Am, ^{109}Cd, ^{60}Co, ^{134}Cs, ^{54}Mn, ^{75}Se, ^{65}Zn	Genta-Jouve et al. (2012)
H. heliophila	PAH	Batista et al. (2013)
A. pigmentifera	Cd, Cu, Pb, Zn	Srikanth and Rao (2014)

et al. 2012; Srikanth and Rao 2014). The ability of sponges to tolerate fluctuations in the environment due to contamination in most cases has helped them to have been proposed for a "Sponge Watch Programme" (Patel et al. 1985; Hansen et al. 1995). Use of sessile organisms such as sponges for biomonitoring metal contamination in the aquatic environment is a suitable tool to monitor the influence of contaminants in aquatic ecosystem (Batista et al. 2013). The study conducted by many researchers clearly revealed that sponges seem to be good biomonitors in aquatic ecosystems since their contaminant levels reflect the local contamination (Table 8.1).

8.5 Conclusion and Environmental Constraints

We conclude that selection of a proper biomonitoring organism for assessing the environmental damage is essential for monitoring the health status of the aquatic ecosystem and accurately reflects environmental levels. The nature of PTEs concentration in sponge species is very important in response to nature management and to find the most ideal species of sponges which actually reflect the environment levels. The rapid industrialisation, urban sprawl and unplanned tourism have caused a negative impact on the positive health of aquatic ecosystem. The contamination of water with PTEs is also transferred through the biological compartment and is seen affecting the food chain. Sponge species show considerable proof for using them as biomonitoring organism. These are benthic and sessile and provide enough tissue for analysis of metals which are accumulated. Sponges which are very massive are much easy to work because there is less risk of contamination with other organisms.

Acknowledgements The authors are thankful to the Department of Biotechnology (DBT), Government of India, for their financial assistance and also thankful to the Director of IICT for providing the facilities and for his constant encouragement. The author KS is thankful to FCT (SFRH/BPD/79490/2011) for providing the fellowship.

References

Adema MM (1978) *Daphnia magna* as a test animal in acute and chronic toxicity tests. Hydrobiologia 59:125–134

Arnoux A, Harmelin JG, Monod JL, Romaña LA, Zibrowius F (1992) Altérations des peuplements benthiques de roches profondes en Méditerranée nord-occidentale: quelques aspects biologiques et molysmologiques. CR Acad Sci Paris 314:219–225

Arsene C, Bougatioti A, Mihalopoulos N (2009) Sources and variability of non-methane hydrocarbons in the Eastern Mediterranean. Glob NEST J 11:333–340

Bargagli R, Nelli L, Ancora S, Focardi S (1996) Elevated cadmium accumulation in marine organisms from Terra Nova Bay (Antarctica). Polar Biol 16:513–520

Barnes PB (2009) Environmental impacts and the ecology of sponges and ascidians in south-eastern Australian coastal lakes and lagoons. Unpublished PhD Thesis, School of Biological Sciences, University of Wollongong, NSW, pp 7–20

Batista D, Tellini K, Nudi AH, Massone TP, Scofield A, Wagener ALR (2013) Marine sponges as bioindicators of oil and combustion derived PAH in coastal waters. Mar Environ Res 92:234–243

Bayen S, Thomas GO, Lee HK, Obbard JP (2004) Organochlorine pesticides and heavy metals in green mussel, *Perna viridis* in Singapore. Water Air Soil Pollut 155:103–116

Berthet B, Mouneyrac C, Perez T, Amirad-Triquet C (2005) Metallothionein concentration in sponges (*Spongia officinalis*) as a biomarker of metal contamination. Comp Biochem Physiol Part C 141:306–313

Cachada A, Dias A, Pato P, Mieiro C, Rocha-Santos T, Pereira M, Ferreira da Silva E, Duarte A (2012) Major inputs and mobility of potentially toxic elements contamination in urban areas. Environ Monit Assess 185:279–294

Capon RJ, Elsbury K, Butler MS, Lu CC, Hooper NAR, Rostas JAP, O'Brien KJ, Mudge LM, Sim ATR (1993) Extraordinary levels of cadmium and zinc in a marine sponge, *Tedania charcoti* Topsent: inorganic chemical defense agents. Experientia 49:263–264

Cebrian E, Martí R, Uriz MJ, Turon X (2003) Sublethal effects of contamination on the Mediterranean sponge *Crambe crambe*: metal accumulation and biological responses. Mar Pollut Bull 46:1273–1284

Cebrian E, Agell G, Marti R, Uriz MJ (2006) Response of the Mediterranean sponge *Chondrosia reniformis* Nardo to copper pollution. Environ Pollut 141:452–458

Cebrian E, Uriz M, Turon X (2007) Sponge as biomonitors of heavy metals in spatial and temporal surveys in northwestern Mediterranean: multispecies comparison. Environ Toxicol Chem 26:2430–2439

Clason B, Langston WJ, Zauke GP (2004) Bioaccumulation of trace metals in the amphipod *Chaetogammarus marinus* (Leach, 1815) from the Avon and Tamar estuaries (UK): comparison of two-compartment and hyperbolic toxicokinetic models. Mar Environ Res 57:171–195

Conti ME, Beatriz Tudino M, Oscar Muse J, Cecchetti GF (2002) Biomonitoring of heavy metals and their species in the marine environment: the contribution of atomic absorption spectroscopy and inductively coupled plasma spectroscopy. Trends Appl Spectrosc 4:295–324

Cubadda F, Conti ME, Campanella L (2001) Side dependent concentration of trace metals in four Mediterranean gastropods. Chemosphere 45:561–569

de Mestre C, Maher W, Roberts D, Broad A, Krikowa F, Davis AR (2012) Sponges as sentinels: patterns of spatial and intra-individual variation in trace metal concentration. Mar Pollut Bull 64(1):80–89

Demirak A, Yilmaz F, Tuna AL, Ozdemir N (2006) Heavy metals in water, sediment and tissues of Leuciscus cephalus from a stream in southwestern Turkey. Chemosphere 63:1451–1458

Desideri D, Meli MA, Roselli C (2010) A biomonitoring study: ^{210}Po and heavy metals in marine organisms from the Adriatic Sea (Italy). J Radioanal Nucl Chem 285:373–382

Efremova SM, Margulis BA, Guzhova IV, Itskovich VB, Lauenroth S, Müller WE, Schröder HC (2002) Heat shock protein Hsp70 expression and DNA damage in Baikalian sponges exposed to model pollutants and wastewater from Baikalsk pulp and paper plant. Aquat Toxicol 57:267–280

Fialkowski W, Fialkowska E, Smith BD, Rainbow PS (2003) Biomonitoring survey of trace metal pollution in streams of a catchment draining a zinc and lead mining area of upper Silesia, Poland using the amphipod *Gammarus fossarum*. Int Rev Hydrobiol 88 (2):187–200

Garg S, Gupta RK, Jain KL (2009) Sublethal effects of heavy metals on biochemical composition and their recovery in Indian major carps. J Hazard Mater 163:1369–1384

Genta-Jouve G, Cachet N, Oberhansli F, Noyer C, Teyssie JL (2012) Comparative bioaccumulation kinetics of trace elements in Mediterranean marine sponges. Chemosphere 89:340–349

Girgin S, Kazancý N, Dügel M (2010) Relationship between aquatic insects and heavy metals in an urban stream using multivariate techniques. Int J Environ Sci Technol 7:653–664

Gochfeld M (2003) Cases of mercury exposure, bioavailability, and adsorption. Ecotoxicol Environ Safety 56:174–179

Hansen IV, Weeks M, Depledge MH (1995) Accumulation of Copper, Zinc, Cadmium and chromium by the marine sponge *Halichondria panacea* Pallas and the implications for biomonitoring. Mar Pollut Bull 31:133–138

Hill M, Stabile C, Steffen LK, Hill A (2002) Toxic effects of endocrine disrupters on freshwater sponges: common developmental abnormalities. Environ Pollut 117:295–300

Horiguchi T, Shiraishi H, Shimizu M, Morita M (1997) Effects of triphenyltin chloride and five other organotin compounds on the development of imposex in the rock shell, Thais clavigera. Environ Pollut 95:85–91

Ilan M, Gugel J, Van Soest RWM (2004) Taxonomy, reproduction and ecology of new and known Red sea sponges. Sarsia 89:410–2004

Kahle J, Zauke GP (2003) Trace metals in Antarctic copepods from the Weddell Sea (Antarctica). Chemosphere 51:409–417

Kaviarasan T, Yogamoorthi A, Siva Sankar R (2012) Heavy metal analysis of three gastropod species in Pondicherry, South eastcoast of India. Int J Curr Res 4:104–106

Kumar NJI, Soni H, Kumar RN (2006) Biomonitoring of selected freshwater macrophytes to assess lake trace element contamination: a case study of Nal Sarovar Bird Sanctuary, Gujarat, India. J Limnol 65:1–8

Ledda ED, Ramoino P, Ferrando S, Gallus L, Bianchini P, Diaspro A, Manconi R (2011) Biomonitoring of coastal areas: cadmium effect on cytoskeleton of the calcisponge *Clathrina clathrus*. Bio Mar Mediterr 18:388–389

Liang LN, He B, Jiang GB, Chen DY, Yao ZW (2004) Evaluation of mollusks as biomonitors to investigate heavy metal contaminations along the Chinese Bohai Sea. Sci Total Environ 324:105–113

Lyons BP, Thain JE, Stentiford GD, Hylland K, Davies IM, Vethaak AD (2010) Using biological effects tools to define Good Environmental Status under the European Union Marine Strategy Framework Directive. Mar Pollut Bull 60:1647–1651

Malik N, Biswas AK, Qureshi TA, Borana K, Virha R (2010) Bioaccumulation of heavy metals in fish tissues of a freshwater lake of Bhopal. Environ Monit Assess 160:267–276

Momoshima N, Toyoshima T, Matsushita R, Fukuda A, Hibino K (2007) Metal concentrations in Japanese medaka, mosquitofish and insect larvae living in uncontaminated rivers in Kumamoto, Japan. J Radioanal Nucl Chem 272:495–499

Negri A, Burns K, Boyle S, Brinkman D, Webster N (2006) Contamination in sediments, bivalves and sponges of McMurdo Sound, Antarctica. Environ Pollut 143:456–467

Nouri J, Khorasani N, Lorestani B, Karami M, Hassani AH, Yousefi N (2009) Accumulation of heavy metals in soil and uptake by plant species with phytoremediation potential. Environ Earth Sci 59:315–323

Olesen TME, Weeks JM (1994) Accumulation of Cd by the marine sponge *Halichondria panicea* Pallas: effects upon filtration rate and its relevance for biomonitoring. Bull Environ Contam Toxicol 52:722–728

Onen SA, Kucuksezgin F, Kocak F (2011) Temporal and spatial biomonitoring of heavy metals in eastern Aegean coastal waters using *Amphibalanus amphitrite*. Mar Pollut Bull 62:2548–2556

Pallela R, Koigoora S, Gunda VG, Sunkara MS, Rao VJ (2011) Comparative morphometry, biochemical and elemental composition of three marine sponges (Petrosiidae) from Gulf of Mannar, India. Chem Spec Bioavailab 23:16–23

Pan K, Lee OO, Qian PY, Wang WX (2011) Sponges and sediments as monitoring tools of metal contamination in the eastern coast of the Red Sea, Saudi Arabia. Mar Pollut Bull 62:1140–1146

Patel B, Balani MC, Patel S (1985) Sponge 'sentinel' of heavy metals. Sci Total Environ 41(2):143–152

Perez T, Wafo E, Fourt M, Vacelet J (2003) Marine sponges as biomonitor of polychlorobiphenyls contamination: concentration and fate of 24 congeners. Environ Sci Technol 37:2152–2158

Perez T, Vacelet J, Rebouillon P (2004) In situ comparative study of several Mediterranean sponges as potential biomonitors of heavy metals. In: Pansini M, Pronzato R, Bavestrello G, Manconi R (eds) Sponge science in the new millennium, vol 68. Bolletino dei Musei e degli Instituti Biologici dell'Universita di Genova, Genova, pp 517–525

Perez T, Longet D, Schembri T, Rebouillon P, Vacelet J (2005) Effects of 12 years operation of a sewage treatment plant on trace metal occurrence within a Mediterranean commercial sponge (*Spongia officinalis*, Demospongiae). Mar Pollut Bull 50:301–309

Philip RB, Leung FY, Bradley C (2003) A comparison of the metal content of some benthic species from coastal waters of the Florida Panhandle using high-resolution inductively coupled plasma mass spectrometry (ICP-MS) analysis. Arch Environ Contam Tox 44:218–223

Rainbow PS (1995) Physiology, physicochemistry and metal uptake – a crustacean perspective. Mar Pollut Bull 31:55–59

Rainbow PS, Blackmore G (2001) Barnacles as biomonitors of trace metal bio availabilities in Hong Kong coastal waters: changes in space and time. Mar Environ Res 51:441–463

Rainbow PS, Phillips DJH (1993) Cosmopolitan biomonitors of trace metals. Mar Pollut Bull 26:593–601

Rainbow PS, Wolowicz M, Fialkowski W, Smith BD, Sokolowski A (2000) Biomonitoring of trace metals in the Gulf of Gdansk, using mussels (*Mytilus trossulus*) and barnacles (*Balanus improvisus*). Wat Res 34:1823–1829

Rainbow PS, Poirier L, Smith BD, Brix KV, Luoma SN (2006) Trophic transfer of trace metals from the polychaete worm *Nereis diversicolor* to the polychaete *N. virens* and the decapod crustacean *Palaemonetes varians*. Mar Ecol Prog Ser 321:167–181

Rao JV, Kavitha P, Reddy NC, Rao TG (2006) *Petrosia testudinaria* as a biomarker for metal contamination at Gulf of Mannar, southeast coast of India. Chemosphere 65(4):634–638

Rao JV, Kavitha P, Srikanth K, Usman P, Rao T (2007) Environmental contamination using accumulation of metals in marine sponge, *Sigmadocia fibulata* inhabiting the coastal waters of Gulf of Mannar, India. Toxicol Environ Chem 89:487–498

Rao JV, Srikanth K, Pallela R, Rao T (2009) The use of marine sponge *Haliclona tenuiramosa* as bioindicator to monitor heavy metal pollution in the coasts of Gulf of Mannar, India. Environ Monit Assess 6:451–459

Richelle-Maurer E, Gomez R, Braekman JC, van de Vyver G, van Soest RWM, Devijver C (2003) Primary cultures from the marine sponge *Xestospongia muta* (Petrosiidae, Haplosclerida). J Biotechnol 100:169–176

Rinderhagen M, Ritterhoff J, Zauke GP (2000) Crustaceans as bioindicators. In: Gerhardt A (ed) Biomonitoring of polluted water-reviews on actual topics. Trans Tech Publications-Scitech Publications, Environmental Research Forum, Vol 9, - Uetikon-Zuerich, pp 161–194

Roberts DE, Davis AR, Cummins SP (2006) Experimental manipulation of shade, silt, nutrients and salinity on the temperate reef sponge *Cymbastela concentrica*. Mar Ecol Prog Ser 307:143–154

Srikanth K, Rao JV (2014) Spatial and seasonal variation of potential toxic elements in *Adocia pigmentifera*, seawater and sediment from Rameswaram, southeast coast of India. Environ Earth Sci 72:2905–2916

Stewart GM, Fowler SW, Teyssie JL, Cotret O, Cochran JK, Fisher NS (2005) Contrasting transfer of polonium—210 and lead-210 across three trophic levels in marine plankton. Mar Ecol Prog Ser 290:27–33

Svecevicius G, Kazlauskiene N (2011) Behavioral responses in Rainbow trout *Oncorhynchus mykiss* as indicators of sublethal exposure to heavy metals. Environmental Engineering. The 8th international conference, May 19–20, 2011, Vilnius

Verdenal B, Diana C, Arnoux A, Vacelet J (1990) Pollutant levels in Mediterranean commercial sponges. In: Riitzler K (ed) New perspective in sponge biology. Smithsonian Institution Press, Washington, DC, pp 516–524

Vogel S (1977) Current-induced flow through living sponges in nature. Proc Natl Acad Sci U S A 74:2069–2071

Weeks JM, Rainbow PS (1993) The relative importance of food and seawater as sources of copper and zinc to talitrid amphipods (Crustacea; Amphipoda; Talitridae). J Appl Ecol 30:722–735

Xu K, Choi JK, Yang EJ, Lee KC, Lei Y (2002) Biomonitoring of coastal pollution status using protozoan communities with a modified PFU method. Mar Pollut Bull 44:877–886

Yuzereroglu TA, Gok G, Cogun HY, Firat O, Aslanyavrusu S, Maruldah O, Kargin F (2010) Heavy metals in *Patella caerulea* (Mollusca, Gastropoda) in polluted and non-polluted areas from the Iskenderun Gulf (Mediterranean Turkey). Environ Monit Assess 167:257–264

Zahn RK, Zahn G, Muller WEG, Kurelec B, Rijavec M, Batel R, Given R (1981) Assessing consequences of marine pollution by hydrocarbons using sponges as model organisms. Sci Total Environ 20:147–169

Zhou Q, Zhang J, Fu J, Shi J, Jiang G (2008) Biomonitoring: an appealing tool for assessment of metal pollution in the aquatic ecosystem. Anal Chim Acta 606:135–150

9 Biologically Active Metabolites from Sponges and Their Activities

A. Shanmugam and S. Vairamani

Abstract

Sponges are mostly marine found distributed right from the intertidal region to the deeper waters of the oceans. Its spatial and temporal distribution is found ubiquitous. Though the sponges have simple morphology and anatomy, they show symbiotic association with several microorganisms, which are the main source of secondary metabolites and are capable of producing many biologically active compounds. So there is a good debate going on among the researchers that the source of such biologically active compounds/substances is either the sponge itself or the microorganism residing in the sponges. But unfortunately most of these symbiotic microorganisms are non-culturable. Anyhow the sponges as a whole are the good source of several substances covering the polyketides, alkaloids, terpenes, etc. This chapter deals with the variety of such chemical substances present in the sponges and their biological activities.

Keywords

Marine sponges • Metabolites • Biological activities

9.1 Introduction

Sponges are simple invertebrates with loose organization. Generally, they have spicules of silica or calcium carbonate embedded in their bodies for support and fibrous skeletons made of a horny substance called spongin; however, either or both of these may be lacking. Because sponges lack a distinct enteron and the germ layers are not well established, the phylum Porifera is sometimes classed in a separate subkingdom, the Parazoa, or the Metazoa.

There are approximately 4000 species of sponges. About 1 % (all members of a single family) inhabits freshwater, 10 % are intertidal, and the remaining is marine or benthic. Sponges

A. Shanmugam (✉) • S. Vairamani
Faculty of Marine Sciences, Centre of Advanced Study in Marine Biology, Annamalai University, Parangipettai 608 502, India
e-mail: shanpappu48@gmail.com; babavairamani@gmail.com

R. Pallela, H. Ehrlich (eds.), *Marine Sponges: Chemicobiological and Biomedical Applications*, DOI 10.1007/978-81-322-2794-6_9

obtain feed by propelling water through tiny pores in the body wall, thus capturing microorganisms and organic detritus that may be present in their body. Further the sponges inhabit several millions of symbiotic organisms, particularly microorganisms which are producing many biologically active substances for their successful survival in sponges which are also taking part in it. Because of this reason, there is a debate among the researchers about the source of these biological substances. Thus the sponge as a whole contributes to show a variety of biological activities, including antimicrobial, anticancer, and also reported to have toxic materials (Table 9.1).

9.2 Biologically Active Metabolites from Sponges

9.2.1 Polyketides

9.2.1.1 Fatty Acid Metabolites

The azacyclopropene, dysidazirine (Fig. 9.1) was isolated from the grey sponge *Dysidea fragilis* that lacks a spicule skeleton; instead it has a network of fibers loaded with sand grains, broken spicules, and other foreign material. It is strongly conulose and forming lobate or digitate cushions and elastic when compressed. It is a common sponge along most coasts of Western Europe. The dysidazirine reported an IC50 value of 0.27 μg/ml against L1210, the mouse lymphocytic leukemia cells (Molinski and Ireland 1988).

Ficulinic acids A (Fig. 9.2) and B (Fig. 9.3) from the sponge *Ficulina ficus* (= *Suberites ficus* Linnaeus 1767) reported inhibition on the growth of the mouse lymphocytic leukemia cells (L1210) with an ID50 value of 10–12 μg/ml (Guyot et al. 1986). It is an orange sponge with big massive lobate, occasionally cylindrical, with one or more conspicuous, large oscules. It has a velvety smooth appearance. It enjoys its distributed in North East Atlantic coast mostly in places with tidal currents.

9.2.1.2 Long-Chain Acetylenes

Numerous aliphatic compounds have been isolated from sponges, and a number of these have been reported to be cytotoxic. Five monoacetylenic alcohols with different reactive groups (Fig. 9.4) from the sponge *Cribrochalina vasculum* collected in Belize were toxic to the mouse P388 cell line (IC_{50} 1.0,1.3, 1.1, 0.2, 0.1 μg/ml, respectively), and they also showed in vitro immunosuppressive activity in lymphocyte reaction tests (Gunasekera and Faircloth 1990). This appears to be the first report of branched-chain aliphatic acetylenic compounds from marine organisms. *C. vasculum* is also called *Cribrochalina infundibulum* (Schmidt 1870). Smooth inverted cones, to ear-shaped or fan-shaped, sometimes torn or crooked by waves or predators; color tan to vinaceous. The skeleton of *Cribrochalina* is made of thick multispicular tracts cemented by spongin and is found distributed in Santa Marta, Colombia (Hallock et al. 1995).

Duryne (Fig. 9.5) that was isolated from the Caribbean sponge *Cribrochalina dura*, was found toxic to murine leukemia cells (IC50 0.07 μg/ml) and also colon, lung and mammary cell lines, with MIC (Minimum Inhibitory Concentration) of 0.1 μg/ml (Wright et al. 1987a).

Petrosia ficiformis is one of the sponges found producing more acetylenes, that have different purposes in industry. One among them is Petrosynol (Fig. 9.6), a polyacetylene of 30 atoms, showed antibiotic activity and was also active in the starfish egg assay at 1 μg/ml (Fusetani et al. 1987). Cimino et al. (1990) have described a number of C46 polyacetylenes that were active in the brine shrimp assay (IC_{50} 0.002–0.12 μg/ml) and also the sea urchin egg assay (IC50 1–50 μg/ml).

P. ficiformis has a compact, hard texture, with spherical oscula irregularly spread over the surface. It is found on the underside of rocks, on overhangs and in caves between 5 m and 70 m depth. The species has been reported at Adriatic Sea, Aegean Sea, Azores, Canaries, Madeira,

Table 9.1 Details of the compound, source and its potential biological activity

S. No	Compound with structure	Source	Bioactivity	Reference
1	Dysidazirine (Fig. 9.1)	*Dysidea fragilis*	Showed inhibition on the growth of the mouse lymphocytic leukemia cells (L1210)	Molinski and Ireland (1988)
2	Ficulinic acid A: n = 7 (Fig. 9.2); Ficulinic acid B: n = 9 (Fig. 9.3)	*Ficulina ficus* .	- Do -	Guyot et al. (1986)
3	Monoacetylenic alcohols (Fig. 9.4)	*Cribrochalina vasculum*	In vitro immunosuppressive activity	Gunasekera and Faircloth (1990)
4	Duryne (Molecular Formula – $C_30H_{48}O_2$) (Fig. 9.5)	*Cribrochalina dura*	Toxic to murine leukemia cells and also colon, lung and mammary cell lines	Wright et al. (1987a)
5	Petrosynol (Fig. 9.6)	*Petrosia ficiformis*	Antibiotic activity and active in the starfish egg assay	Fusetani et al. (1987)
			Active in the brine shrimp assay and also the sea urchin egg assay	Cimino et al. (1990)
6	Xestin A (Fig. 9.7)	*Xestospongia* sp.	Toxic against P388 cells	Quinoa et al. (1986)
	Xestin B (Fig. 9.8)			
7	Cyclic peroxide acids (Fig. 9.9)	*Plakortis angulospiculatis*	Inhibiting the growth of P388 cells	Gunasekera et al. (1990a)
8	Acanthifolicin (Fig. 9.10)	*Pandaros acanthifolium*	strong cytotoxic activity against P388 cells	Schmitz et al. (1981)
9	Okadaic acid (Fig. 9.11)	*Halichondria okadai*	- Do -	Tachibana et al. (1981)
10	Discodermolide (Fig. 9.12)	*Discodermia dissoluta*	Potent inhibitor of tumor cell growth in several MDR cancer cell lines.	Gunasekara et al. (1990b).
			Most potent natural promoters of tubulin assembly.	
11	Fijianolides A (Fig. 9.13) Fijianolides B (Fig. 9.14)	*Spongin mycofijiensis* (= *Leiosella lavis*).	Active against P388 and HT-29 human colon tumor cells	Quinoa et al. (1988)
12	Mycalolides A-C (Figs. 9.15 (1), Fig. 9.16 (2) and Fig. 9.17 (3))	*Mycale*	Highly cytotoxic against B16	Fusetani et al. (1898b)
13	Halichondrins B (R = H) (Fig. 9.18) and C (R = OH) (Fig. 9.19),	*Halichondria kadai*	In vitro activity against B16 melanoma cell lines	Hirata and Uemura (1986)
	Norhalichondrins A (Fig. 9.20) ($R_1 = R_2 = H$, $R_3 = R_4 = OH$), B ($R_1 = R_2 = H$, $R_3 = R_4 = H$) (Fig. 9.21) and C ($R_1 = R_2 = R_3 = H$, $R_4 = OH$) (Fig. 9.22)			
	Homohalichondrins A ($R_1 = R_2 = OH$, $R_3 = H$) (Fig. 9.23), B ($R_1 = R_2 = R_3 = H$) (Fig. 9.24) and C ($R_1 = R_3 = H$, $R_2 = OH$) (Fig. 9.25)			

(continued)

Table 9.1 (continued)

S. No	Compound with structure	Source	Bioactivity	Reference
14	Misakinolide A (Fig. 9.26)	*Theonella swinhoei*	In vitro antiviral and antifungal activity	Sakai et al. (1986)
15	Latrunculin A (Fig. 9.27)	*Latrunculia magnifica*	Disturbing microfilament organization in the cell and thus affects normal functioning of the cell	Amiram Groweiss et al. (1983)
16	Hennoxazoles A (R_1 = OH, R_2 = CH_3), B (R_1 = OH, R_2 = CH_2 CH_3), C (R_1 = OH, R_2 = CH_2 CH_2 CH_2 CH_3) and D (R_1 = H, R_2 = CH_3) (Figs. 9.28, 9.29 and0 9.30)	*Polyfibrospongia* sp	Displaying analgesic activity	Ichiba et al. (1991)
			Strong activity against HSV-1	
17	Curcuphenol (Fig. 9.31)	*Didiscus flavus*	Inhibited the growth of several cell lines such as P388, A549 (lung), HCT-8 (colon) and MDAMB (mammary)	Wright et al. (1987b)
18	Metachromin A (Fig.9.32) Metachromin B (Fig. 9.33)	*Hippospongia cf. metachromia*	Toxic to L1210 cells	Ishibashi et al. (1988)
			Showed coronary vasodilating effects and inhibited potassium chloride-induced contraction of the rabbit isolated coronary artery	
19	Avarol (Fig. 9.34)	*Dysidea avara*	Interferes with the mitotic processes, thus preventing telophase formation	Mueller et al. (1985)
20	Puupehenone (Fig. 9.35) (Molecular formula – $C_{21}H_{28}O_3$)	*Strongylophora hartmani*	Inhibits the growth of a number of tumor cell lines such as P388, A549 human lung, HCT-8 human colon and MCF-7 human mammary	Kohmoto et al. (1987a)
21	Amorphane sesquiterpenes (Figs. 9.36)	*Axinyssa fenestratus*	Anthelmintic activity	Alvi et al. (1991)
22	Axisonitrile-3 (Fig. 9.37)	*Topsentia* sp.	- Do -	Alvi et al. (1991)
23	Manoalide (Fig. 9.38)	*Luffariella variabilis*	Irreversibly inhibits PLA2	Glaser and Jacobs (1986), Jacobson et al. (1990)
			Inhibits arachidonic acid release	De Vries et al. (1988)
			Inhibits 5-lipoxygenase	
24	Luffariellolide (Fig. 9.39) (Molecular formula – $C_{25}H_{38}O_3$)	*- Do -*	Anti-inflammatory activity	Albizati et al. (1987)
			Partially reversible PLA2 inhibitor	
25	Variabilin (Fig. 9.40)	*Ircinia* sp.	Cytotoxic to host BSC cells in an antiviral assay	Barrow et al. (1988)

(continued)

Table 9.1 (continued)

S. No	Compound with structure	Source	Bioactivity	Reference
26	Okinonellin A (Fig. 9.41) Okinonellin B (Fig. 9.42)	*Spongionella* sp.	Inhibit division of fertilized starfish eggs	Kato et al. (1986)
27	Phyllofoliaspongin (Fig. 9.43)	*Phyllospongia foliascens*	Inhibited P388 cell growth.	Kitagawa et al. (1989)
			Showed anti- thrombocytic inhibitory effect on ADP-induced and collagen-induced aggregation of rabbit platelets in vitro	
28	Heteronemin (Fig. 9.44) 12-episcalarin (Fig. 9.45)	*Hyrtios erecta*	In vitro anthelmintic activity	Kazlauskas et al. (1976), Kashman and Rudi (1977), Cimino et al. (1977), Crews and Bescansa (1986)
29	Isocyanine (Fig. 9.46)	*Bubaris*	Antitumor, antiviral and antifungal activities	Wright et al. (1988)
30	Kalihinol Y (Fig. 9.47) Kalihinol J (Fig. 9.48)	*Acanthella cavernosa*	Potent in vitro anthelmintic activity	Omar et al. (1988)
31	Spongiadiol (Fig. 9.49)	*Spongia* sp.	Antiviral activity	Kohmoto et al. (1987a), (1987b)
32	Reiswigin A (R = CH CH $(CH_3)_2$) (Fig. 9.50) Reiswigin B (R = –CH = $C(CH_3)_2$)	*Epipolasis reiswigi*	Antiviral activity	Kashman et al. (1989b)
			Inhibiting HSV-1 completely and A59 virus partially	
33	Pouoside A (Fig. 9.51)	*Asteropus* sp.,	Inhibited P388 cell growth	Ksebati et al. (1988), (1989)
34	Penasterol (Fig. 9.52)	*Penares* sp.	Active against L1210 cells	Cheng et al. (1988a)
35	Sarasinoside A1 (Fig. 9.53)	*Asteropus* sp.	Active against P388 cells	Schmitz et al. (1988)
36	Eryloside A (Fig. 9.54)	*Erylus lendenfeldi*	Showed cytotoxic activity against P388 and antifungal activity	Carmely et al. (1989a)
37	2,6 – dibromo – 4 – acetamido – 4 – hydroxycyclohexadienone (Fig.9.55)	*Verongia cauliformis*	Antibacterial activity	Sharma and Burkholder (1967)
38	Aerothionin (Fig. 9.56)	*Aplysia aerophoba* and *Verongia thiona*	Antibiotic activity	Encarnacion et al. (2000); Thoms et al. (2004)
39	Bastadin series of cyclic amides (Fig. 9.57)	*Iarthella basta*	Inhibit P388 cell growth	Pordesimo and Schmitz (1990)
40	Mycalamide A (R = 4) and B (R = Me) (Fig. 9.58)	*Mycale* sp.	Showed antiviral and cytotoxic activity	Perry et al. (1988a); (1990)
41	Calyculin A (R_1 = CN, $R_2 – R_3$ = H); Calyculin B (R_1 = R_3 = H, R_2 = CN); Calyculin C (R_1 = CN, R_2 = H, R_3 = CH_3); Calyculin D (R_1 = H, R_2 = CN, R_3 = CH_3) (Fig. 9.59)	*Discodermia calyx*	Active against L1210 cells	Kato et al. (1986a, b), (1988b)

(continued)

Table 9.1 (continued)

S. No	Compound with structure	Source	Bioactivity	Reference
			Inhibited cell division of both starfish and sea urchin eggs	Kato et al. (1988a, b)
			Exhibited in vivo activity against Erlich and P388 leukemia in mice (Calyculin A)	
			Inhibited uptake of [^{3H}] thymidine, [^{3H}] uridine and [^{3H}] leucine in L1210 murine leukemia cells (Calyculin A)	
42	Alkaloid (Fig. 9.60)	*Teichaxinella morchella* and *Ptilocaulis walpersi*	Showed mild cytotoxicity to L1210 cells	Wright and Thompson (1987)
43	Girolline (Fig. 9.61)	*Pseudaxinyssa cantharella*	Active against P388	Ahond et al. (1989)
44	Pyronaamide (Fig. 9.62)	*Leucetta*	Toxic to KB cells	Akee et al. (1990)
45	Series of 2-amino imidazole alkaloids Naamidines (e.g. Fig. 9.63)	*Leucetia chagosensis*	Showed cytotoxicity against P388 cells	Carmely et al. (1989b)
46	Horbindole A (R = Me); Horbindole B (R = Et); Horbindole C (R = CH = CH-Et) (Fig. 9.64)	*Axinella* sp	Showed cytotoxicity against KB and found to have fish anti- feedant activity	Herb et al. (1990)
47	Dragmacidin (Fig. 9.65)	*Dragmacidian* sp.	Toxic to P388 cells and also to A549 human-8 human colon and MDAMB human mammary cells	Kohmoto et al. (1988)
48	Dragmacidon A (Fig. 9.66)	*Dragmacidian* sp.	Showed cytotoxicity against L1210 cells	Morris and Andersen (1989)
49	Fascaplysin (Fig. 9.67)	*Fascaplysinopsis* sp.,	killed L1210 cells (LD50 0.2 ug/ml) and also showed antibiotic activity	Roll et al. (1988)
50	Eudistomin K (Fig. 9.68)	*Riterella sigillinoides*	Described in a patent as being "very effective in inhibiting growth of L1210, P388, A549 and HCT-8 cells at varying concentrations"	Blunt et al. (1988)
51	Manzamine A (Fig. 9.69)	*Haliclona* sp.	Active against P388 cells in vitro	Sakai et al. (1986)
52	Theonelladins A (R = H); Theonelladin B (R = CH_3-D); Theonelladin C (R = H); Theonelladin D (R = CH_3) (Fig. 9.70)	*Theonella swinhoei*	Showed the cytotoxicity against L1210 cell lines and KB cells	Kobayashi et al. (1989)
			Reported to be 20 times more than caffeine in causing release of Ca^{2+} from sarcoplasmic reticulum	

(continued)

Table 9.1 (continued)

S. No	Compound with structure	Source	Bioactivity	Reference
53	Niphatyne A (Fig. 9.71)	*Niphates* sp.	Cytotoxic to P388 cells	Quinoa and Crews (1987)
	Niphatyne B (Fig. 9.72)			
54	5-(methoxycarbonyl) tubercidin (R_1 = CO_2Me, R_2 = ribose) and Toyocamycin (R_1 = CN, R_2 = ribose) (Fig. 9.73)	*Jaspis*	Showed activity against L1210	Zabriskie and Ireland (1989)
			In vivo activity against L1210, increasing lifetimes by up to 39 % (5-(methoxycarbonyl) tubercidin)	
55	Arabinosides (Fig. 9.74)	*Cryptotethia crypta*	Antiviral and antitumor activities	De Clercq et al. (1977), Gosselin et al. (1986)
	Ara-A Ara-C Ara-T Ara-U			
56	Doridosine (Fig. 9.75)	*Tedania digitala*	Causes reduced arterial pressure and reduced heart rate in mammalians in a manner that is qualitatively similar to adenosine	Quinu et al. (1980)
			Acts as muscle relaxant and showed hypothermic activity	
57	1-Methylisoguanosine (Fig. 9.76)	*Tedania digitata*	Shows potent muscle relaxant, blood pressure lowering, cardiovascular and anti-inflammatory activity	Jamieson and Davis (1980), Bartlett et al. (1981)
58	Aaptamine (R_1 = CH_3, R_2 = H) (Fig. 9.77) and some of its derivatives (R_1 = H, R_2 = H; 163. R_1 = H, R_2 = CH_3)	*Suberites* sp	Reported to have some in vitro and in vivo cell inhibitory activity when tested for antitumor activity against Ehrlich ascites tumor in mice	Fedoreov et al. (1988)
59	Isobatzellines (Fig. 9.78)	*Batzella* sp.	Showed antifungal activity against *C. albicans*	Sun et al. (1990)
60	Renierol (Fig. 9.79)	*Xestospongia caycedoi*	Inhibited the growth of L1210 cells	McKee and Ireland (1987)
61	Indolizidine stellenamide A (Fig. 9.80)	*Stella* sp.	Antifungal activity and also inhibited K562 epithelium cell growth	Hirota et al. (1990b)
62	Discorhabdin A (Fig. 9.81) Discorhabdin – B (Fig. 9.82)	*Latrunculia brevis* and *Prianos* sp.	Reported the cytotoxicity against P388 cells	Perry et al. (1988b, c)
	Discorhabdin – C (Fig. 9.83) Discorhabdin – D (Fig. 9.84)			
63	Prianosin A (Fig. 9.85) Prianosin B (Fig. 9.86)	*Prianos melanos*	Active against L5178Y cells and KB cells	Kobayashi et al. (1987)
	Prianosin C (R = OH) and D (R = H) (Fig. 9.87)			
64	Dysemenin (Fig. 9.88)	*Dysidea herbacea*	Inhibited iodide transfer in thyroid cells	Van Sande et al. (1990)

(continued)

Table 9.1 (continued)

S. No	Compound with structure	Source	Bioactivity	Reference
65	Amphimedine (Fig. 9.89)	*Amphimedon* sp.	Active against P388 in vitro	Schmitz et al. (1983)
66	Dercitin (Fig. 9.90)	*Descitus* sp.	Showed in vitro and in vivo activity in the P388 model	Gunawardana et al. (1988)
			Immunosuppressive and antiviral activity	Burres et al. (1989)
67	Plakinidine A (R = H); Plakinidine B (R = CH_3); Plakinidine C (R = $H\Delta^9$) (Fig. 9.91)	*Plakortis* sp.	Active against L1210 cells	West et al. (1990)
			Inhibited reverse transcriptase activity (Plakinidine A).	Inman et al. (1990)
68	Latrunculin A (Fig.9.92)	*Spongia mycofijiensis*	Showed excellent in vitro activity at 50ug/ml against *N. brasiliensis*	Kashman et al. (1980)
69	Ptilomycalin A ($R_1 = R_2 = H$, n = 13)	*Ptilocaulis spiculifer* and *Hemimycale* sp.	Activity against HSV and antitumor and antifungal activities (Ptilomycalin A)	Kashman et al. (1989a)
	Crambescidins (Crambescidin 816: $R_1 = R_2 = OH$, n = 13; Crambescidin 830: $R_1 = R_2 = OH$, n = 14; Crambescidin 844: $R_1 = R_2 = OH$, n = 15; Crambescidin 800: $R_1 = H$, $R_2 = OH$, n = 13 (Fig. 9.93)		Activity against HSV-1 and exhibited 98 % inhibition of L1210 cell growth (Crambescidins)	
70	Sceptrin (Fig. 9.94), Ageliferin (Fig. 9.95) and oxysceptrin (Fig. 9.96)	*Agelas conifer*	Active against HSV-1 and VSV Sceptrin and Ageliferin)	Keifer et al. (1991)
			Less active Oxysceptrin	
71	Acarnidine 1 a (R = CO $(CH_2)_{10}CH_3$); Acarnidine – 1 b (R = CO $(CH_2)_3CH = CH$ $(CH_2)_5CH_3(z)$); Acarnidine – 1 c (R = $COC_{13}H_{21}$) (Figs. 9.97 and 9.98)	*Acarnus erithacus*	Antiviral property	Carter and Rinehurt (1978a)
72	Discobahamin A (Fig. 9.99)	*Discodermia* sp.	Antifungal activity	
73	Papuamides A and B (Fig. 9.100)	*Theonella* sp.	Inhibited the infection of human T-lymphoblastoid cells	Ford et al. (1999)
74	Microspinosamide (Fig. 9.101)	*Sidonops microspinosa*	Anti-HIV activity	Rashid et al. (2001)
75	Keramamide (Fig. 9.102 and 9.103),	*Theonella* sp.	Reported cytotoxic effect against P388 murine leukemia cells	Fusetani et al. (1991)
76	Cyclotheonamide (Fig. 9.104)	*Theonella* sp.	Reported as a potent antithrombin cyclic peptide which strongly inhibited various proteinases, particularly thrombin	Fusetani et al. (1990)

(continued)

Table 9.1 (continued)

S. No	Compound with structure	Source	Bioactivity	Reference
77	Theonellamide F (Fig. 9.105)	*Theonella* sp.	Showed activity against L1210 and P388 cells	Matsunaga et al. (1989)
78	Hymenistatin 1 (Fig. 9.106)	*Hymeniacidon* sp.,	Showed both in vitro and in vivo activity against P388 murine leukemia cells	Petit and Zeghloul (1990)
79	Microsclerodermin A (R = OH)- Microsclerodermin B (R = H) (Fig. 9.107)	*Theonella* sp. and *Microscleroderma* sp.	Antifungal activity	Schmidt and Faulkner (1998)
80	Theonegramide (Fig. 9.108)	*Theonella* sp. and *Microscleroderma* sp.	Antifungal activity	Bewley and Faulkner (1994)

Fig. 9.1 Dysidazirine

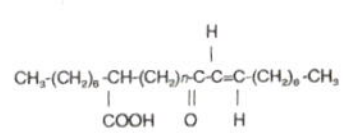

Fig. 9.2 Ficulinic acid A: n = 7

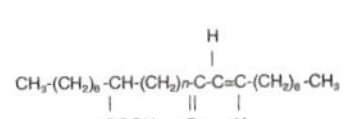

Fig. 9.3 Ficulinic acid B: n = 9

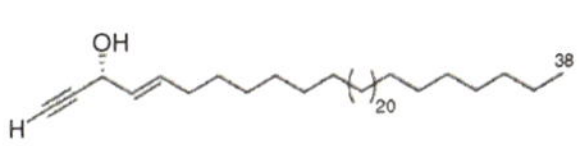

Fig. 9.4 Monoacetylenic alcohols

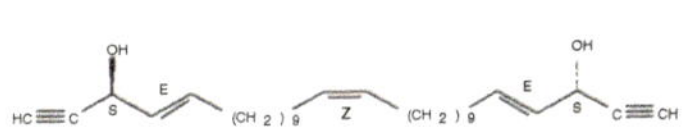

Fig. 9.5 Duryne (Molecular formula – $C_{3}0H_{48}O_{2}$)

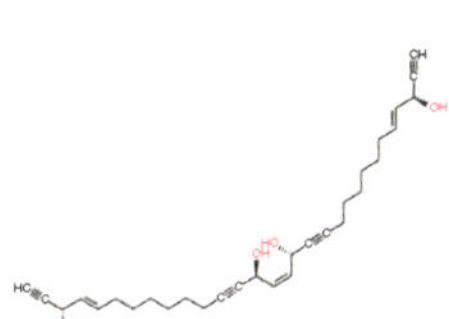

Fig. 9.6 Petrosynol

Cape Verde, Ionian Sea, Levantine Sea, Mediterranean Sea, North Atlantic, Tunisian Plateau/ Gulf of Sidra, West Africa and Western Mediterranean.

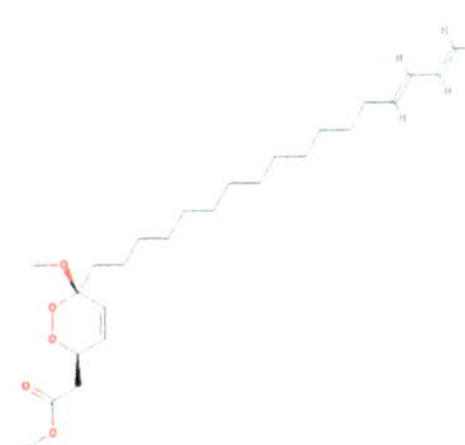

Fig. 9.7 Xestin A

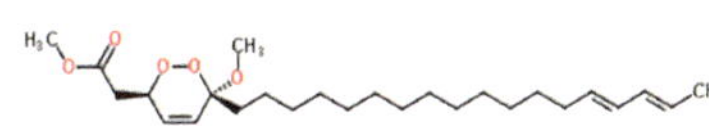

Fig. 9.8 Xestin B

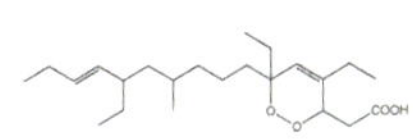

Fig. 9.9 Cyclic peroxide acids

9.2.1.3 Aliphatic Ester Peroxides

Xestins A and B (Figs. 9.7 and 9.8), isolated from *Xestospongia* sp., were found toxic against P388 cells (ID_{50} 0.3 and 3 μg/ml, respectively) (Quinoa et al. 1986). *P. lita* is maroon to pink, with the opening of the barrel pale white. In the intertidal zones, this species ranges from 10 to 20 cm in diameter, and are about 10–20 cm tall. This species is found in the Philippines, Australia, western and central Indian Ocean, Indonesia, Malaya, and New Caledonia.

The cyclic peroxide acids (Fig. 9.9) isolated from the sponge *Plakortis angulospiculatis*, which are much more highly branched were collected in Venezuela, and the esters derived from the sponge were found inhibiting the growth of P388 cells (IC50 0.2–0.9 μg/ml) (Gunasekera et al. 1990b).

9.2.1.4 Complex Polyketides

The polyether carboxylic acids acanthifolicin (Fig. 9.10) and okadaic acid (Fig. 9.11) were initially isolated from sponges – acanthifolicin from the Caribbean sponge *Pandaros acanthifolium* and okadaic acid from *Halichondria okadai* collected in Japan and also from *H. melaodocia* from the Florida Keys, reported strong cytotoxic activity (Schmitz et al. 1981; Tachibana et al. 1981). The ED50 value of 0.0002 and 0.0017 μg/ml against P388 cells were reported by acanthifolicin and okadaic acid, respectively.

The sponge, *P. acanthifolium* (Duchassaing and Michelotti 1864) is erect, dark bushy with flattened branches. Branches up to 25 cm long, 4 cm wide. Oscules inconspicuous. It is a reef dweller and it is found distributed in Florida and the Caribbean.

The polyketide natural product, discodermolide (Fig. 9.12), was isolated from the deep-sea marine sponge *Discodermia dissoluta* in 1990 by Gunasekara et al. (1990a). It was found to be a potent inhibitor of tumor cell growth in several MDR cancer cell lines. Further, it was identified as one of the most potent natural promoters of tubulin assembly.

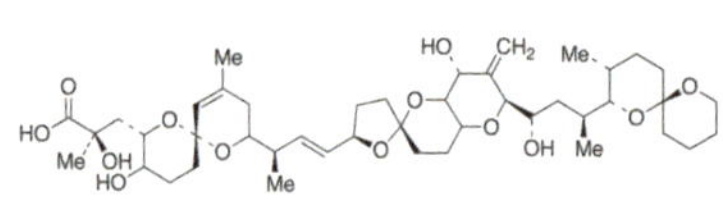

Fig. 9.10 Acanthifolicin

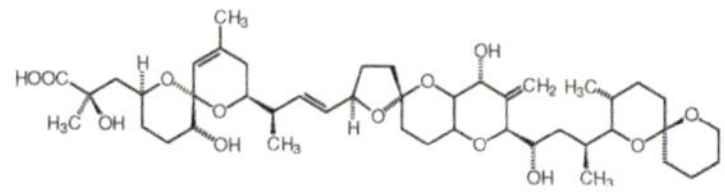

Fig. 9.11 Okadaic acid

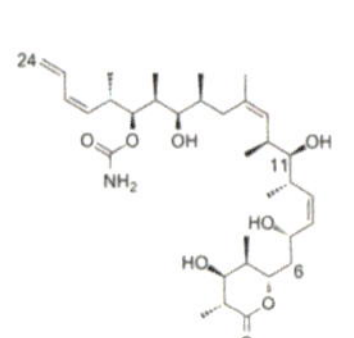

Fig. 9.12 Discodermolide

9.2.1.5 Macrolides

9.2.1.5.1 Assorted Macrolides

Fijianolides A and B (Figs. 9.13 and 9.14) were isolated from the Vanuatuan sponge *Spongin mycofijiensis* (= *Leiosella lavis*) (Quinoa et al. 1988). This sponge is massive, lobate, or tubular, sometimes with a short stalk (2–3 cm). The size varies from 3 to 20 cm in height, and 2–10 cm in diameter. The surface is microconulose, and the texture is compressible and flexible. This species is dark brown/black, in colour, externally and tan inside and is generally found in sheltered reef habitats, under ledges or in caves. It is fairly rare despite its broad range of distribution in the South and Indo Pacific.

Fijianolide A reported the IC50 of 9 μg/ml against P388 and 11 μg/ml vs HT-29 human colon tumor cells. When the diacetate of fijianolide B was tested against the same cells, it reported the IC50 as 6 μg/ml vs P388 and 0.5 μg/ml vs HT-29.

The three other tris-isoxazole containing macrolides, mycalolides A–C (Figs. 9.15, 9.16 and 9.17), were isolated from *Mycale*. Although the above were highly cytotoxic (IC50 0.0005–0.001 μg/ml vs. B16), they have not shown promising results in vivo (Fusetani et al. 1988).

Apart from okadaic acid, the Japanese sponge *H. kadai* was a good source of a group of very

Fig. 9.13 Fijianolides A

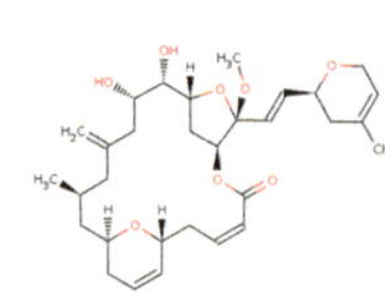

Fig. 9.14 Fijianolides B

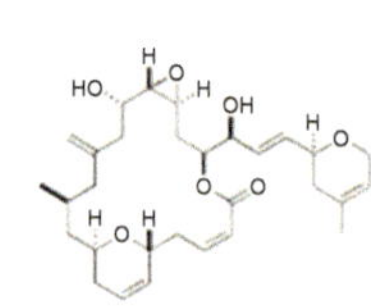

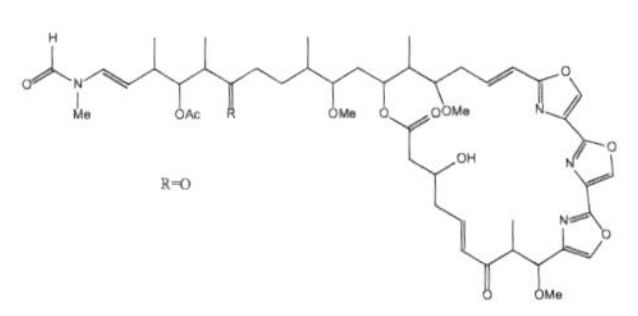

Fig. 9.15 Mycalolides A

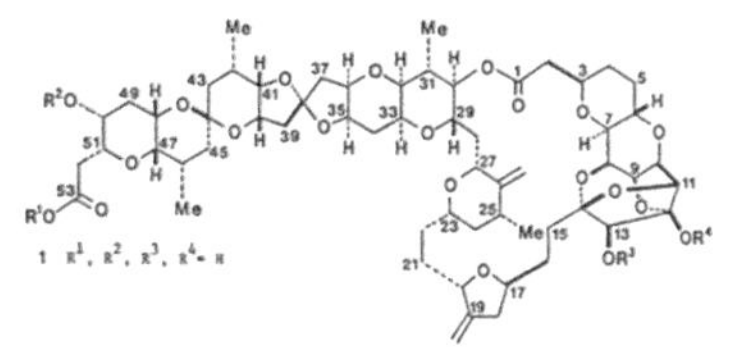

Fig. 9.20 Norhalichondrins A

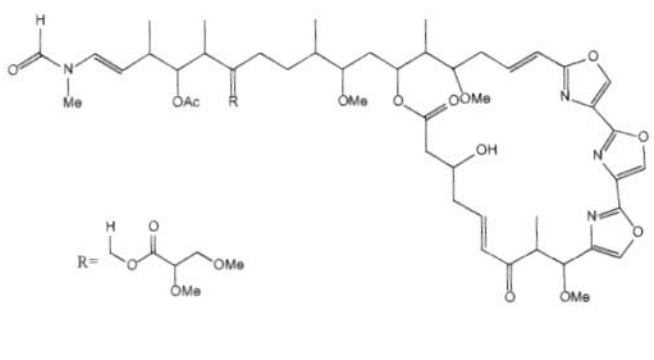

Fig. 9.16 Mycalolides B

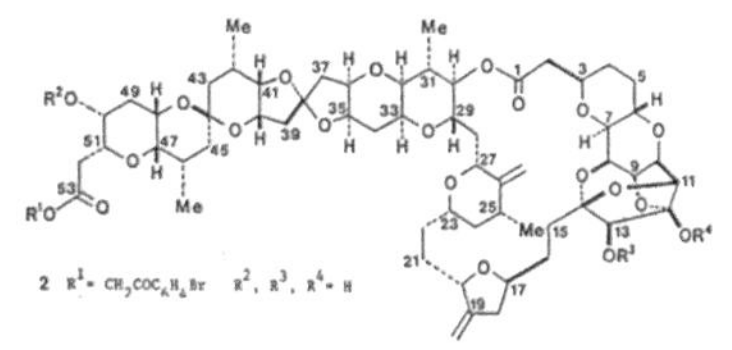

Fig. 9.21 Norhalichondrins B

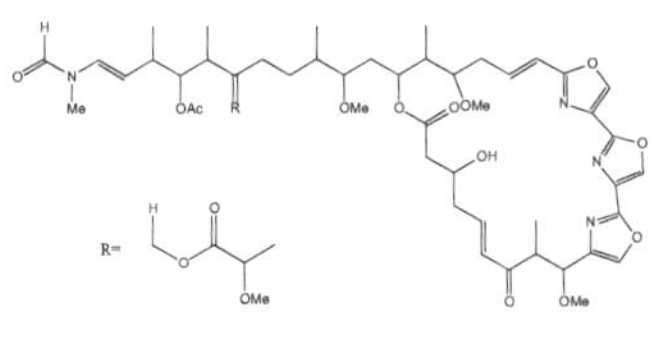

Fig. 9.17 Mycalolides C

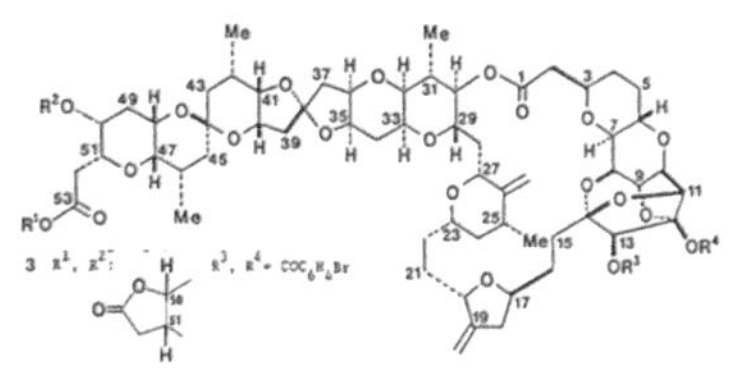

Fig. 9.22 Norhalichondrins C

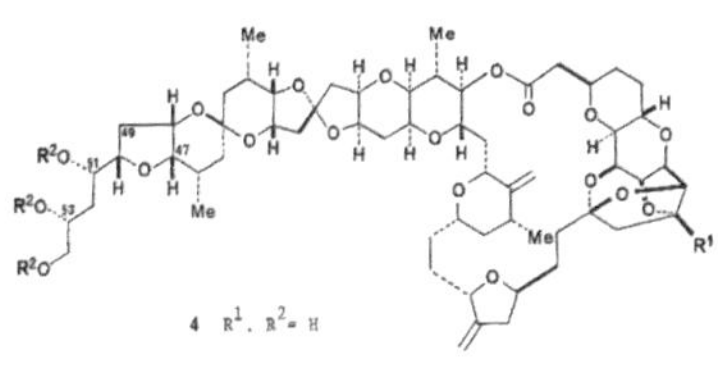

Fig. 9.18 Halichondrins B (R = H)

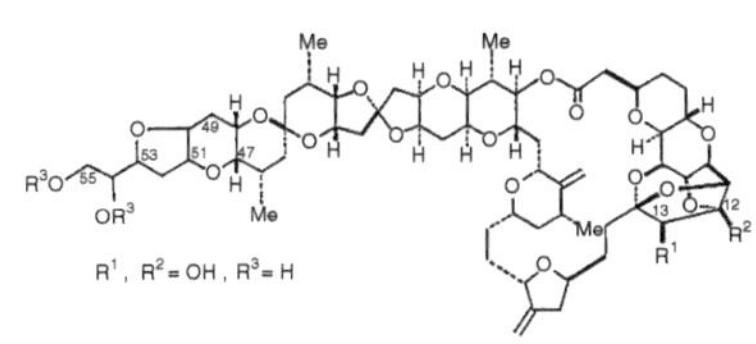

Fig. 9.23 Homohalichondrins A ($R_1 = R_2 = OH$, $R_3 = H$)

$R^1 = OH$, $R^2 = H$

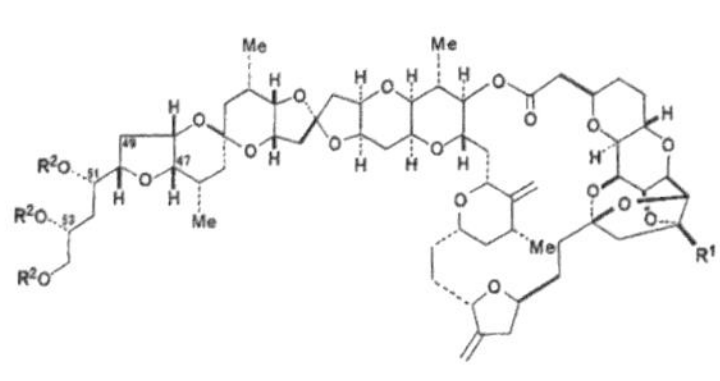

Fig. 9.19 Halichondrins C (R = OH)

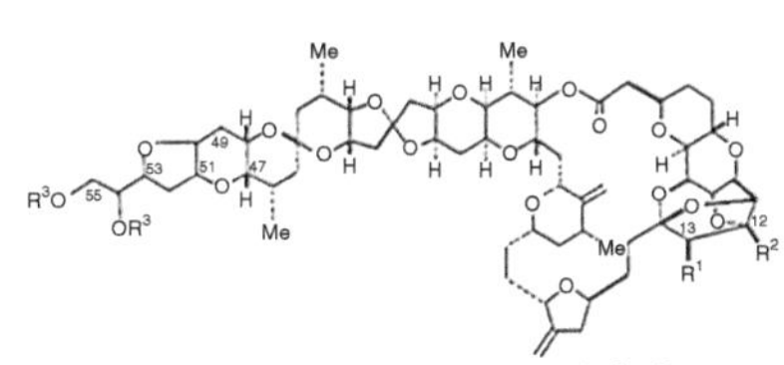

Fig. 9.24 Homohalichondrins B ($R_1 = R_2 = R_3 = H$)

complex and biologically active macrolides, halichondrins B (Fig. 9.18) and C (Fig. 9.19); norhalichondrins A (Fig. 9.20), B (Fig. 9.21), and C (Fig. 9.22) and homohalichondrins A (Fig. 9.23), B (Fig. 9.24), and C (Fig. 9.25) (Hirata and Uemura 1986). The above macrolides showed the following in vitro activity against B16 melanoma cell lines: norhalichondrin A – 0.0052 μg/ml; halichondrin

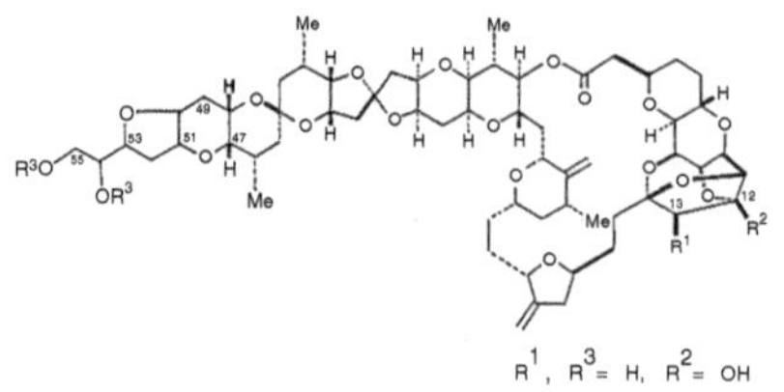

Fig. 9.25 Homohalichondrins C ($R_1 = R_3 = H$, $R_2 = OH$)

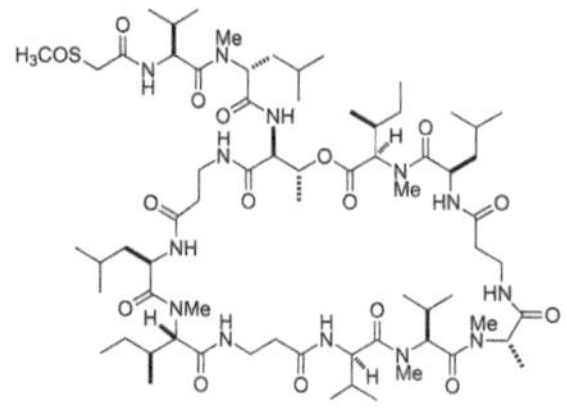

Fig. 9.26 Misakinolide A

B – 0.000093 μg/ml; homohalichondrin A – 0.00026 μg/ml; halichondrin C – 0.00035 μg/ml and homohalichondrin B – 0.0001 μg/ml. Halichondrin B showed good in vivo activity against B16 melanoma in mice (T/C values of 203–244 %, depending on dose (5–20 μg/kg) and regimen), against P388 leukemia in mice (T/C 323 % @ 10 μg/kg), and against L1210 in mice (T/C 207–375 % with doses of 50–100 μg/kg under various injection schedules). From the results, it was concluded that it is important for antitumor activity that the tricyclic ring be relatively lipophilic and that the terminal group have two or more hydroxyls, but not a carboxylate. *Halichondria* are massive, amorphous sponges with clearly separated inner and outer skeletons consisting of bundles of spicules arranged in a seemingly random pattern.

Misakinolide A (Fig. 9.26) isolated from *Theonella* sp. was collected in Okinawas (Sakai et al. 1986), and it showed in vitro antiviral and antifungal activities. *Theonella* sp. is a coral reef sponge (*Theonella swinhoei*), found distributed in the Red Sea and Indian Ocean.

The macrolide latrunculin A (Fig. 9.27) was isolated from the red sea sponge *Latrunculia magnifica*. It binds and stabilizes the globular G-actin in a 1:1 complex, preventing the conversion of globular (monomeric) G-actin into filamentous (polymeric) F-actin, disturbing microfilament organization in the cell. Latruculin A affects normal functioning of the cell by disrupting the polymerization of G-actin and microfilament organization which is essential for the cellular mechanical processes including motility and cytoskeleton scaffolding (Groweiss et al. 1980).

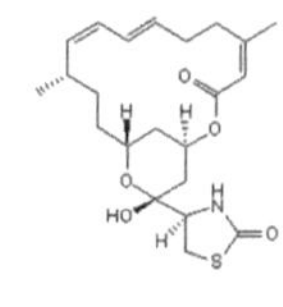

Fig. 9.27 Latrunculin A

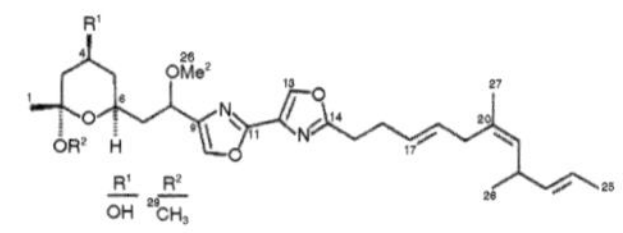

Fig. 9.28 Hennoxazoles A ($R_1 = OH$, $R_2 = CH_3$)

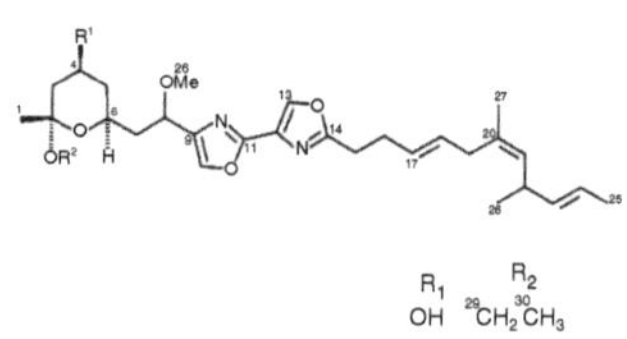

Fig. 9.29 Hennoxazoles B ($R_1 = OH$, $R_2 = CH_2\ CH_3$)

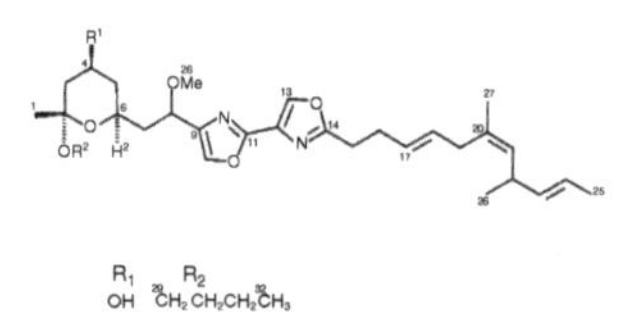

Fig. 9.30 Hennoxazoles C ($R_1 = OH$, $R_2 = CH_2\ CH_2\ CH_2\ CH_3$)

9.2.1.6 Miscellaneous

The sponge, *Polyfibrospongia* sp., collected on the island of Miyako in Okinawa was the source for hennoxazoles A–D (Figs. 9.28, 9.29 and 9.30) (Ichiba et al. 1991). Apart from displaying analgesic activity, hennoxazole A, the major component (0.01 % of wet weight) showed strong activity against HSV-1 (IC50 0.6 μg/ml).

9.3 Terpenes

9.3.1 Sesquiterpenes

Curcuphenol (Fig. 9.31) extracted from the sponge *Didiscus flavus* collected in both shallow and deep waters in the Bahamas and Belize was found inhibiting the growth of several cell lines [IC50 7 μg/ml vs. P388; MIC for human cell lines: A549 (lung) 10 μg/ml; HCT-8 (colon) 0.1 μg/ml; MDAMB (mammary) 0.1 μg/ml] (Wright et al. 1987b).

Metachromins A (Fig. 9.32) and B (Fig. 9.33), isolated from the sponge *Hippospongia cf. metachromia*, were reported to be toxic to L1210 cells (IC50 2.4 and 1.62 μg/ml, respectively) (Ishibashi et al. 1988). Further they also showed coronary vasodilating effects and inhibited potassium chloride-induced contraction of the rabbit isolated coronary artery.

Avarol (Fig. 9.34) from the sponge *Dysidea avara* interferes with the mitotic processes, thus preventing telophase formation which may be due to changes of the intracellular pools and/or alterations of the permeability properties of the cell membranes for the precursors (Mueller et al. 1985).

Fig. 9.31 Curcuphenol

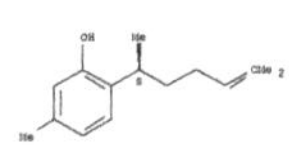

Fig. 9.32 Metachromin A

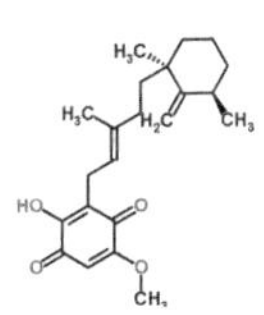

Fig. 9.33 Metachromin B

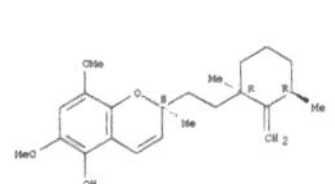

Fig. 9.34 Avarol

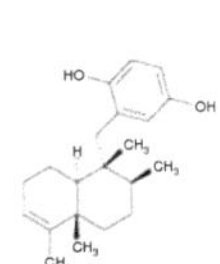

Fig. 9.35 Puupehenone (Molecular formula – $C_{21}H_{28}O_3$)

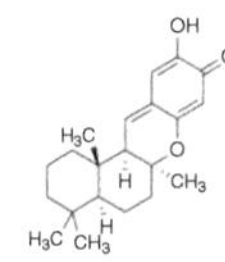

Fig. 9.36 Amorphane sesquiterpenes

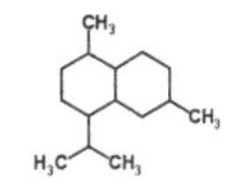

Fig. 9.37 Axisonitrile-3

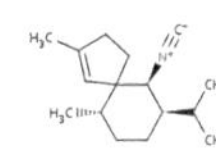

Puupehenone (Fig. 9.35) was isolated from a deep water sponge, *Strongylophora hartmani* by Kohmoto et al. (1987a). It was found to inhibit the growth of a number of tumour cell lines (IC50; P388, 1 μg/ml; A549 human lung, 0.1–1 μg/ml; HCT-8 human colon, 1–10 μg/ml; MCF-7 human mammary, 0.1–1 μg/ml). Besides the above, it also showed very modest in vivo effects on p388 cell lines (19 % increase in lifetime @ 25 mg/kg for 9 days).

Isonitrile, isothiocyanate, and related functionalized terpenes are characteristic metabolites of sponges belonging to the order Halichondida. Four amorphane sesquiterpenes (Fig. 9.36) were isolated from the Fijian sponge *Axinyssa fenestratus* (Alvi et al. 1991) and tested for their anthelmintic activity.

Another sesquiterpene, axisonitrile-3 (Fig. 9.37) (D' Blassio et al. 1976) extracted from *Topsentia* sp. from Thailand (Alvi et al. 1991). Though it reported superior anthelmintic activity in vitro at 50 μg/ml, it was not active in vivo.

9.3.2 Sesterterpenes

A nonsteroidal sesterterpene, manoalide (Fig. 9.38), isolated from the sponge *Luffariella variabilis* (De Silva and Scheuer 1980) has emerged as a potent tool for studying inflammation. It irreversibly inhibited PLA2 (Glaser and Jacobs 1986; Jacobson et al. 1990).

In addition to inhibiting PLA2, manoalide inhibited 5-lipoxygenase (de Vries et al. 1988), leading to speculation that its anti-inflammatory activity of manoalide was attributed to its inhibitory effect on Ca^{2+} channels (Wheeler et al. 1988). Interestingly at low concentrations, manoalide inhibited calcium channels with no effect on phosphor-inositide metabolism. The ability of manoalide to dissect these two components of the inflammation process may prove to be its most useful attribute in studying the role of Ca^{2+} signaling in inflammation and proliferation (Barzaghi et al. 1989).

Another analog of manoalide, luffariellolide (Fig. 9.39), isolated from the same organism, also exhibited anti-inflammatory activity, but it was slightly less potent than manoalide and was a partially reversible PLA2 inhibitor (Albizati et al. 1987).

A number of cytotoxic furanosesterpenes have been obtained from a variety of sponges. Variabilin (Fig. 9.40) and the related sesterpene tetronic acids from a Caribbean *Ircinia* sp. sponge were all described as being cytotoxic to host BSC cells at 2 μg/ml in an antiviral assay (Barrow et al. 1988).

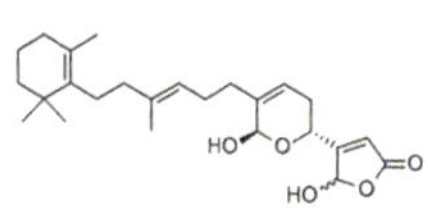

Fig. 9.38 Manoalide

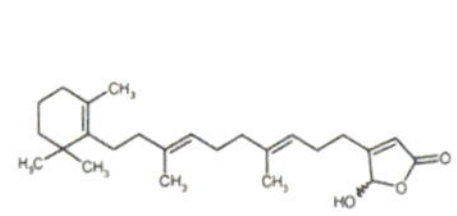

Fig. 9.39 Luffariellolide (Molecular formula – $C_{25}H_{38}O_3$)

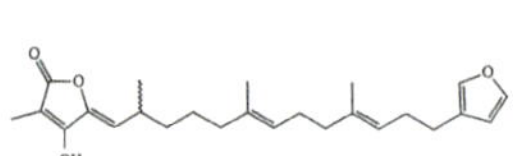

Fig. 9.40 Variabilin

Okinonellins A and B (Figs. 9.41 and 9.42), from *Spongionella* sp., were reported to inhibit division of fertilized starfish eggs at 5 μg/ml (Kato et al. 1986a).

The bishomo scalarene sesterpene phyllofoliaspongin (Fig. 9.43) from *Phyllospongia foliascens* inhibited P388 cell growth at 5 μg/ml (Kitagawa et al. 1989). Another activity noted for this compound was its antithrombocytic inhibitory effect on ADP-induced and collagen-induced aggregation of rabbit platelets in vitro.

Sesterterpenes extracted from the sponge *Hyrtios erecta* showed in vitro anthelmintic activity. Heteronemin (Fig. 9.44) (Kazlauskas et al. 1976; Kashman and Rudi 1977) showed in vitro activity with varying results. Another compound 12-episcalarin (Fig. 9.45) (Cimino

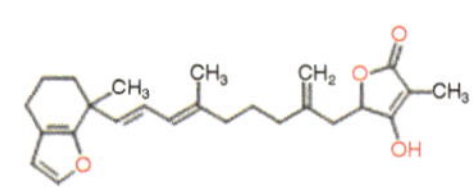

Fig. 9.41 Okinonellin A

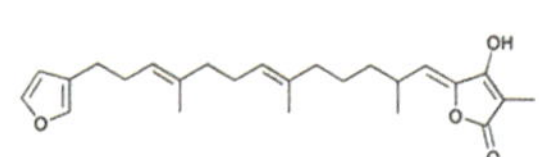

Fig. 9.42 Okinonellin B

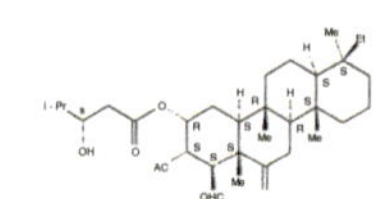

Fig. 9.43 Phyllofoliaspongin

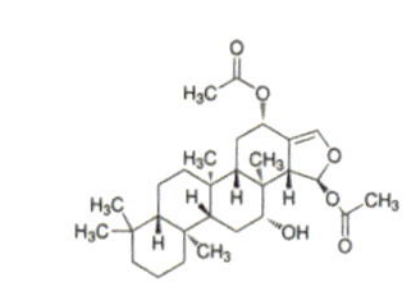

Fig. 9.44 Heteronemin

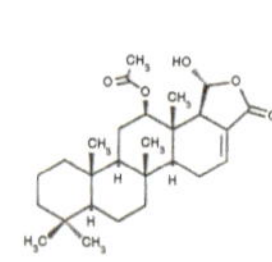

Fig. 9.45 12-episcalarin

et al. 1977; Crews and Bescansa 1986) exhibited moderate in vitro anthelmintic activity.

9.3.3 Sesquiterpenoid Isocyanide

Wright et al. (1988) reported the antitumor, antiviral, and antifungal activities for a sesquiterpenoid isocyanine (Fig. 9.46) isolated from the marine sponge *Bubaris* sp.. At 20 μg/ 0.5 ml, the A59 coronavirus in mouse liver cells was partially inhibited, indicating that the sesquiterpenoid compound is only weakly virucidal.

9.3.4 Diterpenes

Among the various kalihinols extracted from the sponge *Acanthella cavernosa*, Kalihinols Y (Fig. 9.47) and J (Fig. 9.48) reported potent in vitro anthelmintic activity (Chang et al. 1987; Omar et al. 1988; Alvi et al. 1991).

Kohmoto et al. (1987b) isolated spongiadiol (Fig. 9.49), epispongiadiol ($R_1 + R_2 = O$, $R_3 = OH$, $R_4 = H$), and the new isospongiadiol [2∞, 19-dihydroxyspongia – 13(16), 14-dien-3-one] ($R_1 = H$, $R_2 = OH$, $R_3 + R_4 = O$) from the deep-water Caribbean sponge *Spongia* sp.. Both antiviral activity and cytotoxicity were reported for all the three spongiodiols. In vitro assays against HSV-1 revealed a spectrum of activities ranging from the very active spongiadiol (IC50 = 0.25 μg/ml) to the modestly active epispongiadiol (IC50 = 12.5 μg/ml), with isospongiadiol exhibiting intermediate activity (IC50 = 2.0 μg/ml). Further the studies on antitumour and antiviral activities of these three furanoditerpenoids, spongidiol and isospongiadiol gave 100 % inhibition on HSV-1 plaque formation at 20 and 0.5 μg/(6 mm disk), and epispongiadiol gave partial inhibition at 12.5 μg/ml (Kohmoto et al. 1987b).

Kashman et al. (1987) isolated reiswigins A ($R = CH\ CH(CH_3)_2$) and B ($R = -CH = C(CH_3)_2$) (Fig. 9.50), bioactive terpenes from the sponge *Epipolasis reiswigi*. Both reiswigins A and B were found reporting the inhibition of HSV-1 completely at 2 μg and A59 virus partially at 20 μg (+ +). Particularly reiswigin A completely inhibited VSV at 2 μg without accompanying cytotoxicity (Kashman et al. 1989a).

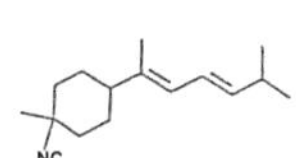

Fig. 9.46 Isocyanine

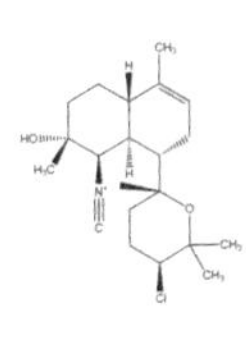

Fig. 9.47 Kalihinol Y

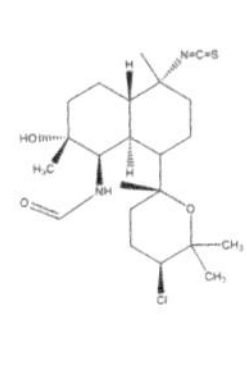

Fig. 9.48 Kalihinol J

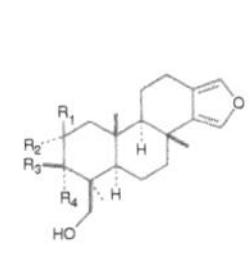

Fig. 9.49 Spongiadiol

9.3.5 Triterpenes

Pouoside A (Fig. 9.51) from *Asteropus* sp., collected in Truk Lagoon, inhibited P388 cell growth with an ED50 of 1.5 μg/ml (Ksebati et al. 1988, 1989). The Okinawan sponge *Penares* sp. was the source of penasterol (Fig. 9.52), which was active against L1210

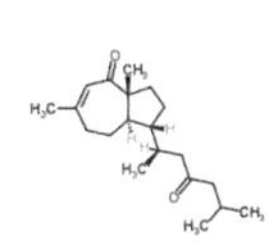

Fig. 9.50 Reiswigin A

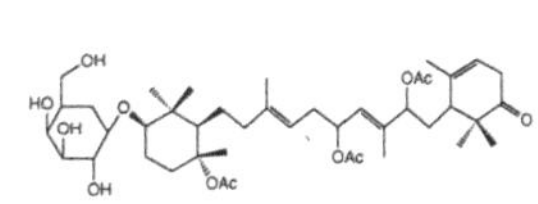

Fig. 9.51 Pouoside A

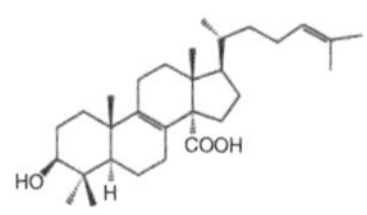

Fig. 9.52 Penasterol

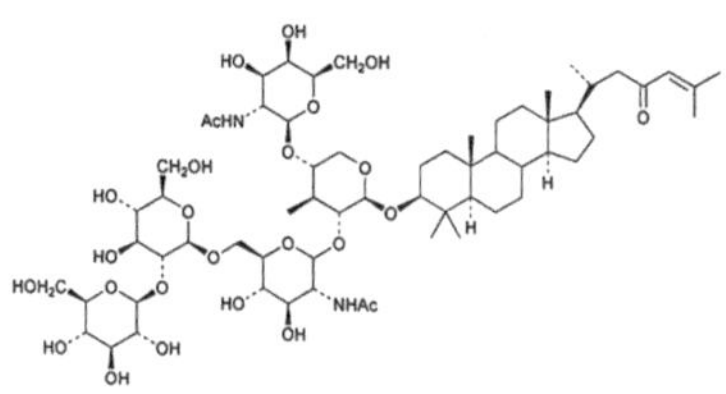

Fig. 9.53 Sarasinoside A1

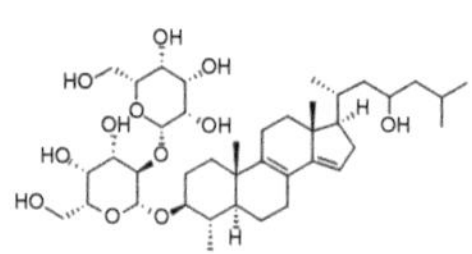

Fig. 9.54 Eryloside A

cells with an ED50 of 3.6 μg/ml (Cheng et al. 1988a).

9.3.6 Sterols

Several polyoxygenated sterols and glycosylated sterols showed cytotoxicity. Sarasinoside A1 (Fig. 9.53), a saponin containing amino sugar, exhibited an ED50 of 2.8 μg/ml against P388 cells. This saponin was isolated by Schmitz et al. (1988) from *Asteropus* sp. from Truk and Guam Islands.

Eryloside A (Fig. 9.54) from the red sea sponge *Erylus lendenfeldi* reported to have both cytotoxic (IC50 4.2 μg/ml vs. P388) and antifungal activity against *Candida albicans* (MIC 15.6 μg/ml) (Carmely et al. 1989a).

9.4 Brominated Compounds

The compound isolated from the marine sponge *Verongia cauliformis* (Sharma and Burkholder 1967) has been characterized as 2,6-dibromo-4-acetamido-4-hydroxycyclohexadienone

Fig. 9.55 2,6-dibromo-4-acetamido-4-hydroxycyclohexadienone

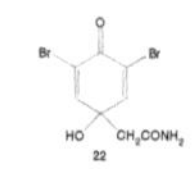

Fig. 9.56 Aerothionin

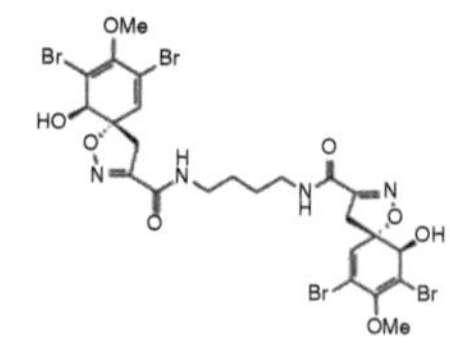

(Fig. 9.55) showed antibacterial activity (Sharma et al. 1970).

Aerothionin (Fig. 9.56) having a spirocyclohexadienylisoxazole skeleton was isolated from two sponges namely *Aplysia aerophoba* and *Verongia thiona* showed antibiotic activity (Encarnacion et al. 2000; Thoms et al. 2004).

9.5 Nitrogen-Containing Compounds

9.5.1 Tyrosine-Based Metabolites

Several members of the bastadin series of cyclic amides (Figs. 9.57) isolated from the sponge *Iarthella basta* were found to inhibit P388 cell growth (ED50 2–4 μg/ml) (Pordesimo and Schmitz 1990).

9.5.2 Other Amines

Mycalamides A (R = 4) and B (R = Me) (Fig. 9.58) obtained from a New Zealand sponge *Mycale* sp. (Perry et al. 1988a, 1990) showed antiviral and cytotoxic activity.

Calyculins A–D (Fig. 9.59) are unusual amines isolated from *Discodermia calyx* (Kato et al. 1986b, c, 1988a, b) that showed the IC50 value of 7.4×10^{-4}, 8.8×10^{-4}, 8.6×10^{-4}, and 1.5×10^{-3} μg/ml respectively against L1210 cells. They also inhibited cell division of both starfish and sea urchin eggs in the 10^{-2} μg/ml range. Further, the calyculin A (Fig. 9.59) exhibited in vivo activity against Erlich and P388 leukemia in mice (T/C 245 and 144 %,

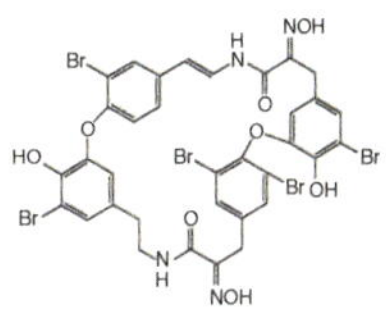

Fig. 9.57 Bastadin series of cyclic amides

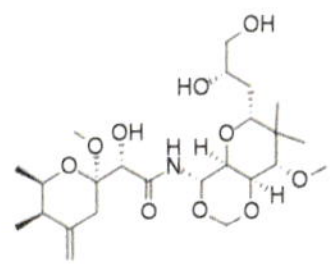

Fig. 9.58 Mycalamide A (R = 4) and B (R = Me)

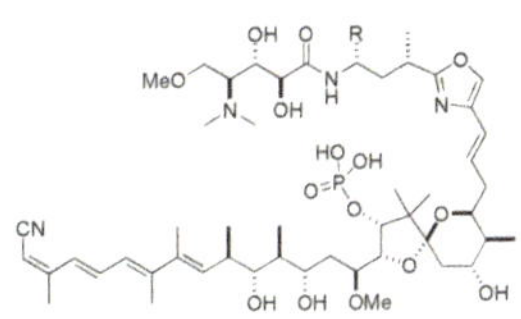

Fig. 9.59 Calyculin A

respectively) apart from inhibiting the uptake of $[^{3H}]$ thymidine, $[^{3H}]$ uridine and $[^{3H}]$ leucine in L1210 murine leukemia cells (Kato et al. 1988a, b).

9.5.3 Pyrroles

The alkaloid 300 (Fig. 9.60) isolated from the sponges *Teichaxinella morchella* and *Ptilocaulis walpersi* reported mild cytotoxicity to L1210 cells (IC50 19 μg/ml) (Wright and Thompson 1987).

9.5.4 Imidazoles

The girolline (Fig. 9.61) extracted from the sponge *Pseudaxinyssa cantharella* was found active against P388 at 0.001–1 μg/ml, and this activity was confirmed in vivo also in mice models (P388 at 1 mg/kg doses (Ahond et al. 1989).

Pyronaamide (Fig. 9.62), obtained from *Leucetta* sponge from Saipan and Guam, was toxic to KB cells (MIC 5 μg/ml) (Akee et al. 1990). A series of 2-amino imidazole

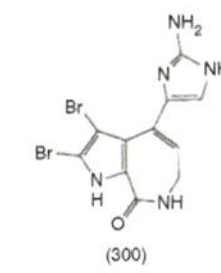

Fig. 9.60 Alkaloid

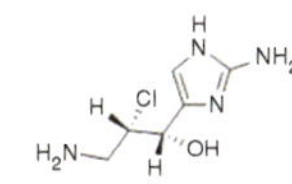

Fig. 9.61 Girolline

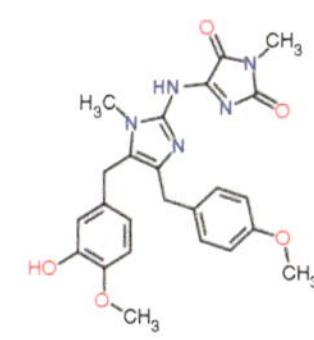

Fig. 9.62 Pyronaamide

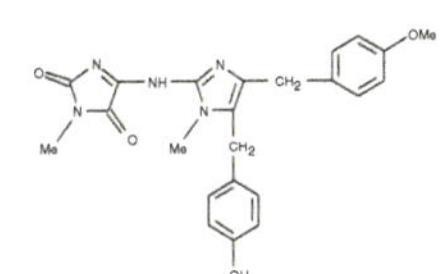

Fig. 9.63 Series of 2-amino imidazole alkaloids Naamidines

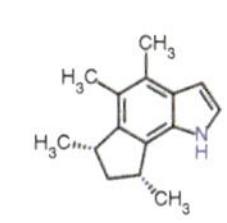

Fig. 9.64 Horbindole A (R = Me); Horbindole B (R = Et); Horbindole C (R = CH = CH-Et)

alkaloids called naamidines (e.g. Fig. 9.63) were obtained by Carmely et al. (1989b) from the marine sponge *Leucetia chagosensis* that showed cytotoxicity at 2–10 μg/ml against P388 cells.

9.5.5 Indoles

Horbindoles A–C (Fig. 9.64) extracted from *Axinella* sp. from western Australia showed cytotoxicity (KB; MIC 5, >10, and 10 μg/ml, respectively) and were also found to have fish antifeedant activity (Herb et al. 1990).

A deep water sponge, *Dragmacidian* sp. was the source for dragmacidin (Fig. 9.65) that was found to be toxic to P388 cells (IC50 15 μg/ml) and also to A549 human-8 human colon, and MDAMB human mammary cells, all with IC50 of 1–10 μg/ml (Kohmoto et al. 1988). Morris and Anderson (1990) isolated the closely related dragmacidon A (Fig. 9.66) which showed cytotoxicity against L1210 cells (ED50 10 μg/ml) similar to that of dragmacidin.

An antimicrobial pigment, fascaplysin (Fig. 9.67), obtained from a Fijian sponge, *Fascaplysinopsis* sp., killed L1210 cells (LD50 0.2 μg/ml) and also showed antibiotic activity against four different microorganisms (Roll et al. 1988).

The other group of indoles, eudistomins were initially reported as antiviral agents, but Eudistomin K (Fig. 9.68), obtained from *Riterella sigillinoides*, is described in a patent as being "very effective in inhibiting growth of L1210, P388, A549 and HCT-8 cells at varying concentrations" (Blunt et al. 1988).

Manzamine A, an alkaloid (Fig. 9.69) was reported as its hydrochloride salt from a *Haliclona* sp. of sponge from Okinawa with as IC50 of 0.07 μg/ml against P388 cells in vitro (Sakai et al. 1986b).

Fig. 9.65 Dragmacidin

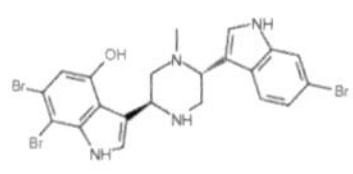

Fig. 9.66 Dragmacidon A

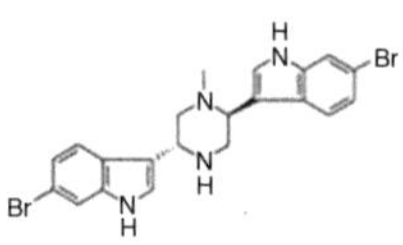

Fig. 9.67 Fascaplysin

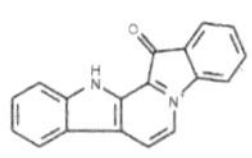

Fig. 9.68 Eudistomin K

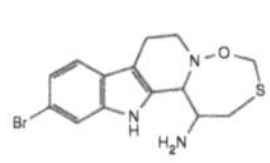

Fig. 9.69 Manzamine A

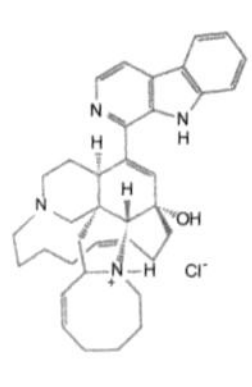

9.5.6 Pyridines

The pyridine alkaloids theonelladins A–D (Fig. 9.70) isolated from the sponge *Theonella swinhoei* (Kobayashi et al. 1989) showed the cytotoxicities of 4.7, 1.0, 3.6, and 1.6 μg/ml (IC50) against L1210 cell lines and 10.0, 3.6, 10.0, and 5.2 μg/ml (ED50) against KB cells. These compounds were also reported to be 20 times more than caffeine in causing release of Ca^{2+} from sarcoplasmic reticulum.

The related pyridine alkaloids niphatynes A (Fig. 9.71) and B (Fig. 9.72), from *Niphates* sp. collected in Fiji, were found cytotoxic to P388 cells (IC50 0.5 μg/ml) (Quinoa and Crews 1987).

9.5.7 Nucleosides

Nucleosides are vital components of all living cells and are involved in several biological processes.

The two cytotoxic nucleosides 5-(methoxycarbonyl) tubercidin (Fig. 9.73) and toyocamycin (Fig. 9.73) isolated from the Fijian sponge *Jaspis* (Zabriskie and Ireland 1989) showed IC50 values of 0.0026 and 0.27 μg/ml,

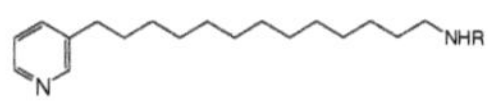

Fig. 9.70 Theonelladins A (R = H); Theonelladin B (R = CH_3-D); Theonelladin C (R = H); Theonelladin D (R = CH_3)

Fig. 9.71 Niphatyne A

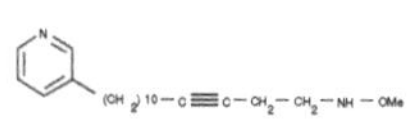

Fig. 9.72 Niphatyne B

respectively, against L1210. The 5-(methoxycarbonyl) tubercidin (Fig. 9.73) also reported earlier to have in vivo activity against L1210, increasing lifetimes by up to 39 %.

The two antiviral and antitumor compounds presently in clinical use as antiviral or antitumor agents (i.e., ara-A, 9-β-D-arabinofuranosyladenine, Fig. 9.74; ara-C, 1-β-D-arabinosylcytosine, Fig. 9.74) were isolated from the marine sponge *Cryptotethia crypta* in the early 1950s (Bergmann and Feeney 1950, 1951). Bergmann collected *C. crypta* in 1945, within next few years he reported the presence of spongothymidine (ara-T, 1-β-D-arabinofuranosylthymidine, Fig. 9.74), spongouridine (ara-U, 1-β-D-arabinofuranosyluracil, Fig. 9.74), and spongosine (1-β-D-arabinofuranosyl-2-methoxyadenine) (Cohen 1966).

The in vitro studies of the arabinosides (Fig. 9.74) showed varying antiviral activity against HSV-1 or HSV-2. Using rabbit kidney and human skin fibroblast cultures, De Clercq et al. (1977) reported MICs (minimum inhibitory concentration) as low as 0.02 and 1 μg/ml for ara-C and ara-A, respectively, against HSV-1; and 200 and 10 μg/ml, respectively, against HSV-2. Besides the above, a significant in vitro activity was also observed for a number of xylofuranonucleosides against three DNA viruses (HSV-1, HSV-2, and vaccinia) and one RNA virus (thinovirus-9) (Gosselin et al. 1986).

Doridosine (Fig. 9.75) (Quinu et al. 1980) was isolated from marine sponge *Tedania digitala* from Australia. It causes reduced arterial pressure and reduced heart rate in mammalians in a manner that is qualitatively similar to adenosine. It also acts as muscle relaxant and showed hypothermic activity.

1-Methylisoguanosine (Fig. 9.76) was isolated from the sponge *Tedania digitata* (Quinu et al. 1980). This nucleoside showed potent muscle relaxant, blood pressure lowering, cardiovascular, and anti-inflammatory activity (Jamieson and Davis 1980).

9.5.8 Quinolines and Isoquinolines

The aaptamine (Fig. 9.77) and some of its derivatives (R_1 = H, R_2 = H; 163. R_1 = H, R_2 = CH_3) were isolated from the sponge *Suberites* sp., which were reported to have some in vitro and in vivo cell inhibitory activity when tested for antitumor activity against Ehrlich ascites tumor in mice. A 95 % inhibition was reported in the case of mice inoculated with Ehrlich ascites tumor cells pretreated with the derivative with R_1 = H, R_2 = H or R_1 = H, and R_2 = CH_3 at 25 μg/ml (Fedoreov et al. 1988).

A series of pyrroloquinoline alkaloids namely isobatzellines A–D (Figs. 9.78) were found in

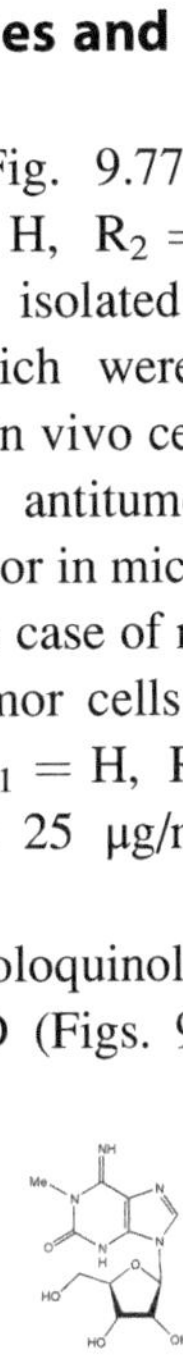

Fig. 9.75 Doridosine

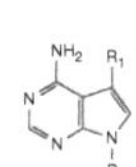

Fig. 9.73 5-(methoxycarbonyl) tubercidin (R_1 = CO_2Me, R_2 = ribose) and Toyocamycin (R_1 = CN, R_2 = ribose)

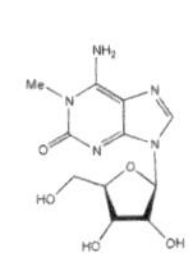

Fig. 9.76 1-Methylisoguanosine

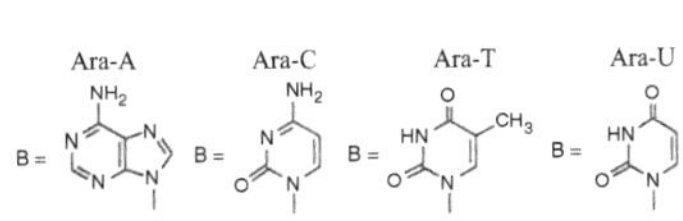

Fig. 9.74 Arabinosides

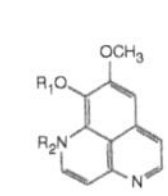

Fig. 9.77 Aaptamine (R_1 = CH_3, R_2 = H) and some of its derivatives (R_1 = H, R_2 = H; 163. R_1 = H, R_2 = CH_3)

extracts of the Caribbean sponge *Batzella* sp. (Sun et al. 1990). These compounds showed antifungal activity against *C. albicans*. Renierol (Fig. 9.79), obtained from the Fijian sponge *Xestospongia caycedoi*, inhibited the growth of L1210 cells (IC50 3 μg/ml) (McKee and Ireland 1987).

9.5.9 Quinilizidines and Indolizidines

The indolizidine stellenamide A (Fig. 9.80) from the sponge *Stella* sp. showed antifungal activity and also inhibited K562 epithelium cell growth (IC50 of 5.1 μg/ml) (Hirota et al. 1990a).

9.5.10 Prianosins/Discorhabdins

The prianosins and discorhabdins, the two closely related sulfur-containing alkaloids, were extracted from *Latrunculia* sp. and *Prianos* sp. The first of these to be reported was discorhabdin C (Fig. 9.83) (Perry et al. 1988a). The remaining discorhabdins (Figs. 9.81, 9.82, 9.83 and 9.84 = discorhabdins A,B,C,D, respectively) were described subsequently (Perry et al. 1988a,b). The discorhabdins A–D isolated from the sponges *Latrunculia brevis* and *Prianos* sp. reported the cytotoxicity (IC50) of 0.05, 0.1, 0.03, and 6.0 μg/ml, respectively, against P388 cells. Only discohabdin D (Fig. 9.84) showed any in vivo activity in the P388 model and that was modest (T/C 132 at 20 mg/kg).

Prianosin A (Fig. 9.85), from the Okinawan sponge *Prianos melanos* (Kobayashi et al. 1987), is the nonprotonated form of discorhabdin A (Fig. 9.81). The remaining prianosins B–C (Figs. 9.86 and 9.87) were reported in 1988 (Cheng et al. 1988b). Prianosin D (Fig. 9.87) and discorhabdin D (Fig. 9.84) are a hydroquinone/quinine pair. The prianosins A–D reported the IC50 of 0.037, 2.0, 0.15, and 0.18 μg/ml against L1210 cells, 0.014, 1.8, 0.024 and 0.048 μg/ml against L5178Y cells and 0.073, >5, 0.57

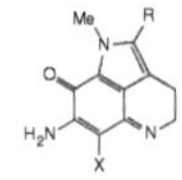

Fig. 9.78 Isobatzellines

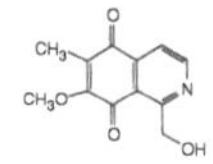

Fig. 9.79 Renierol

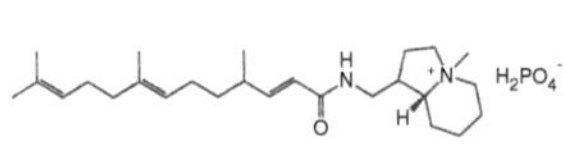

Fig. 9.80 Indolizidine stellenamide A

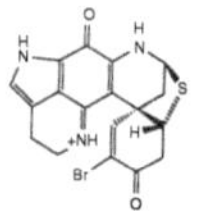

Fig. 9.81 Discorhabdin A

Fig. 9.82 Discorhabdin B

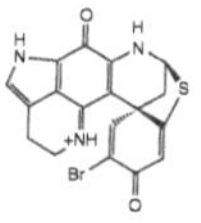

Fig. 9.83 Discorhabdin C

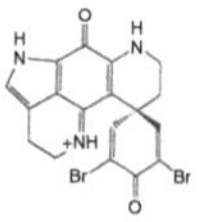

Fig. 9.84 Discorhabdin D

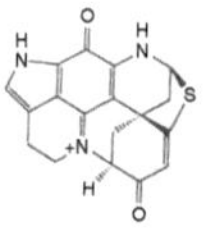

Fig. 9.85 Prianosin A

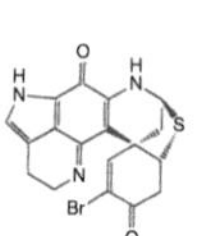

Fig. 9.86 Prianosin B

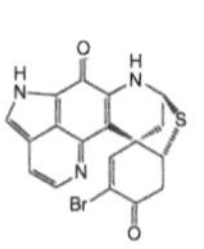

and 0.46 μg/ml against KB cells respectively. In addition to these activities, the prianosin D (Fig. 9.87), but not the others, induced Ca^{2+} release from sarcoplasmic reticulum, with potency ten times than that of caffeine.

9.5.11 Marine Alkaloids

Dysemenin (Fig. 9.88) a hexachlorinated alkaloid isolated from the sponge *Dysidea herbacea* (Charles et al. 1978, 1980; Biskupiak and Ireland 1984) was found inhibiting iodide transfer in thyroid cells. This molecule might provide insight into the mechanism of the elusive "iodide pump" as it inhibits iodine transport by a different mechanism than ouabain (Van Sande et al. 1990).

The polycyclic aromatic alkaloid, amphimedine (Fig. 9.89), isolated from *Amphimedon* sp. was found active against P388 in vitro with an ED50 value of 0.4 μg/ml, but proved inactive in vivo (Schmitz et al. 1983).

Dercitin (Fig. 9.90), from a deepwater sponge *Descitus* sp., showed in vitro and in vivo activity

Fig. 9.87 Prianosin C (R = OH) and D (R = H)

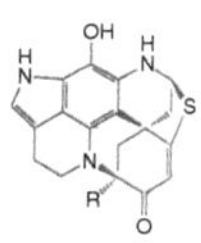

Fig. 9.88 Dysemenin

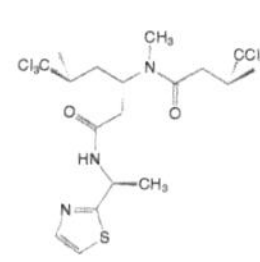

Fig. 9.89 Amphimedine

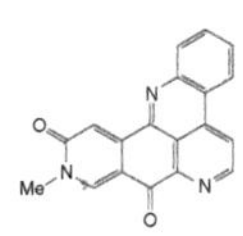

Fig. 9.90 Dercitin

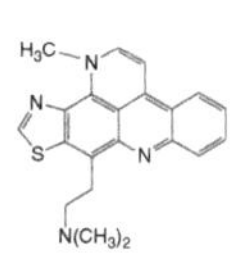

Fig. 9.91 Plakinidine A (R = H); Plakinidine B (R = CH_3); Plakinidine C (R = $H\Delta^9$)

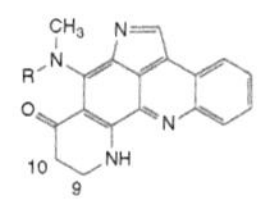

Fig. 9.92 Latrunculin A

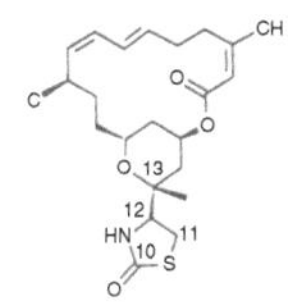

(T/C 170 at 5 mg/kg) in the P388 model (Gunawardana et al. 1988). In addition, dercitin was described as having immunosuppressive and antiviral activity. The dercitin was found disrupting the macromolecular synthesis (DNA, RNA, and protein) in the P388 system by binding to DNA and inhibiting nucleic acid synthesis (Burres et al. 1989).

Inman et al. (1990) isolated plakinidines A (Fig. 9.91) and B (Fig. 9.91), using an antiparasite bioassay, from the fijian sponge *Plakortis* sp. The planinidine A inhibited reverse transcriptase activity at 1 μg/ml. Thereafter West et al. (1990) described plakinidines A, B, and C (Fig. 9.91) from the same Fijian sponge species and reported the IC50 values of 0.1, 0.3 and 0.7 μg/ml, respectively, for these compounds against L1210 cells.

Other anthelmintic-active alkaloids were isolated from a Fijian sponge of the family Spongiidae, originally identified as *Spongia mycofijiensis* (Kakou et al. 1987). This sponge yielded latrunculin A (Fig. 9.92) (Kashman et al. 1980), which showed excellent in vitro activity at 50 μg/ml against *N. brasiliensis*.

9.5.12 Guanidines

Kashman et al. (1989b) isolated ptilomycalin A (Fig. 9.93) from the Caribbean sponge *Ptilocaulis spiculifer* and a red sea sponge *Hemimycale* sp. that reported activity against HSV at a concentration of 0.2 μg/ml (Kashman et al. 1989a). In addition to the high antiviral activity, this compound exhibited antitumor and antifungal activities also.

In later years, Janes Erijman et al. (1991) isolated a series of compounds related to ptilomycalin A from the Mediterranean sponge *Crambe crambe*. The new compounds, the crambescidins (Fig. 9.93) showed activity against HSV-1 at 1.25 μg/ml and exhibited 98 % inhibition of L1210 cell growth at 0.1 μg/ml.

The diacetate salts of the series of bromopyrroles were extracted from the Caribbean sponge *Agelas conifer* (Rinehart 1988; Keifer et al. 1991). Based on spectroscopic comparisons to the known sceptrin (Fig. 9.94) (Walker et al. 1981), as well as on FABMS and NMR data, the structures assigned included the oxysceptrins (Fig. 9.96) and ageliferins (Fig. 9.95). The compounds of the sceptrin and

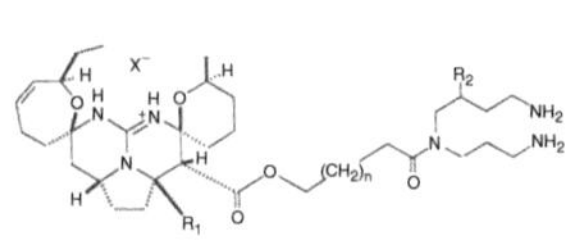

Fig. 9.93 Ptilomycalin A ($R_1 = R_2 = H, n = 13$) Crambescidins (Crambescidin 816: $R_1 = R_2 = OH, n = 13$; Crambescidin 830:$R_1 = R_2 = OH, n = 14$; Crambescidin 844: $R_1 = R_2 = OH, n = 15$; Crambescidin 800: $R_1 = H, R_2 = OH, n = 13$

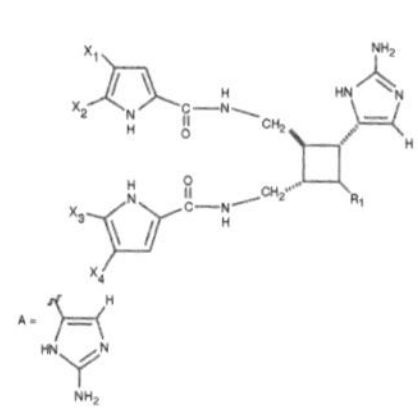

Fig. 9.94 Sceptrin

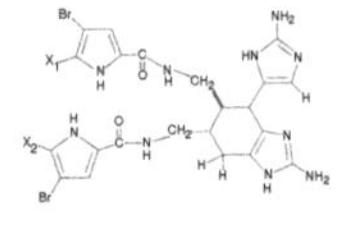

Fig. 9.95 Ageliferin

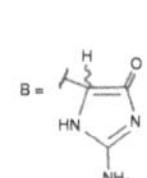

Fig. 9.96 Oxysceptrin

Fig. 9.97 Acarnidine la ($R = CO(CH_2)_{10}CH_3$)

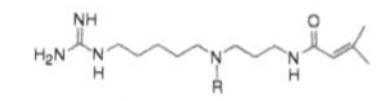

Fig. 9.98 Acarnidine 1b ($R = CO(CH_2)_3CH = CH (CH_2)_5CH_3(z)$)

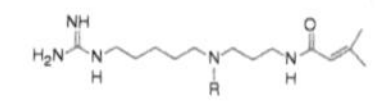

ageliferin groups were found active against HSV-1 at 20 μg/disk and VSV at 100 μg/disk, while the oxysceptrins were less active (Keifer et al. 1991).

Acarnidines la–1c (Figs. 9.97 and 9.98) were isolated from *Acarnus erithacus*, collected from Gulf of California, and were reported to show antiviral property (Carter and Rinehart 1978). The homospermidine skeleton common to these three guanidine compounds was assigned based on GC/MS data, and the compounds were distinguished from one another by their fatty acid constituents. In addition to some antibacterial activity, the activity against HSV-1 was also obtained at 100 μg/disk.

9.5.13 Peptides and Depsipeptides

Sponges are a large and diverse group of colonial organisms that constitute the phylum Porifera with thousands of different species extensively distributed from superficial waters near the sea shores up to deep waters of the ocean. Active peptides from sponges most of them with unique unprecedent structures in comparison with these kind of compounds from other sources are often cyclic or linear peptides containing unusual amino acids which are either rare in terrestrial and microbial systems or even totally novel, and also frequently containing uncommon condensation between amino acids (Aneiros and Garateix 2004).

Discobahamin A (Fig. 9.99) was a bioactive antifungal peptide evaluated as inhibitor of the growth of *Candida albicans* isolated from the Bahamian deep water marine sponge *Discodermia* sp. (Gunasekera et al. 1994; Tohma et al. 2003).

The cyclic depsipeptides papuamides A and B (Fig. 9.100) isolated from sponges of the

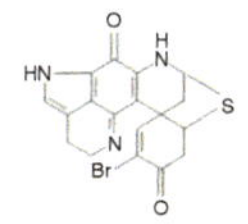

Fig. 9.99 Discobahamin A

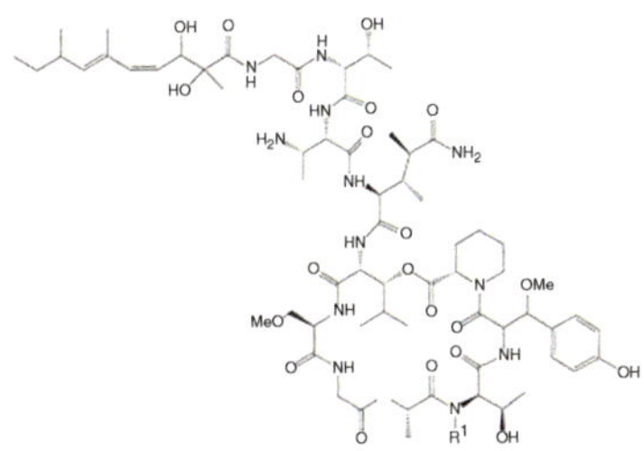

Fig. 9.100 Papuamides A and B

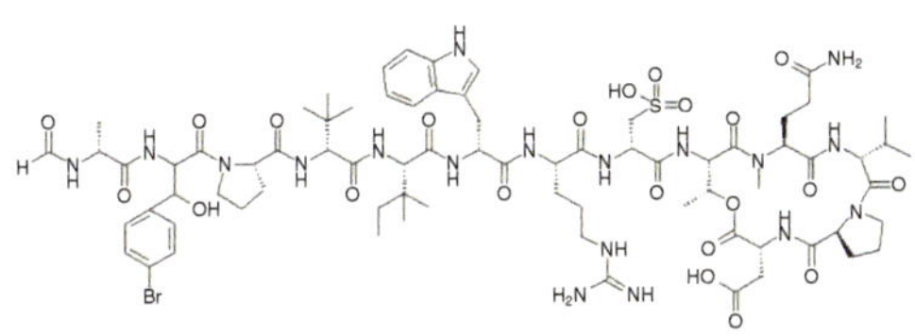

Fig. 9.101 Microspinosamide

genus *Theonella*, containing a number of unusual amino acids are also the first marine-derived peptides reported to contain 3-hydroxyleucine and homoproline residues (Ford et al. 1999). They inhibited the infection of human T-lymphoblastoid cells be HIV-1 sub (RF) in vitro with an EC50 of approximately 4 ng/ml.

Microspinosamide a new cyclic depsipeptide incorporating 12 amino acid residues (Fig. 9.101) from the sponge *Sidonops microspinosa* reported anti-HIV activity (Rashid et al. 2001). It also inhibited the cytopathic effect of HIV-1 infection in an XTT-based in vitro assay.

Another novel peptide, keramamide (Fig. 9.102) (Kobayashi et al. 1991) as well as orbiculamide A (Fig. 9.103) (Fusetani et al. 1991) isolated from the marine sponge *Theonella* sp. reported cytotoxic effect against P388 murine leukemia cells (IC50 = 4.7 ng/ml). The other active peptide Cyclotheonamide (Fig. 9.104) (Fusetani et al. 1990) isolated from the species of the same genus was reported as

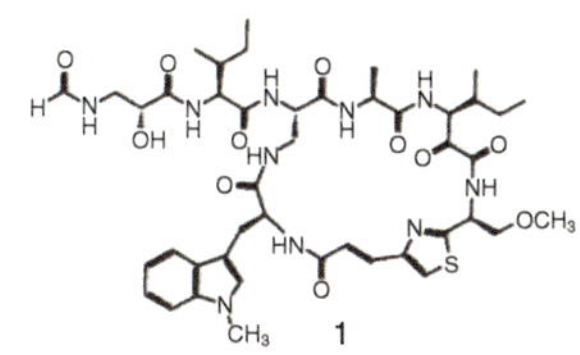

Fig. 9.102 Keramamide

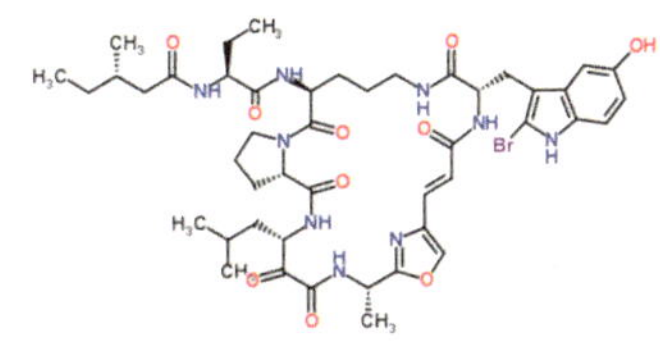

Fig. 9.103 Orbiculamide A

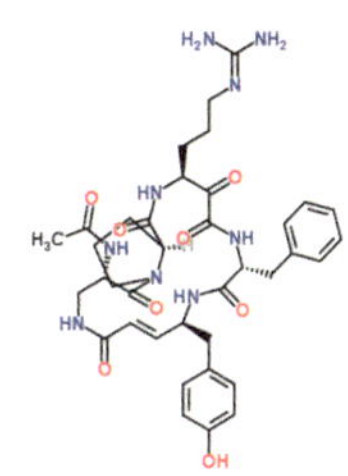

Fig. 9.104 Cyclotheonamide

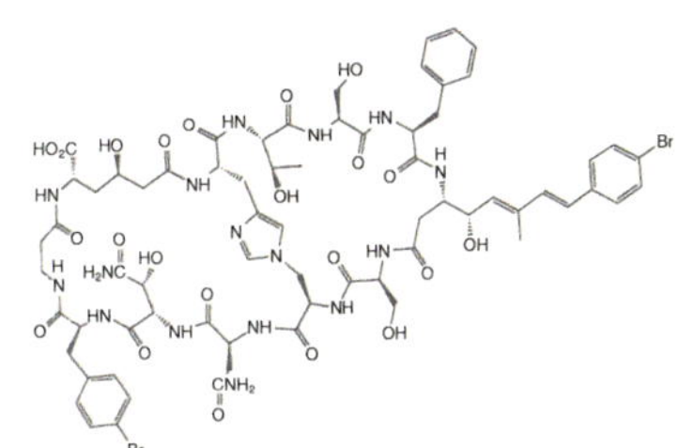

Fig. 9.105 Theonellamide F

a potent antithrombin cyclic peptide which strongly inhibited various proteinases, particularly thrombin.

Theonellamide F (Fig. 9.105) an antifungal peptide isolated from *Theonella* sp. from Japan also showed activity against L1210 and P388 cells (IC50 3.2 and 2.7 μg/ml, respectively) (Matsunaga et al. 1989).

From a western Pacific sponge, *Hymeniacidon* sp., collected at Palau, Pettit et al. (1990) isolated the cyclic octopeptide, hymenistatin 1 (Fig. 9.106)

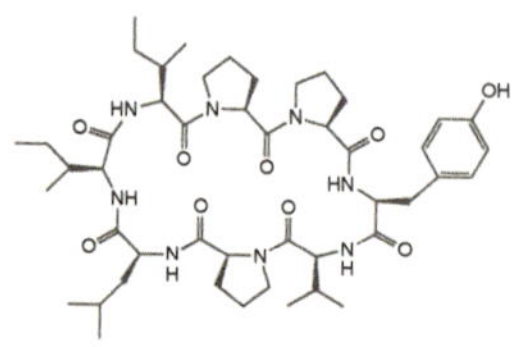

Fig. 9.106 Hymenistatin 1

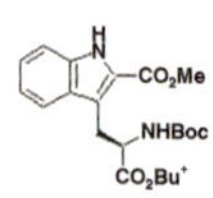

Fig. 9.107 Microsclerodermin A (R = OH) – Microsclerodermin B (R = H)

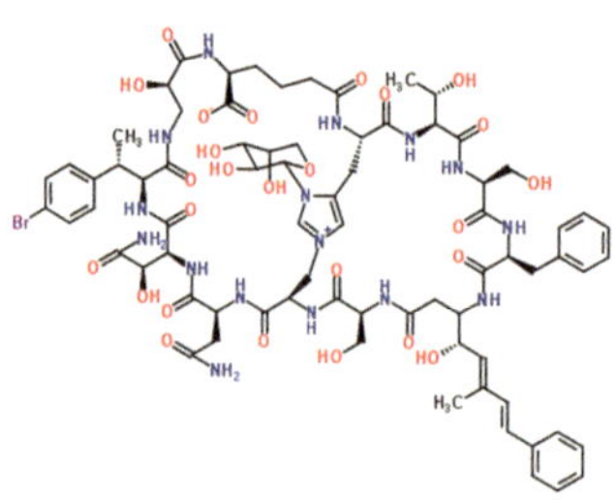

Fig. 9.108 Theonegramide

in which all amino acids therein having the schirality. It showed both in vitro (ED50 3.5 μg/ml) and in vivo activity (T/C 130) against P388 murine leukemia cells.

Three new antifungal cyclic peptides with unprecedented amino acids, microsclerodermins A-B (Figs. 9.107) were isolated from two species of sponges, *Theonella* sp. and *Microscleroderma* sp. from the Philippines (Schmidt and Faulkner 1998). Another antifungal cyclic peptide isolated from the same sponges was the Theonegramide (Fig. 9.108) (Bewley and Faulkner 1994).

9.6 Conclusion

The researchers studying the marine natural products report several substances with interesting pharmacological properties. But only very few of them are available as potent drugs in the market which are being superseded by the synthetic ones. This may be because of the non-availability of source materials for the continuous supply of such biologically active compounds. Further this acts as a limiting factor for the pharmaceutical companies to go for patenting. So the pharmaceutical companies prefer the synthetic compounds to get continuous supply after launching their product in the market. However, it is not that much easy to synthesize, economically, some of the natural products, since they have more complex structure. Hence, further research is needed to find out the ways and means to synthesize the more complex marine natural products.

Above all, the research in the field of marine natural products needs to be encouraged by the funding agencies to get fruitful results in future which need not be immediate as that of synthetic chemistry outcomes.

References

Alvi KA, Tenenbaum L, Crews P (1991) Anthelmintic polyfunctional nitrogen containing terpenes from marine sponges. J Nat Prod 54:71–78

Ahond A, Zurita MB, Collin M, Fizames C, Laboute P, Lavelle F, Laurent D, Poupat C, Pusset J, Pusset M, Thoison 0, Potier P (1989) Girolline, a new antitumoral compound extracted from the sponge *Pseudaxinyss acantharella* (Axinellidae). C R Acad Sci Paris 307:145–148

Akee RK, Carroll TR, Yoshida WY, Scheuer PJ, Stout TJ, Clardy J (1990) 1\vo imidazole alkaloids from a sponge. J Org Chem 55(6):1944–1946

Albizati KF, Holman T, Faulkner DJ, Glaser KB, Jacobs RS (1987) Luffariellolide, an anti-inflammatory sesterterpene from the marine sponge *Luffariella* sp. Experientia 43(949–9):50

Aneiros A, Garateix A (2004) Bioactive peptides from marine sources: pharmacological properties and isolation procedures. J Chromatogr B Analyt Technol Biomed Life Sci 803(1):41–53

Barrow CJ, Blunt JW, Munro MHG, Perry NB (1988) Oxygenated furanosesterterpenetetronic acids from a sponge of the genus *Ircinia*. J Nat Prod 51:1294–1298

Bartlett RT, Cook AF, Holman MJ, Mccomas WW, Nowoswait EF, Poonian MS, Bairdlambert JA, Baldo BA, Marwood JF (1981) Synthesis and pharmacological evaluation of a series of analogues of 1-methylisoguanosine. J Med Chem 24:947–954

Barzaghi G, Sarace HM, Mong S (1989) Platelet-activating factor induced phosphoinositide metabolism in differentiated U-937 cells in culture. J Pharmacol Exp Ther 248:559–566

Bergmann W, Feeney RJ (1950) The isolation of a new thymine pentoside from sponges. J Am Chem Soc 72:2809–2810

Bergmann W, Feeney RJ (1951) Contributions to the study of marine products. XXXII the nucleosides of sponges. J Org Chem 16:981–987

Bewley CA, Faulkner DJ (1994) Theonegramide, an antifungal glycopeptide from the Philippine lithistid sponge *Theonella swinhoei*. J Org Chem 59:4849–4852

Biskupiak JE, Ireland CM (1984) Revised absolute configuration of dysidenin and Isodysidenin. Tetrahedron Lett 25:2935–2936

Blunt JW, Lake RJ, Munro MHG (1988) Antitumor polycyclic compounds from marine *Ritterella sigillinoides* and their use and preparation. WO 8800826 AI 11 February 1988.

Burres NS, Sazesh S, Gunawardana GP, Clement JJ (1989) Antitumor activity and nucleic acid binding properties of dercitin, a new acridine alkaloid isolated from a marine *Dercitus* species sponge. Cancer Res 49:5267–5274

Carmely S, Roll M, Loya Y, Kashman Y (1989a) The structure of Eryloside A, A new antitumor and antifungal 4-methylated steroidal glycoside from the sponge *Erylus lendenfeldi*. J Nat Prod 52 (1):167–170

Carmely S, Ilan M, Schmitz Y (1989b) 2-Amino imidazole alkaloids from the marine sponge *Leucetta chagosensis*. Tetrahedron 45:2193–2200

Carter GT, Rinehart KL Jr (1978) Acarnidines, novel antiviral and antimicrobial compounds from the sponge *Acarnus erithacus* (de Laubenfels). J Am Chem Soc 100:4302–4304

Chang CW, Patra A, Baker JA, Scheuer PJ (1987) Kalihinols, multifunctional diterpenoid antibiotics from marine sponges *Acanthella* sp. J Am Chem Soc 109:6119–6123

Charles C, Braekman JC, Daloze D, Thrsch B, Karlson R (1978) Isodysidenin, a further hexachlorinated metabolite from the sponge *Dysidea herbacea*. Tetrahedron Lett 17:1519–1520

Charles C, Braekman JC, Dalose D, Thrsch B (1980) The relative and absolute configuration of dysidenin. Tetrahedron 36:2133–2135

Cheng J, Kobayashi J, Nakamura H, Ohizumi Y, Hirata Y, Sasaki T (1988a) Penasterol, a novel antileukemic sterol from the Okinawan marine sponge *Penares* sp. J Chem Soc Perkin 1(8):2403–2406

Cheng J, Ohizumi Y, Walchli MR, Nakamura H, Hirata Y, Sasaki T, Kobayashi J (1988b) Prianosins B, C, and D, novel sulfur-containing alkaloids with potent antineoplastic activity from the Okinawan marine sponge *Prianos melanos*. J Org Chem 53(19):4621–4624

Cimino G, De Stefano S, Minale L, Trivellone E et al (1977) 12-Epi-scalarin and 12-epideoxyscalarin, sesterterpenes from the sponge *Spongia nitens*. J Chem Soc Perkin Trans I 1977:1587–1588

Cimino G, De Giulio A, De Rosa S, Di Marzo V (1990) Minor bioactive polyacetylenes from *Petrosia ficiformis*. J Nat Prod 53(2):345–353

Cohen SS (1966) Introduction to the biochemistry of o-arabinosyl nucleosides. In: Davidson JN, Cohn WE (eds) Progress in nucleic acid research and molecular biology, vol 5. Academic, New York, pp 1–88

Crews P, Bescansa P (1986) Sesterterpenes from a common marine sponge, *Hyrtios erecta*. J Nat Prod 49:1041–1052

D' Blassio G, Fattorusso E, Mango S, Mayo L, Pedone C, Santacroce C, Sica D (1976) Axisonitrile-3, axisothiocyanate-3, and axamide-3, sesquiterpenes with a novel spiro [4,5] decane skeleton from the sponge *Axinella cannabina*. Tetrahedron 32:473–478

De Clercq E, Krajewska E, Descamps J, Torrence PF (1977) Anti-herpes activity of deoxythymidine analogues: specific dependence on virus-induced deoxythymidine kinase. Mol Pharmacol 13:980–984

De Silva ED, Scheuer PJ (1980) Manoalide, an antibiotic sesterterpenoid from the marine sponge *Luffariella variabilis* (Polejaeff). Tetrahedron Lett 21:1611–1614

De Vries GW, Amdahl L, Mobasser A, Wenzel M, Wheeler LA (1988) Preferential inhibition of 5-lipoxygenase activity by manoalide. Biochem Pharmacol 37:2899–2905

Encarnacion RD, Sandoval E, Malmstrom J, Christophersen C (2000) One pot spiropyrazoline synthesis via intramolecular cyclization methylation. J Nat Prod 63:874

Fedoreov SA, Prokofeva NG, Denisenko VA, Rebachuk NM (1988) Cytotoxic activity of aaptamines derived from Suberitidae sponges. Khim Farm Zh 22 (8):943–946

Ford PW, Gustafson KR, McKee TC, Shigematsu N, Maurizi LK, Pannell LK (1999) Papuamides A–D, HIV-inhibitory and cytotoxic depsipeptides from the sponges *Theonella mirabilis* and *Theonella swinhoei* collected in Papua New Guinea. J Am Chem Soc 121:5899–5909

Fusetani N, Shiragaki T, Matsunaga S, Hashimoto K (1987) Bioactive marine metabolites XX. Petrosynol and petrosynone, antimicrobial C30 polyacetylenes from the marine sponge *Petrosia* sp: determination of the absolute configuration. Tetrahedron Lett 28 (37):4313–4314

Fusetani N, Yasumuro K, Matsunaga S, Hashimoto K et al (1988) Mycalolides A-C, hybrid macrolides of ulapualides and halichondramide, from a sponge of the genus *Mycale*. Tetrahedron Lett 30 (21):2809–2812

Fusetani N, Matsunaga S, Matsumoto H, Takebayashi Y (1990) Bioactive marine metabolites. 33. Cyclotheonamides, potent thrombin inhibitors, from a marine sponge *Theonella* sp. J Am Chem Soc 112:7053–7054

Fusetani N, Sugawara T, Matsunaga S (1991) Orbiculamide A: a novel cytotoxic cyclic peptide from a marine sponge *Theonella* sp. J Am Chem Soc 113:7811–7812

Glaser KB, Jacobs RS (1986) Molecular pharmacology of manoalide. Biochem Pharmacol 35:449–453

Gosselin G, Bergogne MC, de Rudder J, De Clercq E, Imbach JL (1986) Systematic synthesis and biological evaluation of α- and 3-xylofuranosyl nucleosides of the five naturally occurring bases in nucleic acids and related analogues. J Med Chem 29:203–213

Groweiss A, Kashman Y, Shmueli U (1980) Latrunculin, a new 2-thiazolidinone macrolide from the marine sponge *Latrunculaia magnifica*. Tetrahedron Lett 21:3629–3632

Groweiss A, Shmueli U, Kashman Y (1983) Marine toxins of Latrunculia magnifica. J Org Chem 48 (20):3512–3516

Gunasekera SP, Faircloth GT (1990) New acetylenic alcohols from the sponge *Cribrochalina vasculum*. J Org Chem 55(25):6223–6225

Gunasekera SP, Gunasekera M, Longley RE, Schulte GK (1990a) Discodermolide: a new bioactive polyhydroxylated lactone from the marine sponge *Discodermia dissolute*. J Org Chem 55:4912–4915

Gunasekera S, Gunasekera M, Gunawardana GP, McCarthy P, Burres N (1990b) Two new bioactive cyclic peroxides from the marine sponge *Plakortis angulospiculatis*. J Nat Prod 53(3):669–674

Gunasekera SP, McCarthy PJ, Borges MK (1994) Discobahamins A and B, new peptides from the Bahamian deep water marine sponge *Discodermia* sp. J Nat Prod 57:1437

Gunawardana GP, Kohmoto S, Gunasekera SP, McConnell OJ, Koehn EE (1988) Dercitin, a new biologically active acridine alkaloid from a deep water marine sponge, *Dercitus* sp. J Am Chem Soc 110:4856–4858

Guyot M, Durgeat M, Morel E (1986) Ficulinic acid A and B, two novel cytotoxic straight chain acids from the sponge *Ficulina ficus*. J Nat Prod 49(2):307–309

Hallock YF, Cardellina JH 2nd, Balaschak MS, Alexander MR (1995) Antitumor activity and stereochemistry of acetylenic alcohols from the sponge *Cribrochalina vasculum*. J Nat Prod 58(12):1801–1807

Herb R, Carroll AR, Yoshida WY, Scheuer PJ, Paul VJ (1990) Polyalkylated cyclopentindoles: cytotoxic fish antifeedants from a sponge, *Axinella* sp. Tetrahedron 46(8):3089–3092

Hirata Y, Uemura D (1986) Halichondrins – antitumor polyether macrolides from a marine sponge. Pure Appl Chem 58(5):701–710. doi:10.1351/pac198658050701

Hirota H, Matsunaga S, Fusetani N (1990a) Bioactive marine metabolites. Part 32. Stellettamide A, an antifungal alkaloid from a marine sponge of the genus *Stelletta*. Tetrahedron Lett 31(29):4163–4164

Hirota H, Matsunaga S, Fusetani N (1990b) Bioactive marine metabolites. Part 32. Stellettamide A, an antifungal alkaloid from a marine sponge of the genus *Stelletta*. Tetrahedron Lett 31(29):4163–4164

Ichiba T, Yoshiba WY, Scheuer PJ (1991) Hennoxazoles: bioactive bisoxazoles from a marine sponge. J Am Chem Soc 113:3173–3174

Inman WD, O'Neill-Johnson M, Crews P (1990) Novel marine sponge alkaloids. l. Plakinidine A and B, anthelmintic active alkaloids from a *Plakortis* sponge. J Am Chem Soc 112(1):1–4

Ishibashi M, Ohizumi Y, Cheng JF, Nakamura H, Hirata Y, Sasaki T, Kobayashi J (1988) Metachromins A and B, novel antineoplastic sesquiterpenoids from the Okinawan sponge *Hippospongia* cf. *metachromia*. J Org Chem 53(12):2855–2858

Jacobson PB, Marshall LA, Sunf A, Jacobs RS et al (1990) Inactivation of human sinovial fluid phospholipase by the marine natural product manoalide. Biochem Pharmacol 39:1557–1564

Jamieson D, Davis P (1980) Interactions of the anticonvulsant carbamazepine with adenosine receptors. 2. Pharmacological studies. Eur J Pharmacol 67:295

Janes Erijman EA, Sakai R, Rinehart KL (1991) Crambescidins, new antiviral and cytotoxic compounds from the sponge *Crambe crambe*. J Org Chem 56:5712–5715

Kakou Y, Crews P, Bakus GJ (1987) Dendrolasin and lactrunculin a from the Fijian sponge *Spongia microfijiensis* and an associated nudibranch *Chromodoris lochi*. J Nat Prod 50(3):482–484

Kashman Y, Rudi A (1977) The IJC-NMR spectrum and stereochemistry of heteronemin. Tetrahedron 33:2997–2998

Kashman Y, Groweiss A, Schmueli U (1980) Latrunculin, a new 2-thiazolidinone macrolide from the marine sponge *Latrunculia magnifica*. Tetrahedron Lett 21:3629

Kashman Y, Hirsch S, Koehn F, Cross S (1987) Reiswigins A and B, novel antiviral diterpenes from a deepwater sponge. Tetrahedron Lett 28:5461–5464

Kashman Y, Hirsch S, Cross S, Koeh F (1989a) Antiviral compositions derived from marine sponge *Epipolasis reiswigi* and their methods of use European Patent EP 306,282, 8 March 1989. US Patent 91,078, 31 Aug 1987

Kashman Y, Hirsch S, McConnell OJ, Ohtani I, Kusumi T, Kakisawa H (1989b) Ptilomycalin A: a novel polycyclic guanidine alkaloid of marine origin. J Am Chem Soc 111:8925–8926

Kato T, Yamaguchi Y, Ohnuma S, Uyehara T, Namai T, Kodama M, Shiobara Y (1986) Structure and synthesis of 11,12,13-trihydroxy-9Z,l5Z-octadecadienoic acids from rice plant suffering from rice blast disease. Chern Lett:577–580

Kato Y, Fusetani N, Matsunaga S, Hashimoto K et al (1986a) Okinonellins A and B, two novel furanosesterterpenes which inhibit cell division of fertilized starfish eggs, from the marine sponge *Spongionella* sp. Experientia 42(11–12):1299–1300

Kato Y, Fusetani N, Matsunaga S, Hashimoto K, Fujita S, Furuya T (1986b) Bioactive marine metabolites, Part 16. In: Calyculin A (ed) A novel antitumor metabolite from the marine sponge *Discodermia calyx*. J Am Chem Soc 108(10):2780–2781

Kato Y, Fusetani N, Matsunaga S, Hashimoto K, Fujita S, Furuya T, Koseki K (1986c) Structures of calyculins, novel antitumor substances from the marine sponge *Discodermia calyx*. Tennen Yuki KagobutsuToronkai Koen Yoshishu 28:168–175

Kato Y, Fusetani N, Matsunaga S, Hashimoto K et al (1988a) Calyculins, potent antitumor metabolites from the marine sponge *Discodermia calyx:* biological activities. Drugs Exp Clin Res 14(12):723–728

Kato Y, Fusetani N, Matsunaga S, Hashimoto K, Koseki K (1988b) Bioactive marine metabolites. 24. Isolation and structure elucidation of calyculins B, C, and D, novel antitumor metabolites, from the marine sponge *Discodermia calyx*. J Org Chem 53 (17):3930–3932

Kazlauskas R, Murphy PT, Quinn RJ, Wells RJ (1976) Heteronemin a new scalarin type sesterterpene from the sponge *Heteronema erecta*. Tetrahedron Lett 17:2631–2634

Keifer PA, Schwartz RE, Koker MES, Hughes RG Jr, Rittschof D, Rinehart KL (1991) Bioactive bromopyrrole metabolites from the Caribbean sponge *Agelas conifer*. J Org Chem 56(2):5–2975

Kitagawa I, Kobayashi M, Lee NK, Oyama Y, Kyogoku Y (1989) Marine natural products XX. Bioactive scalarane type bishomosesterterpenes from the Okinawan marine sponge *Phyllospongia foliascens*. Chem Pharm Bull 37(8):2078–2082

Kobayashi J, Cheng JF, Ishibashi M, Nakamura H, Ohizumi Y, Hirata Y, Sasaki T, Lu H, Clardy J (1987) Prianosin A, a novel antileukemic alkaloid from the Okinawan marine sponge *Prianos melanos*. Tetrahedron Lett 28(43):4939–4942

Kobayashi J, Murayama T, Ohizumi Y, Sasaki T, Ohta T, Nozoe S (1989) Theonelladins A.apprx. D, novel antineoplastic pyridine alkaloids from the Okinawan marine sponge *Theonella swinhoei*. Tetrahedron Lett 30(36):4833–4836

Kobayashi J, Tsuda M, Tanabe A, Ishibashi M, Cheng JF, Yamamura S, Sasaki TJ (1991) Cystodytins D-I, new cytotoxic tetracyclic aromatic alkaloids from the Okinawan marine tunicate *Cystodytes dellechiajei*. J Nat Prod 54:1634

Kohmoto S, McConnell OJ, Wright A, Koehn F, Thompson W, Lui M, Snader KM (1987a) Puupehenone, a cytotoxic metabolite from a deep water marine sponge, *Strongylophara hartmani*. J Nat Prod 50(2):336

Kohmoto S, McConnell OJ, Wright A, Cross S (1987b) Isospongiadiol, a cytotoxic and antiviral diterpene from a Caribbean deep water marine sponge, *Spongia* sp. Chern Lett 16(9):1687–1690

Kohmoto S, Kashman Y, McConnell OJ, Rinehart KL, Wright A, Koehn F (1988) Dragmacidin, a new cytotoxic bis(indole) alkaloid from a deep water marine sponge, *Dragmacidon* sp. J Org Chem 53 (13):3116–3118

Ksebati MB, Schmitz FJ, Gunasekera SP (1988) Pouosides A-E, novel triterpene galactosides from a marine sponge, *Asteropus* sp. J Org Chem 53 (17):3917–3921

Ksebati MB, Schmitz FJ, Gunasekera SP (1989) Pouosides A-E, novel, triterpene galactosides from a marine sponge, *Asteropus* sp. J Org Chem 54(8):2026

Matsunaga S, Fusetani N, Hashimoto K, Walchli M (1989) Titeonellarnide F. A novel antifungal bicyclic peptide from a marine sponge *Theonella* sp. J Am Chem Soc 111(7):2582–2588

McKee TC, Ireland CM (1987) Cytotoxic and antimicrobial alkaloids from the Fijian sponge *Xestospongia caycedoi*. J Nat Prod 50(4):754–756

Molinski TF, Ireland CM (1988) Dysidazirine, a cytotoxic azacyclopropene from the marine sponge *Dysidea fragilis*.J. Org Chem 53(9):2103–2105

Morris SA, Anderson RJ (1990) Brominated bis(indole) alkaloids from the marine sponge *Hexadella* SP. Tetrahedron 46:715–764

Mueller WEG, Zahn RK, Gasic MJ, Dogovic N, Maidhof A, Becker C, Diehl-Seifert B, Eich E (1985) Avarol, a cytostatically active compound from the marine sponge *Dysidea avara*. Comp Pharmacol Toxicol 80(1):47–52

Omar S, Albert C, Fanni CP (1988) Polyfunctional diterpene isonitriles from a marine sponge, *Acanthella cavernosa*. J Org Chem 53:5971–5972

Perry NB, Blunt JW, Munro MHG, Pannell LK (1988a) Mycalamide A, an antiviral compound from a New Zealand sponge of the genus *Mycale*. J Am Chem Soc 110:4850–4851

Perry NB, Blunt JW, Munro MHG (1988b) Cytotoxic pigments from New Zealand sponges of the genus *Latrunculia:* discorhabdins A, B, and C. Tetrahedron 44(6):1727–1734

Perry NB, Blunt JW, Munro MHG, Thompson AM et al (1990) Antiviral and antitumor agents from a New Zealand Sponge, *Mycale* sp. 2, Structures and solution conformations of mycalamides A and B. J Org Chem 55:223–227

Pettit GR, Clewlow PJ, Dufresne C, Doubek DL, Cerny RL, Rutzler K (1990) Antineoplastic agents. 193. Isolation and structure of the cyclic peptide hymenistatin I. Can J Chem 68(5):708–711

Petit J, Zeghloul A (1990) Environmental and microstructural influence on fatigue propagation of small surface cracks. In: Lisagor WB, Crooker TW, Leis BN (eds) Environmentally assisted cracking: science and engineering, ASTM STP 1049. American Society for Testing and Materials, Philadelphia, pp 334–346

Pordesimo EO, Schmitz F (1990) Newbastadins from the sponge *Ianthella basta*. J Org Chem 55 (15):4704–4709

Quinoa E, Crews P (1987) Niphatynes, methoxylamine pyridines from the marine sponge *Niphates* sp. Tetrahedron Lett 28(22):2467–2468

Quinoa E, Kho E, Manes LV, Crews P, Sakus G (1986) Heterocycles from the marine sponge *Xestospongia* sp. J Org Chem 51(22):4260–4264

Quinoa E, Kakou Y, Crews P (1988) Fijianolides, polyketide heterocycles from a marine sponge. J Org Chem 53(15):3642–3644

Quinu RJ, Gregson RP, Cool AF, Bartlett RT (1980) Recent developments in natural products: potential impact on antibacterial drug discovery. In: Emerging trends in antibacterial discovery: answering the call to arms. Tetrahedron Lett 21:367

Rashid MA, Gustafson KR, Boyd MR (2001) New cytotoxic N-Methylated β-Carboline alkaloids from the marine ascidian *Eudistomagil boverde*. J Nat Prod 64:1454

Rinehart KL (1988) Bioactive metabolites from the caribbean sponge *Agelas coniferin*. US Patent 4,737,510, 2 Apr 1988

Roll DM, Ireland CM, Lu HSM, Clardy J (1988) Fascaplysin, an unusual antimicrobial pigment from the marine sponge *Fascaplysinopsis* sp. J Org Chem 53:3276–3278

Sakai LY, Keene DR, Engvall E (1986b) Fibrillin, a new glycoprotein, is a component of extracellular microfibrils. J Cell Biol 103:2499–2509

Sakai R, Higa T, Jefford CW, Bernardinelli G (1986) Manzamine A, a novel antitumor alkaloid from a sponge. J Am Chem Soc 108(20):6404–6405

Schmidt EW, Faulkner DJ (1998) Microsclerodermins C – E, antifungal cyclic peptides from the lithistid marine sponges *Theonella* sp. and *Microscleroderma*. Tetrahedron 54:3043–3056

Schmitz FJ, Prasad RS, Yalamanchili G, Hossain MB, van der Helm D, Schmidt P (1981) Acanthifolicin, a new episulfide-containing polyether carboxylic acid from extracts of the marine sponge *Pandaros acanthifolium*. J Am Chem Soc 103:2467–2469

Schmitz FJ, Agarwal SK, Gunasekera SP, Schmidt PG, Shoolery JN (1983) Amphimedine, new aromatic alkaloid from a Pacific sponge, *Amphimedon* sp. carbon connectivity determination from natural abundance 13C-13C coupling constants. J Am Chem Soc 105:4835

Schmitz FJ, Ksebati MB, Gunasekara SP, Agarwal S (1988) Sarasinoside A: a saponin containing amino sugars isolated from a sponge. J Org Chem 53 (25):5941–5947

Sharma GM, Burkholder PR (1967) Studies on the antimicrobial substances of sponges II. Structure and synthesis of a bromine-containing antibacterial, compound from a marine sponge. Tetrahedron Lett 8:4147

Sharma GM, Vig B, Burkholder PR (1970) In: Youngken HW (ed) Food drugs from the sea. Marine Technology Society, Washington, DC, pp 307

Sun HH, Sakemi S, Burres N, McCarthy P (1990) Isobatzellines A, B, C, and D. Cytotoxic and antifungal pyrroloquinoline alkaloids from the marine sponge *Batzella* sp. J Am Chem Soc 55:4964–4966

Tachibana K, Scheuer PJ, Tsukitani Y, Kikuchi H, Van Engen D, Clardy J, Gopichand Y, Schmitz FJ (1981) Okadaic acid, a cytotoxic polyether from two marine sponges of the genus *Halichondria*. J Am Chem Soc 103:2469

Thoms C, Wolff M, Padmakumar K, Ebel R, Proksch PZ (2004) Chemical defense of Mediterranean sponges *Aplysina cavernicola* and *Aplysina aerophoba*. Natur forsch 59:113

Tohma H, Maegawa T, Kita Y (2003) Facile and efficient oxidation of sulfides to sulfoxides in water using hypervalent iodine reagents. ARKIVOC 4:62–70

Van Sande J, Deneubourg F, Beauivens R, Breakman JC, Daloze D, Dumont JE (1990) Inhibition of iodide transport in thyroid cells by dysidenin, a marine toxin, and some of its analogs. Mol Pharmacol 37:583–589

Walker RP, Faulkner DJ, Van Engen D, Clardy J (1981) Sceptrin, an antimicrobial agent from the sponge *Agelas sceptrum*. J Am Chem Soc 103:6772–6773

West RR, Mayne CL, Ireland CM, Brinen LS, Clardy J (1990) Plakinidines: cytotoxic alkaloid pigments from the Fijian sponge *Plakortis* sp. Tetrahedron Lett 31 (23):3271–3274

Wheeler LA, Sachs G, Goodrum D, Amdahl L, Horowitz N, de Vries W (1988) Importance of marine natural products in the study of inflammation and calcium channels. In: Fautin DG (ed) Biomedical importance of marine organisms (Memoirs of the California Academy of Sciences), vol 13. California Academy of Sciences, San Francisco, pp 125–132

Wright AE, Thompson WC (1987) Antitumor compositions containing imidazolyl pyrroloazepines. WO 8707274 A2 3 December 1987 Chem Abstr 110 (17):147852

Wright AE, Mc Connell OJ, Kohmoto S, Lui MS, Thompson W, Snader KM (1987a) Duryne, a new cytotoxic agent from the marine sponge *Cribro chalinodura*. Tetrahedron Lett 28(13):1377–1380

Wright AE, Pomponi SA, McConnell OJ, Kohmoto S, McCarthy PJ (1987b) (+)Curcuphenol and (+)-curcudiol, sesquiterpene phenols from shallow and deep water collections of the marine sponge *Didiscus flavus*. J Nat Prod 50(5):976–978

Wright AE, Mc Carthy P, Cross SS, Rake JB, Mc Connell OJ, (1988) Sesquiterpenoid isocyanide purification from a marine sponge and its use as a neoplasm inhibitor, virucide, and fungicide, european Patent Application EP 285,302, October 5, 1988; US Patent Application 32,289, 30 March, Chem Abstr 111:50414

Zabriskie TM, Ireland CM (1989) The isolation and structure of modified bioactive nucleosides from *Jaspis johnstoni*. J Nat Prod 52(6):1353–1356

10 Bioactive Potential of Sponge Secondary Metabolites

Irudayaraj Rajendran

Abstract

The advent of spongothymidine and spongouridine from the Caribbean sponge *Tethya crypta* turned the attraction of chemists toward sponges for their unique and novel compounds which have a broad spectrum of biological activity and anticancer property in particular. They find the candidates for the development of new drug analogues with the expected biological activity. The sponge compounds are having a series of bioactivity broadly classified into three groups, *viz.*, pharmacological activity, antagonizing activities, and antifouling activity. Each group is dealt with the allied bioactivities which were observed on the range of sponge compounds during evaluation studies. The biological activities are grouped depending on their similarity and type of testing. The structural groups of the compounds include terpenoids, alkaloids, heterocycles, acetylenic compounds, steroids, and nucleosides. The uniqueness of the selected compounds derived from sponges with the observed bioactivity is discussed.

Keywords

Marine natural products • Secondary metabolites • Sponge extracts • Bioprospecting • Pharmacological screening

10.1 Introduction

The oceans and seas constitute more than 70 % of the earth's surface. Marine ecosystem covers 34 of the total 36 phyla out of which 13 are exclusive of marine origin. Thus, such biologically diverse and ecologically complex marine ecosystem comprising both flora and fauna may, therefore, be considered as the largest scope for bioprospecting on the earth. But this rich source has become the subject of systematic investigation for the benefit of human race only around the beginning of the twentieth century. This is in contrast to the terrestrial natural products, which have been exploited almost. With the

I. Rajendran (✉)
Mandapam Regional Centre, Marine Fisheries PO, ICAR-Central Marine Fisheries Research Institute, Ramanathapuram 623 520, Tamil Nadu, India
e-mail: cmfrirajendran@gmail.com

R. Pallela, H. Ehrlich (eds.), *Marine Sponges: Chemicobiological and Biomedical Applications*,
DOI 10.1007/978-81-322-2794-6_10

help of underwater exploration facilities like self-contained underwater breathing apparatus (SCUBA), deep sea laboratory, research submarines, etc., it has become possible to go down deep into the sea of the production-rich continental shelf. Eventually it has been felt that the marine organisms have been found to be the storehouse of compounds of significant bioactivity. They are microorganisms and phytoplankton; green, brown, and red algae; sponges, corals; cnidarians; bryozoans; mollusks; tunicates; echinoderms; mangroves; and other intertidal plants including minor sedentary organisms. Of these, sponges contribute to about one third of the compounds explored. The compounds present in the bycatches of fisheries may be attributed to have been evolved from dietary, commensalic, or endosymbiotic bacteria, fungi, cyanobacteria, etc. (McConnell et al. 1994; Hentschel et al. 2002; Moore 2006; Taylor et al. 2007; Thomas et al. 2010).

10.2 Marine Natural Products

Marine natural products' (MNPs) chemistry has been proliferating, since the pioneering work of Bergmann about 60 years ago with his isolation of spongothymidine and spongouridine (Fig. 10.1) from the Caribbean sponge *Tethya crypta* paved way to the development of new pharmacologically important compounds (Bergmann and Burke 1955). Good accumulation of literature on MNP based on the compound types and the phylum as well started evolving since 1970, as the chemists have been unraveling the fine structures of these organic compounds often termed as the secondary metabolites produced by marine organisms, and the published reports have been regularly reviewed (Munro et al. 1999; Blunt et al. 2010). It has been the subject of increased interest to unravel the mystery behind the role of these chemicals in the marine ecosystem.

Fig. 10.1 R=H, spongouridine; R=CH_3, spongothymidine

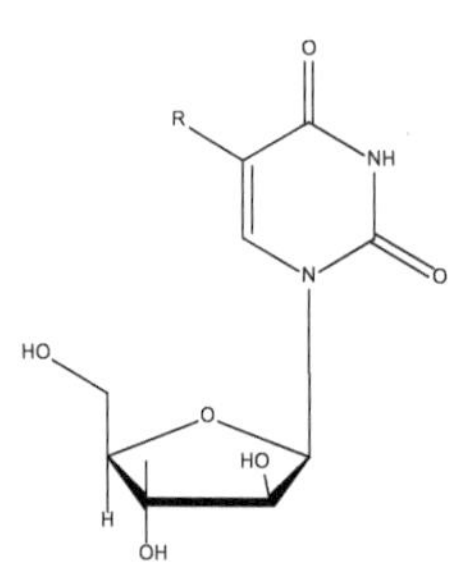

R = H, Spongouridine
R = CH_3, Spongothymidine

It is intriguing to find new compounds arising out of dramatically different marine environment by biosynthetic pathway with new building blocks incorporating unprecedented enzymatic reactions. The novelty of these molecules has attracted synthetic chemists to target new analogues and new synthetic methodologies (Albizati et al. 1990). Though they appear for chemist as a series of new organic compounds, they serve in the effective communication system among marine organisms in the marine ecosystem, in the form of *allomones* (chemicals benefiting the producing organism, e.g., repellents), *kairomones* (benefiting the receiving organism, e.g., toxin-producing bacteria), and *pheromones* (used for intraspecific communication) for their mutual benefits (Brown 1975; Pawlik et al. 1995). They are group specific, coexistent, and coevolution species (Luckner 1983). There has been increased interest in this new field of multidisciplinary nature by chemists, biologists, and pharmacologists as evidenced by the appearance of increased number of patents on methods of isolation and application of natural products and periodic review of the publications in the past few decades. With more than 20,000 compounds reported so far, new compounds are being added up every year showing the pace at which new explorations of marine natural products are going on globally (Sipkema et al. 2005; Blunt et al. 2014). Marine environment accommodates more than 80 % of the earth's phyla. Marine plants, animals, and microorganisms exhibit processes and produce substances unknown in terrestrial organisms. Unique chemical structures, unusually high biological activity, and participation in intra- and interspecific relationships in underwater communities are the reasons for great attention to these substances.

Though the potential of economic and public health benefits of pharmaceuticals, pesticides, hormones, enzymes, and polymers derived from marine organisms is high, this prospective sector is yet to be further exploited. The development of novel products from the sea has the potential to greatly contribute for the treatment of dreaded human diseases such as cancer, AIDS, inflammation diseases, etc. by eliminating drug resistance. Some of the compounds of significant importance in terms of their antitumor, cytotoxic, antiviral, antiparasitic, antimicrobial, receptor-antagonistic, anticancer, antibiotic, and enzyme inhibition activities are on clinical trials. With the development and update of taxonomy of marine organisms in general and sponges in particular (Pattanayak and Buddadeb 2001; Hooper and Van Soest 2002), modern powerful spectroscopic techniques like ^{1}H and ^{13}C NMR, 2D NMR, HRMS, TOF-MALDI, and single-crystal X-ray diffraction (XRD) analysis find it very useful for characterization of compounds of nanoscale yields. As a result, amazing parade of novel structures has been published in the core journals. This situation hardly relies on the older methods of chemical degradations or correlations on this minute quantity of compounds isolated from marine organisms. These compounds with novel structures and pharmacological importance (e.g., prostaglandins) further increased the curiosity of the scientists to explore more marine organisms (Okuda et al. 1982). MNPs have been the continued attraction for drug development and neurophysiological probes, and the marine biomedical research has now become the popular subject as inferred from many recent reviews (Schwartsmann et al. 2003; Tziveleka et al. 2003). The compounds, like tetrodotoxin, however command the field of pharmacology (French et al. 2010).

The main sources of MNPs are marine microorganisms and phytoplankton, green algae, brown algae, red algae, sponges, coelenterates, bryozoans, mollusks, tunicates (ascidians), echinoderms, polychaete worms, fin fishes, and crustaceans. Sponges and coelenterates continue to dominate as source phyla for new compounds. The inheritance of the latest drugs obviously has the origin of the explorations of marine fauna and flora as has been seen in the case of some of the antiviral drug like acyclovir and AZT (Moore 2006; Thomas et al. 2010; Taylor et al. 2007; McConnell et al. 1994). Few others to quote are *Yondelis* better known as ecteinascidin 743 and conus toxin *ziconotide* or *Prialt* in Phase II/III trials against soft tissue sarcoma and Phase III trials for intractable pain, respectively (Newman and Cragg 2004).

10.3 Sponge Metabolites

Sponges are sessile (sedentary) animals and are the storehouse of extraneous marine biota (Moore 2006). As sponges are the filter feeders, they accumulate an array of microorganisms which are in symbiotic relationship on inter- and intracellular mode for nutrition and other mutual benefits as well. Some natural products seem to have originated from sponge since they are located within sponge cells (Uriz et al. 1996). Other compounds are associated with microbial symbionts, suggesting that microbes are the true producers (Farkašovský 2013). The abundance of the source of MNPs from the organisms is approximately in the following order:

Sponge > *Coelenterates* > *Microorganisms* and *Phytoplankton* > *Echinoderms* = *Tunicates* > *Red algae* > *Mollusks* ≥ *Brown algae* ≥ *Green algae* > *Bryozoans*

From the order it is obvious that sponges are the major contributors. The order of abundance in the type of compounds reported may be given as:

Terpenoid > Polyketide > Alkaloid > Peptide > Shikimate

Terpenoids are prevalent among sponges and coelenterates. The symbiotic relation that existed between marine organisms in the marine environment was correlated to the terrestrial insect chemical ecology by the MNP chemists as this association to every organism was inevitable in terms of the organism's defense, camouflage, etc., to protect them from their enemies. Initially MNP chemistry was widely developed in three tracks, viz., marine toxins, marine biomedicines, and marine chemical ecology. Now they have been integrated to give more vigor to this field as they are interdisciplinary in nature. It is difficult to refer to any single compendium for the details of MNP research. The identification of the species and phylum during the chemical work is the integral part enabling one to correlate the result of the work of the significance and status of the particular compound with respect to other members of the family and the marine ecosystem.

Invoking the above sequence it is obvious that sponges are the largest storehouse of novel compounds being explored every year and nearly one third of the reports emerging from marine resources are from sponges. Even though the quantity of the compounds isolated from sponges is very low, they give the gateway for the development of novel compounds with their unique structural features typical of marine origin and also drug candidates (Butler 2004).

As mentioned earlier, sponges are one of the major contributors of novel MNPs. Many of the compounds derived from sponges now have become the precursor for the development of new drugs.

Sponges are one of the sources of potential anticancer compounds. Various screening processes of the extracts/pure compounds are to be conducted to confirm the activity. More studies are being carried out on the extracts/pure compounds for anticancer and antimicrobial properties. Various parameters of the pharmacological activities are used to detect particular biological activity. The most cited tests for sponge products are divided into like-groups, *albeit* there are overlaps of testing between the bioactivity types to confirm the activity of compound tested. The range of sponge products reported for each group is discussed under each group of biological activity. Only products with significant to potent activity are taken. Though there are various types of pharmacological and other bioactivity experiments, the experiments used for MNPs as reported earlier are summarized here.

10.4 Bioactivity Tests Used for Screening the Spongean Products

10.4.1 Pharmacological Activity Tests

10.4.1.1 Antitumor (AT) and Anticancer (AC) Tests

Protein kinase C (PKC) inhibitor, 3-fucosyltransferase (FTase) inhibitor, kinesin motor protein (KMP) inhibitor, stabilization of microtubules (MTS), tubulin polymerization (TP) inhibitor, actin depolymerization (actin DP), topoisomerase (TI) II inhibitor, nitric oxide synthase (NOS) inhibitor, NKT cell activator, reverses drug resistance of cancer cells, v-ATPase inhibitor, Ca^{2+} channel blocker, immunosuppressive (IS), cytotoxic (CTX), P-388 leukemia (P-388L) cells, inhibitor of T-cell proliferation, IL-2 inhibitor, IL-8 inhibitor, histamine release (HR) inhibitor, inhibitors of proton pump activity (PPAI), inhibition of DNA replication (DNAR), lipoxygenase (LP) inhibitor, cysteine protease (CP) inhibitor, phosphatase activity (PA) inhibitor, ovarian human tumor (OHT) cell, inhibitors of serine/threonine (STI)

10.4.1.2 Neurosuppressives (NS) and Muscle Relaxants (MR)

Glutamate receptor antagonist, serotonergic receptor (SR) antagonist, IP3-inhibitor, adrenergic receptor antagonist, actomyosin ATPase inhibitor (antiasthmatic, uterine relaxation), glutamate receptor antagonist (GRA)

10.4.1.3 Anti-inflammatory (AI)

Phospholipase A2 (PLA2) inhibitor

10.4.1.4 Antiviral (AV)

AV (HIV-1 integrase inhibitor), AV (herpes and polio), AV (feline leukemia, mouse influenza, mouse corona), UAG suppressor glutamine tRNA inhibitor, interferon mediator, herpes simplex virus type 1 (HSV-1), HIV reverse transcriptase (HIV RT)

10.4.1.5 Blood-Related Diseases

Serine protease inhibitor (SPI), thrombin receptor antagonist (TRA), VCAM-1 inhibitor, α-glucosidase inhibitor

10.4.1.6 Antibacterial (AB), Antifungal (AF), Antimalarial (AM), Antibiotic (ATB)

Gram-positive: *Bacillus subtilis*, *Staphylococcus aureus*, *Candida albicans*; fungus, *Cladosporium cucumerinum*; malarial parasite, *Plasmodium falciparum*

10.4.1.7 Toxic Activities

(i) Antifouling (AFL) – *repellent*

(ii) Inhibition (INB) – *fertilized sea urchin egg cell division*
(iii) Ichthyotoxic (ITX) – *artemia, fish, feed deterrent*
(iv) Nematocidal

10.5 Compounds Derived from Sponges

(i) Terpenoids
(ii) Alkaloids and heterocycles
(iii) Polyacetylenes, polyethers, polyketides, peptides, macrolides
(iv) Glycosides and nucleosides
(v) Steroids

10.5.1 Spongean Terpenoids

The compounds (Table 10.1) include sesterterpenes (STTPN), triterpenes (TTPN), sesquiterpenes (SQTPN), and diterpenes (DTPN) as per the abundance and their hydroquinone (HQ) derivatives. They inhibit enzymes responsible for cell proliferation, thus controlling tumor or cancer cell formation as tested both for murine and human cell lines. Some directly act on tumor/cancer cells arresting further development of cells. Scalarane and drimane class terpenoids are having more activity as cytotoxic. Muqubilone (Fig. 10.2) (El Sayed et al. 2001b) inhibits herpes simplex virus (HSV) showing importance as antiviral agent. Okinonellin B (Kato et al. 1986) finds extensive screening for human tumor cell line growth inhibition assay.

10.5.2 Spongean Alkaloids and Heterocycles

Spongean alkaloids and heterocycles (Table 10.2) are unique in structural features typical of marine origin. In this way, manzamines are unique marine alkaloids possessing an intricate nitrogen-containing ring system at C-1 of the β-carboline (indole alkaloid with pyridine ring fusion) ring. By the virtue of this structural feature, manzamine derivatives and isomers (Sakai and Higa 1986) (Fig. 10.3) are having a broad spectrum of biological activity like antimalarial, cytotoxicity, insecticidal, and antibacterial including in the control of AIDS opportunistic infections. Plakortamine A (Sandler et al. 2002) (Fig. 10.4) is an indole alkaloid derivative having the range of cytotoxic activity on different cell lines.

Fascaplysin A (JimBnez et al. 1991), an indole alkaloid derivative, has the effect on HIV reverse transcriptase and inhibits cyclin-dependent kinases (CDKs) with surprising selectivity (Shafiq et al. 2012). Batzellines (Chang et al. 2002) and batzelladines (Patil et al. 1996) are pyrroloquinoline (PQ) group having promising activity on HIV-related disease.

Table 10.1 Spongean terpenoids useful in pharmacological activities

Compounds	Compound type	Antagonist activity	Source	Reference
Adociasulfate 1	TTPN HQ	PPA inhibitor	*Adocia* sp.	Kalaitzis et al. (1999)
Agelasidines B and C	Hypotaurocyamine	Na,K-ATPase inhibitor	*Agelas nakamurai*	Nakamura et al. (1985)
Sodwanone M	TTPN	CTX to (P-388L) cells	*Axinella weltneri*	Rudi et al. (1997)
Yardenone A	TTPN	CTX to NSCLC cells	*Axinella* cf. *bidderi*	Carletti et al. (2003)
18-*epi*-scalaradial	Scalarane	CTX to tumor cells	*Cacospongia scalaris*	Rueda et al. (1997)
Metachromin A	SQTPN HQ	CTX to COLO-205 and KB	*Hippospongia metachromia*	Shen et al. (2001)
Sesterstatin 1	Cyclic STTPN	CTX to (P-388L) cells	*Hyrtios erecta*	Pettit et al. (1998)
Honulactone A	Homoscalarane	-do-	*Strepsichordaia aliena*	Jiménez et al. (2000)
Polyfibrospongol A	SQTPN HQ	CTX to human KB-16	*Polyfibrospongia australis*	Shen and Hsieh (1997)
Furan STTPN	–	Inhibition of DNAR	*Psammocinia* sp.	Choi et al. (2004)
Bolinaquinone	Furano STTPN	Inhibited human PLA2	*Dysidea* sp.	Giannini et al. (2001)
Stellettin A	TRTPN	CTX to leukemia	*Stelletta* sp.	McCormick et al. (1996)
Manoalide	Cyclohexane	STTPN A2 inhibitor	*Luffariella variabilis*	Bennet et al. (1987)
Dysidotronic acid	Drimane	SQTPN PLA2 inhibitor	*Dysidea* sp.	Giannini et al. (2000)
Cacospongionolide	BSTTPN lactone	-do-	*Fasciospongia cavernosa*	Garcia et al. (1999)
Ircinins 1 and 2	Acyclic STTPN	PLA2 inhibitor	*Ircinia oros*	Cimino et al. (1972)
Petrosaspongiolides	Cheilantane STTPN	-do-	*Petrosaspongia nigra*	Randazzo et al. (1998)
Jaspaquinol	Benzenoid DTPN	LP inhibitor	*Jaspis splendens*	Carroll et al. (2001)
Subersic acid	-do-	-do-	*Suberea* sp.	[- do -]
Muqubilone	STTPN	(HSV-1)	*Diacarnus erythraeanus*	El Sayed et al. (2001b)
Okinonellin B	Furano STTPN	Muscle relaxant	*Spongionella* sp.	Kato et al. (1986)

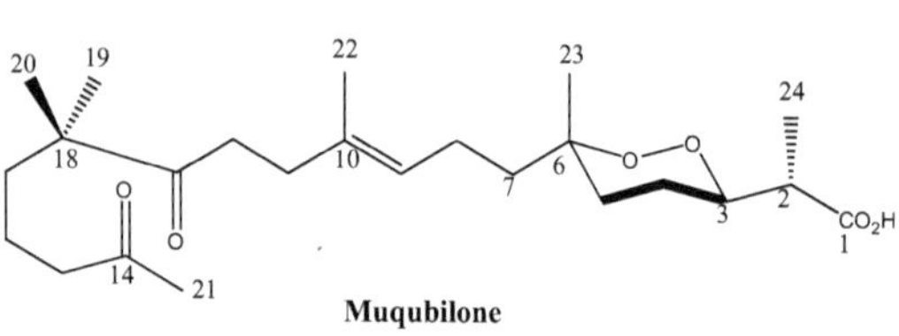

Fig. 10.2 Muqubilone

Hennoxazole (Ichiba et al. 1991) is a bisoxazole derivative having antiviral activity on HSV. Topsentin (Tsujii et al. 1988) is a bis(indolyl) imidazole compound inhibitory to both human and murine tumor cell proliferation apart from being a potent antiviral compound. Isoaaptamine (Fedoreev et al. 1989) is having significant antitumor activity against murine P-388 and human tumor cells including KB16, A549, and HT-29 cell lines. Other sponge alkaloids having imidazole, thiazole, guanidine moieties are having cytotoxicity against the screened cell lines.

10.5.3 Spongean Polyacetylenes, Polyethers, Polyketides, Peptides, Macrolides, etc.

This group comprises macromolecules (Table 10.3) with toxic activities over a range of pathogens, cell lines, and potent antitumor; stabilization of microtubules; and inhibition of enzymes responsible for cell proliferation, inflammation, and HIV. Polyketides are natural products consisting of two principal fatty acid chains with various functionalities such as a

Table 10.2 Spongean alkaloids and heterocycles useful in pharmacological activities

Compounds	Compound type	Specific activity	Source	Reference
1-Carboxymethyl nicotinic acid	–	CP inhibitor	*Anthosigmella* cf. *raromicrosclera*	Matsunaga et al. (1998b)
Discorhabdin S	Pyrroloiminoquinone	CTX	*Batzella*	Gunasekera et al. (2003)
Plakortamine A	β-Carbolines	HCT-116 inhibitor	*Plakortis nigra*	Sandler et al. (2002)
Discorhabdin L	Quinoline deriv.	CTX	*Latrunculia brevis*	Reyes et al. (2004)
Topsentin	Bis(indolyl)imidazoles	AV	*Halichondriidae*	Tsujii et al. (1988)
Secobatzelline A	Batzellines	(PA) inhibitor	*Batzella*	Gunasekera et al. (1999)
Manzamine A	–	P-388 inhibition	–	Sakai and Higa (1986)
Mimosamycin	Isoquinoline	OHT	*Haliclona* sp.	Rashid et al. (2001)
8-Hydroxymanzamine A	–	AIDS infections	*Petrosiidae*	El Sayed et al. (2001a)
Renieramycin J	Isoquinoline	CTX	*Neopetrosia* sp.	Oku et al. (2003)
Dragmacidin E	Bisindole	STI	*Spongosorites* sp.	Capon et al. (1998)
Plakinamine G	Steroidal pyrrole	CTX	*Corticium* sp.	Borbone et al. (2002)
Spongidines A–D	Pyridinium alkaloid	PLA2 inhibitor	*Spongia* sp.	De Marino et al. (2000)
Isoaaptamine	Benzonaphthyridine	PKC inhibitor	*Aaptos aaptos*	Fedoreev et al. (1989)
Debromohymenialdisine	Pyrrole-guanidine	-do-	*Hymeniacidon aldis*	Kitagawa et al. (1983)
Neoamphimedine	Pyridoacridine alkaloid	TI-II inhibitor	*Xestospongia* cf *carbonaria*	De Guzman et al. (1999)
Elenic acid	Alkylphenol	-do-	*Plakinastrella* sp.	Juagdan et al. (1995)
Naamine D	Imidazole alkaloid	NOS inhibitor	*Leucetta* cf. *chagosensis*	Dunbar et al. (2000)
Agelasphin	α-Galactosylceramide	NKT cell activator	*Agelas mauritianus*	Shimosaka (2002)
Taurodispacamide A	Pyrrole-imidazole	IL-2 inhibitor	*Agelas oroides*	Fattorusso and Taglialatela-Scafati (2000)
Pateamine A	Thiazole macrolide	-do-	*Mycale* sp.	Northcote et al. (1991)
Keramadine	Pyrrole-guanidine	SR antagonist	*Agelas* sp.	Nakamura et al. (1984)
Penaresidin A	Azetidine	Actomyosin ATPase inhibitor	*Penares* sp.	Kobayashi et al. (1991)
Variolin B	Pyridopyrrolopyrimidine	AV	*Kirkpatrickia varialosa*	Perry et al. (1994)
Avarol	Hydroquinone	Glutamine tRNA inhibitor	*Dysidea avara*	Muller et al. (1987)
Hennoxazole A	Bisoxazole	AV to HSV	*Polyfibrospongia* sp.	Ichiba et al. (1991)
Crambescidin 816	Polycyclic guanidine	(HSV-1)	*Crambe crambe*	Erijman et al. (1993)
Isobatzelline E	PQ alkaloids	HIV-1 cell fusion	*Zyzzya fuliginosa*	Chang et al. (2002)

(continued)

Table 10.2 (continued)

Compounds	Compound type	Specific activity	Source	Reference
Batzelladine A	-do-	Binding of HIVgp -120 to CD4	*Batzella* sp.	Patil et al. (1996)
Fascaplysin A	–	HIV RT	*Fascaplysinopis reticulate*	JimBnez et al. (1991), Shafiq et al. (2012)
Dragmacidin d	Bisindole alkaloids	P-388 and A549 cells	*Spongosorites*	Wright et al. (1992)]

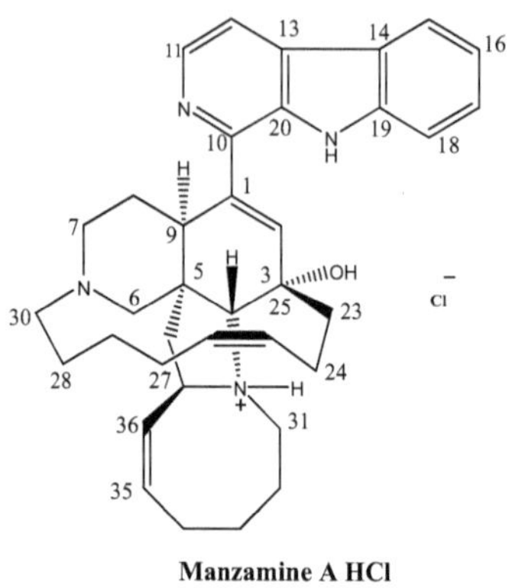

Fig. 10.3 Manzamine A hydrochloride

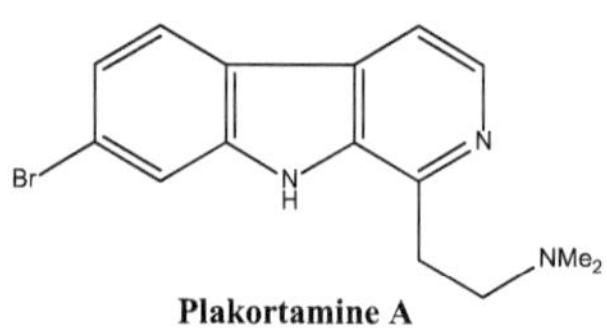

Fig. 10.4 Plakortamine A

sulfate ester, an oxazole, and a thiazole group, constituting a macrocyclic lactone ring bearing a long side chain attached through an amide linkage. Though it is difficult to attribute the correct structure-activity relationship, the stereochemistry of the molecular constituents may be responsible for the activity exhibited by these macromolecules.

The calyculins (Kato et al. 1988) are the unique polyketides bearing nitrogen and phosphorus functions. They are having spiroketal of an unprecedented skeleton bearing phosphate, oxazole, nitrile, and amide functionalities. They are potent antitumor agents inhibiting protein phosphatases 1 and 2A.

Okadaic acid (Tachibana et al. (1981) (Fig. 10.5) is a complex derivative of a C_{38} fatty acid. Though the origin of this compound is indeed dinoflagellate, *Prorocentrum concavum*, it was first isolated from *Halichondria* (Murakami et al. 1982). It is responsible for diarrheal shellfish poisoning (DSP) accumulated in bivalves. It is a potent and selective inhibitor of protein phosphatases. Discodermolide (Gunasekera et al. 1990) (Fig. 10.6) is a polyketide and potent inhibitor of tumor cell growth. It is having broad spectrum of activity as antiproliferative, neuroprotective, etc.

Plakortide M (Jiménez et al. (2003) is having potent cytotoxicity in human cancer cell lines in addition to its antimalarial activity. Plakosides A and B (Costantino et al. 1997) are unique glycosphingolipids (prenylated glycolipid) with strong immunosuppressive activity on activated T cells. Fulvinol is a long-chain acetylene with cytotoxicity against four tumor cell lines (Ortega et al. 1996). Distinctive feature of dysinosin A (Carroll et al. 2002) is the presence of a 5,6-dihydroxy-octahydroindole-2-carboxylic acid, 3-amino-ethyl 1-*N*-amidino-Δ-3-pyrroline, sulfated glyceric acid, and D-leucine, assembled through three peptidic linkages.

Clavosines A, B, and C are closely related to calyculins and calyculinamides. They are potent cytotoxins for tumor cell lines, potent inhibitors of type 1 and 2A serine/threonine protein phosphatases (Fu et al. 1998a). Salicylihalamides A and B (Erickson et al. 1997) are the macrolides with a 12-membered lactone ring with the incorporation of salicylic acid and an enamide side chain. They are highly potent and cytotoxic

Table 10.3 Spongean polyacetylenes, polyethers, polyketides, peptides, macrolides, etc., testified in pharmacological assays

Compounds	Activity and type	Source	Reference
Calyculins B	Potent AT	*Discodermia calyx*	Kato et al. (1988)
Okadaic acid	CTX, polyether	*Halichondria*	Tachibana et al. (1981), Murakami (1982)
Discodermolide	IS and CTX	*Discodermia dissoluta*	Gunasekera et al. (1990)
Polyacetylenic alcohols	CTX	*Petrosia*	Lim et al. (1999)
Clavosines A and B	CTX to 60 tumor cells	*Myriastra clavosa*	Fu et al. (1998a)
Plakortide M	CTX, polyketide	*Plakortis halichondrioides*	Jiménez et al. (2003)
Andavadoic acid	CTX to 13 tumor cells	*Plakortis aff simplex*	Rudi et al. (2003)
Plakosides	IS on T cells	*Plakortis simplex*	Costantino et al. (1997)
Fulvinol	CTX to tumor cells	*Reniera fulva*	Ortega et al. (1996)
Scleritodermin A	CTX to tumor cells	*Scleritoderma nodosum*	Schmidt et al. (2004)
Cyclotheonamide E4	Inhibition on human tryptase	*Ircinia* sp.	Murakami et al. (2002)
Cinachyrolide A	CTX on L1210 leukemia	*Cinachyra* sp.	Fusetani et al. (1993)
Dysinosin A	Anticoagulant	*Dysideidae*	Carroll et al. (2002)
Laulimalide	CTX on the KB cell line	*Hyattella* sp.	Corley et al. (1988), Mooberry et al. (1999)
	MTS	*Cacospongia mycofljiensis*	
Theopederin A	CTX against P-388	*Theonella* sp.	Fusetani et al. (1992)
Peloruside A	-do-	*Mycale hentscheli*	Hood et al. (2002)
Latrunculin A	Actin DP, macrolide	*Latrunculia magnifica*	Kashman et al. (1980)
Thiomycalolide A	CTX against P-388	*Mycale* sp.	Matsunaga et al. (1998a)
Spongistatin 1	Activity against HT	*Spongia*	Pettit et al. (1993)
Cyclotheonamides E	Active against thrombin	*Theonella*	Nakao et al. (1998)
Taurospongin A	HIV RT	*Hippospongia* sp.	Ishiyama et al. (1997)
Mycalamide A	HSV-1 and *polio* viruses	*Mycale*	Perry et al. (1988)
Hemiasterlin	MTS	*Auletta* sp.	Anderson et al. (1997)
Dictyostatin	-do-	*Corallistidae* sp.	Isbrucker et al. (2003)
Halichondrin B	TP inhibitor	*Halichondria okadai*	Hirata and Uemura (1986)
Arenastatin A	-do-	*Dysidea arenaria*	Koiso et al. (1996)
Swinholide A	Actin DP	*Theonella swinhoei*	Bubb et al. (1995)
Salicylihalamide	v-ATPase inhibitor	*Haliclona* sp.	Erickson et al. (1997)
Chondropsins A and B	-do-	*Chondropsis* sp.	Cantrell et al. (2000)
Simplexides	Inhibitor of T cell, glycolipid	*Plakortis simplex*	Costantino et al. (1999)
Pateamine A	IL-2 inhibitor, thiazole	*Mycale* sp.	Northcote et al. (1991)
Eryloside F	TRA, penasterol	*Erylus formosus*	Stead et al. (2000)
Callyspongynic acid	α-Glucosidase inhibitor, polyacetylene	*Callyspongia truncata*	Nakao et al. (2002)
Dysiherbaine	GRA, amino acid	*Dysidea herbacea*	Sakai et al. (1997)
Xestospongin C	IP3-inhibitor, bis-oxaquinolizidine	*Xestospongia* sp.	De et al. (1999)

(continued)

Table 10.3 (continued)

Compounds	Activity and type	Source	Reference
Papuamides C and D	Cyclic peptide, AV (HIV-1)	*Theonella mirabilis*	Ford et al. (1999)
Mololipids	Tyramine lipid	*Verongida*	Ross et al. (2000)
Hamigeran B	AV (herpes and polio), phenolic macrolide	*Hamigera tarangaensis*	Wellington et al. (2000)

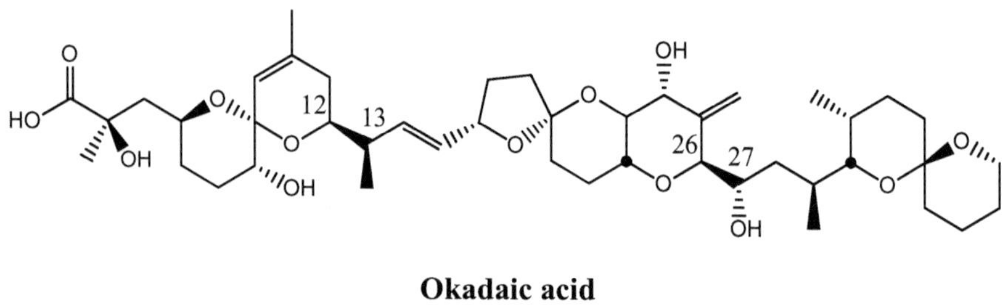

Fig. 10.5 Okadaic acid

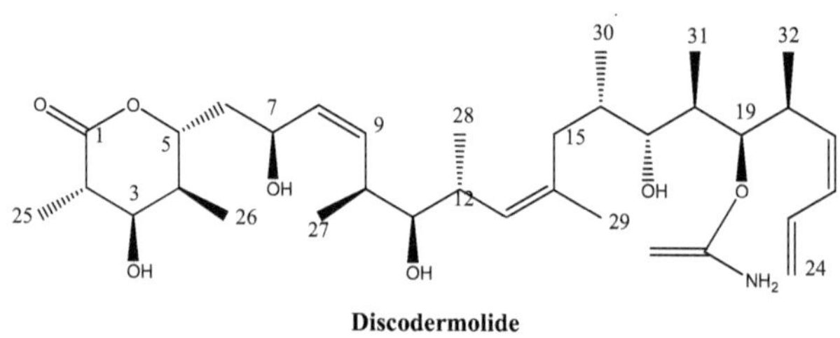

Fig. 10.6 Discodermolide

and also represent a potentially important new class for antitumor lead optimization.

Cyclotheonamides E4 and E5 (Nakao et al. 1998) are cyclic pentapeptides obtainable from *Theonella*. They are potent inhibitors for thrombin, serine protease, and human tryptase and are also useful as a therapeutic agent in the treatment of allergic diseases including asthma. Taurospongin A is an acetylene-containing natural product consisting of taurine and two fatty acid residues. It is a potent inhibitor for DNA polymerase *â* and HIV reverse transcriptase (Ishiyama et al. 1997). Spongistatin 1 is a macrocyclic lactone having extremely potent activity against selected human tumor cell. Scleritodermin A (Schmidt et al. 2004) is a cyclic peptide, inhibited tubulin polymerization and showed significant in vitro cytotoxicity against human tumor cell lines. Cinachyrolide A (Fusetani et al. (1993)), laulimalide, and isolaulimalide (Corley et al. 1988; Mooberry et al. 1999) are macrolides with potent cytotoxicity.

10.5.4 Spongean Sterols

Spongian sterols (Table 10.4) are distinct from the normal sterols with functional groups of sulfate, hydroxyls, unsaturation, etc. Clathriol is a highly oxygenated steroid with unusual 14β configuration, and it has in vitro anti-inflammatory activity against human neutrophil and rat mast cells (Keyzers et al. 2002). Clathsterol is a sulfated sterol (Fig. 10.7), and it is an inhibitor of

Table 10.4 Spongean sterols useful in pharmacological assays

Compounds	Activity and type	Source	Reference
Clathriol	AI activity	*Clathria lissosclera*	Keyzers et al. (2002)
Plakinamine I	CTX	*Corticium niger*	Ridley and Faulkner (2003)
Crellastatin A	CTX	*Crella* sp.	D'Auria et al. (1998)
Aragusterol A	Potent AT	*Xestospongia*	Iguchi et al. (1994)
Oxygenated C29 sterols	AT activity	*Polymastia tenax*	Santafe' et al. (2002)
9,11-secosterols	AT activity	*Spongia agaricina*	Rueda et al. (1998)
Clathsterol	HIV-1 RT	*Clathria* sp.	Rudi et al. (2001)
Callipeltin A	Protect HIV infected cells, cyclic depsidecapeptide	*Callipelta*	Zampella et al. (1996)
Agosterol A	Reverses drug resistancy of cancer cells	*Spongia* sp.	Aoki et al. (1998)
Contignasterol	HR inhibitor	*Petrosia contignata*	Takei et al. (1994)
Xestobergsterols A and B	-do-, pentacyclic steroids	*Xestospongia berquistia*	Shoji et al. (1992)
Haplosamates A and B	AV (HIV-1), sulfamated steroid	*Xestospongia* sp.	Qureshi and Faulkner (1999)
Weinbersterols A and B	AV feline leukemia, sulfated sterol	*Petrosia weinbergi*	Sun et al. (1991)

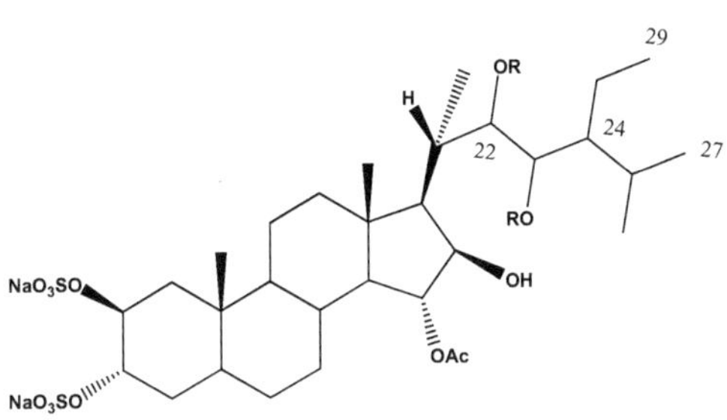

Clathsterol, R=C_3H_7CO

Fig. 10.7 Clathsterol

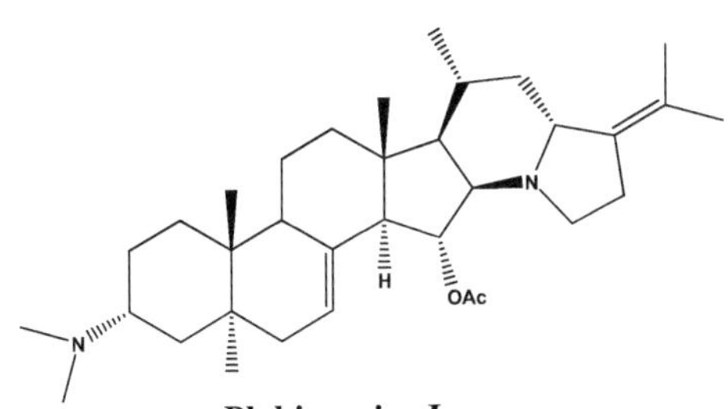

Plakinamine I

Fig. 10.8 Plakinamine I

human deficiency virus type 1 (HIV-1) reverse transcriptase (RT) (Rudi et al. 2001). Plakinamines I–K (Fig. 10.8) are steroidal alkaloids that exhibited significant in vitro cytotoxicity (Ridley and Faulkner 2003). Potent antitumor category sterols include aragusterol A (Iguchi et al. 1994); $5\alpha,6\alpha$-epoxy-24R^*-ethylcholest-8(14)-en-3β,7α-diol [Santafe' et al. 2002); 3-*O*-deacetylluffasterol B; and 9,11-secosterol sesterterpenoids (Rueda et al. 1998) with cell lines A-549, HT-29, H-116, MS-1, and PC-3 tested. Xestobergsterol A (Shoji et al. 1992) is a potent inhibitor of histamine release from rat mast cells induced by anti-IgE.

10.6 Anti(bacterial)microbial, Antifungal, Antimalarial, Antibiotic Products

The products obtained from sponges are also having properties against pathogens harmful to humans. Both simple and complex molecules were found to have this property when tested. Some compounds were having broad-spectrum activity with their unique structural features typical to MNP.

10.6.1 Spongean Terpenoids (Table 10.5)

Agelasine (Fu et al. 1998b) is a diterpene possessing a 9-methyladeninium substituent with antimicrobial activity. Kalihinols (Fig. 10.9) are the diterpene with functional groups of formamide (kalihipyran derivatives) and triisocyano groups obtainable from *Acanthella* spp. They have been found to have antibiotic activity (Patra et al. 1984) and also antifouling activity (Chang et al. 1984) against larvae of the barnacle *Balanus amphitrite*. Halisulfates are the sulfated sesterterpene hydroquinones with inhibition activity over the growth of *Staphylococcus aureus* and *Candida albicans* (Satitpatipan and Suwanborirux 2004). Methanol adduct of puupehenone (Fig. 10.10) is both antimicrobial and antifungal active (Nasu et al. 1995; Kondracki et al. 1999).

Table 10.5 Sponge products useful as antibacterial, antifungal, antimalarial, antibiotic, etc.

Compounds	Compound type	Specific activity	Source	Reference
Terpenes				
Agelasine	DTPN	AB	*Agelas*	Fu et al. (1998b)
Membranolide C	DTPN	Gram-negative AB	*Dendrilla membranosa*	Ankisetty et al. (2004)
Kalihinols	DTPN	Gram-positive AB	*Acanthella* sp.,	Patra et al. (1984)
			Acanthella cavernosa	Chang et al. (1984)
Cacospongionolide F	STTPN	AB and IT	*Fasciospongia cavernosa*	De Rosa et al. (1999)
Nitrogenous Germacranes	SQTPN	AB	*Axinyssa* n. sp.	Satitpatipan and Suwanborirux (2004)
Halisulfate 1	STTPN	AB 1	*Halichondriidae*	Kernan and Faulkner (1988)
Puupehenones	–	AB and AF	*Hyrtios* sp.	Nasu et al. (1995), Kondracki et al. (1999)
Homofascaplysin	STTPN	AB	*Hyrtios* cf. *erecta*	Kirsch et al. (2000)
Sigmosceptrellin B	Nor STTPN	AM	*Diacarnus erythraeanus*	El Sayed et al. (2001b)
Diterpenes	–	AM	*Cymbastela hooperi*	König et al. (1996)
Axisonitrile-3	STTPN	AM	*Acanthella klethra*	Angerhofer et al. (1992)
Arenosclerins A	Alkylpiperidine	AB	*Arenosclera brasiliensis*	Torres et al. (2002)

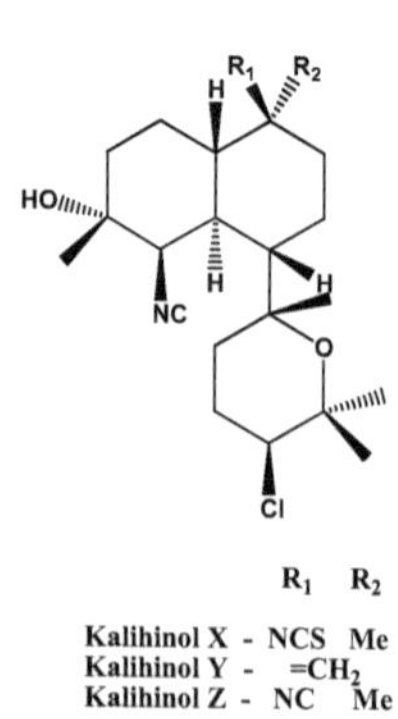

Fig. 10.9 Kalihinols

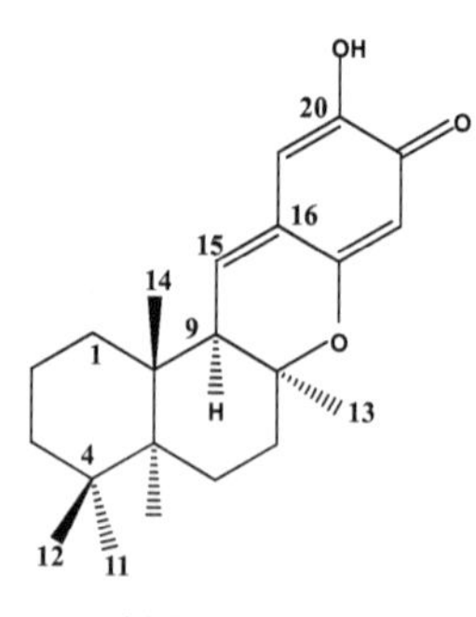

Fig. 10.10 (+)Puupehenone

Homofascaplysin A is potently active in vitro against the malarial parasite, *Plasmodium falciparum* (Kirsch et al. 2000). *Hyrtios* sp. is a good source of these two types of compounds. The presence of isocyanate, isothiocyanate and isonitrile functional groups in the compounds isolated from *Cymbastela hooperi* makes them significant and selective for in vitro antimalarial activity test (König et al. (1996).

10.6.2 Spongean Alkaloids and Heterocycles (Table 10.6)

Papuamine is a pentacyclic alkaloid having antifungal property (Baker et al. 1988). Chelonin A is an aromatic alkaloid with tryptophan and tyrosine unit and found to have multifunctional antimicrobial activity against *Bacillus subtilis* and anti-inflammatory (Bobzi and Faulkner 1991). A range of bioactive compounds were isolated from *Smenospongia* sp. with indole, pyrroloiminoquinone, and tryptamine units having significant antimalarial and antimycobacterial activity (Djura et al. 1980; Hu et al. 2002). Stellettazole B is a geranylgeranyl derivative with antibacterial activity against *Escherichia coli* (Matsunaga et al. 1999).

Palau'amine (Fig. 10.11) is a hexacyclic bisguanidine derivative having broad-spectrum activity against both gram-negative and gram-positive organisms and also resistant to fungal growth (Kinnel et al. 1993), and an isoquinoline quinine derivative, obtained from *Xestospongia* sp., 6-dimethyl-7-methoxy-5,8-dihydroisoquinoline-5,8-dione, was found to be active against the gram-positive bacteria *Bacillus subtilis* and *Staphylococcus aureus* and fungus *Cladosporium cucumerinum* (Edrada et al. 1996a, b). Manzamine class of alkaloids is important compounds with complex pentacyclic diamine linked to C-1 of a β-carboline moiety. The derivatives obtained from *Xestospongia*

Table 10.6 Spongean alkaloid products useful as antibacterial, antifungal, antimalarial, antibiotic, etc.

Compounds	Compound type	Specific activity	Source	Reference
Sceptrin	Bromopyrrole	ATB	*Agelas nakamurai*	Eder et al. (1999)
Papuamine	Pentacyclic	AF	*Haliclona* sp.	Baker et al. (1988)
Chelonin A	Aromatic alkaloid	AB	*Chelonaplysilla* sp.	Bobzin and Faulkner (1991)
Manzamine A	Cyclic β-carboline	AM	*Petrosiidae*	El Sayed et al. (2001a)
6-Bromoaplysinopsin	Indole alkaloid	AM	*Smenospongia aurea*	Djura et al. (1980), Hu et al. (2002)
Plakortide M	Polyketide	AM	*Plakortis halichondrioides*	Jiménez et al. (2003)
Naamine A	Imidazole alkaloids	AF	*Leucetta chagosensis*	Hassan et al. (2004)
3-Bromomaleimide	Pyrrole alkaloids	AF	*Axinella brevistyla*	Tsukamoto et al. (2001)
Aureol *N*,*N*-dimethylthiocarbamate	–	AF and AM	*Smenospongia aurea*	Hu et al. (2002)
Stellettazole B	–	AB	*Stelletta* sp.	Matsunaga et al. (1999)
Palau'amine	Bisguanidine	Gram-negative and gram-positive	*Stylotella agminata*	Kinnel et al. (1993)
Isoquinoline quinone	–	Gram-positive	*Xestospongia*	Edrada et al. (1996a)
Manzamine congeners	–	Insecticidal and gram-positive	*Xestospongia ashmorica*	Edrada et al. (1996b)
Axinellamines B–D	Imidazo-azolo-imidazole	AB	Axinella sp.	Urban et al. (1999)

ashmorica are both insecticidal as well as cytotoxic when studied in vitro against L1578 mouse lymphoma cells (Edrada et al. 1996a, b).

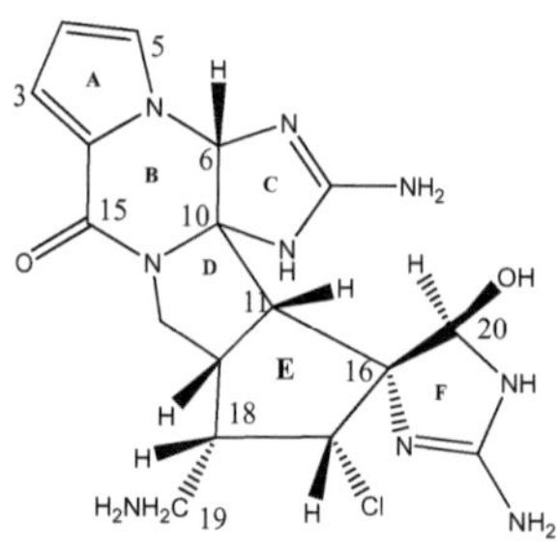

Palau'amine, R_1=R_2=R_3=H

Fig. 10.11 Palau'amine

10.6.3 Spongean Acetylenes, Polyethers, Polyketides, Macrolides

These groups include complex molecules having antimicrobial properties when tested against various pathogens (Table 10.7). Melophlin C (Fig. 10.12) is a tetramic acid derivative isolated from the sponge, *Melophlus sarassinorum* (Wang et al. 2003). It has pronounced antibacterial activity against gram-positive pathogens of *Bacillus subtilis*, *Staphylococcus aureus*, and *Candida albicans*. Aciculitins A–C are bicyclic peptides with unusual histidino-tyrosine bridge attached to a bicyclic peptide along with C_{13}-C_{15} 2,3-dihydroxy-4,6-dienoic acids bearing D-lyxose at the 3-position, obtained from *Aciculites orientalis*. They inhibited the growth of *Candida albicans* and

Table 10.7 Spongean acetylenes, polyethers, polyketides, macrolides, glycosides, nucleosides, hydroquinone derivatives, and sterols useful as antibacterial, antifungal, antimalarial, antibiotic, etc.

Compounds	Compound type	Specific activity	Source	Reference
Acetylenes, polyethers, polyketides, macrolides				
Melophlin C	Tetramic acid	AB	*Melophlus sarassinorum*	Wang et al. (2003)
Celenamide E	Tripeptide alkaloid	ATB	*Cliona chilensis*	Palermo et al. (1998)
Aciculitins 1–3	Cyclic peptides	Gram-positive	*Aciculites orientalis*	Bewley et al. (1996)
Phorboxazoles A and B	Macrolides	AF	*Phorbas* sp.	Searle and Molinski (1995)
Jaspamide	Peptide	AF	*Jaspis*	Zabriskie et al. (1986)
Swinhoeiamide A	Calyculin derivative	AF	*Theonella swinhoei*	Edrada et al. (2002)
Discodermins B, C, and D	Cyclic peptide	AB	*Discodermia kiiensis*	Matsunaga et al. (1985)
Spongistatin	Polyether macrolide	AF	*Hyrtios erecta*	Pettit et al. (1998)
Leucascandrolide A	Polyether macrolide	AF	*Leucascandra caveolata*	D'Ambrosio et al. (1996)
Glycosides, nucleosides, hydroquinone derivatives				
Oceanapiside	Bis-α,ω-amino alcohol glycoside	AF	*Oceanapia phillipensis*	Nicholas et al. (1999)
Aurantosides	Tetramic acid glycosides	AF	*Siliquariaspongia japonica*	Sata et al. (1999)
Isoquinoline quinones	–	Gram-positive	*Xestospongia*	Edrada et al. (1996a, b)
Sterols				
Topsentiasterol sulfate A	Sulfated sterol	AB and AF	*Topsentia* sp.	Fusetani et al. (1994)
Acanthosterols I and J	Sulfated sterol	AF	*Acanthodendrilla* sp.	Tsukamoto et al. (1998)

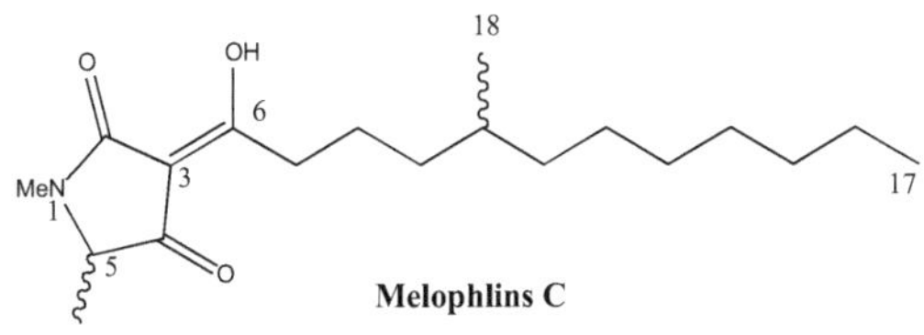

Fig. 10.12 Melophlin C

were cytotoxic toward the HCT-116 cell line (Bewley et al. 1996). Jaspamide is a mixed peptide with potent *insecticidal* activity against *Heliothis virescens* and antifungal activity against *Candida albicans* (Zabriskie et al. 1986). Macrocyclic lactone polyether, Spongistatin, isolated from Hyrtios erecta, was a broad spectrum antifungal compound (Petit et al. 1997). Swinhoeiamide A is a calyculinamide-related congener with insecticidal activity toward neonate larvae of the polyphagous pest *Spodoptera littoralis* when incorporated in an artificial diet and also fungicidal against *Candida albicans* and *Aspergillus fumigates* (Edrada et al. 2002).

10.6.4 Spongean Glycosides, Nucleosides, Hydroquinone, Sterols

Oceanapiside is an α,ω-bis-aminohydroxylipid glycoside having significant antifungal activity against the pathogenic, fluconazole-resistant yeast, *Candida glabrata* (Nicholas et al. 1999) and was obtained from *Oceanapia phillipensis*.

10.7 Antifouling, Inhibition of Sea Urchin Egg Cell Division, Toxicity for Artemia, Feed Deterrent, Repellent, Ichthyotoxic, Nematocidal Products

10.7.1 Spongean Terpenes (Table 10.8)

Apart from the range of bioactivities we have seen so far, sponge compounds are also having deterrent activity to marine macroorganisms paving the way to develop the probable products which may find similar uses under controlled conditions. The toxicity of these compounds present in sponges saves them from their predators leaving them unharmed and better survival in the competitive marine environment.

Cacospongionolide F is a diterpenoid having lethal activity on brine shrimp, *Artemia salina*, and fish (De Rosa et al. 1999). Puupehenone derivative, 15-oxopuupehenone obtained from *Hyrtios* spp., is having both antitumor and antimalarial activity (Nasu et al. 1995; Kondracki et al. 1999). Strongylophorine dimer obtained from *Strongylophora* sp. is a meroditerpenoid having activity against *Micrococcus luteus*, *Salmonella typhi*, and phytopathogenic fungus *Cladosporium cucumerinum* and also against the neonate larvae of the polyphagous pest insect *Spodoptera littoralis* (Oliveros et al. 1998).

10.7.2 Spongean Alkaloids and Heterocycles (Table 10.9)

Bengazole A (Fig. 10.13) is a bisoxazole that has anthelmintic activity (Adamczeski et al. 1988). Mauritiamine is an oroidin dimer found to inhibit larval metamorphosis of the barnacle *Balanus amphitrite* but promoted larval metamorphosis of the ascidian *Ciona savignyi* (Tsukamoto et al. 1996b).

10.7.3 Spongean Acetylenes, Polyethers, Polyketides, Macrolides

Callytriols are polyacetylenic compounds with potent metamorphosis inducing activity on the ascidian *Halocynthia roretzi* larvae and antifouling activity against the barnacle *Balanus amphitrite* larvae (Tsukamoto et al. 1997). Spongiadioxin C and its methyl ether are polybrominated diphenyl ethers that inhibited the cell division of fertilized sea urchin eggs (Utkina et al. 2002). Amphilactams A–D are macrocyclic lactone/lactams having potent in vitro nematocidal properties, obtained from *Amphimedon* spp. (Ovenden et al. 1999). Cyclotheonamide (Fig. 10.14) is having potent

Table 10.8 Sponge terpene and alkaloid products useful as antifouling, repellent, etc.

Compounds	Compound type	Specific activity	Source	Reference
Terpenes				
Kalihinene X	IsocyanoTPN	AFL	*Acanthella cavernosa*	Okino et al. (1995)
Kalihipyrans	DTPN formamides	AFL to *Balanus amphitrite*	*Acanthella cavernosa*	Okino et al. (1996)
Cacofurans 1 and 2	Labdane-class DTPN	INB	*Cacospongia* sp.	Tanaka et al. (2001)
Furanosesquiterpene	–	AFL *Mytilus edulis galloprovincialis*	*Dysidea herbacea*	Sera et al. (1999a)
Cavernosolide	STTPN	ITX	*Fasciospongia cavernosa*	De et al. (1997)
Cacospongionolide F	STTPN	ITX	*Fasciospongia cavernosa*	De Rosa et al. (1997)
15-Oxopuupehenol	–	AT and AM	*Hyrtios* spp.	Nasu et al. (1995)
Strongylophorine dimer	mero DTPN	ITX	*Strongylophora*	Oliveros et al. (1998)
Alkaloids				
E/*Z* bromoindole ethyl esters	–	nematocidal	*Hymeniacidon* sp.	Capon et al. (2002)
Sventrin	Bromopyrrole alkaloid	ITX	*Agelas sventres*	Assmann et al. (2001)
Bengazole A	Oxazoles	Anthelmintic	*Jaspidae*	Adamczeski et al. (1988)

Table 10.9 Sponge *heterocycles, acetylenes, polyethers, and steroids* useful as antifouling, repellent, etc.

Compounds	Compound type	Specific activity	Source	Reference
Heterocycles				
Mauritiamine	Oroidin dimer	AF	*Agelas mauritiana*	Tsukamoto et al. (1996b)
Pseudoceratidine 2	Dibromopyrrolo spermidine	AF	*Pseudoceratina purpurea*	Tsukamoto et al. (1996c)
Ceratinamides A and B	Bromotyrosine deriv	AF	*Pseudoceratina purpurea*	Tsukamoto et al. (1996d)
Acetylenes, polyethers, etc				
Callytriols	Polyacetylene	AF	*Callyspongia truncata*	Tsukamoto et al. (1997)
Spongiadioxins	Polybrominated dibenzo-*p*-dioxins	INB	*Dysidea dendyi*	Utkina et al. (2002)
Geodin A Mg salt	Macrocyclic polyketide	Larval nematocidal	*Geodia*	Capon et al. (1999)
Callyspongins A and B	Polyacetylene Sulfates	INB starfish gametes	*Callyspongia truncata*	Uno et al. (1996)
Ceratinamine	Cyanoformamide	AF	*Pseudoceratina purpurea*	Tsukamoto et al. (Tsukamoto et al. 1996a, b, c, d)
Amphilactams A–D	Macrocyclic lactone	Larval nematocidal	*Amphimedon*	Ovenden et al. (1999)
Cyclotheonamide E4	Cyclic peptide	Treatment of asthma	*Ircinia* sp.	Murakami et al. (2002)
Waiakeamide	Hexapeptide	AFL	*Haliclona* sp.	Sera et al. (2003)
Steroids				
Epidioxy sterols	–	AFL	*Lendenfeldia chondrodes*	Sera et al. (1999b)
Mycalosides A	Steroidal oligoglycosides	INB	*Mycale laxissima*	Antonov et al. (2003)

inhibitory activity against human tryptase and also useful as a therapeutic agent in the treatment of allergic diseases including asthma (Murakami et al. 2002).

10.7.4 Spongean Steroids

Epidioxy sterols showed repellent activity against the blue mussel *Mytilus edulis gallоprovincialis* (Sera et al. 1999a). Steroidal oligoglycosides, mycalosides B–I, inhibit the fertilization of eggs by sperm of the sea urchin *Strongylocentrotus nudus* preincubated with these compounds (Antonov et al. 2003).

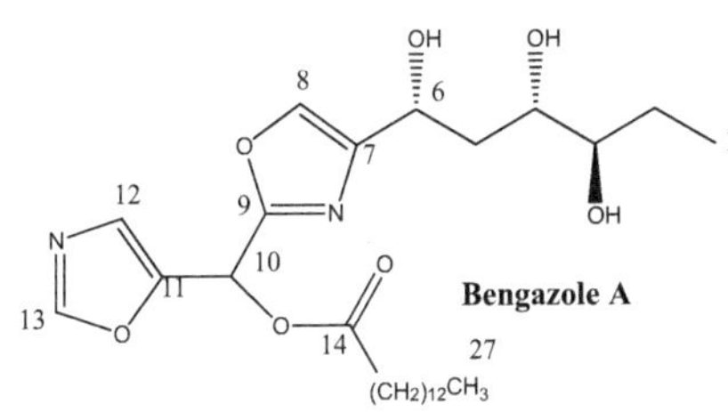

Fig. 10.13 Bengazole A

10.8 Conclusion

The biotechnology revolution has impacted the diverse fields of science and many sectors of the economy. In the environmental arena, application of molecular technologies has brought new ways to identify and mitigate ecological stresses and may hold the keys to remediation. Sales of products developed through biotechnology have been increased by 17 % in 1998 to $13 billion – a figure with the potential to reach about $30 billion in the coming years. Remarkably, these developments have been largely based upon the molecular genetic characterization of terrestrial organisms, even though more than 80 % of all the earth's phyla are found only in the sea. Studies show that the extension of biotechnology to the marine environment is few despite numerous, compelling incentives. Marine plants, animals, and microorganisms exhibit processes and produce substances unknown in terrestrial organisms. The potential economic and public health benefits of pharmaceuticals, pesticides, hormones, enzymes, and polymers derived from marine organisms are high and yet to be exploited further. If the mankind is to realize

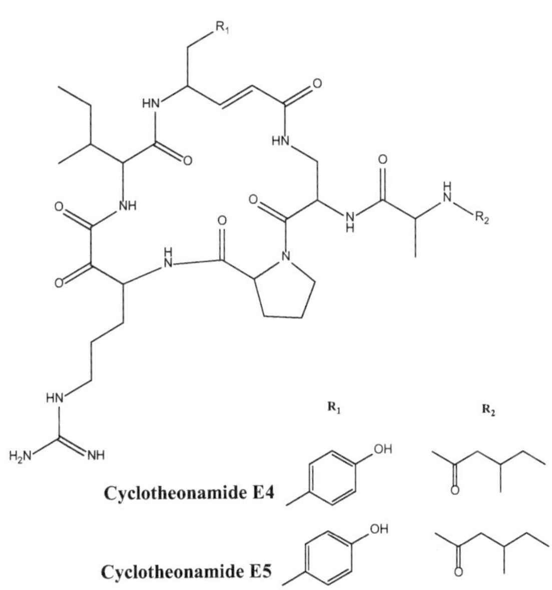

Fig. 10.14 Cyclotheonamides

the benefits to be derived from marine organisms including myriads of beneficial microorganisms as sources of new products and processes and develop viable strategies to conserve them, an increased investment and attention in marine biotechnology is essential.

Recent advances in molecular genetics, sensor biology, environmental remediation, and bioengineering have greatly expanded the ability to find, manipulate, and utilize marine organisms in a sustainable manner. Recognizing the potential of marine biotechnology as the "greatest remaining technology and industrial frontier," it is expected in the future to unravel the seas with more investment to have the sustainability of the nature and to realize the benefits of nature for the mankind.

Acknowledgment I am grateful to Dr. A. Gopalakrishnan, Director, ICAR-Central Marine Fisheries Research Institute, Kochi for his support and permission to publish this work.

References

Adamczeski M, Quiñoá E, Crews P (1988) Unusual anthelminthic oxazoles from a marine sponge. J Am Chem Soc 110:1598–1602

Albizati KF, Martin VA, Agharahimi MR, Stolze DA (1990) In: Scheuer PJ (ed) Bioorganic marine chemistry, vol 5. Springer, Berlin/Heidelberg

Anderson HJ, Coleman JE, Andersen RJ, Roberge M (1997) Cytotoxic peptides hemiasterlin, hemiasterlin A and hemiasterlin B induce mitotic arrest and abnormal spindle formation. Cancer Chemother Pharmacol 39:223–226

Angerhofer CK, Pezzuto JM, Konig GM, Wright AD, Sticher O (1992) Antimalarial activity of sesquiterpenes from the marine sponge Acanthella klethra. J Nat Prod 55:1787–1789

Ankisetty S, Amsler CD, McClintock JB, Baker BJ (2004) Further membranolide diterpenes from the Antarctic sponge *Dendrilla membranosa*. J Nat Prod 67:1172–1174

Antonov AS, Afiyatullov SS, Kalinovsky AI, Ponomarenko LP, Dmitrenok PS, Aminin DL, Agafonova IG, Stonik VA (2003) Mycalosides B-I, eight New spermostatic steroid oligoglycosides from the sponge *Mycale laxissima*. J Nat Prod 66:1082–1088

Aoki S, Yoshioka Y, Miyamoto Y, Higuchi K, Setiawan A, Murakami N, Chen ZS, Sumizawa T, Akiyama S, Kobayashi M (1998) Agosterol A, a novel polyhydroxylated sterol acetate reversing multidrug resistance from a marine sponge *Spongia* sp. Tetrahedron Lett 39:6303–6306

Assmann M, Zea S, Köck M (2001) Sventrin, a New bromopyrrole alkaloid from the Caribbean sponge *Agelas sventres*. J Nat Prod 64:1593–1595

Baker BJ, Scheuer PJ, Shoolery JN (1988) Papuamine, an antifungal pentacyclic alkaloid from a marine sponge, *Haliclona* sp. J Am Chem Soc 110:965–966

Bennet CF, Mong S, Clark MA, Kruse, LI, Crooke ST (1987) Differential effects of manoalide on secreted intracellular phospholipases. Biochem Pharmacol 36: 2079–2086; ED de Silva, PJ Scheuer *Tetrahedron Lett* 1980 21:1611

Bergmann W, Burke DC (1955) Contributions to the study of marine products. XXXIX. The nucleosides of sponges. III. Spongothymidine and spongouridine. J Org Chem 20:1501–1507

Bewley CA, He H, Williams DH, Faulkner DJ (1996) Aciculitins A–C: cytotoxic and antifungal cyclic peptides from the lithistid sponge *Aciculites orientalis*. J Am Chem Soc 118:4314–4321

Blunt JW, Copp BR, Munro MHG, Northcotec PT, Prinsep MR (2010) Marine natural products. Nat Prod Rep 27:165–237

Blunt JW, Copp BR, Keyzers RA, Munro MHG, Prinsep MR (2014) Marine natural products. Nat Prod Rep 31:160–258

Bobzin SC, Faulkner DJ (1991) Aromatic alkaloids from the marine sponge *Chelonaplysilla* sp. J Org Chem 56:4403–4407

Borbone N, Marino SD, Iorizzi M, Zollo F, Debitus C, Esposito G, Iuvone T (2002) Minor steroidal alkaloids from the marine sponge *Corticium* sp. J Nat Prod 65:1206–1209

Brown JL (1975) The evolution of behavior. Norton, New York, pp 761

Bubb MR, Spector I, Bershadsky AD, Korn ED (1995) Swinholide A is a microfilament disrupting marine toxin that stabilizes actin dimers and severs actin filaments. J Biol Chem 270:3463–3466

Butler MS (2004) The role of natural product chemistry in drug discovery. J Nat Prod 67(12):2141–2153

Cantrell CL, Gustafson KR, Cecere MR, Pannell LK, Boyd MR (2000) Chondropsins A and B: novel antitumor cell growth-inhibitory macrolide lactams from the marine sponge *Chondropsis* sp. J Am Chem Soc 122:8825–8829

Capon RJ, Rooney F, Murray LM, Collins E, Sim ATR, Rostas JAP, Butler MS, Carroll AR (1998) Dragmacidins: new protein phosphatase inhibitors from a Southern Australian deep-water marine sponge, *Spongosorites* sp. J Nat Prod 61:660–662

Capon RJ, Skene C, Lacey E, Gill JH, Wadsworth D, Friedel T (1999) Geodin A magnesium salt: a novel nematocide from a southern Australian marine sponge, *Geodia*. J Nat Prod 62:1256–1259

Capon RJ, Skene C, Vuong D, Lacey E, Gill JH, Heiland K, Friedel T (2002) Equilibrating isomers: bromoindoles and a seco-xanthine encountered during a study of nematocides from the southern Australian marine sponge *Hymeniacidon* sp. J Nat Prod 65:368–370

Carletti I, Long C, Funel C, Amade P (2003) Yardenone A and B: new cytotoxic triterpenes from the Indian ocean sponge *Axinella* cf. *bidderi*. J Nat Prod 66:25–29

Carroll J, Johnsson EN, Ebel R, Hartman MS, Holman TR, Crews P (2001) Probing sponge-derived terpenoids for human 15-L-lipoxygenase inhibitors. J Org Chem 66:6847–6851

Carroll AR, Pierens GK, Fechner G (2002) Dysinosin a: a novel inhibitor of factor VIIa and thrombin from a new genus and species of Australian sponge of the family *Dysideidae*. J Am Chem Soc 124:13340–13341

Chang CWJ, Patra A, Roll DM, Scheuer PJ, Matsumoto GK, Clardy J (1984) Kalihinol-A, a highly functionalized diisocyano diterpenoid antibiotic from a sponge. J Am Chem Soc 106:4644–4646

Chang LC, Quintero SO, Hooper JNA, Bewley CA (2002) Batzelline D and Isobatzelline E from the Indopacific Sponge *Zyzzya fuliginosa*. J Nat Prod 65:776–778

Choi K, Hong J, Lee CO, Kim DK, Sim CJ, Im KS, Jung JH (2004) Cytotoxic furanosesterterpenes from a marine sponge *Psammocinia* sp. J Nat Pro 67:1186–1189

Cimino G, De Stefano S, Minale L, Fattorusso E (1972) Ircinin 1 and 2, linear sesterterpenes from the marine sponge *Ircinia oros*. Tetrahedron 28:333–341

Corley DG, Moore RHE, Scheuer PJ (1988) Laulimalides: new potent cytotoxic macrolides from a marine sponge and a nudibranch predator. J Org Chem 53:3644–3646

Costantino V, Fattorusso E, Mangoni A, Di Rosa M, Ianaro A (1997) Glycolipids from sponges. Plakoside A and B, two unique prenylated glycosphingolipids with immunosuppressive activity from the marine sponge *Plakortis simplex*. J Am Chem Soc 119:12465–12470

Costantino V, Fattorusso E, Mangoni A, Di Rosa M, Ianaro A (1999) Glycolipids from sponges, VII: simplexides, novel immunosuppressive glycolipids from the Caribbean sponge *Plakortis simplex*. Bioorg Med Chem Lett 9:271–276

D'Ambrosio M, Guerriero A, Debitus C, Pietra F (1996) Leucascandrolide A, a new type of macrolide: the first powerfully bioactive metabolite of calcareous sponges (*Leucascandra caveolata*, a new genus from the coral sea). Helv Chim Acta 79:51–60

D'Auria MV, Giannini C, Zampella A, Minale L, Debitus C, Roussakis C (1998) Crellastatin a: a cytotoxic bis-steroid sulfate from the Vanuatu marine sponge *Crella* sp. J Org Chem 63:7382–7388

De Guzman FS, Carte B, Troupe N, Faulkner DJ, Harper MK (1999) Neoamphimedine: a new pyridoacridine topoisomerase II inhibitor which catenates DNA. J Org Chem 64:1400–1402

De Marino S, Iorizzi M, Zollo F, Debitus C, Menou JL, Ospina LF, Alcaraz MJ, Payá M (2000) New pyridinium alkaloids from a marine sponge of the genus *Spongia* with a human phospholipase A2 inhibitor profile. J Nat Prod 63:322–326

De Rosa S, Crispino A, De Giulioc A, Iodice C, Tommonaro G (1997) Cavernosolide, a new sesterterpene from a Tyrrhenian sponge. J Nat Prod 60:844–846

De Rosa S, Crispino A, De Giulio A, Iodice C, Amodeo P, Tancredi T (1999) A New cacospongionolide derivative from the sponge *Fasciospongia cavernosa*. J Nat Prod 62:1316–1318

De Smet P, Parys JB, Callewaert G, Weidema AF, Hill E, De Smedt H, Erneux Cn Sorrentino V, Missianen L (1999) Xestospongin C is an equally potent inhibitor of the inositol 1,4,5-triphosphate receptor and the endoplasmic-reticulum Ca2+ pumps. Cell Calcium 26:9–13

Djura P, Stierle DB, Sullivan B (1980) Some metabolites of the marine sponges *Smenospongia aurea* and *Smenospongia* (*Polyfibrospongia*) *echina*. J Org Chem 45:1435–1441

Dunbar DC, Rimoldi JM, Clark AM, Kelly M, Hamann MT (2000) Anti-cryptococcal and nitric oxide synthase inhibitory imidazole alkaloids from the calcareous sponge *Leucetta cf. chagosensis*. Tetrahedron 56:8795–8798

Eder C, Proksch P, Wray V, Van Soest RWM, Ferdinandus E, Pattisina LA, Sudarsono (1999) New bromopyrrole alkaloids from the indopacific sponge *Agelas nakamurai*. J Nat Prod 62:1295–1297

Edrada RA, Proksch P, Wray V, Christ R, Witte L, Van Soest RWM (1996a) Bioactive isoquinoline quinone from an undescribed Philippine marine sponge of the genus *Xestospongia*. J Nat Prod 59:973–976

Edrada RA, Proksch P, Wray V, Witte L, Müller WEG, Van Soest RWM (1996b) Four new bioactive manzamine-type alkaloids from the Philippine marine sponge *Xestospongia ashmorica*. J Nat Prod 59:1056–1060

Edrada RA, Ebel R, Supriyono A, Wray V, Schupp P, Steube K, Van Soest R, Proksch P (2002) Swinhoeiamide A, a new highly active calyculin derivative from the marine sponge *Theonella swinhoei*. J Nat Prod 65:1168–1172

El Sayed KA, Kelly M, Kara UAK, Ang KKH, Katsuyama DC, Dunbar AA, Khan AA, Hamann MT (2001a) New manzamine alkaloids with potent activity against infectious diseases. J Am Chem Soc 123:1804–1808

El Sayed KA, Hamann MT, Hashish NE, Shier WT, Kelly M, Khan AA (2001b) Antimalarial, antiviral, and antitoxoplasmosis norsesterterpene peroxide acids from the Red Sea sponge *Diacarnus erythraeanus*. J Nat Prod 64:522–524

Erickson KL, Beutler JA, Cardellina JH, Boyd MR (1997) Salicylihalamides A and B, novel cytotoxic macrolides from the marine sponge *Haliclona* sp. J Org Chem 62:8188–8192

Erijman EAJ, Ingrum AL, Carney JR, Rinehart KL, Sakai R (1993) Polycyclic guanidine-containing compounds from the Mediterranean sponge *Crambe crambe*: the structure of 13,14,15-isocrambescidin 800 and the absolute stereochemistry of the pentacyclic guanidine moieties of the crambescidins. J Org Chem 58:4805–4808

Farkašovský JNM (2013) Bioprospecting microbial metagenome for natural products. Biol Sect Cel Mol Biol 68(6):1079–1086

Fattorusso E, Taglialatela-Scafati O (2000) Two novel pyrrole-imidazole alkaloids from the Mediterranean sponge *Agelas oroides*. Tetrahedron Lett 41:9917–9922

Fedoreev SA, Prokof_eva NG, Denisenko VA, Rebachuk NM (1989) Cytotoxic activity of aaptamines from suberitid marine sponges. Pharm Chem J 22:615–618

Ford PW, Gustafson KR, McKee TC, Shigematsu N, Maurizi LK, Pannell LK, Williams DE, De Silva ED, Lassota P, Allen TM, Van Soest R, Andersen RJ, Boyd MR (1999) Papuamides A–D, HIV-inhibitory and cytotoxic depsipeptides from the sponges *Theonella mirabilis* and *Theonella swinhoei* collected in Papua New Guinea. J Am Chem Soc 121:5899–590

French RJ, Yoshikami D, Sheets MF, Olivera BM (2010) The tetrodotoxin receptor of voltage-gated sodium channels – perspectives from interactions with μ-conotoxins. Mar Drugs 8:2153–2161

Fu X, Schmitz FJ, Borges MK, McCready TL, Holmes CFB (1998a) Clavosines A-C from the marine sponge *Myriastra clavosa*: potent cytotoxins and inhibitors of protein phosphatases 1 and 2A. J Org Chem 63:7957–7963

Fu X, Schmitz FJ, Tanner RS, Kelly-Borges M (1998b) Agelasines H and I, 9-methyladenine-containing diterpenoids from an *Agelas* sponge. J Nat Prod 61:548–550

Fusetani N, Sugawara T, Mataunaga S (1992) Theopederins A–E, potent antitumor metabolites from a marine sponge *Theonella* sp. J Org Chem 57:3828–3832

Fusetani N, Shinoda K, Matsunaga S (1993) A potent cytotoxic macrolide possessing two Spiro ketals from marine sponge *Cinachyra* sp. J Am Chem Soc 115:3917–3981

Fusetani N, Takahashi M, Matsunaga S (1994) Topsentiasterol sulfates, antimicrobial sterol sulfates possessing novel side chains, from a marine sponge, *Topsentia* sp. Tetrahedron 50:7765–7770

Garcia Pastor P, De Rosa S, De Giulio A, Payá M, Alcaraz MJ (1999) Modulation of acute and chronic inflammatory processes by cacospongionolide B, a novel inhibitor of human synovial phospholipase A2. Br J Pharmacol 126:301–311

Giannini C, Debitus C, Posadas I, Payá M, D'Auria MV (2000) Dysidotronic acid, a new and selective human phospholipase A2 inhibitor from the sponge *Dysidea* sp. Tetrahedron Lett 41:3257–3260

Giannini C, Debitus C, Lucas R, Ubeda A, Paya M, Hooper JNA (2001) New sesquiterpene derivatives from the sponge *Dysidea* species with a selective inhibitor profile against human phospholipase A2 and other leukocyte functions. J Nat Prod 64:612–615

Gunasekera SP, Gunasekera M, Longley RE, Schulte GK (1990) Discodermolide: a new bioactive polyhydroxylated lactone from the marine sponge *Discodermia dissoluta*. J Org Chem 55:4912–4915

Gunasekera SP, McCarthy PJ, Longley RE, Pomponi SA, Wright AE (1999) Secobatzellines A and B, two new enzyme inhibitors from a deep-water Caribbean sponge of the genus *Batzella*. J Nat Prod 62:1208–1211

Gunasekera SP, Zuleta IA, Longley RE, Wright AE, Pomponi SA (2003) Discorhabdins S, T, and U, new cytotoxic pyrroloiminoquinones from a deep-water Caribbean sponge of the genus *Batzella*. J Nat Prod 66:1615–1617

Hassan W, Edrada RA, Ebel R, Wray V, Berg A, Van Soest R, Wiryowidagdo S, Proksch P (2004) New imidazole alkaloids from the Indonesian sponge *Leucetta chagosensis*. J Nat Prod 67:817–822

Hentschel U, Hopke J, Horn M, Friedrich AB, Wagner M, Hacker J, Moore BS (2002) Molecular evidence for a uniform microbial community in sponges from different oceans. Appl Environ Microbiol 68:4431–4440

Hirata Y, Uemura D (1986) Halichondrins – antitumor polyether macrolides from a marine sponge. Pure Appl Chem 58:701–710

Hood KA, West LM, Rouwe´ B, Northcote PT, Berridge MV, Wakefield SJ, Miller JH (2002) Peloruside A, a novel antimitotic agent with paclitaxel-like microtubule-stabilizing activity. Cancer Res 62:3356–3360

Hooper JNA, Van Soest RWM (eds) (2002) Systema Porifera: a guide to the classification of sponges. Kluwer Academic/Plenum Publishers, New York

Hu JF, Schetz JA, Kelly M, Peng JN, Ang KK, Flotow H, Leong CY, Nq SB, Buss AD, Wilkins SP, Hamann MT (2002) New antiinfective and human 5-HT2 receptor binding natural and semisynthetic compounds from the Jamaican sponge *Smenospongia aurea*. J Nat Prod 65:476–480

Ichiba T, Yoshida WY, Scheuer PJ (1991) Hennoxazoles, bioactive bisoxazoles from a marine sponge. J Am Chem Soc 113:3173–3174

Iguchi K, Shimura H, Taira S, Yokoo C, Matsumoto K, Yamada Y (1994) Aragusterol B and D, New 26,27-cyclosterols from the Okinawan marine sponge of the genus *Xestospongia*. J Org Chem 59:7499–7502

Isbrucker RA, Cummins J, Pomponi SA, Longley RE, Wright AE (2003) Tubulin polymerizing activity of dictyostatin 1, a polyketide of marine sponge origin. Biochem Pharmacol 66:75–82

Ishiyama H, Ishibashi M, Ogawa A, Yoshida S, Kobayashi J (1997) Taurospongin a, a novel acetylenic fatty acid derivative inhibiting DNA polymerase β and HIV reverse transcriptase from sponge *Hippospongia* sp. J Org Chem 62:3831–3836

JimBnez C, Quiiiol E, Adamczeski M et al (1991) Novel sponge-derived amino acids. 12. Tryptophan-derived pigments and accompanying sesterterpenes from

Fascaplysinopis reticulata. J Org Chem 56:3403–3410

Jiménez JI, Yoshida WY, Scheuer PJ, Lobkovsky E, Clardy J, Kelly M (2000) Honulactones: new bishomoscalarane sesterterpenes from the Indonesian sponge *Strepsichordaia aliena*. J Org Chem 65:6837–6840

Jiménez MS, Garzòn SP, Rodrı´guez AD (2003) Plakortides M and N, bioactive polyketide endoperoxides from the Caribbean marine sponge *Plakortis halichondrioides*. J Nat Prod 66:655–661

Juagdan EG, Kalindindi RS, Scheuer PJ, Kelly-Borges M (1995) Elenic acid, an inhibitor of topoisomerase II, from a sponge, *Plakinastrella* sp. Tetrahedron Lett 36:2905–2908

Kalaitzis JA, Leone PA, Harris L, Butler MS, Ngo A, Hooper JNA, Quinn RJ (1999) Adociasulfates 1, 7, and 8: new bioactive hexaprenoid hydroquinones from the marine Sponge *Adocia* sp. J Org Chem 64:5571–5574

Kashman Y, Groweiss A, Shmueli U (1980) Latrunculin, a new 2-thiazolidinone macrolide from the marine sponge *Latrunculia magnifica*. Tetrahedron Lett 21:3629–3632

Kato Y, Fusetani N, Matsunaga S, Hashimoto K (1986) Okinonellins A and B, two novel furanosesterterpenes, which inhibit cell division of fertilized starfish eggs, from the marine sponge *Spongionella* sp. Experientia 42:1299–1300

Kato Y, Fusetani N, Matsunaga S, Hashimoto K, Koseki K (1988) Isolation and structure elucidation of calyculins B, C, and D, novel antitumor metabolites, from the marine sponge *Discodermia calyx*. J Org Chem 53:3930–3932

Kernan MR, Faulkner DJ (1988) Sesterterpene sulfates from a sponge of the family *Halichondriidae*. J Org Chem 53:4574–4578

Keyzers RA, Northcote PT, Webb V (2002) Clathriol, a novel polyoxygenated 14*β* steroid isolated from the New Zealand marine sponge *Clathria lissosclera*. J Nat Prod 65:598–600

Kinnel RB, Gehrken HP, Scheuer PJ (1993) Palau'amine: a cytotoxic and immunosuppressive hexacyclic bisguanidine antibiotic from the sponge *Stylotella agminata*. J Am Chem Soc 115:3376–3377

Kirsch G, Köng GM, Wright AD, Kaminsky R (2000) A new bioactive sesterterpene and antiplasmodial alkaloids from the marine sponge *Hyrtios* cf. *erecta*. J Nat Prod 63:825–829

Kitagawa I, Kobayashi M, Kitanaka K, Kido M, Kyogoku Y (1983) Marine natural products, XII: on the chemical constituents of the Okinawan marine sponge *Hymeniacidon aldis*. Chem Pharm Bull 31:2321–2328

Kobayashi J, Cheng JF, Ishibashi M, Walchli MR, Yamamura S, Ohizumi Y (1991) Penaresidin A and B, two novel azetidine alkaloids with potent actomyosin ATPase activating activity from the Okinawan marine sponge *Penares* sp. J Chem Soc Perkin Trans 1:1135–1138

Koiso Y, Morita K, Kobayashi M, Wang W, Ohyabu N, Iwasaki S (1996) Effects of arenastatin A and its synthetic analogs on microtubule assembly. Chemico-Biol Interact 102:183–191

Kondracki MLB, Lacombe F, Guyot M (1999) Methanol adduct of puupehenone, a biologically active derivative from the marine sponge *Hyrtios* species. J Nat Prod 62:1304–1305

König GM, Wright AD, Angerhofer CK (1996) Novel potent antimalarial diterpene isocyanates, isothiocyanates, and isonitriles from the tropical marine sponge *Cymbastela hooperi*. J Org Chem 61:3259–3267

Lim YJ, Kim JS, Im KS, Jung JH, Lee CO, Hong J, Kim DK (1999) New cytotoxic polyacetylenes from the marine sponge *Petrosia*. J Nat Prod 62 (9):1215–1217

Luckner M (1983) Secondary metabolites in microorganisms, plants and animals. Berlin, Springer. In: Torssell KBG (ed) (1984) Natural product chemistry. A mechanistic and biosynthetic approach to secondary metabolism, 2nd edn. John Wiley, New York

Matsunaga S, Fusetani N, Konosu S (1985) Bioactive marine metabolites, VII: structures of discodermins B, C, and D, antimicrobial peptides from the marine sponge *Discodermia kiiensis*. Tetrahedron Lett 26:855–856

Matsunaga S, Nogata Y, Fusetani N (1998a) Thiomycalolides: new cytotoxic trisoxazole-containing macrolides isolated from a marine sponge *Mycale* sp. J Nat Prod 61:663–666

Matsunaga S, Kamimura T, Fusetani N (1998b) Isolation of 1-carboxymethylnicotinic acid from the marine sponge *Anthosigmella* cf. *raromicrosclera* as a cysteine protease inhibitor. J Nat Prod 61:671–672

Matsunaga S, Yamashita T, Tsukamoto S, Fusetani N (1999) Three new antibacterial alkaloids from a marine sponge *Stelletta* species. J Nat Prod 62:1202–1204

McConnell O, Longley RE, Koehn FE (1994) In: Gullo VP (ed) In the discovery of natural products with therapeutic potential. Butterworth-Heinemann, Boston, pp 109–174

McCormick JL, McKee TC, Cardellina JH, Leid M, Boyd MR (1996) Cytotoxic triterpenes from a marine sponge, *Stelletta* sp. J Nat Prod 59:1047–1050

Mooberry SL, Tien G, Hernandez AH, Plubrukarn A, Davidson BS (1999) Laulimalide and isolaulimalide, new paclitaxel-like microtubule-stabilizing agents. Cancer Res 59:653–660

Moore BS (2006) Biosynthesis of marine natural products: macroorganisms (Part B). Nat Prod Rep 23:615–629

Muller WEG, Sobel C, Diehl-Seifert B, Maidhof A, Schroder HC (1987) Influence of the antileukemic and anti-human immunodeficiency virus agent avarol on selected immune responses in vitro and in vivo. Biochem Pharmacol 36:1489–1494

Munro MHG, Blunt JW, Dumdei EJ, Hickford SJH, Lill RE, Li S, Battershill CN, Duckworth AR (1999) The discovery and development of marine compounds with pharmaceutical potential. J Biotechnol 70:15–25

Murakami Y, Oshima Y, Yasumoto T (1982) Identification of okadaic acid as a toxic component of a marine dinoflagellate *Prorocentrum lima*. Bull Jpn Soc Sci Fish 48:69–72

Murakami Y, Takei M, Shindo K, Kitazume C, Tanaka J, Higa T, Fukamachi H (2002) Cyclotheonamide E4 and E5, New Potent Tryptase Inhibitors from an *Ircinia* species of sponge. J Nat Prod 65:259–261

Nakamura H, Ohizumi Y, Kaboyashi J (1984) Keramadine, a novel antagonist of serotonergic receptors isolated from the Okinawan sea sponge *Agelas* sp. Tetrahedron Lett 25:2475–2478

Nakamura H, Wu H, Kobayashi J, Kobayashi M, Ohizumi Y, Hirata Y (1985) Agelasidines. Novel hypotaurocyamine derivatives from the Okinawan Sea sponge *Agelas nakamurai* Hoshino. J Org Chem 50(14):2494–2497

Nakao Y, Oku N, Matsunaga S, Fusetani N (1998) Cyclotheonamides E2 and E3, new potent serine protease inhibitors from the marine sponge of the genus *Theonella*. J Nat Prod 61:667–670

Nakao Y, Uehara T, Matsunaga S, Fusetani N, Van Soest RWM (2002) Callyspongynic acid, a polyacetylenic acid which inhibits a-glucosidase, from the marine sponge *Callyspongia truncata*. J Nat Prod 65:922–924

Nasu SS, Yeung BKS, Hamann MT, Scheuer PJ, Kelly-Borges M, Goins K (1995) Puupehenone-related metabolites from two Hawaiian sponges, *Hyrtios* spp. J Org Chem 60:7290–7292

Newman DJ, Cragg GM (2004) Marine natural products and related compounds in clinical and advanced preclinical trials. J Nat Prod 67(8):1216–1238

Nicholas GM, Hong TW, Molinski TF, Lerch ML, Cancilla MT, Lebrilla CB (1999) Oceanapiside, an antifungal Bis-α, ω-amino alcohol glycoside from the marine sponge *Oceanapia phillipensis*. J Nat Prod 62:1678–1681

Northcote PT, Blunt JW, Munro MHG (1991) Pateamine: a potent cytotoxin from the New Zealand marine sponge, *Mycale* sp. Tetrahedron Lett 32:6411–6414

Okino T, Yoshimura E, Hirota H, Fusetani N (1995) Antifouling kalihinenes from the marine sponge Acanthella cavernosa. Tetrahedron Lett 36:8637–8640

Okino T, Yoshimura E, Hirota H (1996) New antifouling Kalihipyrans from the marine sponge *Acanthella cavernosa*. J Nat Prod 59:1081–1083

Oku N, Matsunaga S, van Soest RWM, Fusetani N (2003) Renieramycin J, a highly cytotoxic tetrahydroisoquinoline alkaloid, from a marine sponge *Neopetrosia* sp. J Nat Prod 66:1136–1139

Okuda RK, Klein D, Kinnel RB, Li M, Scheuer PJ (1982) Marine natural products: the past twenty years and beyond. Pure Appl Chem 54(10):1907–1914

Oliveros MB, Edrada RA, Proksch P, Wray V, Witte Nad L, Van Soest RW (1998) A new meroditerpenoid dimer from an undescribed Philippine marine sponge of the genus *Strongylophora*. J Nat Prod 61:948–952

Ortega MJ, Zubía E, Carballo JL, Salva J (1996) Fulvinol, a new long-chain diacetylenic metabolite from the sponge *Reniera fulva*. J Nat Prod 59:1069–1071

Ovenden SPB, Capon RJ, Lacey E, Gill JH, Friedel T, Wadsworth D (1999) Amphilactams A–D: novel nematocides from southern Australian marine sponges of the genus *Amphimedon*. J Org Chem 64:1140–1144

Palermo JA, Brasco MFR, Cabezas E, Balzaretti V, Seldes AM (1998) Celenamide E, a tripeptide alkaloid from the Patagonian sponge *Cliona chilensis*. J Nat Prod 61:488–490

Patil AD, Kumar NV, Kokke WC, Bean MF, Freyer AJ, De Brosse C, Mai S, Truneh A, Faulkner DJ, Carte B, Breen AL, Hertzberg RP, Johnson RK, Westley JW, Potts BCM (1996) Novel alkaloids from the Sponge *Batzella* sp.: inhibitors of HIV gpl2O-Human CD4 Binding. J Org Chem 60:1182–1188

Patra A, Chang CWJ, Scheuer PJ, Van Duyne GD, Matsumoto GK, Clardy J (1984) An unprecedented triisocyano diterpenoid antibiotic from a sponge. J Am Chem Soc 106:7981–7983

Pattanayak JG, Buddadeb M (2001) Distribution of marine sponges (Porifera) in India. Proc Zool Soc Calcutta 54:73–101

Pawlik JR, Chanas B, Toonen RJ, Fenical W (1995) Defenses of Caribbean sponges against predatory reef fish.1. Chemical deterrency. Mar Ecol Prog Ser 127:183–194

Perry NB, Blunt JW, Munro MHG, Pannell LK (1988) Mycalamide a, an antiviral compound from a New Zealand sponge of the genus *Mycale*. J Am Chem Soc 110:4850–4851

Perry NB, Ettouati L, Litaudon M, Blunt JW, Munro MHG, Parkin S, Hope H (1994) Alkaloids from the antarctic sponge Kirkpatrickia varialosa, part 1: variolin B, a new antitumour and antiviral compound. Tetrahedron 50:3987–3992

Pettit GR, Cichacz ZA, Gao F, Herald CL, Boyd MR, Schmidt JM, Hooper JNA (1993) Isolation and structure of spongistatin 1. J Org Chem 58:1302–1304

Pettit RK, McAllister SC, Pettit GR, Herald CL, Johnson JM, Cichacz ZA (1997) A broad-spectrum antifungal from the marine sponge *Hyrtios erecta*. Int J Antimicrob Agents 9:147–152

Pettit GR, Cichacz ZA, Tan R, Hoard MS, Melody N, Pettit RK (1998) Antineoplastic agents. 386. Isolation of sesterstatins 1–3 from the marine sponge *Hyrtios erecta*. J Nat Prod 61:13–16

Qureshi A, Faulkner DJ (1999) Haplosamates A and B: new steroidal sulfamate esters from two haplosclerid sponges. Tetrahedron 55:8323–8330

Randazzo A, Debitus C, Minale L, Pastor PG, Alcaraz MJ, Payá M, Gomez-Paloma L (1998) Petrosaspongiolides M-R: new potent and selective phospholipase A2 inhibitors from the New Caledonian marine sponge *Petrosaspongia nigra*. J Nat Prod 61:571–575

Rashid MA, Gustafson KR, Boyd MR (2001) A new isoquinoline alkaloid from the marine sponge *Haliclona* species. J Nat Prod 64:1249–1250

Reyes F, Martín R, Rueda A, Fernández R, Montalvo D, Gómez C, Sánchez-Puelles JM (2004) Discorhabdins I and L, cytotoxic alkaloids from the sponge *Latrunculia brevis*. J Nat Prod 67:463–465

Ridley CP, Faulkner DJ (2003) New cytotoxic steroidal alkaloids from the Philippine sponge *Corticium niger*. J Nat Prod 66:1536–1539

Ross SA, Weete JD, Schinazi RF, Wirtz SS, Tharnish P, Scheuer PJ, Hamann MT (2000) Mololipids, a new series of anti-HIV bromotyramine-derived compounds from a sponge of the order Verongida. J Nat Prod 63:501–503

Rudi A, Aknin M, Gaydou EM, Kashman Y (1997) Sodwanones K, L, and M; new triterpenes from the marine sponge, *Axinella weltneri*. J Nat Prod 60:700–703

Rudi A, Yosief T, Loya S, Hizi A, Schleyer M, Kashman Y (2001) Clathsterol, a novel anti-HIV-1 RT sulfated sterol from the sponge *Clathria* species. J Nat Prod 64:1451–1453

Rudi A, Afanii R, Gravalos LG, Aknin M, Gaydou E, Vacelet J, Kashman Y (2003) Three new cyclic peroxides from the marine sponge *Plakortis aff* simplex. J Nat Prod 66:682–685

Rueda A, Zubía E, Ortega MJ, Carballo JL, Salva J (1997) New cytotoxic metabolites from the sponge *Cacospongia scalaris*. J Org Chem 62:1481–1485

Rueda A, Zubía E, Ortega MJ, Carballo JL, Salva J (1998) New metabolites from the sponge *Spongia agaricina*. J Nat Prod 61:258–261

Sakai R, Higa T (1986) Manzamine A, a novel antitumor alkaloid from a sponge. J Am Chem Soc 108:6404–6405

Sakai R, Kamiya H, Murata M, Shimamoto K (1997) A new neurotoxic amino acid from the Micronesian marine sponge *Dysidea herbacea*. J Am Chem Soc 119:4112–4116

Sandler JS, Colin PL, Hooper JNA, Faulkner DJ (2002) Cytotoxic β-carbolines and cyclic peroxides from the Palauan sponge *Plakortis nigra*. J Nat Prod 65:1258–1261

Santafé G, Paz V, Rodríguez J, Jimenez C (2002) Novel cytotoxic oxygenated C29 sterols from the Colombian marine sponge *Polymastia tenax*. J Nat Prod 65:1161–1164

Sata NU, Matsunaga S, Fusetani N, Van Soest RWM (1999) Aurantosides D, E, and F: new antifungal tetramic acid glycosides from the marine sponge *Siliquariaspongia japonica*. J Nat Prod 62:969–971

Satitpatipan V, Suwanborirux K (2004) New nitrogenous germacranes from a Thai marine sponge, *Axinyssa* n. sp. J Nat Prod 67:503–505

Schmidt EW, Suarez CR, Bifano M, Menendez AT, Fairchild CR, Faulkner DJ (2004) Scleritodermin A, a cytotoxic cyclic peptide from the lithistid sponge *Scleritoderma nodosum*. J Nat Prod 67:475–478

Schwartsmann G, Da Rocha AB, Mattei J, Lopes R (2003) Marine-derived anticancer agents in clinical trials. Expert Opin Investig Drugs 12(8):1367–1383

Searle PA, Molinski TF (1995) Phorboxazoles A and B: potent cytostatic macrolides from marine sponge *Phorbas* Sp. J Am Chem Soc 117:8126–8131

Sera Y, Adachi K, Shizuri Y (1999a) A new epidioxy sterol as an antifouling substance from a Palauan marine sponge, *Lendenfeldia chondrodes*. J Nat Prod 62:152–154

Sera Y, Adachi K, Nishida F, Shizuri Y (1999b) A new sesquiterpene as an antifouling substance from a Palauan marine sponge, *Dysidea herbacea*. J Nat Prod 62:395–396

Sera Y, Adachi K, Fujii K, Shizuri Y (2003) A new antifouling hexapeptide from a Palauan sponge, *Haliclona* sp. J Nat Prod 66:719–721

Shafiq MI, Steinbrecher T, Schmid R (2012) FASCAPLYSIN as a specific inhibitor for CDK4: insights from molecular modelling. PLoS One 7(8): e42612. doi:10.1371/journal.pone.0042612

Shen YC, Hsieh PW (1997) New sesquiterpene hydroquinones from a Taiwanese marine sponge *Polyfibrospongia australis*. J Nat Prod 60:93–97

Shen YC, Chen CY, Kuo YH (2001) New sesquiterpene hydroquinones from a Taiwanese marine sponge, *Hippospongia metachromia*. J Nat Prod 64:801–803

Shimosaka A (2002) Role of NKT cells and a-galactosyl ceramide. Int J Hematol 76:277–279

Shoji N, Umeyama A, Shin K, Takeda K, Arihara S, Kobayashi J, Takei M (1992) Two unique pentacyclic steroids with *cis* C/D ring junction from *Xestospongia bergquistia* fromont, powerful inhibitors of histamine release. J Org Chem 57:2996–2997

Sipkema D, Franssen MCR, Osinga R, Tramper J, Wijffels RH (2005) Marine sponges as pharmacy. Marine Biotechnol 7:142–162

Stead P, Hiscox S, Robinson PS, Pike NB, Sidebottom PJ, Roberts AD, Taylor NL, Wright AE, Pomponi SA, Langley D (2000) Eryloside F, a novel penasterol disaccharide possessing potent thrombin receptor antagonist activity. Bioorg Med Chem Lett 10:661–664

Sun HH, Cross SS, Gunasekera M, Koehn FE (1991) Weinbersteroldisulfates A and B, antiviral steroid sulfates from the sponge *Petrosia weinbergi*. Tetrahedron 47:1185–1190

Tachibana K, Scheuer PJ, Tsukitani Y, Kikuchi H, Van Engen D, Clardy J, Gopichand Y, Schmitz FJ (1981) Okadaic acid, a cytotoxic polyether from two marine sponges of the genus *Halichondria*. J Am Chem Soc 103:2469–2471

Takei M, Burgoyne DL, Andersen RJ (1994) Effect of contignasterol on histamine release induced by antiimmunoglobulin E from rat peritoneal mast cells. J Pharm Sci 83:1234–1235

Tanaka J, Marriott G, Higa T, Higa T (2001) Cacofurans A and B, new furanoditerpenes from a marine sponge. J Nat Prod 64:1468–1470

Taylor MW, Radax R, Steger D, Wagner M (2007) Sponge associated microorganisms: evolution, ecology and biotechnological potential. Microbiol Mol Biol Rev 71:295–347

Thomas TRA, Kavlekar DP, LokaBharathi PA (2010) Marine drugs from sponge-microbe association – a review. Mar Drugs 8:1417–1468. doi:10.3390/md8041417

Torres YR, Berlinck RGS, Nascimento GGF, Fortier SC, Pessoa C, de Moraes MO (2002) Antibacterial activity against resistant bacteria and cytotoxicity of four alkaloid toxins isolated from the marine sponge *Arenosclera brasiliensis*. Toxicon 40:885–891

Tsujii S, Rinehart KL, Gunasekera SP, Kashman Y, Cross SS, Lui MS, Pomponi SA, Diaz MC (1988) Topsentin, bromotopsentin, and dihydrodeoxybromotopsentin: antiviral and antitumor Bis(indolyl)imidazoles from Caribbean deep-sea sponges of the family Halichondriidae. Structural and synthetic studies. J Org Chem 53:5446–5453

Tsukamoto S, Kato H, Hirota H, Fusetani N (1996a) Ceratinamine: an unprecedented antifouling cyanoformamide from the marine sponge *Pseudoceratina purpurea*. J Org Chem 61:2936–2937

Tsukamoto S, Kato H, Hirota H, Fusetani N (1996b) Mauritiamine, a new antifouling oroidin dimer from the marine sponge *Agelas mauritiana*. J Nat Prod 59:501–503

Tsukamoto S, Kato H, Hirota H, Fusetani N (1996c) Ceratinamides A and B: new antifouling dibromotyrosine derivatives from the marine sponge *Pseudoceratina purpurea*. Tetrahedron 52:8181–4186

Tsukamoto S, Kato H, Hirota H, Fusetani N (1996d) Pseudoceratidine: a new antifouling spermidine derivative from the marine sponge *Pseudoceratina purpurea*. Tetrahedron Lett 37:1439–1440

Tsukamoto S, Kato H, Hirota H, Fusetani N (1997) Seven new polyacetylene derivatives, showing both potent metamorphosis-inducing activity in ascidian larvae and antifouling activity against barnacle larvae, from the marine sponge *Callyspongia truncata*. J Nat Prod 60:126–130

Tsukamoto S, Matsunaga S, Fusetani N, Van Soest RWM (1998) Acanthosterol sulfates A–J: ten new antifungal steroidal sulfates from a marine sponge *Acanthodendrilla* sp. J Nat Prod 61:1374–1378

Tsukamoto S, Tane K, Ohta T, Matsunaga S, Fusetani N, Van Soest RWM (2001) Four new bioactive pyrrole-derived alkaloids from the marine sponge *Axinella brevistyla*. J Nat Prod 64:1576–1578

Tziveleka LA, Vagias C, Roussis V (2003) Natural products with anti-HIV activity from marine organisms. Curr Top Med Chem 3(13):1512–1535

Uno M, Otha S, Otha E, Ikegami S (1996) Callyspongins a and B: novel polyacetylene sulfates from the marine sponge *Callyspongia truncata* that inhibit fertilization of starfish gametes. J Nat Prod 59:1146–1148

Urban S, De Almeida LP, Carroll AR, Fechner GA, Smith J, Hooper JNA, Quinn RJ (1999) Axinellamines A–D, novel imidazo-azolo-imidazole alkaloids from the Australian marine sponge *Axinella* sp. J Org Chem 64:731–735

Uriz MJ, Turon X, Galera J, Tur JM (1996) New light on the cell location of avarol within the sponge *Dysidea avara* (Dendroceratida). Cell Tissue Res 285:519–527

Utkina NK, Denisenko VA, Virovaya MV, Scholokova OV, Prokofeva NG (2002) Two new minor polybrominated Dibenzo-p-dioxins from the marine sponge *Dysidea dendyi*. J Nat Prod 65:1213–1215

Wang CY, Wang BG, Wiryowidagdo S, Wray V, Van Soest R, Steube KG, Guan HS, Proksch P, Ebel R (2003) Melophlins C-O, thirteen novel tetramic acids from the marine sponge *Melophlus sarassinorum*. J Nat Prod 66:51–56

Wellington KD, Cambie RC, Rutledge PS, Bergquist PR (2000) Chemistry of sponges, 19: novel bioactive metabolites from *Hamigera tarangaensis*. J Nat Prod 63:79–85

Wright AE, Pomponi SA, Cross SS, McCarthy P (1992) A new Bis(indole) alkaloid from a deep-water marine sponge of the genus *Spongosorites*. J Org Chem 57:4772–4775

Zabriskie TM, Klocke JA, Ireland CM, Marcus AH, Molinski TF, Faulkner DJ, Xu C, Clardy JC (1986) Jaspamide, a modified peptide from a *Jaspis* sponge, with insecticidal and antifungal activity. J Am Chem Soc 108:3123–3124

Zampella A, D'Auria MV, Paloma LG, Casapullo A, Minale L, Debitus C, Henin Y (1996) Callipeltin A, an anti-HIV cyclic depsipeptide from the New Caledonian Lithistida sponge *Callipelta* sp. J Am Chem Soc 118:6202–6209

11 Typification of Chemical Compounds of Marine Sponge Metabolites

Irudayaraj Rajendran

Abstract

Sponges are the storehouse of extraneous marine biota paving way for the formation of unique structured organic compounds with interesting biological activities either by themselves or by their symbionts in the form of secondary metabolites for their defense from their enemies. Some compounds are specific to the genus and some are dependent on the location. The range of compounds includes major classes *viz*., terpenoids, alkaloids, heterocycles, steroids, polyacetylenes, peptides, polyethers, polyketides, macrolides, glycosides, and nucleosides. Major compounds under each class with reported biological activities are discussed. Compounds with unique structural features and chemical interest are taken into account.

Keywords

Marine natural products • Sponge extracts • Bioprospecting • Bioactive compounds

11.1 Introduction

Sponges are wealthy reservoirs of compounds with novel structural features found typical of marine origin. Many are having definite biological activity with unique structural features mainly of chemical interest. Any report of the new product from a sponge is followed by its synthesis reported by another group elsewhere. The compounds are the result of either sponge metabolism or of the symbiotic microbiota present in inter- or intraspecific mode in the cells of the sponge (Schmitz et al. 1984; Ishibashi et al. 1986; Kobayashi et al. 1988). Some compounds are indeed specific to particular genus useful for chemotaxonomical classification. However the incidence of the products is location specific with the presence of type of microbes present in them as have been ascertained from the reports.

On the basis of novelty in the structures, they are classified under the major groups of:

I. Rajendran (✉)
Mandapam Regional Centre, Marine Fisheries PO, ICAR-Central Marine Fisheries Research Institute, Ramanathapuram 623 520, Tamil Nadu, India
e-mail: cmfrirajendran@gmail.com

R. Pallela, H. Ehrlich (eds.), *Marine Sponges: Chemicobiological and Biomedical Applications*, DOI 10.1007/978-81-322-2794-6_11

1. Terpenoids
2. Alkaloids
3. Heterocycles
4. Steroids
5. Polyacetylenes, peptides, polyethers, polyketides, and macrolides
6. Glycosides and nucleosides

Major compounds under each group with the reported biological activities are taken primarily for discussion. Compounds with unique structural features and chemical interest are discussed. Various types of compounds reported under each group in this chapter.

11.2 Terpenoids

Terpenes are the major class of compounds found among the sponge secondary metabolites. Natural compounds, artifacts, analogues, and functional derivatives are among the range of the terpenoids identified from sponge extracts. The terpenes isolated and characterized from sponges generally include sesquiterpenes (C_{15}), diterpenes (C_{20}), sesterterpenes (C_{25}), and triterpenes (C_{30}), with functional groups comprising formamide, hydroquinone, epoxy, halogen-substituted carbonimides, peroxides, isocyano, furan, sulfate, keto, aldehyde, hydroxyl, acetoxy, aromatic, isonitrile, pyrrole, amino, guanidine, adenine, pyran etc. Devoid of one or two carbons result in the norterpenes. Stereo isomers include either enantiomers or diastereoisomers. The compounds are cyclic and/or acyclic and rearranged form. The vernacular names of the compounds are arrived at with the incorporation of the genus of sponge source or person or place of the sponge source. The major terpenoids and their derivatives with reported biological activity are discussed. The repeated structural features are omitted among the compounds identified, for the knowledge of novel structures.

11.2.1 Sesquiterpenes

The sesquiterpenes having unusual carbocyclic skeleton, are present in *Cymbastela hooperi*. The compounds are (1*R**,2*S**,5*R**,6*R**,7*S**,8*R**)-1,5-dimethyl-7-(1′-methylethenyl)-tricyclo [6.2.0.3] decane (kelsoene, Fig. 11.1), (1*R**,2*S**,5*R**,6*R**,7*R**,8*S**)-1,5-dimethyl-8-(1′-methylethenyl)-tricyclo [5.3.0.2] decane (prespatane, Fig. 11.2), (1*R**,4*R**,7*S**,10*S**)-4,10-dimethyl -7-(1′-methylethenyl)-bicyclo [5.3.0]dec-5-ene (epi-γ- gurjunene, Fig. 11.3),

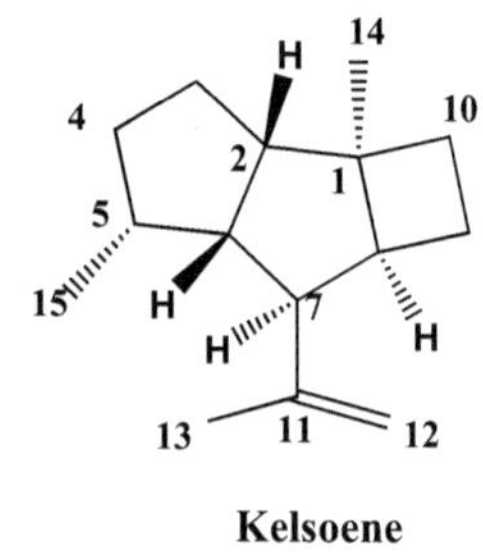

Fig. 11.1 Kelsoene

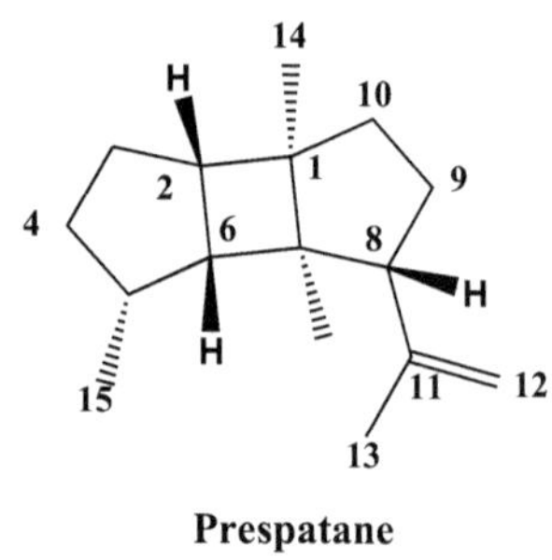

Fig. 11.2 Prespatane

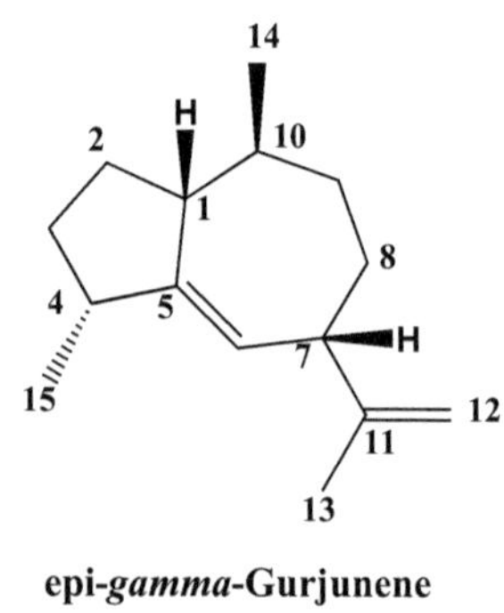

Fig. 11.3 Epi-γ-Gurjunene

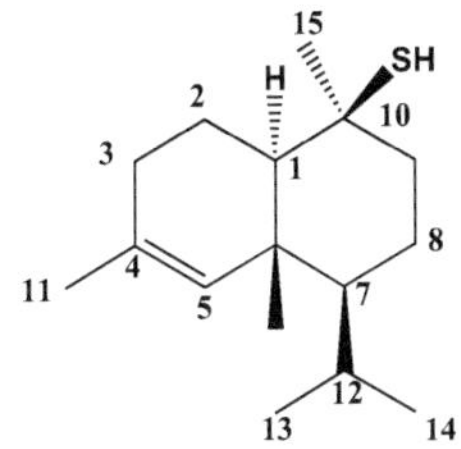

T-cadinthiol

Fig. 11.4 T-cadinthiol

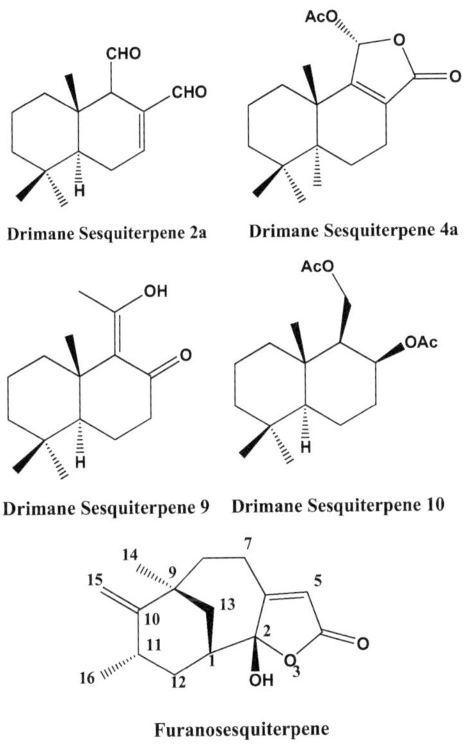

Fig. 11.5 Drimanes

and (1*R**,6*R**,7*S**,10*S**)-4,10-dimethyl-7-(1′-methylethyl)-10-mercaptobicyclo [4.4.0]-dec-4-ene (T-cadinthiol, Fig. 11.4) (König and Wright 1997). Sesquiterpenes and 12- norsesquiterpenoids with drimane skeleton are found in *Dysidea* sp. (Paul et al. 1997). Sesquiterpene with incorporated furanone moiety has been reported from sponge, *Dysidea herbacea* (Sera et al. 1999a, Fig. 11.5).

The compounds melemeleone A, an 18-methoxyavarone derivative (Fig. 11.6), and popolohuanone-C (Fig. 11.7) having quinone moieties with additional heteroatoms were isolated from *Dysidea avara*. *Bilosespens* A and B (Fig. 11.8), *Dysidea cinerea* (Alvi et al. 1992; Rudi et al. 1999) respectively.

Eudesm-11-en-4-ylamine hydrochloride, axinyssamine hydrochloride, 4-isocyanatoeudesm-11-ene, and formamidoeudesm-11-ene, *Axinyssa ambrosia* (Petrichtcheva et al. 2002, Fig. 11.9).

Germacrane sesquiterpenes (Fig. 11.10): (1Z,4Z)-7R*H*-11-aminogermacra-1(10),4-diene, *N*, *N*-11-bis[(1Z,4Z)-7*αH*-germacra-1(10),4-dienyl] urea, *Axinyssa* n. sp. (Satitpatipan and Suwanborirux 2004); sesquiterpene hydroquinones (Fig. 11.11), metachromins A and B and hippochromins A and B, *Hippospongia metachromia* (Ishibashi et al. 1988; Shen et al. 2001).

Puupehenones (a unique class of merosesquiterpenes, Fig. 11.12), 21-chloropuupehenol, 16-oxopuupehenol, and molokinenone were isolated from *Hyrtios* sp. (Nasu et al. 1995). Similar structural analogues were also isolated along

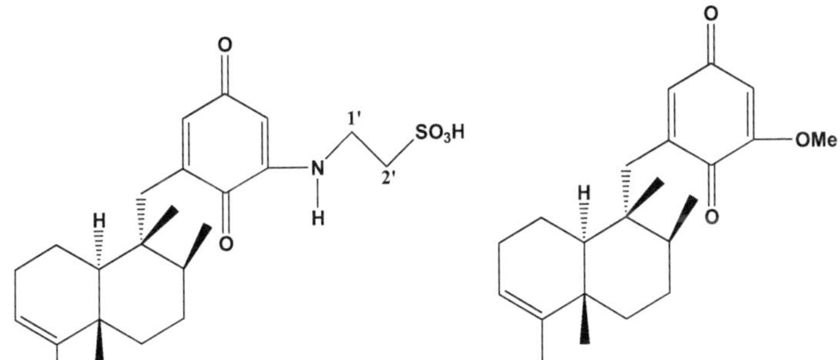

Melemeleone A   18-Methoxyavarone

Fig. 11.6 Melemeleone A, 18-methoxyavarone

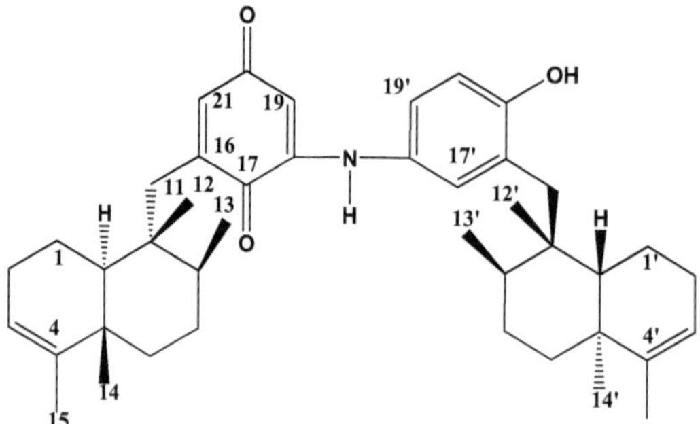

Popolohuanone-C

Fig. 11.7 Popolohuanone-C

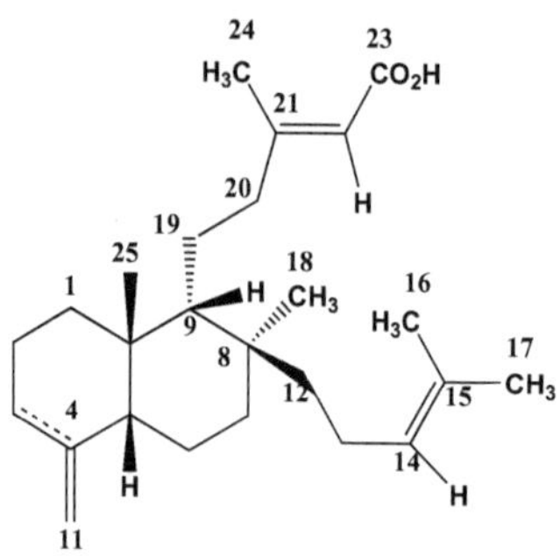

Bilosespene A
Bilosespene B, 3 ene

Fig. 11.8 Bilosespens A and B

with these compounds. The formation of the artifacts from the parent compound puupehenone by methanol adduct is discussed (Kondracki et al. 1999). *Hyrtios* sp. (puupehenone congeners), (+)-(5*S*,8*S*,9*R*,10*S*)-20-methoxy puupehenone and (+)-(5*S*,8*S*,9*R*,10*S*)-15,20-dimethoxy-puupehenol (Fig. 11.13, Piña et al. 2003). Ilimaquinone derivatives (sesquiterpene hydroquinones, Fig. 11.14): polyfibrospongols A and B, dictyoceratin A, and ilimaquinone analogues, *Polyfibrospongia australis* (Shen and Hsieh 1997); ilimaquinone and 5-epi-ilimaquinone, *Fenestraspongia* (Salmoun et al. 2000; Goclik et al. 2000). Parahigginols A–D and parahigginic acid, *Parahigginsia* sp. (Chen et al. 1999); 5-isothiocyanatopupukeanane, *Axinyssa* (Marcus et al. 1989).

Sesquiterpene cyclopentenones, sesquiterpene aminoquinone (rearranged drimane skeleton, Fig. 11.15): Dysidenones A and B and dysidine, bolinaquinone, *Dysidea* sp. (Giannini et al. 2001). Sesquiterpene amide of 1,4-diguanidinobutane and bistellettadines A and B, *Stelletta* sp. (Tsukamoto et al. 1999).

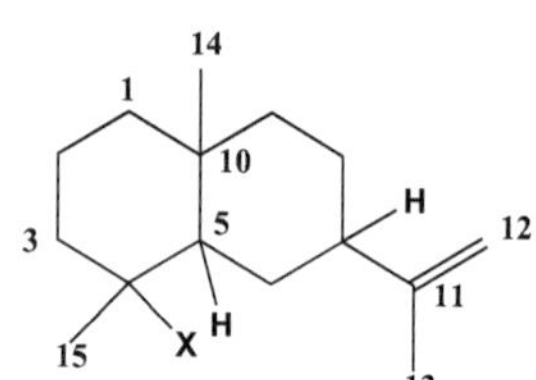

(4R*,5R*,7S*,10R*)-Eudesm-11-en-4-ylamine hydrochloride, X=NH$_2$HCl
(4R*,5R*,7S*,10R*)-4-isocyanatoeudesm-11-ene, X=NC
(4R*,5R*,7S*,10R*)-formamidoeudesm-11-ene, X=NHCHO

Fig. 11.9 Eudesmenes

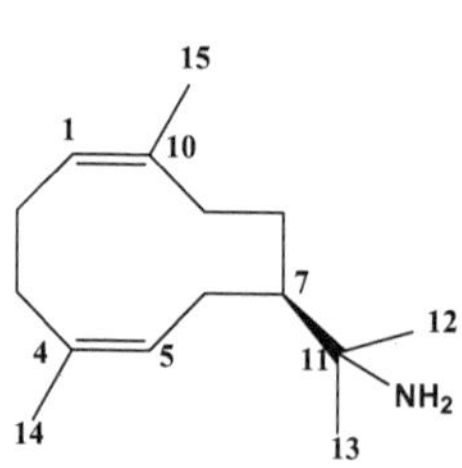

(1Z,4Z)-7$_{alpha}$*H*-11-amino-germacra-1(10),4-diene

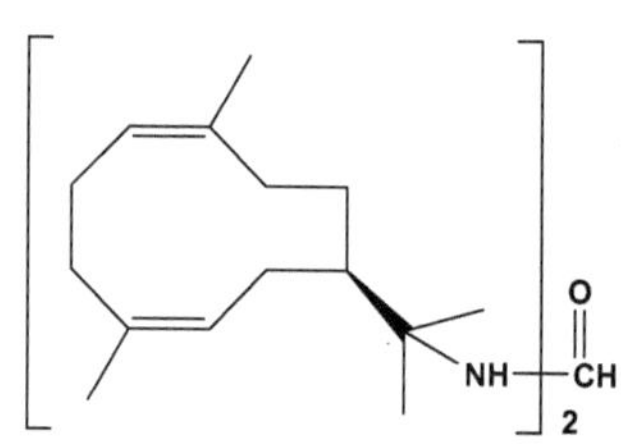

***N,N*-11-bis[(1Z,4Z)-7$_{alpha}$*H*-germacra-1(10),4-dienyl]urea**

Fig. 11.10 Germacrane sesquiterpenes

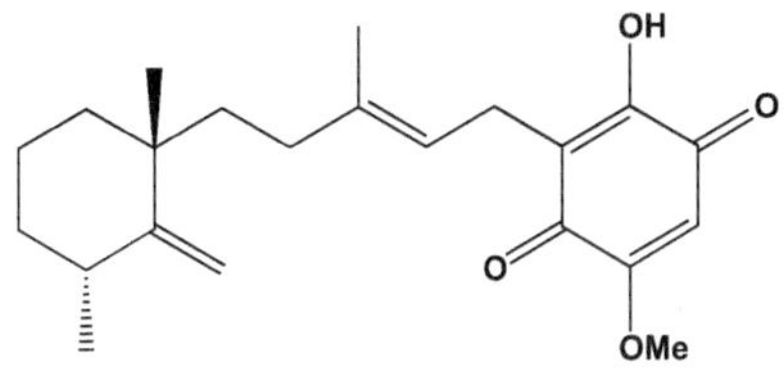

Metachromin A

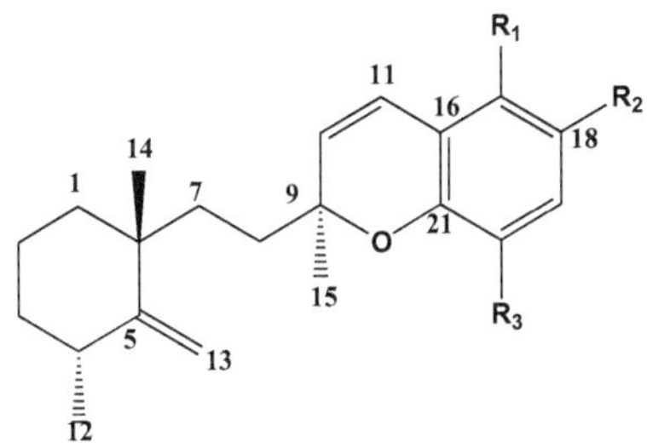

Metachromin B, R_1=OH, R_2=R_3=OMe
Hippochromin A, R_1= R_2=OH, R_3=OMe
Hippochromin B, R_1= R_3=OH, R_2=OMe

Fig. 11.11 Metachromins A and B, hippochromins A and B

11.2.2 Diterpenes

They include agelasine, kalihinane, nakamurol, cacofurans, dendrillolides, muqubilin, dolabellanes, polones, ambliols, spongines, phorbasins, and strongylophorines. Only novel structures have been depicted omitting the repeated structural feature with different functional groups.

Agelasins They were isolated from *Agelas* sp. and the extract contained both diterpene and sesquiterpenes. They are quaternary 9-methyladenine salt of diterpenes (Capon and Faulkner 1984; Fu et al. 1998a, Fig. 11.16). *Agelasidines A* and *B* (Fig. 11.17): they are hypotaurocyamine derivative from a Pacific sponge *Agelas* sp. (Capon and Faulkner 1984): *Agelas nakamurai* (Nakamura et al. 1985).

Kalihinenes Diterpene formamides and kalihipyran derivatives were isolated to charac-

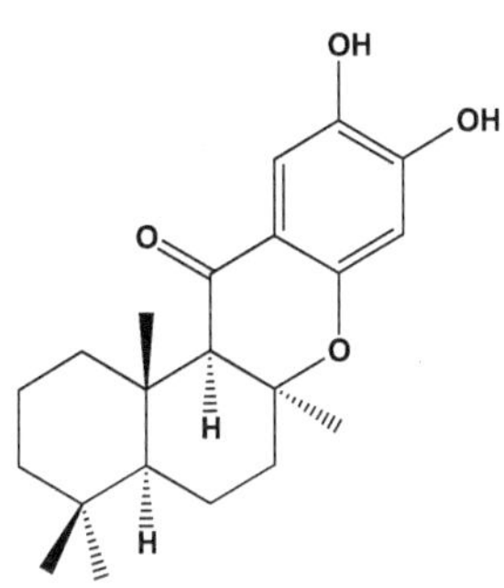

16-Oxopuupehenol

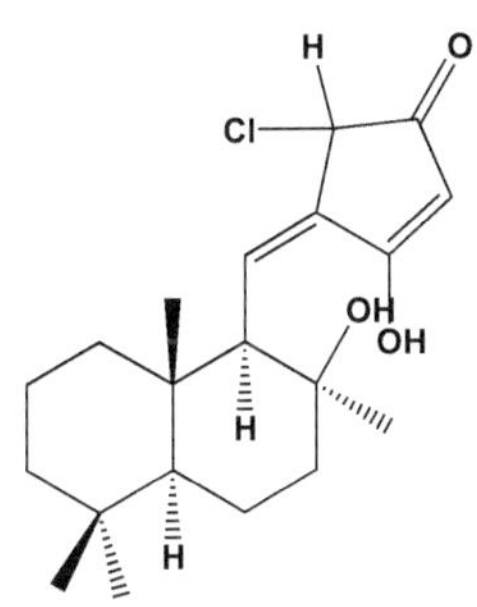

Molokinenone

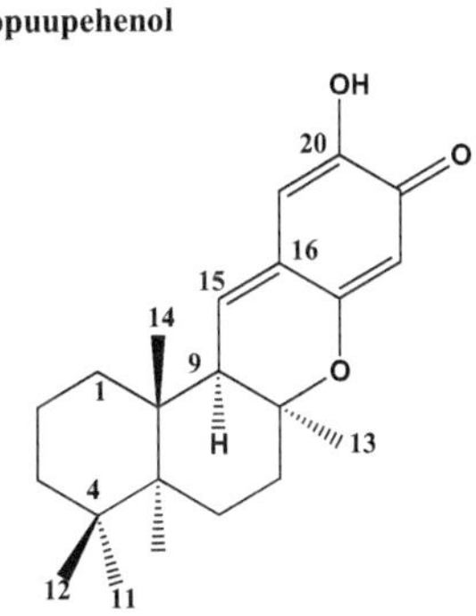

(+)-Puupehenone

Fig. 11.12 Puupehenones

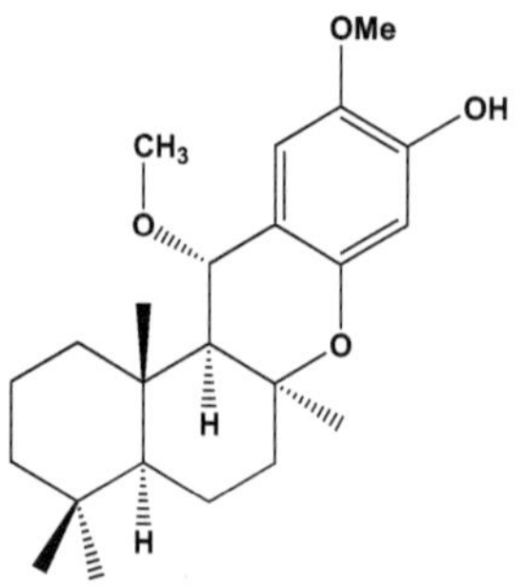

(+)-(5S,8S,9R,10S)-15,20-Dimethoxypuupehenol

Fig. 11.13 Dimethoxy-puupehenol

terize the compounds as kalihinenes X–Z, kalihipyrans A and B, kalihinol A, 10-formamidokalihinene, 15-formamidokalihinene, and biflora-4,9,15-triene, *Acanthella cavernosa* (Okino et al. 1996) (Fig. 11.18).

Kalihinols (Triisocyano Diterpenoid) Kalihinols A, E, and F, unprecedented triisocyano diterpenoid, *Acanthella* sp. (Chang et al. 1984; Patra et al. 1984). Seven new diterpene isonitriles and isothiocyanates were isolated from the sponge *Phakellia pulcherrima* along with eight known ones. Six of the new

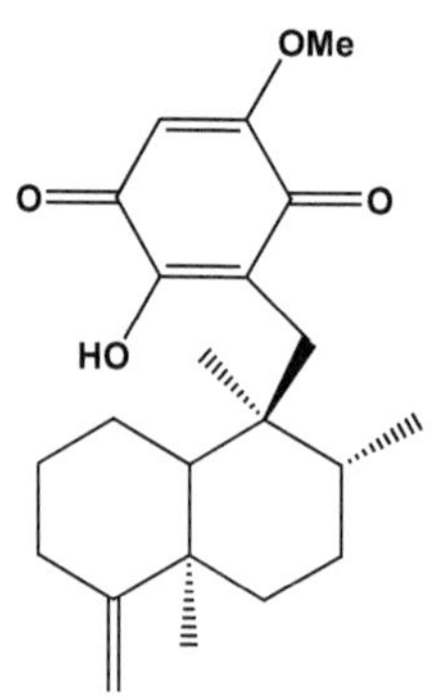

Ilimaquinone

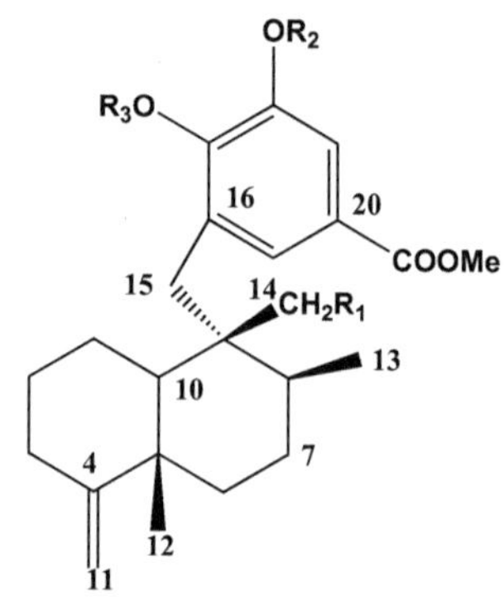

Polyfibrospongol A, R_1=R_3=H, R_2=Me
Polyfibrospongol B, R_1=OH,R_2=Me,R_3=H
Dictyoceratin A, R_1=R_2=R_3=H

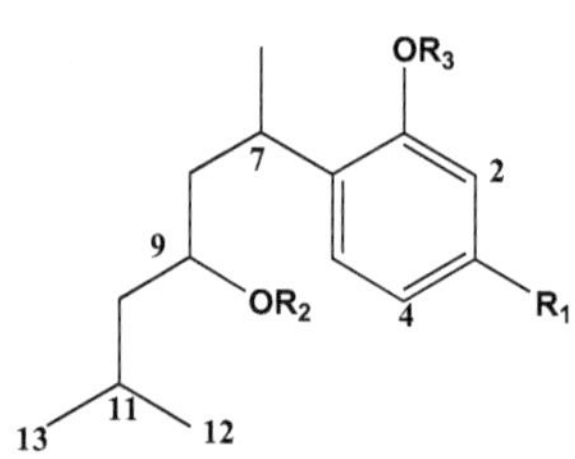

Parahigginols A, R_1=CH_3, R_2=R_3=H
Parahigginols B, R_1=CHO, R_2=Ac, R_3=H
Parahigginols C, R_1=CH_3, R_2=Ac, R_3=H

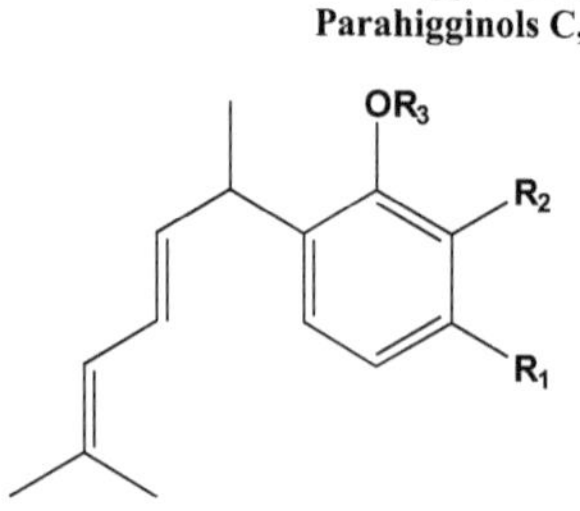

Parahigginols A, R_1=CHO, R_2=OH, R_3=H
Parahigginic acid, R_1=CO_2H, R_2=H, R_3=CH_3

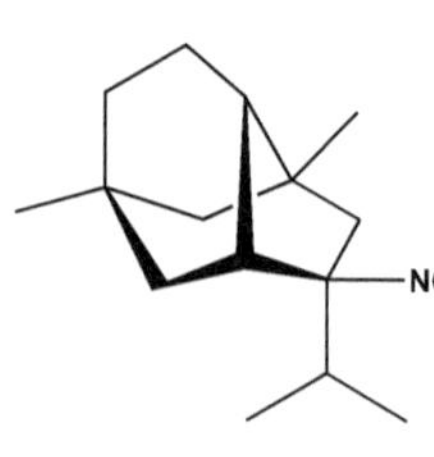

5-isothiocyanatopupukeanane

Fig. 11.14 Ilimaquinone derivatives

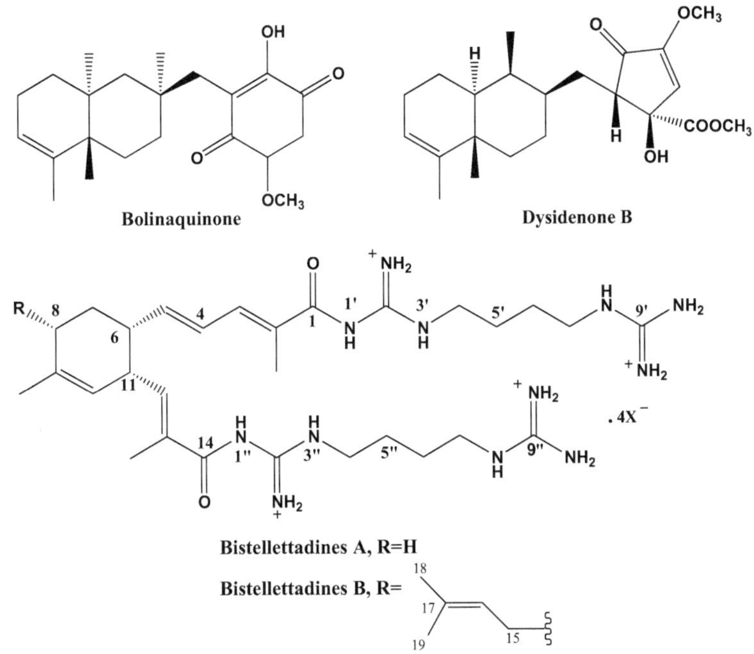

Fig. 11.15 Rearranged drimane skeletons

compounds, 9–14, and the eight known ones, 1–8, belong to the kalihinol family of diterpenes (Wolf and Schmitz 1998) (Fig. 11.19).

Nakamurols A–D (Fig. 11.20), *Agelas nakamurai* (Shoji et al. 1996); cacofurans A and B (Fig. 11.21): They are labdane-class diterpenes, *Cacospongia* sp. (Tanaka et al. 2001). Membranolides B–D (Fig. 11.22), *Dendrilla membranosa* (Ankisetty et al. 2004).

Nuapapuins A and B (norditerpene peroxides) (Fig. 11.23), *Diacarnus* cf. *spinopoculum* [Sperry et al. (1998)]; aikupikoxide A, aikupikoxides B–D (Fig. 11.24), muqubilin, nuapapuin A methyl ester, and *O*-methyl guaianediol, *Diacarnus erythraenus* (Youssef et al. 2001).

Strongylophorines (Fig. 11.25): They are meroditerpenoids possessing a hydroquinone situated on an isocopalane-type diterpene skeleton, strongylophorines 3, 4, 5, 6, and 8, strongylophorine dimer, *Strongylophora durissima*, *Petrosia* (*Strongylophora*) *corticata* (Salvlf and Faulkner 1990; Oliveros et al. 1998; Hoshino et al. 2003).

11.2.3 Sesterterpenes

Cavernosolide, *Fasciospongia cavernosa* (De Rosa et al. 1997); cacospongionolide E, *Fasciospongia cavernosa* (De Rosa et al. 1998, 1999); furospongins, furospongin-5, cyclofurospongin-2, and demethylfurospongin-4, *Spongia officinalis* (Garrido et al. 1997) (Fig. 11.26).

Sesterterpene sulfates: Halisulfates 1–3 sulfated sesterterpene hydroquinone, family *Halichondriidae* (Kernan and Faulkner 1988); hipposulfates A and B, *Hippospongia* cf. *metachromia* (Musman et al. 2001) (Fig. 11.27); sesterstatins 1–3, *Hyrtios erecta*

Fig. 11.16 Quaternary 9-methyladenine salt of diterpenes

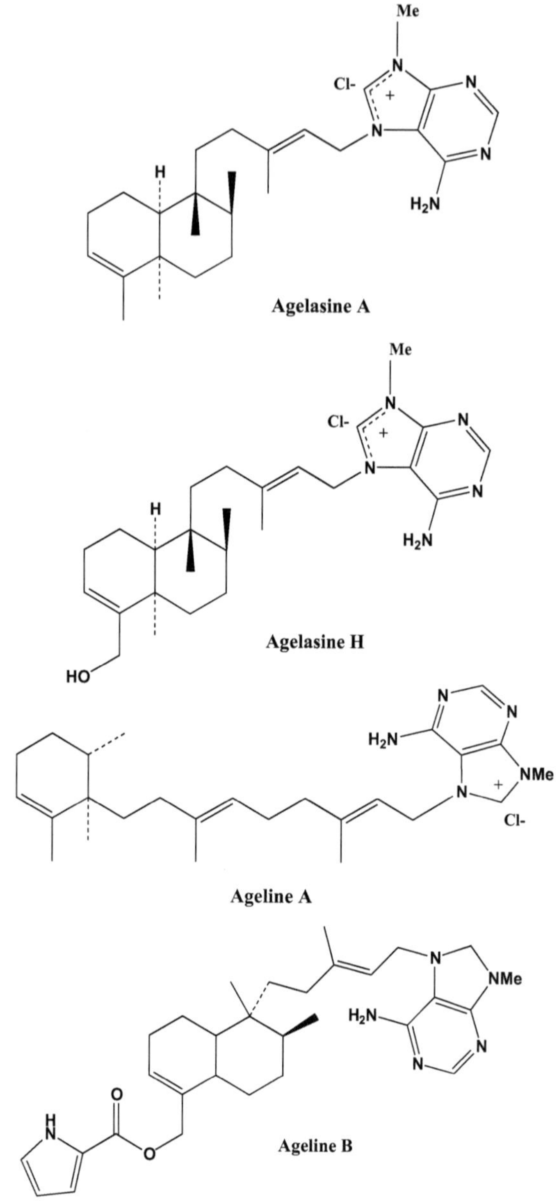

(Pettit et al. 1998); *Hyrtios* cf. *erectus*, 16-*O*-deacetyl-16-*epi*-scalarolbutenolide, 12-*O*-acetyl- 16-*O*-deacetyl-16-*epi*-scalarolbutenolide, and 12-deacetoxy-21acetoxyscalarin (Ryu et al. 1996a); isodehydroluffariellolide, homofascaplysin A, and fascaplysin (Fig. 11.28), *Hyrtios* cf. *erecta* (Kirsch et al. 2000); hyrtiolide, 16-hydroxyscalarolide, and 12-deacetyl-Δ^{17}-hyrtial, *Hyrtios erectus* (Miyaoka et al. 2000); furanosesterterpenes

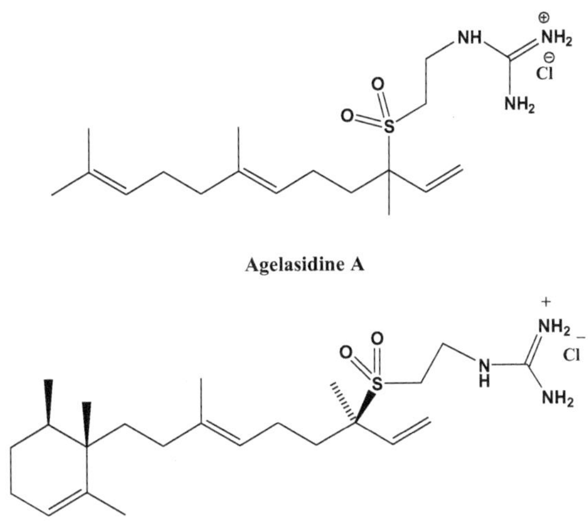

Fig. 11.17 Agelasidines A and B

containing a tetronic acid, norsesterterpenes, hippospongin 1, and untenic acid, *Ircinia* sp. (Issa et al. 2003) (Fig. 11.29).

Dysidiolide: Sesterterpene γ-hydroxybutenolide (C_{25} isoprenoid), *Dysidea etheria* (Gunasekera et al. 1996a); honulactones A–L (20,24-bishomoscalarane sesterterpenes), *Strepsichordaia aliena* (Jiménez et al. 2000a); hyrtiolides, hyrtiosal, *Hyrtios erectus* (Iguchi et al. 1992) (Fig. 11.30).

Scalarane Group

New scalarane sestertepenes (Fig. 11.31) together with three uncommon norscalaranes are reported from sponge *Cacospongia scalaris* (Rueda et al. 1997). The structures **a–d** are correlated with aragusterols A, *Hyrtios erecta* (Tsuchiya et al. 1998; Kobayashi et al. 1996); *Spongia* (Tsukamoto et al. 2003a).

Six new 20,24-bishomoscalarane sesterterpenes, honu'enone, phyllofolactones H–K, and phyllofenone C., *Phyllospongia foliascens*; phyllofolactone C (homoscalarane sesterterpenes), *Strepsichordaia aliena* (Jiménez et al. 2000b; Fu et al. 1999) (Fig. 11.32).

Suberitenones A and B, *Suberites* sp. (Shin et al. 1996); luffariolides A–E, *Luffariella variabilis* (Tsuda et al. 1992); mycaperoxide H, cyclic norsesterterpene peroxide, *Mycale* sp. (Phuwapraisirisan et al. 2003) (Fig. 11.33).

Muqubilone (norsesterterpene acid), *Diacarnus erythraeanus* (El Sayed et al. 2001a); tasnemoxides A (cyclic norsestertepene peroxides), *Diacarnus erythraenus* (Youssef 2004); petrosaspongiolides M, N, and R, *Petrosaspongia nigra* (Randazzo et al. 1998) (Fig. 11.34).

Salmahyrtisols A, B, and C, 3-acetyl sesterstatin 1, 19-acetyl sesterstatin 3, hyrtiosal, and scalarolide, *Hyrtios erecta* (Youssef et al. 2002); 12-*O*-desacetylfuroscalar-16-one, *Cacospongia* sp. (Cambie et al. 1998) (Fig. 11.35).

Furanosesterterpenes Strobilinin, felixinin furanosesterterpenes, *Psammocinia* sp. (mixture of compounds **7** and **8** displayed significant inhibition of DNA replication and moderate antioxidant profile (Choi et al. 2004); *furanosesterterpene tetronic acids*, sarcotins B and D; ircinin-1, *Sarcotragus* sp. (Liu et al. 2001) (Fig. 11.36).

Pyrrolo- and Furanosesterterpenoids Sarcotrins A–C, including two trinorsesterterpenes, two diterpenes, *Sarcotragus* sp. (Liu et al. 2002, 2003) (Fig. 11.37).

Fig. 11.18 Kalihinenes

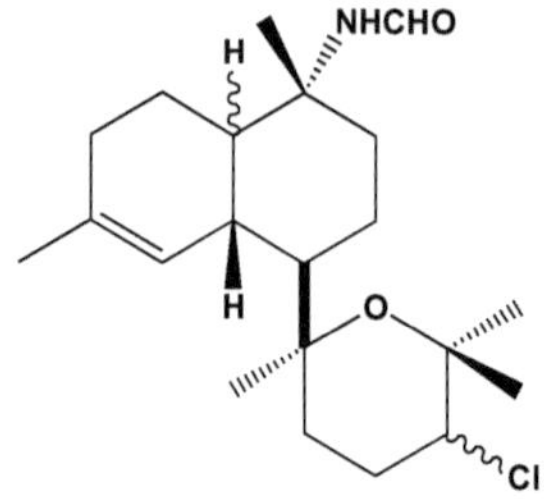

Kalihinene X - 1beta H, 14 alpha Cl
Kalihinene Y - 1alpha H,14 alpha Cl
Kalihinene Y - 1alpha H, 14 alpha Cl

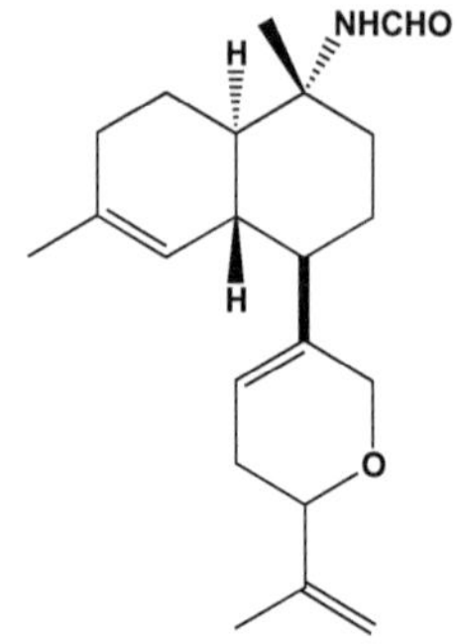

Kalihipyran A

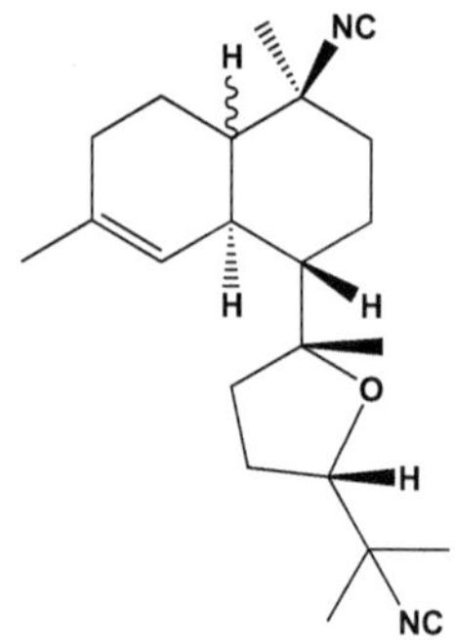

Kalihinene - 1H_{alpha}
1-*epi*-Kalihinene - 1H_{beta}

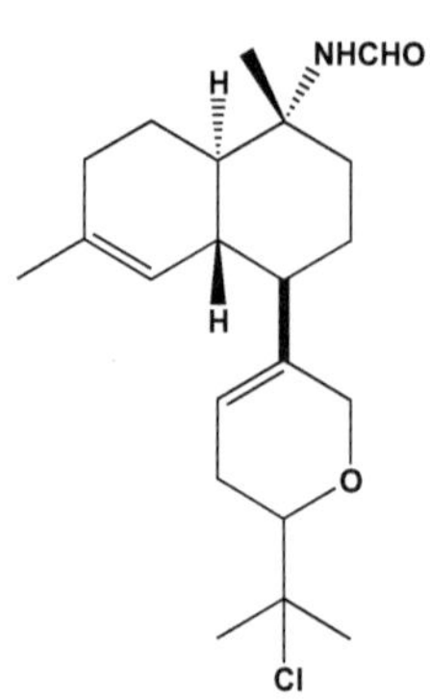

Kalihipyran B

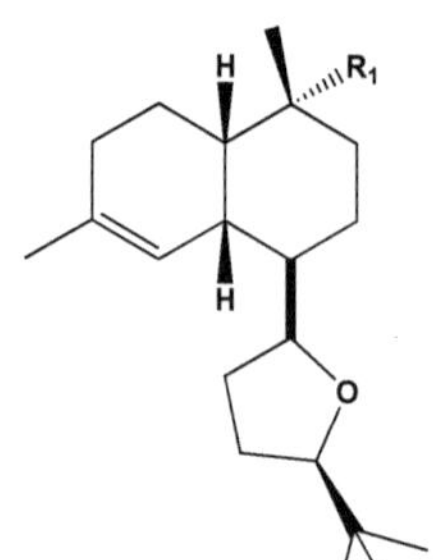

10-Formamidokalihinene,
R_1=NHCHO, R_2=NC
15-Formamidokalihinene,
R_1=NC, R_2=NHCHO

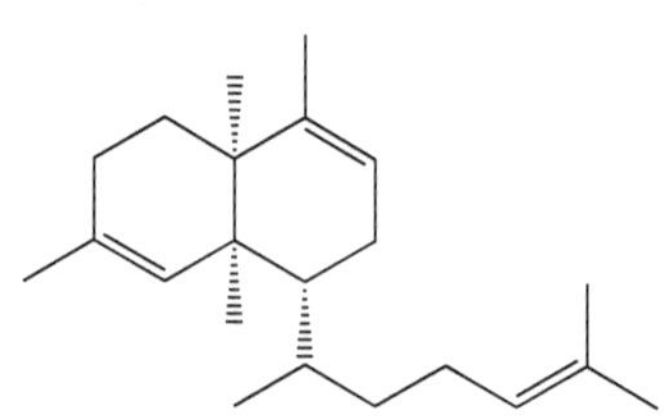

Biflora-4,9,15-triene

Fig. 11.19 Kalihinols

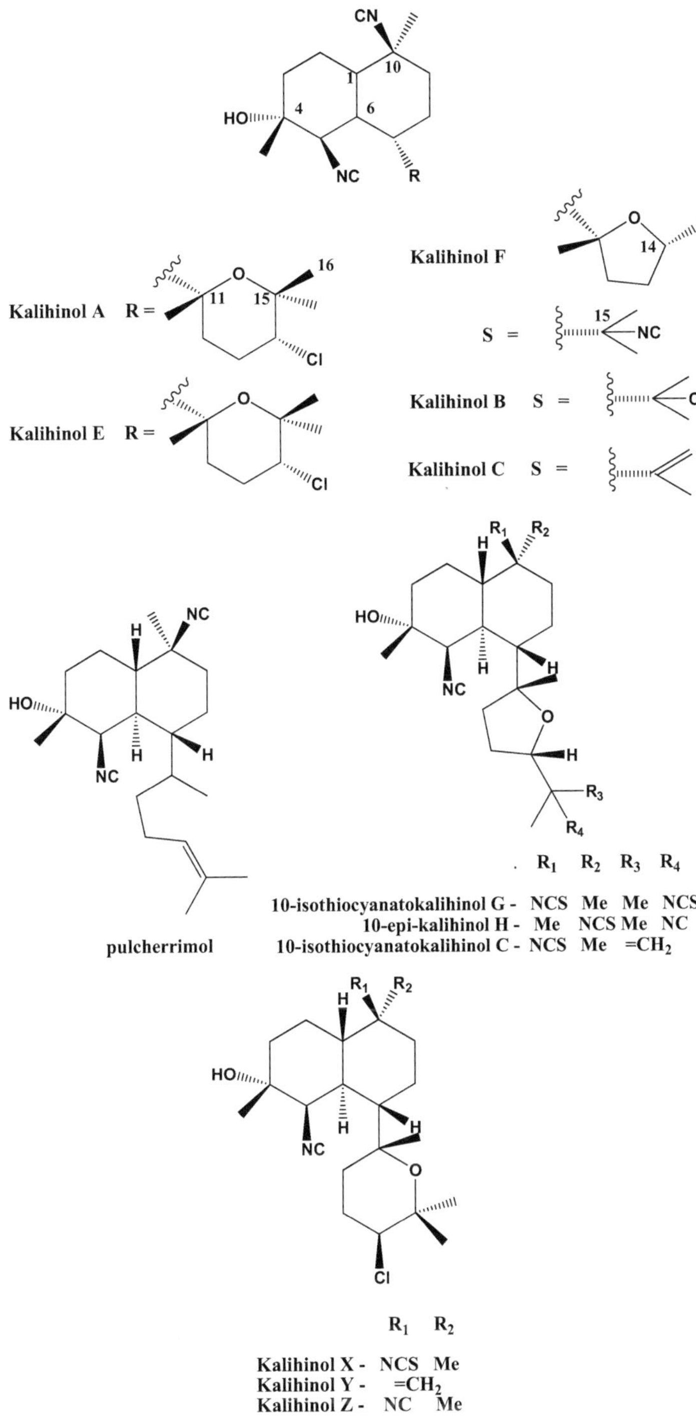

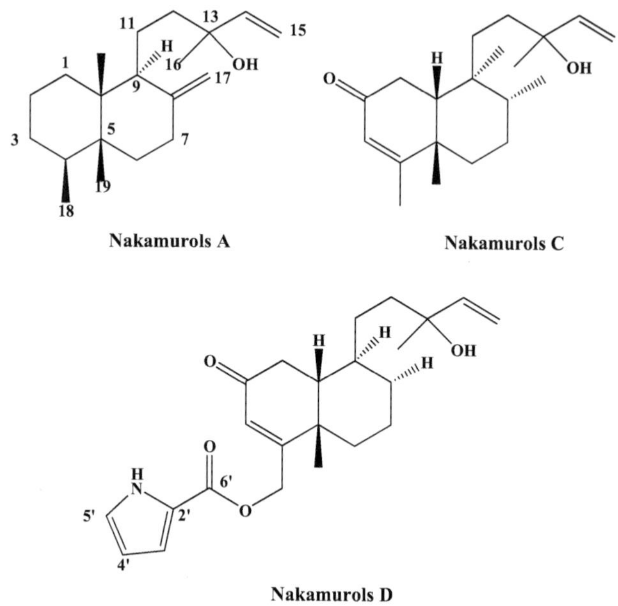

Fig. 11.20 Nakamurols A–D

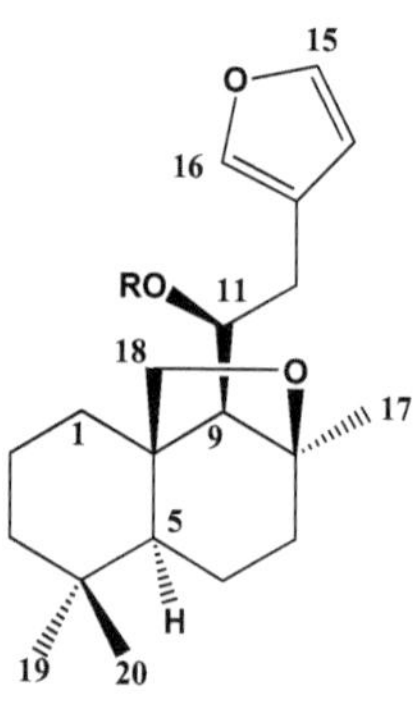

Cacofuran A, R=Ac
Cacofuran B, R=H

Fig. 11.21 Cacofurans A and B

11.2.4 Triterpenes

Sodwanones K and L, *Axinella weltneri* (Rudi et al. 1997); yardenone, yardenone A, *Axinella* cf. *bidderi* (Carletti et al. 2003) (Fig. 11.38); *stellettins*, isomalabaricane triterpenes, stellettins A, C, E, H, and I and rhabdastrellic acid A, *Rhabdastrella globostellata* (Tasdemir et al. 2002a) (Fig. 11.39); 29-hydroxystelliferins A and E, *Jaspis* sp. (Meragelman et al. 2001); globostellatic acids A and B, *Stelletta globostellata*, *Stelletta* sp. (Ryu et al. 1996b; McCormick et al. 1996) (Fig. 11.40).

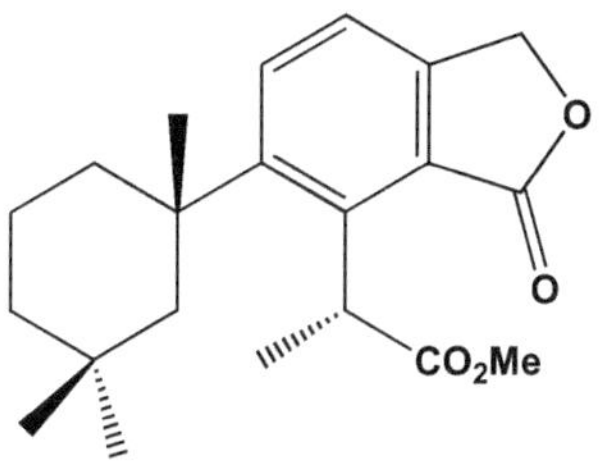

Membranolide

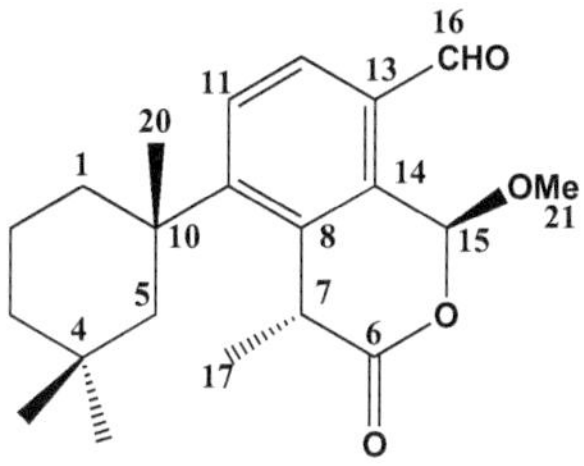

Membranolide B

Fig. 11.22 Membranolides B–D

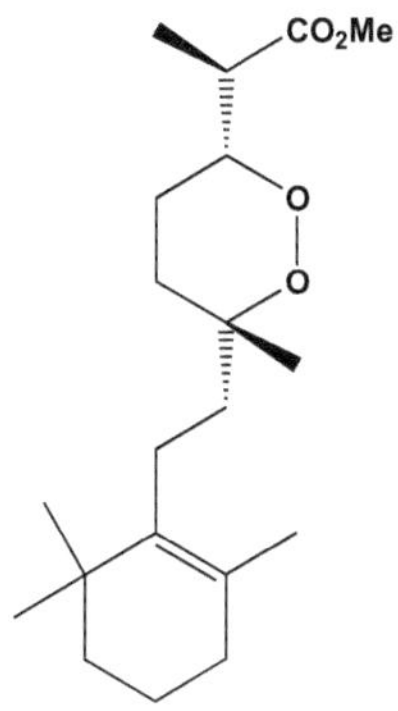

Nuapapuin A methyl ester

Fig. 11.23 Nuapapuins A and B (norditerpene peroxides)

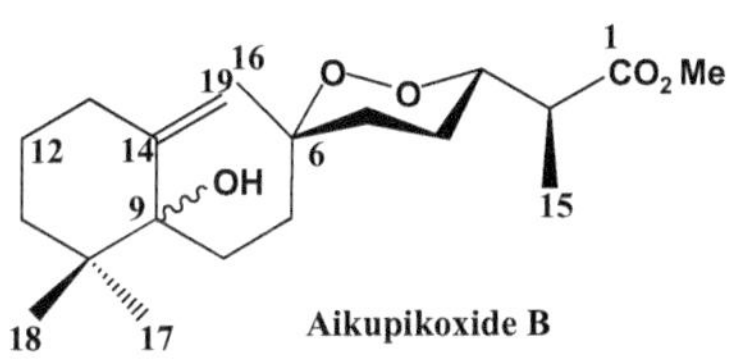

Aikupikoxide B

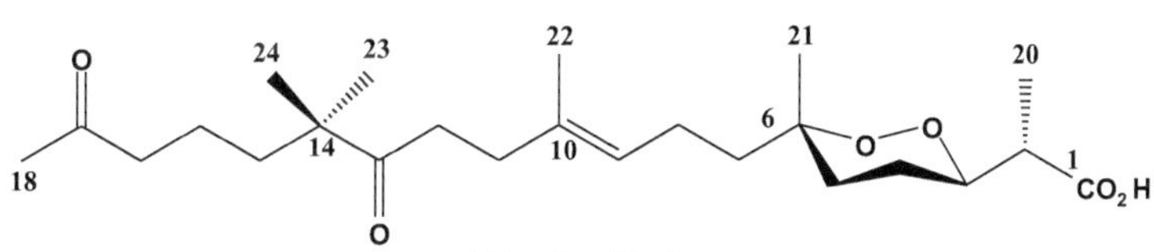

Aikupikoxide A

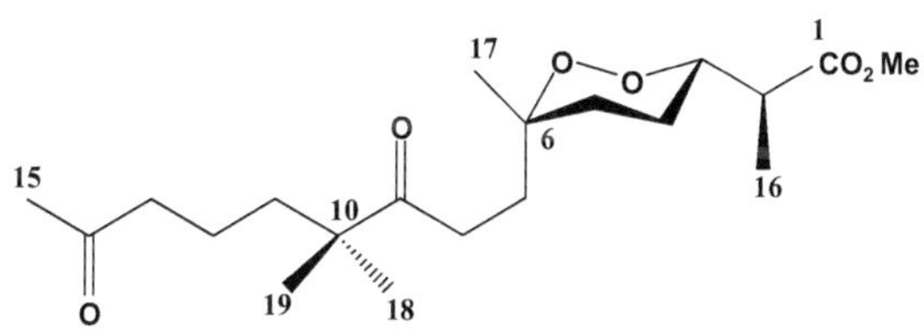

Aikupikoxide C

Fig. 11.24 Aikupikoxides B–D

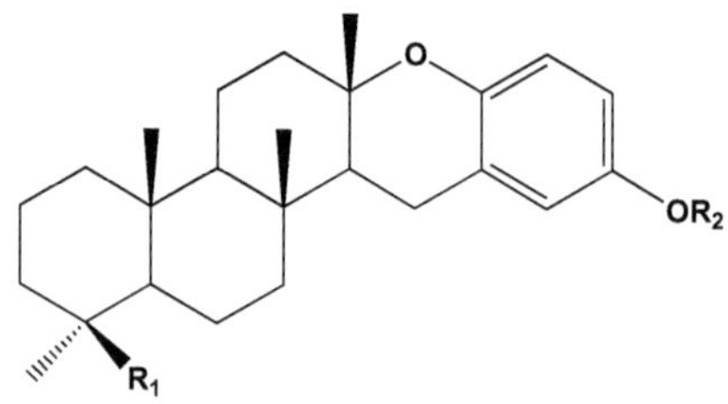

Strongylophorine 3, R_1=COOH, R_2=H
Strongylophorine 4 R_1=CHO, R_2=H
Strongylophorine 5, R_1=CH_2OH, R_2=H

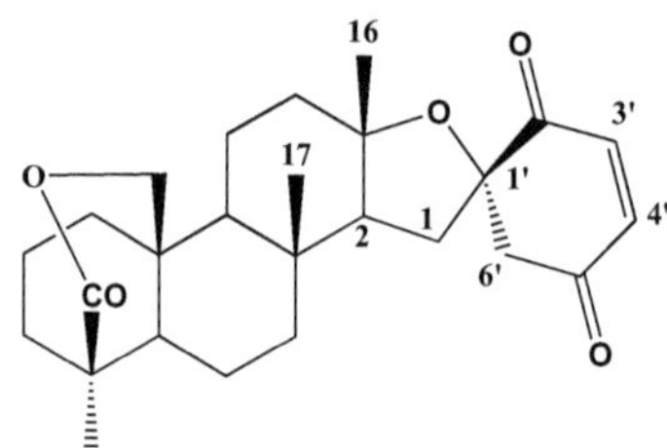

Strongylophorine 6

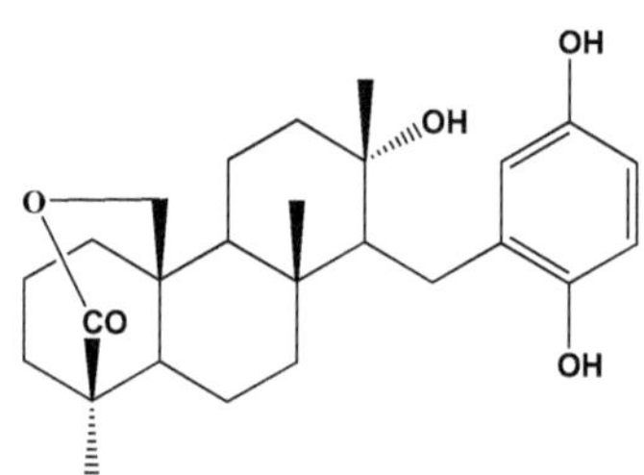

Strongylophorine 8

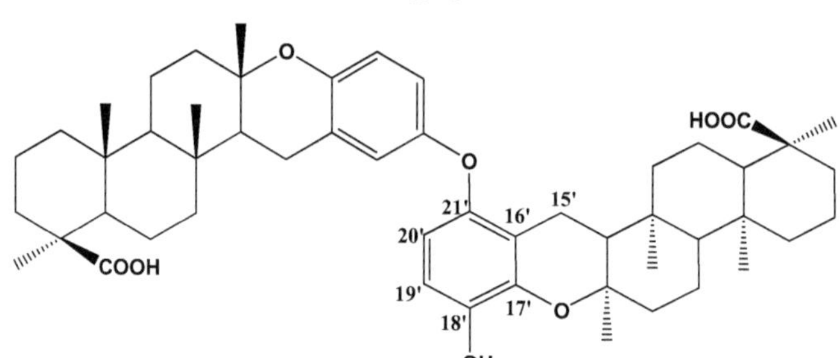

Strongylophorine dimer

Fig. 11.25 Strongylophorines

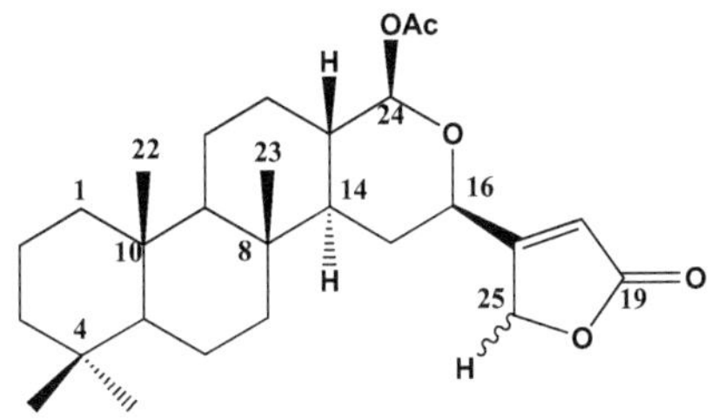

Cavernosolide

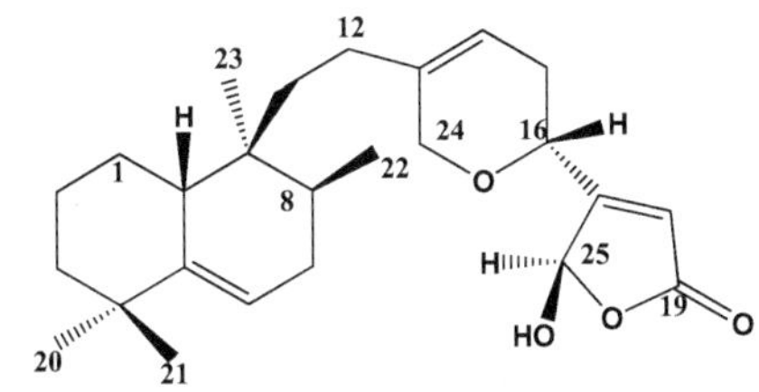

Cacospongionolide F

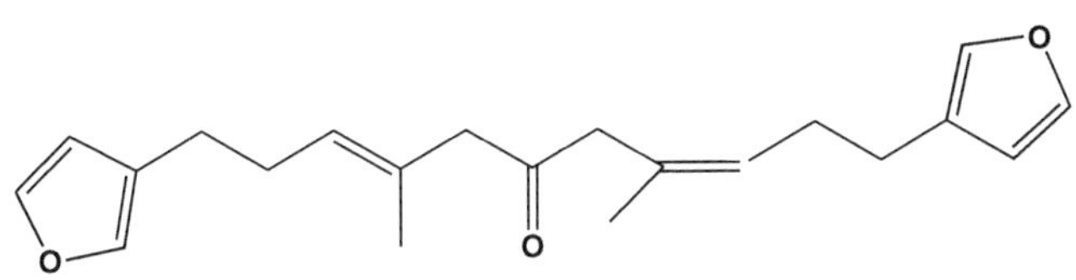

Furospongin-5

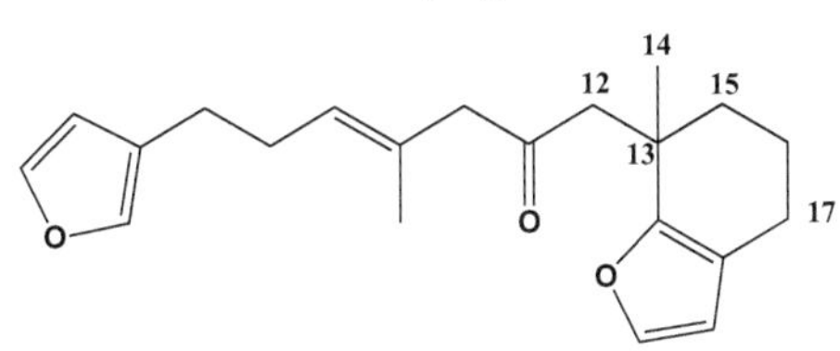

Cyclofurospongin-2

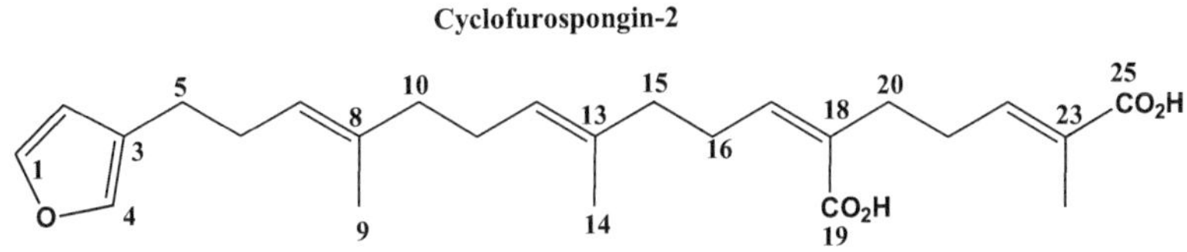

Demethylfurospongin-4

Fig. 11.26 Cavernosolide, cacospongionolide, furospongins

Fig. 11.27 Halisulfates, hipposulfates

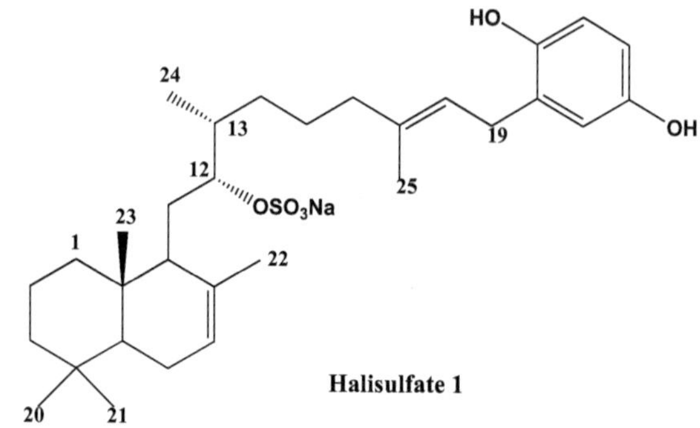

Halisulfate 1

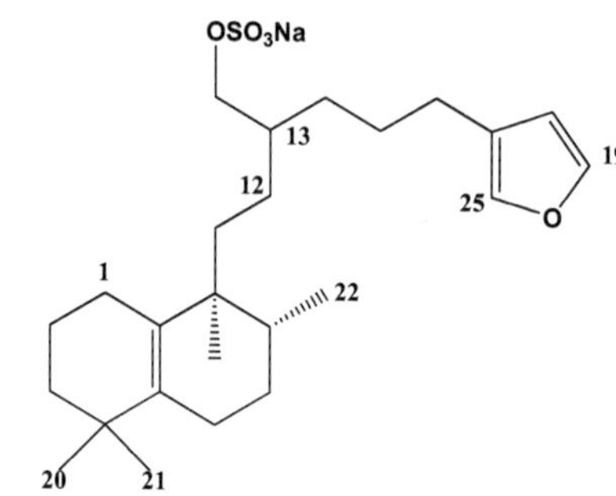

Halisulfate 3

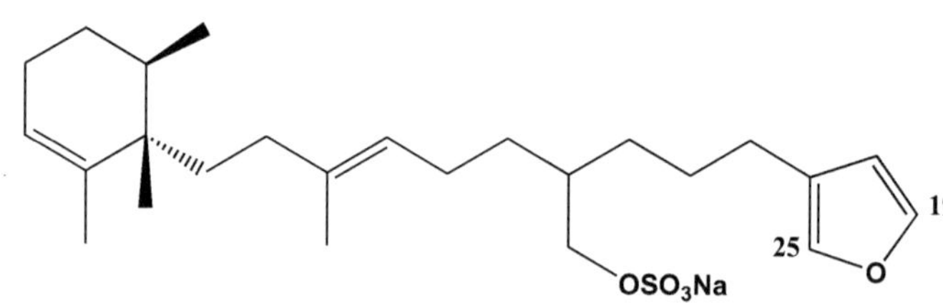

Halisulfate 2

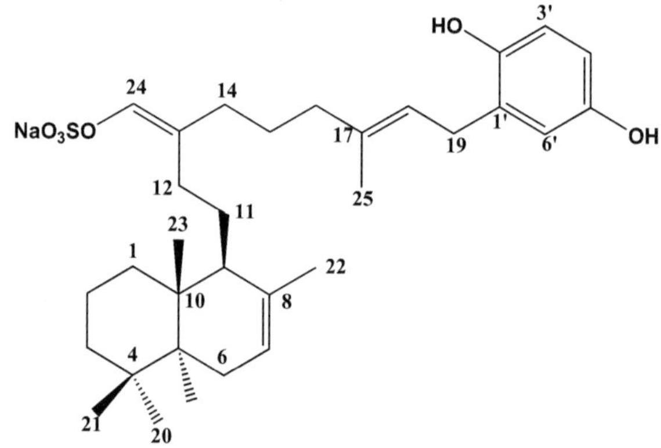

Hipposulphate A

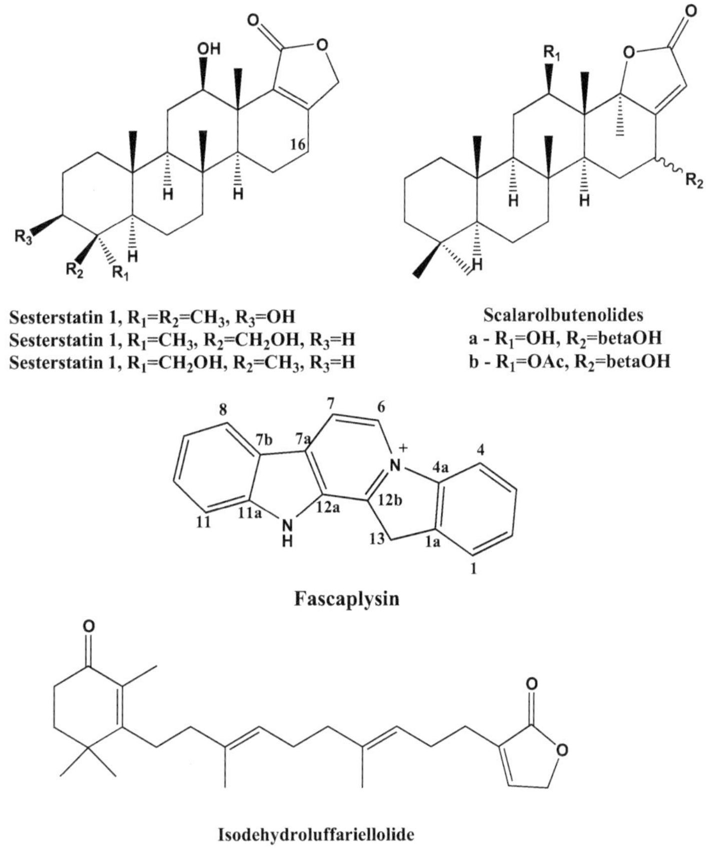

Fig. 11.28 Sesterstatins, isodehydroluffariellolide, homofascaplysin A, and fascaplysin

Adociasulfates Novel hexaprenoid hydroquinone sulfates, Adociasulfates, *Adocia aculeate*, *Adocia* sp., *Haliclona* (*aka Adocia*) sp. (Kalaitzis et al. 1999a, b; Blackburn et al. 1999) (Fig. 11.41) tyrosine, guanidine, isoquinoline, pyridine, purine, etc. The biologically active and structurally novel alkaloids have been taken for discussion.

11.3 Spongean Alkaloids

Spongean alkaloids are having unique structures different from that of terrestrial origins. They have heterocyclic structural units of bromopyrrole, pyrroloquinoline, pyrroloiminoquinone, bromoindole, cyclic amine linked to a β-carboline, imidazole, oxazoles, tryptophan,

11.3.1 Pyrroles and Bromopyrroles

Longamide B: Clathramides C and D (Fig. 11.42) were isolated from *Agelas dispar* (Cafieri et al. 1998). Debromosceptrin, *Agelas conifera* (Shen et al. 1998); nakamuric acid, 5-bromopyrrole-2-carboamide, 5-bromopyrrole-2-(*N*-methoxymethyl) carboxamide, isomer of

Fig. 11.29 Hyrtiolide, hydroxyscalarolide, hyrtial, tetronic acids, norsesterterpenes

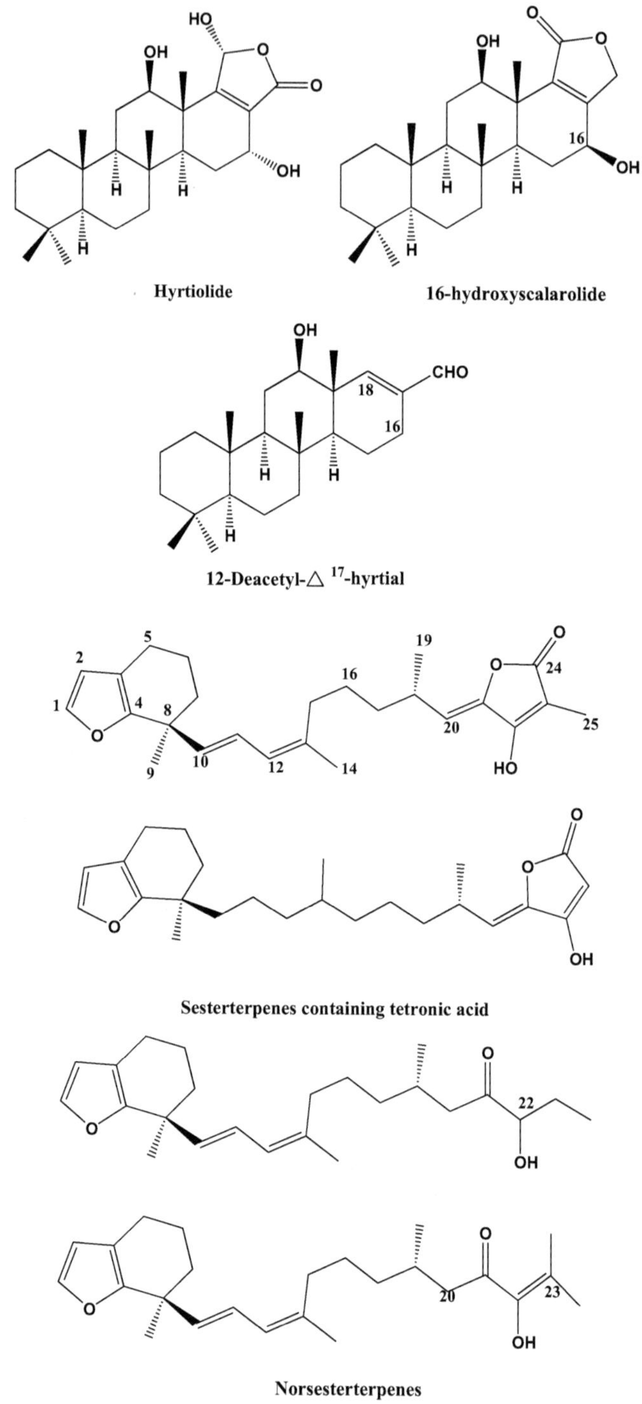

agelasine B, *Agelas nakamurai* (Eder et al. 1999; Iwagawa et al. 1998) (Fig. 11.43); 3-bromomaleimide, 3,4-dibromomaleimide, 12-chloro-11-hydroxydibromoisophakellin, and *N*-methylmanzacidin C, *Axinella brevistyla* (Tsukamoto et al. 2001); sventrin, *Agelas sventres* (Assmann et al. 2001) (Fig. 11.44).

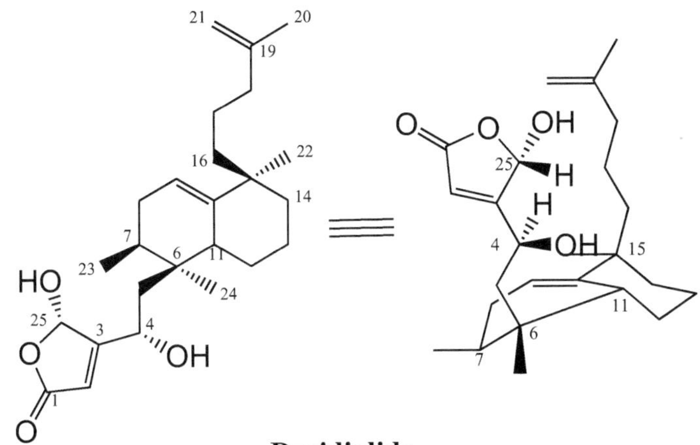

Dysidiolide

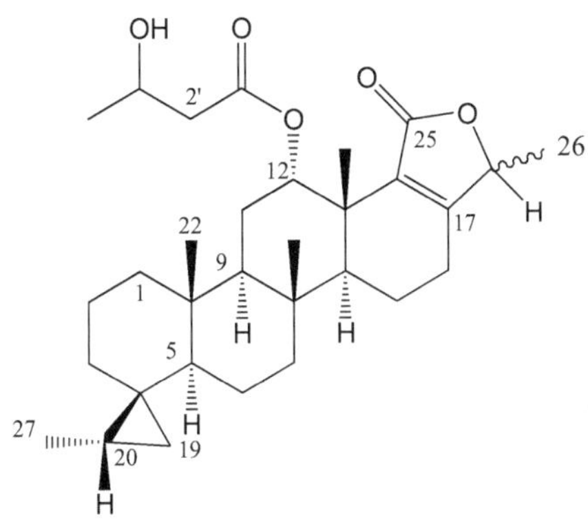

Honulactone A, 26_{beta}Me
Honulactone B, 26_{alpha} Me

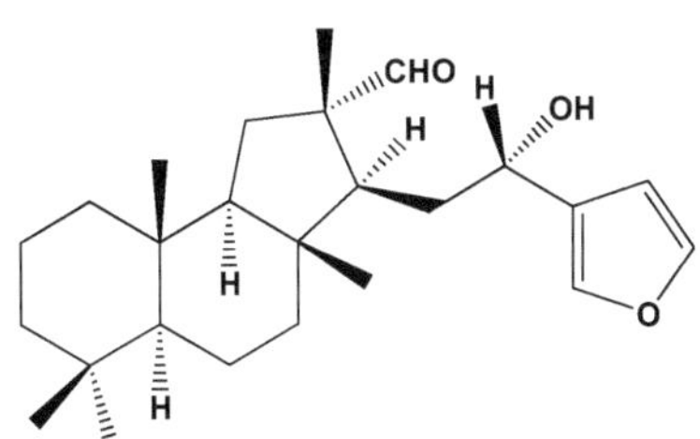

Hyrtiosal

Fig. 11.30 Dysidiolide, honulactones, hyrtiosal

11.3.2 Pyrroloquinoline and Pyrroloiminoquinone Alkaloids

Pyrroloquinoline alkaloids: Isobatzellines A, B, and C, *Zyzzya fuliginosa* (Venables et al. 1997); *Batzella* sp. (Gunasekera et al. 2003; Sun et al. 1990); batzellines C and D (Chang et al. 2002) (Fig. 11.45); discorhabdins L, P, and Q (spirocyclohexadienone skeleton) (Fig. 11.46) *Batzella* (Gunasekera et al. 1999a); *Latrunculia purpurea*, *Zyzzya massalis*, and *Zyzzya* spp. (Dijoux et al. 1999); *Negombata* and *Latrunculia* (Ford and Capon 2000; Copp et al. 1994); *Latrunculia brevis* (Reyes et al. 2004); secobatzellines A and B, *Batzella* (Gunasekera et al. 1999b).

Pyrroloiminoquinone alkaloids: Makaluvamines (Fig. 11.47), *Zyzzya* cf. *fuliginosa* (Casapullo et al. 2001); makaluvamine O, *Smenospongia* sp. (Tasdemir et al. 2002b); makaluvamines D, J, K, N, O, and P (Chang et al. 2002).

11.3.3 Bromoindoles

E/*Z* bromoindole esters, hymeniacidin (seco-xanthine formamides), *Hymeniacidon* sp. (Capon et al. 2002); bromoindole sulfonic acids, echinosulfonic acids A–C (Fig. 11.48), *Echinodictyum* sp. (Ovenden and Capon 1999); bis(indolyl)imidazoles, topsentin, bromotopsentin, *Spongosorites* sp. (Tsujii et al. 1988), *Rhaphisia lacazei* (Casapullo et al. 2000), nortopsentins A, B, and C, *Spongosorites ruetzleri* (Shinichi Sakemit et al. 1991; Carletti et al. 2000); brominooxindole alkaloid, *Iotrochota purpurea* (Hassan et al. 2004); *Spongosorites genitrix* (Shin et al. 1999) (Fig. 11.49).

Indole and pyrroloiminoquinone alkaloid: 5-bromo-l-tryptophan, 5-bromoabrine (5-bromo-N-methyl-l-tryptophan), and 5,6-dibromoabrine (5-bromo-N-methyl-l-tryptophan (Tasdemir et al. 2002b), *Smenospongia aurea* and *Smenospongia echina* (Djura et al. 1980); bis(indole) alkaloids, dragmacidin d, *Spongosorites* sp. (Wright et al. 1992); 6-bromo-2′-de-*N*-methylaplysinopsin, 6-bromoaplysinopsin and *N*-3′-ethylaplysinopsin, *Smenospongia aurea* (Hu et al. 2002); dragmacidins D and E, indole alkaloids, *Spongosorites* sp. (Capon et al. 1998); plakortamines A–D (Sandler et al. 2002); (Figs. 11.50 and 11.51).

11.3.4 Cyclic Amine Linked to β-Carbolines

This group includes important biopotent manzamine group of compounds which are the complex pentacyclic diamine linked to C-1 of a *β*-carboline moiety. Manzamine A hydrochloride, 32,33-dihydro-31-hydroxymanzamine A, 32,33-

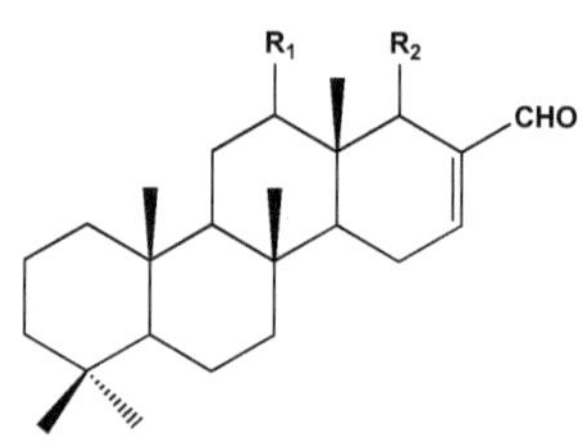

	R_1	R_2
Scalaradial -	12_{alpha}OAc	18_{beta}CHO
12-deacetoxyscalaradial -	H	18_{beta}CHO
18-epi-scalaradial -	12_{alpha}OAc	18_{alpha}CHO
19-dihydroscalaradial -	12_{alpha}OAc	$18_{beta}CH_2OH$

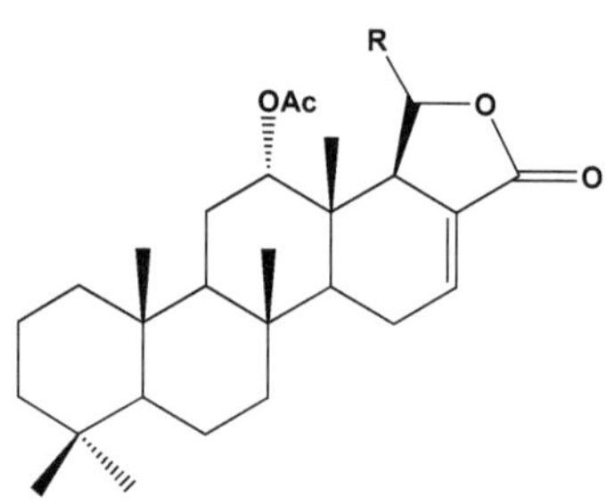

Scalarin - R = 12_{alpha}OH
19-deoxyscalarin- R = H

Fig. 11.31 Scalarane sestertepenes

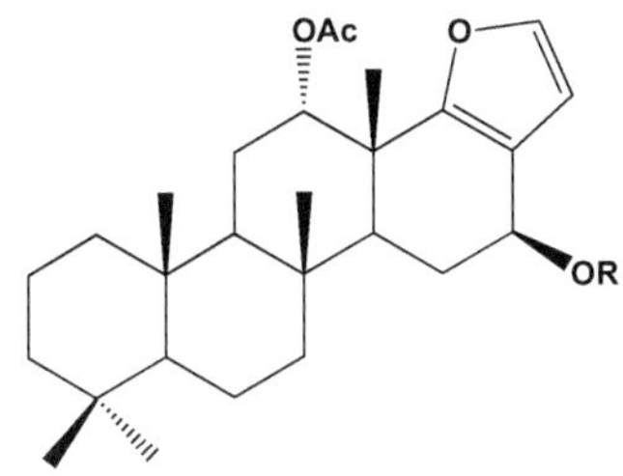

Furoscalarol - R = H
16-acetylfuroscalarol - R = Ac

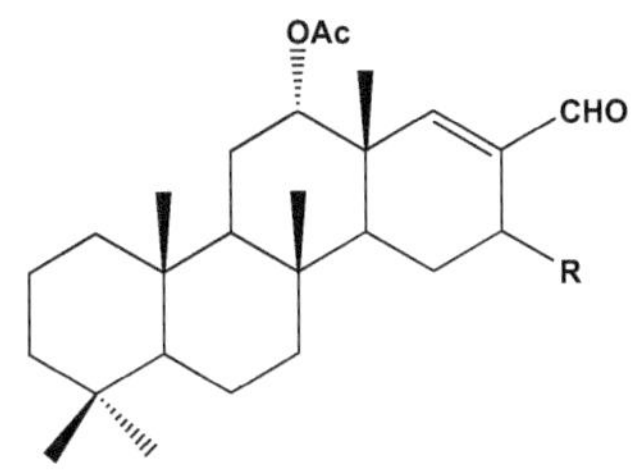

Norscalaral A- R = 12 $_{beta}$OH
Norscalaral B- R = 12 $_{alpha}$OH

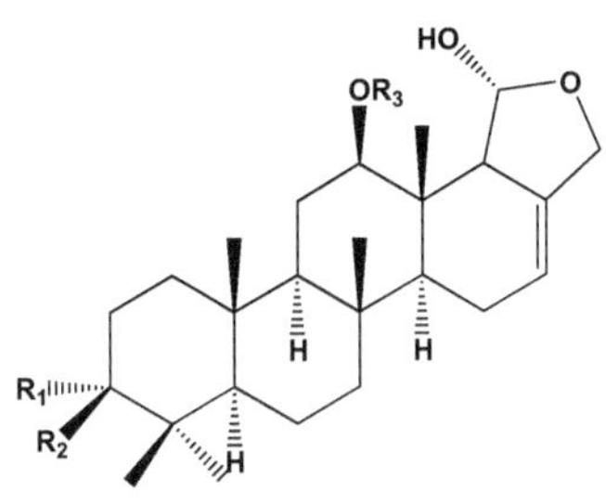

Scalarane sesterterpenes
a - R_1, R_2=O; R_3=H
b - R_1=H; R_2=OH
c - R_1=R_2=R_3=H

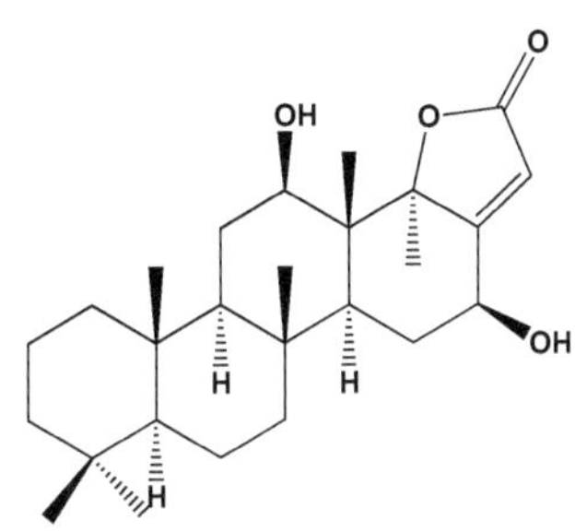

Scalarane sesterterpene d

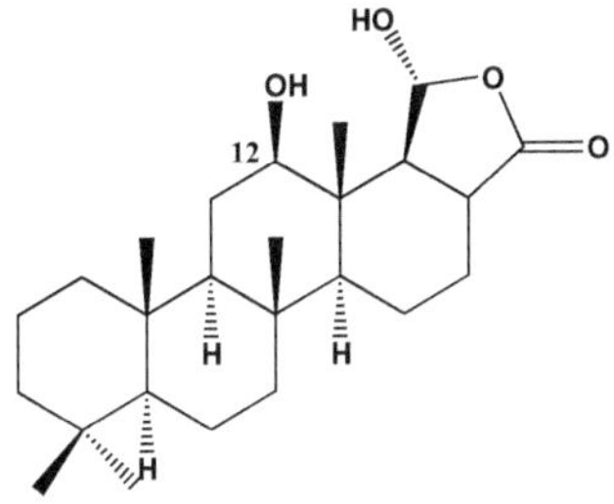

12-*O*-Deacetyl-12-*epi*-scalarin

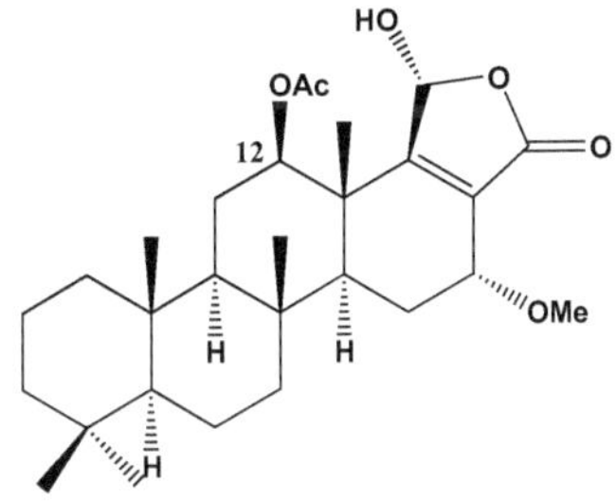

12-*O*-Acetyl-16-*O*-methylhyrtiolide

Fig. 11.31 (continued)

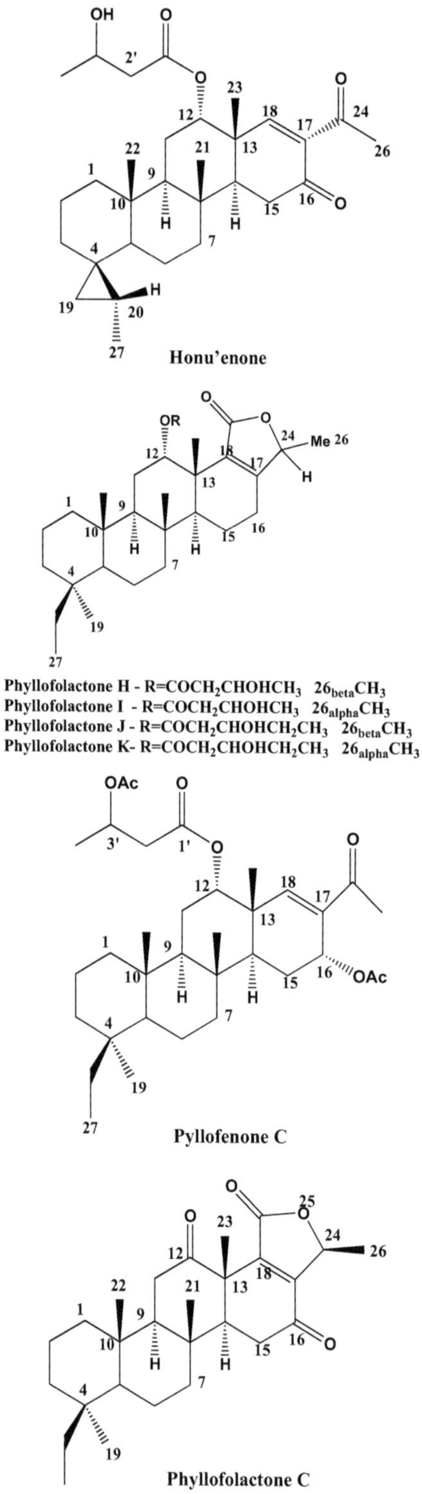

Fig. 11.32 Honu'enone, phyllofolactones

dihydro-6-hydroxymanzamine A-35-one, des-*N*-methylxestomanzamine A, *Haliclona*, *Prianos*, *Pachypellina*; *Xestospongia ashmorica* (Sakai and Higa 1986; Edrada et al. 1996a; El Sayed et al. 2001b; Rao et al. 2003); ircinal A, precursors of the manzamine alkaloids, *Ircinia* sp. (Kazuhiko Kondo et al. 1992); (Figs. 11.52 and 11.53).

11.3.5 Imidazoles and Oxazoles

Pyrroloaminopropylimidazole: Agelastatin A, *Cymbastela* sp. (Hong et al. 1998); acridine Alkaloid, dercitin (fused pentacyclic aromatic), *Dercitus* sp. (Gunawardana et al. 1988); bis-oxazoles, bengazole A, *Jaspidae* (Adamczeski et al. 1988); (Fig. 11.54).

11.3.5.1 Imidazole Alkaloids

Isonaamidine E, 5-{[4-(3,4-dimethoxybenzyl)-1-(4-methoxybenzyl)-1*H*-imidazol-2-yl]imino}-3-methyl-2,4-imidazolidinedione, naamine E, 5-{[2-imino-4-(4-methoxybenzyl)-1-methyl-1,2-dihydro-1*H*-imidazol -5-yl]methyl}-2-methoxy-1,3-benzenediol (Gross et al. 2002); (+)-calcaridine A, (−)-spirocalcaridine A (Fig. 11.55), *Leucetta* sp. (Edrada et al. 2003); kealiinine A, *Leucetta chagosensis* (Hassan et al. 2004); (2*E*,9*E*)-pyronaamidine 9-(*N*-methylimine), *Leucetta* sp. cf. *chagosensis* (Plubrukarn et al. 1997); pyronaamidine, kealiiquinone, *Leucetta* sp. (Akee et al. 1990) (Fig. 11.56).

11.3.6 Tryptophan and Tyrosine Alkaloids

Chelonins A and C, bromochelonin B, *Chelonaplysilla* sp. [Bobzin and Faulkner (1991)]; purealidin S, purpuramine J, *Druinella* sp. (Tabudravu and Jaspars 2002) (Fig. 11.57).

11.3.7 Isoquinoline Alkaloids

N-Formyl-1,2-dihydro-5-hydroxy-7-methoxyisoquinoline derivative, *O*-demethyl renierol

Fig. 11.33 Suberitenones, luffariolides mycaperoxide H

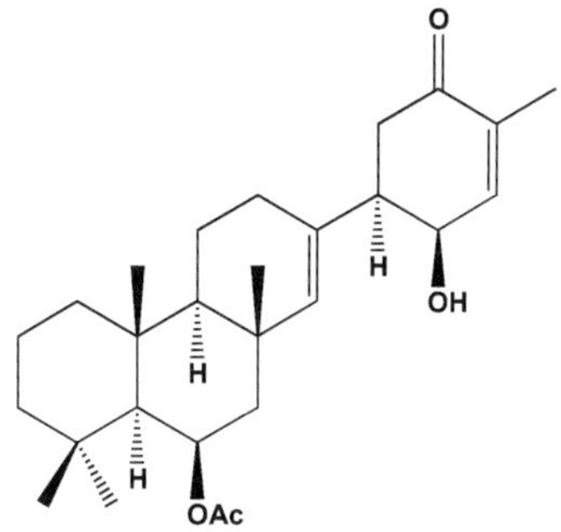

Suberitenone A

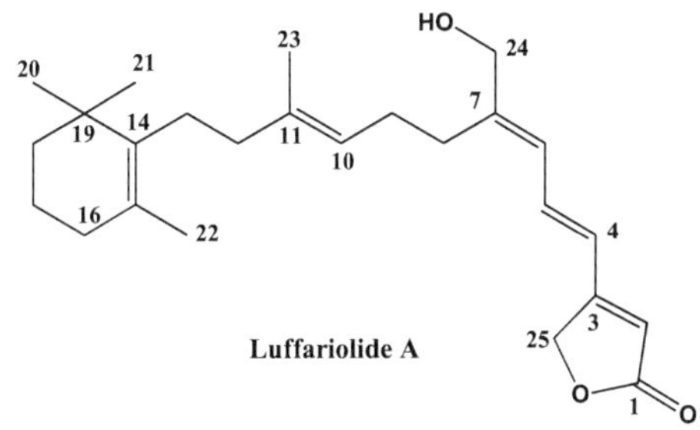

Luffariolide A

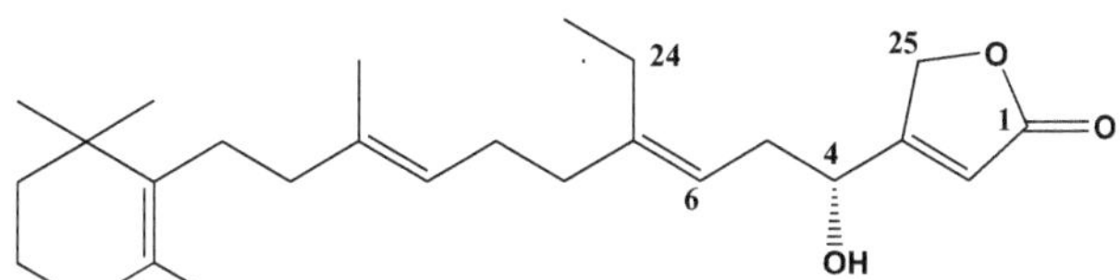

Luffariolide B

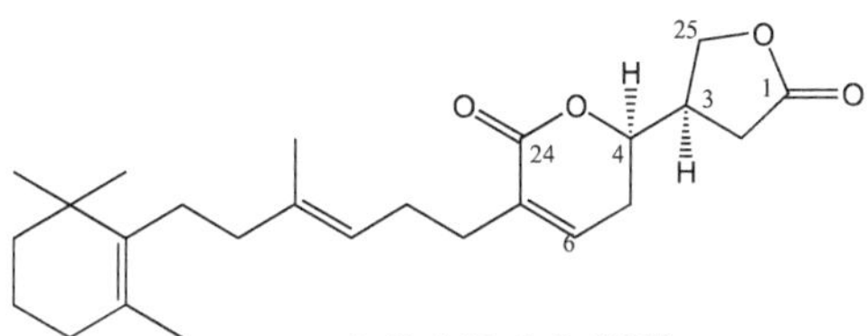

Luffariolide D, R=H,OH
Luffariolide E, R=O

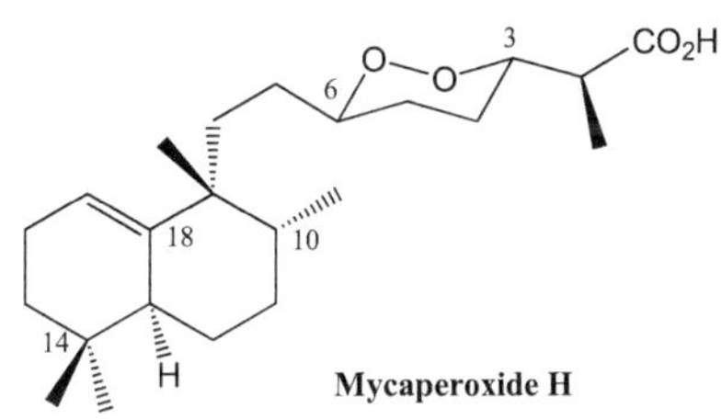

Mycaperoxide H

Fig. 11.34 Muqubilone, tasnemoxide A, petrosaspongiolides

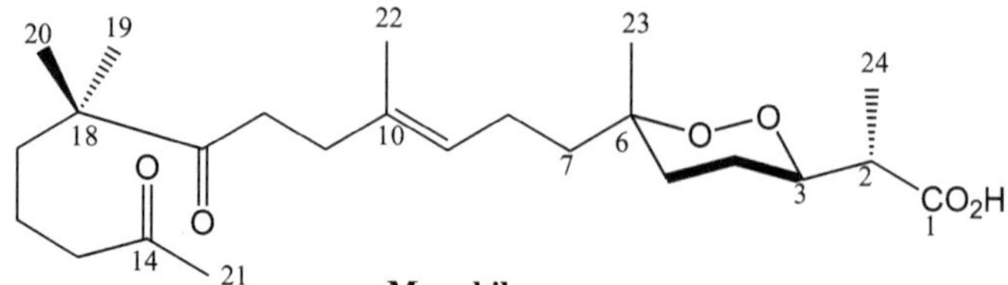

Muqubilone

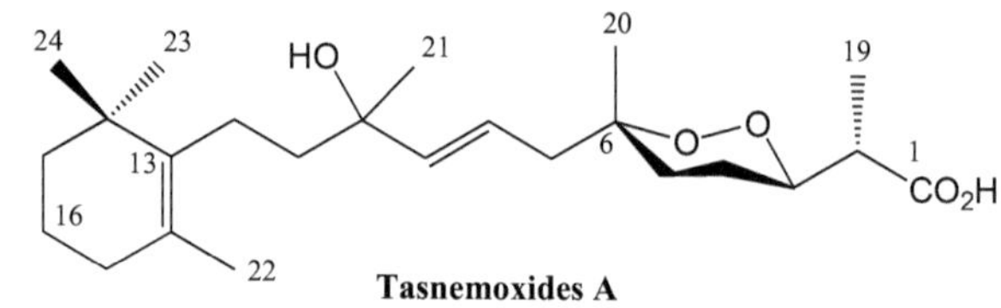

Tasnemoxides A

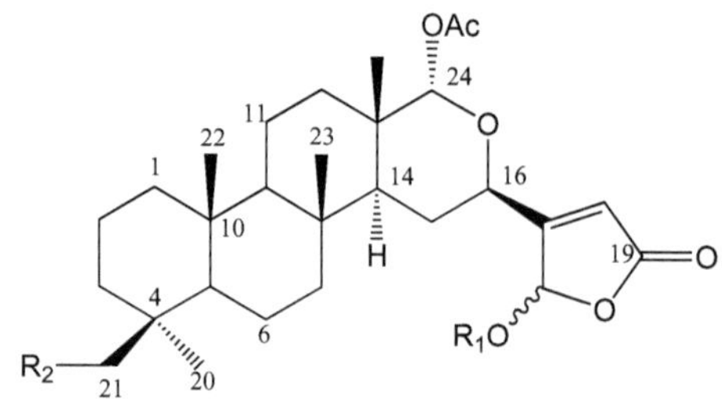

Petrosaspongiolide M, R_1,R_2=H
Petrosaspongiolide N, R_1,R_2=H

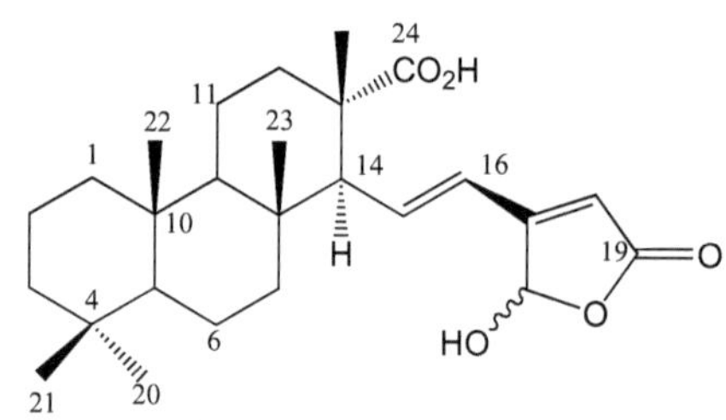

Petrosaspongiolide R

acetate, *Petrosia similis* (Ramesh et al. 1999); *Haliclona* sp. (Rashid et al. 2001); renieramycin J, tetrahydroisoquinoline alkaloid, potent cytotoxin, *Neopetrosia* sp. (Oku et al. 2003); isoquinoline quinines, renierone, *N*-ethylene methyl ketone derivative of renierone, 1,6-dimethyl-7-methoxy-5,8-dihydroisoquinoline-5,8-dione, mimosamycin, *Xestospongia* (Edrada et al. 1996b); 4.2 isoquinoline derivatives, cribrostatins 3, 4, 5, and 6 in 10^{-5}–10^{-7} % of the wet weight, *Cribrochalina* sp. (Pettit et al. 2000, 2003) (Figs. 11.58 and 11.59).

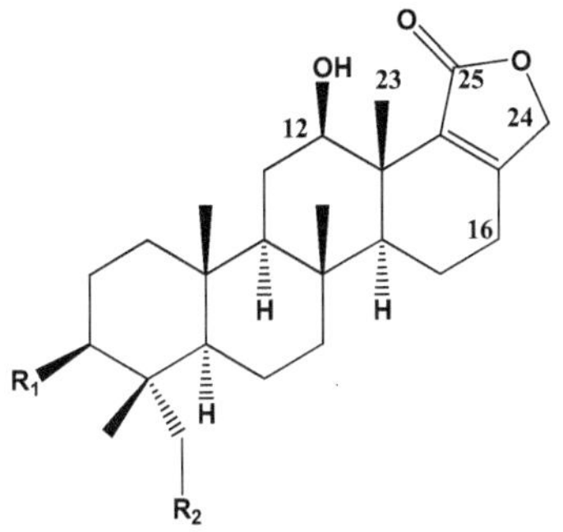

3-acetyl sesterstatin 1, R_1=H,R_2=OAc
19-acetyl sesterstatin 3, R_1=OAc,R_2=H
Scalarolide, R_1=H,R_2=H

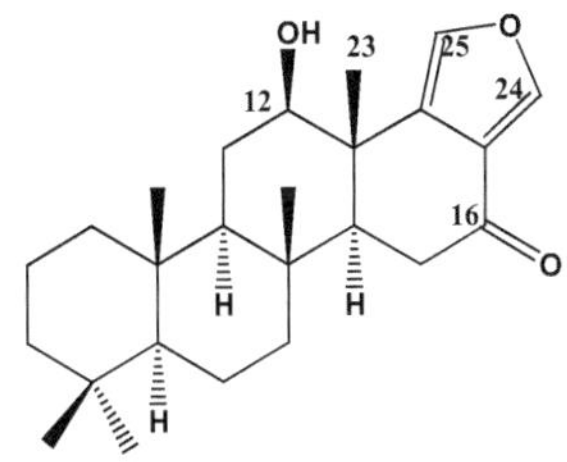

Salmahyrtisol B

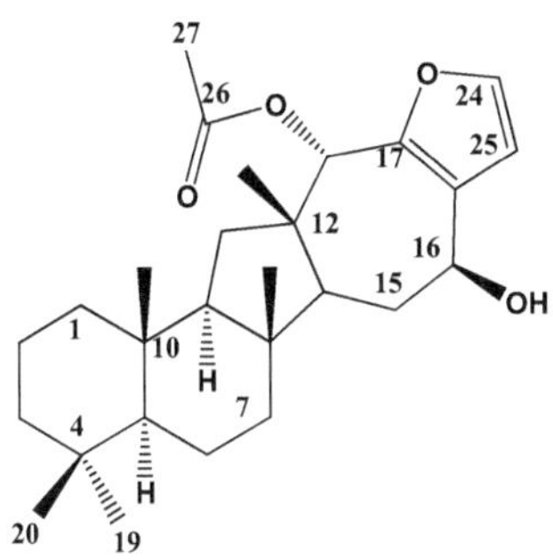

Salmahyrtisol A

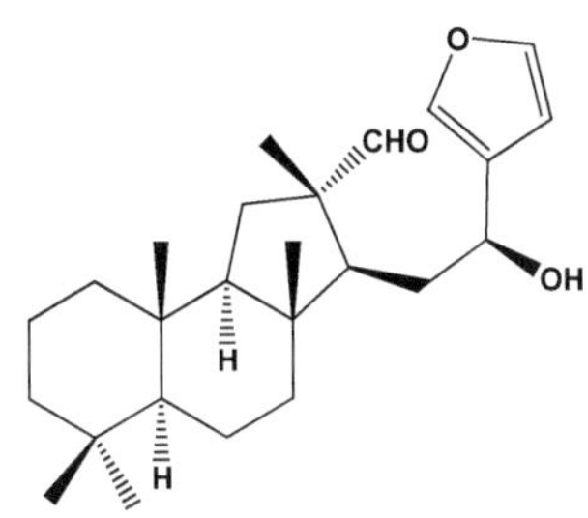

Hyrtiosal

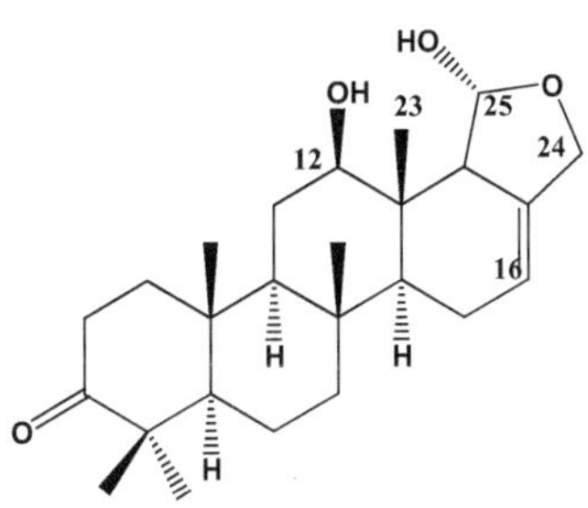

Salmahyrtisol C

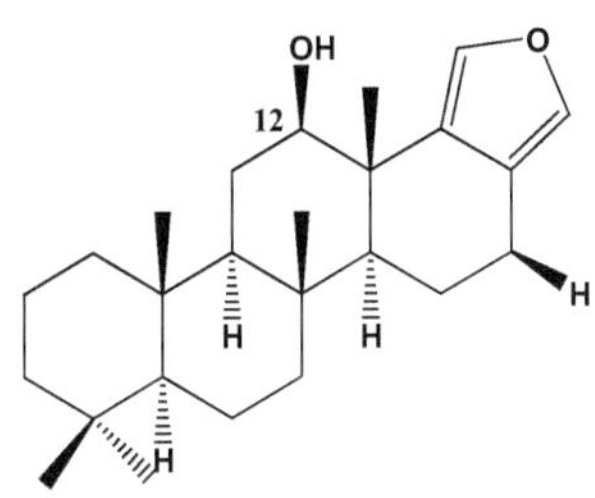

12-*O*-Deacetylscalarafuran

Fig. 11.35 Salmahyrtisol B, acetyl sesterstatins, hyrtiosal, 12-*O*-desacetylfuroscalar-16-one

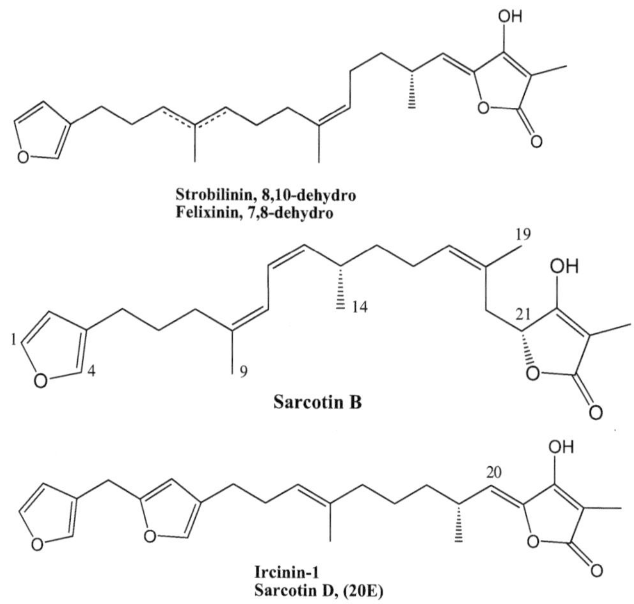

Fig. 11.36 Furanosesterterpenes

11.3.8 Guanidine Alkaloids

13,14,15-Isocrambescidin800 (pentacyclic guanidine), *Crambe crambe* (Jares-Erijman et al. 1993); batzelladines A, *Batzella* sp. (Patil et al. 1995); mirabilin G, *Clathria* sp. (Capon et al. 2001) (Fig. 11.60); hexacyclic bisguanidine, palau'amine, 4-bromo, and 4,5-dibromo derivatives of palau'amine, *Stylotella agminata* (Kinnel et al. 1993); styloguanidine which also has similar structural feature of palau'amine and its bromo derivatives, *Stylotella aurantium* (Kinnel et al. 1998); batzelladines F–I, *Batzella* sp. (Patil et al. 1997) (Fig. 11.61).

11.3.9 Sulfamate Indoles

Ancorinolates A–C, *Ancorina* sp. (Meragelman et al. 2002); 1-carboxymethylnicotinic acid, *Anthosigmella* cf. *raromicrosclera* (Matsunaga et al. 1998a); pyridoacridine alkaloids and kuanoniamines C and D, *Oceanapia* sp. Eder et al. 1998) (Fig. 11.62).

11.3.10 Indolizidine Alkaloid

Stellettamide B, stellettadine A, *Stelletta* sp. (Shin et al. 1997); geranylgeranyl moiety, stellettazole B, *Stelletta* (Matsunaga et al. 1999) (Fig. 11.63).

11.3.11 Steroidal Alkaloids

Plakinamines G and H, *Corticium* sp. (Borbone et al. 2002); lokysterolamine A, plakinamine E (Lee et al. 2001); motuporamines A and C, saturated 15-membered cyclic amine, *Xestospongia exigua* (Williams et al. 2002); spermidine, motuporamines A and C, petrosin, and xestospongin/araguspongine class of 3-alkylpiperidine alkaloids, *Xestospongia exigua* (Williams et al. 1998) (Fig. 11.64).

Fig. 11.37 Pyrrolo- and furanosesterterpenoids

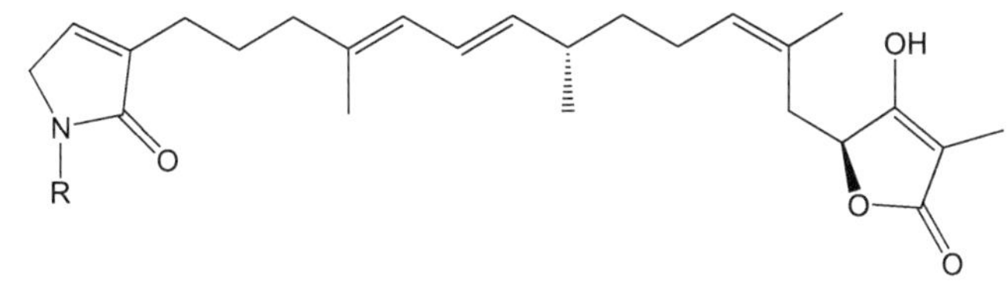

Sarcotrine A, (21S), R= 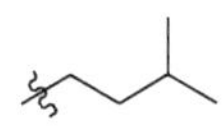 ***epi*-Sarcotrine A (21R), R=** 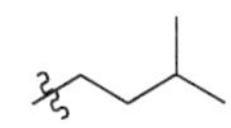

Sarcotrine B, (21S), R= 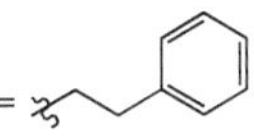 ***epi*-Sarcotrine B (21R), R=** 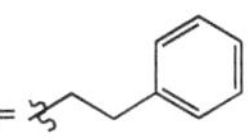

Sarcotrine C, (21S), R= 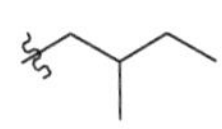 ***epi*-Sarcotrine B (21R), R=** 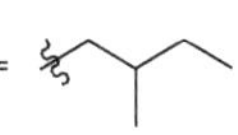

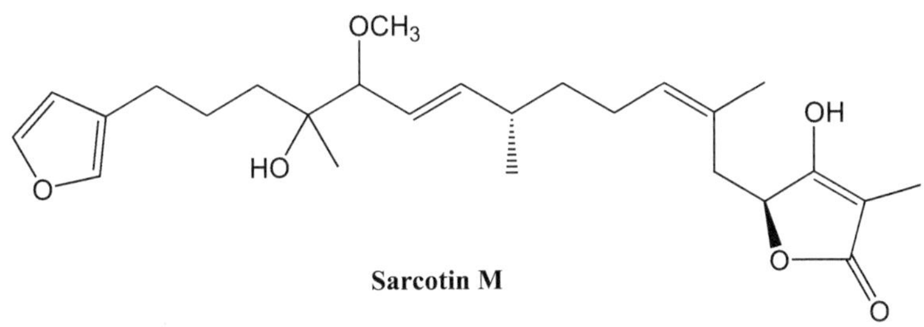

Sarcotin M

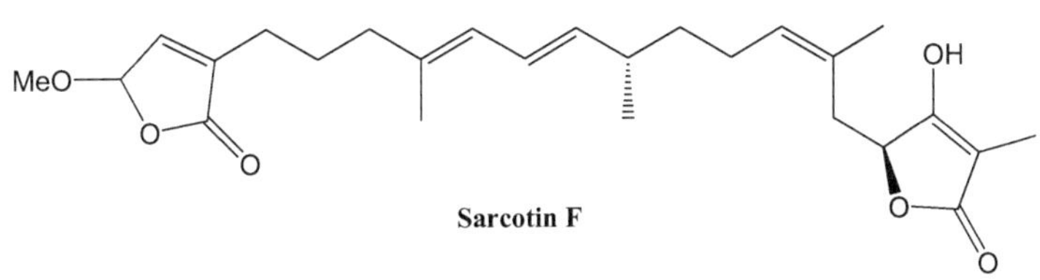

Sarcotin F

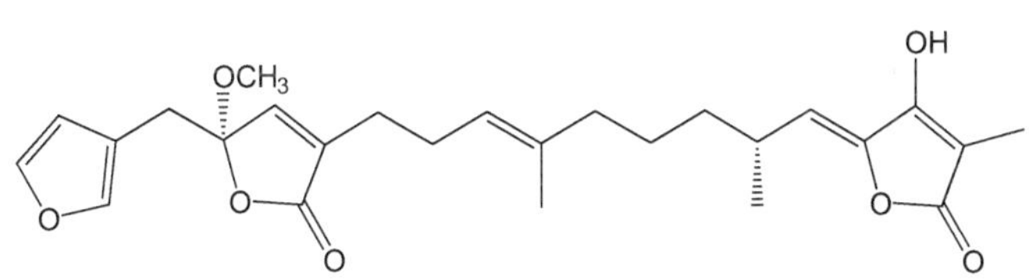

Sarcotin G

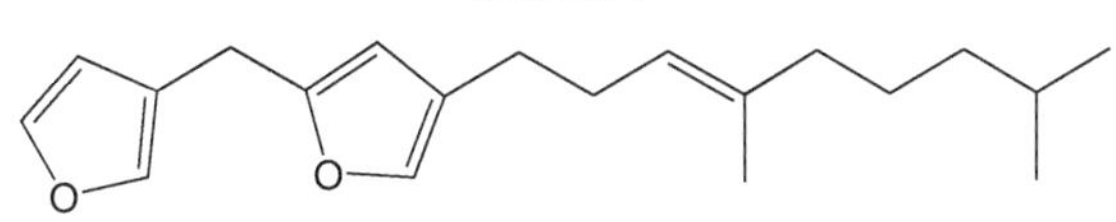

Sarcotin K

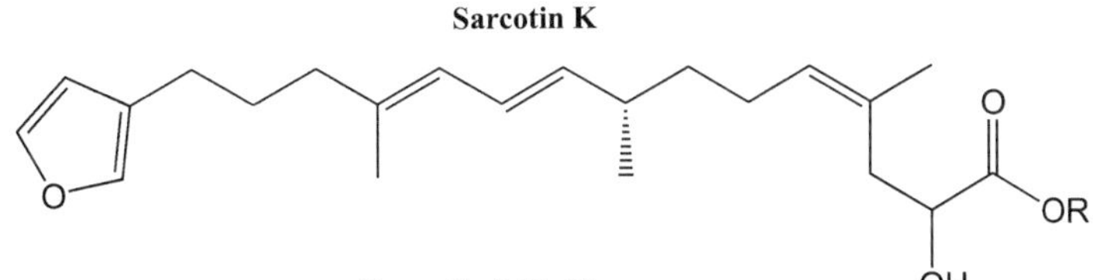

Sarcotin I, R=Na
Sarcotin J, R=CH_3

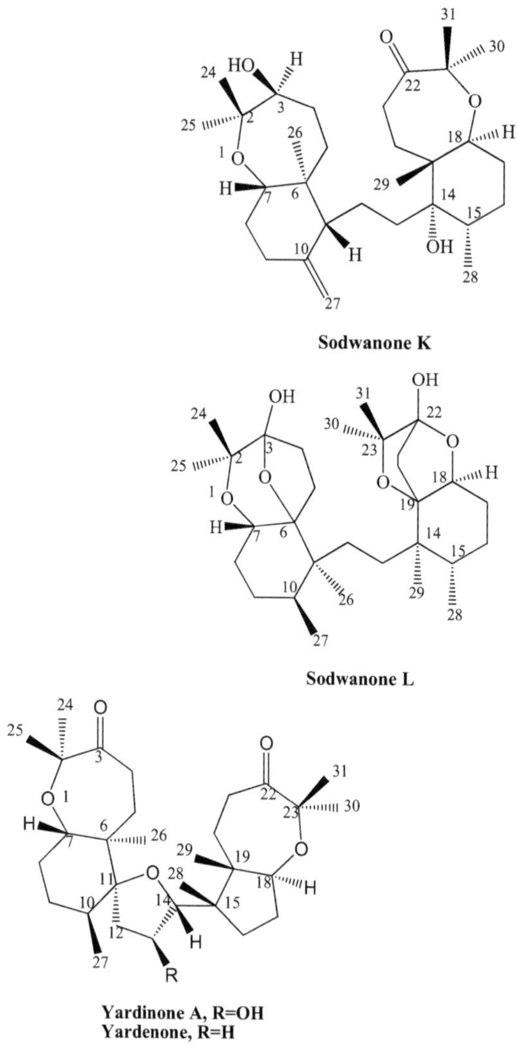

Fig. 11.38 Sodwanone, yardenone

11.3.12 3-Alkylpyridine Alkaloids

Hachijodines C, E, and G, *Xestospongia* and *Amphimedon* (Tsukamoto et al. 2000); amphimedine, pyridoacridines, *Xestospongia* sp., *Xestospongia* cf. *carbonaria* (de Guzman et al. 1999) (Fig. 11.65).

11.4 Heterocycles

11.4.1 Pyrrole Derivatives

Mauritiamine, oroidin, and 4,5-dibromopyrrole-2-carbamide, *Agelas mauritiana* (Tsukamoto et al. 1996).

11.4.2 Bengamides

Bengamide L, *Pachastrissa* sp. [146]; bengamides Y and Z, *Jaspis* sp. (Groweiss et al. 1999); (Fig. 11.66).

11.4.3 Purine and Nucleoside Metabolites

Erinacea and *p*-hydroxybenzaldehyde, *Isodictya erinacea* (Moon et al. 1998).

11.4.4 Aaptamines

Demethyloxyaaptamine and aaptamine, *Hymeniacidon* sp. (Pettit et al. 2004) (Fig. 11.67).

11.4.5 Asmarines

A–F, methyl 3-oxo-cholan-24-oate, *Raspailia* sp. (Yosief et al. 2000); fijianolides A and B, heterocyclic macrocyclic lactones (polyketide), *Spongia mycofijiensis*, 10 (Quiñoá et al. 1988);

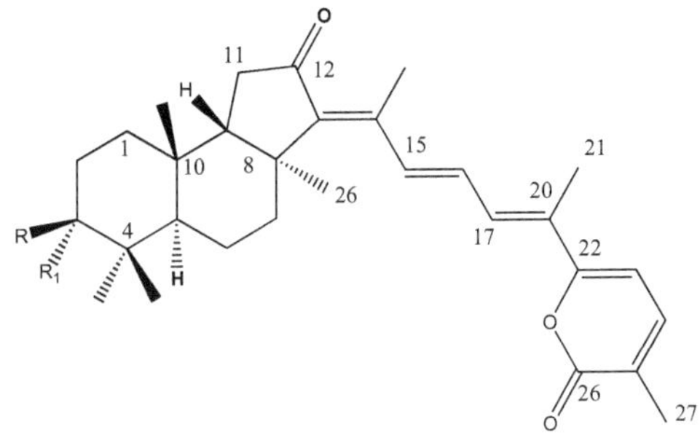

Stellettin A, R_1,R_2=O
Stellettin C, R=OAc, R_1=H

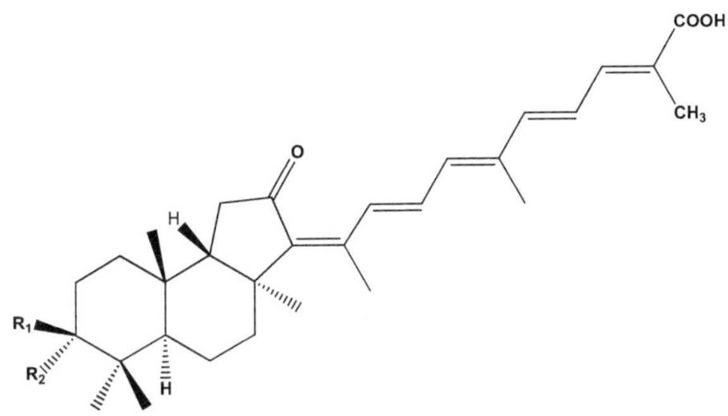

Stellettin E, R_1&R_2=O
Stellettin I, R_1=OAc, R_2=H

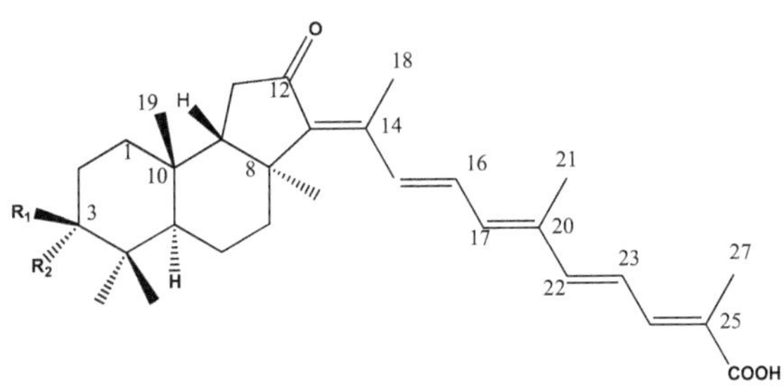

Stellettin H, R_1=OCOCH$_3$, R2=H
Rhabdastrellic acid-A, R_1=R_2=O

Fig. 11.39 Stellettins, rhabdastrellic acid

fascaplysin, *Fascaplysinopsis* sp. (Roll et al. 1988) (Fig. 11.68).

11.4.6 Bengazoles

Bengazoles are homologous fatty acid esters of a heterocyclic nucleus comprised of a bis(oxazolyl)-methanol further substituted with a hexanetetrol side chain, reminiscent of a carbohydrate analogue. Bengazoles (C–G), known bengazoles A and B, comprise a homologous series of *n*, *iso*, and *anteiso* fatty acid esters (C13–C16) of the same heterocyclic bis-(oxazolyl)-methanol parent, *Jaspis* sp., 11 (Searle et al. 1996); bengazoles (1–6), bengamide L, *Pachastrissa* sp. (Fernández et al. 1999) Microxine, purine derivative, *Microxina* sp., 12 (Killday et al. 2001) (Fig. 11.69).

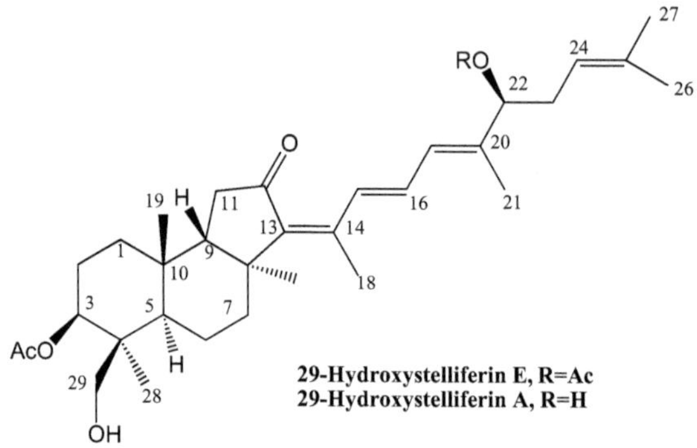

29-Hydroxystelliferin E, R=Ac
29-Hydroxystelliferin A, R=H

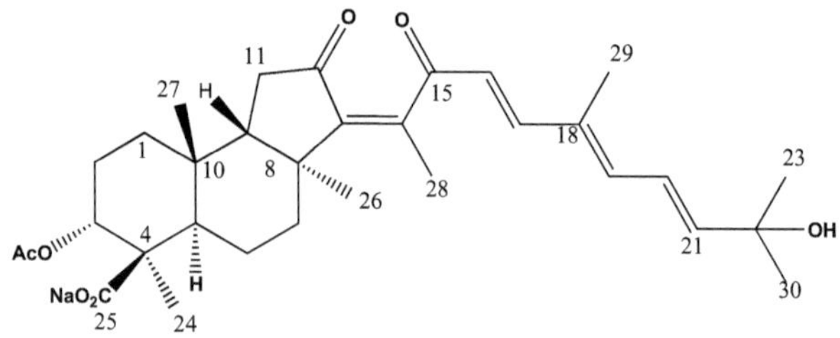

Globostellatic acid A

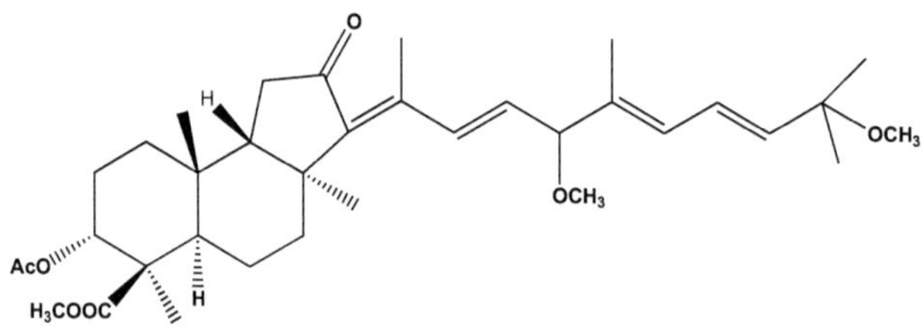

Globostellatic acid E methyl ester

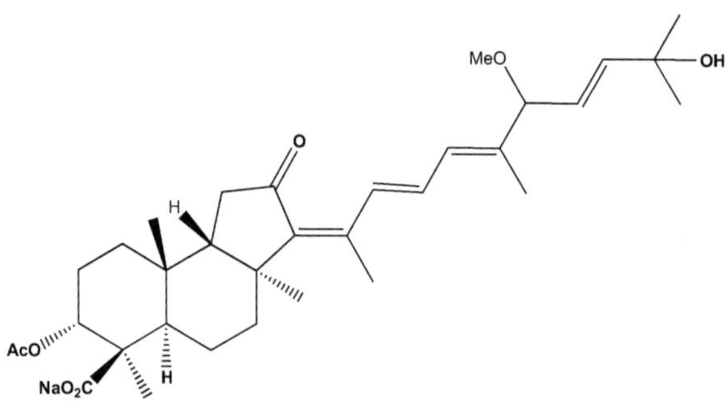

Globostellatic acid B

Fig. 11.40 Hydroxystelliferin, globostellatic acids

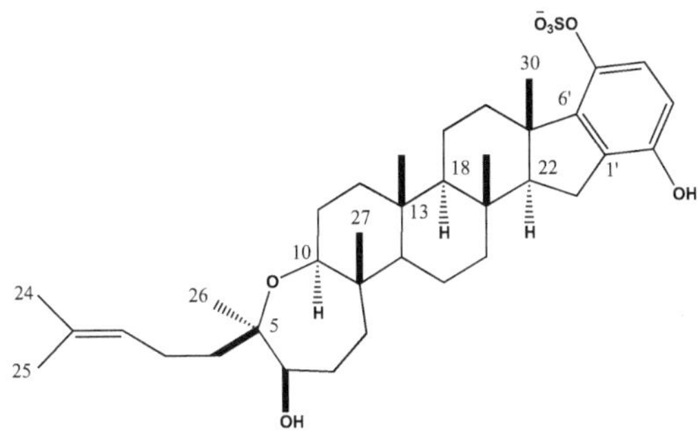

Adociasulfate-9

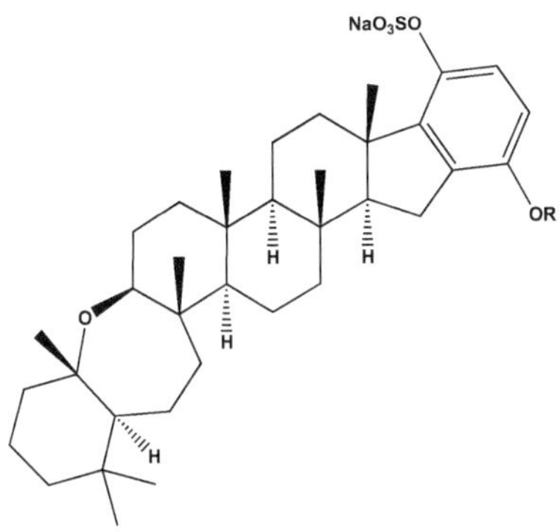

Adociasulfate-2, R=SO_3Na
Adociasulfate-6, R=H

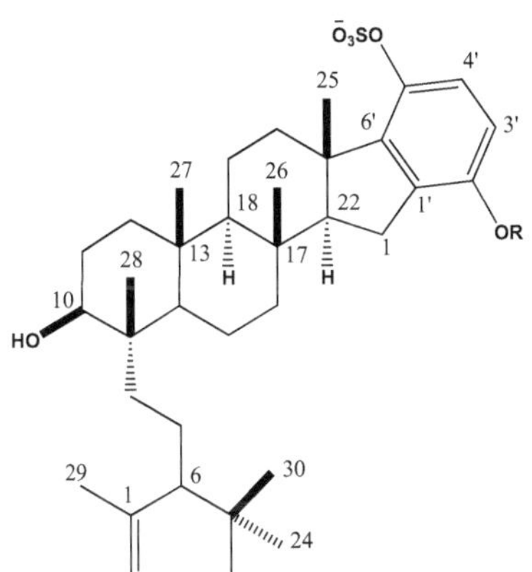

Adociasulfate-1, R=SO_3Na
Adociasulfate-5, R=H

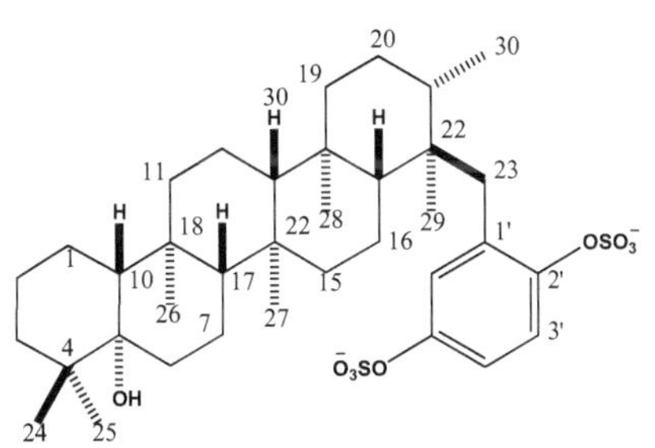

Adociasulfate-8

Fig. 11.41 Adociasulfates

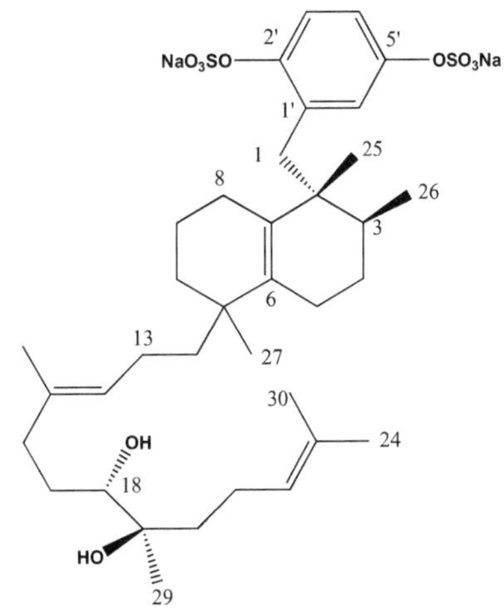

Adociasulfate 3

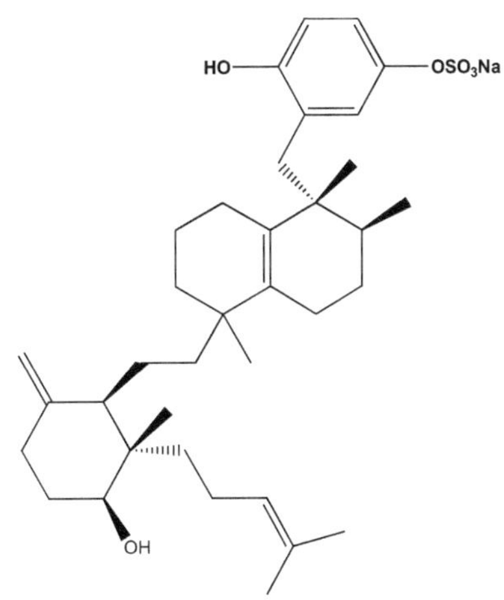

Adociasulfate 4

Fig. 11.41 (continued)

11.4.7 Nitrogen Heterocyclic Constituents

Dibromophakellstatin, dibromophakellin, thymidine, debromohymenialosine, deoxyuridine, and thymine. *Phakellia mauritiana*, 13 (Pettit et al. 1997) (Fig. 11.70).

11.5 Polyacetylenes, Polyethers, Polyketides, Peptides, and Macrolides

This group includes long-chain polyacetylenes, polyethers, macrolides, peptides, and polyketides with multiple functional group substitutions.

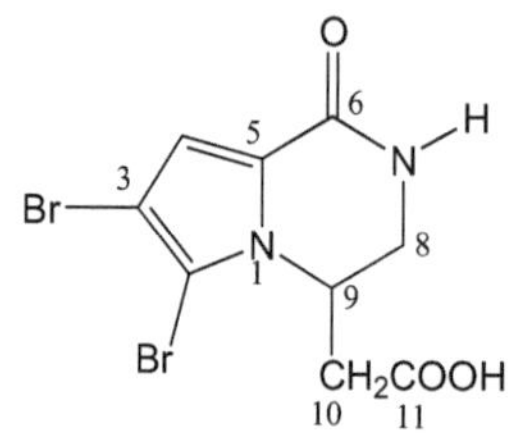

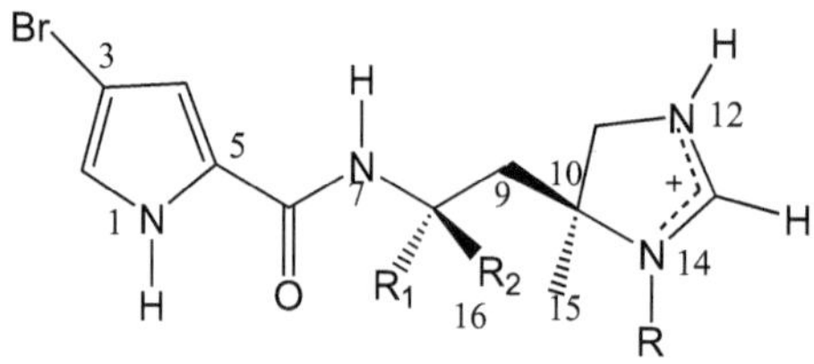

Fig. 11.42 Longamide B, clathramides C and D

Only compounds with tested activity are covered. Representative compounds with novel structural feature under each type are given with the sponge sources.

11.5.1 Polyacetylenes

Polyacetylenes include long-chain fatty compounds with alcohol functional group.

Callyspongenol A, C_{22}-polyacetylenic alcohol, *Callyspongia* sp. [Youssef et al. (2003)]; callypentayne, *Callyspongia truncata* (Sachiko Tsukamoto et al. 1997); vasculyne, C_{43} acetylenic alcohol, *Cribrochalina vasculum* (Dai et al. 1996) (Fig. 11.71).

Diplynes A and B, diplyne A 1-sulfate, *Diplastrella* sp. (Lerch 2003); durissimol B, *Strongylophora durissima* (Shen and Prakash 2000); C_{46} polyacetylenic alcohols, petrocortyne A, *Petrosia* sp. (Kim et al. 1999) (Fig. 11.72).

C_{14} acetylenic acid, *Oceanapia* sp. (Matsunaga et al. 2000); brominated acetylenic fatty acid, sterol esters, *Xestospongia testudinaria* (Pham et al. 1999); polyacetylene sulfates, callyspongins A and B, *Callyspongia truncata* (Uno et al. 1996); fulvinol, *Reniera fulva* (Ortega et al. 1996) (Fig. 11.73).

11.5.2 Calyculinamide-Related Compounds

Geometricin A, *Luffariella geometrica* [Kehraus et al. (2002a)]; 1-methylherbipoline salts of halisulfate-1 and of suvanine, *Coscinoderma mathewsi* (Kimura et al. 1998) (Fig. 11.74).

11.5.3 Polybrominated Diphenyl Ethers

Phyllospongia dendyi (Liu et al. 2004). The effect of polybrominated ethers on cell division of the fertilized eggs of marine organisms was studied. Some of the organisms are sea urchins and star fish (Fig. 11.75).

11.5.4 Calyculins

Calyculin J: It is a spiroketal of an unprecedented skeleton bearing phosphate, oxazole, nitrile, and amide functionalities, *Discodermia calyx* (Matsunaga et al. 1997); *Hamigera tarangaensis* (Wellington et al. 2000) (Fig. 11.76).

11.5.5 Taurospongin A

It is an acetylene-containing natural product consisting of a taurine and two fatty acid residues, *Hippospongia* sp. (Ishiyama et al. 1997).

11.5.6 Butenolides

It is a cyclopentenone derivative, *Homaxinella* sp. (Mansoor et al. 2004) (Fig. 11.77).

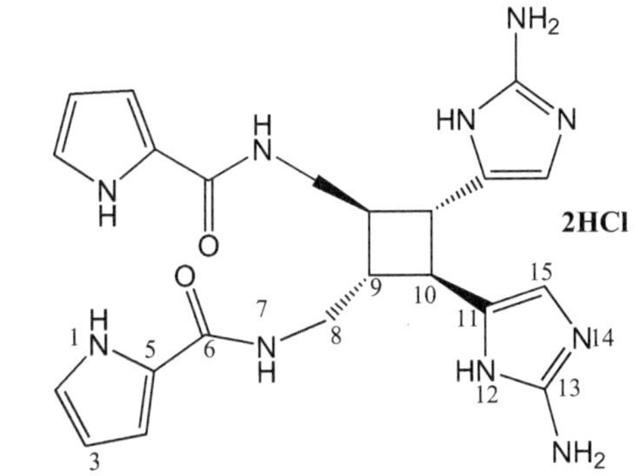

Debromosceptrin

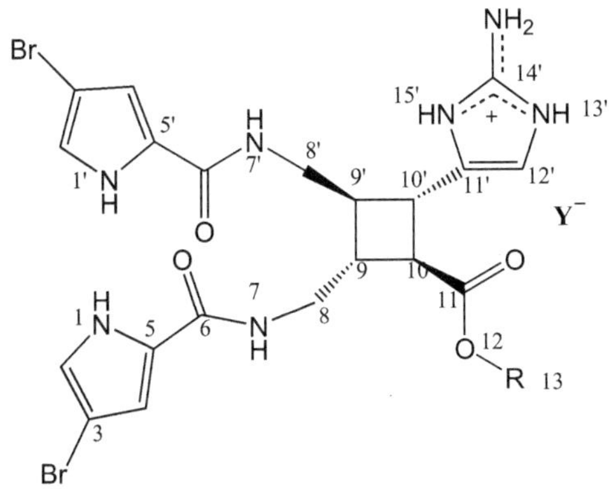

Nakamuric acid, R=H, Y=CF_3COO^-
Nakamuric acid Meester, R =CH_3, Y=CF_3COO^-

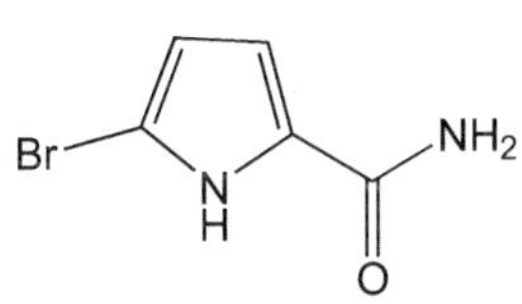

5-Bromopyrrole-2-carboamide

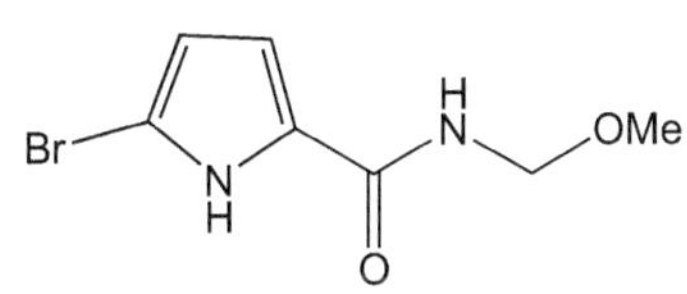

5-Bromopyrrole-2-(N-methoxymethyl) carboxamide

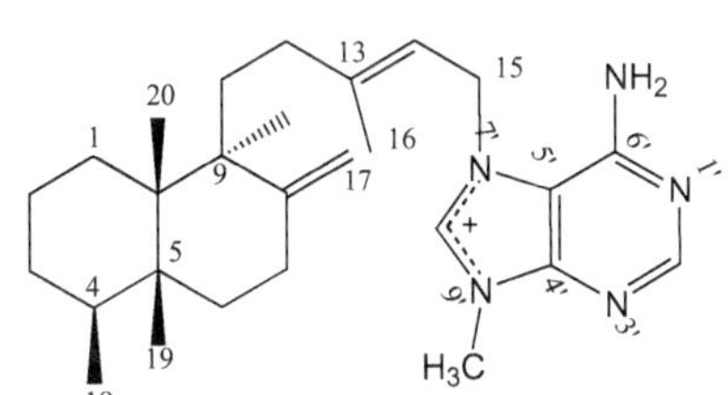

Agelasine B isomer

Fig. 11.43 Debromosceptrin, agelasine

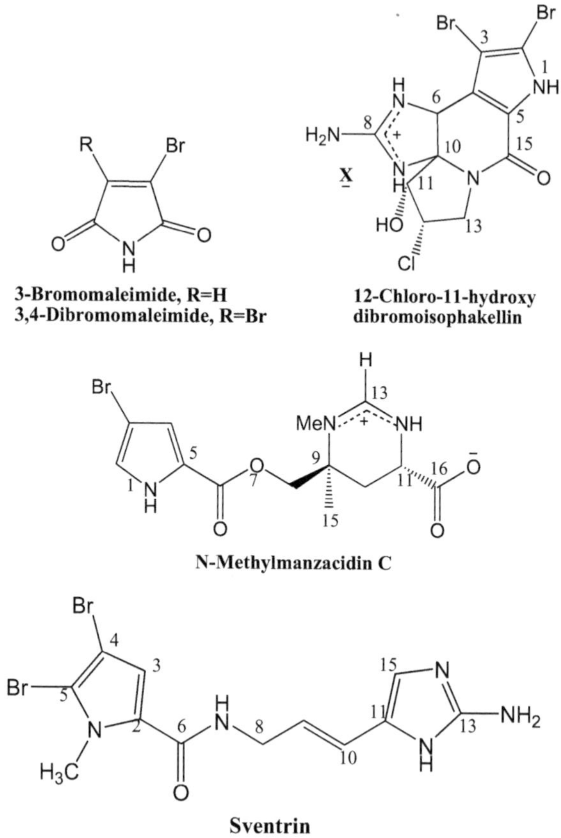

Fig. 11.44 3-bromomaleimide, 3,4-dibromomaleimide, 12-chloro-11-hydroxydibromoisophakellin, *N*-methylmanzacidin C, sventrin

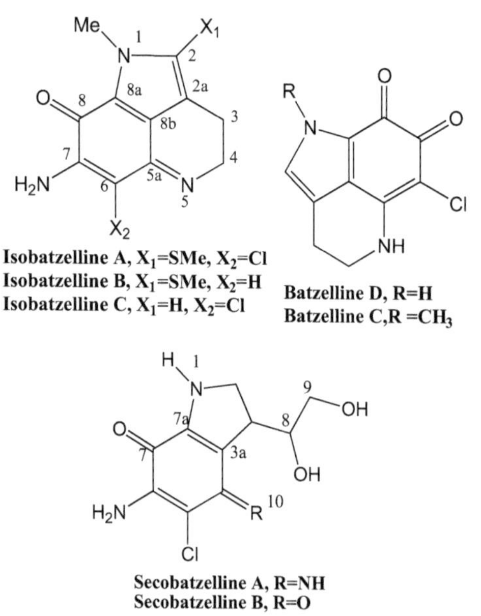

Fig. 11.45 Isobatzellines, batzellines

11.5.7 Dysiherbaine

It is a *cis*-fused hexahydrofuro[3,2-*b*]pyran ring substituted with a 3-[2-aminopropanoic acid] side chain, *Dysidea herbacea* (Sakai et al. 1997); halenaquinone, pentacyclic polyketide, *Xestospongia exigua* (Roll et al. 1983) (Fig. 11.78).

11.5.8 Discodermolide

It is a polyhydroxylated lactone, *Discodermia dissoluta* (Gunasekera et al. 1990); acetylenic enol ethers of glycerols of the yne-diene series, linear acetylenic alcohol, *Petrosia* (Seo et al. 1999) (Fig. 11.79).

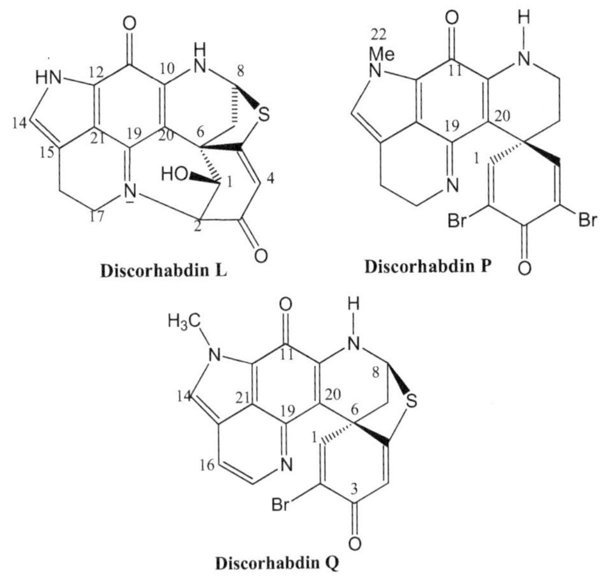

Fig. 11.46 Discorhabdins

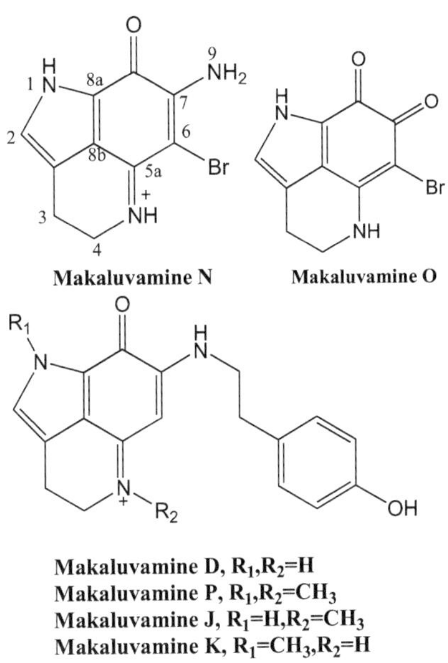

Fig. 11.47 Makaluvamines

11.5.9 Polyether Macrolide

Homohalichondrin B, axinastatin 1 (Fig. 11.80), cycloheptapeptides, *Axinella* sp. (Pettit et al. 1994); geodin A Mg salt, macrocyclic polyketide lactam tetramic acid magnesium salt, *Geodia* (Capon et al. 1999); mycalamide D, *Stylinos* n. sp. (Simpson et al. 2000; West et al. 2000) (Fig. 11.81).

Homo-plakotenin, *Plakortis lita* (Qureshi et al. 1999); tridecanoate and pentadecanoate analogues of bengamides A and B, *Jaspis carteri* (D'Auria et al. 1997); melophlins C, tetramic acid derivatives, *Melophlus sarassinorum* (Wang et al. 2003) (Fig. 11.82).

Fig. 11.48 *E*/*Z* bromoindole esters, *Hymeniacidon*, bromoindole sulfonic acids: echinosulfonic acids

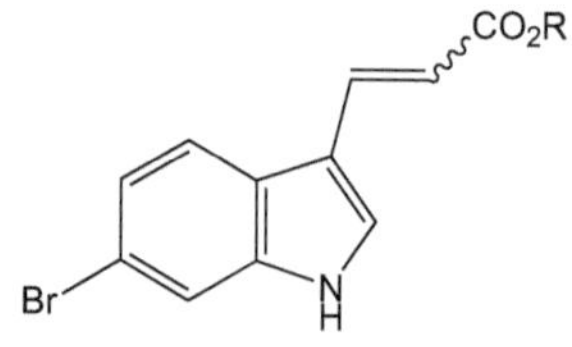

E-Bromoindole ethylester, R=Et
Z-Bromoindole ethylester, R=Et
E-Bromoindole methyl ester, R=Me
Z-Bromoindole methyl ester, R=Me

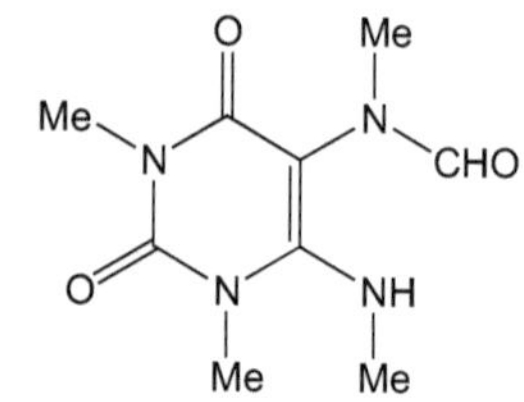

Hymeniacidin

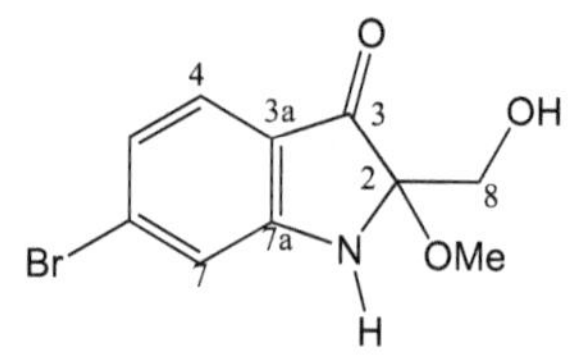

6-bromo-2-methoxy-2-hydroxy-methane-3-indolinone

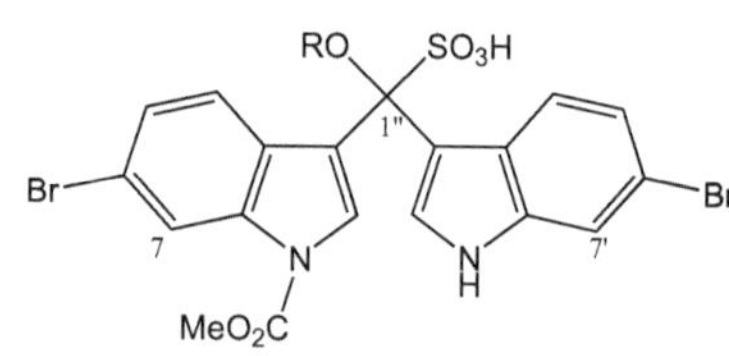

Echinosulfonic acid A, R=Et
Echinosulfonic acid B, R=Me
Echinosulfonic acid C, R=H

11.5.10 Clavosines

Clavosines A and B are closely related to calyculins and calyculinamides, *Myriastra clavosa* (Fu et al. 1998b); plakortides N, polyketide endoperoxides, *Plakortis halichondrioides* (Jiménez et al. 2003) (Fig. 11.83).

Andavadoic acid, *Plakortis aff simplex* (Rudi et al. 2003); plakoside B: It is a unique glycosphingolipid (prenylated glycolipid), *Plakortis simplex* (Costantino et al. 1997). Ethyl didehydroplakortide Z, *Plakortis lita* (Blaine Harrison and Phillip Crews 1998); spongiadioxins A, tetrabromodibenzo-*p*-dioxins, *Dysidea dendyi* (Utkina et al. 2001); (Fig. 11.84). Bitungolides A–D, polyketides, *Theonella* cf. *swinhoei* (Sirirath et al. 2002).

11.5.11 Macrocyclic Lactone/Lactams

Celenamide E, *Amphimedon* spp. (Ovenden et al. 1999); scleritodermin A, cyclic peptide, *Scleritoderma nodosum* (Schmidt et al. 2004) (Figs. 11.85 and 11.86).

11.5.12 Bicyclic Peptides

Aciculitins A–C contain an unusual histidino-tyrosine bridge with attachment to the bicyclic peptide. They are C_{13}-C_{15} 2,3-dihydroxy-4,6-dienoic acids bearing D-lyxose at the 3-position, *Aciculites orientalis* (Bewley et al. 1996); cyclotheonamides E4 and E5, *Ircinia* (Murakami et al. 2002); dysinosin A: Distinctive

Fig. 11.49 Topsentin, bromotopsentin, nortopsentins

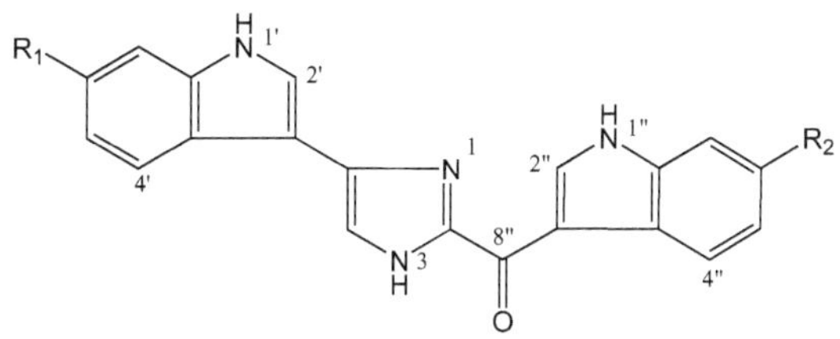

Topsentin A, R_1=H, R_2=H
Topsentin B1, R_1=H, R_2=OH
TopsentinB2, R $_1$=Br, R_2=H

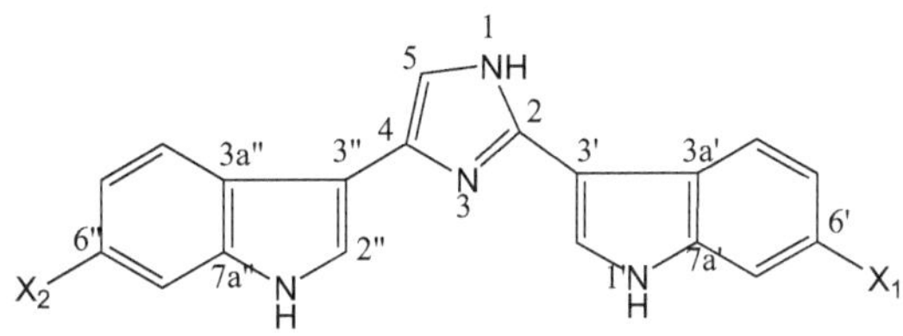

Nortopsentin A,X $_1$=X_2=Br
Nortopsentin B, X_1 =Br, X_2=H
Nortopsentin C,X $_1$=H, X_2=Br

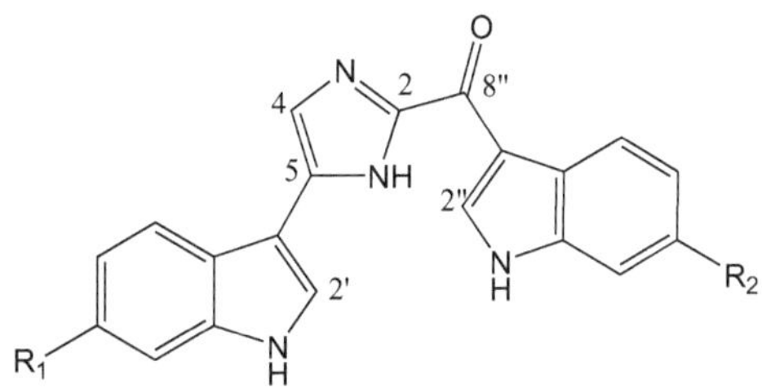

Bromodeoxytopsentin, R_1=Br, R_2=H
Isobromodeoxytopsentin, R_1=H, R_2=Br
Deoxytopsentin, R_1= R_2=H
Bromotopsentin, R_1=Br, R_2=OH

features of dysinosin A are the presence of a 5,6-dihydroxyoctahydro-indole -2-carboxylic acid, 3-amino-ethyl 1-*N*-amidino-Δ-3-pyrroline, a sulfated glyceric acid, and D-leucine, assembled through three peptidic linkages (Carroll et al. 2002); geodiamolides A and B (Fig. 11.87). They are cyclodepsipeptides, *Geodia* sp. They contain tripeptide unit of two (S)-alanines and a (R)-3-halotyrosine joined to a polypropionate unit in an 18-membered ring (Chan et al. 1987) (Fig. 11.88).

Halicylindramides D is a tridecapeptide with the N-terminus blocked by a formyl group and

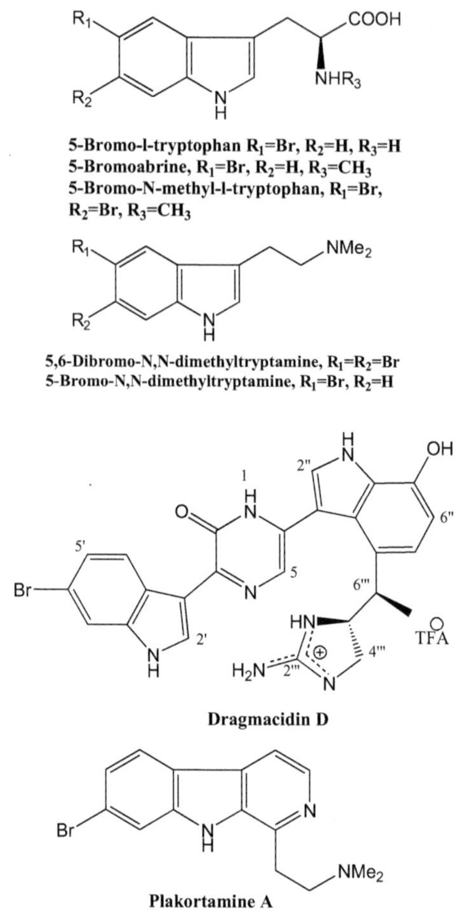

Fig. 11.50 Tryptophans, tryptamines, dragmacidin, plakortamine

the C-terminus lactonized with a threonine residue, *Halichondria cylindrata* (Li et al. 1996) (Fig. 11.89).

11.5.13 Macrolide

Salicylihalamides A (Fig. 11.90) has salicylic acid to a 12-membered lactone ring and an enamide side chain, *Haliclona* sp. (Erickson et al. 1997).

11.5.14 Macrolactone

5-desacetylaltohyrtin A, 42-membered ring, two spiroketals, two tetrahydro-pyranes, and a halogen atom, *Hyrtios altum* (Aoki et al. 2001; Kobayashi et al. 1993) (Fig. 11.91).

Phorboxazoles A and B and laulimalide (Fig. 11.92): They are macrolides, isolated from *Phorbas* sp. (Searle and Molinski 1995) and *Hyattella* sp. (Corley et al. 1988), respectively. Cyclotheonamides A and B: Cyclic peptides,

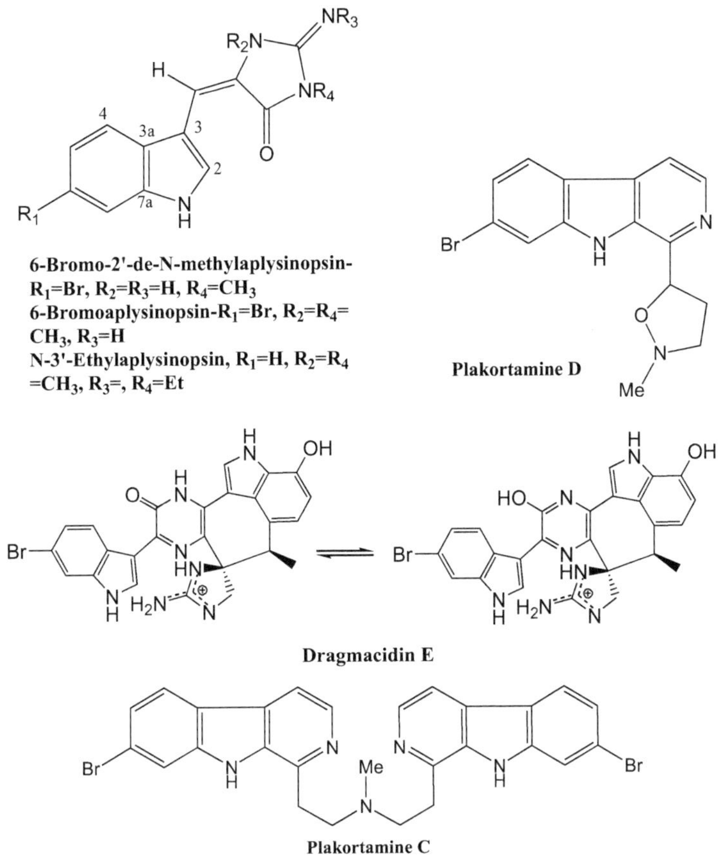

Fig. 11.51 Plakortamines, aplysinopsins, dragmacidins

Theonella (Fusetani and Matsunaga 1990). Stylopeptide 1: Cycloheptapeptide, with Pro-Leu-Ile-Phe-Ser-Pro-Ile amino acid units, *Stylotella* sp. and *Phakellia costata* (Pettit et al. 1995); theopederins A and B: Pederins are highly cytotoxic against P388 murine leukemia cells with promising antitumor activity, *Theonella* sp. (Fusemi et al. 1992) (Fig. 11.93).

Keramamide F: It is a thiazole-containing peptide, *Theonella*. It contains unusual amino acids such as (O-methylseryl) thiazole, α,-β-dehydrotryptophan, isoserine, 2,3 diaminopropionic acid, and 3-amino-4-methyl-2-oxohexanoic acid (Itagaki et al. 1992). Leucamide A (Fig. 11.94), cyclic heptapeptide, *Leucetta microraphis*. It contains a unique mixed

Fig. 11.52 Manzamines

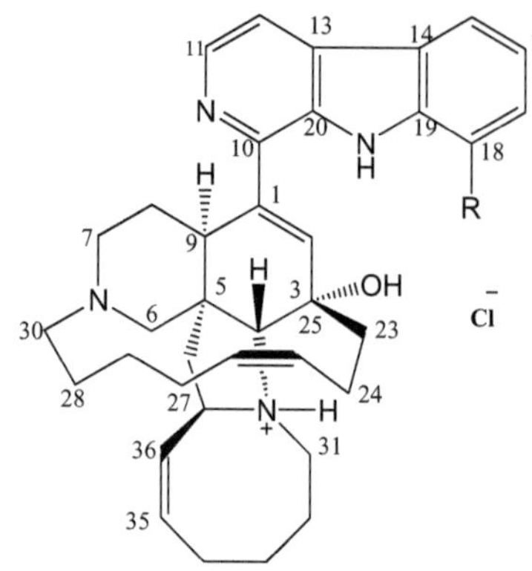

Manzamine A HCl, R=H
-18-Hydroxymanzamine A HCl, R=OH

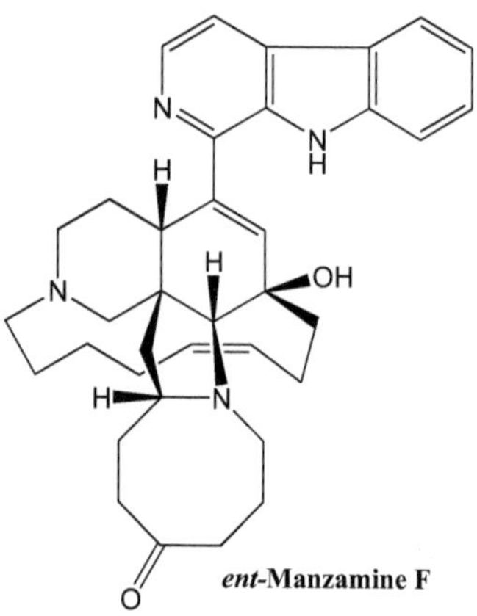

***ent*-Manzamine F**

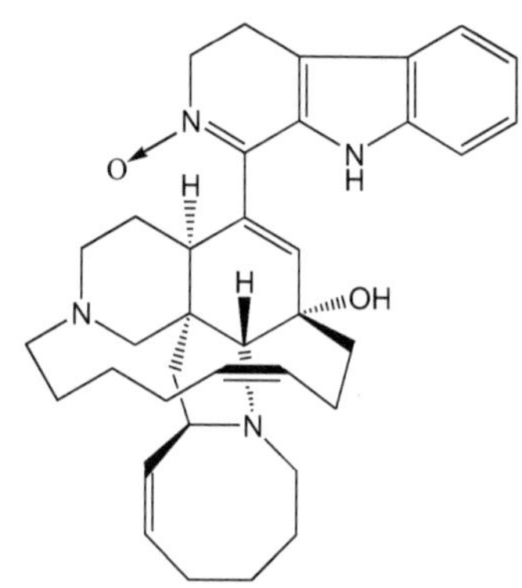

3,4-Dihydromanzamine A *N*-oxide

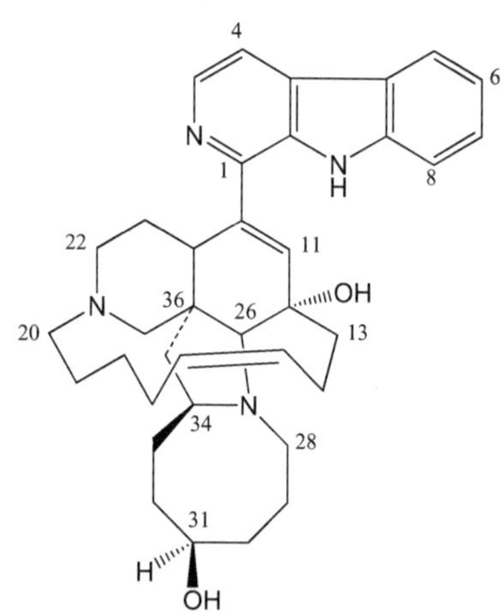

32,33-Dihydro-31-hydroxymanzamine A

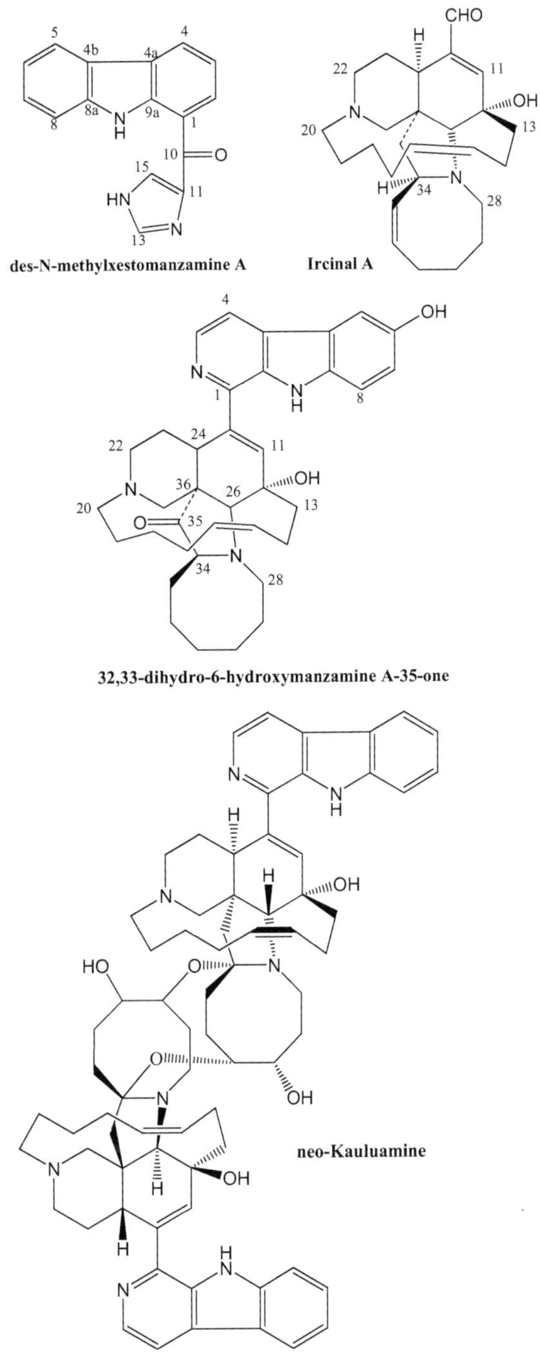

Fig. 11.53 Manzamine derivatives

Fig. 11.54 Agelastatin, dercitin

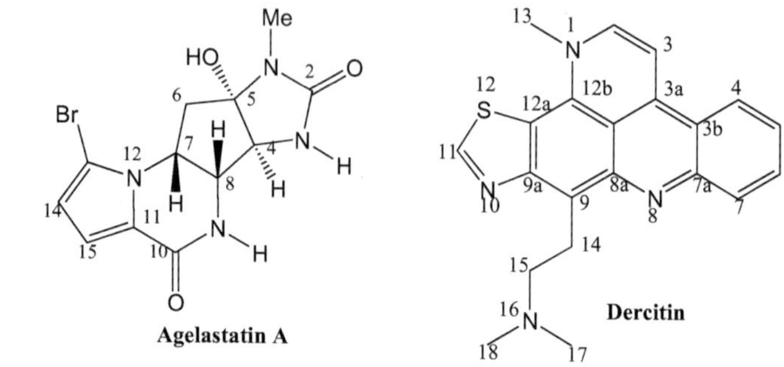

Fig. 11.55 Naamines, spirocalcaridine A

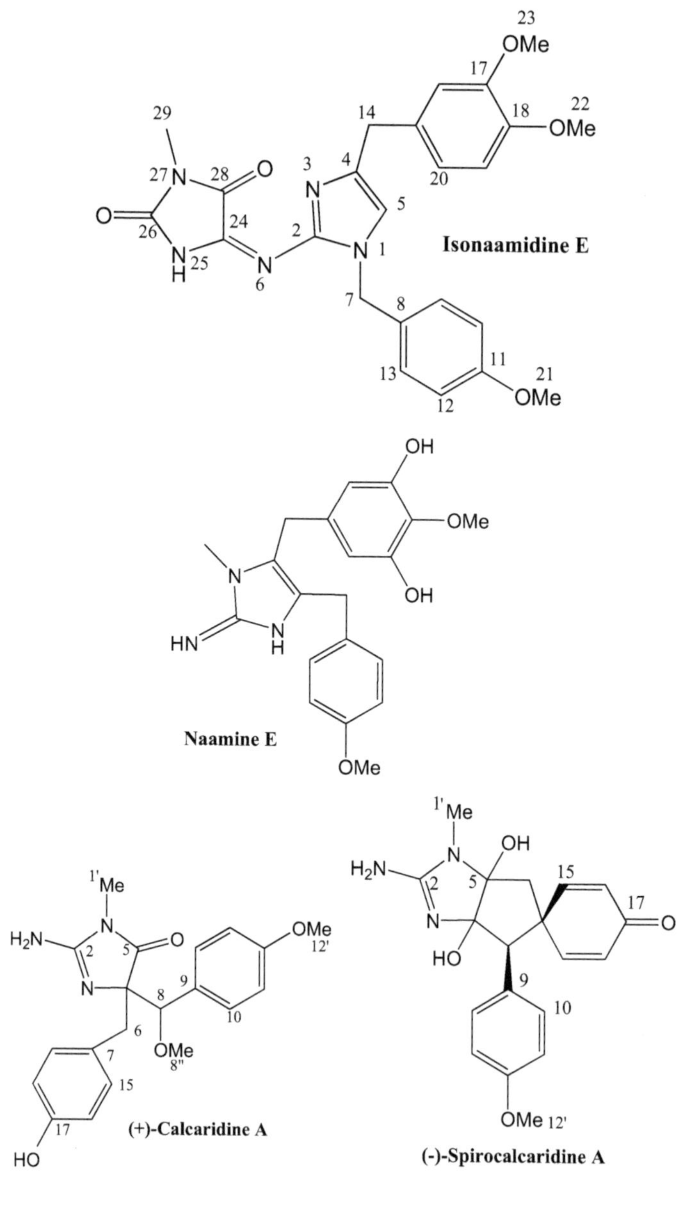

4,2-bisheterocycle tandem pair consisting of a methyloxazole and thiazole subunit (Kehraus et al. 2002b).

Thiomycalolides A and B: Trisoxazole macrolides, *Mycale* sp. (Matsunaga et al. 1998b); spongistatin 1 (Fig. 11.95), macrocyclic lactone, *Spongia* (Pettit et al. 1993); cyclotheonamides E2 and E3 (Fig. 11.96): They are cyclic pentapeptide containing unusual amino acid residues, *i.e.*, vinylogous tyrosine (VTyr), α-ketohomoarginine (K-Arg), and

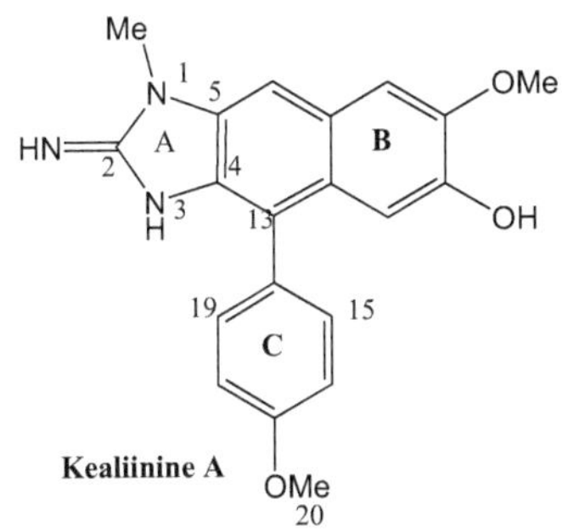

Kealiinine A

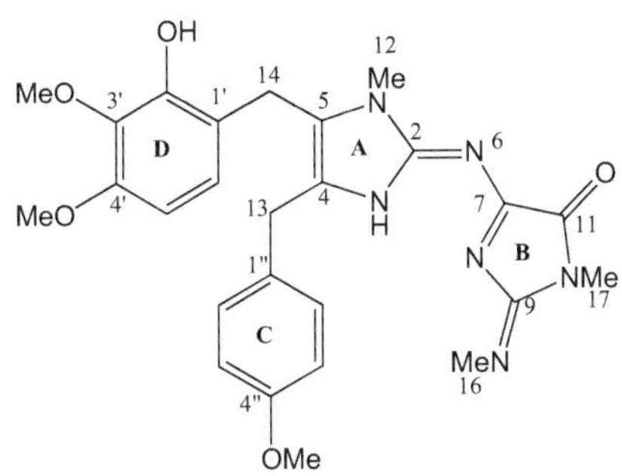

(2E,9E)-Pyronaamidine 9-(N-methylimine)

Fig. 11.56 Kealiinine A, (2*E*,9*E*)-pyronaamidine 9-(*N*-methylimine), pyronaamidine, kealiiquinone

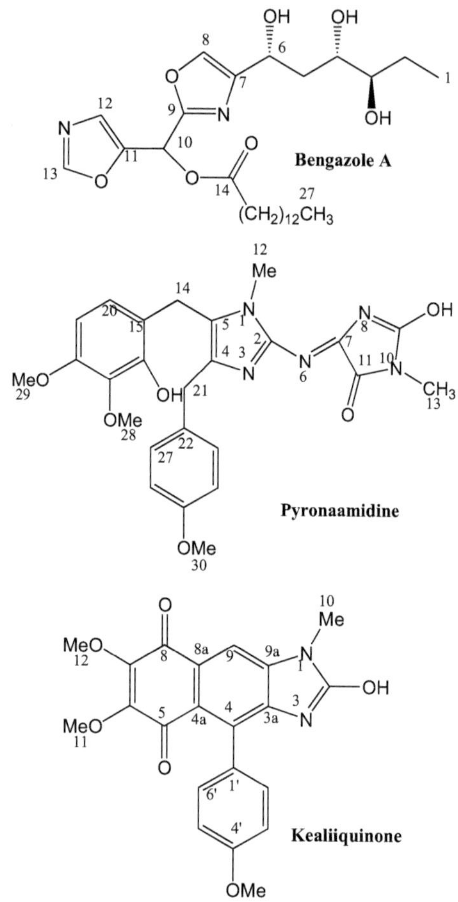

Fig. 11.56 (continued)

Fig. 11.57 Chelonin A and C, bromochelonin B, purealidin S, purpuramine J

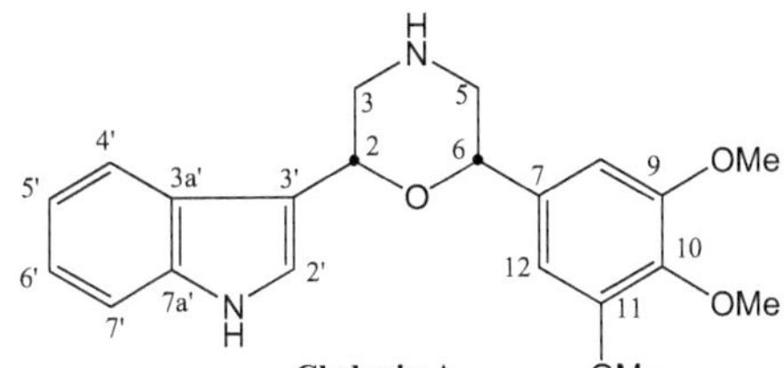

Chelonin A

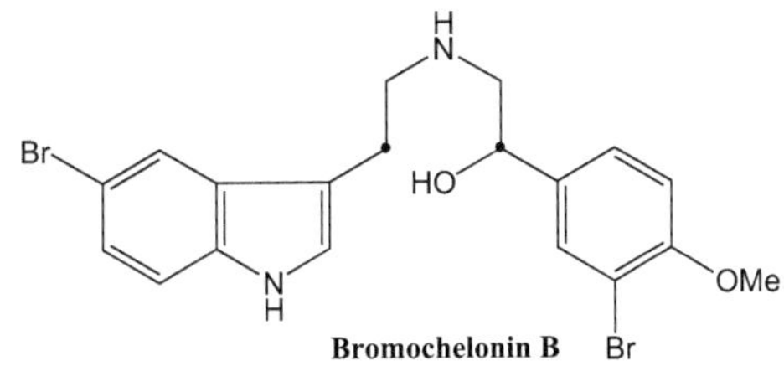

Bromochelonin B

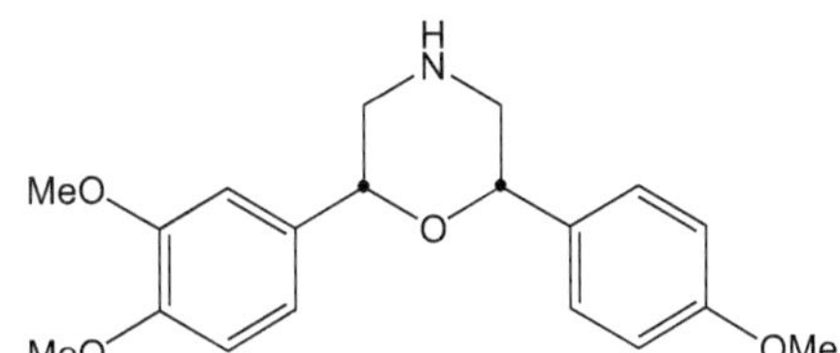

Chelonin C

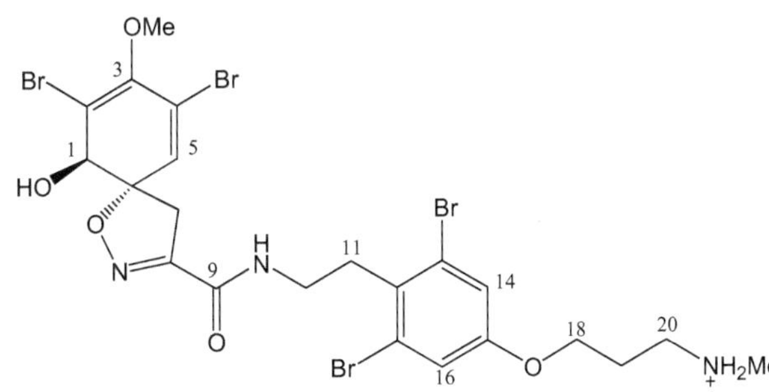

Purealidin S

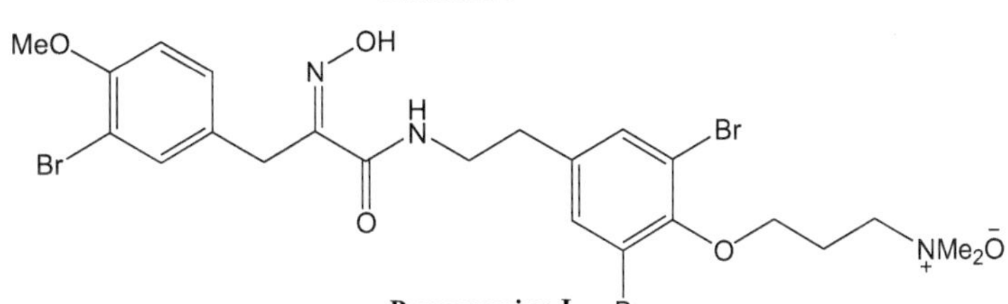

Purpuramine J

Fig. 11.58 Cribrostatins

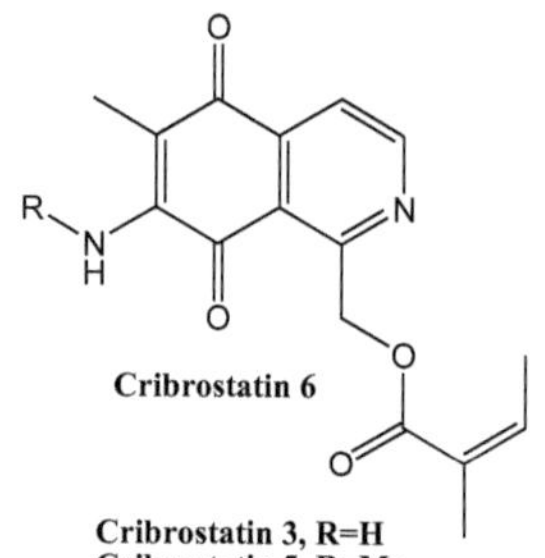

Cribrostatin 3, R=H
Cribrostatin 5, R=Me

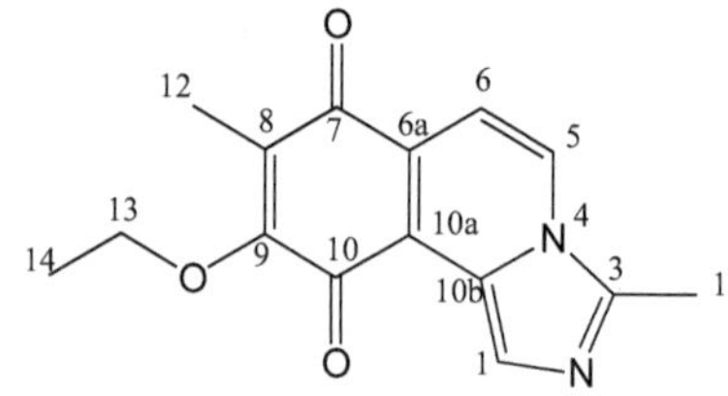

Cribrostatin 6

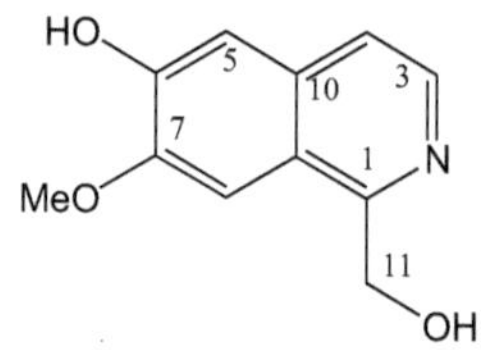

1-Hydroxymethyl-7-methoxyisoquinolin-6-ol

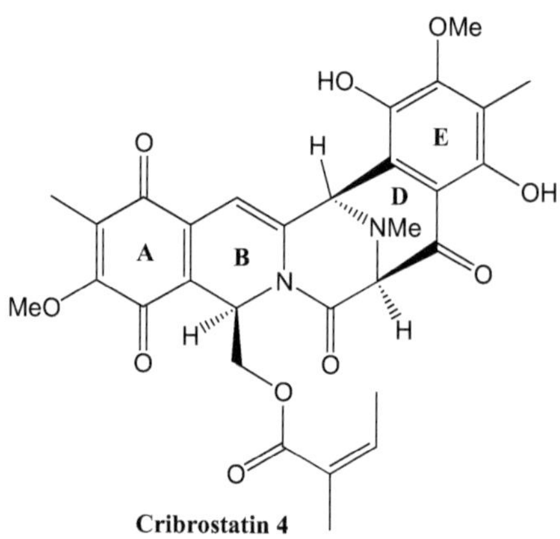

Cribrostatin 4

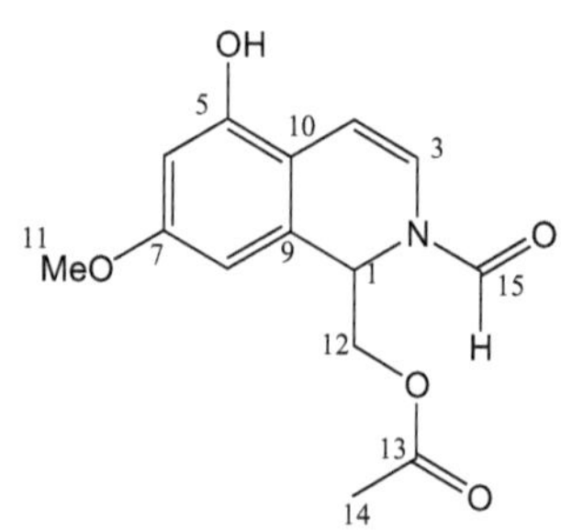

N-Formyl-1,2-dihydro isoquinoline derivative

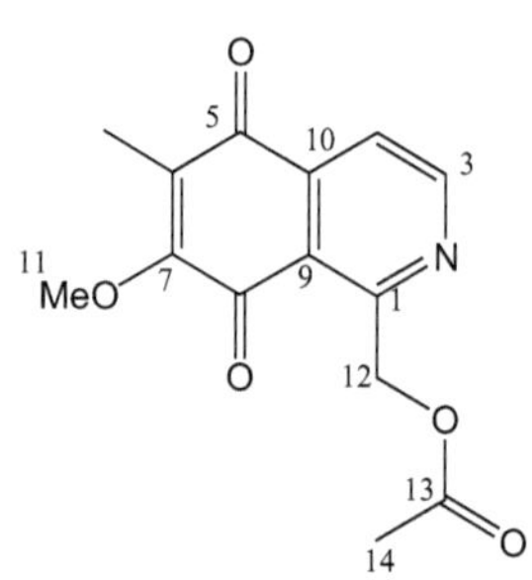

***O*-Demethyl renierol acetate**

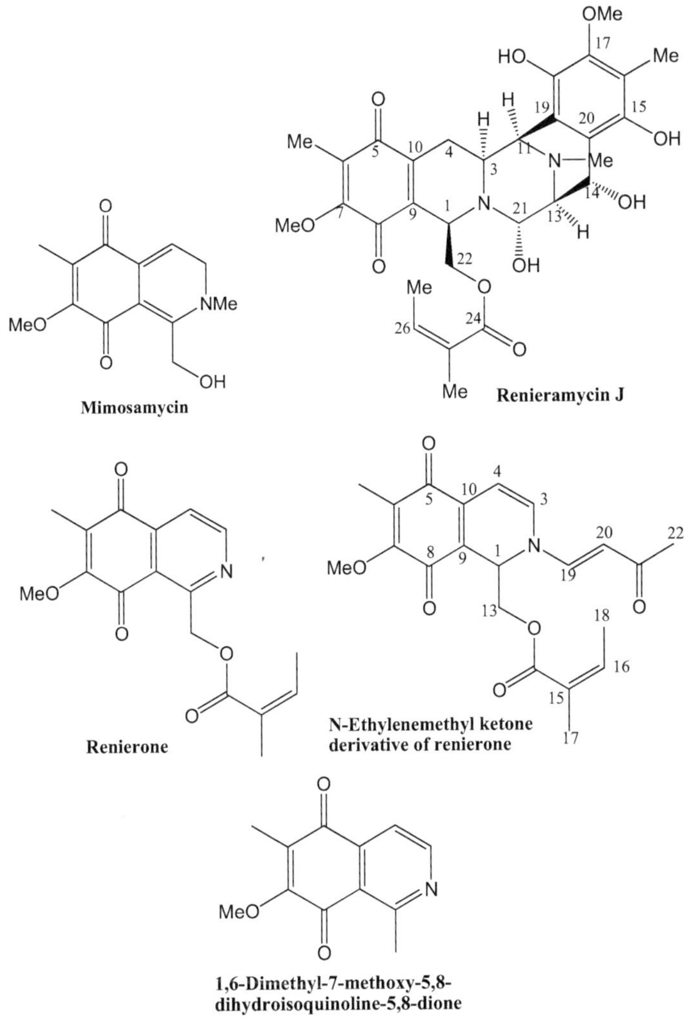

Fig. 11.59 Renieramycin, renierones, mimosamycin, cribrostatins

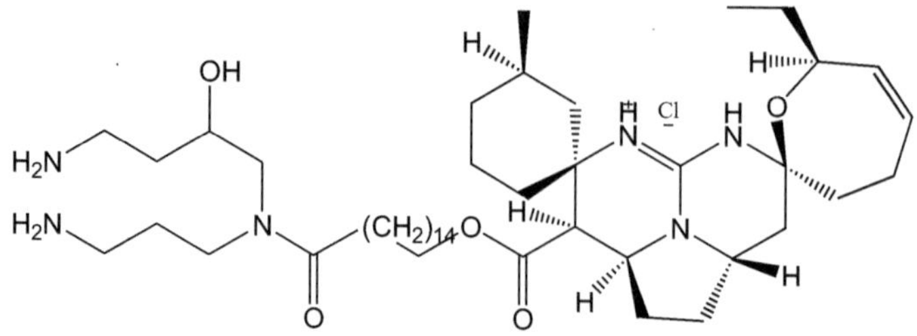

13,14,15-Isocrambescidin800

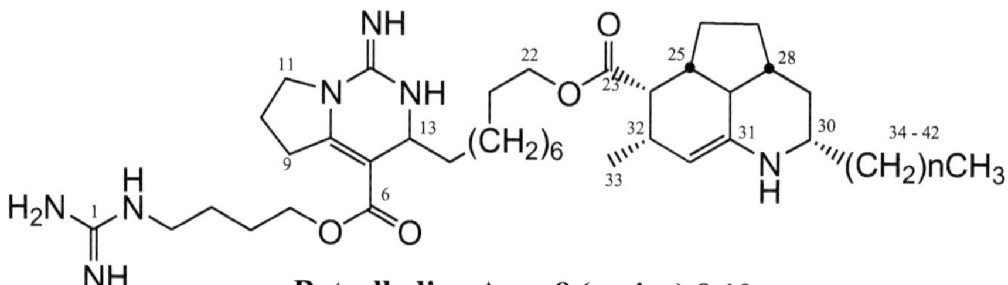

Batzelladine A, n=8 (major),9,10

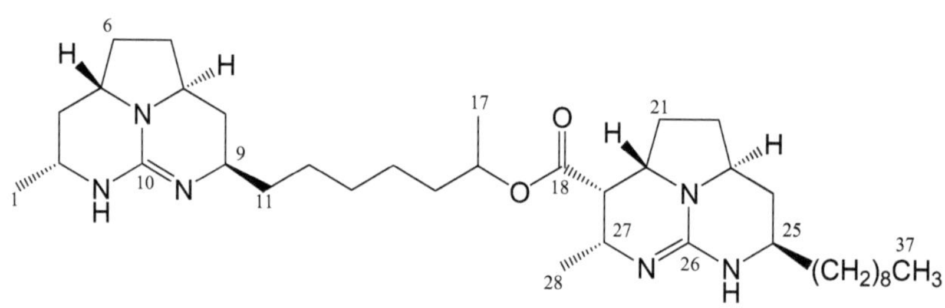

Batzelladine F

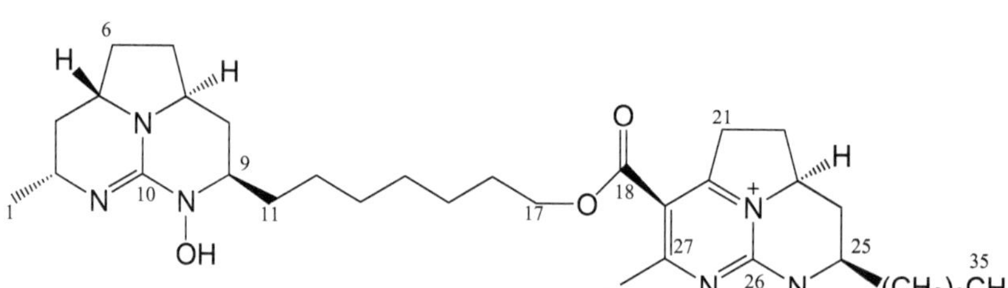

Batzelladine I

Fig. 11.60 Isocrambescidin, batzelladines

Fig. 11.61 Mirabilins, palau'amines

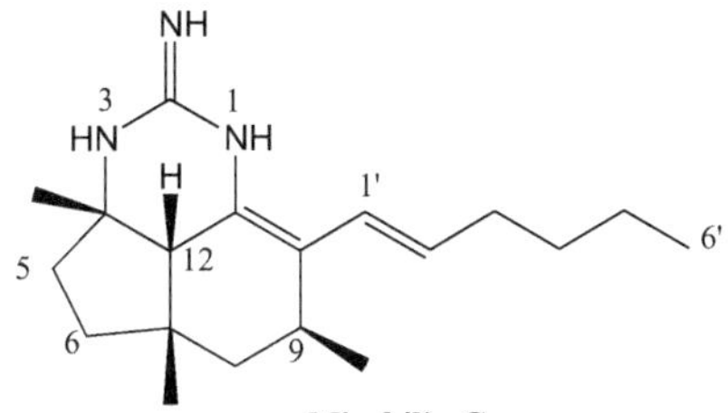

Mirabilin G

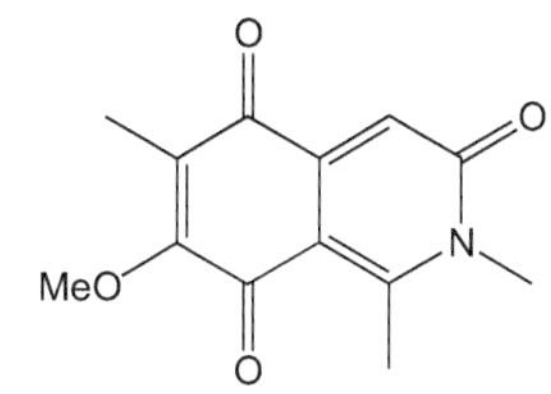

Mimosamycin

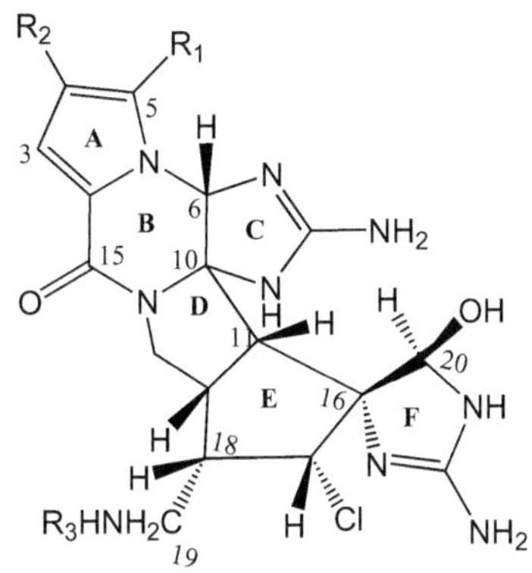

Palau'amine, R_1=R_2=R_3=H
4-Bromopalau'amine, R_1, R_3=H, R_2=Br
4,5-Dibrompalau'amine, R_1, R_2=Br, R_3=H,

Fig. 11.62 Ancorinolates, 1-carboxymethylnicotinic acid, Kuanoniamine

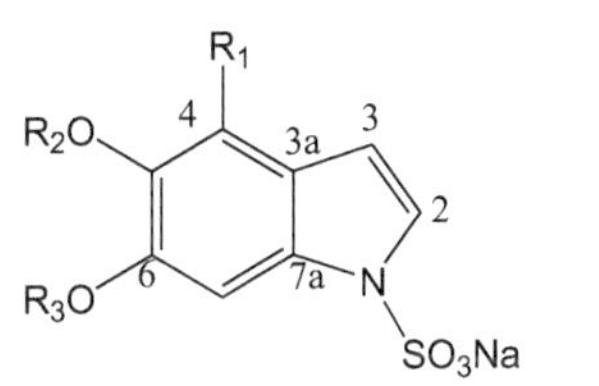

Ancorinolate A, R_1=Cl, R_2=H, R_3=SO_3Na
Ancorinolate C, R_1=Cl, R_2=H, R_3=H

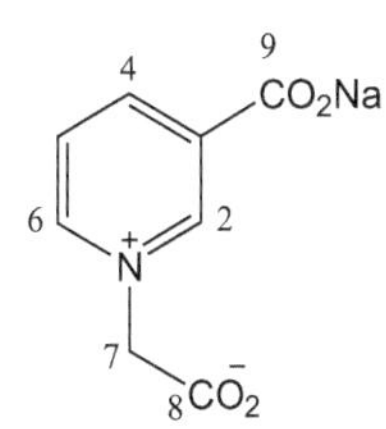

1-Carboxymethyl-nicotinic acid

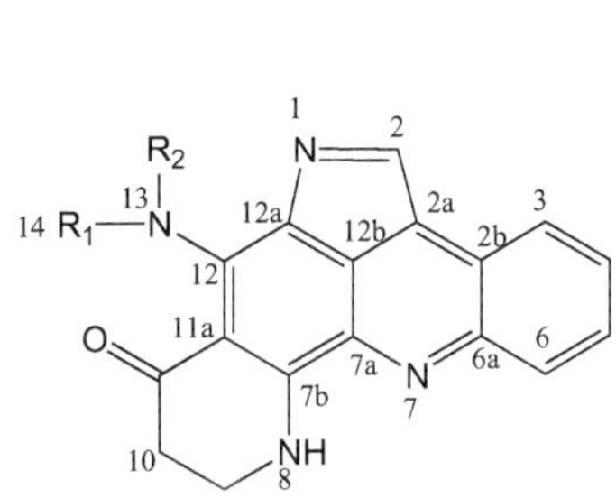

Plakinidine A, R_1=H, R2=CH_3
Plakinidine B, R_1, R2=CH_3

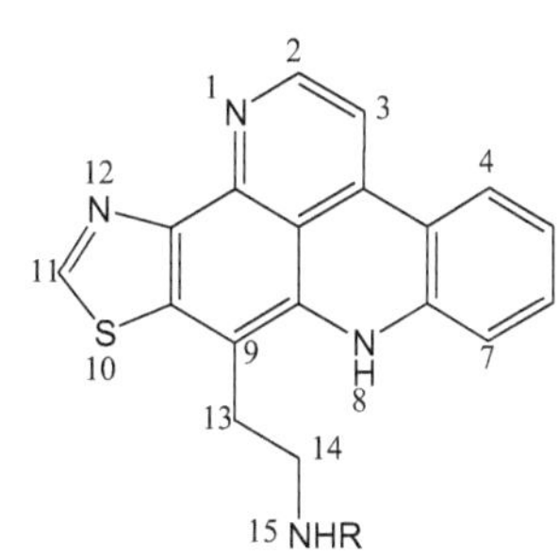

Kuanoniamine C, R=COEt
Kuanoniamine D, R=COMe
N-deacyl derivative, R=H

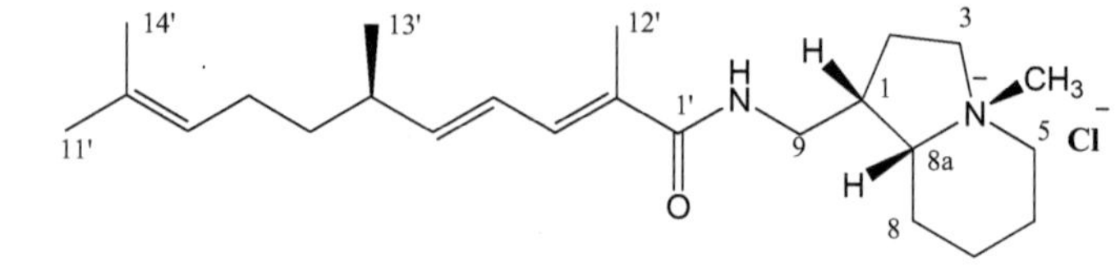

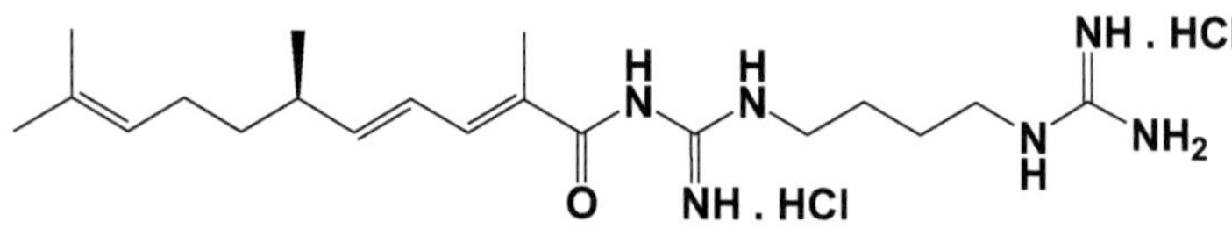

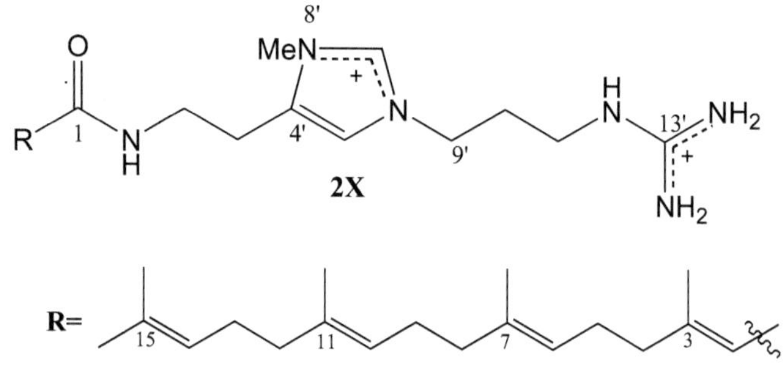

Fig. 11.63 Stellettamides, stellettadine, stellettazole

Fig. 11.64 Plakinamines, motuporamines

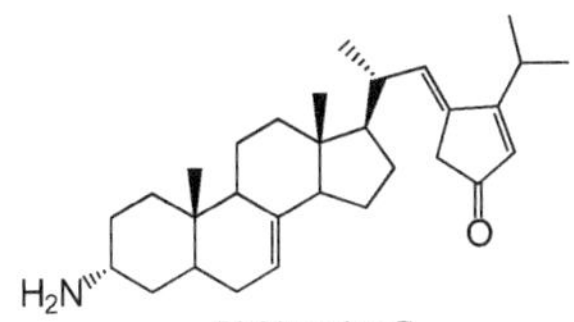

Plakinamine G

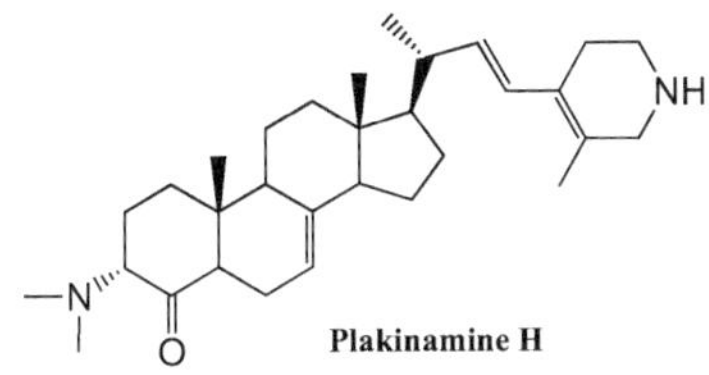

Plakinamine H

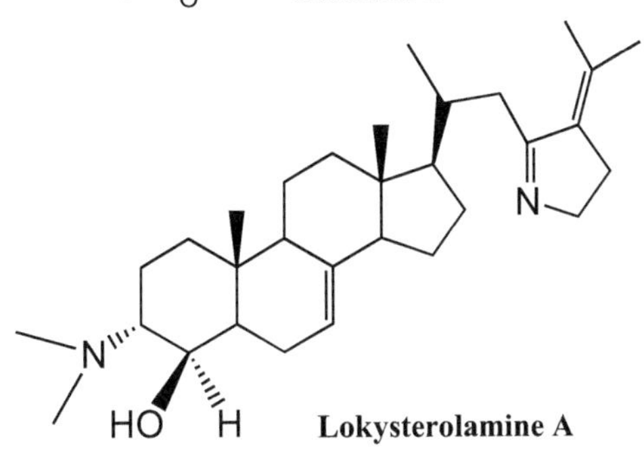

Lokysterolamine A

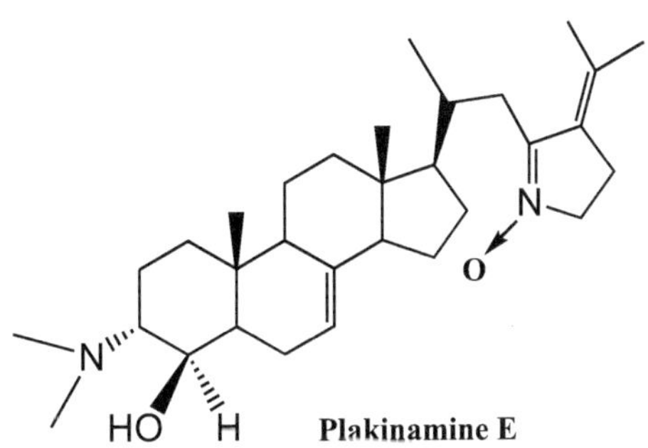

Plakinamine E

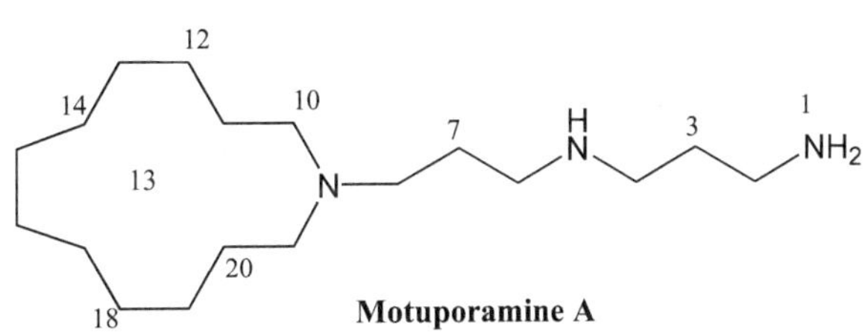

Motuporamine A

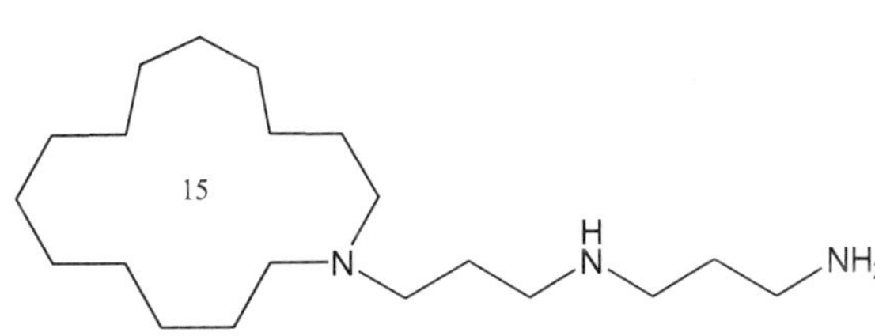

Motuporamine C

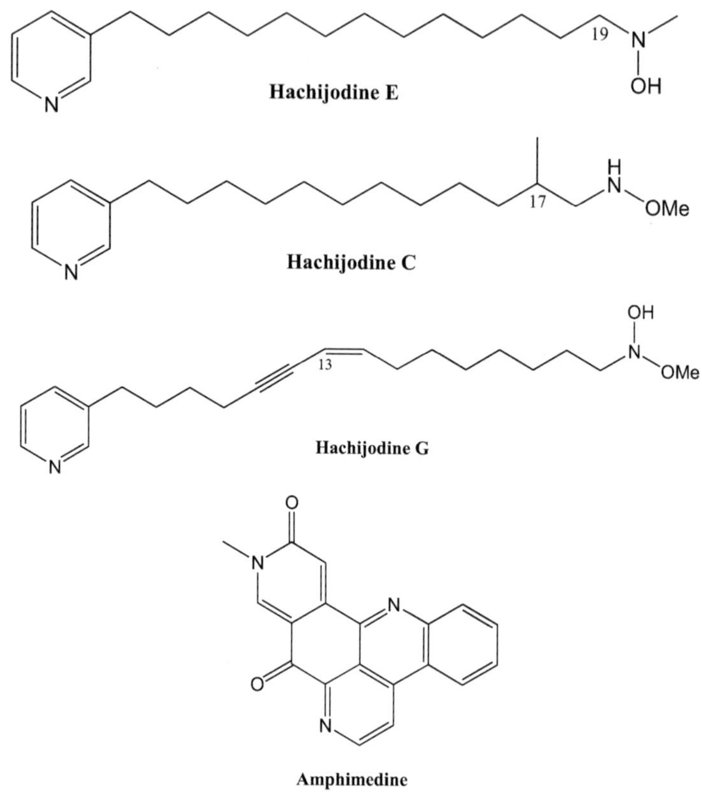

Fig. 11.65 Hachijodines, amphimedine

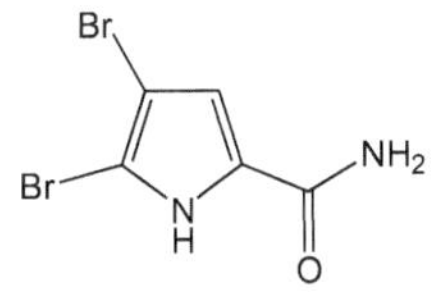

4,5-Dibromopyrrole-2-carbamide

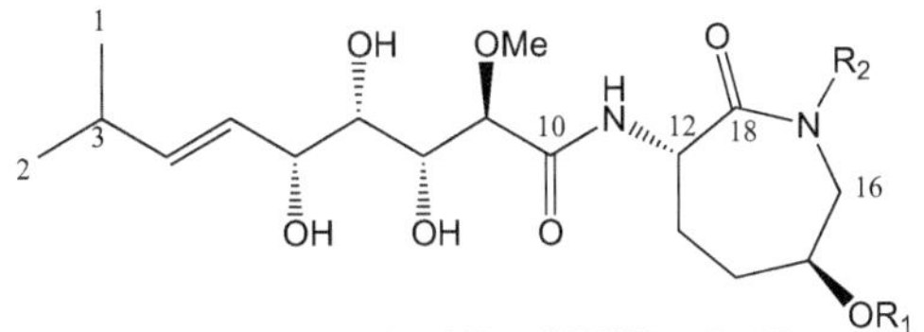

Bengamide L, R_1=CO(CH_2)$_{11}$CH(CH_3)$_2$, R_2=H

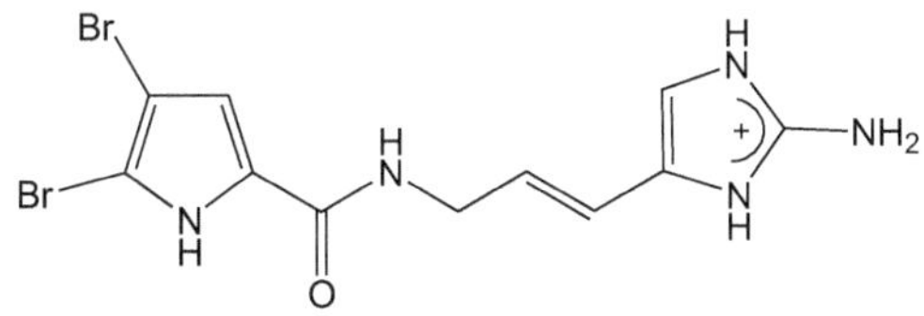

Oroidin

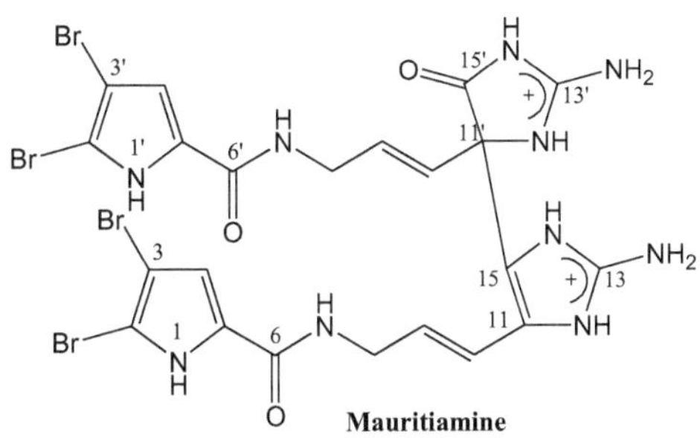

Mauritiamine

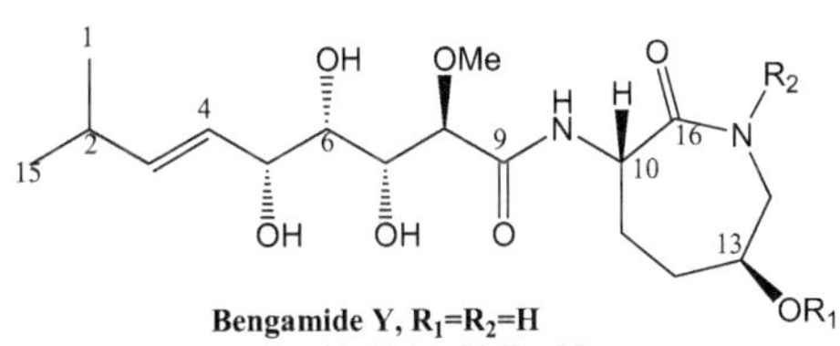

Bengamide Y, R_1=R_2=H
Bengamide Z, R_1=H, R_2=Me

Fig. 11.66 Mauritiamine, oroidin, bengamides

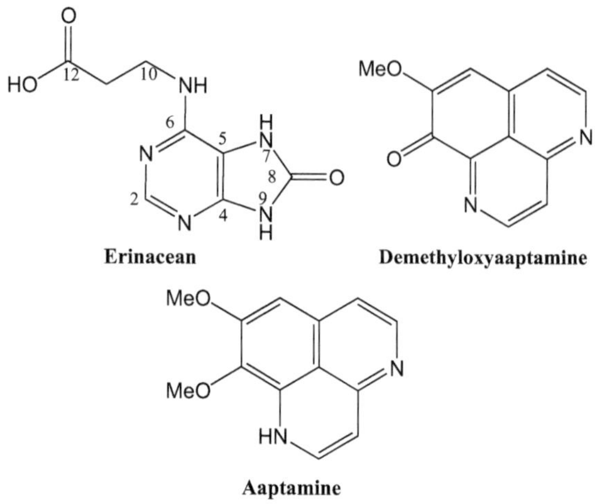

Fig. 11.67 *Erinacea*, demethyloxyaaptamine, aaptamine

β-linked diaminopropionic acid (Dpr), *Theonella* (Nakao et al. 1998).

11.6 Spongean Sterols, Glycosides, and Nucleoside Derivatives

11.6.1 Spongean Sterols

Agosterol C and acanthosterol sulfate J (Fig. 11.97), polyhydroxylated sterol, *Acanthodendrilla* sp. (Tsukamoto et al. 2003b, 1998).

Clathriol is a highly oxygenated steroid with the unusual *cis* C/D ring fusion with 14β configuration, *Clathria lissosclera* (Keyzers et al. 2002). Clathsterol, sterol sulfate, *Clathria* sp. (Rudi et al. 2001); plakinamines I and J, steroidal alkaloids, *Corticium niger* (Ridley and Faulkner 2003) (Fig. 11.98).

Crellastatin A: Nonsymmetric dimeric steroid, *Crella* sp. (D'Auria et al. 1998); aragusterol D (Fig. 11.99), *Xestospongia* (Iguchi et al. 1994);

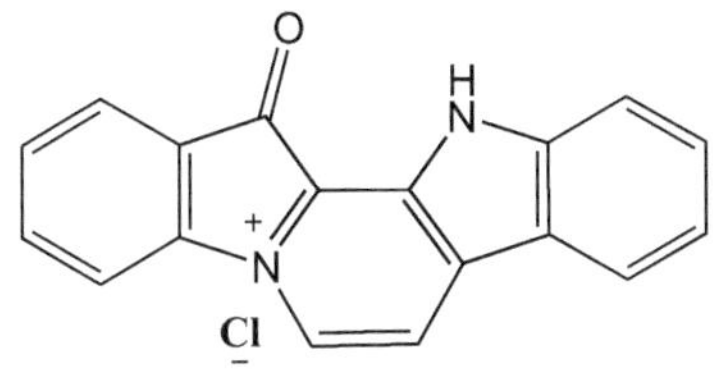

Fascaplysin

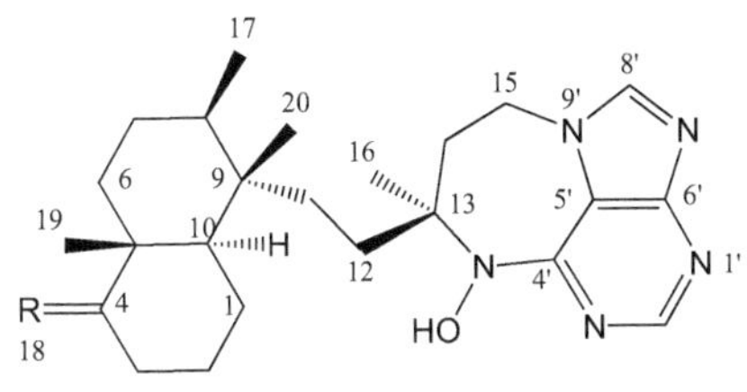

Asmarine A, R=CH_2
Asmarine B, 5 *epi* R=CH_2

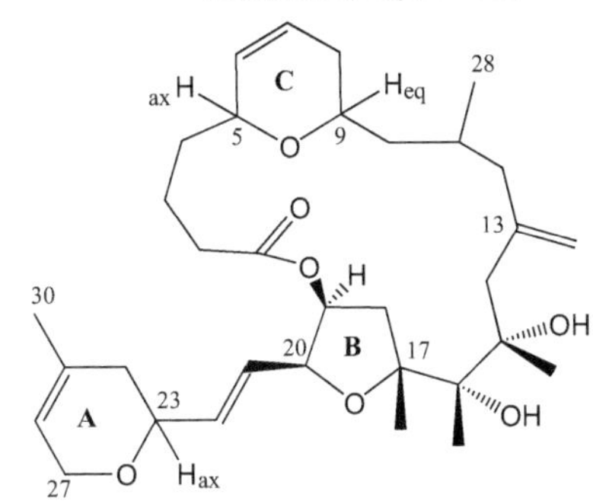

Fijianolide A

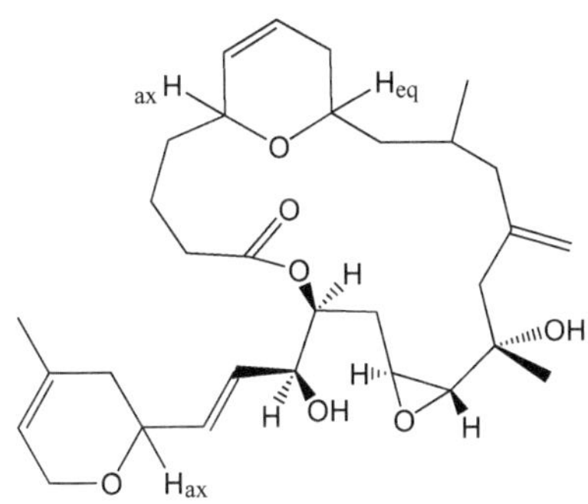

Fijianolide B

Fig. 11.68 Asmarines, fijianolides, fascaplysin

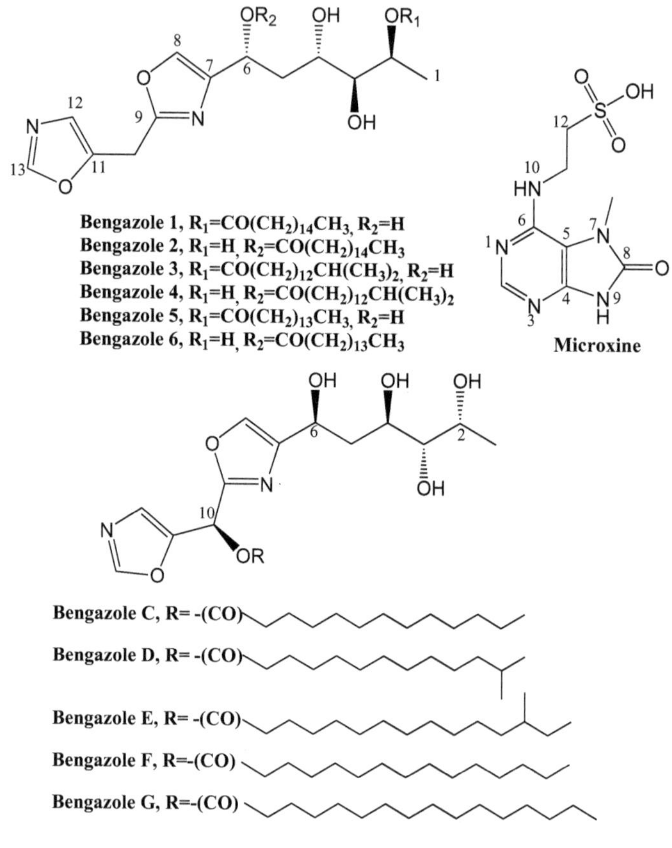

Fig. 11.69 Bengazoles, microxine

polyoxygenated sterols, *Dysidea* (Leone et al. 2000); epidioxy sterols **1a** and **1b**, **2**, *Lendenfeldia chondrodes* (Sera et al. 1999b) (Fig. 11.100).

11.6.2 Steroidal Oligoglycosides

Mycaloside A (Fig. 11.101), *Mycale laxissima* (Antonov et al. 2003); $5\alpha,6\alpha$-epoxy-$24R^*$-ethylcholest-8(14)-en-$3\beta,7\alpha$-diol, *Polymastia*

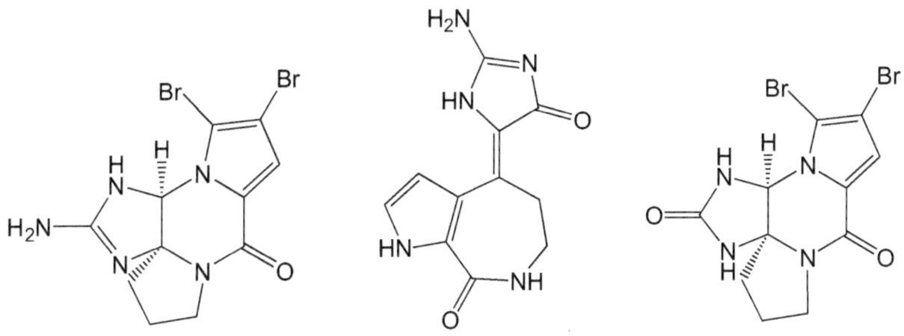

Dibromophakellin debromohymenialdisine Dibromophakellstatin

Fig. 11.70 Dibromophakellstatin, debromohymenialosine, dibromophakellin

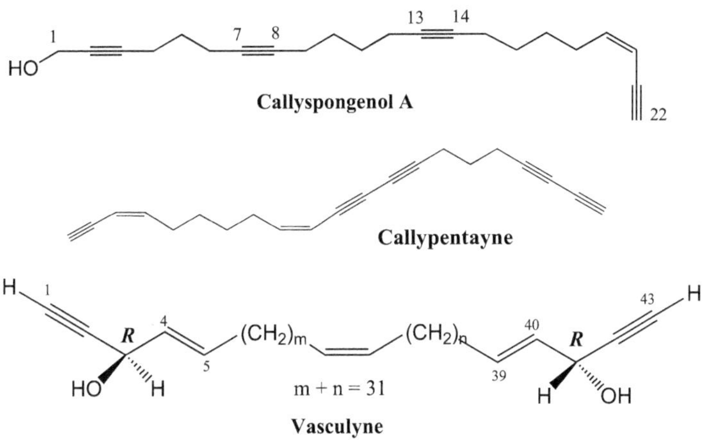

Fig. 11.71 Callyspongenol, callypentayne, vasculyne

tenax (Santafé et al. 2002); 24(*R*)-methyl-5-α-cholest-7-enyl 3*β*-methoxymethyl ether (sterol ether), *Scleritoderma* sp. cf. *paccardi* (Gunasekera et al. 1996b); 3-*O*-deacetylluffasterol B and 3-*O*-deacetyl-22,23-dihydro-24,28-dehydroluffasterol B, 9,11-secosterols, *Spongia agaricina* (Rueda et al. 1998) (Fig. 11.102).

Phorbasterones A–D: Ring A-contracted steroids, *Phorbas amaranthus* (Masuno et al.

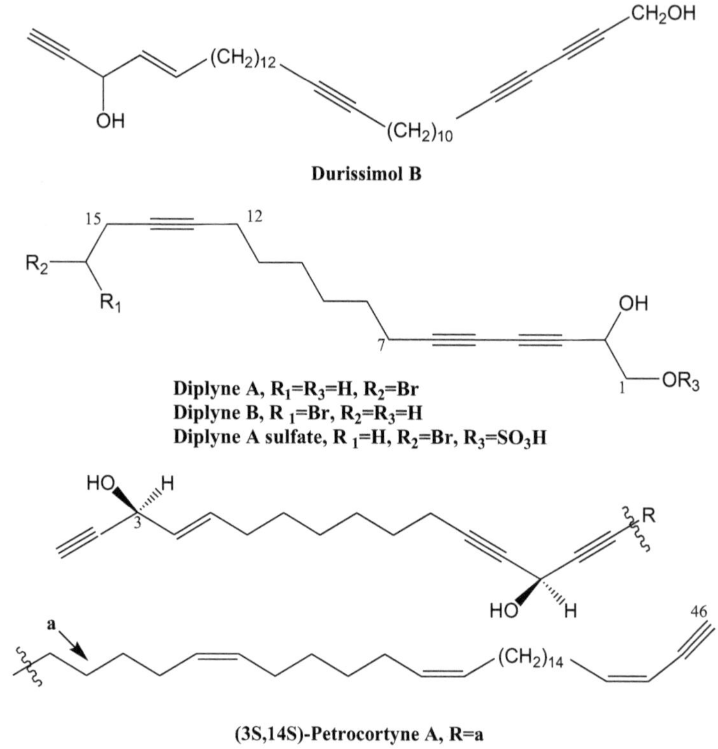

Fig. 11.72 Durissimol, diplynes, petrocortyne

2004; Shoji et al. 1992); xestobergsterol A (Fig. 11.103) (23S-16β,23-cyclo-3α,6α,7β, 23-tetrahydroxy-5α,l4β-cholestan-15-one and B (2) (23S-16β,23-cyclo-lβ,2β,3α, 6α, 7β,23-hexahydroxy-5α,l4β-cholestan -15-one), *Xestospongia bergquistia* (Nicholas et al. 1999).

11.6.3 Glycosides and Nucleoside Derivatives

Oceanapiside: It is a *α,ω-bis*-aminohydroxylipid glycoside and highly polar, *Oceanapia philippensis* (Mitchell et al. 1997);

Fig. 11.73 C_{14} acetylenic acid, callyspongins, brominated acetylenic fatty acid, fulvinol

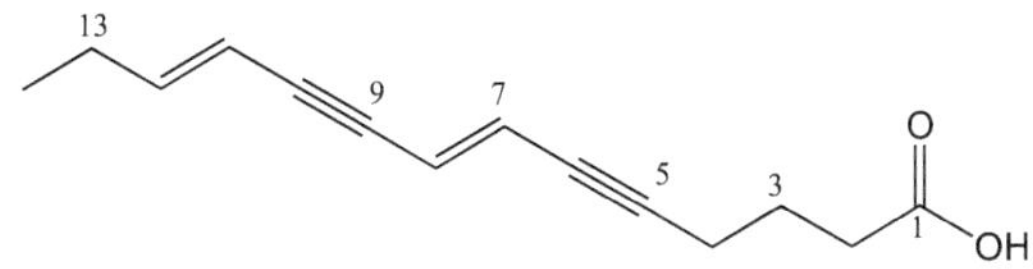

C_{14} Acetylenic acid

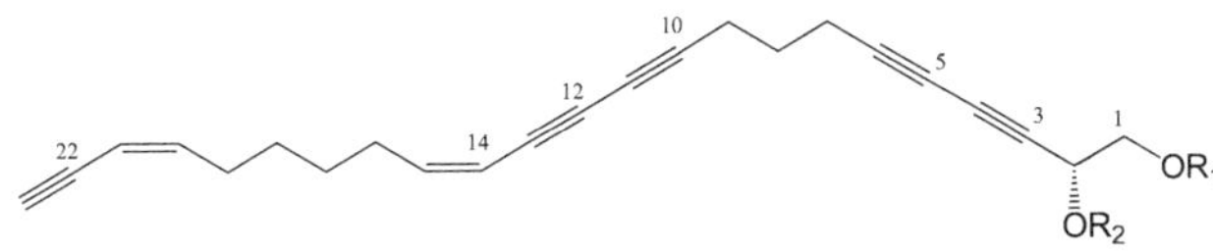

Callyspongin A, R_1=R_2=SO_3Na
Callyspongin B, R_1=SO_3Na, R_2=H

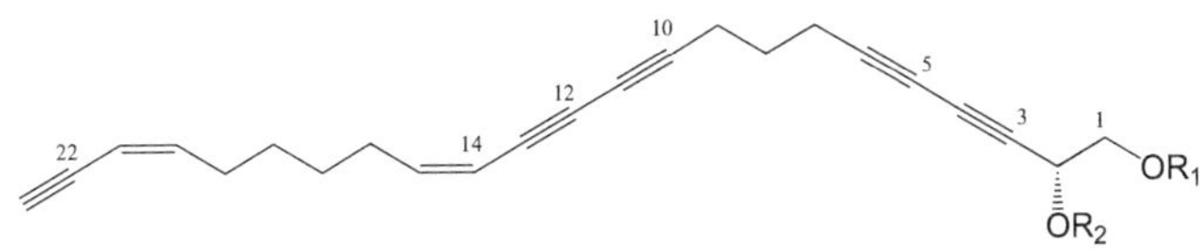

Callyspongin A, R_1=R_2=SO_3Na
Callyspongin B, R_1=SO_3Na, R_2=H

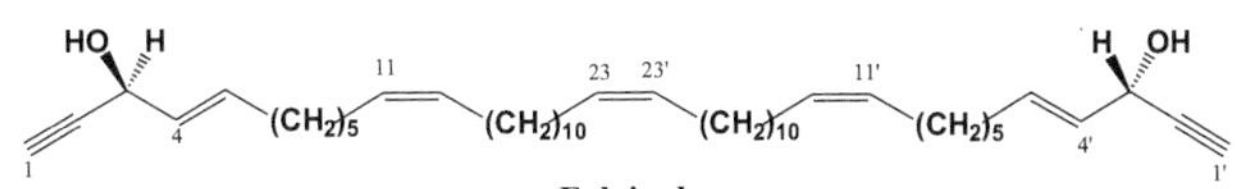

Fulvinol

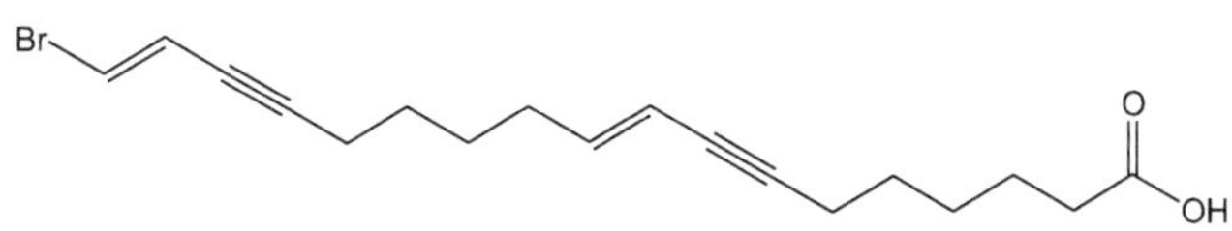

18-Bromooctadeca-(9E,17E)-diene-7,15-diynoic acid

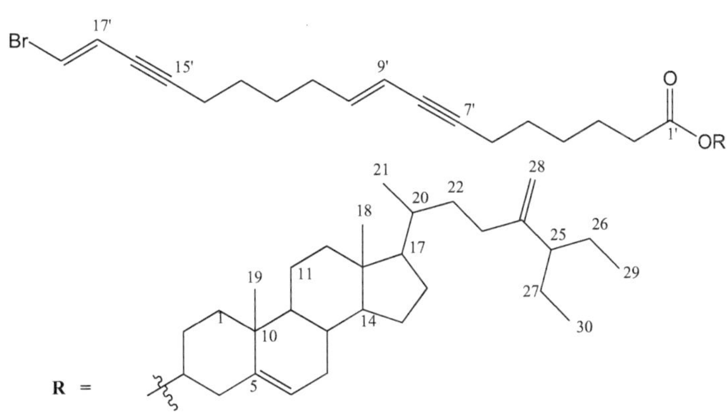

Xestosterol ester of 18-bromooctadeca-(9*E*,17*E*)-diene-7,15-diynoic acid

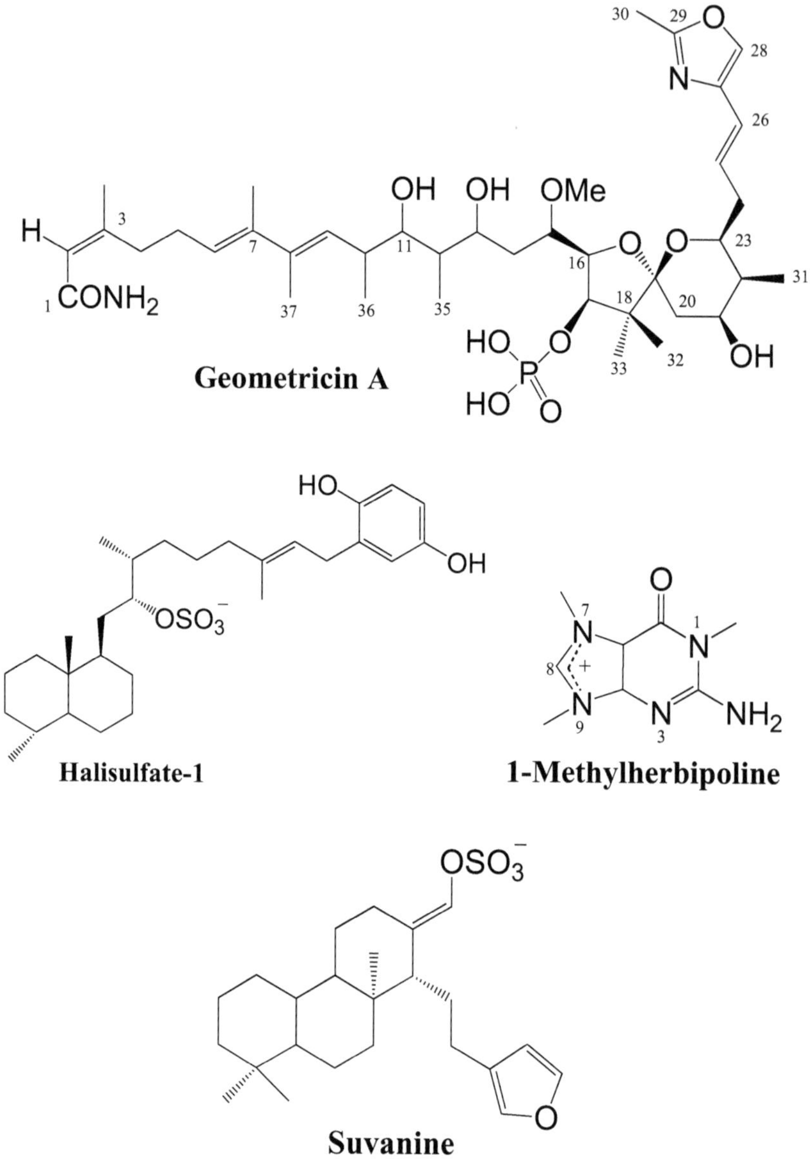

Fig. 11.74 Geometricin A, 1-methylherbipoline salts of halisulfate-1, suvanine

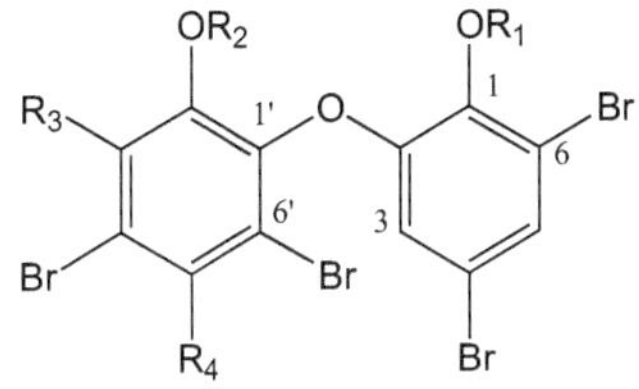

Polybrominated diphenyl ethers
3 = R_1=R_2=H, R_3=R_4=Br
7 = R_1=R_2=H, R_3=Br, R_4=H

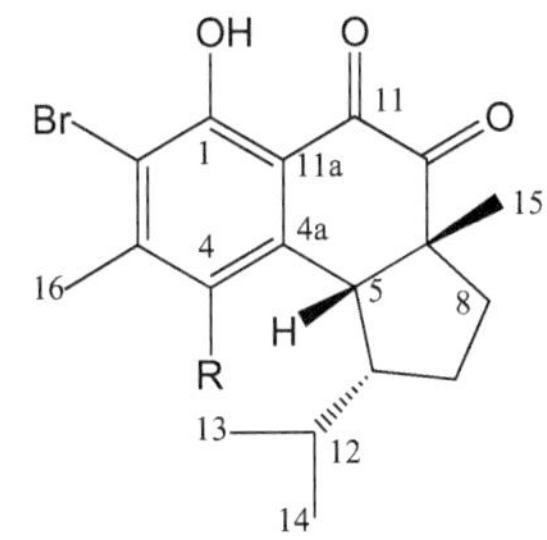

Hamigeran B, R=H
4-Bromohamigeran B, R=Br

Fig. 11.75 Polybrominated diphenyl ethers

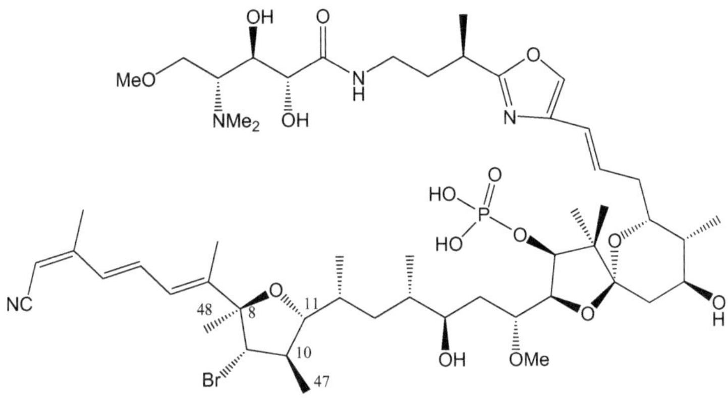

Calyculin J

Fig. 11.76 Calyculin

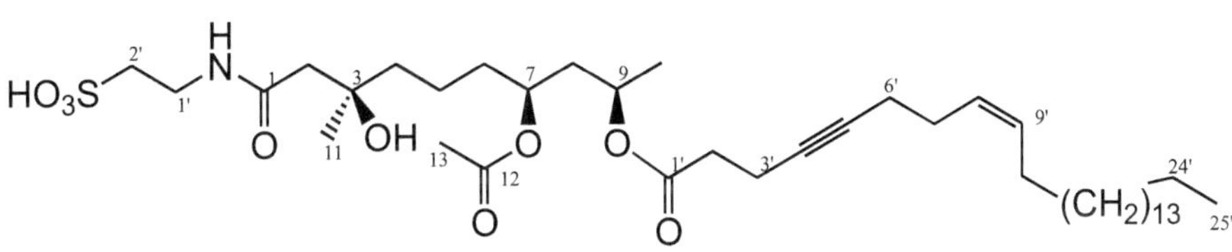

Taurospongin A

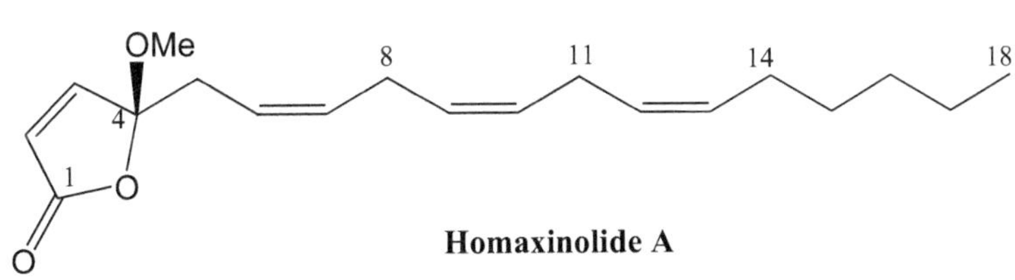

Homaxinolide A

Fig. 11.77 Taurospongin A, butenolides

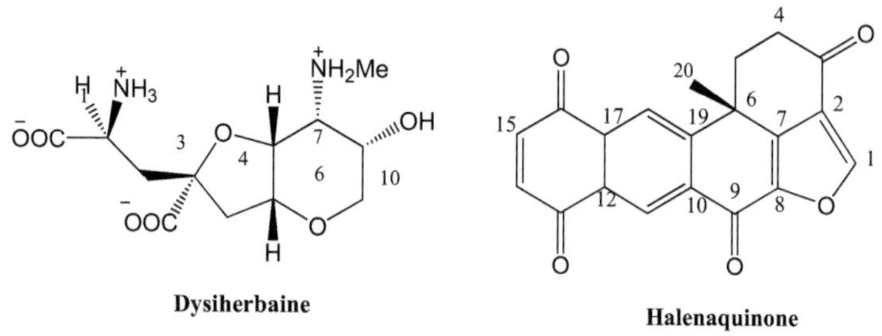

Fig. 11.78 Dysiherbaine, halenaquinone

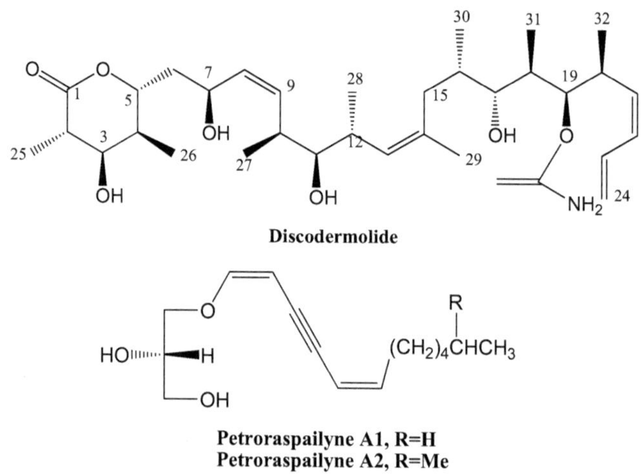

Fig. 11.79 Discodermolide, acetylenic enol ethers

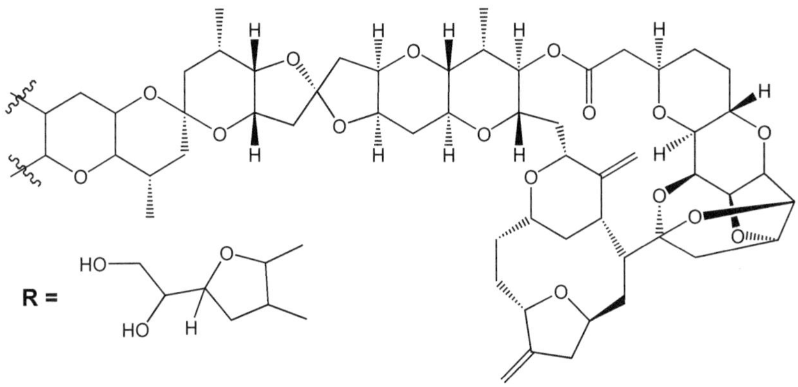

Homohalichondrin B

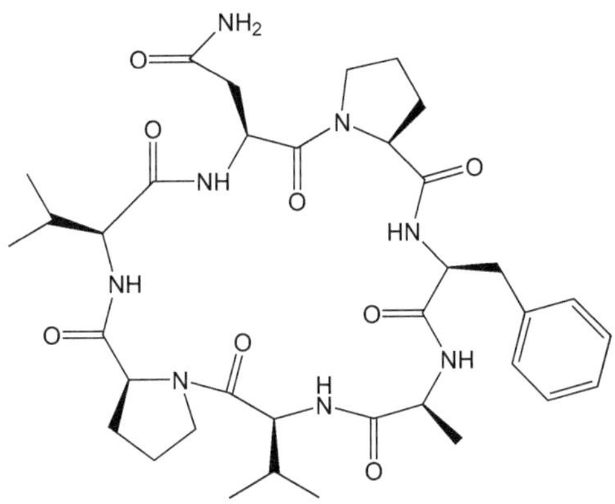

Axinastatin 1

Fig. 11.80 Homohalichondrin, axinastatin

Fig. 11.81 Macrocyclic polyketide, mycalamide

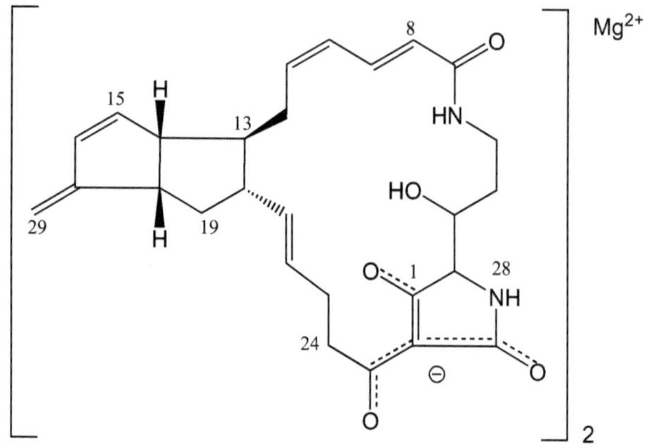

Geodin A Mg salt

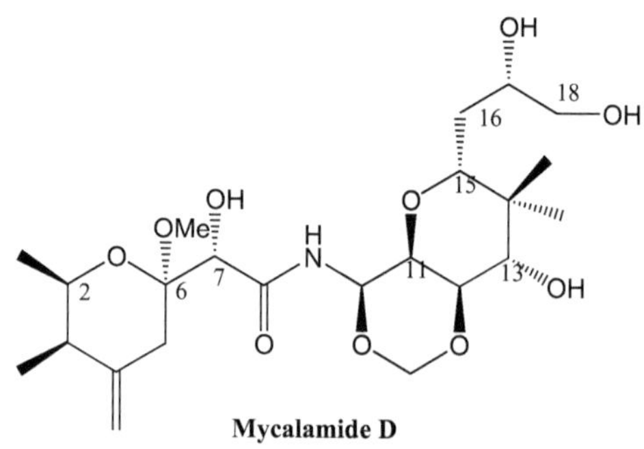

Mycalamide D

Fig. 11.82 Plakotenin, bengamides, melophlins

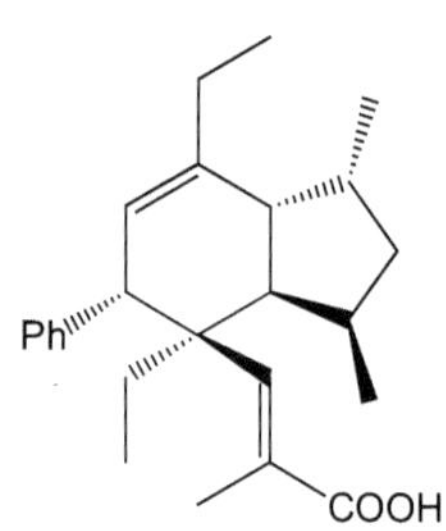

Homo-plakotenin

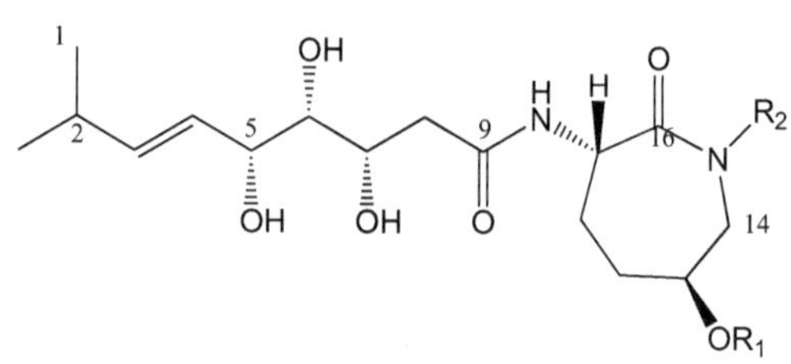

Bengamide G, R_1= -CO(CH$_2$)$_{11}$CH$_3$, R_2=H
Bengamide H, R_1= -CO(CH$_2$)$_{11}$CH$_3$, R_2=Me
Bengamide I, R_1= -CO(CH$_2$)$_{13}$CH$_3$, R_2=H
Bengamide J, R_1= -CO(CH$_2$)$_{13}$CH$_3$, R_2=Me

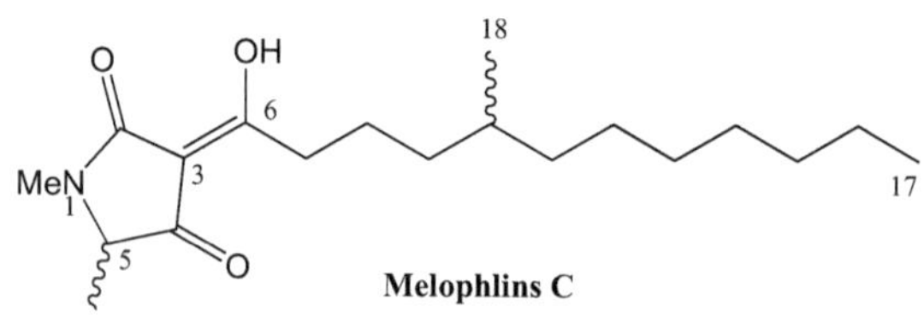

Melophlins C

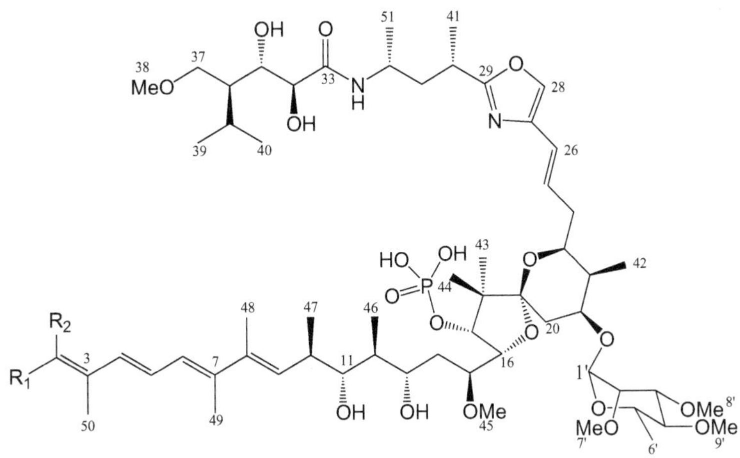

Clavosine A, R_1=H, R_2=$CONH_2$
Clavosine B, R_1=$CONH_2$ R_2=H

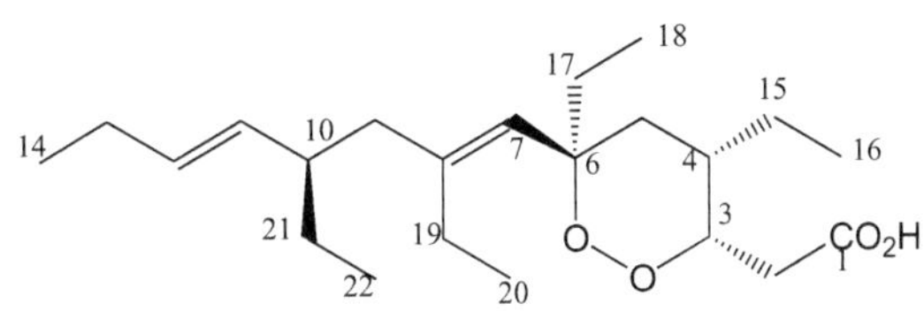

Plakortide N

Fig. 11.83 Clavosines, plakortides

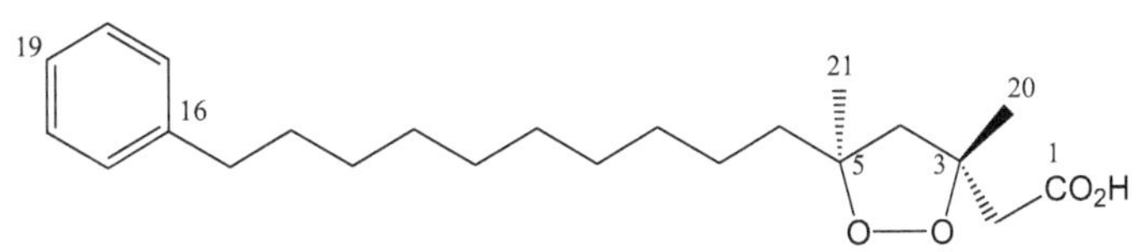

Andavadoic acid

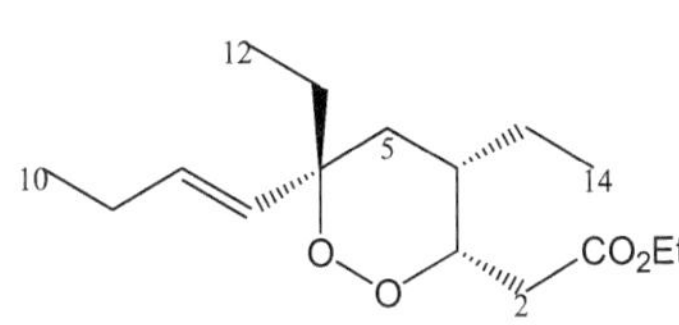

Ethyl didehydroplakortide Z

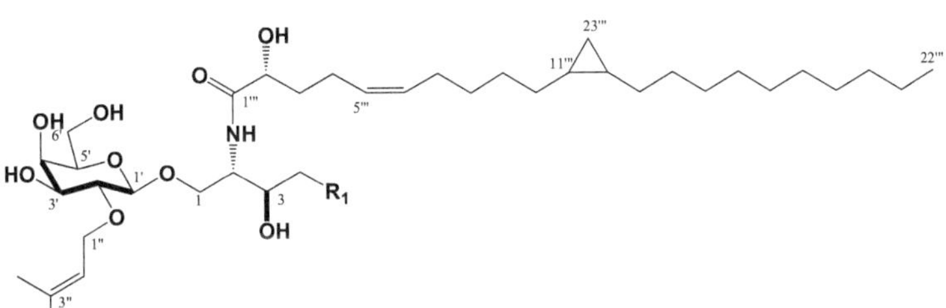

Plakoside B, R_1 = 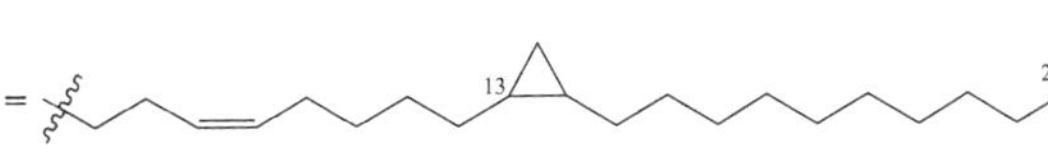

Fig. 11.84 Andavadoic acid, ethyl didehydroplakortide, plakoside

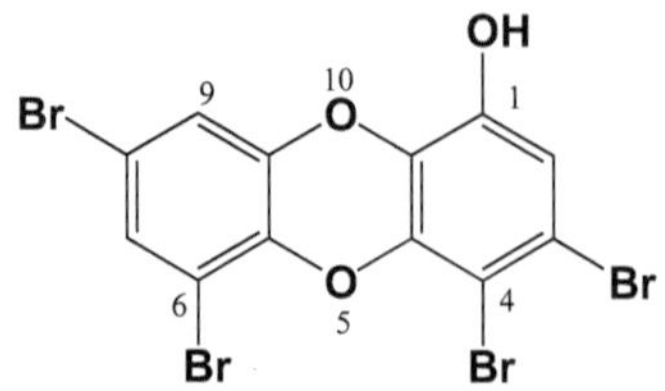

Spongiadioxin A

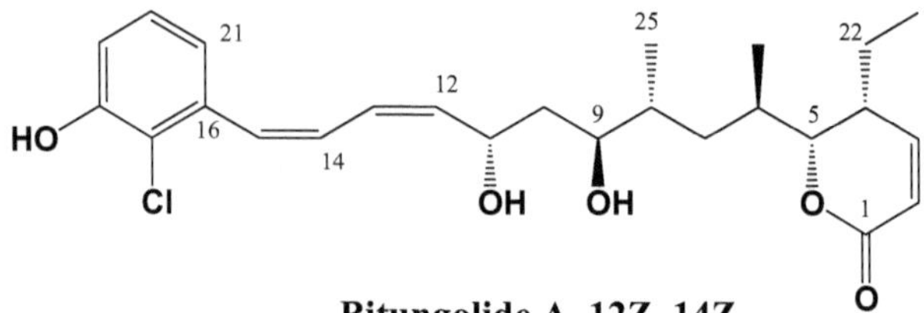

Bitungolide A, 12Z, 14Z
Bitungolide B, 12E, 14E
Bitungolide C, 12Z, 14E
Bitungolide D, 12E, 14Z

Fig. 11.85 Spongiadioxins, bitungolides

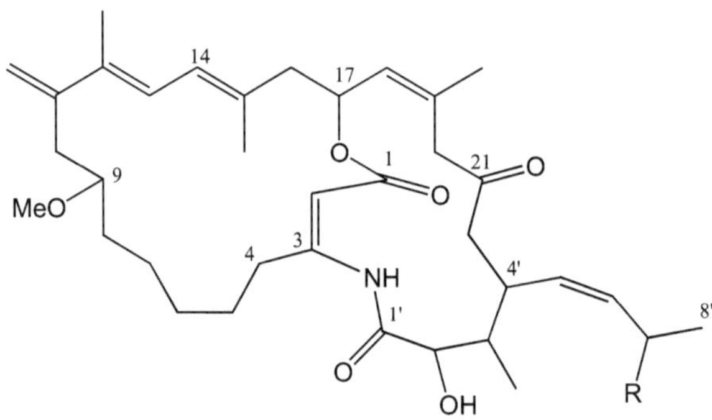

Amphilactam A, R=H
Amphilactam B, R=Me

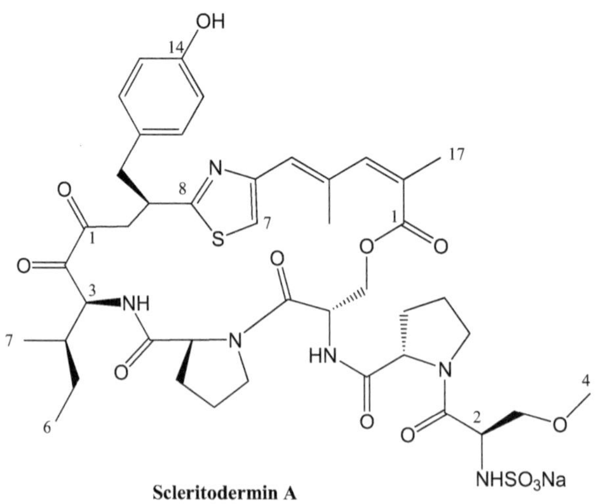

Scleritodermin A

Fig. 11.86 Macrocyclic lactone/lactams

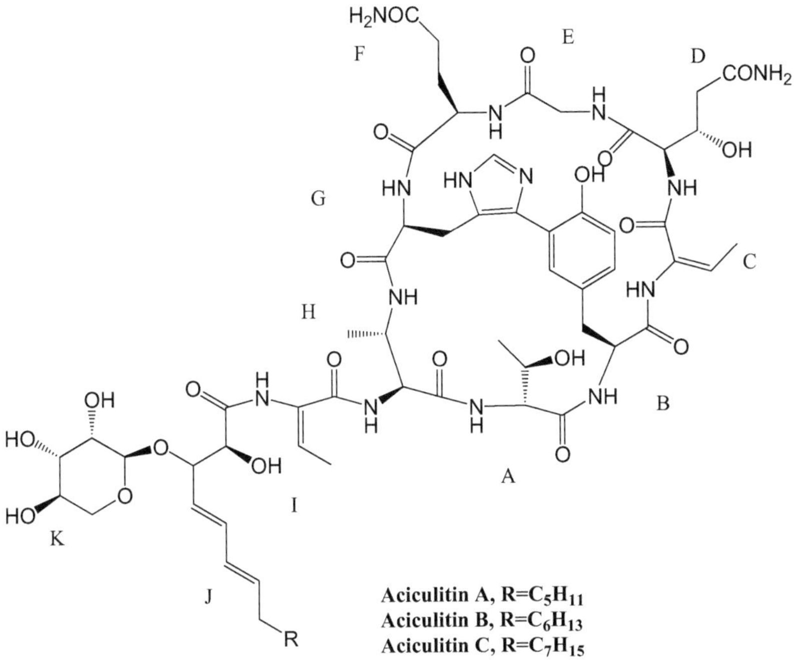

Aciculitin A, R=C_5H_{11}
Aciculitin B, R=C_6H_{13}
Aciculitin C, R=C_7H_{15}

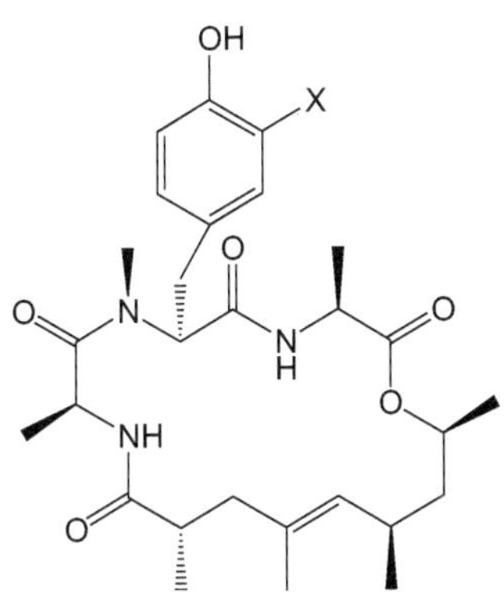

Geodiamolides A, X=I
Geodiamolides B, X=Br

Fig. 11.87 Aciculitins, geodiamolides

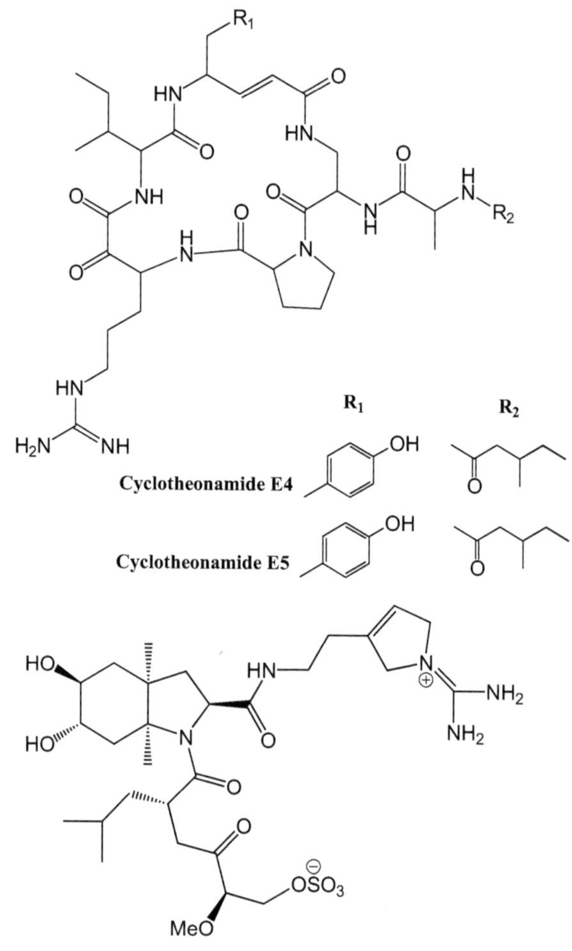

Dysinosin A

Fig. 11.88 Cyclodepsipeptides

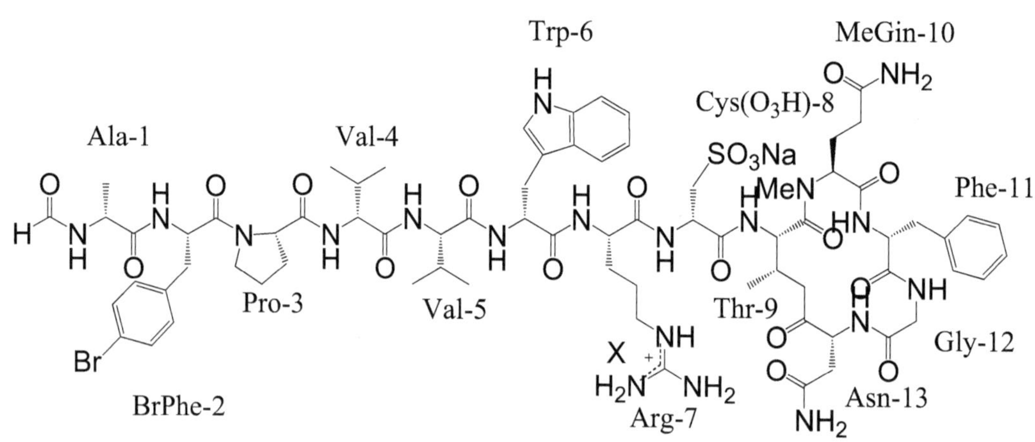

Halicylindramide D

Fig. 11.89 Halicylindramides

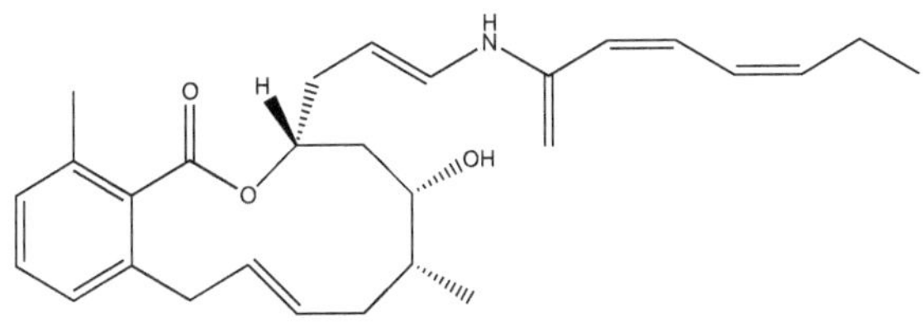

Salicylihalamide A

Fig. 11.90 Salicylihalamides

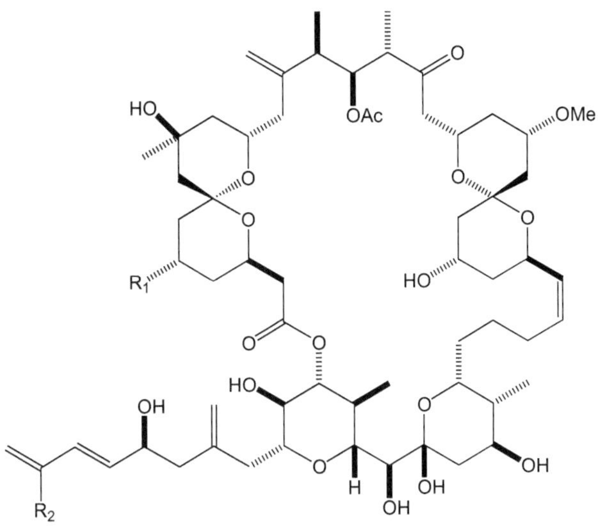

5-Desacetylaltohyrtin A, R1=OH, R2=Cl

Fig. 11.91 5-desacetylaltohyrtin A

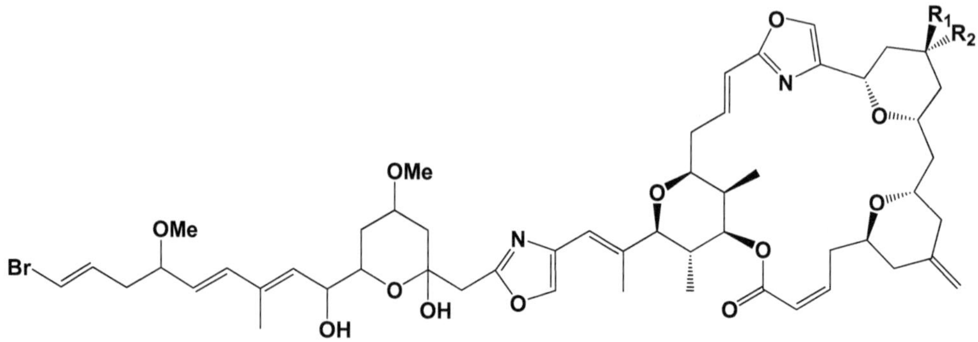

Phorboxazole A, R_1=OH, R_2=H
Phorboxazole B, R_1=H, R_2=OH

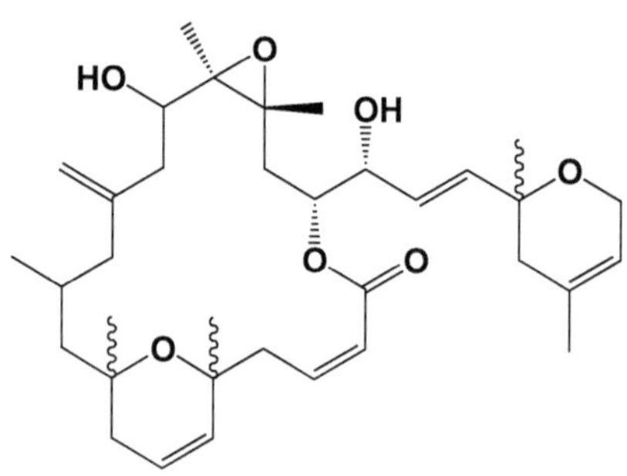

Laulimalide

Fig. 11.92 Phorboxazoles A and B and laulimalide

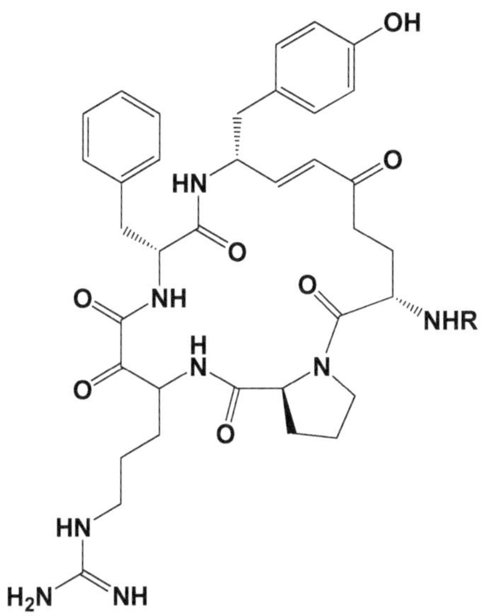

Cyclotheonamide A, R=CHO
Cyclotheonamide B, R=Ac

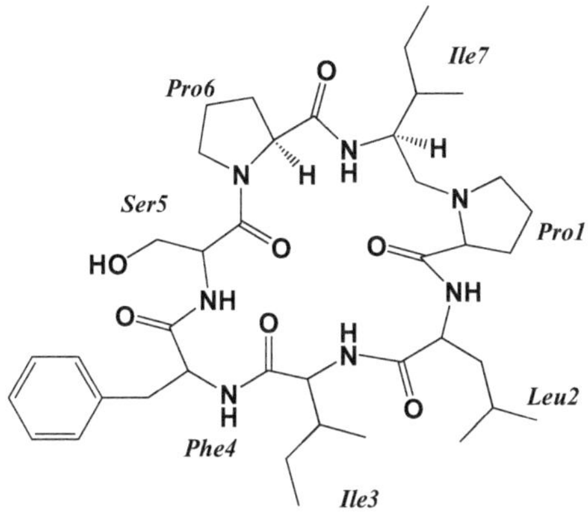

Stylopeptide1

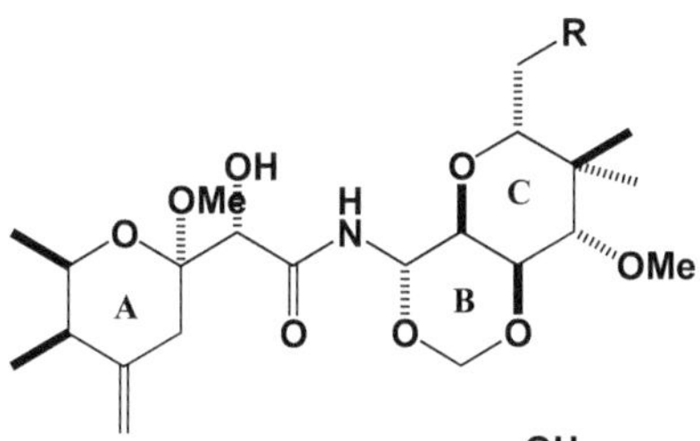

Theopederins A, R= 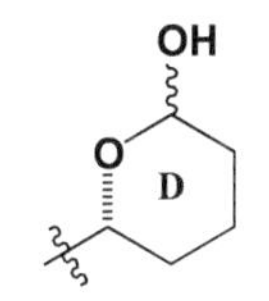

Theopederins B, R= 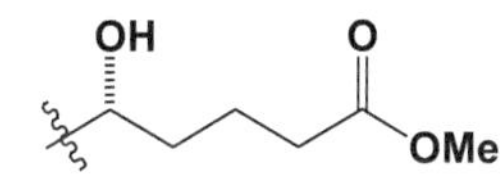

Fig. 11.93 Cyclotheonamides, stylopeptide, theopederin

Fig. 11.94 Keramamide, leucamide

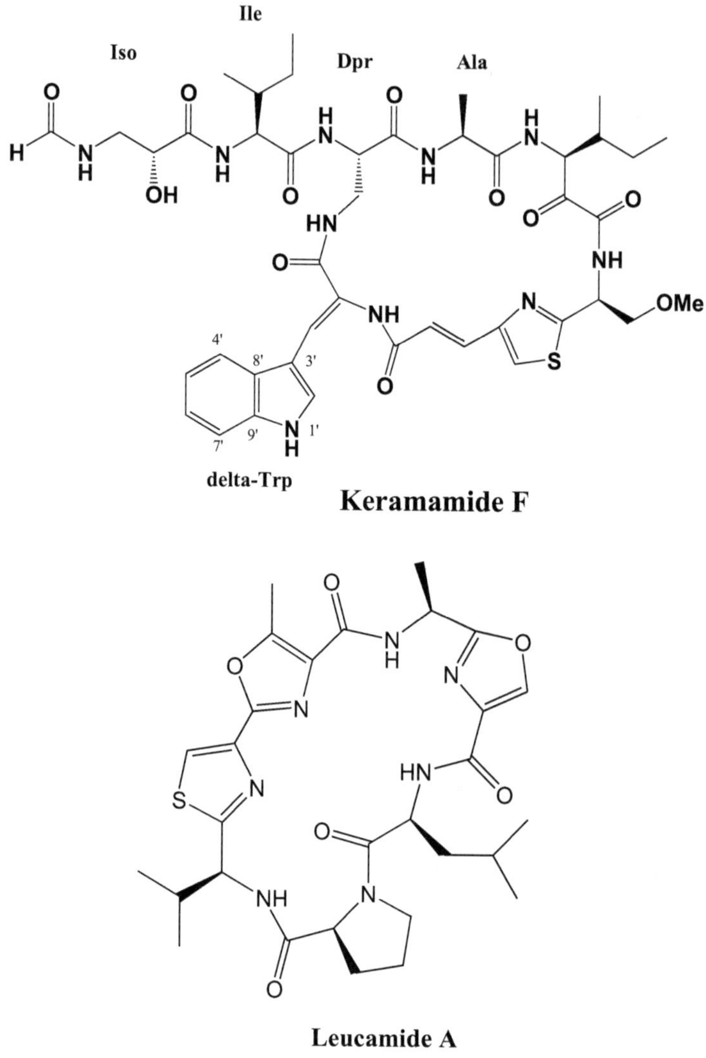

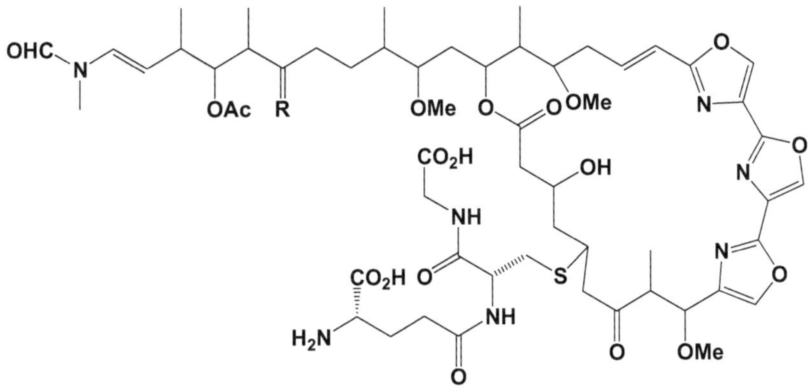

Thiomycalolide A, R=O

Thiomycalolide B, R= 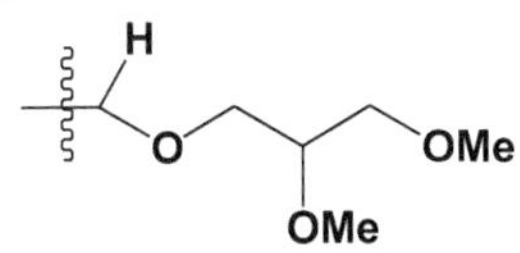

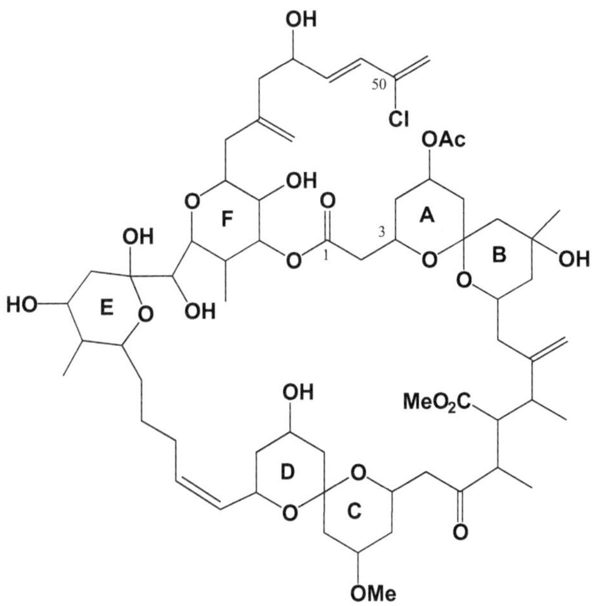

Spongistatin 1

Fig. 11.95 Thiomycalolides, spongistatin

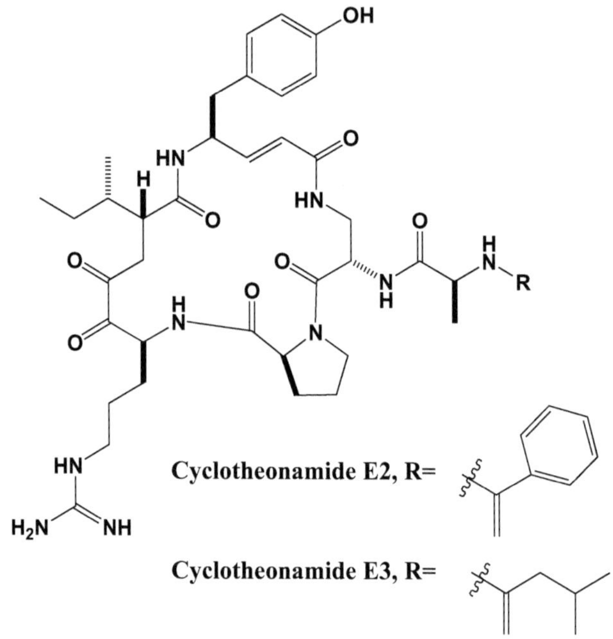

Fig. 11.96 Cyclotheonamides E2 and E3

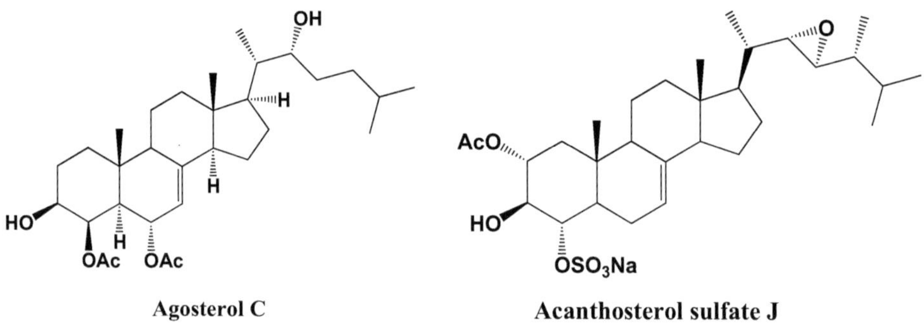

Fig. 11.97 Agosterol C, acanthosterol sulfate J

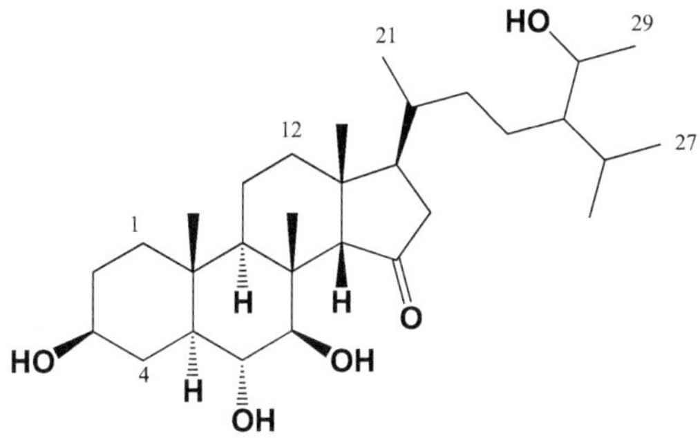

Clathriol

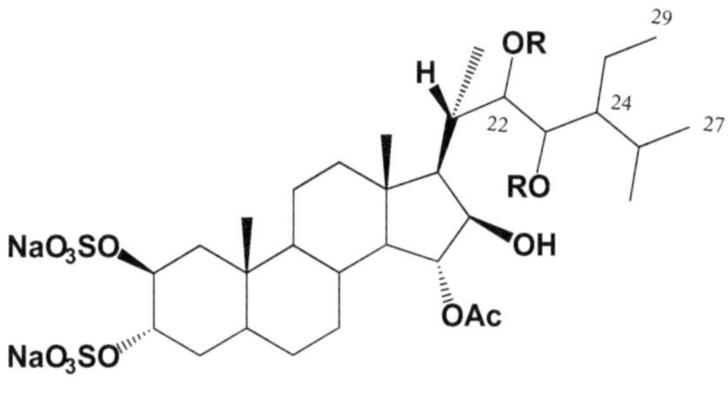

Clathsterol, $R=C_3H_7CO$

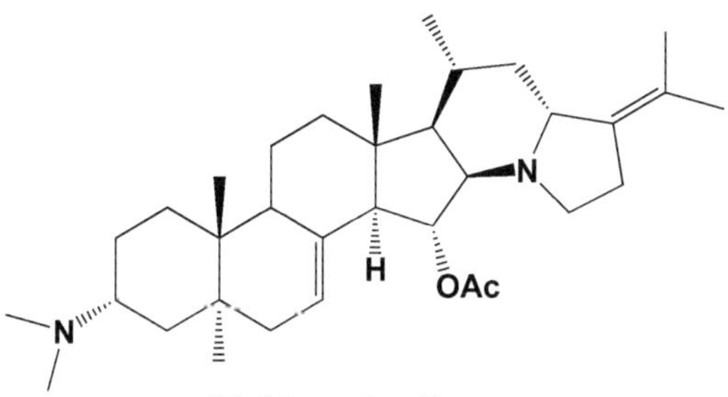

Plakinamine I

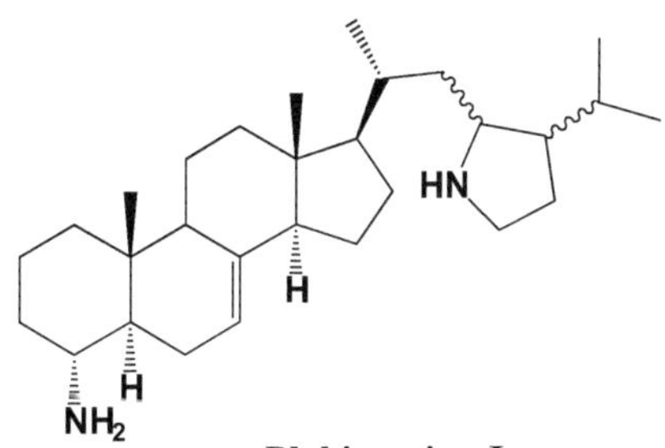

Plakinamine J

Fig. 11.98 Clathriol, clathsterol, plakinamines

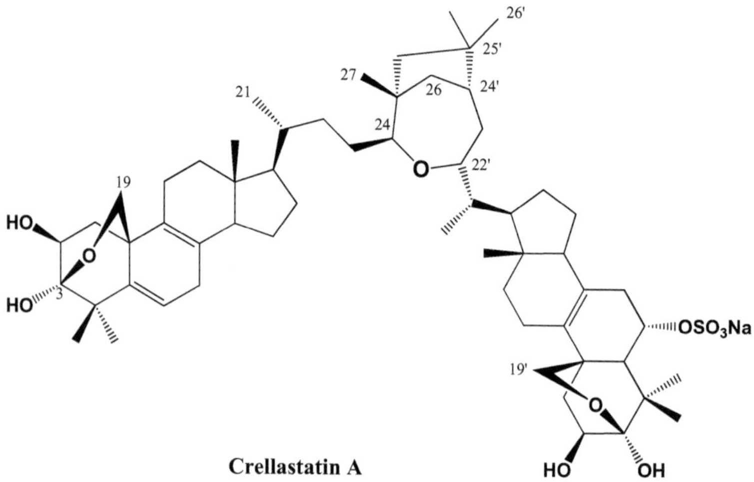

Crellastatin A

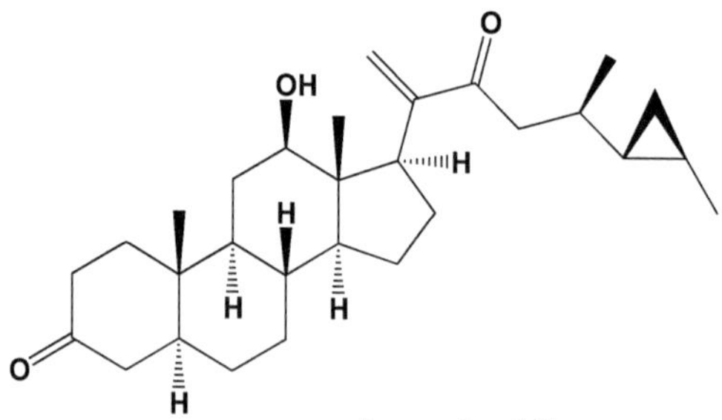

Aragusterol D

Fig. 11.99 Crellastatin, aragusterol

1,3-dimethylisoguanine, *Amphimedon viridis* (Ohta et al. 1997); ancorinoside A (Fig. 11.104), tetramic acid glycoside, *Ancorina* sp. (Sata et al. 1999a); rubrosides A, tetramic acid glycoside, *Siliquariaspongia japonica* (Sata et al. 1999b); aurantosides E and F, polyene tetramic acids comprising an *N*-trisaccharide unit, *Siliquariaspongia japonica* (Sata et al. 1999a) (Fig. 11.105).

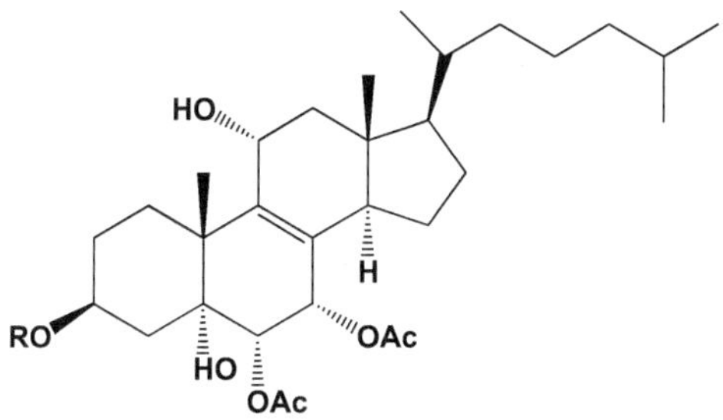

Cholest-8-ene-3$_{beta}$,5$_{alfa}$,6$_{alfa}$,7$_{alfa}$,
10$_{alfa}$-pentol 6,7-diacetate, R=H
Cholest-8-ene-3$_{beta}$,5$_{alfa}$,6$_{alfa}$,7$_{alfa}$,
10$_{alfa}$-pentol 3,6,7-triacetate, R=Ac

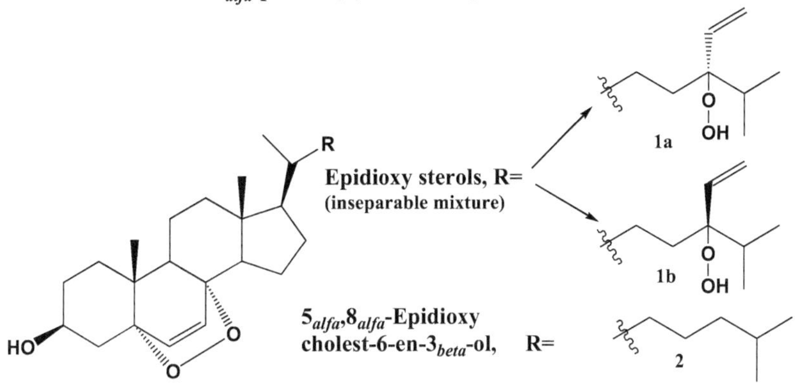

Fig. 11.100 Polyoxygenated sterols, epidioxy sterols

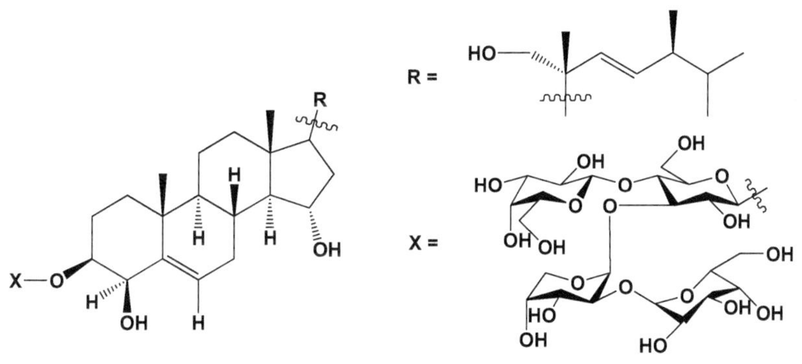

Mycaloside A

Fig. 11.101 Mycaloside A

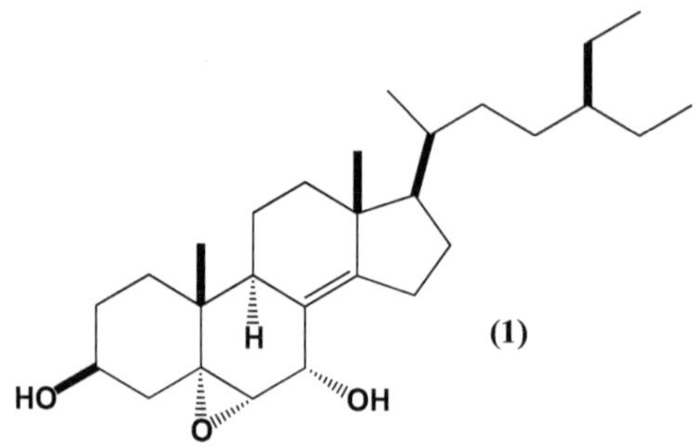

(1)

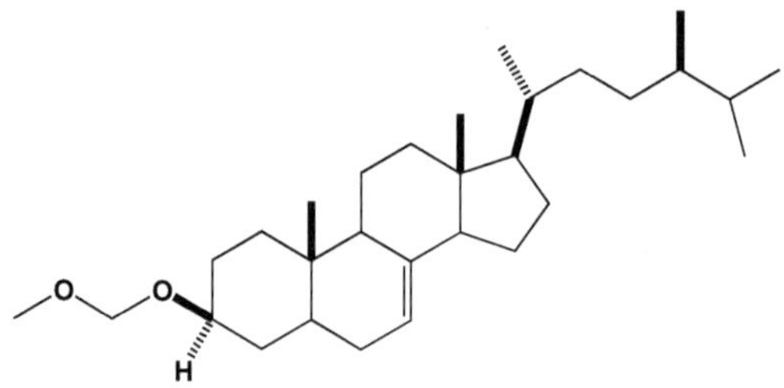

Sterol ether

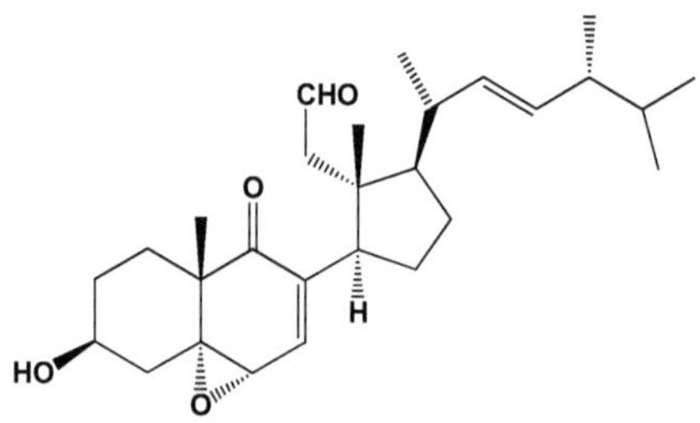

3-*O*-Deacetylluffasterol B

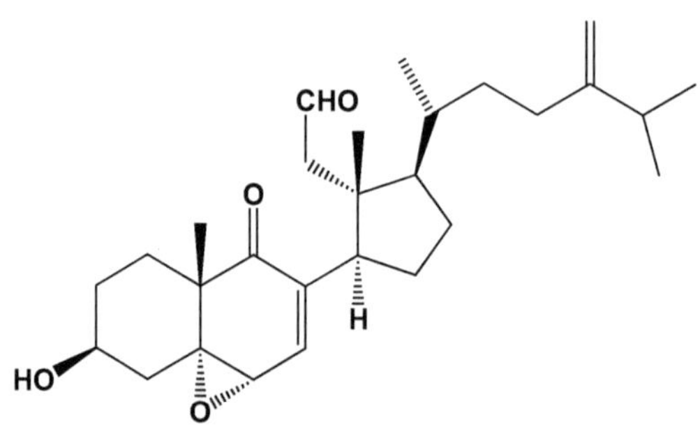

3-*O*-Deacetyl-22,23-dihydro-24,28-dehydroluffasterol B

Fig. 11.102 Sterol ether, luffasterol derivatives

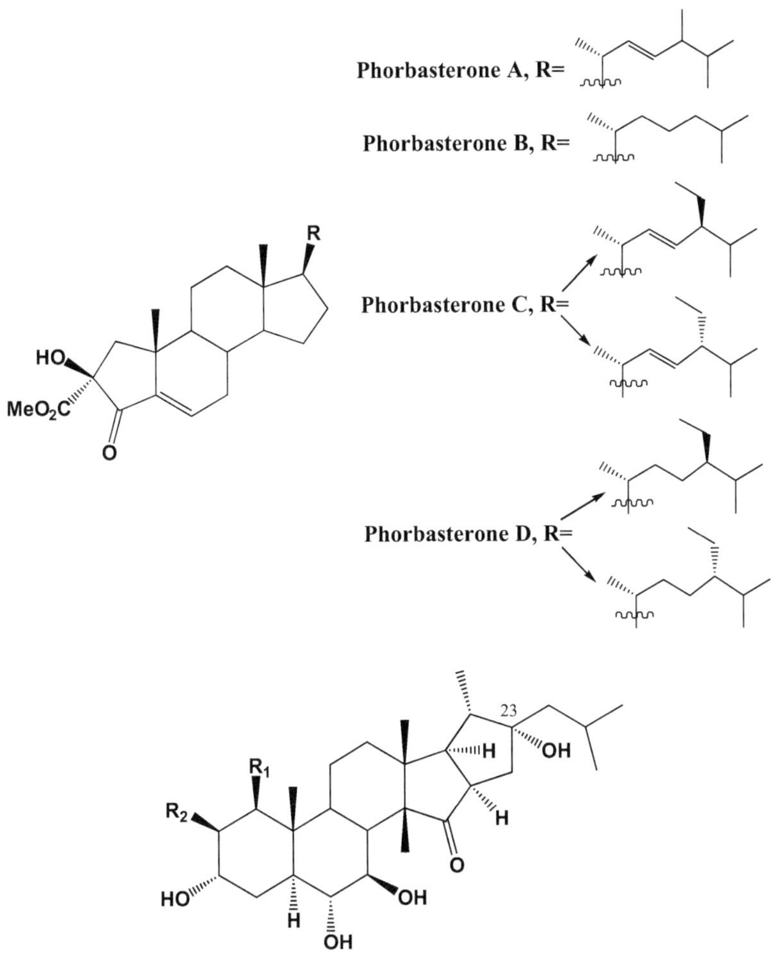

Xestobergsterol A, R_1, R_2=H
Xestobergsterol B, R_1,R_2=OH

Fig. 11.103 Phorbasterones, xestobergsterols

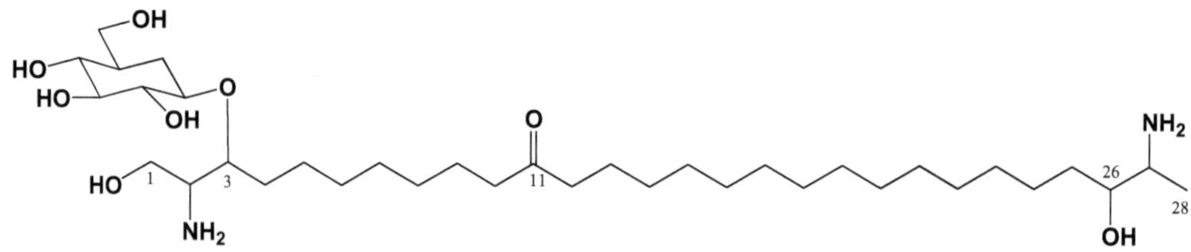

(-)-Oceanapiside

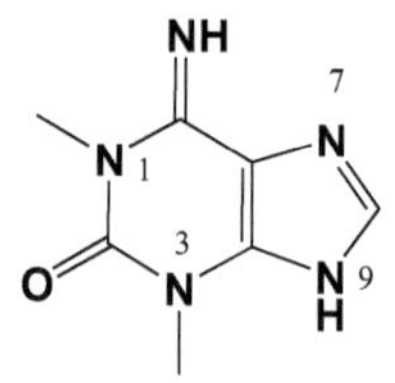

1,3-Dimethylisoguanine

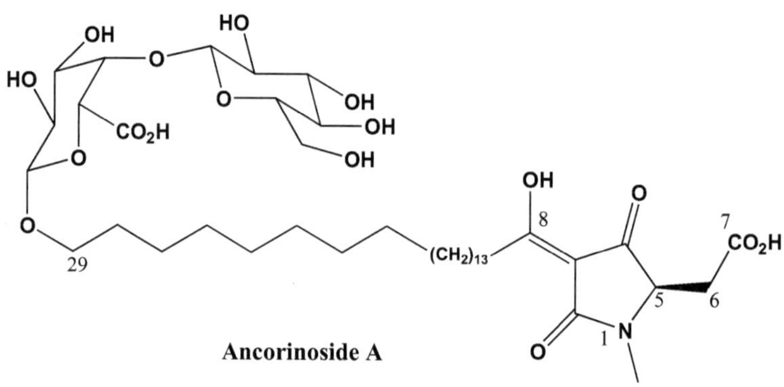

Ancorinoside A

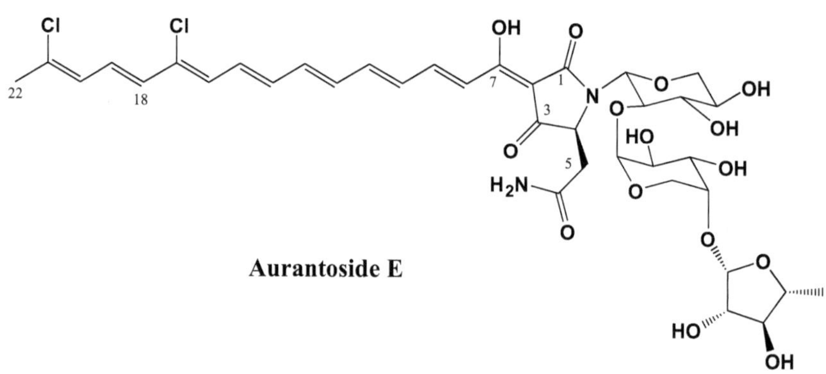

Aurantoside E

Fig. 11.104 Oceanapiside, 1,3-dimethylisoguanine, ancorinoside A

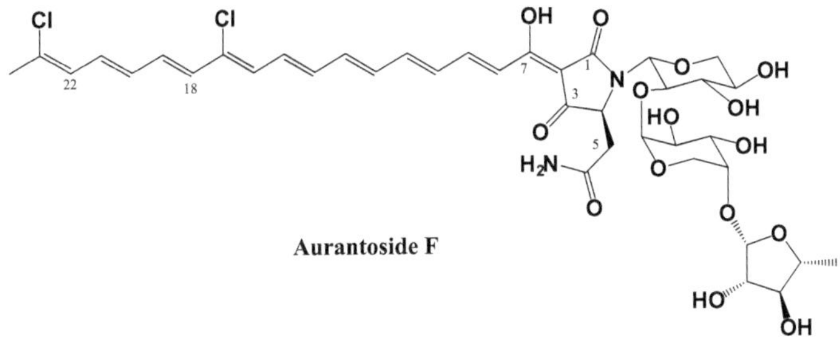

Aurantoside F

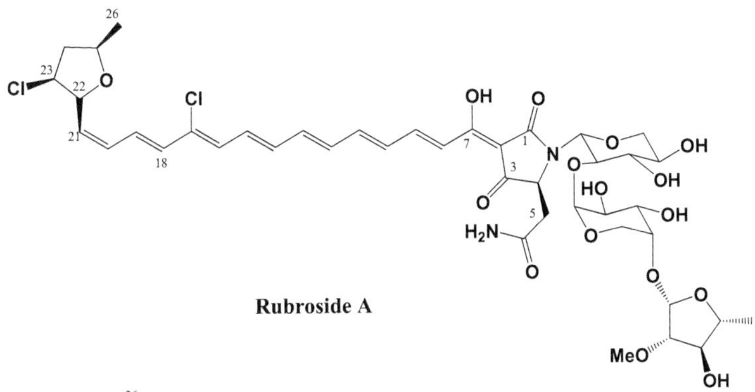

Rubroside A

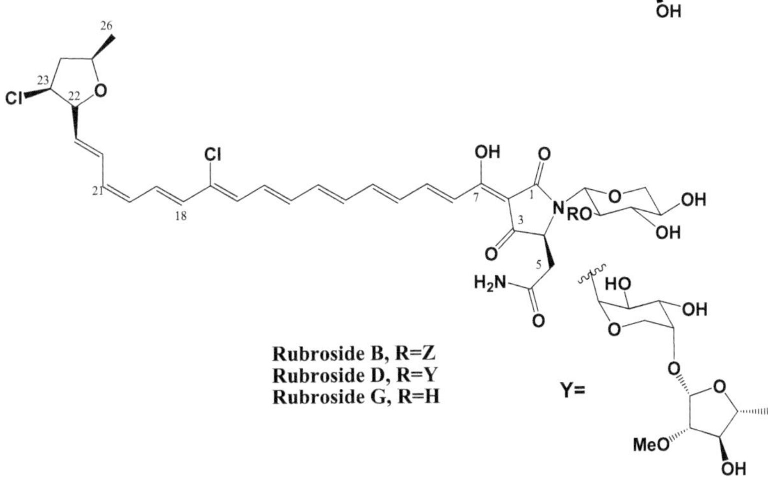

Rubroside B, R=Z
Rubroside D, R=Y
Rubroside G, R=H

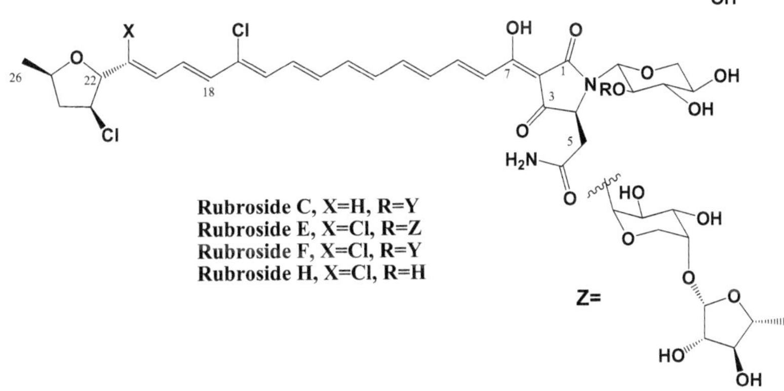

Rubroside C, X=H, R=Y
Rubroside E, X=Cl, R=Z
Rubroside F, X=Cl, R=Y
Rubroside H, X=Cl, R=H

Fig. 11.105 Aurantosides, rubrosides

11.7 Conclusion

Sponges harbouring microbiota, have been the fascinating animals for researchers to have unique molecules. The associated microbes are a prime role in the formation of these secondary metabolites. The recovery of the sponge products is not easily reproducible from the cultured sponges unlike that of wild ones. So this requires attention by the multidisciplinary work involving marine biologists, chemists, pharmacologists, and statisticians to arrive at a holistic view about the product retrieval in an economical way when culture of sponges is taken in captivity. Modern techniques of biotechnology are increasingly essential for genetically modified organisms to get the desired product.

References

Adamczeski M, Quiñoá E, Crews P (1988) Unusual anthelminthic oxazoles from a marine sponge. J Am Chem Soc 110:1598–1602

Akee RK, Carroll TR, Yoshida WY, Scheuer PJ (1990) Two imidazole alkaloids from a sponge. J Org Chem 55:1944–1946

Alvi KA, Diaz MC, Crews P, Slates DL, Lee RH, Moretti R (1992) Evaluation of New sesquiterpene quinones from two *Dysidea* sponge species as inhibitors of protein tyrosine kinase. J Org Chem 57:6604–6607

Ankisetty S, Amsler CD, McClintock JB, Baker BJ (2004) Further membranolide diterpenes from the antarctic sponge *Dendrilla membranosa*. J Nat Prod 67:1172–1174

Antonov AS, Afiyatullov SS, Kalinovsky AI, Ponomarenko MP, Dmitrenok PS, Aminin DL, Agafonova IG, Stonik VA (2003) Mycalosides B–I, eight new spermostatic steroid oligoglycosides from the sponge *Mycale laxissima*. J Nat Prod 66:1082–1088

Aoki S, Nemoto N, Kobayashi Y, Kobayashi M, Kitagawa I (2001) Molecular modeling of 5-desacetylaltohyrtin A, a spongean cytotoxic macrolide. Tetrahedron 57:2289–2292

Assmann M, Zea S, Köck M (2001) Sventrin, a new bromopyrrole alkaloid from the Caribbean sponge *Agelas sventres*. J Nat Prod 64:1593–1595

Bewley CA, He H, Williams DH, Faulkner DJ (1996) Aciculitins A–C: cytotoxic and antifungal cyclic peptides from the lithistid sponge *Aciculites orientalis*. J Am Chem Soc 118:4314–4321

Blackburn CL, Hopmann C, Sakowicz R, Berdelis MS, Goldstein LSB, Faulkner DJ (1999) Adociasulfates 1–6, inhibitors of kinesin motor proteins from the sponge *Haliclona* (aka *Adocia*) sp. J Org Chem 64:5565–5570

Blaine Harrison B, Phillip Crews P (1998) Cyclic polyketide peroxides and acyclic diol analogues from the sponge *Plakortis lita*. J Nat Prod 61:1033–1037

Bobzin SC, Faulkner DJ (1991) Aromatic alkaloids from the marine sponge *Chelonaplysilla* sp. J Org Chem 56:4403–4407

Borbone N, De Marino S, Iorizzi M, Zollo F, Debitus C, Esposito G, Iuvone T (2002) Minor steroidal alkaloids from the marine sponge *Corticium* sp. J Nat Prod 65:1206–1209

Cafieri F, Fattorusso E, Scafati OY (1998) Novel bromopyrrole alkaloids from the sponge *Agelas dispar*. J Nat Prod 61:122–125

Cambie RC, Rutledge PS, Yang XS, Bergquist PR (1998) Chemistry of sponges. 18.1 12-desacetylfuroscalar-16-one, a new sesterterpene from a *Cacospongia* sp. J Nat Prod 61:1416–1417

Capon RJ, Faulkner DJ (1984) Antimicrobial metabolites from a Pacific sponge *Agelas* sp. J Am Chem Soc 106:1819–1822

Capon RJ, Rooney F, Murray LM, Collins E, Sim ATR, Rostas JAP, Butler MS, Carrol AR (1998) Dragmacidins: new protein phosphatase inhibitors from a Southern Australian deep-water marine sponge, *Spongosorites* sp. J Nat Prod 61:660–662

Capon RJ, Skene C, Lacey E, Gill JH, Wadsworth D, Friedel T (1999) Geodin A magnesium salt: a novel nematocide from a Southern Australian marine sponge, *Geodia*. J Nat Prod 62:1256–1259

Capon RJ, Miller M, Rooney F (2001) Mirabilin G: a new alkaloid from a Southern Australian marine sponge, *Clathria* species. J Nat Prod 64:643–644

Capon RJ, Skene C, Vuong D, Lacey E, Gill JH, Heiland K, Friedel T (2002) Equilibrating isomers: bromoindoles and a seco-xanthine encountered during a study of nematocides from the Southern Australian marine sponge *Hymeniacidon* sp. J Nat Prod 65:368–370

Carletti I, Banaigs B, Amade P (2000) Matemone, a new bioactive bromine-containing oxindole alkaloid from the Indian Ocean sponge *Iotrochota purpurea*. J Nat Prod 63:981–983

Carletti I, Long C, Funel C, Amade P (2003) Sodwanones K, L, and M; new triterpenes from the marine sponge *Axinella Weltneri*. J Nat Prod 66:25–29

Carroll AR, Pierens GK, Fechner G, de Almeida LP, Ngo A, Simpson M, Hyde E, Hooper JNA, Boström SL, Musil D, Quinn RJ (2002) Dysinosin A: a novel inhibitor of factor VIIa and thrombin from a new genus and species of Australian sponge of the Family Dysideidae. J Am Chem Soc 124:13340–13341

Casapullo A, Bifulco G, Bruno I, Riccio R (2000) New bisindole alkaloids of the topsentin and hamacanthin classes from the Mediterranean marine sponge *Rhaphisia lacazei*. J Nat Prod 63:447–451

Casapullo A, Cutignano A, Bruno I, Bifulco G, Debitus C, Gomez-Paloma L, Riccio R (2001) Makaluvamine P, a new cytotoxic pyrroloiminoquinone from *Zyzzya* cf. *fuliginosa*. J Nat Prod 64:1354–1356

Chan WR, Tinto WF, Manchand PS, Todaro LJ (1987) Stereostructures of Geodiamolides A and B, novel cyclodepsipeptides from the marine sponge *Geodia sp*. J Org Chem 52:3091–3093

Chang CWJ, Patra A, Roll DM, Scheuer PJ, Matsumoto GK, Clardy J (1984) Kalihinol-A, a highly functionalized diisocyano diterpenoid antibiotic from a sponge. J Am Chem Soc 106:4644–4646

Chang LC, Quintero SO, Hooper JNA, Bewley CA (2002) Batzelline D and Isobatzelline E from the Indopacific sponge *Zyzzya fuliginosa*. J Nat Prod 65:776–778

Chen C, Shen Y, Chen Y, Sheu J, Duh C (1999) Bioactive sesquiterpenes from a Taiwanese marine sponge *Parahigginsia* sp. J Nat Prod 62:573–576

Choi K, Hong J, Lee CO, Kim DK, Sim CJ, Im KS, Jung JH (2004) Cytotoxic furanosesterterpenes from a marine sponge *Psammocinia* sp. J Nat Prod 67:1186–1189

Copp BR, Fulton KF, Perry NB, Blunt JW, Munro MHG (1994) Natural and synthetic derivatives of Discorhabdin C, a cytotoxic pigment from the New Zealand Sponge *Latrunculia* cf. bocagei. J Org Chem 59:8233–8238

Corley DG, Moore RHRE, Scheuer PJ (1988) Laulimalides: new potent cytotoxic macrolides from a marine Sponge and a Nudibranch Predator. J Org Chem 53:3644–3646

Costantino V, Fattorusso E, Mangoni A, Di Rosa M, Ianaro A (1997) Glycolipids from sponges. 6.1 Plakoside A and B, two unique prenylated glycosphingolipids with immunosuppressive activity from the marine sponge *Plakortis simplex*. J Am Chem Soc 119:12465–12470

D'Auria MV, Giannini C, Minale L, Zampella A, Debitus C, Frostin M (1997) Bengamides and related new amino acid derivatives from the New Caledonian marine sponge *Jaspis carteri*. J Nat Prod 60:814–816

D'Auria MV, Giannini C, Zampella A, Minale L, Debitus C, Roussakis C (1998) Crellastatin A: a cytotoxic Bis-Steroid sulfate from the Vanuatu marine sponge *Crella* sp. J Org Chem 63:7382–7388

Dai JR, Hallock YF, Cardellina JH, Gray GN, Boyd MR (1996) Vasculyne, a new cytotoxic acetylenic alcohol from the marine sponge *Cribrochalina vasculum*. J Nat Prod 59:88–89

de Guzman FS, Carte B, Troupe N, Faulkner DJ, Harper MK, Conception GP, Mangalindan GC, Matsumoto SS, Barrows LR, Ireland CM (1999) Neoamphimedine: a new pyridoacridine topoisomerase II inhibitor which catenates DNA. J Org Chem 64:1400–1402

De Rosa S, Crispino A, De Giulio A, Iodice C, Pronzato R, Zavodnik N (1997) Cavernosolide, a new sesterterpene from a Tyrrhenian sponge. J Nat Prod 60:844–846

De Rosa S, Crispino A, De Giulio A, Iodice C, Benrezzouk R, Terencio MC, Ferrándiz ML, Alcaraz MJ, Paya MA (1998) A new cacospongionolide inhibitor of human secretory phospholipase A2 from the Tyrrhenian sponge *Fasciospongia cavernosa* and absolute configuration of Cacospongionolides. J Nat Prod 61:931–935

De Rosa S, Crispino A, De Giulio A, Iodice C, Amodeo P, Tancredi T (1999) A new cacospongionolide derivative from the sponge *Fasciospongia cavernosa*. J Nat Prod 62:1316–1318

Dijoux MG, Gamble WR, Hallock YF, Cardellina JH, Van Soest R, Boyd MR (1999) A new discorhabdin from two sponge genera. J Nat Prod 62:636–637

Djura P, Stierle DB, Sullivan B, Faulkner DJ, Arnold EV, Clardy J (1980) Some metabolites of the marine sponges *Smenospongia aurea* and *Smenospongia* (= *Polyfibrospongia*) *echina*. J Org Chem 45:1435–1441

Eder C, Schupp P, Proksch P, Wray V, Steube K, Müller C, van Soest RWM (1998) Bioactive pyridoacridine alkaloids from the Micronesian sponge *Oceanapia* sp. J Nat Prod 61:301–305

Eder C, Proksch P, Wray V, Van Soest RWM, Ferdinandus E, Pattisina LA, Sudarsono (1999) New bromopyrrole alkaloids from the indopacific sponge *Agelas nakamurai*. J Nat Prod 62:1295–1297

Edrada RA, Proksch P, Wray V, Christ R, Witte L, van Soest RWM (1996a) Bioactive isoquinoline quinone from an undescribed Philippine marine sponge of the genus *Xestospongia*. J Nat Prod 59:973–976

Edrada RA, Proksch P, Wray V, Witte L, Muller WE, Van Soest RW (1996b) Four new bioactive manzamine-type alkaloids from the Philippine marine sponge *Xestospongia ashmorica*. J Nat Prod 59:1056–1060

Edrada RA, Stessman CC, Crews P (2003) Uniquely modified imidazole alkaloids from a calcareous *Leucetta* sponge. J Nat Prod 66:939–942

El Sayed KA, Kelly M, Kara UAK, Ang KKH, Katsuyama DC, Dunbar AA, Khan AA, Hamann MT (2001a) New manzamine alkaloids with potent activity against infectious diseases. J Am Chem Soc 123:1804–1808

El Sayed KA, Hamann MT, Hashish NE, Shier WT, Kelly M, Khan AA (2001b) Antimalarial, antiviral, and antitoxoplasmosis norsesterterpene peroxide acids from the Red Sea sponge *Diacarnus erythraeanus*. J Nat Prod 64:522–524

Erickson KL, Beutler JA, Cardellina JH, Boyd MR (1997) Salicylihalamides A and B, novel cytotoxic macrolides from the marine sponge *Haliclona* sp. J Org Chem 62:8188–8192

Fernández R, Dherbomez M, Letourneux Y, Nabil M, Verbist JF, Biard JF (1999) Antifungal metabolites from the Marine sponge *Pachastrissa* sp.: new Bengamide and Bengazole derivatives. J Nat Prod 62:678–680

Ford J, Capon RJ (2000) Discorhabdin R: a new antibacterial pyrroloiminoquinone from two

Latrunculiid marine sponges, *Latrunculia* sp. and *Negombata* sp. J Nat Prod 63:1527–1528

Fu X, Schmitz FJ, Tanner RS, Kelly-Borges M (1998a) Agelasines H and I, 9-Methyladenine-containing diterpenoids from an *Agelas* sponge. J Nat Prod 61:548–550

Fu X, Schmitz FJ, Borges MK, McCready TL, Holmes CFB (1998b) Clavosines A–C from the marine sponge *Myriastra clavosa*: potent cytotoxins and inhibitors of protein phosphatases 1 and 2A. J Org Chem 63:7957–7963

Fu X, Zeng L, Su J, Schmitz FJ (1999) Phyllofolactones C and D, two new minor Homoscalarane Sesterterpenes from the Chinese sponge *Phyllospongia foliascens*. J Nat Prod 62:644–646

Fusemi N, Sugawara T, Mataunaga S (1992) Theopederins A–E, potent antitumor metabolites from a marine sponge, *Theonella* sp. J Org Chem 57:3828–3832

Fusetani N, Matsunaga S (1990) Cyclotheonamides, potent thrombin inhibitors, from a marine sponge theonella sp. J Am Chem Soc 112:7053–7054

Garrido L, Zubía E, Ortega MJ, Salva J (1997) New furanoterpenoids from the sponge *Spongia officinalis*. J Nat Prod 60:794–797

Giannini C, Debitus C, Lucas R, Ubeda A, Paya M, Hooper JNA (2001) New sesquiterpene derivatives from the sponge *Dysidea* species with a selective inhibitor profile against human phospholipase A2 and other leukocyte functions. J Nat Prod 64:612–615

Goclik E, König GM, Wright AD, Kaminsky R (2000) Pelorol from the tropical marine sponge *Dactylospongia elegans*. J Nat Prod 63:1150–1152

Gross H, Kehraus S, König GM, Woerheide G, Wright AD (2002) New and biologically active imidazole alkaloids from two sponges of the genus *Leucetta*. J Nat Prod 65:1190–1193

Groweiss A, Newcomer JJ, O'Keefe BR, Blackman A, Boyd MR (1999) Cytotoxic metabolites from an Australian collection of the sponge *Jaspis* species. J Nat Prod 62:1691–1693

Gunasekera SP, Gunasekera M, Longley RE, Schulte GK (1990) Discodermolide: a new bioactive polyhydroxylated lactone from the marine sponge *Discodermia dissoluta*. J Org Chem 55:4912–4915

Gunasekera SP, Borges MK, Longley RE (1996a) A new cytotoxic sterol methoxymethyl ether from a deep water marine sponge *Scleritoderma* sp. cf. *paccardi*. J Nat Prod 59:161–162

Gunasekera SP, McCarthy PJ, Borges MK, Lobkovsky E, Clardy J (1996b) Dysidiolide: a novel protein phosphatase inhibitor from the Caribbean sponge *Dysidea etheria* de Laubenfels. J Am Chem Soc 118:8759–8760

Gunasekera SP, McCarthy PJ, Longley RE, Pomponi SA, Wright AE, Lobkovsky E, Clardy J (1999a) Discorhabdin P, a new enzyme inhibitor from a deep-water Caribbean sponge of the genus *Batzella*. J Nat Prod 62:173–175

Gunasekera SP, McCarthy PJ, Longley RE, Pomponi SA, Wright AE (1999b) Secobatzellines A and B, two new enzyme inhibitors from a deep-water Caribbean sponge of the genus *Batzella*. J Nat Prod 62:1208–1211

Gunasekera SP, Zuleta IA, Longley RE, Wright AE, Pomponi SA (2003) Discorhabdins S, T, and U, new cytotoxic pyrroloiminoquinones from a deep-water Caribbean sponge of the genus *Batzella*. J Nat Prod 66:1615–1617

Gunawardana GP, Kohmoto S, Gunasekera SP, McConnell OJ, Koehn FE (1988) Dercitin, a new biologically active acridine alkaloid from a deep water marine sponge, *Dercitus* sp. J Am Chem Soc 110:4856–4858

Hassan W, Edrada RA, Ebel R, Wray V, Berg A, Van Soest RWM, Wiryowidagdon S, Proksch P (2004) New imidazole alkaloids from the Indonesian sponge *Leucetta chagosensis*. J Nat Prod 67:817–822

Hong TW, Jímenez DR, Molinski TF (1998) Agelastatins C and D, new pentacyclic bromopyrroles from the sponge *Cymbastela* sp., and potent arthropod toxicity of (−)-agelastatin A. J Nat Prod 61:158–161

Hoshino A, Mitome H, Miyaoka H, Shintani A, Yamada Y, Van Soest RWM (2003) New strongylophorines from the Okinawan marine sponge *Petrosia (Strongylophora) corticata*. J Nat Prod 66:1600–1605

Hu JF, Schetz JA, Kelly M, Peng JN, Ang KK, Flotow H, Leong CY, Ng SB, Buss AD, Wilkins SP, Hamann MT (2002) New antiinfective and human 5-HT2 receptor binding natural and semisynthetic compounds from the Jamaican sponge *Smenospongia aurea*. J Nat Prod 65:476–480

Iguchi K, Shimada Y, Yamada Y (1992) Hyrtiosal, a New sesterterpenoid with a novel carbon skeleton from the Okinawa marine sponge *Hyrtios erectus*. J Org Chem 57:522–524

Iguchi K, Shimura H, Taira S, Yokoo C, Matsumoto K, Yamada Y (1994) Aragusterol B and D, new 26,27-cyclosterols from the okinawan marine sponge of the genus *Xestospongia*. J Org Chem 59:7499–7502

Ishibashi M, Moore RE, Patterson GML, Xu C, Clardy J (1986) Scytophycins, cytotoxic and antimycotic agents from the cyanophyte Scytonema pseudohofmanni. J Org Chem 51:5300–5306

Ishibashi M, Ohizumi Y, Cheng J, Nakamura H, Hirata Y, Sasaki T, Kobayashi J (1988) Metachromins A and B, novel antineoplastic sesquiterpenoids from the Okinawan sponge *Hippospongia* c f. *Metachromia*. J Org Chem 53:2855–2858

Ishiyama H, Ishibashi M, Ogawa A, Yoshida S, Kobayashi J (1997) Taurospongin A, a novel acetylenic fatty acid derivative inhibiting DNA polymerase *â* and HIV reverse transcriptase from sponge *Hippospongia* sp. J Org Chem 62:3831–3836

Issa HH, Tanaka J, Higa T (2003) New cytotoxic furanosesterterpenes from an Okinawan marine sponge, *Ircinia* sp. J Nat Prod 66:251–254

Itagaki F, Shigemori H, Ishibashi M, Nakamura T, Sasaki T, Kobayashi J (1992) Keramamide F, a new thiazole-containing peptide from the Okinawan marine sponge *Theonella* sp. J Org Chem 57:5540–5542

Iwagawa T, Kaneko M, Okamura H, Nakatani M, Van Soest RWM (1998) New alkaloids from the Papua New Guinean aponge *Agelas nakamurai*. J Nat Prod 61:1310–1312

Jares-Erijman EA, Ingrum AL, Carney JR, Rinehart KL, Sakai R (1993) Polycyclic guanidine-containing compounds from the Mediterranean sponge *Crambe crambe*: the structure of 13,14,15-isocrambescidin 800 and the absolute stereochemistry of the pentacyclic guanidine moieties of the crambescidins. J Org Chem 58:4805–4808

Jiménez JI, Yoshida WY, Scheuer PJ, Kelly M (2000a) Scalarane-based sesterterpenes from an Indonesian sponge *Strepsichordaia aliena*. J Nat Prod 63:1388–1392

Jiménez JI, Yoshida WY, Scheuer PJ, Lobkovsky E, Clardy J, Kelly M (2000b) Honulactones: new bishomoscalarane sesterterpenes from the Indonesian sponge *Strepsichordaia aliena*. J Org Chem 65:6837–6840

Jiménez MS, Garzón SP, Rodríguez AD (2003) Plakortides M and N, bioactive polyketide endoperoxides from the Caribbean marine sponge *Plakortis halichondrioides*. J Nat Prod 66:655–661

Kalaitzis JA, Quinn RJ (1999a) Adociasulfate-9, a new hexaprenoid hydroquinone from the Great Barrier Reef sponge *Adocia aculeata*. J Nat Prod 62:1682–1684

Kalaitzis JA, Leone PA, Harris L et al (1999b) Adociasulfates 1, 7, and 8: new bioactive hexaprenoid hydroquinones from the marine sponge *Adocia* sp. J Org Chem 64:5571–5574

Kazuhiko K, Shigemori H, Kikuchi Y, Masami I, Takuma S, Kobayashi S (1992) Ircinals A and B from the Okinawan marine sponge *Ircinia* sp.: plausible biogenetic precursors of manzamine alkaloids. J Org Chem 57:2480–2483

Kehraus S, König GM, Wright AD, Woerhelde G (2002a) Leucamide a: a new cytotoxic heptapeptide from the Australian sponge *Leucetta microraphis*. J Org Chem 67:4989–4992

Kehraus S, König GM, Wright AD (2002b) A new cytotoxic calyculinamide derivative, geometricin a, from the Australian sponge *Luffariella geometrica*. J Nat Prod 65:1056–1058

Kernan MR, Faulkner DJ (1988) Sesterterpene sulfates from a sponge of the family Halichondriidae. J Org Chem 53:4574–4578

Keyzers RA, Northcote PT, Webb V (2002) Clathriol, a novel polyoxygenated 14α steroid isolated from the New Zealand marine sponge *Clathria lissosclera*. J Nat Prod 65:598–600

Killday KB, Yarwood D, Sills MA, Murphy PT, Hooper JN, Wright AE (2001) Microxine, a new cdc2 kinase inhibitor from the Australian marine sponge *Microxina* species. J Nat Prod 64:525–526

Kim JS, Lim YJ, Im KS, Jung JH, Shim CJ, Lee CO, Hong J, Lee H (1999) Cytotoxic polyacetylenes from the marine sponge *Petrosia* sp. J Nat Prod 62:554–559

Kimura J, Ishizuka E, Nakao Y, Yoshida WY, Scheuer PJ, Kelly-Borges M (1998) Isolation of 1-methylherbipoline salts of halisulfate-1 and of suvanine as serine protease inhibitors from a marine sponge, *Coscinoderma mathewsi*. J Nat Prod 61:248–250

Kinnel RB, Gehrken HP, Scheuer PJ (1993) Palau'amine: a cytotoxic and immunosuppressive hexacyclic bisguanidine antibiotic from the sponge *Stylotella agminata*. J Am Chem Soc 115:3376–3377

Kinnel RB, Gehrken HP, Swali R, Skoropowski G, Scheuer PJ (1998) Palau'amine and its congeners: a family of bioactive bisguanidines from the marine sponge *Stylotella aurantium*. J Org Chem 63:3281–3286

Kirsch G, Köng GM, Wright AD, Kaminsky R (2000) A new bioactive sesterterpene and antiplasmodial alkaloids from the marine sponge *Hyrtios* cf. *Erecta*. J Nat Prod 63:825–829

Kobayashi J, Ishibashi M, Walchili MR, Nakamura H, Hirata Y, Sasaki T, Ohizuni Y (1988) Amphidinolide C: the first 25-membered macrocyclic lactone with potent antineoplastic activity from the cultured dinoflagellate *Amphidinium* sp. J Am Chem Soc 110:490–494

Kobayashi M, Aoki S, Sakai H, Kihara N, Sasaki T, Kitagawa I (1993) Altohyrtin A, a potent anti-tumor macrolide from the Okinawan marine sponge *Hyrtios altum*. Tetrahedron Lett 34:2795–2798

Kobayashi M, Chen Higuchi YJK, Aoki S (1996) Marine natural products. XXXVII. Aragusteroketals A and C, two novel cytotoxic steroids from a marine sponge of *Xestospongia* sp. Chem Pharm Bull 44:1840–1842

Kondracki MLB, Lacombe F, Guyot M (1999) Methanol adduct of puupehenone, a biologically active derivative from the marine sponge *Hyrtios* sp. J Nat Prod 62:1304–1305

König GM, Wright AD (1997) New and unusual sesquiterpenes: kelsoene, prespatane, *epi*-γ-gurjunene, and T-cadinthiol, from the tropical marine sponge *Cymbastela hooperi*. J Org Chem 62:3837–3840

Lee HS, Seo Y, Rho JR, Shin J, Paul VJ (2001) New steroidal alkaloids from an undescribed sponge of the genus *Corticium*. J Nat Prod 64:1474–1476

Leone PA, Redburn J, Hooper JNA, Quinn RJ (2000) Polyoxygenated *Dysidea* sterols that inhibit the binding of [I125] IL-8 to the human recombinant IL-8 receptor type A. J Nat Prod 63:694–697

Lerch ML, Harper MK, Faulkner DJ (2003) Brominated polyacetylenes from the Philippines sponge *Diplastrella* sp. J Nat Prod 66:667–670

Li H, Matsunaga S, Fusetani N (1996) Halicylindramides D and E, antifungal peptides from the marine sponge *Halichondria cylindrata*. J Nat Prod 59:163–166

Liu Y, Bae BH, Alam N, Hong J, Sim C, Lee C, Im KS, Jung JH (2001) New cytotoxic sesterterpenes from the sponge *Sarcotragus* species. J Nat Prod 64:1301–1304

Liu Y, Hong J, Lee CO, Im KS, Kim ND, Choi JS, Jung JH (2002) Cytotoxic pyrrolo- and furanoterpenoids from the sponge *Sarcotragus* species. J Nat Prod 65:1307–1314

Liu Y, Mansoor TA, Hong J, Lee CO, Sim CJ, Im KS, Kim ND, Jung JH (2003) New cytotoxic sesterterpenoids and norsesterterpenoids from two sponges of the genus *Sarcotragus*. J Nat Prod 66:1451–1456

Liu H, Namikoshi M, Meguro S, Nagai H, Kobayashi H, Yao X (2004) Isolation and characterization of polybrominated diphenyl ethers as inhibitors of microtubule assembly from the marine sponge *Phyllospongia dendyi* collected at Palau. J Nat Prod 67:472–474

Mansoor TA, Hong J, Lee CO (2004) New cytotoxic metabolites from a marine sponge *Homaxinella* sp. J Nat Prod 67:721–724

Marcus AH, Molinski TF, Fahy E, Faulkner DJ, Xu CF, Clardy J (1989) 5-Isothiocyanatopupukeanane from a sponge of the genus *Axinyssa*. J Org Chem 54:5184–5186

Masuno MN, Pawlik JR, Molinski TF (2004) Phorbasterones A–D, cytotoxic *Nor*-ring a steroids from the sponge *Phorbas amaranthus*. J Nat Prod 67:731–733

Matsunaga S, Wakimoto T, Fusetani N (1997) Isolation of four new calyculins from the marine sponge *Discodermia calyx*. J Org Chem 62:2640–2642

Matsunaga S, Nogata Y, Fusetani N (1998a) Thiomycalolides: new cytotoxic trisoxazole-containing macrolides isolated from a marine sponge *Mycale* sp. J Nat Prod 61:663–666

Matsunaga S, Kamimura T, Fusetani N (1998b) Isolation of 1-Carboxymethylnicotinic acid from the Marine sponge *Anthosigmella* cf. *raromicrosclera* as a cysteine protease inhibitor. J Nat Prod 61:671–672

Matsunaga S, Yamashita T, Tsukamoto S, Fusetani N (1999) Three new antibacterial alkaloids from a marine sponge *Stelletta* species. J Nat Prod 62:1202–1204

Matsunaga S, Okada Y, Fusetani N, Van Soest RWM (2000) An antimicrobial C14 acetylenic acid from a marine sponge *Oceanapia* species. J Nat Prod 63:690–691

McCormick JL, McKee TC, Cardellina JH, Leid M, Boyd MR (1996) Cytotoxic triterpenes from a marine sponge, *Stelletta* sp. J Nat Prod 59:1047–1050

Meragelman KM, McKee TC, Boyd MR (2001) New cytotoxic isomalabaricane triterpenes from the sponge *Jaspis* species. J Nat Prod 64:389–392

Meragelman KM, West LM, Northcote PT, Pannell LK, McKee TC, Boyd MR (2002) Unusual sulfamate indoles and a novel Indolo[3,2-*a*]carbazole from *Ancorina* sp. J Org Chem 67:6671–6677

Mitchell SS, Whitehill AB, Rosenthal HGT, Ireland CM (1997) Isolation and characterization of 1,3-dimethylisoguanine from the Bermudian sponge *Amphimedon viridis*. J Nat Prod 60:727–728

Miyaoka H, Nishijima S, Mitome H, Yamada Y (2000) Three new scalarane sesterterpenoids from the Okinawan sponge *Hyrtios erectus*. J Nat Prod 63:1369–1372

Moon B, Baker BJ, McClintock JB (1998) Purine and nucleoside metabolites from the antarctic sponge *Isodictya erinacea*. J Nat Prod 61:116–118

Murakami Y, Takei M, Shindo K, Kitazume C, Tanaka J, Higa T, Fukamachi H (2002) Cyclotheonamide E4 and E5, new potent tryptase inhibitors from an *Ircinia* species of sponge. J Nat Prod 65:259–261

Musman M, Ohtani II, Nagaoka D, Tanaka J, Higa T (2001) Hipposulfates A and B, new sesterterpene sulfates from an Okinawan Sponge, *Hippospongia* cf. *Metachromia*. J Nat Prod 64:350–352

Nakamura H, Wu H, Kobayashi J, Kobayashi M, Ohizumi Y, Hirata Y (1985) Agelasidines. Novel hypotaurocyamine derivatives from the Okinawan sea sponge *Agelas nakamurai* Hoshino. J Org Chem 50:2494–2497

Nakao Y, Oku N, Matsunaga S, Fusetani N (1998) Cyclotheonamides E2 and E3, new potent serine protease inhibitors from the Marine sponge of the genus *Theonella*. J Nat Prod 61:667–670

Nasu SS, Yeung BKS, Hamann MT, Scheuer PJ, Kelly-Borges M, Goins K (1995) Puupehenone-related metabolites from two Hawaiian sponges, *Hyrtios* spp. J Org Chem 60:7290–7292

Nicholas GM, Hong TW, Molinski TF, Lerch ML, Cancilla MT, Lebrilla CB (1999) Oceanapiside, an antifungal Bis-r, ö-amino alcohol glycoside from the Marine sponge *Oceanapia phillipensis*. J Nat Prod 62:1678–1681

Ohta S, Ohta E, Ikegami S (1997) Ancorinoside A: a novel tetramic acid glycoside from the Marine sponge, *Ancorina* sp. Which specifically inhibits blastulation of starfish embryos. J Org Chem 62:6452–6453

Okino T, Yoshimura E, Hirota H, Fusetani N (1996) New antifouling kalihipyrans from the Marine sponge *Acanthella cavernosa*. J Nat Prod 59:1081–1083

Oku N, Matsunaga S, van Soest RWM, Fusetani N (2003) Renieramycin J, a highly cytotoxic tetrahydroisoquinoline alkaloid, from a Marine sponge *Neopetrosia* sp. J Nat Prod 66:1136–1139

Oliveros MB, Edrada RA, Proksch P (1998) A new meroditerpenoid dimer from an undescribed Philippine marine sponge of the genus *Strongylophora*. J Nat Prod 61:948–952

Ortega MJ, Zubía E, Carballo JL, Salva J (1996) Fulvinol, a new long-chain diacetylenic metabolite from the sponge *Reniera fulva*. J Nat Prod 59:1069–1071

Ovenden SPB, Capon RJ (1999) Echinosulfonic acids A-C and echinosulfone A: novel bromoindole sulfonic acids and a sulfone from a Southern Australian marine sponge, *Echinodictyum*. J Nat Prod 62:1246–1249

Ovenden SPB, Capon RJ, Lacey E, Gill JH, Friedel T, Wadsworth D (1999) Amphilactams A-D: novel nematocides from Southern Australian marine sponges of the genus *Amphimedon*. J Org Chem 64:1140–1144

Patil AD, Kumar NV, Kokke WC, Bean MF, Freyer AJ, De Brosse C, Mai S, Truneh A, Faulkner DJ, Carte B, Breen AL, Hertzberg RP, Johnson RK, Westley JW, Potts BCM (1995) Novel alkaloids from the sponge *Batzella* sp.: inhibitors of HIV gpl20-human CD4 binding. J Org Chem 60:1182–1188

Patil AD, Freyer AJ, Taylor PB, Carté B, Zuber G, Johnson RK, Faulkner DJ (1997) Batzelladines F-I, novel alkaloids from the sponge *Batzella* sp.: inducers of p56 ^{lck}-CD4 dissociation. J Org Chem 62:1814–1819

Patra A, Chang CWJ, Scheuer PJ, Van Duyne GD, Matsumoto GK, Clardy J (1984) An unprecedented triisocyano diterpenoid antibiotic from a sponge. J Am Chem Soc 106:7981–7983

Paul VJ, Seo Y, Cho KW, Rjo JR, Shin J, Berquist PR (1997) Sesquiterpenoids of the drimane class from a sponge of the genus *Dysidea*. J Nat Prod 60:1115–1120

Petrichtcheva NV, Duque C, Dueñas A, Zea S, Hara N, Fujimoto Y (2002) New nitrogenous eudesmane-type compounds isolated from the Caribbean sponge *Axinyssa ambrosia*. J Nat Prod 65:851–855

Pettit GR, Cichacz ZA, Gao F, Herald CL, Boyd MR, Schmidt JM, Hooper JNA (1993) Isolation and structure of spongistatin 1. J Org Chem 58:1302–1304

Pettit GR, Gao F, Cerny RL, Doubek DL, Tackett LP, Schmidt JM, Chapuis J (1994) Isolation and structure of the cell growth inhibitory constituents from the Western Pacific marine sponge *axinella* sp. J Med Chem 37:1165–1168

Pettit GR, Srirangam JK, Herald DL, Xu JP, Boyd MR, Cichacz Z, Kamano Y, Schmidt JM, Erickson KL (1995) Isolation and crystal structure of stylopeptide 1, a new marine porifera cycloheptapeptide. J Org Chem 60:8257–8261

Pettit GR, McNulty J, Herald DL, Doubek DL, Chapuis JC, Schmidt JM, Tackett L, Boyd MR (1997) Antineoplastic agents. 362. Isolation and x-ray crystal structure of dibromophakellstatin from the Indian Ocean sponge *Phakellia mauritiana*. J Nat Prod 60:180–183

Pettit GR, Cichacz ZA, Tan R, Hoard MS, Melody N, Pettit RK (1998) Antineoplastic agents. 386. Isolation of sesterstatins 1-3 from the Marine sponge *Hyrtios erecta*. J Nat Prod 61:13–16

Pettit GR, Knight JC, Collins JC, Herald DL, Pettit RK, Boyd MR, Young VG (2000) Antineoplastic agents 430. Isolation and structure of cribrostatins 3, 4, and 5 from the Republic of Maldives *Cribrochalina* species. J Nat Prod 63:793–798

Pettit GR, Collins JC, Knight JC, Herald DL, Nieman RA, Williams MD, Pettit RK (2003) Antineoplastic agents. 485. Isolation and structure of cribrostatin 6, a dark blue cancer cell growth inhibitor from the Marine sponge *Cribrochalina* sp. J Nat Prod 66:544–547

Pettit GR, Hoffmann H, McNulty J, Higgs KC, Murphy A, Molloy DJ, Herald DL, Williams MD, Pettit RK, Doubek DL, Hooper JNA, Albright L, Schmidt JM, Chapuis JC, Tackett LP (2004) Antineoplastic agents. 380. Isolation and x-ray crystal structure determination of isoaaptamine from the Republic of Singapore *Hymeniacidon* sp. and conversion to the phosphate prodrug hystatin 1. J Nat Prod 67:506–509

Pham NB, Butler MS, Hooper JNA, Moni RW, Quinn RJ (1999) Isolation of xestosterol esters of brominated acetylenic fatty acids from the Marine sponge *Xestospongia testudinaria*. J Nat Prod 62:1439–1442

Phuwapraisirisan P, Matsunaga S, Fusetani N, Chaitanawisuti N, Kritsanapuntu S, Menasveta P (2003) Mycaperoxide H, a new cytotoxic norsesterterpene peroxide from a Thai Marine sponge *Mycale* sp. J Nat Prod 66:289–291

Piña IC, Sanders ML, Crews P (2003) Puupehenone congeners from an indo-pacific *Hyrtios* sponge. J Nat Prod 66:2–6

Plubrukarn A, Smith DW, Cramer RE, Davidson BS (1997) (2*E*,9*E*)-Pyronaamidine 9-(*N*-Methylimine), a new imidazole alkaloid from the Northern Mariana Islands sponge *Leucetta* sp. cf. *chagosensis*. J Nat Prod 60:712–715

Quiñoá E, Kakou Y, Crews P (1988) Fijianolides, polyketide heterocycles from a marine sponge. J Org Chem 53:3642–3644

Qureshi A, Stevenson CS, Albert CL, Jacobs RS, Faulkner DJ (1999) *Homo*- and *Nor*-Plakotenin, new carboxylic acids from the Palauan sponge *Plakortis lita*. J Nat Prod 62:1205–1207

Ramesh P, Reddy NS, Venkateswarlu Y (1999) A new 1,2-dihydroisoquinoline from the sponge *Petrosia similis*. J Nat Prod 62:780–781

Randazzo A, Debitus C, Minale L, Pastor PG, Alcaraz MJ, Payá M, Gomez-Paloma L (1998) Petrosaspongiolides M-R: new potent and selective phospholipase A2 inhibitors from the new caledonian marine sponge *Petrosaspongia nigra*. J Nat Prod 61:571–575

Rao KV, Santarsiero BD, Mesecar AD, Schinazi RF, Tekwani BL, Hamann MT (2003) New manzamine alkaloids with activity against infectious and tropical parasitic diseases from an Indonesian sponge. J Nat Prod 66:823–828

Rashid MA, Gustafson KR, Boyd MR (2001) A new isoquinoline alkaloid from the marine sponge *Haliclona* species. J Nat Prod 64:1249–1250

Reyes F, Martín R, Rueda A, Fernández R, Montalvo D, Gómez C, Sánchez-Puelles JM (2004) Discorhabdins I and L, cytotoxic alkaloids from the sponge *Latrunculia brevis*. J Nat Prod 67:463–465

Ridley CP, Faulkner DJ (2003) New cytotoxic steroidal alkaloids from the Philippine sponge *Corticium niger*. J Nat Prod 66:1536–1539

Roll DM, Scheuer PJ, Matsumoto GK, Clardy J (1983) Halenaquinone, a pentacyclic polyketide from a marine sponge. J Am Chem Soc 105:6177–6178

Roll DM, Ireland CM, Lu HS, Clardy J (1988) Fascaplysin, an unusual antimicrobial pigment from the marine sponge *Fascaplysinopsis* sp. J Org Chem 53:3276–3278

Rudi A, Aknin M, Gaydou EM, Kashman Y (1997) J Nat Prod 60:700–703

Rudi A, Yosief T, Schleyer M, Kashman Y (1999) Bilosespens A and B: two novel cytotoxic sesterpenes from the marine sponge *Dysidea cinerea*. Org Lett 1 (3):471–472

Rudi A, Yosief T, Loya S, Hizi A, Schleyer M, Kashman Y (2001) Clathsterol, a novel anti-HIV-1 RT sulfated sterol from the sponge *Clathria* species. J Nat Prod 64:1451–1453

Rudi A, Afanii R, Gravalos LG, Aknin M, Gaydou E, Vacelet J, Kashman Y (2003) Three new cyclic peroxides from the marine sponge *Plakortis aff simplex*. J Nat Prod 66:682–685

Rueda A, Zubía E, Ortega MJ, Carballo JL, Salva J (1997) New cytotoxic metabolites from the sponge *Cacospongia scalaris*. J Org Chem 62:1481–1485

Rueda A, Zubía E, Ortega MJ, Carballo JL, Salva J (1998) New metabolites from the sponge *Spongia agaricina*. J Nat Prod 61:258–261

Ryu G, Matsunaga S, Fusetani N (1996a) Globostellatic acids A-D, new cytotoxic isomalabaricane triterpenes from the marine sponge *Stelletta globostellata*. J Nat Prod 59:512–514

Ryu G, Matsunaga S, Fusetani N (1996b) Three new cytotoxic sesterterpenes from the marine sponge *Hyrtios* cf. *Erectus*. J Nat Prod 59:515–517

Sakai R, Higa T (1986) Manzamine A, a novel antitumor alkaloid from a sponge. J Am Chem Soc 108:6404–6405

Sakai R, Kamiya H, Murata M, Shimamoto K (1997) Dysiherbaine: a new neurotoxic amino acid from the micronesian marine sponge *Dysidea herbacea*. J Am Chem Soc 119:4112–4116

Salmoun M, Devijver C, Daloze D, Braekman JC, Gomez R, De Kluijver M, Van Soest RWM (2000) New sesquiterpene/quinones from two sponges of the genus *Hyrtios*. J Nat Prod 63:452–456

Salvlf J, Faulkner DJ (1990) Metabolites of the sponge *Strongylophora durissima* from Maricabiin Island, Philippines. J Org Chem 55:1941–1943

Sandler JS, Colin PL, Hooper JNA, Faulkner DJ (2002) Cytotoxic β-carbolines and cyclic peroxides from the Palauan sponge *Plakortis nigra*. J Nat Prod 65:1258–1261

Santafé G, Paz V, Rodríguez J, Jimenez C (2002) Novel cytotoxic oxygenated C29 sterols from the Colombian marine sponge *Polymastia tenax*. J Nat Prod 65:1161–1164

Sata NU, Matsunaga S, Fusetani N, Van Soest RWM (1999a) Aurantosides D, E, and F: new antifungal tetramic acid glycosides from the Marine sponge *Siliquariaspongia japonica*. J Nat Prod 62:969–971

Sata NU, Wada SI, Matsunaga S, Watabe S, Van Soest RWM, Fusteani N (1999b) Rubrosides A-H, new bioactive tetramic acid glycosides from the Marine sponge *Siliquariaspongia japonica*. J Org Chem 64:2331–2339

Satitpatipan V, Suwanborirux K (2004) New nitrogenous germacranes from a Thai marine sponge, *Axinyssa* n sp. J Nat Prod 67:503–505

Schmidt EW, Suarez CR, Bifano M et al (2004) Scleritodermin A, a cytotoxic cyclic peptide from the lithistid sponge Scleritoderma nodosum. J Nat Prod 67:475–478

Schmitz FJ, Gunasekera SP, Yalamanchili G, Hossain MB, Van der Helm D (1984) Tedanolid: a potent cytotoxic macrolide from the Caribbean sponge *Tedania ignis*. J Am Chem Soc 106:7251–7252

Searle PA, Molinski TF (1995) Phorboxazoles A and B: potent cytostatic macrolides from marine sponge *Phorbas* Sp. J Am Chem Soc 117:8126–8131

Seo Y, Cho KW, Lee HS, Rho JR, Shin J (1999) New acetylenic enol ethers of glycerol from the sponge *Petrosia* sp. J Nat Prod 62:122–126

Sera Y, Adachi K, Shizuri Y (1999a) A new epidioxy sterol as an antifouling substance from a Palauan marine sponge, *Lendenfeldia chondrodes*. J Nat Prod 62:152–154

Sera Y, Adachi K, Nishida F, Shizuri Y (1999b) A new sesquiterpene as an antifouling substance from a Palauan marine sponge, *Dysidea herbacea*. J Nat Prod 62:395–396

Shen YC, Hsieh PW (1997) New sesquiterpene hydroquinones from a Taiwanese marine sponge *Polyfibrospongia australis*. J Nat Prod 60:93–97

Shen YC, Prakash CVS (2000) Two new acetylenic derivatives and a new meroditerpenoid from a taiwanese marine sponge *Strongylophora durissima*. J Nat Prod 63:1686–1688

Shen X, Perry TL, Dunbar CD, Kelly-Borges M, Hamann MT (1998) Debromosceptrin, an alkaloid from the Caribbean sponge *Agelas conifera*. J Nat Prod 61:1302–1303

Shen YC, Chen CY, Kuo YH (2001) New sesquiterpene hydroquinones from a Taiwanese marine sponge, *Hippospongia metachromia*. J Nat Prod 64:801–803

Shin J, Seo Y, Rho JR, Baek E, Kwon HJ, Jeong TS, Bok SH (1996) Suberitenones A and B: sesterterpenoids of an unprecedented skeletal class from the Antarctic sponge *Suberites* sp. J Org Chem 60:7582–7588

Shin J, Seo Y, Cho KW, Rho JR, Sim CJ (1997) Stellettamide B, a new indolizidine alkaloid from a sponge of the genus *Stelletta*. J Nat Prod 60:611–613

Shin J, Seo Y, Cho KW, Rho JR, Sim CJ (1999) New Bis (Indole) alkaloids of the topsentin class from the sponge *Spongosorites genitrix*. J Nat Prod 62:647–649

Shinichi Sakemit S, Hao H, Sun HH (1991) Nortopsentins A, B, and C. Cytotoxic and antifungal imidazolediylbis[indoles] from the sponge *Spongosorites ruetzleri*. J Org Chem 56:4304–4307

Shoji N, Umeyama A, Shin K et al (1992) Two unique pentacyclic steroids with Cis C/D ring junction from *Xestospongia bergquistia* fromont, powerful inhibitors of histamine release. J Org Chem 57:2996–2997

Shoji N, Umeyama A, Teranaka M, Arihara S (1996) Four novel diterpenoids, including nakamurol A with a unique thelepogane skeleton, from the marine sponge *Agelas nakamurai*. J Nat Prod 59:448–450

Simpson JS, Garson MJ, Blunt JW, Munro MHG, Hooper JNA (2000) Mycalamides C and D, cytotoxic compounds from the marine sponge *Stylinos* n. species. J Nat Prod 63:704–706

Sirirath S, Tanaka J, Ohtani II, Ichiba T, Rachmat R, Ueda K, Usui T, Osada H, Higa T (2002) Bitungolides A-F, new polyketides from the Indonesian sponge *Theonella* cf. *swinhoei*. J Nat Prod 65:1820–1823

Sperry S, Valeriote FA, Corbett TH, Crews P (1998) Isolation and cytotoxic evaluation of marine sponge-derived norterpene peroxides. J Nat Prod 61:241–247

Sun HH, Sakemi S, Burres N, McCarthy P (1990) Isobatzellines A, B, C, and D. Cytotoxic and antifungal pyrroloquinoline alkaloids from the marine sponge *Batzella* sp. J Org Chem 55:4964–4966

Tabudravu JN, Jaspars M (2002) Purealidin S and Purpuramine J, bromotyrosine alkaloids from the Fijian marine sponge *Druinella* sp. J Nat Prod 65:1798–1801

Tanaka J, Marriott G, Higa T, Higa T (2001) Cacofurans A and B, new furanoditerpenes from a marine sponge. J Nat Prod 64:1468–1470

Tasdemir D, Mangalindan GC, Concepción GP, Verbitski SM, Rabindran S, Miranda M, Greenstein M, Hooper JNA, Harper MK, Ireland CM (2002a) Bioactive isomalabaricane triterpenes from the marine sponge *Rhabdastrella Globostellata*. J Nat Prod 65:210–214

Tasdemir D, Bugni TS, Mangalindan GC, Concepción GP, Harper MK, Ireland CM (2002b) Cytotoxic bromoindole derivatives and terpenes from the Philippine marine sponge *Smenospongia* sp. Z Naturforsch 57c:914–922

Tsuchiya N, Sato A, Hata T, Sato N, Sasagawa K, Kobayashi T (1998) Cytotoxic scalarane sesterterpenes from a sponge, *Hyrtios erecta*. J Nat Prod 61:468–473

Tsuda M, Shigemori H, Ishibashi M, Sasaki T, Kobayashi J (1992) Luffariolides A-E, new cytotoxic sesterterpenes from the Okinawan marine sponge *Luffariella* sp. J Org Chem 57:3503–3507

Tsuji S, Rinehart KL, Gunasekera SP, Kashman Y, Cross SS, Lui MS, Pomponi SA, Diaz MC (1988) Topsentin, bromotopsentin, and dihydrodeoxybromotopsentin: antiviral and antitumor Bis(indolyl)imidazoles from Caribbean deep-sea sponges of the family *Halichondriidae*. Structural and synthetic studies. J Org Chem 53:5446–5453

Tsukamoto S, Kato H, Hirota H, Fusetani N (1996) Mauritiamine, a new antifouling oroidin dimer from the marine sponge *Agelas mauritiana*. J Nat Prod 59:501–503

Tsukamoto S, Kato H, Hirota H, Fusetani N, Fusetani N (1997) Seven new polyacetylene derivatives, showing both potent metamorphosis-inducing activity in ascidian larvae and antifouling activity against barnacle larvae, from the marine sponge *Callyspongia truncata*. J Nat Prod 60:126–130

Tsukamoto S, Matsunaga S, Fusetani N (1998) Acanthosterol sulfates A-J: ten new antifungal steroidal sulfates from a marine sponge *Acanthodendrilla* sp. J Nat Prod 61:1374–1378

Tsukamoto S, Yamashita T, Matsunaga S, Fusetani N (1999) Bistellettadines A and B: two bioactive dimeric stellettadines from a marine sponge *Stelletta* sp. J Org Chem 64:3794–3795

Tsukamoto S, Takahashi M, Matsunaga S, Fusetani N, Van Soest RWM (2000) Hachijodines A-G: seven new cytotoxic 3-alkylpyridine alkaloids from two marine sponges of the genera *Xestospongia* and *Amphimedon*. J Nat Prod 63:682–684

Tsukamoto S, Tane K, Otha T, Matsunaga S, Fusetani N, Van Soest RWM (2001) Four new bioactive pyrrole-derived alkaloids from the marine sponge *Axinella brevistyla*. J Nat Prod 64:1576–1578

Tsukamoto S, Miura S, van Soest WM, Ohta T (2003a) Three new cytotoxic sesterterpenes from a marine sponge *Spongia* sp. J Nat Prod 66:438–440

Tsukamoto S, Tatsuno M, van Soest RWM, Yokosawa H, Ohta T (2003b) New polyhydroxy sterols: proteasome inhibitors from a marine sponge *Acanthodendrilla* sp. J Nat Prod 66:1181–1185

Uno M, Otha S, Otha E, Ikegami S (1996) Callyspongins A and B: novel polyacetylene sulfates from the marine sponge *Callyspongia truncata* that inhibit fertilization of starfish gametes. J Nat Prod 59:1146–1148

Utkina NK, Denisenko VA, Scholokova OV, Virovaya MV, Gerasimenko AV, Popov DY, Krasokhin VB, Popov AM (2001) Spongiadioxins A and B, two new polybrominated dibenzo-*p*-dioxins from an Australian marine sponge *Dysidea dendyi*. J Nat Prod 64:151–153

Venables DA, Concepción GP, Matsumoto SS, Barrows LR, Ireland CM (1997) Makaluvamine N: a new pyrroloiminoquinone from *Zyzzya fuliginosa*. J Nat Prod 60:408–410

Wang CY, Wang BG, Wiryowidagdo S, Wray V, Van Soest R, Steube KG, Guan HS, Proksch P, Ebel R (2003) Melophlins C-O, thirteen novel tetramic acids from the marine sponge *Melophlus sarassinorum*. J Nat Prod 66:51–56

Wellington KD, Cambie RC, Rutledge PS, Bergquist PR (2000) Chemistry of Sponges. 19. Novel bioactive metabolites from *Hamigera tarangaensis*. J Nat Prod 63:79–85

West LM, Northcote PT, Hood KA, Miller JH, Page MJ (2000) Mycalamide D, a new cytotoxic amide from the New Zealand marine sponge *Mycale* Species. J Nat Prod 63:707–709

Williams DE, Lassota P, Andersen RJ (1998) Motuporamines A–C, cytotoxic alkaloids isolated from the marine sponge *Xestospongia exigua* (Kirkpatrick). J Org Chem 63:4838–4841

Williams DE, Craig KS, Patrick B, McHardy LM, Van Soest R, Roberge M, Andersen RJ (2002) Motuporamines, anti-invasion and anti-angiogenic alkaloids from the marine sponge *Xestospongia exigua* (Kirkpatrick): isolation, structure elucidation, analogue synthesis, and conformational analysis. J Org Chem 67:245–258

Wolf D, Schmitz FJ (1998) New diterpene isonitriles from the sponge *Phakellia pulcherrima*. J Nat Prod 61:1524–1527

Wright AE, Pomponi SA, Cross SS, McCarthy P (1992) A new Bis(indole) alkaloid from a deep-water marine sponge of the genus *Spongosorites*. J Org Chem 57:4772–4775

Yosief T, Rudi A, Kashman Y (2000) Asmarines A–F, novel cytotoxic compounds from the marine sponge *Raspailia* species. J Nat Prod 63:299–304

Youssef DTA (2004) Tasnemoxides A-C, new cytotoxic cyclic norsesterterpene peroxides from the red sea sponge *Diacarnus erythraenus*. J Nat Prod 67:112–114

Youssef DTA, Yoshida WY, Kelly M, Scheuer PJ (2001) Cytotoxic cyclic norterpene peroxides from a red sea sponge *Diacarnus erythraenus*. J Nat Prod 64:1332–1335

Youssef DTA, Yamaki RK, Kelly M, Scheuer PJ (2002) Salmahyrtisol A, a novel cytotoxic sesterterpene from the red sea sponge *Hyrtios erecta*. J Nat Prod 65:2–6

Youssef DTA, van Soest RWM, Fusetani N (2003) Callyspongenols A-C, new cytotoxic C22-polyacetylenic alcohols from a red sea sponge, *Callyspongia* species. J Nat Prod 66:679–681

Bioactive Alkaloids from Marine Sponges

12

Keisham S. Singh and Mahesh S. Majik

Abstract

Marine sponges are considered to be a rich source of biologically active secondary metabolites with unique and diverse chemical structures. They constitute nearly one third of the secondary metabolites isolated from marine organisms. Chemicals obtained from marine sponges find a wide range of pharmaceutical values, and as a result of these properties, isolation and identification of lead molecules from marine sponges continued to play a leading role in drug discovery research. Some of the molecules obtained from marine sponges have entered in market, while many are under clinical and preclinical trials. There is convincing report about the role of ecology on the production of these valuable secondary metabolites by marine organisms including sponges. The unique body structure of marine sponges which can filter and absorb nutrients from surrounding environment and unique adaptation to variable conditions lead sponges as a major source of bioactive metabolites among the marine organisms. Alkaloids constitute one of the main classes of secondary metabolites isolated from marine sponges. They have wide range of chemical structures and exist in derivatives of several heterocyclic rings. Alkaloids were found almost in all marine sponges and exhibited a wide range of biological activities. This chapter reviews on the various alkaloids, viz., pyridoacridine, indole, isoquinoline, pyridine, piperidine, quinolizidine, steroidal, and bromotyrosine alkaloid isolated from various marine sponges. A brief review on these alkaloids with their diverse structures available in each class along with their biological significance has been

K.S. Singh (✉)
Bioorganic Chemistry Laboratory, CSIR-National Institute of Oceanography, Dona Paula, Goa 403004, India
e-mail: keisham@nio.org

M.S. Majik
Department of Chemistry, Goa University, Taligao, Goa 403206, India

R. Pallela, H. Ehrlich (eds.), *Marine Sponges: Chemicobiological and Biomedical Applications*, DOI 10.1007/978-81-322-2794-6_12

presented. The class of alkaloid along with the name of sponge from which the alkaloids were isolated and chemical structures of these alkaloids are presented.

Keywords
Marine sponges • Pyridoacridine • Bioactive alkaloids • Quinolizidine alkaloids • Alkyl pyridine alkaloids • Bromotyrosine alkaloids

12.1 Alkaloids in Marine Sponges

Marine life represents a uniquely adapted reservoir of bioactive secondary metabolites due to their special environmental and oceanographic condition. Combination of knowledge of multidisciplinary sciences such as natural product chemistry, ecology, biology, and medicinal chemistry has inspired researchers for the development of many of the most successful medicines in particular from marine resources. In ocean, water pressure, temperature, light salt contents, etc., play an important role in adaptation of flora and fauna. As a result, species inhabiting these depths adapt their biochemical machinery to cope such varying pressures. Theses adaptations of marine organisms to deep-sea life and their effect on gene regulation and primary and secondary metabolic pathways gave rise to a wealth of interesting new marine natural products. Among the marine invertebrates, sponges have been considered as the most prolific phylum and prolific source of natural products with more novel compounds isolated from this taxon than from any other marine taxon (Blunt et al. 2011).

Many sponge-derived secondary metabolites possess a unique structural motif and pharmacological activities, thus making them highly desirable drug candidates for the treatment of a wide range of diseases. It has been known from the very early time that marine sponges contain bioactive compounds that are of potential medicinal value. Sponges are simple, multicellular sessile animals with no true tissue layers or organs and habitat every type of marine environment, from polar seas to temperate and tropical waters. Some species of sponges has the capacity of filtering out several tons of water to get nutrition. As a consequence of this, marine sponges are exposed to vast number of pathogenic and nonpathogenic microorganisms. In order to cope up with these microorganisms, sponges have developed strong immune system and they have possessed efficient chemical defense mechanism against the predators. There are more than 5000 (Whitehead 1999) species of marine sponges and many of these organisms have been investigated for their chemical and biological activities.

It is estimated that more than 10,000 bioactive molecules have been discovered from marine sources. In marine environment, this leading source has been taken by invertebrates such as sponges, tunicates, and bryozoans, mostly lacking morphological defense structure. They have developed the largest number of marine-derived secondary metabolites including some of most promising drug candidates (Newman and Cragg 2004). Indeed, out of 13 marine natural products that are currently under clinical trials as new drug candidates, 12 are derived from marine invertebrates (Proksch et al. 2003). As per review of literature on marine natural products, Blunt et al. (2004) described that sponges constitute nearly 40 % of the total secondary metabolites so far discovered from marine organisms. In the early 1950, spongouridine and spongothymidine, the first bioactive compounds from marine organisms, were isolated from the Caribbean sponge, *Cryptotethya crypta* (Bergmann and Feeney

Table 12.1 Different alkaloids with their biological activities obtained from various marine sponges

Class of alkaloids	Compound name	Biological activities	Name of sponge	References
Alkyl piperidine	Arenosclerins A, B, and C	Antibacterial	*Arenosclera brasiliensis*/ Haplosclerida	Torres et al. (2002)
Fused pyrrolo-phenanthroline	Discorhabdin D	Antitumor	*Latrunculia brevis*/ *Prianos* sp.	Perry et al. (1988)
Pyrrole guanidine	Isoaaptamine	Antitumor	*Aaptos aaptos*	Kitagawa et al. (1983)
	Debromohymenialdisine		*Hymeniacidon aldis*	
Pyrrole guanidine	Keramadine	Neurosuppressives	*Agelas* sp.	Nakamura et al. (1984)
Pyrrole imidazole	Taurodispacamide A	Immunosuppressive	*Agelas oroides*	Fattorusso and Taglialatela-Scafati (2000)
Indole	Dragmacidin F	Antiviral	*Halicortex* sp.	Cutignan et al. (2000)
Bisindole	Bromotopsentin	Neurosuppressives	*Spongosorites* sp./ Halichondria	Phife et al. (1996)
Pyridoacridine	Neoamphimedine	Antitumor	*Xestospongia* cf. *carbonaris*	Guzman et al. (1999)
Imidazole	Naamine D	Antitumor	*Leucetta* cf. *chagosensis*	Dunbar et al. (2000)
Azetidine	Penaresidin A	Neurosuppressives	*Penares* sp.	Kobayashi et al. (1991)
Bis-oxa-quinolizidine	Xestospongin-C	Neurosuppressives	*Xestospongia* sp.	De Smet et al. (1999)
Pyridopyrrolo pyrimidine	Variolin B	Antiviral	*Kirkpatrickia varialosa*	Perry et al. (1994)
Manzamine	Manzamine A	Antimalarial	*Haliclona* sp.	Ang et al. (2000)
Imidazo-azolo-imidazole	Axinellamines B–D	Antibacterial and antifungal	*Axinella* sp.	Urban et al. (1999)

1950, 1951). They were approved as anticancer (cytosine arabinoside Ara-C) and antiviral compounds (adenine arabinoside Ara-A), respectively, 15 years later (Jimino et al. 2004). Sponge chemistry is dominated by the presence of nitrogenous metabolites which could be basically divided into two structural type-based groups, peptides and polycyclic aromatic alkaloids. Alkaloid class isolated from sponge indeed includes a large variety of structures, ranging from very complex pyridoacridines and tyrosine-derived alkaloids to simple protoalkaloids. Alkaloids isolated from marine sponges comprise a vast structural diversity and possess several biological properties. Some of the alkaloids isolated from marine sponges along with their biological properties are presented in Table 12.1. This chapter reviews a brief discussion on alkaloids isolated from marine sponges and discussed in terms of their occurrence, structural type, and reported pharmacological activity. The chapter summarizes the recent development in the area of marine alkaloids, viz., pyridoacridine, indole, isoquinoline, alkyl pyridine, piperidine, quinolizidine, steroidal, and bromotyrosine alkaloids with few selected examples.

12.2 Pyridoacridine Alkaloids

Pyridoacridines are highly colored marine natural products having polycyclic planar heteroaromatic 11H-pyrido [4,3,2, mn] acridine systems (Patterson et al. 1960). Pyridoacridines are the largest group of marine alkaloids mostly isolated from sponges and tunicates. A first review on marine pyridoacridines has been published by Molinski (1993) and in later years, by Ding et al. (1999). Schmitz and Shoolery research groups reported the structure of first marine pyridoacridine alkaloids, amphimedine (**1**) (Schmitz et al. 1983); since then over 40 additional examples have been published. Although similar alkaloids containing isomeric ring systems have been found in terrestrial plants, namely, eupomatidine from angiosperm *Eupomatia bennettii*, the pyridoacridines [4,3,2-mnn], carbon skeleton is exclusive to marine invertebrates. Pyridoacridine alkaloids show various biological properties including cytotoxicity and certain other specific biological properties, viz., fungicidal and bactericidal properties, inhibition of topoisomerase II, anti-HIV, intercalation of DNA property, Ca^{+2}-releasing activity, and production of reactive oxygen species (Taraporewala et al. 1992). Pyridoacridines are pH indicator, and the indicator property is correlated with the presence of at least two basic electronic perturbations and extended chromophore with charge-transfer properties. Some other quaternary alkaline solution of pyridoacridine free base generally appeared orange or red, while, in acid solution, they are green to purple. However, simple indicator properties are absent in the less basic iminoquinones, such as cystodytin and diplamine. Pyridoacridine alkaloids have been isolated from several marine sponges, viz., *Oceanapia* sp., *Xestospongia* cf. *carbonaria* (Guzman et al. 1999), *Petrosia* sp. (Molinski et al. 1988), *Dercitus* sp. (Gunawardana et al. 1988), *Stelleta* sp. (Gunawardana et al. 1992), etc.

Hooper and coworkers isolated petrosamine B (**2**) alkaloids from the Australian sponge *Oceanapia* sp. (Carroll et al. 2005). The methanolic solution of the sponge sample imparted green-blue color, but when extract was diluted with water, the color changed to purple. Correlation of solvent-dependent changes in the UV spectrum and NMR spectra suggested that the remarkable color changes observed by varying solvent polarity were associated with shifts in the position of keto-enol equilibrium favoring the enol form. Petrosamine B alkaloid was found to be an inhibitor of the *Helicobacter pylori* enzyme aspartyl semialdehyde dehydrogenase (Carroll et al. 2005). Petrosamine B (**2**) was obtained as optically inactive blue solid and it is isomeric with petrosamine (**3**), isolated from the marine sponge *Petrosia* sp. with the only difference the position of bromine atom (Molinski et al. 1988). Notably, pyridoacridine alkaloids are grouped by total ring counts, viz., tetracyclic, pentacyclic, hexacyclic, heptacyclic, and octacyclic alkaloids. Soest's *group* isolated bioactive pyridoacridine alkaloids, kuanoniamine C (**4**), kuanoniamine D (**5**), and deacyl kuanoniamine derivative (**6**) from Micronesian sponge *Oceanapia* sp. (Eder et al. 1998). Kuanoniamines C and D isolated from the Marine sponge *Oceanapia sagittaria* were studied for anticancer activities, and it was found that kuanoniamine A is a potent growth inhibitor of all the tumor and nontumor cell lines, while kuanoniamine C was less potent but showed high selectivity toward the estrogen-dependent breast cancer cell line (Kijjoa et al. 2007). Recently, Davis and coworkers reported two new cytotoxicity pyridoacridine alkaloids, viz., ecionines A and B from the Australian marine sponge *Ecionemia geodides* (Barnes et al. 2010). Ecionines A and B (**7–8**) are imine-substituted pyridoacridine alkaloids, a very uncommon pyridoacridine family, and so far there are only three alkaloids of these classes available in literature. Wei et al. isolated 1-hydroxydeoxyamphimedine (**10**), 3-hydroxydeoxyamphimedine (**11**), and debromopetrosamine (**12**) along with the known neoamphimedine (**9**) and amphimedine (**1**) from the sponge *Xestospongia* cf. *carbonaria* (Wei et al. 2010) (Fig. 12.1).

In general, pyridoacridine alkaloids show significant biological activity such as cytotoxic, potent antiviral, antifungal, antibacterial, antitumor, and antiparasitic activity (Marshall and Barrows 2004). In fact, the crucial structural features of these alkaloids are the core of a planar iminoquinone moiety which can intercalate into DNA and cleave the DNA double helix or inhibit the action of TOPO II. As a consequence, there

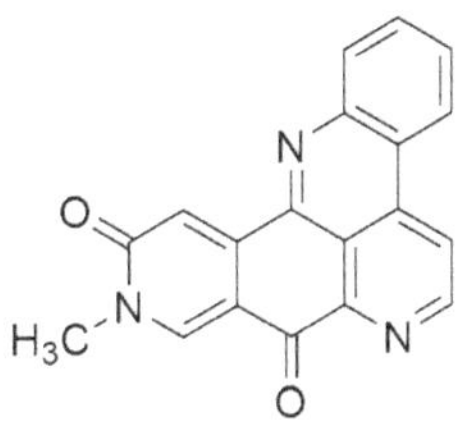

1 Amphimedine

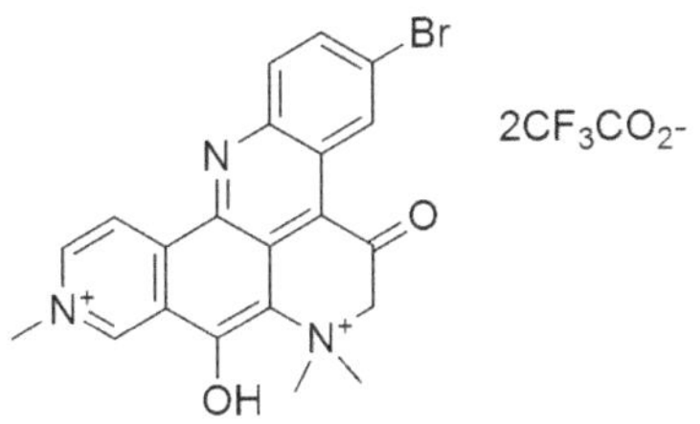

2 Petrosamine B

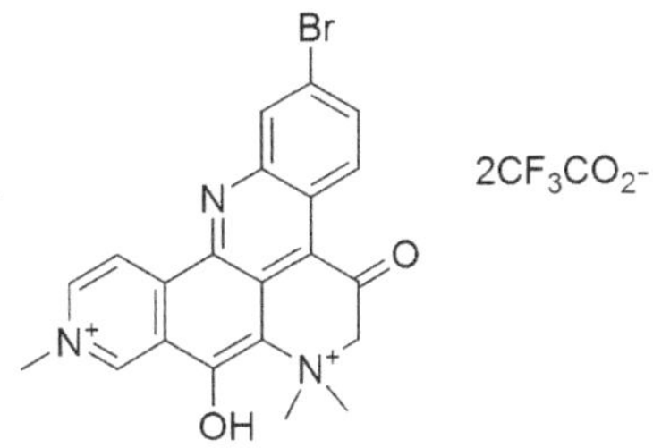

3 Petrosamine

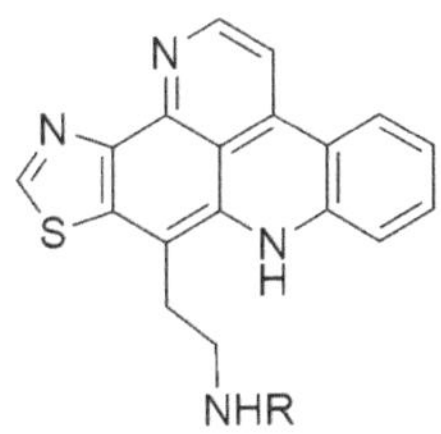

4 R = COC_2H_5 : Kuanoniamine C
5 R = $COCH_3$: Kuanoniamine D
6 R = H : Deacyl kuanoniamine

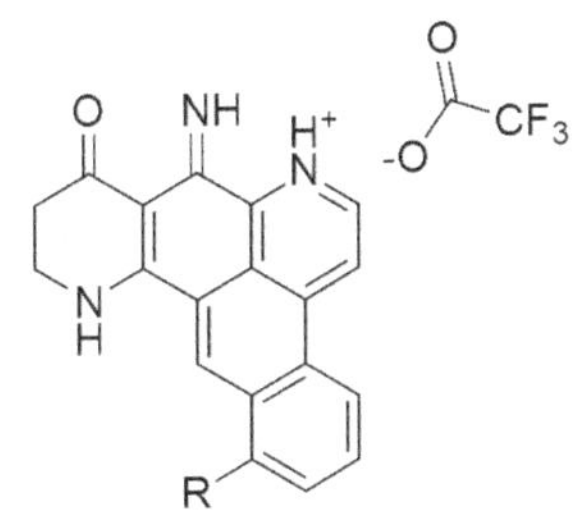

7 R = H: Ecionine A
8 R = OH: Ecoinine B

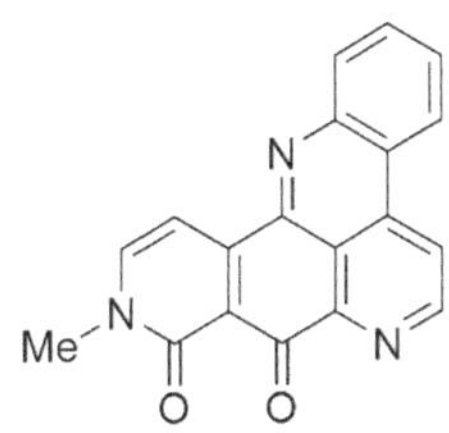

9 Neoamphimedine

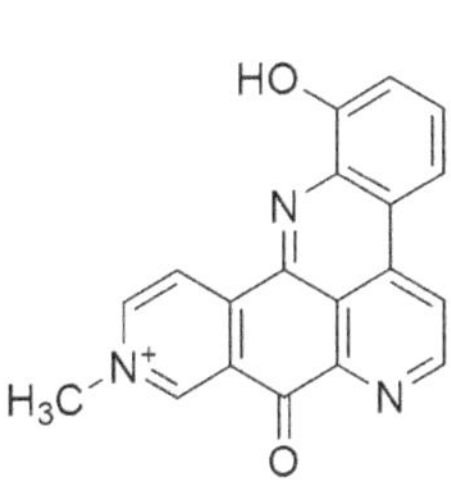

10 1-hydroxy-deoxyamphimedine

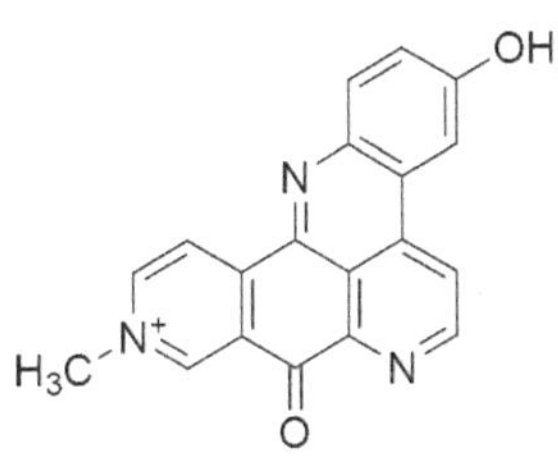

11 3-hydroxy-deoxyamphimedine

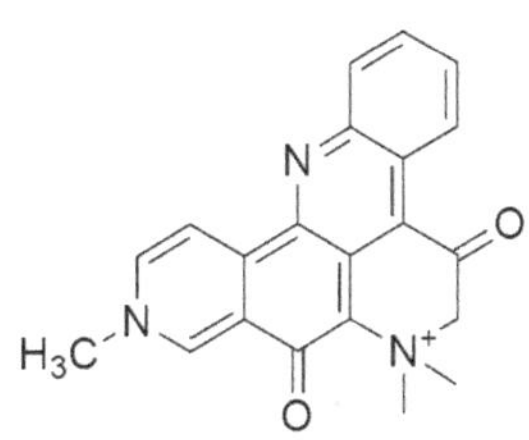

12 debromopetrosamine

Fig. 12.1 All compounds are cited (figure is just for reference)

have been considerable demands for these compounds as antitumor agents (Delfourne and Bastide 2003). Many of these compounds have generated interest as challenging problems both for structure elucidation and synthetic target and for their biological activities (Schmitz et al. 1983; Gunawardana et al. 1992). The red sponge *Plakortis*, collected by Inman and coworkers from different marine sources, led to the isolation of two novel alkaloids, namely, plakinidine-A (**13**) and plakinidine-B (**14**) (Inman et al. 1990), which contain a pyrrolo [2,3,4-kl] acridine fused-ring skeleton representing a new structural variation within polycyclic aromatic alkaloids from marine organisms. The discorhabdin C (**15**) was isolated from both *Latrunculia brevis*, from New Zealand, and *Prianos* sp. from Okinawa (Perry et al. 1988). Cheng et al. have isolated sulfur-containing alkaloids, prianosins A–D (**16–19**), from the green sponge *Prianos melanos* which showed cytotoxicity against L1210 murine leukemia cells (Cheng et al. 1988). The sponge *Bratzella* sp. has also furnished four additional pyrroloacridine alkaloids, namely, isobatzellines

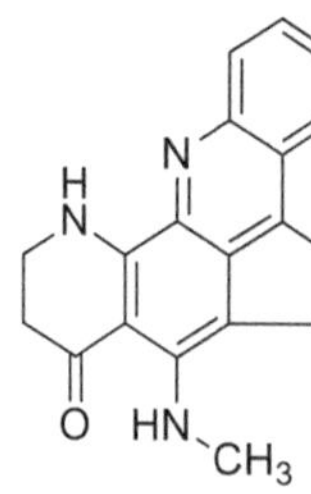

13 Plakinidine A

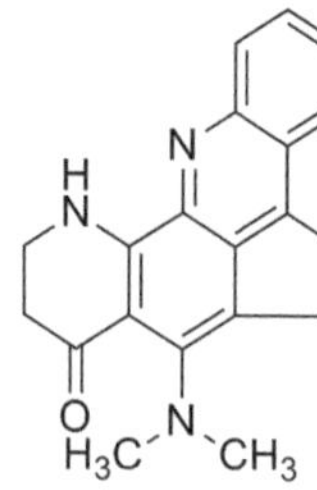

14 Plakinidine B

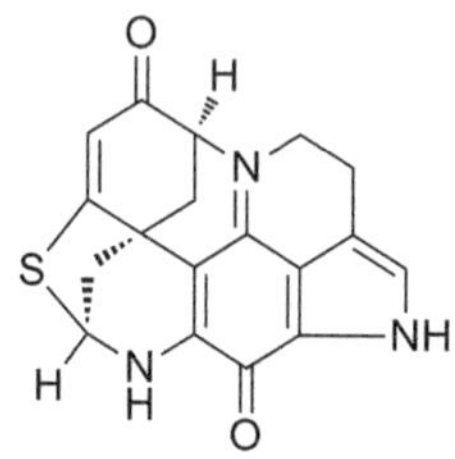

15 Discorhabdin C

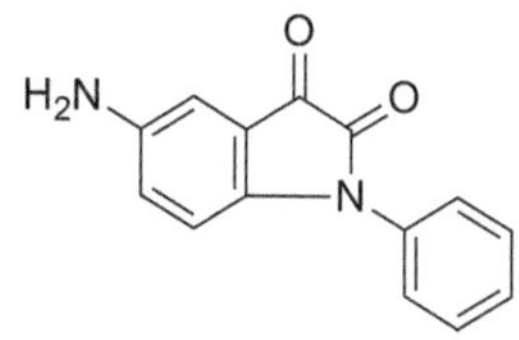

16 Prianosin A

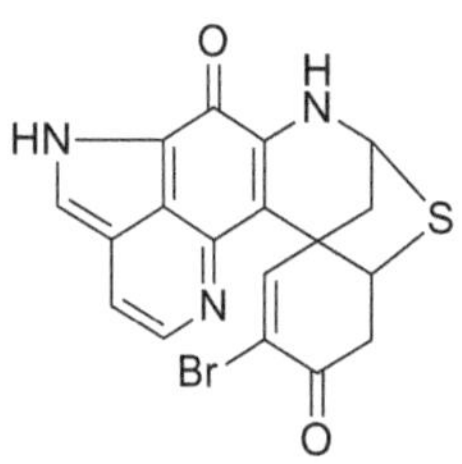

17 Prianosin B

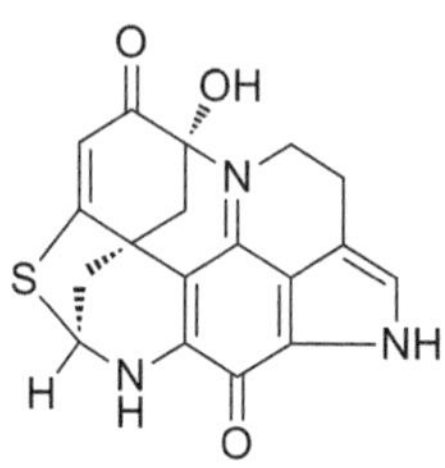

18 Prianosin C

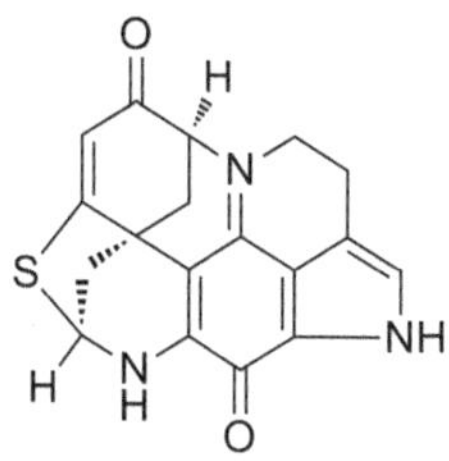

19 Prianosin D

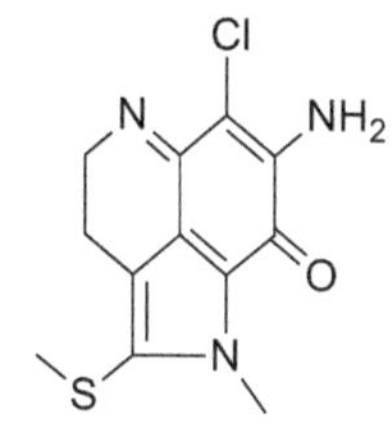

20 Isobatzelline A

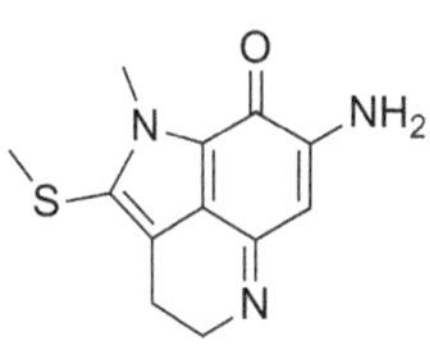

21 Isobatzelline B

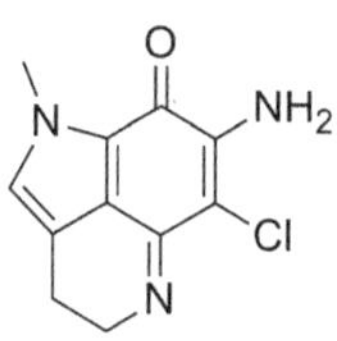

22 Isobatzelline C

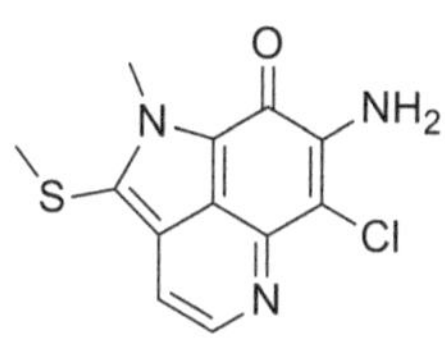

23 Isobatzelline D

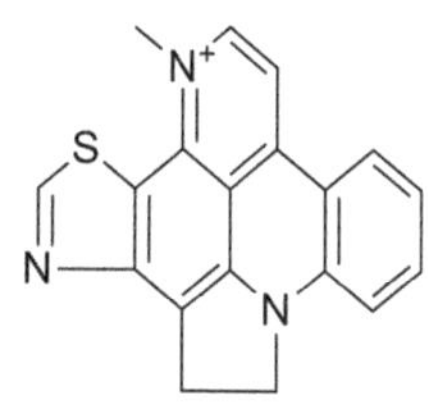

24 Cyclodercitin

Fig. 12.2 All compounds are cited (figure is just for reference). NB: Compounds **25–33** are cited in Table 12.1

A–D (**20–23**) (Sun et al. 1990). In 1975, hexacyclic alkaloids, cyclodercitin (**24**), have been reported from the deep-water sponges *Dercitus* sp. and in *Stelleta* sp. (Gray 1975) (Fig. 12.2).

Pyridoacridines is vast class of alkaloid which varies from each other structurally by attachment of different side chains or fusion of different rings to ring C of the basic structure and sometimes to the acridine nitrogen. Based on the structure, pyridoacridines are divided into tetracyclic, pentacyclic, hexacyclic, heptacyclic, and octacyclic alkaloids (Kumar and Rawat 2011). They show significant biological activity primarily cytotoxicity and certain specific biological properties like fungicidal and bactericidal properties, inhibition of topoisomerase II, anti-HIV, and intercalation of DNA (McCarthy et al. 1992; Kobayashi et al. 1988). A few selected pyridoacridines (**25–33**) showing interesting biological activities along with their source have been depicted in Table 12.2.

12.3 Indole Alkaloids

Indole-containing alkaloids have frequently been isolated from diverse marine invertebrates including bryozoans, coelenterates, sponges,

Table 12.2 Some pyridoacridines: source of bioactive alkaloids

Pyridoacridines	Source	Structures	References
Labuanine A (**25**)	*Biemna fortis* sponge (Indonesia)	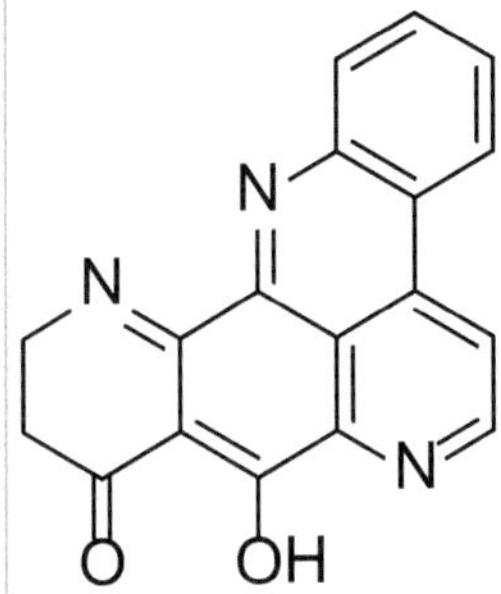	Aoki et al. (2003)
Sagitol (**26**)	*Oceanapia sagittaria* sponge (Palau)	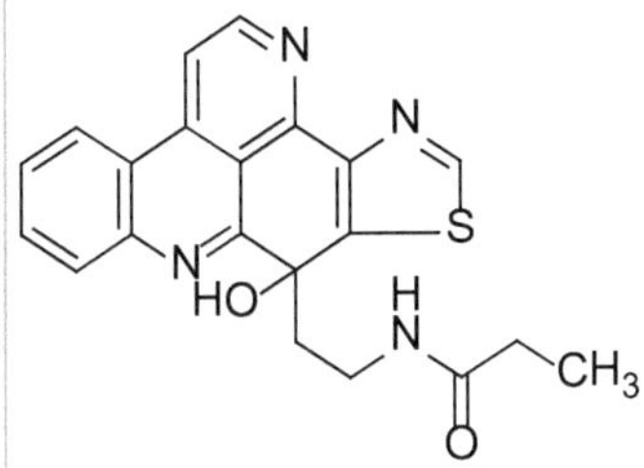	Salomon and Faulkner (1996)
Biemnadin (**27**)	*Biemna fortis* sponge (Indonesia)	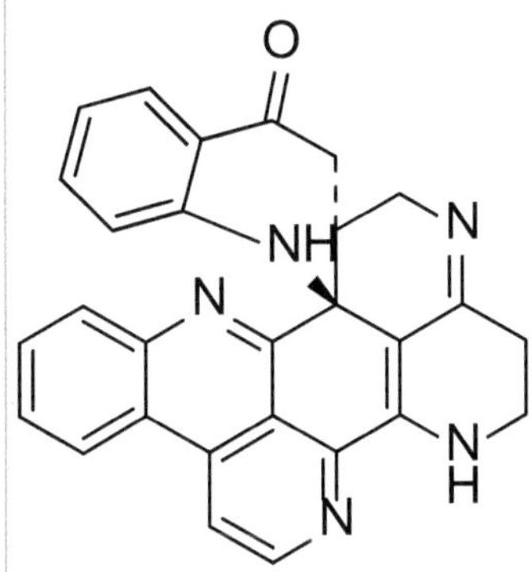	Kumar and Rawat (2011)
Neoamphimedine (**28**)	*Xestospongia* sp. sponge (Philippines) *Xestospongia* cf. *carbonaria* (Micronesia) *Xestospongia* c *carbonaria*, X. cf. *exigua* (Indo-Pacific)	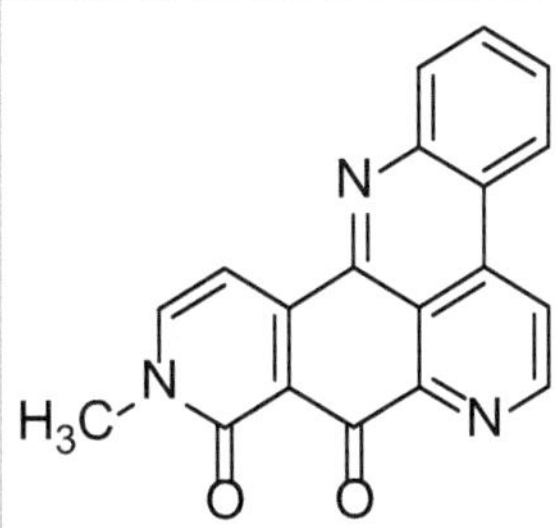	Rodriguez et al. (1993), Kong et al. (1994), and Tasdemir et al. (2001)
Neoamphimedine Y (**29**)	*Xestospongia* c *carbonaria*, X. cf. *exigua* (Indo-Pacific)	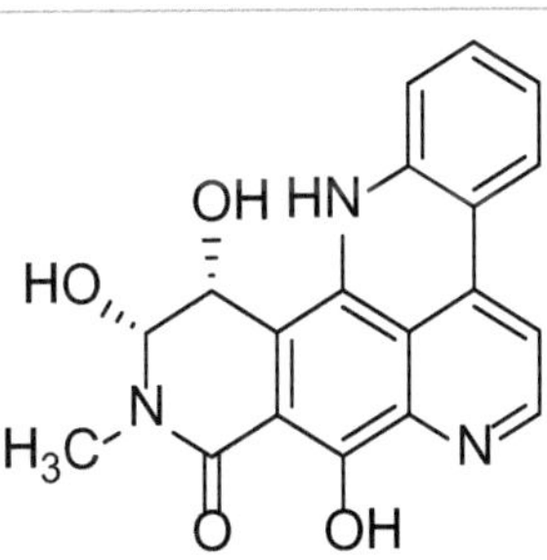	Tsotinis et al. (1996)

(continued)

Table 12.2 (continued)

Pyridoacridines	Source	Structures	References
Neoamphimedine Z (**30**)	*Xestospongia* cf. *carbonaria*, X. cf. *exigua* (Indo-Pacific)	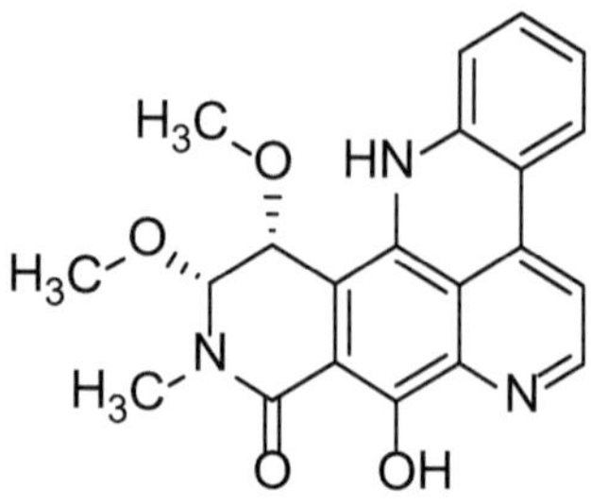	Schmitz et al. (1983)
Nordercitin (**31**)	*Stelletta* sp. sponge *Derdtus* sp. sponge (Bahamas)	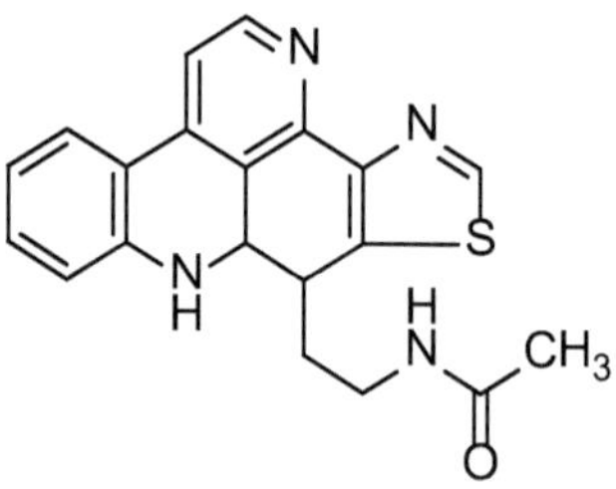	Gunawardana et al. (1992)
Stellettamine (**32**)	*Stelletta* sp. sponge	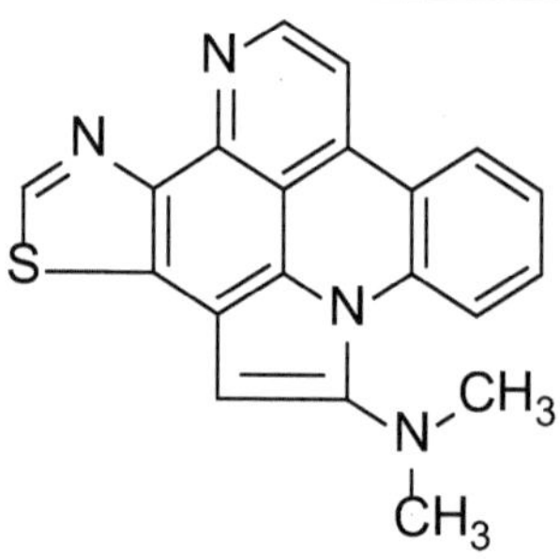	Shin et al. (1997)
Dercitamine (**33**)	*Stelleta* sp. sponge, *Dercitus* sp. sponge (Bahamas)	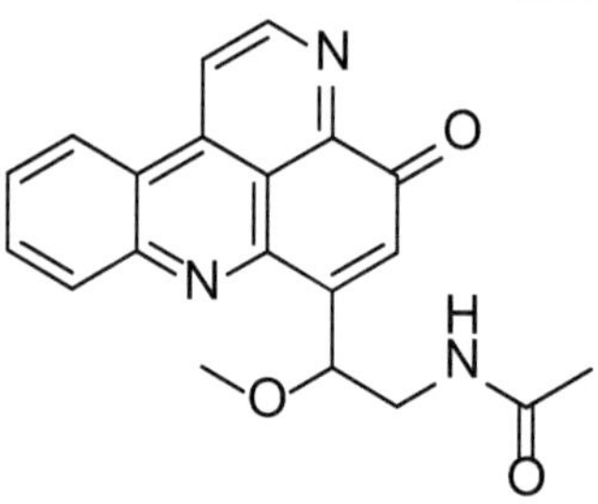	Djura and Faulkner (1980)

tunicates, algae, symbiotic bacteria, and fungi (Moriarty et al. 1987; Tanaka et al. 1988). Moreover, they show interesting biological activities such as cytotoxic, antitumor, antiviral, antimicrobial, etc. Corresponding to their unique structural features and impressive biological activities, the indole series have become attractive targets for the development of new pharmacological lead compounds. Indole alkaloids are distributed in many marine sponges, viz., sponge *Smenospongia* sp., *Topsentia genitrix*, *Dictyodendrilla* sp., *Spongosorites* sp., and *Hyrtios* sp. (Sauleau et al. 2006). Kazlauskas et al. isolated for the first time a novel indole alkaloid, aplysinopsin (**34**), from Indo-Pacific sponge species (Kazlauskas et al. 1977) which are the representatives of the genus *Thorecta* (later assigned as the separate *Aplysinopsis* genus). Since that time, aplysinopsin and its derivatives have been reported in many

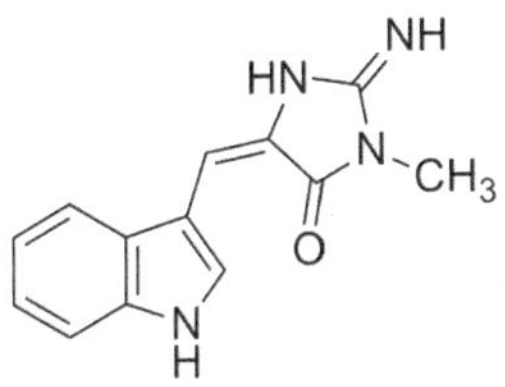

34 Aplysinopsin

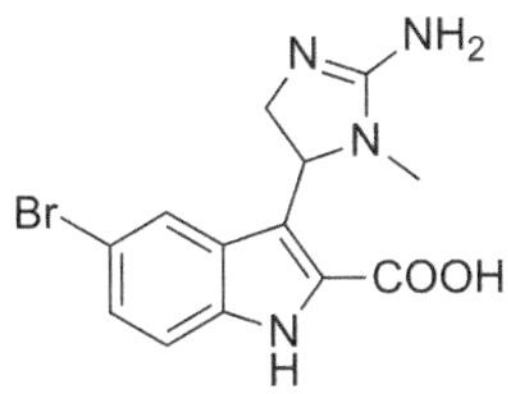

35 Trachycladindole A

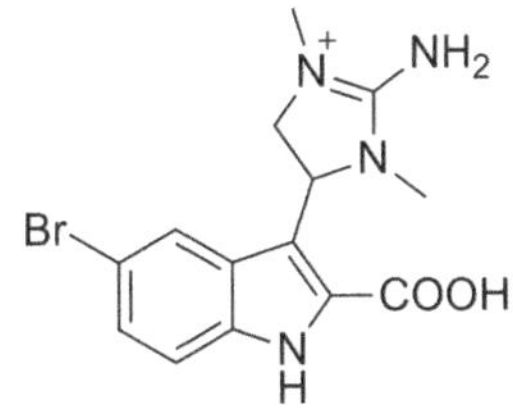

36 Trachycladindole B

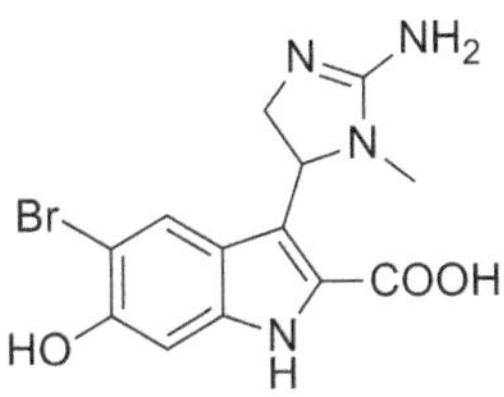

37 Trachycladindole C

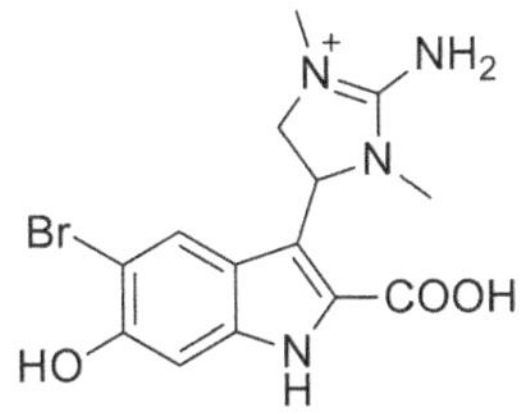

38 Trachycladindole D

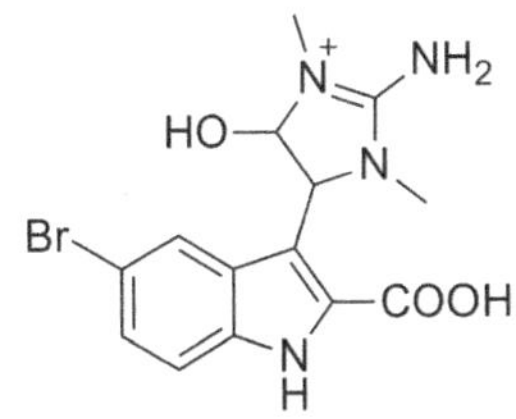

39 Trachycladindole E

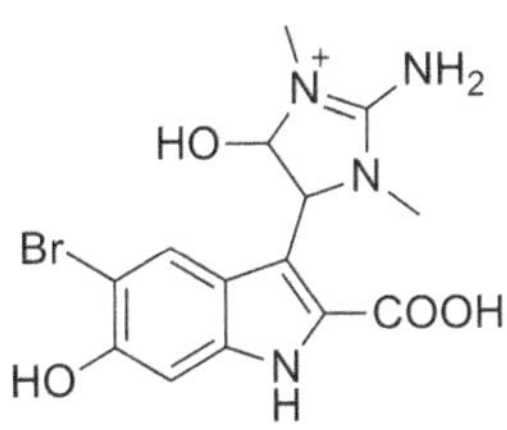

40 Trachycladindole F

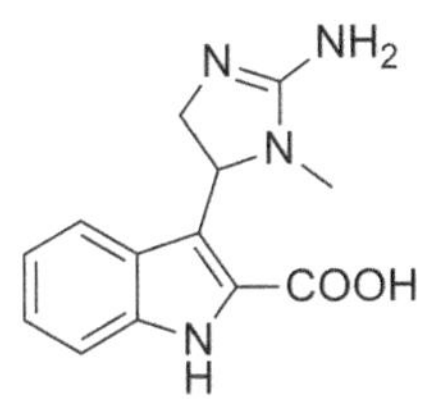

41 Trachycladindole G

Fig. 12.3 All compounds are cited (figure is just for reference)

other marine organisms. Aplysinopsin-type compounds have been found in sponges of the Caribbean, *Verongia spengelli* (Hollenbeak and Schmitz 1977), *Dercitus* sp. (Djura and Faulkner 1980), *Smenospongia aurea* (Djura et al. 1980), and *Verongula rigida* (Kochanowska et al. 2008); the Mediterranean Sea, *Dictyoceratida* sp. (Bergquist and Wells 1983); as well as in the Indo-Pacific region, *Aplysinopsis reticulata* (Kazlauskas et al. 1977; Baker and Wells 1981), *Aplysina* sp. (Kondo et al. 1994), *Hyrtios erecta* (Aoki et al. 2001), *Smenospongia* sp., and *Thorectandra* sp. (Segrave and Crews 2005). In 2008, Capon et al. (2008) have reported the cytotoxic agent trachycladindoles A–G (**35**–**41**) from southern Australian marine sponge, *Trachycladus laevispirulifer*. Excitingly, it displayed promising selective cytotoxicity against a panel of human cancer cell lines (Fig. 12.3).

12.3.1 Bisindole Alkaloids

Bisindole alkaloids, consisting of two indole moieties connected to each other via heterocyclic units, have been particularly abundant within marine sponges. Isolation of bis(indolyl)imidazole, topsentin A (**42**) or topsentin B1 (**43**), was reported from the sponge *Topsentia genitrix* (*Spongosorites genitrix*) (Blunt et al. 2004). Metabolites containing bis(indole) moiety have been found with various carbon skeletons and functionalities (Shin et al. 1999; Casapullo et al. 2000). These compounds exhibited a wide spectrum of pharmacological activities such as cytotoxic, antiviral, antimicrobial, and anti-inflammatory activities. As consequence, bis (indole) alkaloids is considered as an attractive targets for biomedical and synthetic studies (Bao et al. 2005). Topsentins A (**42**), B1 (**43**), and B2 (**44**) were isolated from marine sponge *Rhaphisia lacazei* and showed antiproliferative activity against human bronchopulmonary cancer cells (NSCLC-N6) (Casapullo et al. 2000). In 1992, Wright et al. collected the Pacific sponge *Hexadella* sp. from the coast of British Columbia which led to the identification of dragmacidin A (**45**) as potent cytotoxic compound (Fig. 12.4). Related bis-(indole)-alkaloid, dragmacidin D (**46**), has been isolated from another marine sponge of the genus *Spongosorites* (Wright et al. 1992). This compound inhibited the growth of the feline leukemia virus, the opportunistic fungal pathogens *Candida albicans* and *Cryptococcus neoformans*, and the growth of P388 and A549 tumor cell lines (Wright et al. 1992). Dragmacidins, member of a bis(indole) alkaloids, were isolated from a variety of marine sponges. This alkaloid family showed a wide range of biological activities such as inhibitors of protein phosphatase and anticancer. Two types of sponges, *Coscinoderm lanuga* and *Ircinia felix*, have proved as the major source of various new dragmacidins or other bis(indole) alkaloids (Crook et al. 2009; Davis-McGibony and Pletcher 2006).

A dipyrroloquinone, zyzzyanone A (**47**) (having a pyrrolo [3,2-f] indole-4,8(1H,7H)-dione skeleton), was isolated from the Australian marine sponge *Zyzzya fuliginosa*, exhibiting moderate cytotoxic activity against mouse Ehrlich carcinoma cells (Utkina et al. 2005). Hyrtimomines A–E (**48–52**) were isolated from an Okinawan marine sponge *Hyrtios* sp. (Tanaka et al. 2013). Later they isolated other hyrtimomines F–K (**53–58**) from the same marine sponge (Tanaka et al. 2014). Hyrtimomines A (**48**) and B (**49**) are heteroaromatic alkaloids possessing a fused hexacyclic 6/5/6/6/7/5 ring system, while hyrtimomine C (**50**) is an alkaloid consisting of hydroxyindole and azepino-hydroxyindole moieties (Fig. 12.5).

Hyrtimomines A–C (**48–50**) and hyrtimomines F–K (**53–58**) were studied for

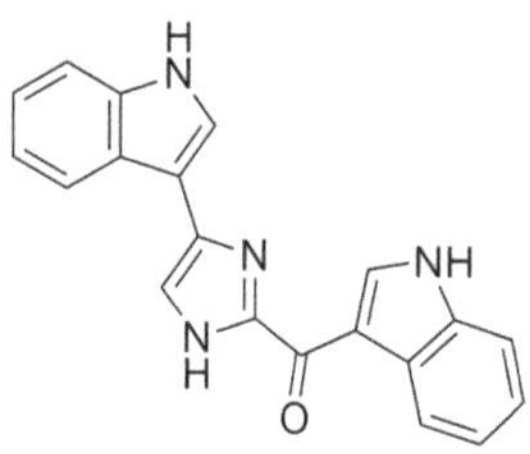

42 Topsentin A

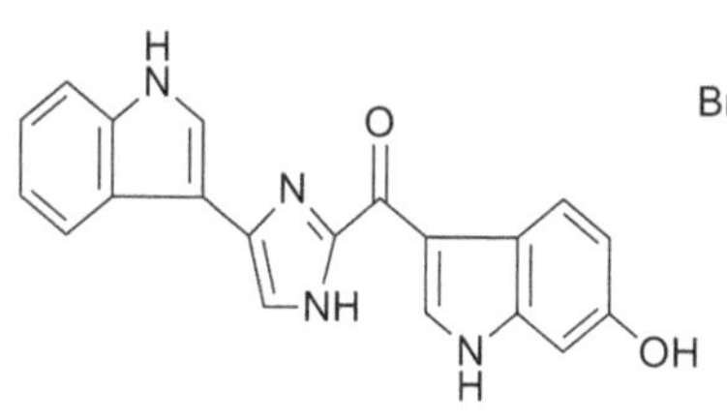

43 Topsentin B1

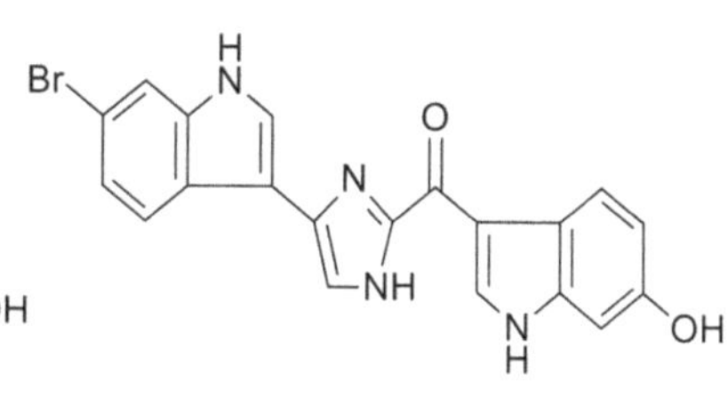

44 Topsentin B2

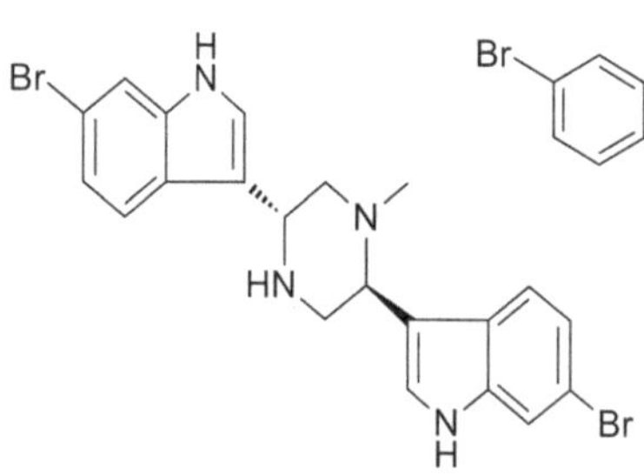

45 dragmacidin A

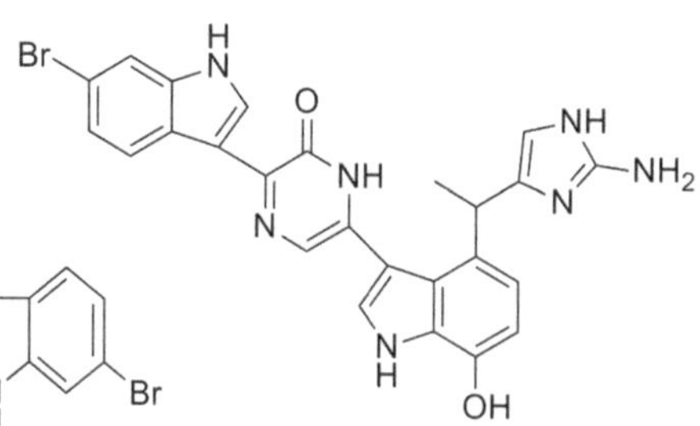

46 Dragmacidin D

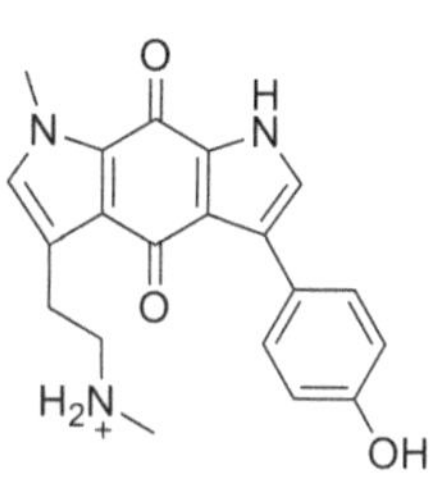

47 Zyzzyanone A

Fig. 12.4 All compounds are cited (figure is just for reference)

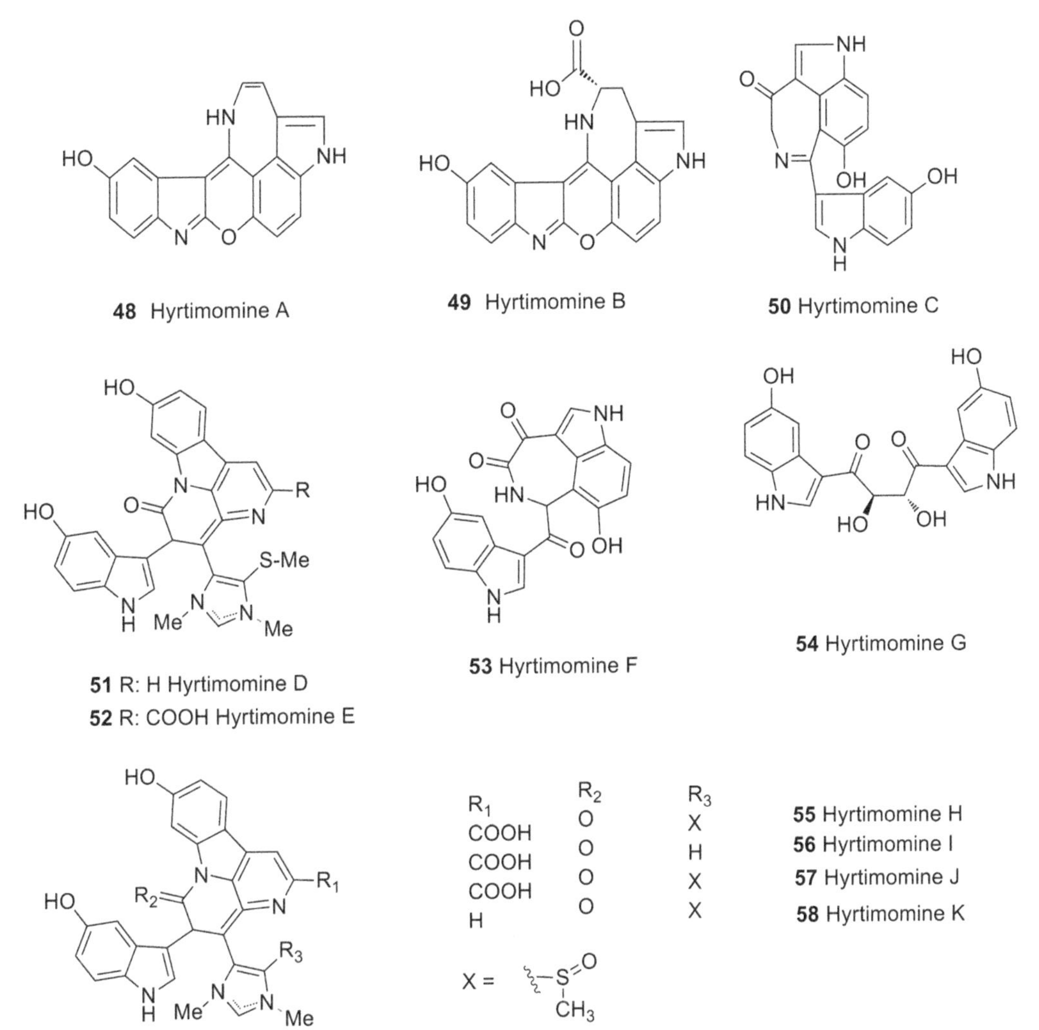

Fig. 12.5 All compounds are cited (figure is just for reference)

antimicrobial activities. Hyrtimomines F (**53**), G (**54**), and I (**56**) exhibited inhibitory effects against *Aspergillus niger*, while hyrtimomine I (**56**) showed inhibitory effect against *Cryptococcus neoformans*. Hyrtimomines A (**48**) and B (**49**) showed antimicrobial activities against *Candida albicans* and *C. neoformans*, while hyrtimomine A (**48**) exhibited an inhibitory activity against *A. niger* (Tanaka et al. 2014). Recently, Kobayashi's groups have shown cytotoxicity activity of hyrtimomine A (**48**) against KB and L1210 cells (Momose et al. 2013) (Figs. 12.5).

12.3.2 Trisindole Alkaloids

Trisindole alkaloids were rarely found in sponges. Bifulco et al. (1994) isolated trisindole alkaloids gelliusines A (**59**) and B (**60**) from deep-water Caledonian sponge *Gellius* or *Orina* sp. possessing cytotoxicity against KB, P-388, P-388/dox, HT-29, and NSCLC-N6 cell lines. The structural feature of gelliusines A and B (**59, 60**) is that the two 6-bromo tryptamine units are linked through their aliphatic chains to the C-2 and C-6 position of a central serotonin moiety, whereas the coupling of the indole unit

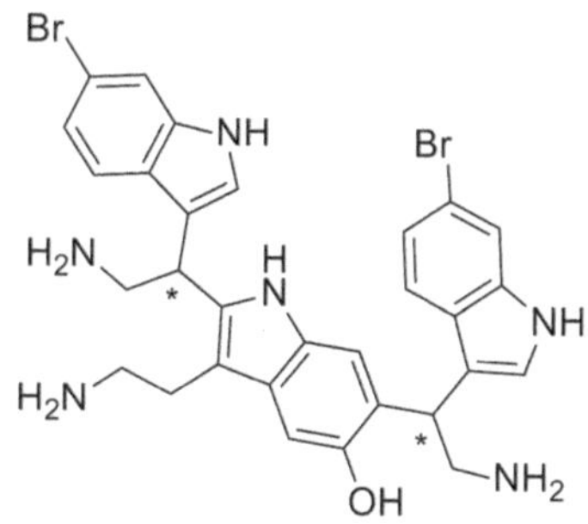

59 Gelliusine A

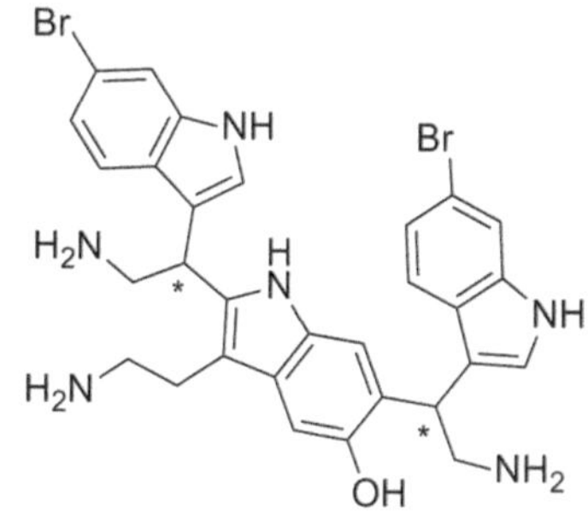

60 Gelliusine B

Fig. 12.6 All compounds are cited (figures are not cited; instead compound's number are cited; it is just for reference)

appears to be non-stereoselective giving two enantiomeric pairs (Fig. 12.6).

12.4 Isoquinoline Alkaloids

Marine sponges of genera *Reniera* and *Xestospongia* are rich in isoquinoline alkaloids. Several isoquinolinequinones have been isolated from blue species of the sponge. Mimosamycin (Kobayashi et al. 1994) and renierol (Mckee and Ireland 1987) are frequently isolated isoquinoline alkaloids and they have been reported from various marine sponges. Mimosamycin (**61**), 4-hydroxymimosamycin (**62**), 1,4-dihydroxymimosamycin (**63**), and O-demethylrenierone (**64**) were isolated from *Haliclona cribricutis* (Parameswaran et al. 1998). They isolated renieramycins H–I (**65–66**), a novel isoquinolinequinone alkaloid from the same sponge (Parameswaran et al. 1998). Isolation of renieramycin M, a *bis*-tetrahydroisoquinoline quinine alkaloid from the Thailand blue sponge *Xestospongia* sp., was reported by Saito and coworkers (Suwanborirux et al. 2013). Renieramycin M exhibited anticancer activity, and it induces human non-small cell lung cancer *H460 cells apoptosis*. The anticancer activity of renieramycin M against human lung carcinoma H460 cells was investigated by incubating the cells in the presence of renieramycin M (0–40 μM) for 24 h, and cell viability was analyzed using MTT assay (Halmi et al. 2011).

Isoquinolinequinones alkaloids, cribrostatins 1 (**68**) and 2 (**69**), were isolated from a deep blue-colored sponge *Cribrochalina* sp. (Pettit et al. 1992) and were found to be active against lymphocytic leukemia cell line (P-388). In 2000, Pettit et al. explored the same sponge *Cribrochalina* sp. which was found to contain other members of this family such as cribrostatins 3 (**70**), 4 (**71**) and 5 (**72**) (Pettit et al. 2000). These compounds (**70–71**) were active against mouse leukemia P-388 cell line. Structurally related alkaloid, cribrostatin 6 (**73**), was also isolated from the same marine sponge *Cribrochalina* sp. (Pettit et al. 2003) and was found to inhibit the growth of murine P-388 lymphocytic leukemia and a panel of human cancer cell lines (Fig. 12.7).

12.5 Pyridine Alkaloids

The sponge of order Haplosclerida are considered the richest source of pyridine alkaloids with diverse carbon skeleton. Several 3-alkyl pyridine alkaloids have been isolated from marine sponges (Faukner 1999). Cytotoxic bis-pyridine alkaloids, pyrinadine A and cribochalines A and B, were isolated from the marine sponge *Cribrochalina* sp. (Kariya et al. 2006). Cribochaline A displayed antifungal activity against both antibiotic-sensitive

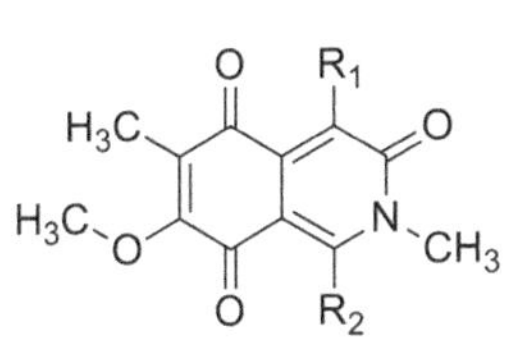

61 Mimosamycin: $R_1 = R_2 = H$
62 4-hydroxymimosamycin: $R_1 = OH$; $R_2 = H$
63 1,4-dihydroxymimosamycin: $R_1 = R_2 = OH$

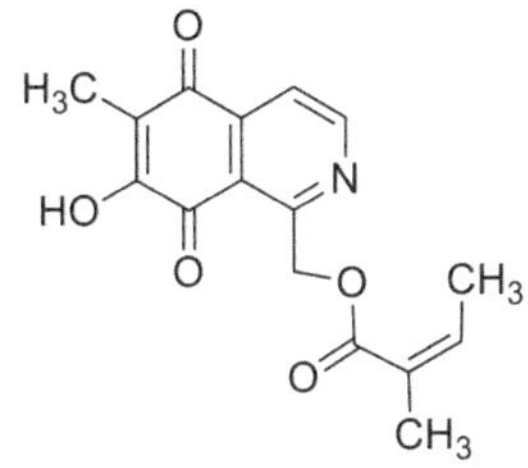

64 O-Demethylrenierone

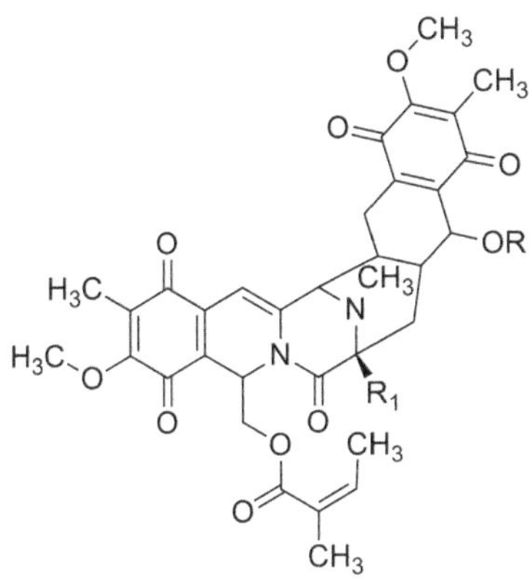

65 Renieramycin H: $R = H$, $R_1 = OH$

66 Renieramycin I: $R = CH_3$, $R_1 = H$

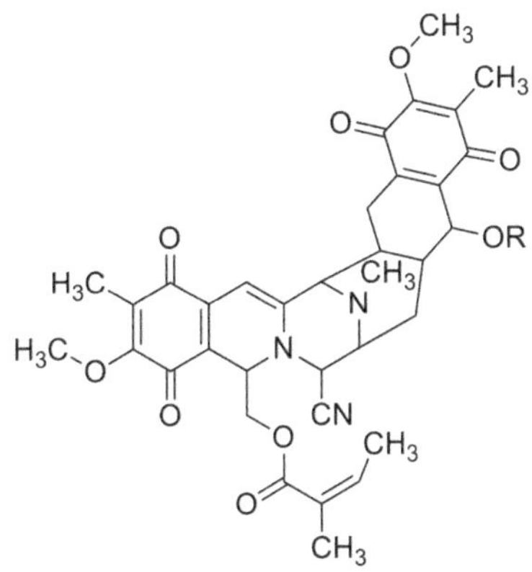

67 Renieramycin M

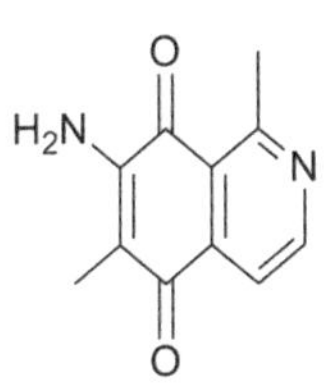

68 Cribrostatin 1

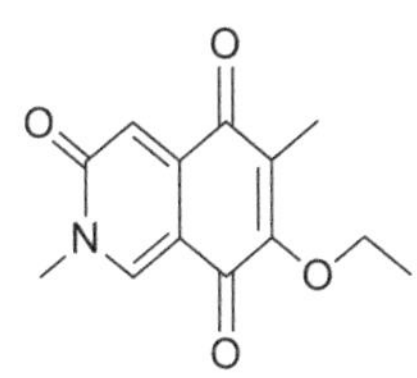

69 Cribrostatin 2

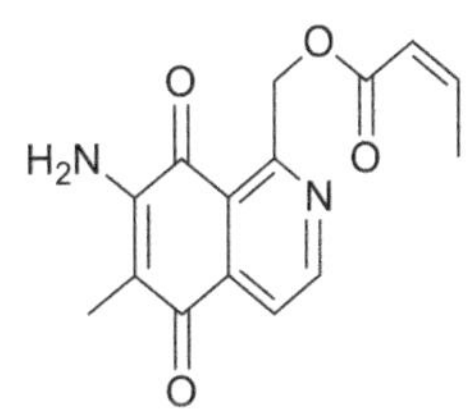

70 Cribrostatin 3

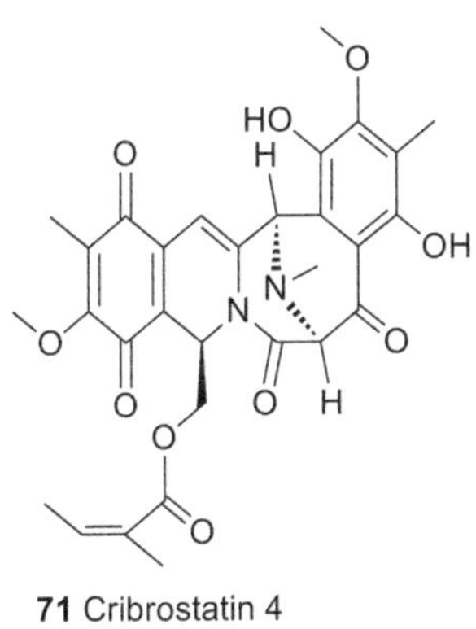

71 Cribrostatin 4

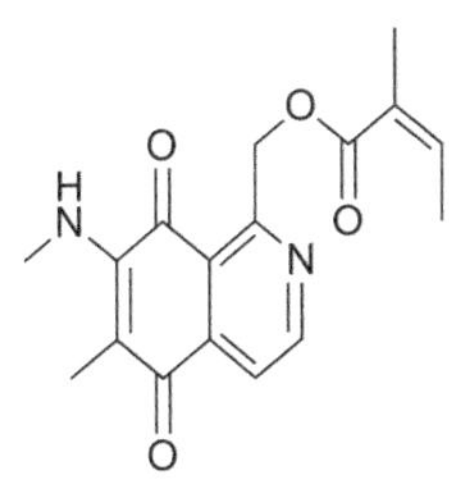

72 Cribrostatin 5

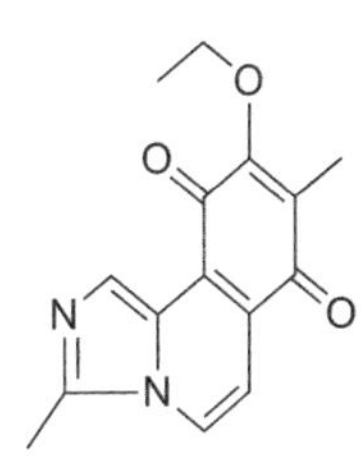

73 Cribrostatin 6

Fig. 12.7 All compounds are cited (figures are not cited; instead compound's number are cited; it is just for reference)

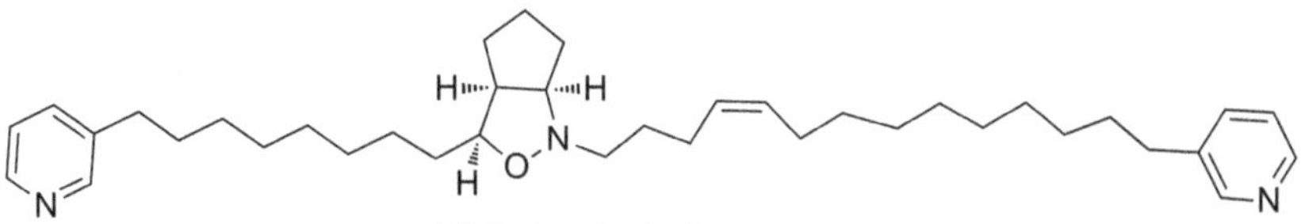

74 Pyrinodemin A

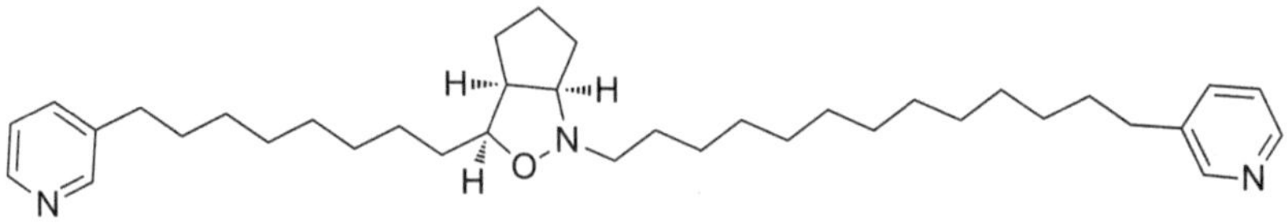

75 Pyrinodemin B

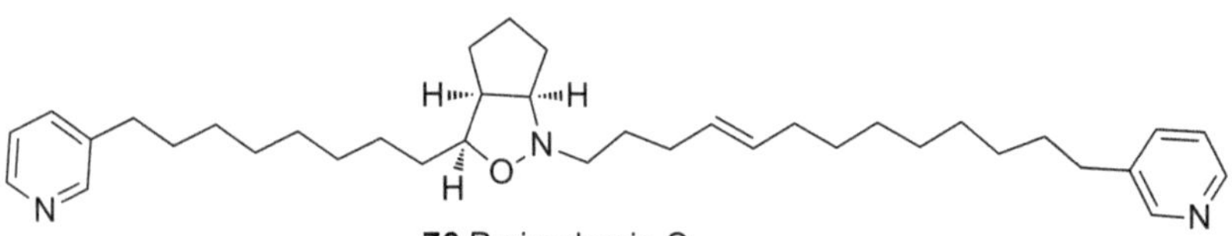

76 Pyrinodemin C

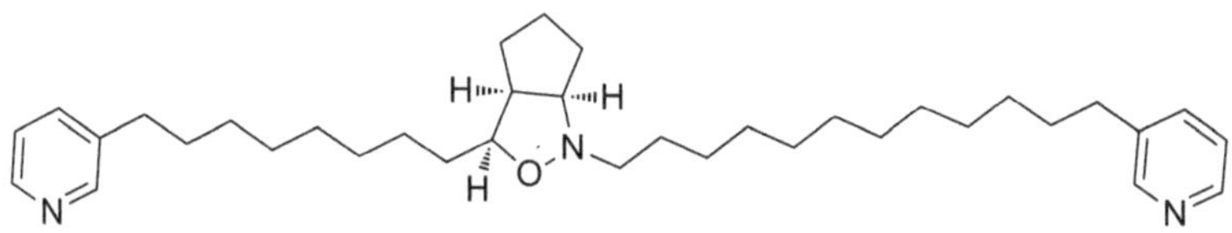

77 Pyrinodemin D

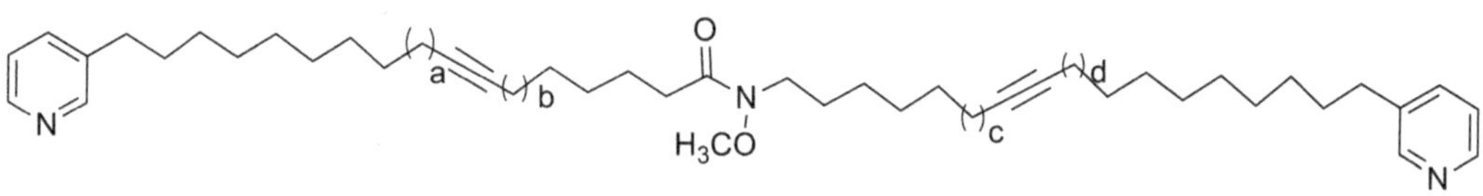

78 Pyrinodemine G(a + b = 1, c + d = 1)

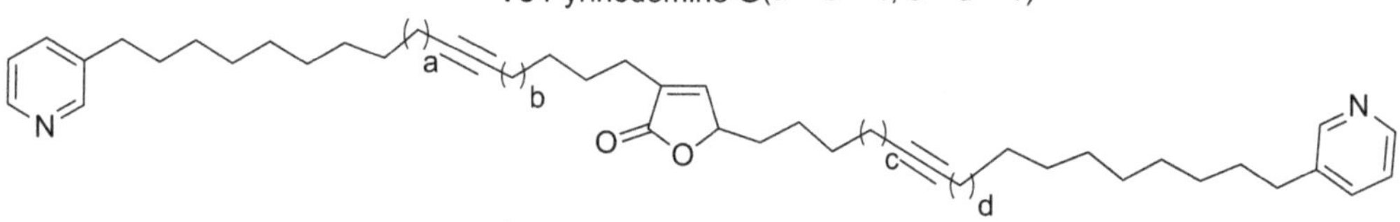

79 Pyrinodemine H (a + b = 1, c + d = 1)

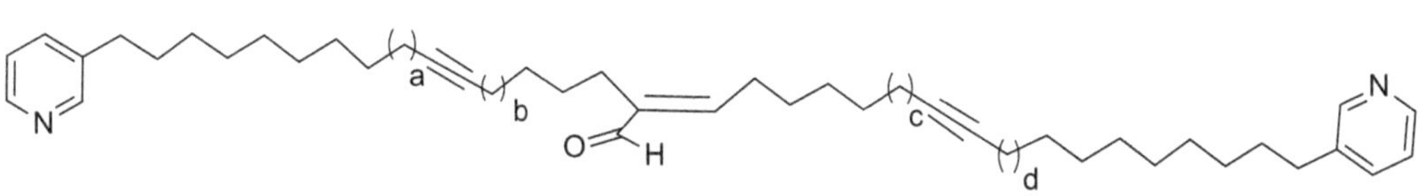

80 Pyrinodemine I (a + b = 1, c + d = 1)

Fig. 12.8 All compounds are cited (figures are not cited; instead compound's number are cited; it is just for reference)

and antibiotic-resistant strains of *Candida* sp. (Nicholas and Molinski 2000). Kobayashi's group have isolated pyrinodemins A–D (**74–77**) (Fig. 12.8) potent cytotoxic bis-pyridine alkaloids with a cis-cyclopent[c]isoxazolidine moiety, from the Okinawan marine sponge *Amphimedon* sp. (Tsuda et al. 1999; Hirano et al. 2000). In the later years, they have isolated several other pyrinodemins, viz., pyrinodemins G–I (**78–80**), bis-3-alkyl pyridine from the same sponge (Kubota et al. 2013) (Fig. 12.8).

Niphatesine F (**81**) was isolated from the Okinawan marine sponge *Niphates* sp. (Kobayashi et al. 1992), while untenines A–C (**82–84**) (Fig. 12.9) were isolated from the Okinawan marine sponge *Callyspongia* sp. (Wang et al. 1996). Cyclic *bis*-pyridine alkaloids, cyclostellettamine alkaloids (**85–93**), were obtained from the sponge *Pachychalina* sp. and the alkaloids exhibited antimicrobial and antimycobacterial activity (De Oliveira et al. 2006). Cytotoxic tripyridine alkaloids, niphatoxins A and B (**94–95**), have been isolated by Kobayashi's group from the Red Sea sponge *Niphates* sp. (Talpira et al. 1992), while nitroalkyl pyridine alkaloids with antimicrofouling properties were isolated from the Okinawan marine sponge *Callyspongia* sp. (Wang et al. 1996). Theonelladins A–D (**96–99**), antineoplastic pyridine alkaloids, were isolated from the marine sponges *Theonella swinhoei* (Kobayashi et al. 1989a). Kitamura et al. isolated echinoclathrines A–C (**100–102**), a new class of pyridine alkaloids having 4-aryl-2-methylpyridine unit from an Okinawan sponge, *Echinoclathria* sp. (Kitamura et al. 1999). Echinoclathrine A (**100**) exhibited a weak cytotoxicity (IC50 = 10 μg/mL) against P-388, A-549, and HT-29 cell lines, while other alkaloids were found to be inactive (Fig. 12.9).

12.6 Piperidine Alkaloids

Piperidines are heterocyclic amines consisting of a six-membered ring containing five methylene bridges (-CH_2-) and one amine bridge (-NH-). Marine sponges belonging to the order Haplosclerida are considered the richest source of alkyl piperidine alkaloids. 3-Alkyl piperidine alkaloid which is a very common piperidine alkaloid includes a variety of metabolites ranging from monomeric 3-alkyl pyridines to condensed bis-alkyl piperidines of the manzamine class. These alkaloids show a wide range of biological activities, viz., antimicrobial, antiviral, and cytotoxic (Schmitz et al. 1978), antimalarial (Ang et al. 2000), and antifouling (Faimali et al. 2003). Unusual oligomeric pyridinium alkaloids, namely, cyclohaliclonamines (Teruya et al. 2006) and viscosamine (Volk et al. 2004), were isolated from *Haliclona* sp. and *Haliclona viscosa*, respectively. A macrocyclic dimeric haliclamines and the linear trimeric viscosaline were also isolated from *H. viscosa* (Volk and Köck 2004).

Fusetani and coworkers have reported piperidine alkaloids, namely, halicyclamine A (**103**), tetradehydrohalicyclamine A (**104**), and 22-hydroxyhalicyclamine A (**105**) from a marine sponge *Amphimedon* sp. (Takekawa, et al. 2006). These halicyclamine piperidine alkaloids (**103–105**) exhibited cytotoxicity against P388 cells with IC_{50} values of 0.45, 2.2, and 0.45 μg/mL, respectively. A new piperidine alkaloid plakoridine C (**106**) has been isolated by Kobayashi's group from an Okinawan marine sponge *Plakortis* sp., and the structure was elucidated from spectroscopic data (Ishiguro et al. 2009). Plakoridine C (**106**) is a new alkaloid possessing a piperidine ring connected to a β-keto-γ-lactone through a double bond. Bis-piperidine alkaloids, madangamine F (**107**), haliclonacyclamine F (**108**), and arenosclerins D (**109**) and E (**110**), have been isolated from the marine sponge *Pachychalina alcaloidifera* and the structures were identified by the analysis of spectroscopic data. The alkaloids displayed cytotoxic activity against different cancer cell lines (Fig. 12.10).

12.7 Quinolizidine Alkaloids

Quinolizidine alkaloids are distinct from other alkaloids in that they contained at least one quinolizidine ring system. They exhibited significant coronary vasodilative effects as well as modest murine leukemia cell growth inhibition and antimicrobial activity (Quirion et al. 1992). Quinolizidine family, namely, 1-oxa-quinolizidine and bis-1-oxa-quinolizidines, is common in marine sponges. The first four "l-oxa-quinolizidines" were isolated from the Australian sponge *Xestospongia exigua*, designated as xestospongins A–D (**111–114**) with the structure of (-)-xestospongin-C (**113**)

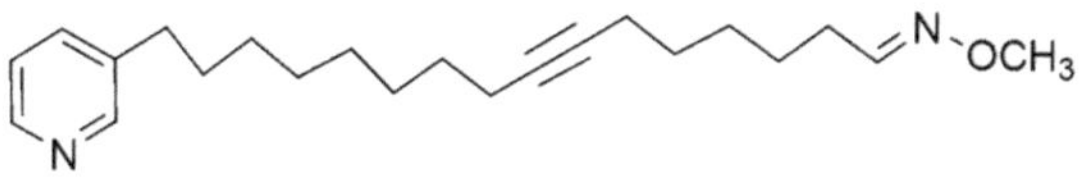

81 Niphatesine F

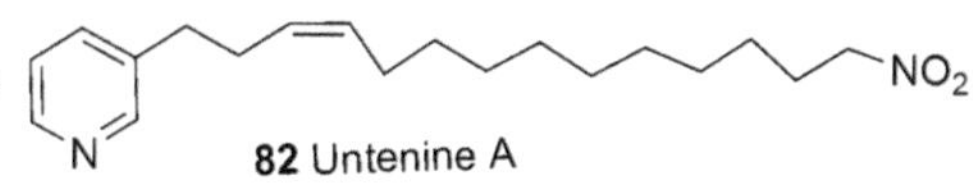

82 Untenine A

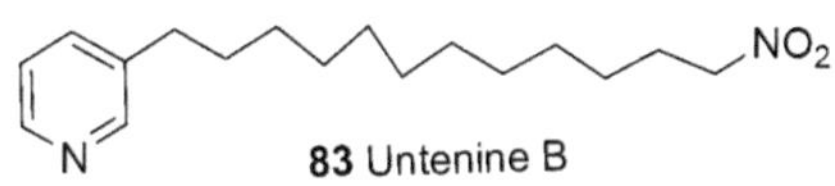

83 Untenine B

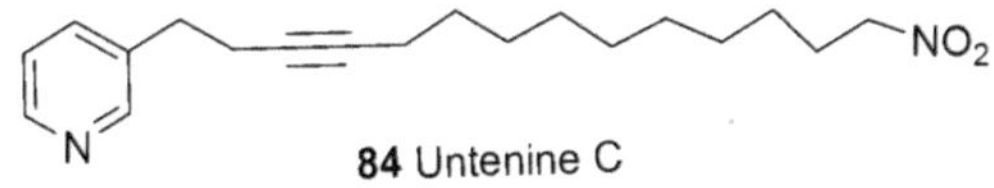

84 Untenine C

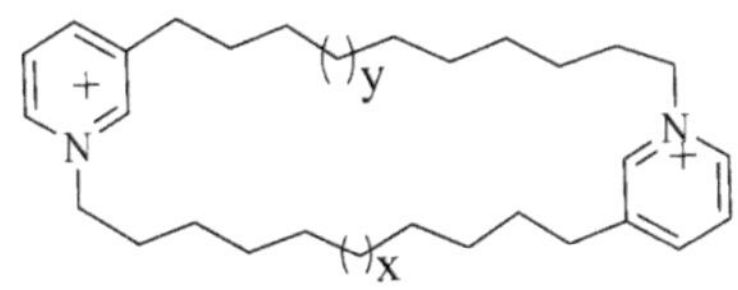

85-87 cyclostellattamines A-C : x =3, y = 3; x = 3, y =4; x = 4, y = 4

88-90 cyclostellattamines D-F: x= 3, y = 5; x =4, y = 5; x= 5, y = 5

91-93 cyclostellattamines G-I: x= 1, y = 3; x = 2, y = 3; x = 1, y = 4

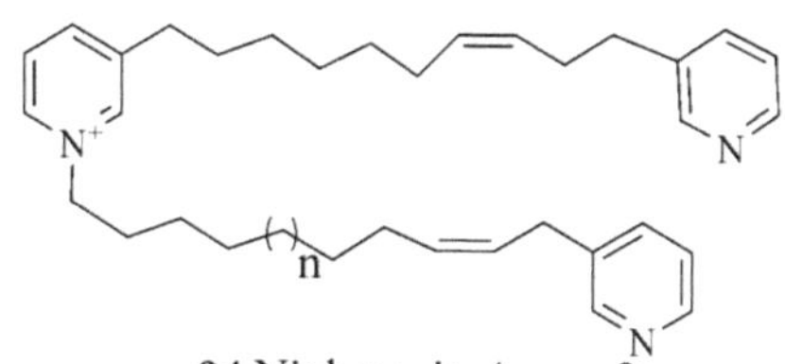

94 Niphatoxin A: n = 2

95 Niphatoxin B: n = 3

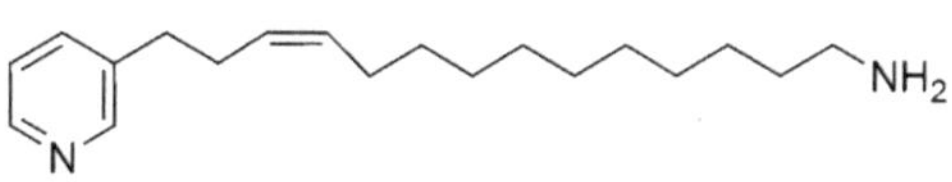

96 Thionelladins A

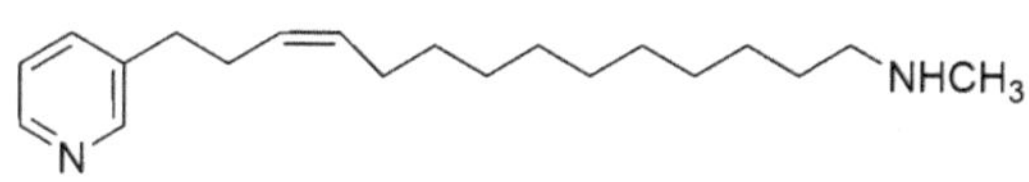

97 Thionelladins B

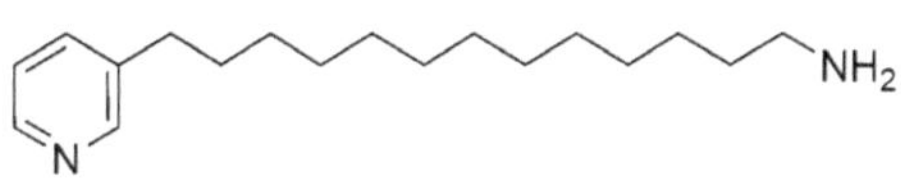

98 Thionelladins C

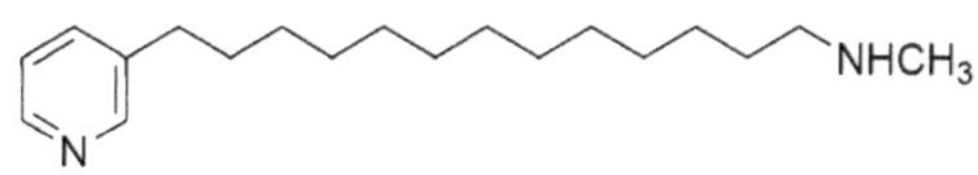

99 Thionelladins D

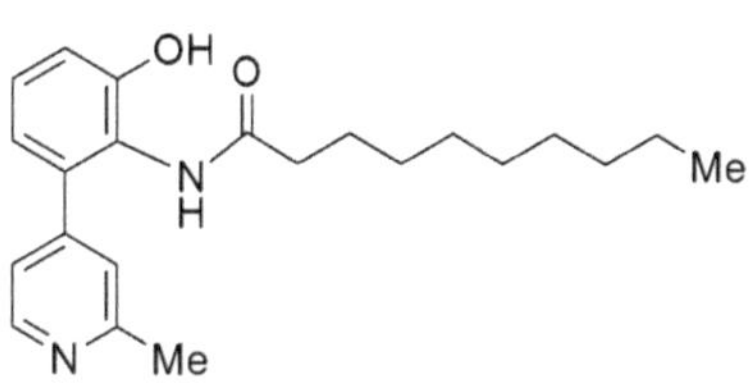

100 Echinoclathrines A

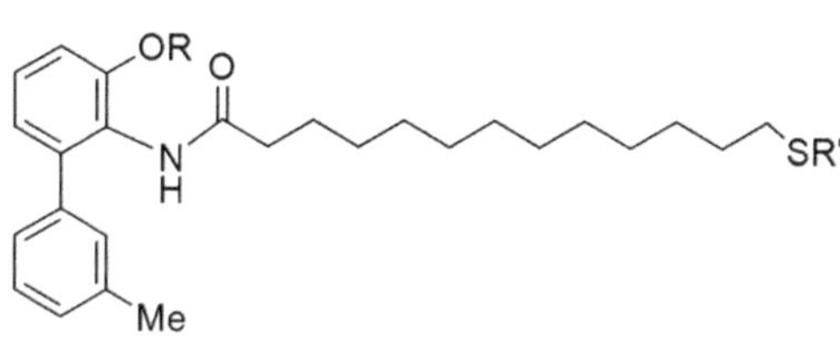

101 Echinoclathrines B R = H, R' = Ac

102 Echinoclathrines C R = R' = H

Fig. 12.9 All compounds are cited (figures are not cited; instead compound's number are cited; it is just for reference)

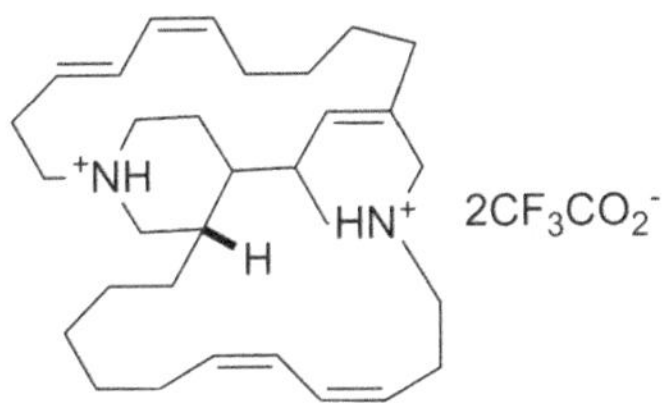

103 Halicyclamine A

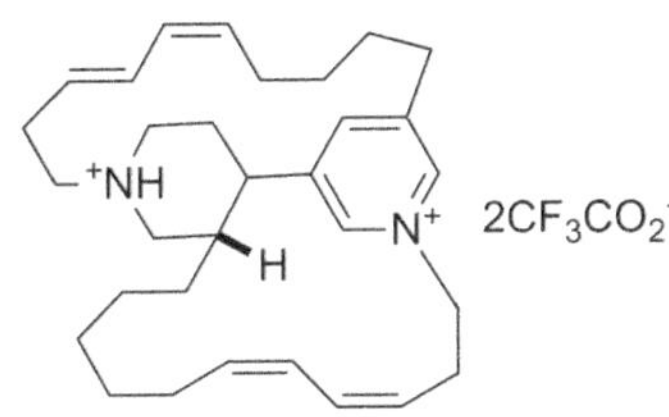

104 Tetradehydrohalicyclamine A

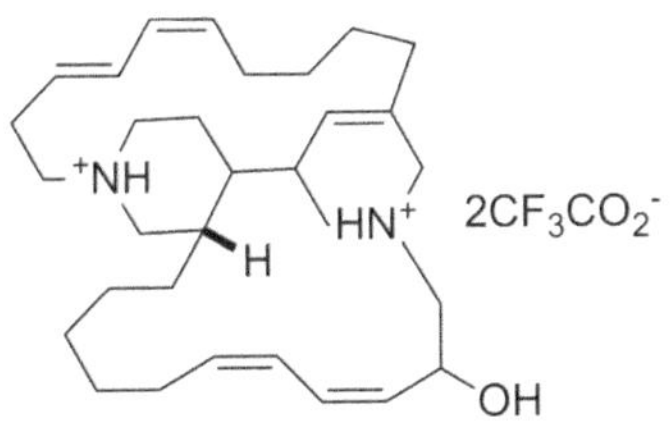

105 22-hydroxyhalicyclamine A

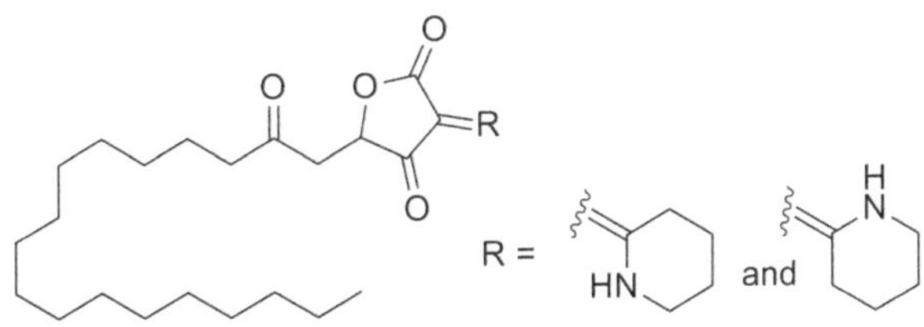

106 Plakoridine C

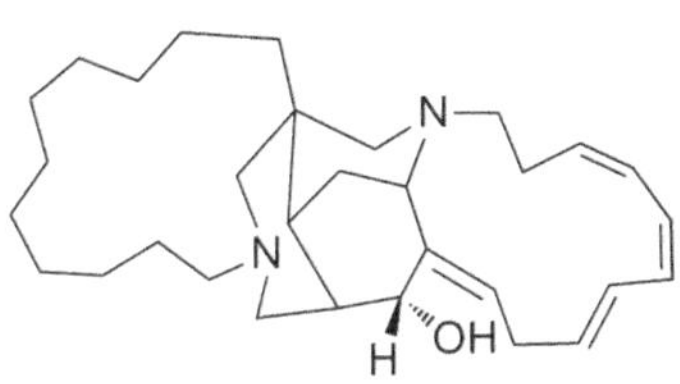

107 madagamine F

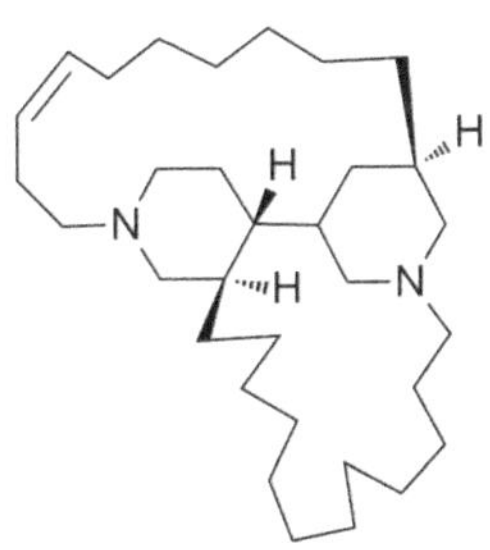

108 halicloacyclamine F

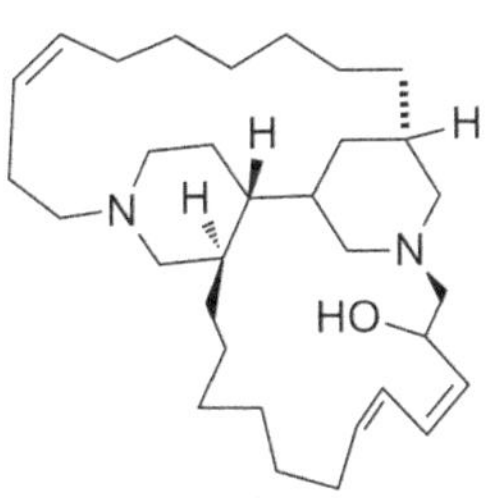

109 arenosclerin D

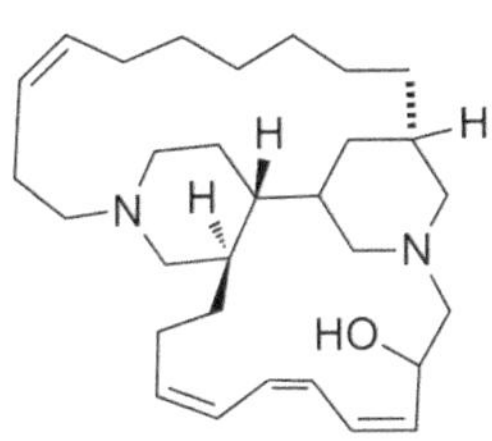

110 arenosclerin E

Fig. 12.10 All compounds are cited (figures are not cited; instead compound's number are cited; it is just for reference)

determined by X-ray techniques (Nakagawa et al. 1984). Later these oxa-quinolizidine and bis-quinolizidine families have also been isolated from several other marine sponges, viz., *Oceanapia* sp. (Singh et al. 2011), *Petrosia similis* (Goud et al. 2003), and *Haliclona exigua* (Venkateswarlu et al. 1994). The family of xestospongin/araguspongine alkaloids comprises

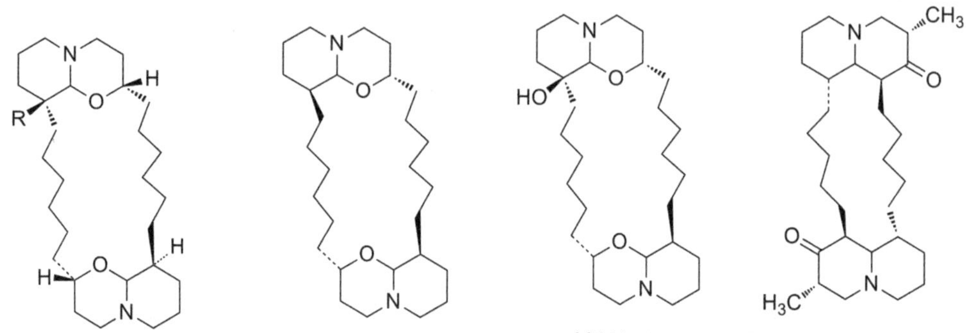

111 R = H (-)-Xestospongin A
112 R = OH (+)-Xestospongin B
113 Xestospongin C
114 Xestospongin D
115 Petrosin

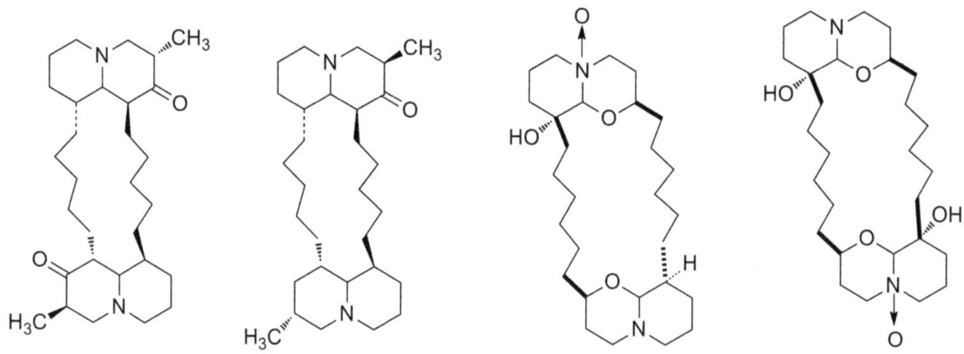

116 Petrosin A
117 Aragupetrosin A
118 Araguspongine K
119 Araguspongine L

Fig. 12.11 All compounds are cited (figures are not cited; instead compound's number are cited; it is just for reference)

of 13 members (Moon et al. 2002; Reddy and Faulkner 1997), and chemically, they are dimeric 2,9-disubstituted 1-oxa-quinolizidines. Braekman et al. reported petrosin (**115**), a bis-quinolizidine alkaloid from the sponge *Petrosia seriata* (Braekman et al. 1982). They have established that petrosin might exist in two isomers in solution; the structure of petrosin was characterized by spectroscopic data and solid-state structure was determined by X-ray diffraction analysis (Braekman et al. 1982). A racemic xestospongin alkaloid (±) xestospongin D (**114**) was isolated from the Singapore marine sponge *Niphates* sp. (Pettit et al. 1996). The absolute stereochemistry at the six chiral centers for this enantiomer was assigned by X-ray analysis. This racemic (±) xestospongin D (**114**) showed several activities including antimicrobial and modest growth inhibitory against a number of tumor cell lines (Pettit et al. 1996). Petrosin A (**116**) vasodilative macrocyclic quinolizidine alkaloid, aragupetrosine A (**117**), and several araguspongin alkaloids have been reported by Kobayashi's group from an Okinawan marine sponge, *Xestospongia* sp. (Kobayashi et al. 1989b). Unique bis-1-oxa-quinolizidine N-oxide alkaloids, araguspongins K (**118**) and L (**119**), were also reported by Orabi et al. from red sponge *Xestospongia exigua* (Orabi et al. 2002) (Fig. 12.11).

12.8 Steroidal Alkaloids

In 2002, Borbone et al. demonstrated the isolation of four steroidal alkaloids, plakinamines G (**120**), H (**121**), and L (**122**) and tetrahydroplakinamine A (**123**) from the marine sponge *Corticium* sp. (Borbone et al. 2002). Among these series, plakinamine G (**120**) and tetrahydroplakinamine A (**123**) were most active against C6 cells, whereas plakinamine H (**121**) and plakinamine L (**122**) were cytotoxic against C6 cells and RAW-264 cell lines. In 2007, three more steroidal alkaloids, cortistatins J–L (**124–126**), were isolated from the Indonesian marine sponge *Corticium simplex* (Aoki et al. 2007). Cortistatin J (**124**) demonstrated potent cytostatic antiproliferative activity against human umbilical vein endothelial cells (HUVEC) and also inhibited migration and tubular formation of HUVEC induced by VEGF or bFGF, whereas cortistatins K (**125**) and L (**126**) were less potent than cortistatin J (**124**). Steroidal alkaloids plakinamine I–K (**127–129**) and dihydroplakinamine K (**130**) were isolated from sponge *Corticium niger* (Ridley and Faulkner 2003) and were tested for cytotoxicity against the human colon tumor cell line (HCT-116). Compounds plakinamine K (**129**) and dihydroplakinamine K (**130**) were found to be the most active in terms of potency, while plakinamines I and J (**127 & 128**) were moderately active (Fig. 12.12).

12.9 Bromotyrosine Alkaloids

Marine sponges from the order Verongida are rich source of bromotyrosine-derived alkaloids (Bergquist 1983; Gribble 1998). Sponges in this order have been reported to show unusual biochemical profiles characterized by the absence of terpenes and the production of sterols and brominated compounds biogenetically tyrosine (Kochanowska et al. 2008). Several bromotyrosine alkaloids, viz., purealin (Tsuda et al. 1992), lipopurealins A–E (Wu et al. 1986; Kobayashi et al. 1995), purealidins A–S (Ishibashi et al. 1991; Kobayashi et al. 1991), psammaplysins A–B (Roll et al. 1985), purpuramines A–J (Tabudravu and Jaspars 2002; Yagi et al. 1993), aplysamines 2–5 (Jurek et al. 1993), and macrocyclic peptides bastadins (Carney et al. 1993; Aoki et al. 2006), have been isolated from this marine sponge order of Verongida. Due to the occurrence of bromotyrosine alkaloids in practically all Verongida marine sponges so far chemically investigated, these alkaloids and their derivatives have been considered as chemotaxonomic markers for sponges of this order (Harper et al. 2001). However, the recent isolation of bromotyrosine-derived compounds from sponges belonging to other distinct taxa, such as *Agelas oroides* (König and Wright 1993), *Oceanapia* sp. (Nicholas et al. 2001), and *Poecillastra wondoensis* (Park et al. 2003), indicated that these compounds are not specific chemotaxonomic markers for marine sponges of Verongida (Erpenbeck and van Soest 2007). Bromotyrosine alkaloids exhibited potent antibacterial (Tsukamoto et al. 1996a, 1996b; Matsunaga et al. 2005), anti-HIV (Ross et al. 2000), antimalarial (Xu et al. 2011), and cytotoxic (Tabudravu and Jaspars 2002) activities.

Purealidin S and purpuramine J were isolated from the Fijian marine sponge *Druinella* sp. (Tabudravu and Jaspars 2002). Fujiwara et al. isolated a new bromotyrosine alkaloid JBIR-44 (**131**) from *Psammaplysilla purpurea*. JBIR-44 (**131**) showed cytotoxic effects against human cervical carcinoma HeLa cells (Fujiwara et al. 2009). Bromotyrosine-derived metabolites purpuramines A–I were isolated from the marine sponge *Psammaplysilla purpurea* (Jurek et al. 1993). Purpuramines A (**132**) and C (**133**) differ only at amine substituent at the aromatic ring.

A novel dibromotyrosine derivative, Aplysfistularine (**134**), was isolated from the marine sponge *Aplysina fistularis* (Lira et al. 2012). This species have been well documented for the presence of a large number of brominated metabolites including fistularines, aerothionines, ceratinamines, aplysamines, anamonianes, and psammaplysines (Ciminiello et al. 1994; Thoms et al. 2005; Saeki et al. 2002). Purealidins B–C (**135–136**) (Kobayashi et al. 1991) and lipopurealins D–E (**137–138**) (Kobayashi et al., 1995) were isolated from the Okinawan marine sponge *Psammaplysilla purea* (Fig. 12.13).

Yin et al. isolated pseudoceramines A–D (**139–142**), a series of antibacterial bromotyrosine alkaloids from the marine sponge *Pseudoceratina* sp. of Erskine Is., Great Barrier

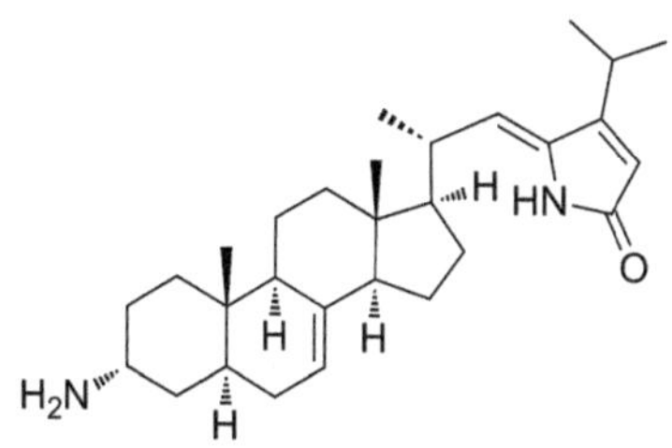

120 Plakinamine G

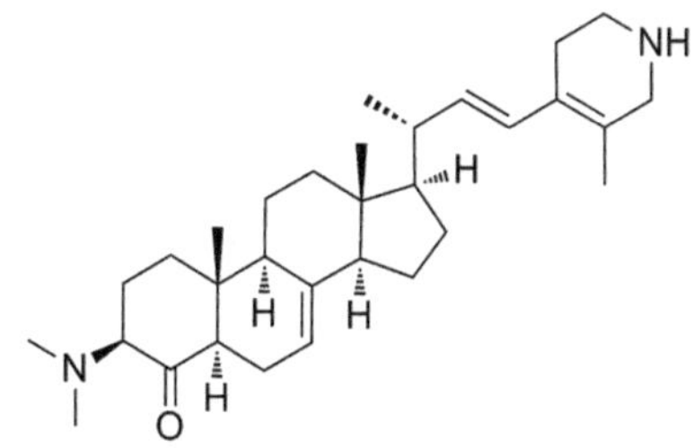

121 Plakinamine H

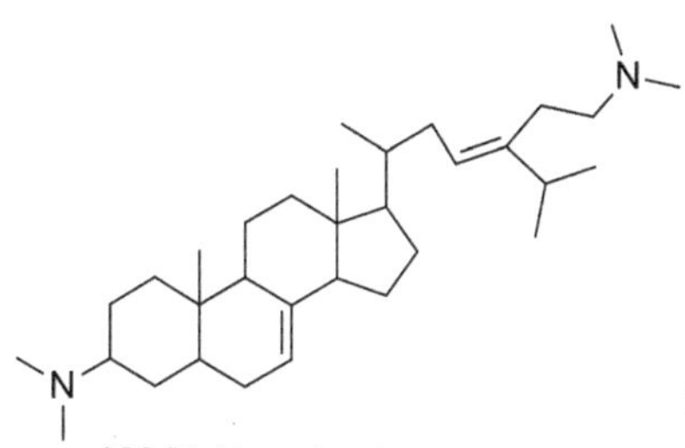

122 Plakinamine L

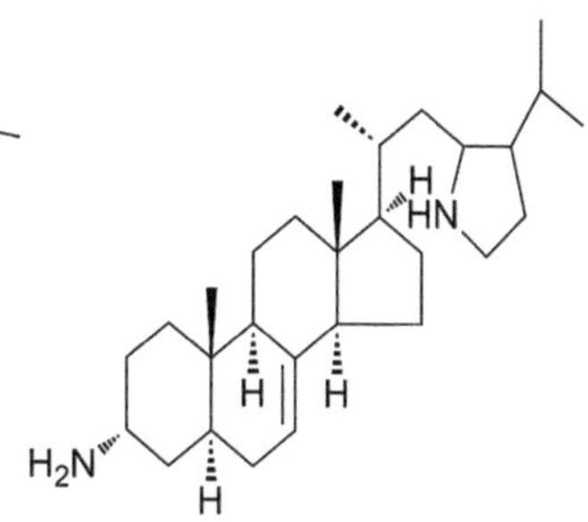

123 Tetrahydroplakinamine A

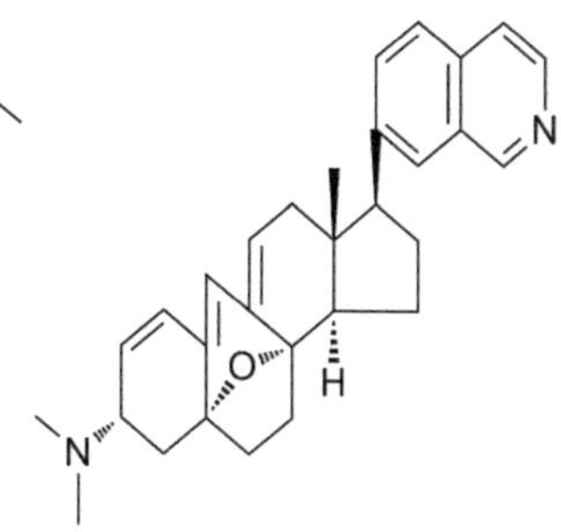

124 Cortistatin J

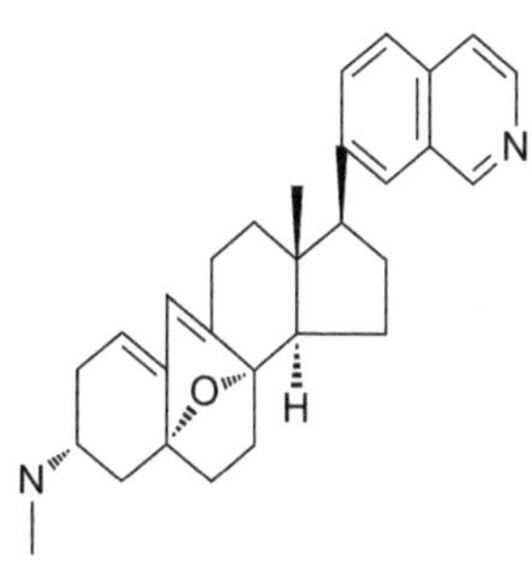

125 Cortistatin K

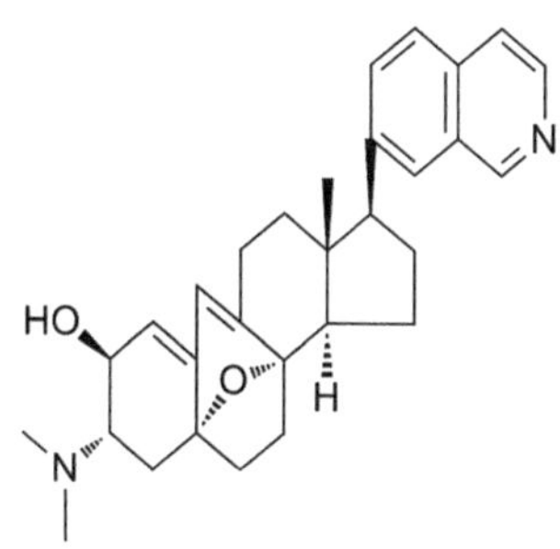

126 Cortistatin L

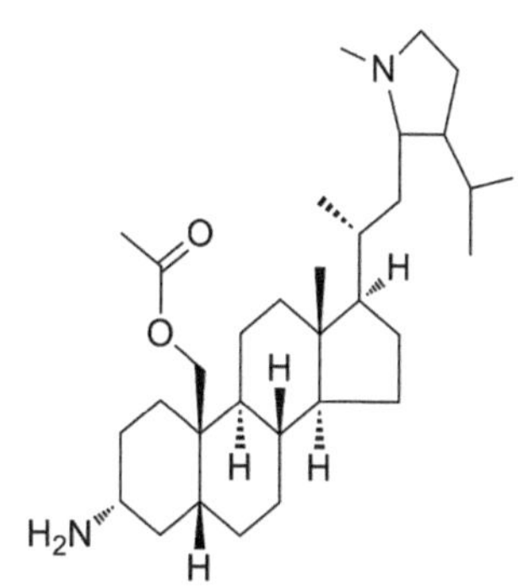

127 Plakinamine I

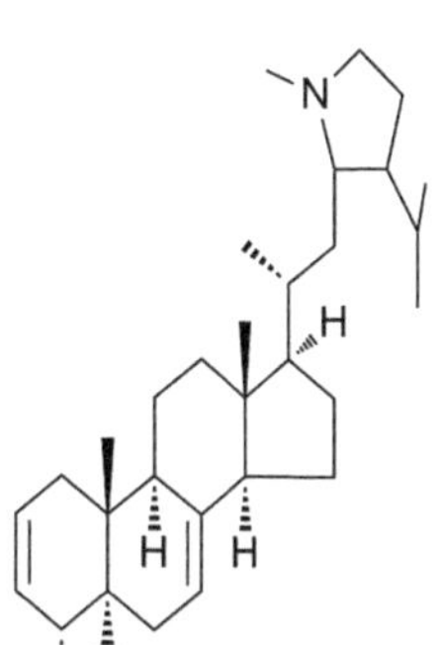

128 Plakinamine J

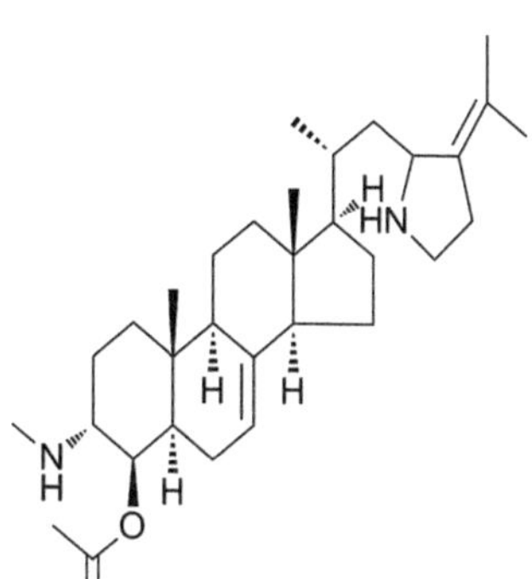

129 Plakinamine K

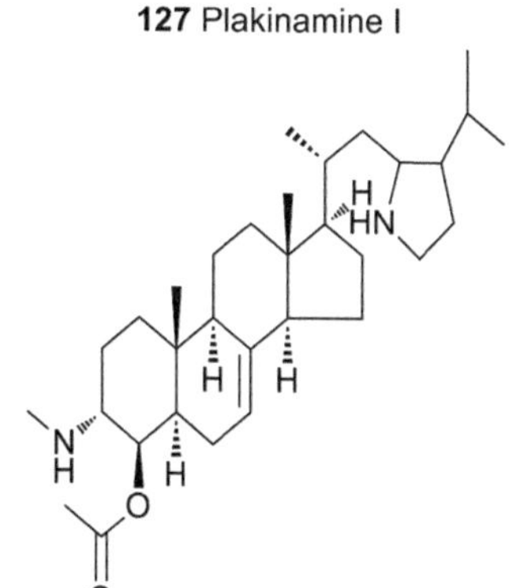

130 Dihydroplakinamine K

Fig. 12.12 All compounds are cited (figures are not cited; instead compound's number are cited; it is just for reference)

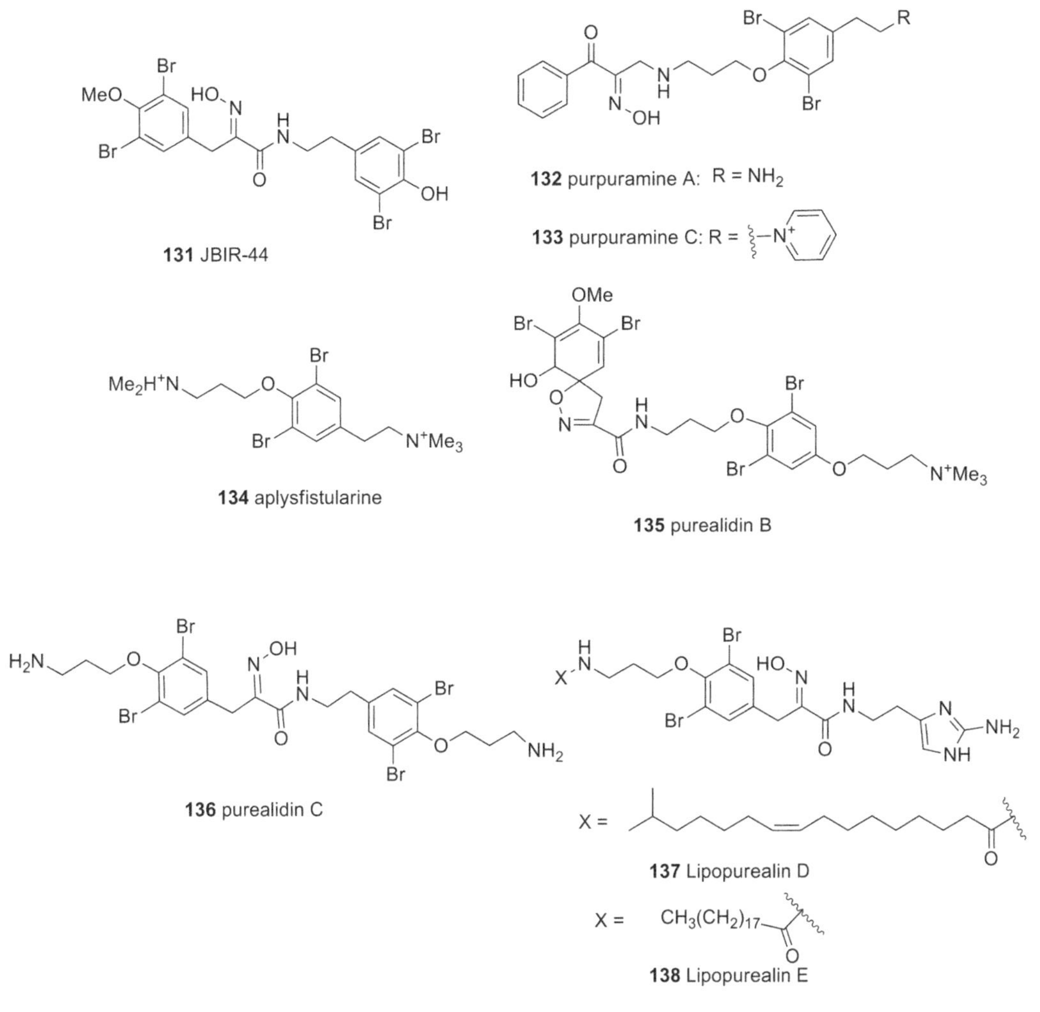

Fig. 12.13 All compounds are cited (figures are not cited; instead compound's number are cited; it is just for reference)

Reef (Yin et al. 2011). They have reported that pseudoceramine C (**141**) was a cleavage derivative of spermatinamine (**143**). Pseudoceramine B (**140**) inhibits secretion of the virulence factor Yersinia outer protein E (Yin et al. 2011). Bromotyrosine-derived alkaloids, purealidin-L (**144**), aerophobin-1 (**145**) and aerophobin-2 (**146**) (Cimino et al. 1983), and isofistularin-3 (**147**), were isolated from several marine sponges (Gopichand and Schmitz 1979) (Fig. 12.14).

Kobayashi's group isolated purealidin-L (**144**) (Kobayashi et al. 1995) from *Psammaplysilla purea*, and tyrokeradines A and B (**148–149**) were isolated from Okinawan marine sponge of order Verongida (Mukai et al. 2009). In later years, they isolated other related bromotyrosine alkaloids tyrokeradines C (**150**) from the same sponge (Kubota et al. 2012). His group also isolated ceratinadins A–C (**151–153**) from Okinawan marine sponge *Pseudoceratina* sp. (Kona et al. 2010). Aplysamine-4 (**154**), a bromotyrosine-derived alkaloid, was isolated from the sponge *Psammaplysilla purpurea* (Jurek et al. 1993). Proksch's group has isolated a new bromotyrosine alkaloid N-methyl-aerophobin-2 (**155**) along with known bromotyrosine alkaloids, purealidin-L (**144**), aerophobin-1(**145**), and aerophobin-2 (**146**),

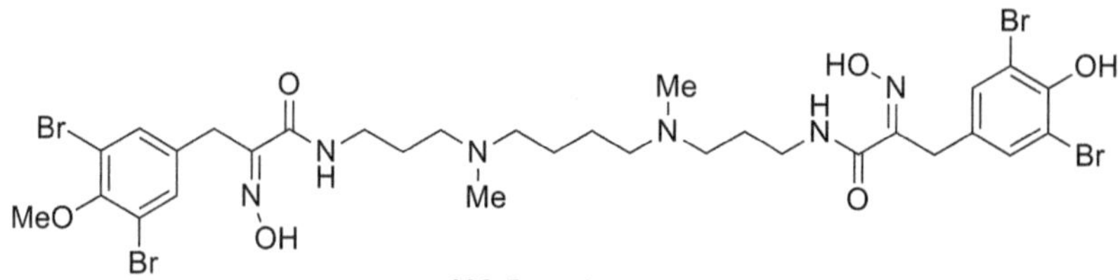

139 Pseudoceramine A

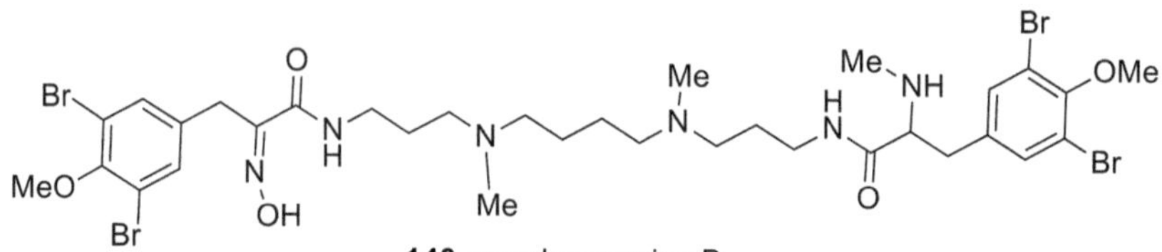

140 pseudoceramine B

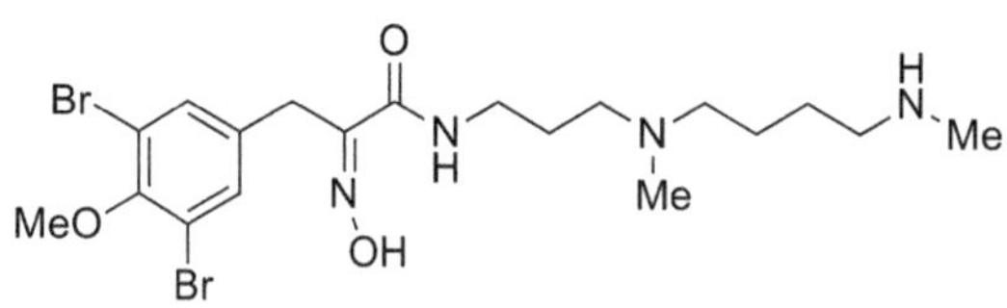

141 pseudoceramine C

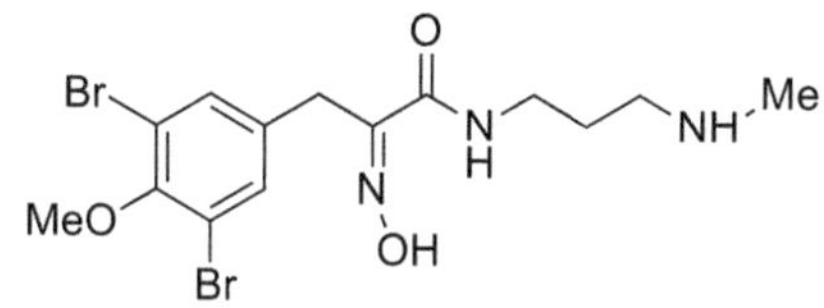

142 pseudoceramine D

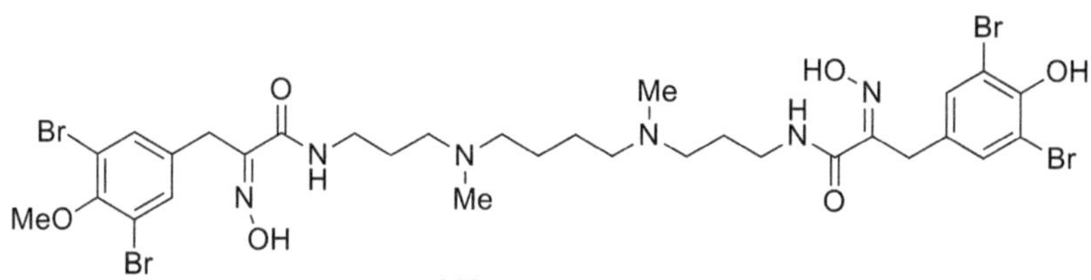

143 spermatinamine

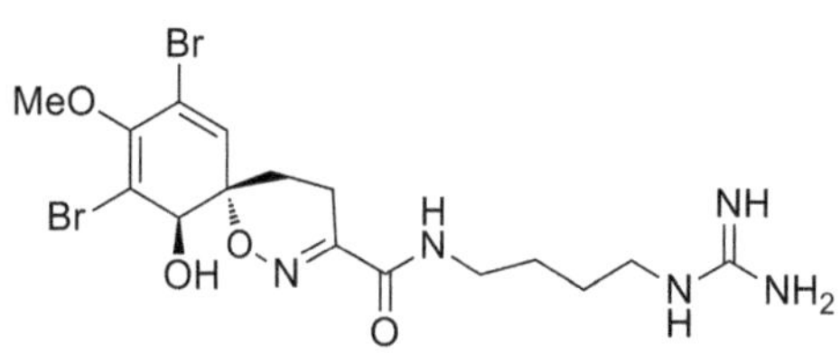

144 purealidin L

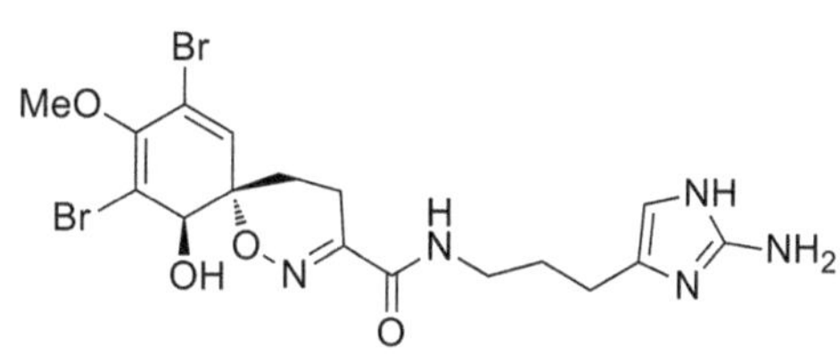

145 aerophobin-1

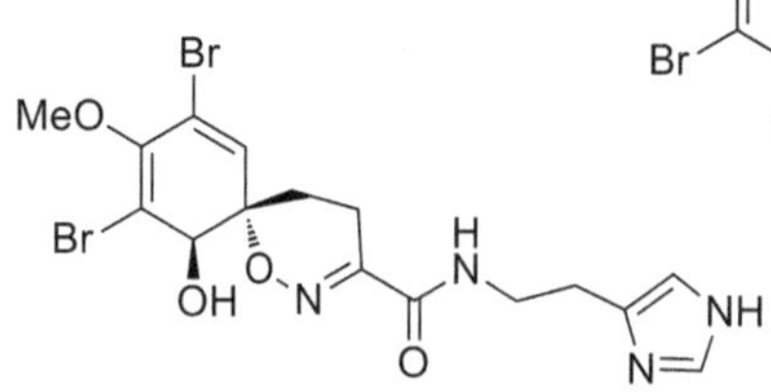

146 aerophobin-2

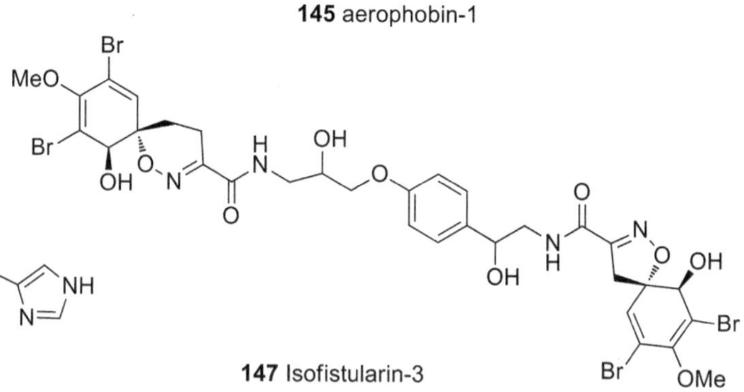

147 Isofistularin-3

Fig. 12.14 All compounds are cited (figures are not cited; instead compound's number are cited; it is just for reference)

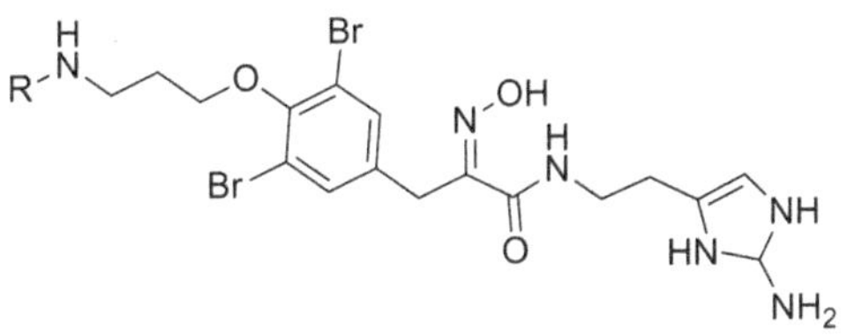

148 tyrokeradine A: R = $N^{+}Me_3$
149 tyrokeradine B: R = $N^{+}H_3$
150 tyrokeradine C: R = NHMe

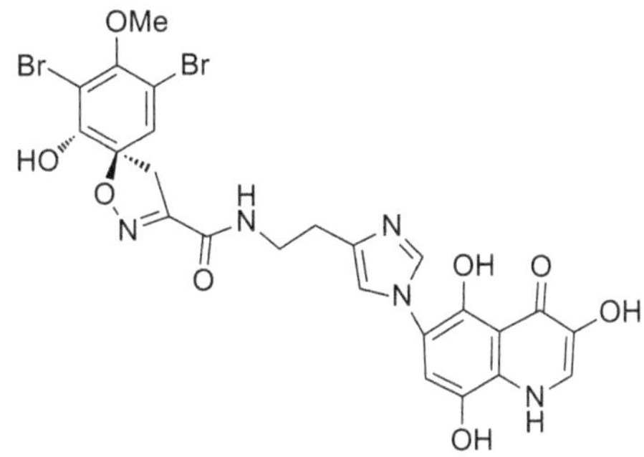

151 ceratinadin A

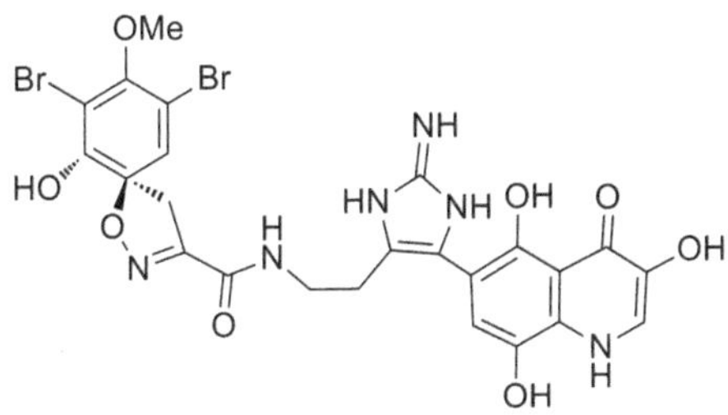

152 ceratinadin B

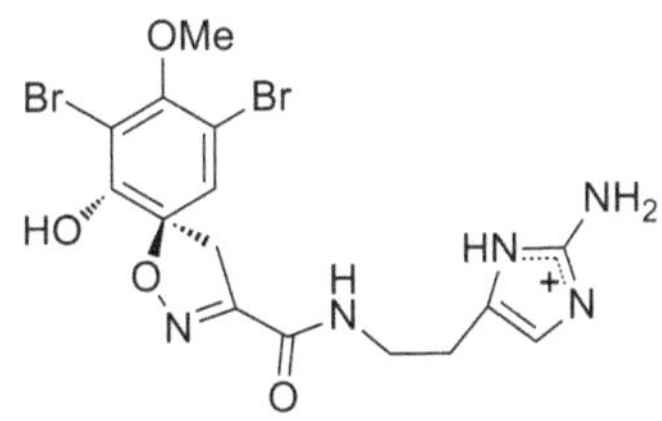

153 ceratinadin C

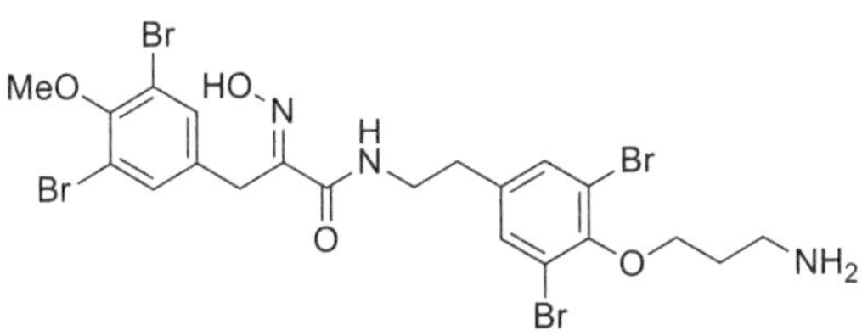

154 apylysamine 4

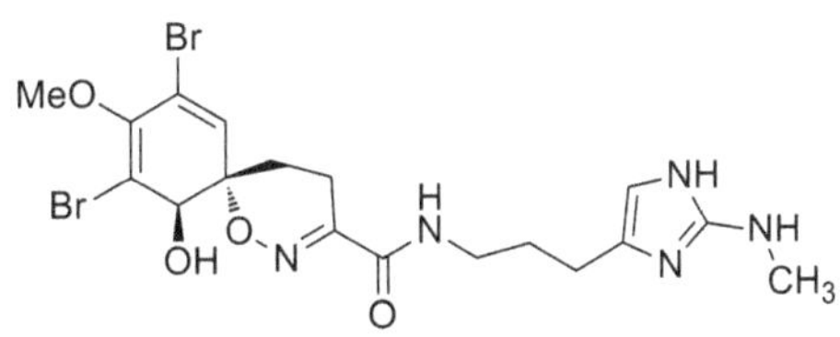

155 N-Methyl aerophobin-2

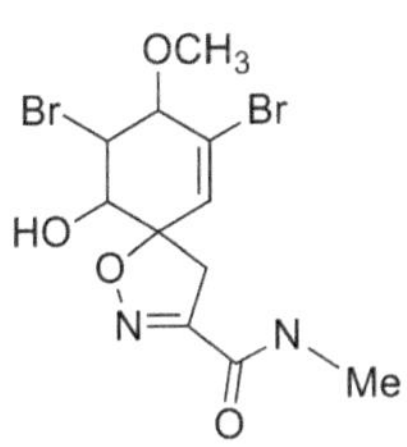

156 purprealidin A

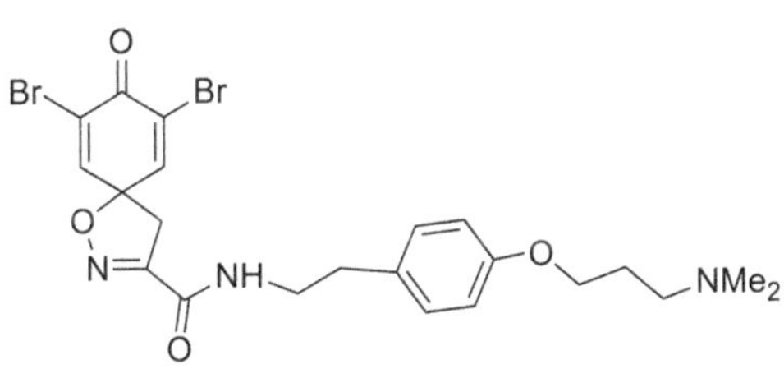

157 purprealidin B

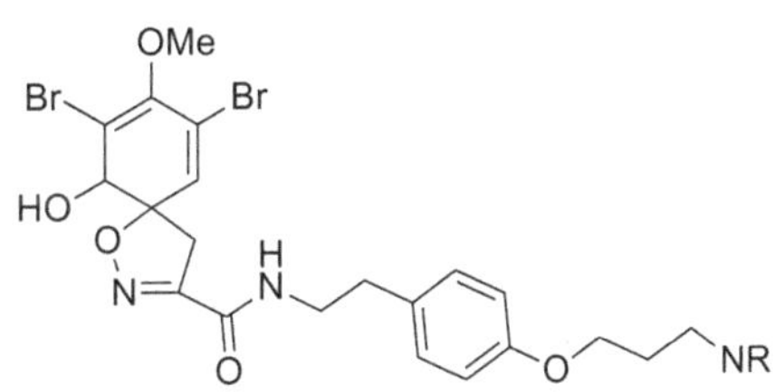

158 purpurealidin C: R = $CO(CH_2)_{11}CH(CH_3)_2$

159 purpurealidin D: R = $CO(CH_2)_{12}CH_2CH_2CH_3$

Fig. 12.15 All compounds are cited (figures are not cited; instead compound's number are cited; it is just for reference)

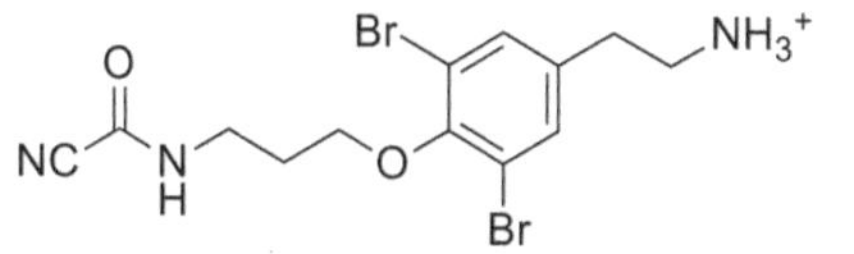

160 ceratinamine

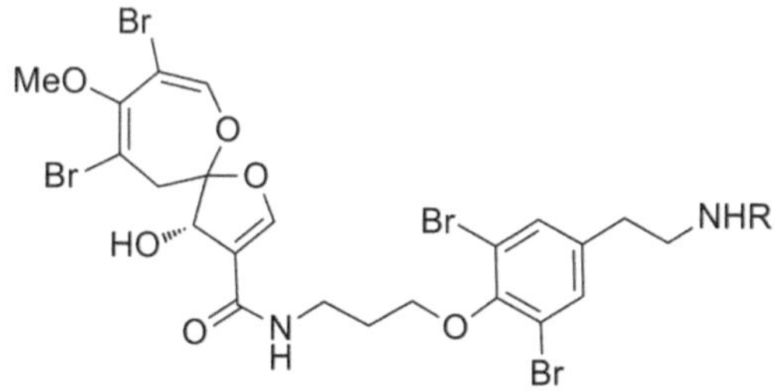

161 ceratinamide A: R = CHO
162 ceratinamide B: R = $CO(CH_2)_{11}CH(CH_3)_2$

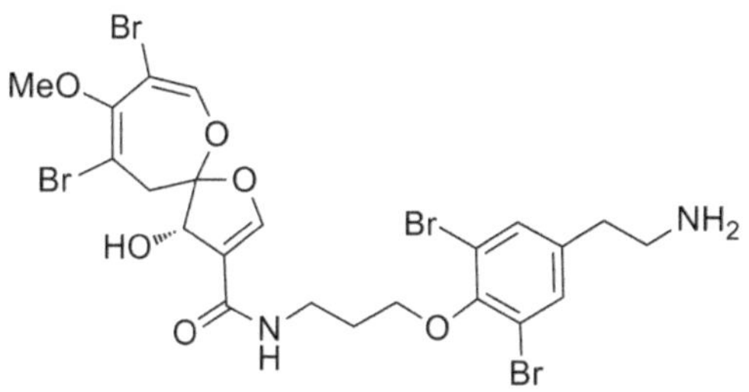

163 psammalysin A

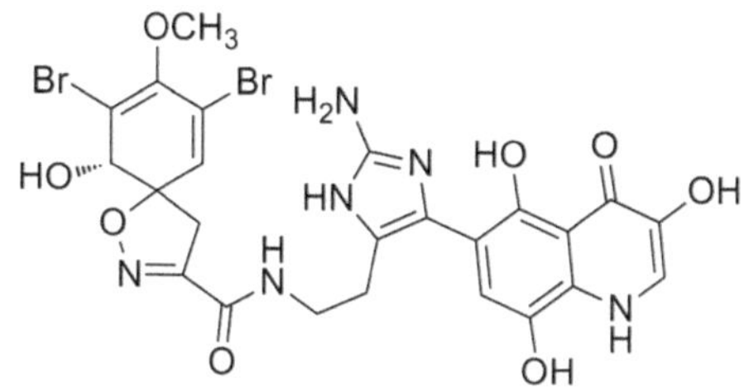

164

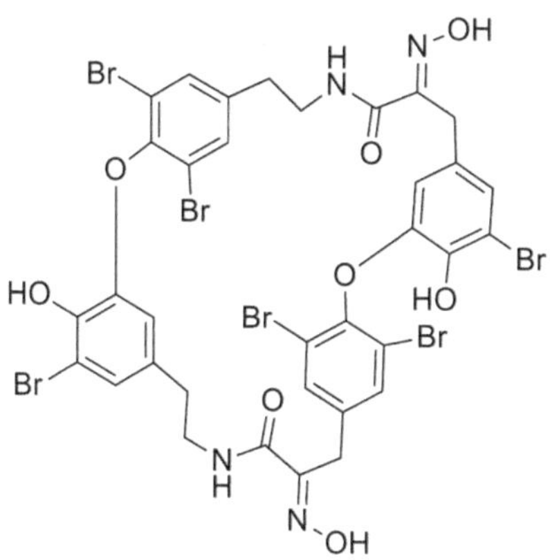

165 bastadin 6

Fig. 12.16 All compounds are cited (figures are not cited; instead compound's number are cited; it is just for reference)

from the Caribbean marine sponge *Aiolochroia crassa* (Assmann et al. 1998). A series of purpurealidins A–D (**156–159**) were isolated by Tilvi et al., from the Indian marine sponge *Psammaplysilla purpurea* (Tilvi et al. 2004) (Fig. 12.15).

Bromotyrosine alkaloids with antifouling activities were reported from *P. purpurea* collected in various locations of Japan, among which the most interesting is ceratinamine (**160**) which contains a cyanoformamide functionality, unprecedented in natural products (Tsukamoto et al. 1996a). Ceratinamine showed potent antifouling activities against barnacle larvae with an EC_{50} value of 5.0 μg mL^{-1}. Other bromotyrosine-derived alkaloids such as ceratinamides A (**161**) and B (**162**) and psammaplysin A (**163**) exhibited potent activity with EC_{50} values of 0.10, 2.40, and 0.27 μg mL^{-1}, respectively (Tsukamoto et al. 1996b).

Bewley's research group isolated a novel bromotyrosine alkaloid (**164**), which inhibits mycothiol S-conjugate amidase (MCA) from marine sponge *Oceanapia* species (Nicholas et al. 2001). Macrocyclic bromotyrosine alkaloids, bastadins, were isolated from several marine sponges, such as *Psammaplysilla purpurea* (Carney et al. 1993) and *Ianthella basta* (Aoki et al. 2006). Bastadin-6 (**165**) exhibited antiproliferative activities against endothelial cells (Aoki et al. 2006) (Fig. 12.16).

12.10 Conclusion

This chapter presents the various alkaloids isolated from marine sponges and discusses their biological properties. In order to simplify to general readers, the chapter presents different class of alkaloids isolated from various marine sponges with their selected chemical structures in each separate section. The source of sponge from which they are isolated and their bioactivities have been discussed. The chapter reviews on alkaloids, viz., pyridoacridines, alkyl pyridine, piperidine, indole, quinolizidine, isoquinoline, steroidal, and bromotyrosine alkaloids and their derivatives isolated from various marine sponges. Since there are several alkaloids of marine sponge origin, it is not possible to include all alkaloids isolated from them. We highlighted only selected alkaloids of marine sponge and discussed their potential biological properties. We believe that this chapter may find interest to general readers and researchers working in natural product sciences both from the academic and industries. We also acknowledged that several published works on the topic which deserved to be cited have been excluded due to page limitation.

Acknowledgments We are grateful to the Council of Scientific and Industrial Research (CSIR), India, for providing financial support. MSM thanks to DST-SERB, India for young scientist award. Thanks to the Director of CSIR-NIO for the constant encouragement.

References

Ang KKH, Holmes MJ, Higa T, Hamann MT, Kara UAK (2000) In vivo antimalarial activity of the b carboline alkaloid manzamine A. Antimicrob Agents Chemother 44(6):1645–1649

Aoki S, Ye Y, Higuchi K, Takashima A, Tanaka Y, Kitagawa I, Kobayashi M (2001) Novel neuronal nitric oxide synthase (nNOS) selective inhibitor, aplysinopsin-type indole alkaloid, from marine sponge *Hyrtios erecta*. Chem Pharm Bull 49 (10):1372–1374

Aoki S, Wei H, Matsui K, Rachmat R, Kobayashi M (2003) Pyridoacridine alkaloids inducing neuronal differentiation in a neuroblastoma cell line, from marine sponge *Biemna fortis*. Bioorg Med Chem 11 (9):1969–1973

Aoki S, Cho S-H, Ono M, Kuwano T, Nakao S, Kuwano M, Nakagawa S, Gao J-Q, Mayumi T, Shibuya M, Kobayashi M (2006) Bastadin 6, a spongean brominated tyrosine derivative, inhibits tumor angiogenesis by inducing selective apoptosis to endothelial cells. Anti-Cancer Drugs 17 (3):269–278

Aoki S, Watanabe Y, Tanabe D, Setiawan A, Arai M, Kobayashi M (2007) Cortistatins J K L novel abeo-9 (10-19)-androstane-type steroidal alkaloids with isoquinoline unit, from marine sponge *Corticium simplex*. Tetrahedron Lett 48(26):4485–4488

Assmann M, Wraay V, van Soest RWM, Proksch P (1998) A new bromotyrosine alkaloid from Caribbean sponge Aiolochroia crassa. Z Naturforsch 53c:398–401

Baker JT, Wells RJ (1981) Biological active substances from Australian marine organisms. In: Beal JL, Reinhard E (eds) Natural products as medicinal agents. Hipocrates Verlag, Stuttgart, pp 281–318

Bao B, Sun Q, Yao X, Hong J, Lee C-O, Sim CJ, Im KS, Jung JH (2005) Cytotoxic bisindole alkaloids from a marine sponge *Spongosorites* sp. J Nat Prod 68 (5):711–715

Barnes EC, Said NABM, Williams ED, Hooper JNA, Davis RA (2010) Ecionines A and B, two new cytotoxic pyridoacridine alkaloids from the Australian marine sponge, *Ecionemia geodides*. Tetrahedron 66 (1):283–287

Bergmann W, Feeney RJ (1950) The isolation of a new thymine pentoside from sponge. J Am Chem Soc 72 (6):2809–2810

Bergmann W, Feeney RJ (1951) Contributions to the study of marine natural products. XXXII. The nucleoside of sponge. J Org Chem 16(6):981–987

Bergquist PR, Wells RJ (1983) Chemotaxonomy of the porifera: the development and current status of the field. In: Scheuer PJ (ed) Marine natural products: chemical and biological perspectives, vol 5. Academic, New York, pp 1–50

Bifulco G, Bruno I, Minale L, Riccio R, Calignano A, Debitus C (1994) (±)-Gelliusines a and B, two diastereomeric brominated tris-indole alkaloids from a deep water New Caledonian marine sponge (*Gellius or Orina* sp.). J Nat Prod 57(9):1294–1299

Blunt JW, Copp BR, Murno MHG, Northcote PT, Prinsep MR (2004) Marine natural products. Nat Prod Rep 21 (1):1–49

Blunt JW, Copp BR, Murray HG, Munro MH, Northcote PT, Prinsep MR (2011) Marine natural products. Nat Prod Rep 28(2):196–268

Borbone N, De Marino S, Iorizzi M, Zollo F, Debitus C, Esposito G, Iuvone T (2002) Minor steroidal alkaloids from the marine sponge *Corticium* sp. J Nat Prod 65:1206–1209

Braekman JC, Daloze D, Abreu PMD, Leopardi CP, Germain G, Meerssche MV (1982) A novel type of bisquinolizidine alkaloids from the sponge *Petrosia Sertata*. Tetrahedron Lett 23(41):4277–4280

Capon RJ, Peng C, Dooms C (2008) Trachycladindoles A-G: cytotoxic heterocycles from an Australian marine sponge, *Trachycladus laevispirulifer*. Org Biomol Chem 6(15):2765–2771

Carney JR, Scheuer PJ, Kelly-Borges M (1993) A new Bastadin from the sponge *Psammaplysilla purpurea*. J Nat Prod 56(1):153–157

Carroll AR, Ngo A, Quinn RJ, Redburn J, Hooper JNA (2005) Petrosamine B, an inhibitor of the Helicobacter pylori enzyme aspartyl semialdehyde dehydrogenase from the Australian sponge *Oceanapia* sp. J Nat Prod 68(5):804–806

Casapullo A, Bifulco G, Bruno I, Riccio R (2000) New bisindole alkaloids of the topsentin and hamacanthin classes from the Mediterranean marine sponge *Rhaphisia lacazei*. J Nat Prod 63(4):447–451

Cheng JF, Ohizumi Y, Walchli MR, Nakamura H, Hirata Y, Sasaki T, Kobayashi J (1988) Prianosins B, C, and D, novel sulfur-containing alkaloids with potent antineoplastic activity from the Okinawan marine sponge Prianos melanos. J Org Chem 53(19):4621–4624

Ciminiello P, Constantino V, Fattorusso E, Magno S, Mangoni A, Pansi M (1994) Chemistry of verongida sponges II constituents of the Caribbean sponge *Aplysina fistularis forma fulva*. J Nat Prod 57 (6):705–712

Cimino G, Rosa SD, Stefano SD, Self R, Sodano G (1983) The bromo compounds of the true sponge *Verongia aerophoba*. Tetrahedron Lett 24(29):3029–3032

Crook S, Davis-McGibony M, Whitelock C (2009) 3,6-Bis(5-bromo-3′-indolyl)-1,4-dimethylpiperazine-2,5-dione. Mobank, M627

Cutignan A, Bifulco G, Bruno I, Casapullo A, Gomez-Paloma L, Riccio R (2000) Dragmacidin F: a new antiviral bromoindole alkaloid from the Mediterranean sponge *Halicortex* sp. Tetrahedron 56(23):3743–3748

Davis-McGibony CM, Pletcher PC (2006) Isolation and characterization of novel(bis)indole alkaloids from local marine sponges. Am Chem Soc 231st ACS National meeting CHED 739

Delfourne E, Bastide J (2003) Marine pyridoacridine alkaloids and synthetic analogues as antitumour agents. Med Res Rev 23(2):234–252

De Oliveira JHHL, Seleghim MHR, Timm C, Grube A, Köck M, Nascimento GGF, Martins ACT, Silva EGO, De Souza AO, Minarini PRR, Galetti FCS, Silva CL, Hajdu E, Berlinck RGS (2006) Antimicrobial and antimycobacterial activity of cyclostellettamine alkaloids from sponge *Pachychalina* sp. Mar Drugs 4 (1):1–8

De Smet P, Parys JB, Callewaert G, Weidema AF, Hill E, Smedt HD, Erneux C, Sorrentino V, Missiaen L (1999) Xestospongin C is an equally potent inhibitor of the inositol 1,4,5-triphosphate receptor and the endoplasmic-reticulum Ca^{2+} pumps. Cell Calcium 26 (1–2):9–13

Ding Q, Chichak K, Lown JW (1999) Pyrroloquinoline and pyridoacridine alkaloids from marine source. Curr Med Chem 6(1):1–27

Djura P, Faulkner DJ (1980) Metabolites of the marine sponge *Dercitus species*. J Org Chem 45 (4):735–737

Djura P, Stierle DB, Sullivan B, Faulkner DJ (1980) Some metabolites of the marine sponges, *Smenospongia aurea* and *Smenospongia* (*Polyfibrospongia*) *echina*. J Org Chem 45(8):1435–1441

Dunbar DC, Rimoldi JM, Clark AM, Kelly M, Hamann MT (2000) Anti-cryptococcal and nitric oxide synthase inhibitory imidazole alkaloids from the calcareous sponge *Leucetta cf. chagosensis*. Tetrahedron 56(45):8795–8798

Eder C, Schupp P, Proksch P, Wray V, Steube K, Muller CE, Frobenius W, Herderich M, van Soest RWM (1998) Bioactive pyridoacridine alkaloids from the Micronesian sponge *Oceanapia* sp. J Nat Prod 61 (2):301–305

Erpenbeck D, van Soest RWM (2007) Status and perspective of sponge chemosystematics. Mar Biotechnol 9 (1):2–19

Faimali M, Sepcic K, Turk T, Geraci S (2003) Non-toxic antifouling activity of polymeric 3-alkylpyridinium salts from the Mediterranean sponge *Reniera sarai* (Pulitzer-Finali). Biofouling 19(1):47–56

Fattorusso E, Taglialatela-Scafati O (2000) Two novel pyrrole-imidazole alkaloids from the Mediterranean sponge *Agelas oroides*. Tetrahedron Lett 41 (50):9917–9922

Faukner DJ (1999) Marine natural products. Nat Prod Rep 16(2):155–198

Fujiwara T, Hwang J-H, Kanamoto A, Nagai H, Takagi M, Shinya K (2009) JBIR-44, a new bromotyrosine compound from a marine sponge *Psammaplysilla purpurea*. J Antibiot 62:393–395

Gopichand Y, Schmitz FJ (1979) Marine natural product: fistularin-1,-2 and -3 from the sponge *Aplysina fistularis forma fulva*. Tetrahedron Lett 20(41):3921–3924

Goud TV, Reddy NS, Swamy NR, RAM TS, Venkateswarlu Y (2003) Anti-HIV active petrosins from the marine sponge *Petrosia similis*. Biol Pharm Bull 26(10):1498–1501

Gray GD (1975) Ara-C and derivatives as examples of immunosuppressive nucleoside analogs. Ann N Y Acad Sci 255:372–379

Gribble GW (1998) The diversity of naturally occurring organobromine compounds. Acc Chem Res 31 (3):141–150

Gunawardana GP, Kohmoto S, Gunasekera SP, McConnell OJ, Koehn FE (1988) Dercitine, a new biologically active acridine alkaloid from a deep water marine sponge, *Dercitus* sp. J Am Chem Soc 110(14):4856–4858

Gunawardana GP, Koehn FE, Lee AY, Clardy J, He HY, Faulkner DJ (1992) Pyridoacridine alkaloids from deep-water marine sponges of the family Pachastrellidae: structure revision of dercitin and related compounds and correlation with the kuanoniamines. Org Chem 57 (5):1523–1526

Guzman FSD, Carte B, Troupe N, Faulkner DJ, Harper MK, Conception GP, Mangalindan GC, Matsumoto SS, Barrows LR, Ireland CM (1999) Neoamphimedine: a new pyridoacridine topoisomerase II inhibitor which catenates DNA. J Org Chem 64(4):1400–1402

Halmi H, Chunhacha P, Suwanborirux K, Chanvorachote P (2011) Anticancer and antimetastatic activities of renieramycin M, a marine tetrahydroisoquinoline alkaloid, in human non-small cell lung cancer cells. Anticancer Res 31(1):193–201

Harper MK, Bugni TS, Copp BR, James RD, Lindsay BS, Richardson AD, Schnabel PC, Tasdemir D, Van Wagoner RM, Verbitski SM, Ireland CM (2001) Introduction to the chemical ecology of marine natural products. In: McClintock JB, Baker BJ (eds) Marine chemical ecology. CRC Press, Boca Raton, pp 3–69

Hirano K, Kubota T, Tsuda M, Mikami Y, Kobayashi J (2000) Pyrinodemins B-D, potent cytotoxic bis-pyridine alkaloids from marine sponge *Amphimedon* sp. Chem Pharm Bull 48(7):974–977

Hollenbeak KH, Schmitz FJ (1977) Aplysinopsin: antineoplastic tryptophan derivative from marine sponge *Verongia spengelii*. Lloydia 40(5):479–481

Inman WD, O'Neill-Johnson M, Crews P (1990) Novel marine sponge alkaloids. 1. Plakinidine A and B, anthelmintic active alkaloids from a Plakortis sponge. J Am Chem Soc 112(1):1–4

Ishibashi M, Tsuda M, Ohizumi Y, Sasaki T, Kobayashi J (1991) Purealidin A, a new cytotoxic bromotyrosine-derived alkaloid from the Okinawan marine sponge *Psammaplysilla purea*. Experientia 47(3):299–300

Ishiguro Y, Kubota T, Ishiuchi K, Fromont J, Kobayashi J (2009) Plakoridine C, a novel piperidine alkaloid from an Okinawan marine sponge *Plakortis* sp. Tetrahedron Lett 50(26):3202–3204

Jimino J, Faircloth G, Fernandez JM S-F, Scheuer P, Rinehart K (2004) New marine derived anticancer Therapeutics-a journey from sea to clinical trials. Mar Drugs 2(1):14–29

Jurek J, Yoshida WY, Scheuer PJ, Kelly-Borges M (1993) Three new bromotyrosine-derived metabolites of the sponge *Psammaplysilla purpurea*. J Nat Prod 56 (9):1609–1612

Kariya Y, Kubota T, Fromont J, Kobayashi J (2006) Pyrinadine A, a novel pyridine alkaloid with an azoxy moiety from sponge *Cribrochalina* sp. Tetrahedron Lett 47(6):997–998

Kazlauskas R, Murphy PT, Quinn RJ, Wells RJ (1977) Aplysinopsin, a new tryptophan derivative from a sponge. Tetrahedron Lett 18(1):61–64

Kijjoa A, Wattanadilok R, Campos N, Nascimento NSJ, Pinto M, Herz W (2007) Anticancer activity evaluation of kuanoniamines a and C isolated from the marine sponge *Oceanapia sagittaria*, collected from the Gulf of Thailand. Mar Drugs 5(2):6–22

Kitagawa I, Kobayashi M, Kitanaka K, Kido M, Kyogoku Y (1983) Marine natural products, XII: on the chemical constituents of the Okinawan marine sponge *Hymeniacidon aldis*. Chem Pharm Bull 31 (7):2321–2328

Kitamura A, Tanaka J, Ohtani II, Higa T (1999) Echinoclathrines A-C: a new class of pyridine alkaloids from an Okinawan sponge, *Echinoclathria* sp. Tetrahedron 55(9):2487–2492

Kobayashi J, Cheng J, Walchli MR, Nakamura H, Hirata Y, Sasaki T, Ohizumi Y (1988) Cystodytins A, B, and C, novel tetracyclic aromatic alkaloids with potent antineoplastic activity from the Okinawan tunicate *Cystodytes dellechiajei*. J Org Chem 53(8):1800–1804

Kobayashi J, Murayama T, Ohizumi Y, Sasaki T, Ohta T, Nozoe S (1989a) Theonelladins A~D, novel antineoplastic pyridine alkaloids from the Okinawan marine sponge *Theonella swinhoei*. Tetrahedron Lett 30 (36):4833–4836

Kobayashi M, Kawazoe K, Kitagawa I (1989b) Araupetrosine A, a new vasodilative macrocyclic quinolizidine alkaloid from an Okinawan marine sponge *Xestospongia* sp. Tetrahedron Lett 30 (31):4149–4152

Kobayashi J, Cheng JF, Ishibashi M, Walchli MR, Yamamura S, Ohizumi Y (1991a) Penaresidin A and B, two novel azetidine alkaloids with potent actomyosin ATPase activating activity from the Okinawan marine sponge *Penares* sp. J Chem Soc Perkin Trans 1 (5):1135–1137

Kobayashi J, Tsuda M, Agemi K, Shigemori H, Ishibashi M, Sasaki T, Mikami Y (1991b) Purealidins B and C, new bromotyrosine alkaloids from the okinawan marine sponge *psammaplysilla purea*. Tetrahedron 47(33):6617–6622

Kobayashi J, Zeng C-M, Ishibashi M, Shigemori H, Sasaki T, Mikami Y (1992) Niphatesines E-H, new pyridine alkaloids from the Okinawan marine sponge *Niphates* sp. J Chem Soc Perkin Trans 1(11):1291–1294

Kobayashi M, Rao SR, Chavakula R, Sarma NS (1994) Mimosamycin, 4-aminomimosamycin and 7-amino-7-demethoxymimosamycin from the sponge *Petrosia* sp. J Chem Res (S) 282–283

Kobayashi J, Honma K, Tsuda M, Kosaka T (1995a) Lipopurealins D and E and purealidin H, new bromotyrosine alkaloids from the Okinawan marine sponge *Psammaplysilla purea*. J Nat Prod 58 (3):467–470

Kobayashi J, Honma K, Sasaki T, Tsuda M (1995b) Purealidins J-R, new bromotyrosine alkaloids from the Okinawan marine sponge *Psammaplysilla purea*. Chem Pharm Bull 43(3):403–407

Kochanowska AJ, Rao KV, Childress S, El-Alfy A, Matsumoto RR, Kelly M, Stewart GS, Sufka KJ, Hamann MT (2008) Secondary metabolites from three Florida sponges with antidepressant activity. J Nat Prod 71(2):186–189

Kona Y, Kubota T, Shibazaki A, Gonoi T, Kobayashi J (2010) Ceratinadins A-C, new bromotyrosine alkaloids from an Okinawan marine sponge *Pseudoceratina* sp. Bioorg Med Chem Lett 20 (15):4569–4572

Kondo K, Nishi J, Ishibashi M, Kobayashi J (1994) Two new tryptophan-derived alkaloids from the Okinawan marine sponge *Aplysina* sp. J Nat Prod 57 (7):1008–1011

Kong F, Andersen RJ, Allen TM (1994) Madangamine A, a novel cytotoxic alkaloid from the marine sponge Xestospongia ingens. J Am Chem Soc 116 (13):6007–6008

König GM, Wright AD (1993) Agelorin-A and agelorin-B, and epi-11-fistularin-3, three new antibacterial fistularin-3 derivatives from the tropical marine sponge *Agelas oroides*. Heterocycles 36:1351–1358

Kubota T, Watase S, Mukai H, Fromont J, Kobayashi J (2012) Tyrokeradines C-F, new bromotyrosine alkaloids from the Verongid sponges. Chem Pharm Bull 60(12):1599–1601

Kubota T, Kura K, Fromont J, Kobayashi J (2013) Pyrinodemins G-I new bis-3-alkylpyridine alkaloids from a marine sponge *Amphimedon* sp. Tetrahedron 69(1):96–100

Kumar D, Rawat DS (2011) Marine natural alkaloids as anticancer agents. Opportunity, challenge and scope of natural products in medicinal chemistry. Research Signpost, Trivandrum, pp 213–268

Lira NS, Monte-Neto RL, Marchi JGB, da Silva Lins AC, Tavares JF, da Silva MS, Barbosa-Filho CDSDJM, dos Santos CF, Leitao da Cunha EV, dos Santos Pinheiro U, Braz-Filho R (2012) Aplysfistularine: novel dibromotyrosine derivative isolated from *Aplysina fistularis*. Quim Nova 35(11):2189–2193

Marshall KM, Barrows LR (2004) Biological activities of pyridoacridines. Nat Prod Rep 21(6):731–751

Matsunaga S, Kobayashi H, van Soest RWM, Fusetani N (2005) Novel bromotyrosine derivatives that inhibit growth of the fish pathogenic bacterium aeromonas hydrophila, from a marine sponge *Hexadella* sp. J Org Chem 70(5):1893–1896

McCarthy PJ, Pitts TP, Gunawardana GP, Kelly-Borges-M, Pomponi SA (1992) Antifungal activity of meridine, a natural product from the marine sponge *Corticium* sp. J Nat Prod 55(11):1664–1668

Mckee TC, Ireland CM (1987) Cytotoxic and antimicrobial alkaloids from the Fijian sponge, *Xestospongia caycedoi*. J Nat Prod 50(4):754–756

Molinski TF (1993) Marine pyridoacridine alkaloids: structure, synthesis and biological chemistry. Chem Rev 93(5):1825–1838

Molinski TF, Fahy E, Faulkner DJ, Van Duyne GD, Clardy J (1988) Petrosamine, a novel pigment from the marine sponge *Petrosia* sp. J Org Chem 53 (6):1340–1341

Momose R, Tanaka N, Fromont J, Kobayashi J (2013) Hyrtimomines A-C, new heteroaromatic alkaloids from a sponge *Hyrtios* sp. Org Lett 15(8):2010–2013

Moon S, MacMillan J, Olmstead M, Ta T, Pessah I, Molinski T (2002) (+)-7S-hydroxyxestospongin a from the marine sponge *Xestospongia* sp. and absolute configuration of (+)-xestospongin D. J Nat Prod 65 (3):249–254

Moriarty RM, Roll DM, Ku YY, Nelson C, Ireland CM (1987) A revised structure for the marine bromoindole derivative citorellamine. Tetrahedron Lett 28 (7):749–752

Mukai H, Kubota T, Aoyama K, Mikami Y, Fromont J, Kobayashi J (2009) Tyrokeradine A and B: new bromotyrosine alkaloids with an imidazolyl-quinolinone moiety from a Verongid sponge. Bioorg Chem Lett 19(5):1337–1339

Nakagawa NN, Endo M, Tanaka N, Gen-Pei L (1984) Structures of Xestospongin A, B, C and D, Novel vasodilative compounds from marine sponge *Xestospongia Exigua*. Tetrahedron Lett 25 (30):3227–3230

Nakamura H, Ohizumi Y, Kaboyashi J (1984) Keramadine, a novel antagonist of serotonergic receptors isolated from the Okinawan sea sponge *Agelas* sp. Tetrahedron Lett 25(23):2475–2478

Newman DJ, Cragg GM (2004) Marine natural product and related compounds in clinical and preclinical trials. J Nat Prod 67(8):1216–1238

Nicholas GM, Molinski TF (2000) Structures of cribochalines a and B, branched-chain methoxylaminoalkyl pyridines from the Micronesian sponge, *Cribochalina* sp. Absolute configuration and enantiomeric purity of related O-methyl oximes. Tetrahedron 56(19):2921–2927

Nicholas GM, Newton GL, Fahey RC, Bewley CA (2001) Novel bromotyrosine alkaloids: inhibitors of mycothiol S-conjugate amidase. Org Lett 3(19):1543–1545

Orabi KY, El Sayed KA, Hamann MT, Dunbar DC, Al-Said MS, Higa T, Kelly M (2002) Araguspongines K and L, new bioactive Bis-1-oxaquinolizidine N-oxide alkaloids from Red Sea specimens of *Xestospongia exigua*. J Nat Prod 65(12):1782–1785

Parameswaran PS, Naik CG, Kamat SY, Pathak BN (1998) Renieramycins H, I, two novel alkaloids from the sponge *Haliclona cribricutis Dendy*. Ind J Chem 37B:1258–1263

Park Y, Liu Y, Hong J, Lee CO, Cho H, Kim DK, Im KS, Jung JH (2003) New bromotyrosine derivatives from an association of two sponges, *Jaspis wondoensis* and *Poecillastra wondoensis*. J Nat Prod 66 (11):1495–1498

Patterson AM, Capell LT, Walker DF (1960) The ring index, 2nd edn. American Chemical Society, Washington, DC

Perry NB, Blunt JW, Munro MHG, Higa T, Sakai R (1988) Discorhabdin D an antitumor alkaloid from the sponges, Latrunculia brevis and *Prianos* sp. J Org Chem 53(17):4127–4128

Perry NB, Ettouati L, Litaudon M, Blunt JW, Munro MHG (1994) Alkaloids from the antarctic sponge Kirkpatrickia varialosa, part 1: variolin B, a new antitumour and antiviral compound. Tetrahedron 50 (13):3987–3992

Pettit GR, Collins JC, Herald DL, Doubek DL, Boyd MR, Schmidt JM, Hooper JNA, Tackett LP (1992) Isolation and structure of cribrostatins 1 and 2 from the blue marine sponge *Cribrochalina* sp. Can J Chem 70:1170–1175

Pettit GR, Orr B, Herrald DL, Doubek DL, Tackett L, Schmidt JM, Boyd MR, Pettit RK, Hooper JNA (1996) Isolation and X-Ray structure of racemic xestospongin D from the Singapore marine sponge *Niphates* sp. Bioorg Med Chem Lett 6(12):1313–1318

Pettit GR, Knight JC, Collins JC, Herald DL, Pettit RK, Boyd MR, Young VG (2000) Antineoplastic agents 430. Isolation and structure of cribrostatins 3, 4, and 5 from the Republic of Maldives Cribrochalina species. J Nat Prod 63(6):793–798

Pettit GR, Collins JC, Knight JC, Herald DL, Nieman RA, Williams MD, Pettit RK (2003) Antineoplastic agents. 485. Isolation and structure of cribrostatin 6, a dark blue cancer cell growth inhibitor from the marine sponge *Cribrochalina* sp. J Nat Prod 66(4):544–547

Phife DW, Ramos RA, Feng M, King I, Gunasekera SP, Wright A, Patel M, Pachter JA, Coval SJ (1996) Marine sponge bis(indole) alkaloids that displace ligand binding to alpha-1-adrenergic receptors. Bioorg Med Chem Lett 6(17):2103–2106

Proksch P, Ebel R, Edrada RA, Wray V, Steube K (2003) In: Müller WEG (ed) Marine molecular biotechnology. Springer, Berlin, pp 117–143

Quirion JC, Sevenet T, Husson H-P, Weiger B, Debitus C (1992) Two new alkaloids from *Xestospongia* sp. a new Caledonian sponge. J Nat Prod 55 (10):1505–1508

Reddy M, Faulkner DJ (1997) 3β, 3′β-dimethylxestospongin C, a new Bis-1-oxaquinolizidine alkaloid from the Palauan sponge *Xestospongia* sp. Nat Prod Lett 11:53–59

Ridley CP, Faulkner DJ (2003) New cytotoxic steroidal alkaloids from the Philippine sponge *Corticium niger*. J Nat Prod 66(12):1536–1539

Rodriguez J, Peters BM, Kurz L, Schatzman RC, McCarley D, Lou L, Crews P (1993) An alkaloid protein kinase C inhibitor, xestocyclamine A, from the marine sponge *Xestospongia* sp. J Am Chem Soc 115(22):10436–10437

Roll DM, Chang CWJ, Scheuer PJ, Gray GA, Shoolery JN, Matsumoto GK, Duyne GDV, Clardy J (1985) Structure of the psammaplysins. J Am Chem Soc 107(10):2916–2920

Ross SA, Weete JD, Schinazi RF, Wirtz SS, Tharnish P, Scheuer PJ, Hamann MT (2000) Mololipids, a new series of anti-HIV bromotyramine-derived compounds from a sponge of the order verongida. J Nat Prod 63 (4):501–503

Saeki BM, Granato AC, Berlink RGS, Magalhães A, Schefer AB, Ferreira AG, Pinheiro US, Hajdu E (2002) Two unprecedented dibromotyrosine-derived alkaloids from the Brazilian endemic marine sponge *Aplysina caissara*. J Nat Prod 65(5):796–799

Salomon CE, Faulkner DJ (1996) Sagitol, a pyridoacridine alkaloid from the sponge Oceanapia sagittaria. Tetrahedron Lett 37(51):9147–9148

Sauleau P, Martin M-T, Dau M-ETH, Youssef DTA, Bourguet-Kondracki M-L (2006) Hyrtiazepine, an azepino-indole-type alkaloid from the Red Sea marine sponge *Hyrtios erectus*. J Nat Prod 69(12):1676–1679

Schmitz FJ, Hollenbeak KH, Campbell DC (1978) Marine natural products: halitoxin, toxic complex of several marine sponges of the genus Haliclona. J Org Chem 43(20):3916–3922

Schmitz FJ, Agarwal SK, Gunasekera SP, Schmidt PG, Shoolery JN (1983) Amphimedine, new aromatic alkaloid from a pacific sponge, *Amphimedon* sp. Carbon connectivity determination from natural abundance ^{13}C-^{13}C coupling constants. J Am Chem Soc 105(14):4835–4836

Segrave NL, Crews P (2005) Investigation of brominated tryptophan alkaloids from two Thorectidae sponges: *Thorectandra* and *Smenospongia*. J Nat Prod 68 (10):1484–1488

Shin J, Seo Y, Cho KW, Rho JR, Sim CJ (1997) Stellettamide B, a new indolizidine alkaloid from a sponge of the *Genus stelletta*. J Nat Prod 60 (6):611–613

Shin J, Seo Y, Cho KW, Rho J-R, Jim SCJ (1999) New bis (indole) alkaloids of the topsentin class from the sponge *Spongosorites genitrix*. J Nat Prod 62 (4):647–649

Singh KS, Das B, Naik CG (2011) Quinolizidines alkaloids: petrosin and xestospongins from the sponge *Oceanapia* sp. J Chem Sci 123(5):601–607

Sun HH, Sakemi S, Burres N, McCarthy P (1990) Isobatzellines A, B, C, and D. Cytotoxic and antifungal pyrroloquinoline alkaloids from the marine sponge *Batzella* sp. J Org Chem 55(16):4964–4966

Suwanborirux K, Amnuoypol S, Plubrukarn A, Pummangura S, Kubo A, Tanaka C, Saito NJ (2013) Chemistry of renieramycins. Part 3. Isolation and structure of stabilized renieramycin type derivatives possessing antitumor activity from Thai sponge *Xestospongia* species, pretreated with potassium cyanide. J Nat Prod 66(11):1441–1446

Tabudravu JN, Jaspars M (2002) Purealidin S and purpuramine J, bromotyrosine alkaloids from the Fijian marine sponge *Druinella* sp. J Nat Prod 65 (12):1798–1801

Takekawa Y, Matsunaga S, van Soest RWM, Fusetani N (2006) Amphimedosides, 3-alkylpyridine glycosides from a marine sponge *Amphimedon* sp. J Nat Prod 69 (10):1503–1505

Talpira R, Rudia A, Ilanb M, Kashman Y (1992) Niphatoxin A and B; two new ichthyo- and cytotoxic tripyridine alkaloids from a marine sponge. Tetrahedron Lett 33(21):3033–3034

Tanaka J, Higa T, Bernardinelli G, Jefford CW (1988) Itomanindoles A and B. Methylsulfinylindoles from *Laurencia brongniartii*. Tetrahedron Lett 29 (47):6091–1694

Tanaka N, Momose R, Takahashi Y, Kubota T, Takahashi-Nakaguchi T, Gonoi J, Fromont J, Kobayashi J (2013) Hyrtimomines D and E, bisindole alkaloids from a marine sponge *Hyrtios* sp. Tetrahedron Lett 54(31):4038–4040

Tanaka N, Momose R, Takahashi-Nakaguchi A, Gonoi T, Fromont J, Kobayashi J (2014) Hyrtimomines, indole alkaloids from Okinawan marine sponges *Hyrtios spp*. Tetrahedron 70(4):832–837

Taraporewala IB, Cessac JW, Chanh TC, Delgado AV, Schinazi RF (1992) HIV-1 neutralization and tumor cell proliferation inhibition in vitro by simplified analogs of pyrido[4,3,2-mn]thiazolo[5,4-b]acridine marine alkaloids. J Med Chem 35(15):2744–2752

Tasdemir D, Marshall KM, Mangalindan GC, Concepcion GP, Barrows LR, Harper MK, Ireland CM (2001) Deoxyamphimedine, a new pryridoacridine alkaloid from two tropical xestospongia sponges. J Org Chem 66(9):3246–3248

Teruya T, Kobayashi K, Suenaga K, Kigoshi H (2006) Cyclohaliclonamines A-E: dimeric, trimeric, tetrameric, pentameric, and hexameric 3-alkyl pyridinium alkaloids from a marine sponge *Haliclona* sp. J Nat Prod 69(1):135–137

Thoms C, Ebel R, Proksch P (2005) Activated chemical defense in aplysina sponges revisited. J Chem Ecol 32 (1):97–123

Tilvi S, Rodrigues C, Naik CG, Parameswaran PS, Wahidhulla S (2004) New bromotyrosine alkaloids from the marine sponge *Psammaplysilla purpurea*. Tetrahedron 60(45):10207–10215

Torres YR, Berlinck RGS, Nascimento GGF, Fortier SC, Pessoa C, De Moraes MO (2002) Antibacterial activity against resistant bacteria and cytotoxicity of four alkaloid toxins isolated from the marine sponge *Arenosclera brasiliensis*. Toxicon 40(17):885–891

Tsotinis A, Calogeropoulou T, Koufaki M, Souli C, Balzarini J, Clercq ED, Makriyannis A (1996) Synthesis and antiretroviral evaluation of new alkoxy and aryloxy phosphate derivatives of 3-azido-3-deoxythymidine. J Med Chem 39(17):3418–3422

Tsuda M, Shigemori H, Ishibashi M, Kobayashi J (1992) Purealidines E-G, new bromotyrosine alkaloids from the Okinawan marine sponge *Psammaplysilla Purea*. J Nat Prod 55(9):1325–1327

Tsuda M, Hirano K, Kubota T, Kobayashi J (1999) Pyrinodemin A, a cytotoxic pyridine alkaloid with an isoxazolidine moiety from sponge *Amphimedon* sp. Tertahedron Lett 40(26):4819–4820

Tsukamoto S, Kato H, Hirota H, Fusetani N (1996a) Ceratinamine: an unprecedented antifouling cyanoformamide from the marine sponge *Pseudoceratina purpurea*. J Org Chem 61 (9):2936–2937

Tsukamoto S, Kato H, Hirota H, Fusetani N (1996b) Ceratinamides A and B: new antifouling dibromotyrosine derivatives from the marine sponge *Pseudoceratina purpurea*. Tetrahedron 52 (24):8181–8186

Urban S, Almeida LPD, Carroll AR, Fechner GA, Smith J, Hooper JNA, Quinn RJ (1999) Axinellamines A-D, novel imidazo-azolo-imidazole alkaloids from the Australian marine sponge *Axinella* sp. J Org Chem 64(3):731–735

Utkina NK, Makarchenko AE, Denisenko VA (2005) Zyzzyanones B-D, dipyrroloquinones from the marine sponge *Zyzzya fuliginosa*. J Nat Prod 68(9):1424–1427

Venkateswarlu Y, Reddy MVR, Rao JV (1994) Bis-1-oxaquinolizidines *from Haliclona Exigua*. J Nat Prod 57(9):1283–1285

Volk CA, Köck M (2004) Viscosaline: new 3-alkyl pyridinium alkaloid from the Arctic sponge *Haliclona viscosa*. Org Biomol Chem 2(13):1827–1830

Volk CA, Lippert H, Lichte E, Koeck M (2004) Two new haliclamines from the arctic sponge *Haliclona viscosa*. Eur J Org Chem 14:3154–3158

Wang G-Y-S, Kuramoto M, Uemura D (1996) Three novel antimicrofouling nitroalkylpyridine alkaloids from the Okinawan marine sponge *Callyspongia* sp. Tetrahedron Lett 37(11):1813–1816

Wei X, Bugni TS, Harper MK, Sandoval IT, Manos EJ, Swift J, Wagoner RMV, Jones DA, Ireland CM (2010) Evaluation of pyridoacridine alkaloids in a zebrafish phenotypic assay. Mar Drugs 8(6):1769–1778

Whitehead R (1999) Natural product chemistry. Annu Rep Prog Chem Sect B 95:183–205

Wright AE, Pomponi SA, Cross SS, McCarthy P (1992) A new bis-(indole) alkaloid from a deep-water marine sponge of the genus *Spongosorites*. J Org Chem 57 (17):4772–4775

Wu H, Nakamura H, Kobayashi J, Ohizumi Y, Hirata Y (1986) Lipopurealins, novel bromotyrosine derivatives with long chain acyl groups, from the marine sponge *Psammaplysilla purea*. Experientia 42 (7):855–856

Xu M, Andrews KT, Birrell GW, Tran TL, Camp D, Davis RA, Quinn RJ (2011) Psammaplysin H, a new antimalarial bromotyrosine alkaloid from a marine sponge of the genus P*seudoceratina*. Bioorg Med Chem Lett 21(2):846–848

Yagi H, Matsunaga S, Fusetani N (1993) Purpuramines A-I, new bromotyrosine-derived metabolites from the marine sponge *Psammaplysilla purpurea*. Tetrahedron 49(18):3749–3754

Yin S, Davis RA, Shelper T, Sykes ML, Avery VM, Elofsson M, Sundin C, Quinn RJ (2011) Pseudoceramines A-D, new antibacterial bromotyrosine alkaloids from the marine sponge *Pseudoceratina* sp. Org Biomol Chem 9 (19):6755–6760

Proteoglycans from Marine Sponges and Their Biomedical Applications

13

Ramachandran Karthik and Ramachandran Saravanan

Abstract

Marine sponges are the simplest and earliest multicellular organisms, proteoglycans originating from the extracellular milieu fastened to the cell membrane. The extracellular space in the tissues of multicellular creatures is blocked through a gel-like substance, called the extracellular matrix, or the ground substance, which grasps the cells collectively and affords a permeable pathway for the dissemination of nutrients and oxygen to individual cells. The extracellular matrix is compiled of an intermingling network of heteropolysaccharides and fibrous connective tissue proteins such as collagen, elastin, fibronectin, and laminin. The glycosaminoglycans (GAG) are a family of linear polymeric heteropolysaccharides composed of duplicating disaccharide units. To investigate the isolation of proteoglycans and their structure has been intended, by means of chromatography, to mimic the function of proteoglycans in the multicellular adhesion of the marine sponge. The interaction of proteoglycans with GAG is not based on electrostatic communication. In addition, the interaction of proteins with GAG may have potential significant implications for biomedical roles including anticoagulant, antibacterial, antiviral, anti-inflammatory, and so on, and as an alternative therapeutic agent in the field of biochemical/pharmacological/microbial/molecular biology.

R. Karthik
Department of Medical Biotechnology, Faculty of Allied Health Sciences, Chettinad Academy of Research and Education, Kelambakkam 603 103, Tamil Nadu, India

R. Saravanan (✉)
Department of Marine Pharmacology, Faculty of Allied Health Sciences, Chettinad Academy of Research and Education, Kelambakkam 603 103, Tamil Nadu, India
e-mail: saran_prp@yahoo.com

R. Pallela, H. Ehrlich (eds.), *Marine Sponges: Chemicobiological and Biomedical Applications*, DOI 10.1007/978-81-322-2794-6_13

Keywords
Marine sponge • Proteoglycans • GAG • Heparin • Chondroitin sulfate • Hyaluronic acid • Chromatography

13.1 Introduction

Sponges are animals of unusual nature and are of ancient heritage. They are filter-feeding benthonic animals that have endured almost morphologically impassive since the greater Cambrian (509 million years ago; Hooper and Van Soest 2002). Sponges are the simplest multicellular lifeform present nowadays in our world, similar, in evolutionary terms, to prehistoric multicellular organisms (Misevic et al. 2007). Sponges are the mainly primordial animal form living today and hence are considered living fossils (Li et al. 1998) for newer energy in drug production. These animals are so peculiar that, in the Metazoa kingdom, an entire phylum is devoted to them, viz. Phylum Porifera (Hooper 1995). There are 8365 well-known sponge species (Van Soest et al. 2008) and, of these, around 98 % live in marine environments. They remain fastened to the marine substratum in many different rocky bottoms, shores, temperatures, salinities, and light conditions (Rutzler et al. 2004) and continue living in a massive amount of colors and silhouettes.

The liaison sandwiched between sponges and remedies goes back to Alexandrian physicians and is meticulously described by the Roman historian Plinius. The general practitioners make use of sponges that are saturated with iodine to induce blood coagulation, or with bioactive tissue extracts to sedate patients in clinics. Pharmaceutical awareness of sponges has been awakened since the 1950s by the breakthrough of a marine-derived anticancer agent and antiviral drug (Sipkema et al. 2005). The sticking together of marine sponges involves the calcium-independent adherence of proteoglycan-like molecules, called aggregation factors, to the cell surface, and self-association of calcium-dependent aggregation factors. The calcium-dependent affair is species-specific, as demonstrated by the hasty self-association and categorization, on the addition of calcium ions, of a mixture of colored (pink, yellow, and white) proteoglycan-coated beads, each color corresponding to a different species. Monoclonal antibodies are raised in opposition to purified adhesion proteoglycans of *Microciona prolifera* chunked the self-relationship, for which the distinguished epitopes are identified as squat carbohydrate units: the sulfated disaccharide 1 and a pyruvylated trisaccharide (Haseley et al. 2001).

Proteoglycans are proteins with a modified form of carbohydrate moiety such as glucose/galactose/xylose/mannose. The fundamental proteoglycan unit consists of a "core protein" with one or more covalently attached at specific glycosaminoglycan (GAG) chain(s) (Meisenberg and Simmons 2006). The point of attachment is a serine residue, where the GAG joins through a tetrasaccharide bridge (e.g., heparan sulfate/dermatan sulfate/karatan sulfate/chondroitin sulfate-protein). The serine residue is generally in the sequence –Ser–Gly–X–Gly– (where X can be any amino acid residue, but proline), even though not every protein with this sequence has been attached to GAG. Under physiological conditions, the chains are elongated and the linear heteropolysaccharide polymers are negatively charged, because of uronic acid and sulfate groups. Proteoglycans can be categorized based on the nature of their GAG chains. Proteoglycans can also be categorized by size, typically high molecular weight heteropolysaccharides that consist of a backbone of repeating disaccharide units with an assimilation of amino sugar and a uronic acid in atomic mass units or kilo Daltons (kDa), analogous to proteins (Table 13.1).

Table 13.1 Various types of GAG and their proteoglycans

GAG	Small proteoglycans	Large proteoglycans
Heparan sulfate	Testican, 44 kDa	Perlecan, 400–470 kDa
Chondroitin sulfate	Decorin – 36 kDa	Versican –260–370 kDa, present in many adult tissues including blood vessels and skin
Dermatan sulfate	Biglycan – 38 kDa	Versican –260–370 kDa, present in many adult tissues including blood vessels and skin
Keratan sulfate	Fibromodulin – 42 kDa Lumican – 38 kDa	

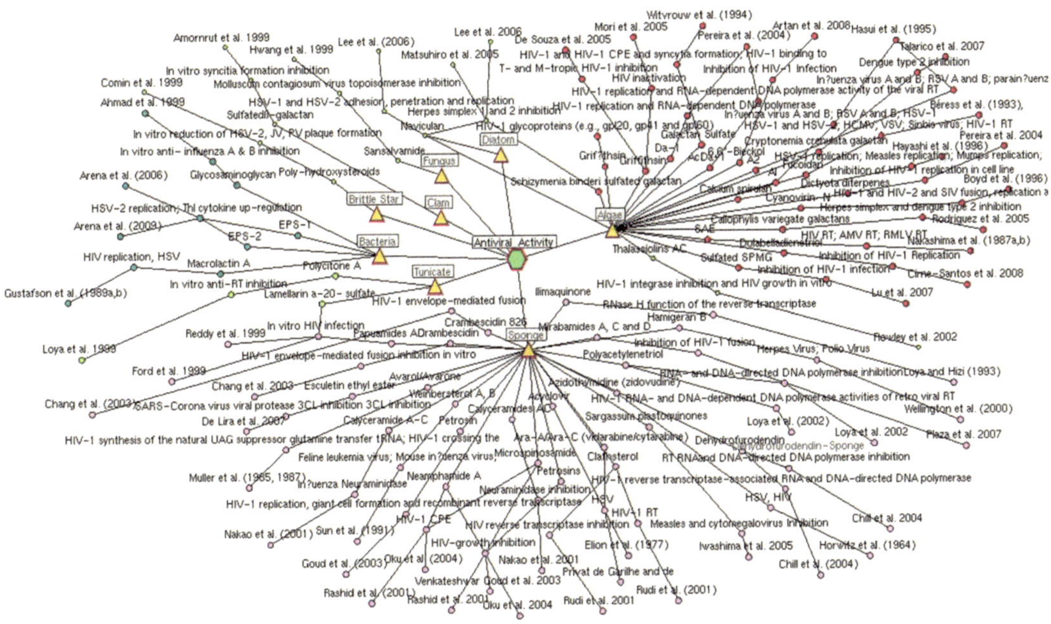

Fig. 13.1 A representational network of natural products extracted from various marine organisms exhibiting antiviral activity

Marine proteoglycans are extremely significant natural macromolecules that are generally present in marine organisms. Marine-sulfated polysaccharides, principally GAG, present a massive array of structures and are still underexploited, thus they should be considered as a novel source of natural compounds for newer drug discovery (Laurienzo 2010). Marine GAG can be divided into three different types, such as animal GAG, plant GAG, and microbial GAG according to their different resources. Marine-derived GAG have been shown to have a variety of bioactivities such as antioxidant, anticoagulant, antitumor, antiviral, immuno-inflammatory, and other medicinal properties, which have less adverse effect, not found in terrestrial sources. In particular, the studies on the antiviral actions of marine GAG and their oligosaccharide derivatives are attracting increasing interest, and marine GAG are the cobblestones of the way for a new trend in antiviral drugs.

From Fig. 13.1 it is evident that marine resources have contributed a great deal to the recognition of lead compounds towards antiviral research. Marine sponges and algae have been the sources of the majority of antiviral drugs during the specified duration (1999–2008), contributing ~75 % of the total compounds identified. With growing time, newer viral mechanisms inside host systems such as viral duplication through topoisomerase binding, replication mechanisms, HIV-1 tat-induced angiogenesis, and so on have been studied and looked at as possible targets of therapy and treatment. The scientific community has been in search of a wide variety of biomolecules such as flavones, alkaloids, sulfated polysaccharides, peptides, and terpenes, among others, which vary in their mechanical mode of action, binding to various cellular targets, to impart antiviral activity. The marine bioactive compounds thus have given the scope of unraveling novel target mechanisms such as inhibition of viral binding and penetration, interaction with HIV-1 glycoproteins, and so on. Drugs such as clathsterol, crambescidin, and dehydrofurodendin have consequently made known potent activities against the viral agents acting through distinct mechanisms. With the emerging antiviral resistance against the ever-increasing barrage of antiviral agents, sponges with their vast diversity and complex structural and efficient mechanisms hold immense promise in countering one of the greatest threats facing civilization.

13.1.1 Types of GAG

Generally GAGs are classified into five types, which are bound to maximum and minimum concentrations of proteoglycans (Table 13.1). GAGs are long, linear, disaccharide repeats of hexosamine and highly sulfated galactose or hexuronic acids, and are usually found bound covalently to a protein "core" to form proteoglycans (Kjellen and Lindahl 1991; Silbert et al. 1997). GAGs are an extremely heterogeneous group of molecules that can be divided into several different general classes, such as heparan sulphate (and its model analogue, heparin), chondroitin sulphate, dermatan sulphate, and hyaluronic acid, depending on the composition of the sugar backbone and the degree of sugar modification. Proteoglycans/GAGs are expressed by all nucleated cells and several bacterial pathogens, such as *Bordetella pertussis*, *Mycobacterium* spp. (Menozzi 1994), *Listeria monocytogenes* (Alvarez-Dominguez et al. 1997), and *Neisseria gonorrhoeae* (Putten and Cole 1998) encode surface proteins that recognize GAGs, that is, GAG-binding adhesins (Rostand 1997).

Jackson et al. (1991) found the GAG, molecular properties, protein interactions, and their role in physiological processes. An overview of low molecular weight heparin and heparinoid (hepain-like substances) basic clinical aspects was given by Hirsh (1992). Volpi (1993) studied the "fast moving" and "slow moving" heparins, dermatan sulfate, and chondroitin sulfate: qualitative and quantitative analysis by agarose-gel electrophoresis. The sequence analysis of heparan sulfate proteoglycans and the identification of variable and constant oligosaccharide regions in eight heparan sulfate proteoglycans from different sources are reported by Tersariol et al. (1994).

13.2 Structures of GAG

13.2.1 Heparin and Heparan Sulfate

The most common disaccharide unit is composed of GlcNAc 1, 4-linked to GlcA. The structure of heparin and heparan sulfate is very similar (Fig. 13.2); however, heparin contains more N-sulfate groups than N-acetyl groups, and the concentration of O-sulfate groups exceeds that of N-sulfate (Saravanan and Shanmugam 2011).

13.2.2 Chondroitin Sulfate

Chondroitin sulfate is another GAG holding opposing views from hyaluronate in two respects: in general it has greatly shorter polymers and is covalently bonded to specific proteins such as proteoglycans. Chondroitin sulfate (Greek *chondros*, cartilage) provides tensile

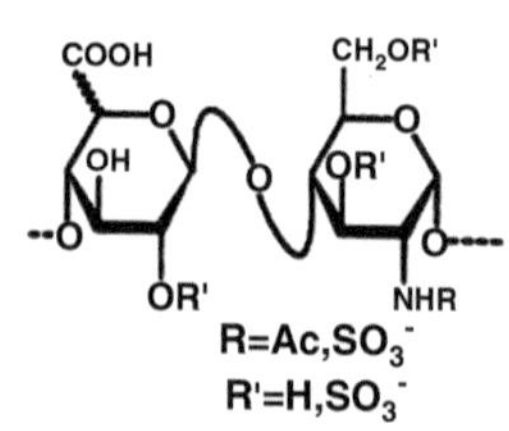

Fig. 13.2 Structure of heparin (Adapted from Falshaw et al. 1999)

strength to the cartilage, tendons, ligaments, and walls of the aorta. It consists of repeating units of sulfated GlcA-GalNAc disaccharides, polymerized into long chains (Fig. 13.3) that can be easily identified using bacterial chondroitin lyases (chondroitinases). In vertebrates, sulfation in chondroitin sulfate is very complex, involving several sulfo-transferases that add sulfate groups at carbon 4 or 6 on the GalNAc residues, and also at carbon 2 on the IdoA residues in dermatan sulphate. There are three types of chondroitin sulfate, namely chondroitin sulfate A, B, and C (Figs. 13.3, 13.4, and 13.5): chondroitin sulfate A carbon 4 of the N-acetylgalactosamine (GalNAc) sugar chondroitin-4-sulfate, chondroitin sulfate C carbon 6 of the GalNAc sugar chondroitin-6-sulfate, chondroitin sulfate D carbon 2 of the glucuronic acid and 6 of the GalNAc sugar chondroitin-2, 6-sulfate, and chondroitin sulfate E carbons 4 and 6 of the GalNAc sugar chondroitin-4,6-sulfate.

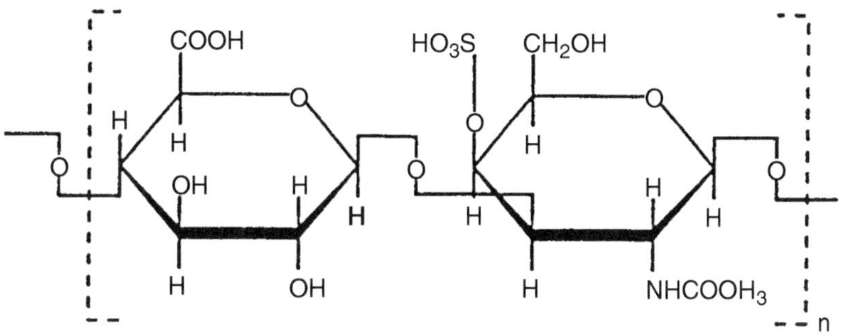

Fig. 13.3 Structure of chondroitin sulfate A (Adapted from Falshaw et al. 1999)

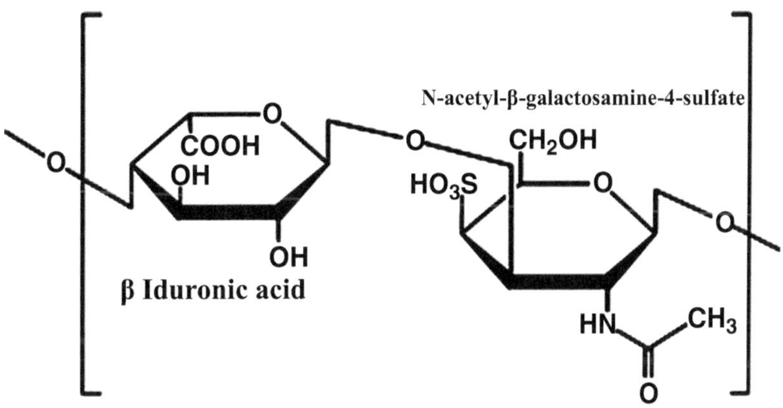

Fig. 13.4 Structure of chondroitin sulfate B (Adapted from Falshaw et al. 1999)

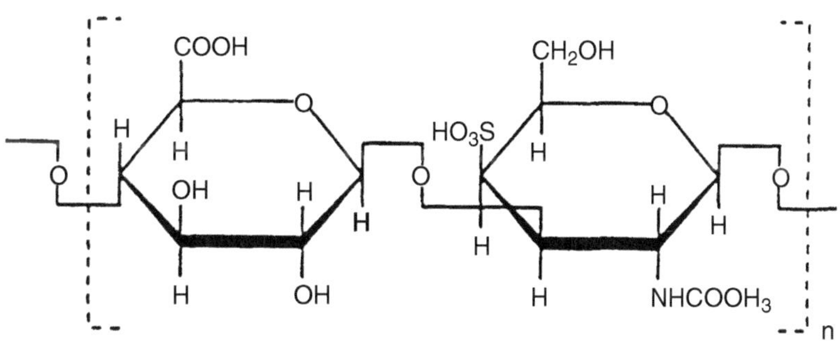

Fig. 13.5 Structure of chondroitin sulfate C (Adapted from Falshaw et al. 1999)

13.2.3 Dermatan Sulfate

Dermatan sulfate (Greek *derma*, skin) contributes to the flexibility of skin and is also present in blood vessels and heart valves. In this polymer, many of the glucuronate (GlcA) residues present in chondroitin sulfate are replaced by their epimer, iduronate (IdoA), via 1, 3 linked to N-acetyl glucosamine. Differences in the degree of sulfation on both hexuronic acid (2-O-sulfated) and GalNAc (4-O- or/and 6-O-sulfated) are responsible for the extensive heterogeneity of this polymer (Rudd et al. 2010). As opposed to chondroitin sulfate, dermatan sulfate refers to a glycan that contains one or more IdoA disaccharides. It is also referred to as chondroitin sulfate-B, although it is no longer classified as a form of chondroitin sulfate, which is represented in Fig. 13.6 (Varki et al. 1999).

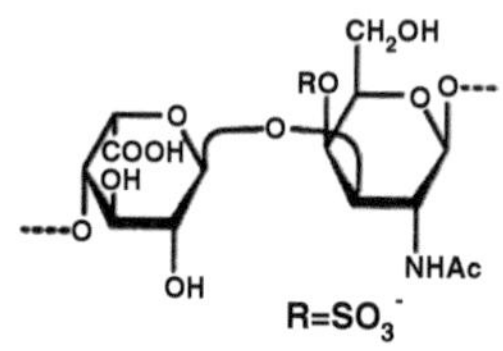

Fig. 13.6 Structure of dermatan sulfate (Adapted from Falshaw et al. 1999)

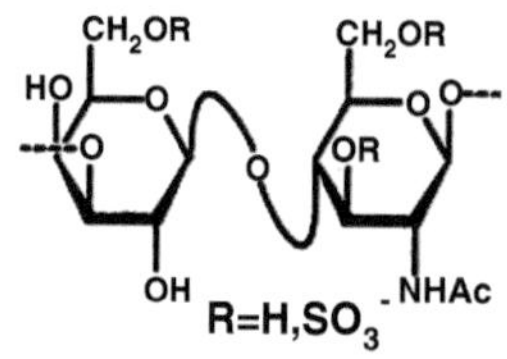

Fig. 13.7 Structure of keratan sulfate (Adapted from Falshaw et al. 1999)

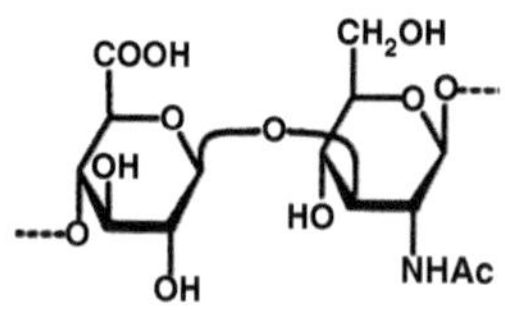

Fig. 13.8 Structure of hyaluronic acid (Adapted from Falshaw et al. 1999)

13.2.4 Keratan Sulfate

Keratan sulfates (Greek *keras*, horn) have no uronic acid and their sulfate content is inconsistent. They are found in cornea, cartilage, bone, and a variety of horny structures formed of dead cells: claws, hair, horn, hoofs, and nails. Keratan is a sulfated polylactosamine chain and contains a mixture of nonsulfated, monosulfated, and disulfated disaccharides (Fig. 13.7). The basic repeating disaccharide unit is →3Gal-β-1→4GlcNAc-β→1. It can be sulfated at carbon 6 of both Gal and GlcNAc monosaccharides. Bacterial keratanase and chondroitinase ABC degrade keratan sulfates at specific positions (Cooper et al. 2002).

13.2.5 Hyaluronic Acid

Hyaluronic acid (hyaluronate at physiological pH) contains alternating residues of D-glucuronic acid and *N*-acetylglucosamine linked via alternating β-1→4 and β-1→3 glycosidic bonds (Fig. 13.8). Several types of hyaluronidases, enzymes that degrade HA, are known to generate either tetrasaccharides or disaccharides as end products (Varki et al. 1999). Hyaluronate is also an important component of the extracellular matrix of cartilage and tendons, to which it donates tensile strength and elasticity as a result of its strong interactions with other components of the matrix. Hyaluronidase, an enzyme produced by some pathogenic bacteria, can hydrolyze the glycosidic linkages of hyaluronate, rendering tissues more susceptible to bacterial offensive action. In many organisms, an analogous enzyme in sperm hydrolyzes an outer GAG coat around the ovum, allowing sperm dissemination.

13.3 Antibodies

Antibodies (immunoglobulins-Ig) are also referred to as glycoproteins (glycoproteins are proteins that carry covalently bonded sugar units); antibodies are antigen-binding proteins present on the B-cell membrane and secreted by plasma cells. Secreted antibodies in circulation serve as effectors of humoral immunity by searching out and neutralizing antigens or scratching them for riddance. All antibodies possess similar structural features, bind to antigen, and take part in associated effector functions. Antibody molecules have a common structure of four peptide chains (Fig. 13.9). This structure consists of two identical

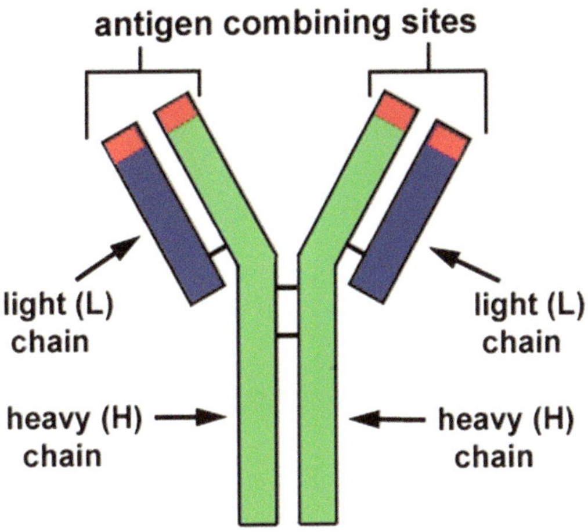

Fig. 13.9 Structure of antibody

light (L) chains, polypeptides of about 23 kDa molecular weight, and two identical heavy (H) chains, larger polypeptides of molecular weight 50–60 kDa or more. As are the antibody molecules they constitute, H and L chains are also called immunoglobulins.

The H-chains and L-chains are connected by interchain disulfide bridges. This prototype structure is common for all monomeric immunoglobulin molecules. Polymeric antibodies of higher molecular weight are formed by 2–6 four-chain subunits, similar to the monomeric immunoglobulin molecule. They possess one or two supplementary peptide chains that are essential for the formation and stabilization of antibody polymers (Vilela-Silva et al. 2001). The immunoglobulin molecules are found at the surface of B-lymphocytes or soluble in the blood and lymph. In general, antibodies are divided into five classes, IgG, IgM, IgA, IgD, and IgE, based on the numbers of prototype structures and the type of heavy-chain polypeptides (Harlow and Lane 1998).

13.3.1 Production of Antibodies

Antibodies are generated by immunization of appropriate animals or by production of hybridomas, using myeloma cells (plasma cell tumor) and activated B-lymphocytes from mouse spleen. The myeloma cells habitually produce a large amount of a single type of abnormal antibody, which is called a para-protein or M protein. It often reduces the production of normal antibodies as it cannot fight infection effectively. However, the methods used for production of antibodies are different. An animal injected with a suitable antigen generates multiple antibodies against different epitopes, whereas the hybridome produces an antibody against a single epitope. A single clone of B-lymphocytes produces antibodies against only a single specific epitope. Because the B-lymphocyte can produce only one type of antibody and a huge number of different types are needed, it divides and gives rise to many B-cells, all producing antibodies against a specific epitope, called a lymphocyte B-cell clone. The ability to stimulate the production of antibodies varies among different molecules. Certain parts of the same molecules are better antigens than others (Goding 1993).

13.3.2 Production of Antibodies Against GAGs and Proteoglycan Epitopes

Carbohydrates are the most abundant and structurally diverse organic molecules in nature that provide energy to living cells. The location of carbohydrates on the outer cell surface enables them to interact effectively with the immune system, which consists of various types of cells and molecules that specifically interact with each other to initiate the host defense mechanism. Their role as cell antigenic determinants has been firmly established and termed glycoimmunology (Guo and Boons 2009). Normally, low molecular weight carbohydrates are not immunogenic except combined with an immunogenic carrier, a hapten. The antigenic mechanism of carbohydrate antigens is somewhat similar to that of proteins. In general, glycans activate B lymphocyte cells in a thymus-independent type-2 response, and polysaccharide antigens induce a T-cell independent response. Thus, the first signal is again the B-cell receptor binding to the antigen. The second signal in this case is either provided by a receptor of the innate immune system, such as a toll-like receptor or by extensive crosslinking of the surface antibody by an antigen with repeating epitopes. Because the carbohydrates typically stimulate the B-cells in a T-cell–independent manner, IgM is the predominant immunoglobulin produced. In addition, affinity maturation and development of memory cells are barely present (Goding 1993; Guo and Boons 2009).

13.4 Types of GAG and Proteoglycan Antibodies

Antibodies or immunoglobulins that are on familiar terms with intact GAG chains or unambiguous epitopes produced by GAG-lyases are commercially accessible. In general, anti-GAG monoclonal antibodies are generated in the experimental animal, against intact human GAG chains, and these antibodies are made to cross-react with homologous epitopes from other species such as monkey, cat, rat, and mouse, but not with other types of GAG. Some of these antibodies categorize the integral GAG chains, but others can only identify a disaccharide succession or "stub" of GAG. The "stub" antibodies are recognized as a specific GAG after extensive digestion using a defined enzyme. Generally, the antibody specificity is according either to the sulfated position such as 4S, 6S or unsulfated position. This is because the self-association surrounded by the GAGs and their interaction with other constituents of the extracellular matrix may be influenced by the meticulous charge on the GAG molecule. Consequently, these antibodies make available an important device to learn the allocation of specific types of some GAG chains that appear to show significant tissue specificity Westerlo et al. (2011).

13.5 Extraction of GAG

13.5.1 Collection of Marine Sponges

The suitable marine sponges are collected from the sea, brought to the laboratory, washed with tap water, and then distilled water. Sand, debris, and tiny particles are completely removed. The whole body tissue is cut into small pieces, ground, defatted with suitable organic (acetone/petroleum ether/chloroform) solvent, and used for further extraction. Otherwise, the mucus of the sponge is filtered and can be considered for separation of GAG.

13.5.2 Extraction of GAGs

The procedure of Holick et al. (1985) is adopted for the extraction of GAGs from the marine sponges. The defatted tissues are ground and mixed with 0.4 M Na_2 SO_4. The mixture is incubated at 55°C for 1 h and 30 min (pH 11.5) after incubation Al_2 $(SO_4)_3$ crystals are added to reduce the pH 7.7 and again incubated to 95 °C for 1 h and centrifuged (2500 × g) for 1 h and 30 min at 4 °C. The retentate is recovered and cetylpyridinium chloride (CPC) is added (0.1 % w/v) and the mixture is allowed to stand for 3 h at 4 °C; centrifugation is performed (2500 × g) at 4 °C for 15 min and the precipitate is recovered and washed two times with 0.1 % CPC solution and recovered each time by centrifugation. Finally, the recovered precipitate is dissolved in 2.5 M NaCl and the crude GAG is recovered by methanol (85 % v/v) precipitation. After standing overnight at 4 °C, the crude GAG precipitate is recovered by centrifugation (2500 × g) at 4 °C for 15 min.

13.6 Fractionation of GAGs

13.6.1 Ion-Exchange Chromatography

The GAGs extracted from the tissue of the marine sponges (10 mg) is then subjected to ion-exchange column chromatography using DEAE-cellulose. The column is eluted with two dissimilar molar concentrations of NaCl (1.0–2.0 M), the flow rate of the column is 8 ml/h, and the active fractions are gathered (Pavao et al. 1998). Both elutes are combined, dialyzed (for low molecular weight GAG—cut off dialysis membrane range is 12–14 kDa), and freeze-dried, which is used for molecular weight determination. Then the fractionation of GAGs (1.0 g) is also done using anionic resin on a column of Amberlite IRA-900 (Cl^-; Saravanan and Shanmugam 2010). Then the sample is recovered by stepwise elution with 0.4 M NaCl and 0.8 M NaCl and the flow rate of the column is 1 ml/min. Both elutes are combined, dialyzed, and freeze-dried.

13.6.2 Conversion of GAGs as GAG Sodium Salts

The freeze-dried GAGs are renovated into GAG sodium salts by using a cationic resin (Amberlilte IR-120 in Na^+) column (Volpi 1994). Then the elute is gathered by precipitation with 2.0 volume of acetone; the collected precipitate is dried under vacuum. The recuperated white powder of GAG complex is used for further investigation.

13.6.3 Chemical Characterization

An Azure-A assay is performed (confirmation test) to estimate the level of sulfo group replacement of the fractionated and purified GAGs. An Azure-B assay is used for the confirmation of nucleic acid presence. Metachromasia of the blue dye on addition of negatively charged GAG results in a concentration-dependent increase in absorbance at 530 nm (Grant et al. 1984); porcine intestinal heparan sulfate is used as a standard.

13.6.4 Purification of GAGs by Gel Chromatography

GAGs are purified on a suitable column of Sephadex G-100. The activities of all the fractions are tested through metachromatic assay. The active fractions are pooled and expansively dialyzed against distilled water and freeze-dried (Laurent et al. 1978).

13.7 Biomedical Applications of Proteoglycans from Marine Sponges

The sulfated polysaccharides are an inherent part of the sponge, characterized by species-specific cell–cell interaction giving rise to a framework intertwined to form the canal system, which performs the numerous physiological functions of the animal. The sponges are characterized by distinct populations of sulfated polysaccharides and with variations in degrees of sulfation. The interactions are mediated by calcium and facilitated in a species-specific manner.

13.7.1 Clinical Significances of Proteoglycans and GAG

Proteoglycans in addition to GAGs act upon copious imperative functions surrounded by the body, several of which still remain to be studied. One of the well-known functions of the GAG heparin is its role in preventing coagulation of the blood by inhibiting the activity of serine protease. Heparin is plentiful in granules of mast cells that line blood vessels. The discharge of heparin from these granules, in reaction to injury, and its consequent doorway into the serum show the ways to an inhibition of blood clotting in the following manner. Free heparin composites with and stimulates antithrombin III, which in turn restrains all the serine proteases of the coagulation cascade. This phenomenon has been clinically exploited in the use of heparin injections for anticoagulation remedies.

13.7.2 Anticoagulant Activity

The anticoagulant activity of enzyme hydrolysis heparan sulfate and chromatography purified GAG of marine scallop are depicted in Table 13.2. The activated partial thromboplastin time (APTT) and prothrombin time (PT) activity of enzyme hydrolysis heparan sulfate of a marrine scallop sample is recorded as 135 and

Table 13.2 APTT and PT activity of fractionated and purified GAG extracted from marine scallop

Source	Type of purification	APTT activity (IU/mg)	PT activity (IU/mg)
Human blood	DEAE-cellulose	72	41
	Amberlite IRA-900	84	57
	Sephadex G-100	95	63
	Enzyme hydrolysis	135	100

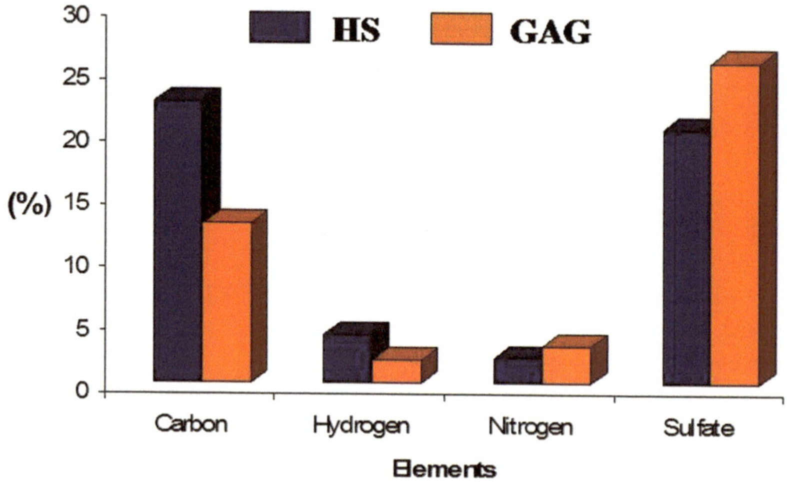

Fig. 13.10 Elements content of HS and GAG of marine scallop (*HS* heparan sulfate)

100 IU/mg, respectively. In the case of APTT and PT activities of fractionated and purified GAG of marine scallop samples are recorded as 72 and 41 IU/mg (DEAE-cellulose), 84 and 57 IU/mg (Amberlite IRA-900), and 95 and 63 IU/mg (Sephadex G-100), correspondingly.

The anticoagulant activity of GAG is low when compared to the heparan sulfate of the marine scallop; the lower anticoagulant activity can be related to a lower degree of sulfation of the polysaccharide and, in particular, to a decrease of the percentage of the trisulfated disaccharides (Volpi 2005). The sequential precipitation and purification of heparin, dermatan sulfate, and chondroitin sulfate from mixtures with various organic solvents are reported by Volpi (1996). This method is routinely used for purification of specific components from the mixture of GAG, by selecting various concentrations of methanol and sodium chloride as eluants. Saravanan and Shanmugam (2010, 2011) made a slight modification in this method for isolation of low molecular weight heparin derivative from GAG of marine animals. They have reported elements such as carbon, hydrogen, nitrogen, and sulfate contents of the marine scallop by using the enyzyme hyrodrolysis method and ion-exchange and gel filtration chromatography. According to them the carbon, hydrogen, nitrogen, and sulfate contents of the heparan sulfate and GAG samples showed their values as 22.80 %, 3.08 %, 2.08 %, and 20.4 % and 12.8 %, 1.88 %, 3.08 %, and 25.76 %, respectively (Fig. 13.10). The increased percentage contents in heparan sulfate are due to the interference of other sulfated GAGs such as chondroitin sulfate and dermatan sulfate in the chromatography purification method. In the GAGs of *H. pugilinus* the sulfate content is 9.91 % and uronic acid 26.9 %. Whereas in the GAGs of heparan sulfate from the snail *Helix aspersa*, the uronic acid and sulfate contents are found varying from 22–53 % to 7–10 %, respectively (Hovingh and Linker 1998).

By degradation with heparitinases and heparinase from *Flavobacterium heparinum* as

Table 13.3 Chemical differences among the sulfated polysaccharides from marine sponges

Species	Sulfated polysaccharides	References
Aplysina fulva	Hex UA, Glu (sulfated)	Zierer and Mourao (2000)
Chondrilla nucula	Hex UA, Ara, Gal, Fuc (sulfated)	Zierer and Mourao (2000)
Cliona celata	Sulfated Hex Nac, Ara, Fuc	Guerardel et al. (2004)
Dysidea robusta	Hex UA, Ara, Gal, Fuc 4-*O*-sulfated	Zierer and Mourao (2000)
Halichondria panicea	Gal Py (4,6), Fuc, GlcNac *N*-sulfated	Guerardel et al. (2004)
Hymeniacidon heliophila	Hex UA, Gal, Fuc (sulfated)	Zierer and Mourao (2000)
Microciona prolifera	Gal, Fuc, Gal Py (4,6), GlcNac *N*-sulfated	Guerardel et al. (2004)
Myxilla rosacea	Glc 4,6-disulfated, Fuc 2,4-disulfated	Cimino et al. (2001)
Ophlithaspongia tenius	HexUA, Glc, GlcNac *N*-sulfated	Parrish et al. (1991)
Suberites ficus	Hex UA, GlcNac, Fuc, Man, Gal (sulfated)	Bucior and Burger (2004)

well as electrophoretic migration in different buffer systems of the sulfated polysaccharides extracted from 22 species of the main classes of invertebrates, it is suggested that heparan sulfate-like and/or heparin-like compounds are present in all tissue-organized species analyzed (Cassaro and Dietrich 1977). In a more recent survey of more than 50 invertebrates from different classes using the same methodology, it is shown that heparan sulfate is a ubiquitous compound. Other authors have also reported the presence of sulfated GAG-like compounds in some species of invertebrates (Hovingh and Linker 1982).

Cosmi et al. (1997) found the effect of nonspecific binding of unfractionated heparin to plasma proteins most likely contributes to the variable anticoagulant-IIa response to unfractionated heparin in patients with thromboembolic disease. Although dermatan sulfate also binds to plasma proteins, the clinical significance of the phenomenon is unclear. In contrast, because low molecular weight heparin does not bind to plasma proteins, the anticoagulant factor-IIa activity of low molecular weight heparin should be just as predictable as its anticoagulant factor – Xa activity.

The mammalian matrix proteoglycans comprise tiny interstitial proteoglycans (biglycan, decorin, fibromodulin), a proteoglycan form of type IX collagen, constituents of the agrecan family unit of proteoglycans (agrecan, brevican, neurocan, or versican). A number of these proteoglycans enclose only one GAG string (e.g., decorin), whereas others have more than 100 chains (e.g., agrecan). Matrix proteoglycans normally surround GAGs of the chondroitin and dermatan sulfate. Nevertheless, the heparan-sulfate–containing proteoglycans from the perlecan and agrin families abound in basement membranes. In addition, heparin proteoglycans (otherwise called serglycin) are found in intracellular granules of immune cells (Varki et al. 1999).

The cellular adhesion and recognition of marine sponges are mediated by proteoglycan-like molecules, also called aggregation factors, spongicans, or glyconectins (Guerardel et al. 2004). These proteoglycan-like molecules are composed of a protein core attached to several sulfated polysaccharide units (Jarchow et al. 2000). The sulfated polysaccharide units of glyconectins are responsible for the cell–cell recognition and adhesion in sponges (Bucior and Burger 2004). The interaction between the sulfated polysaccharides of adjacent sponge cells is calcium dependent and a highly species-specific event (Bucior and Burger 2004; Misevic et al. 2004).

The species-specific interaction of the sulfated polysaccharides from glyconectins is demonstrated by the selective and homophilic aggregation of beads coated with sulfated polysaccharides from different sponges (Popescu and Misevic 1997; Misevic et al. 2004). Other evidence for the species-specificity of sulfated polysaccharides from Porifera species entails their chemical and structural diversity (Zierer and Mourao 2000; Guerardel et al. 2004). These sulfated polysaccharides are highly complex and all the species previously studied showed polymers with different structures and/or sugar and sulfate content (Table 13.3).

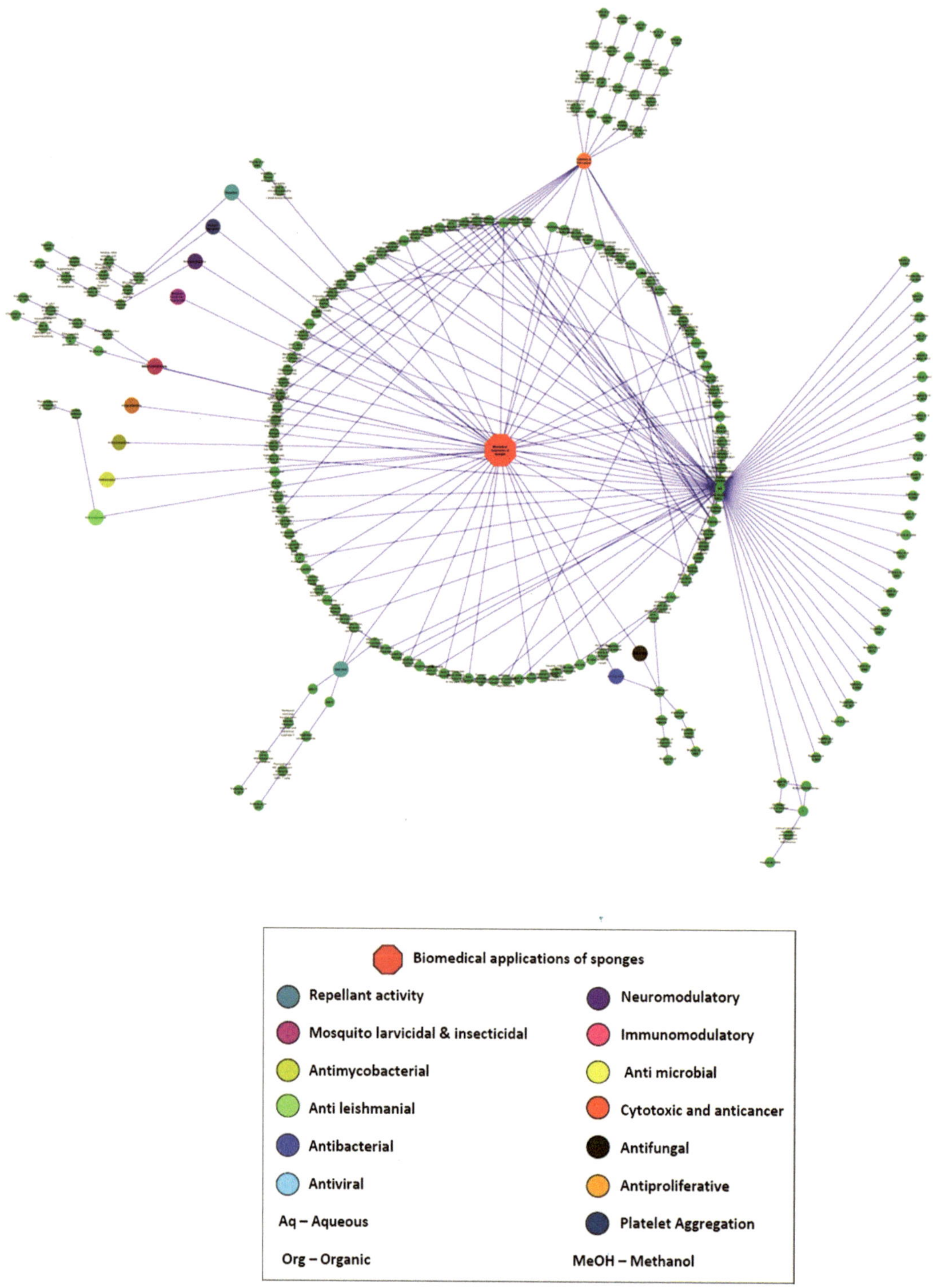

Fig. 13.11 Network of biomedical importance of proteoglycans from marine sponges

13.8 Conclusion

The meshwork (Fig. 13.11) gives a clear picture of the biomedical potential of the diversified applications of marine sponges from the ocean waters around the world. With distinctly complex structural and functional organization, the sponge bioactive compounds exhibit widely varying mechanisms towards exerting the wide range of activities. With comparatively higher concentrations of sulfated polysaccharides and associated carbohydrates such as proteoglycans, GAG, and the like, the sponges are starting to provide a new dimension to antiviral drug research. Novel carbohydrates including halistanol sulfate and other heteropolysaccharides isolated from these sponges have been found to exert significant inhibitions against HIV, herpes simplex virus (HSV), and herpetic viruses among others. Higher heteropolysaccharide content of the sponges has been of great interest to the scientific community towards the development of a potential alternative source of anticoagulant, through improved platelet aggregation activities. Novel proteoglycans such as pachymatismin, derived from sponges, have been found to exhibit cytotoxicity against a wide range of cancer cell lines, portraying potentials of alternate sources of anticancer agents. Sugar moieties in sponges, such as fucans, glycans, and the like, bound to protein molecules have been found to exert significant inhibitory activities against a wide range of gram positive and negative, resistant and nonresistant strains of bacteria. They were also found to suppress the growth of a range of fungal strains. Thus, these molecules from sponges exhibit immense qualities of antimicrobial agents in comparison to antibiotics. With widespread biomedical applications and therapeutic opportunities, the sponge community offers great scope for marine biologists, scientists, and clinicians of the future towards the discovery of drugs for disease and infection and restoring health to mankind and other living organisms.

Acknowledgments The author (RK) is gratefully acknowledged to Chettinad Academy of Research and Education for providing the JRF. The authors are thankful to Mr. Alexander Vincent, JRF, Chettinad Academy of Research & Education, Chennai, for the preparation of network model.

References

Ahmad AS, Matsuda M, Shigeta S, Okutani K (1999) Revelation of antiviral activities by artificial sulfation of a glycosaminoglycan from a marine *Pseudomonas*. Mar Biotechnol 6(1):102–106

Alvarez-Dominguez CVBJ, Marin CE, Mato LP, Cobian FL (1997) Host cell heparin sulfate proteoglycans mediate attachement and entry of *Listeria monocytogenes*, and the listerial surface protein Act A is involved in heparin sulfate receptor recognition. Infect Immunol 65(1):78–88

Amornrut C, Toida T, Imanari T, Woo ER, Park H, Linhardt R, Wu SJ, Kim YS (1999) A new sulfated beta-galactan from clams with anti- HIV activity. Carbohydr Res 321(1–2):121–127

Arena A, Maugeri TL, Pavone B, Iannello D, Gugliandolo C, Bisignano G (2006) Antiviral and immunomodulatory effect of a novel exopolysaccharide from a marine thermotolerant *Bacillus licheniformis*. Int Immunopharmacol 6(1):8–13

Arena A, Gugliandol C, Stassi G, Pavone B, Iannello D, Bisignano G, Maugeri TL (2009) An exopolysaccharide produced by *Geobacillus thermodenitrificans* strain B3–72: antiviral activity on immunocompetent cells. Immunol Lett 123(2): 132–137

Artan M, Li Y, Karadeniz F, Lee SH, Kim MM, Kim S-K (2008) Anti-HIV-1 activity of phloroglucinol derivative, 6, 6′-bieckol, from *Ecklonia cava*. Bioorg Med Chem 16(17):7921–7926

Azevedoa LA, Perazaa GG, Lernerb C, Soaresc A, Murciad N, Baisch ALM (2008) Investigation of the anti-inflammatory and analgesic effects from an extract of *Aplysina caissara*, a marine sponge. Fundam Clin Pharmacol 22(5):549–556

Beress A, Wassermann O, Bruhn T, Beres L, Kraiselburd EN, Gonzalez LV, de Motta GE, Chavez PI (1993) A new procedure for the isolation of anti-HIV compounds (polysaccharides and polyphenols) from the marine alga *Fucus vesiculosus*. J Nat Prod 56(4):478–488

Bhakuni DS, Rawat DS (2005) Bioactive marine natural products. Springer, New York, p 114

Boyd MR, Gustafson K, McMahon J, Shoemaker R (1996) Discovery of cyanovirin-N, a novel HIV-inactivating protein from *Nostoc ellipsosporum* that targets viral gp120. Int Conf AIDS 11:71

Bucior I, Burger MM (2004) Carbohydrate-carbohydrate interaction as a major force initiating cell-cell recognition. Glycoconj J 21(3–4):111–123

Cassaro CM, Dietrich CP (1977) Distribution of sulfated mucopolysaccharides in invertebrates. J Biol Chem 252(7):2254–2261

Chairman K, Jeyamala M, Sankar S, Murugan A, Singh R (2013) Immunomodulating properties of bioactive compounds present in Aurora globostellata. Int J Mar Sci 3(19):151–157

Chang L, Whittaker NF, Bewley CA (2003) Crambescidin 826 and dehydrocrambine A: new polycyclic guanidine alkaloids from the marine sponge *Monanchora* sp. that inhibit HIV-1 fusion. J Nat Prod 66(11):1490–1494

Chill L, Rudi A, Aknin M, Loya S, Hizi A, Kashman Y (2004) New sesterterpenes from Madagascan Lendenfeldia sponges. Tetrahedron 60(47): 10619–10626

Cimino P, Bifulco G, Casapullo A, Gomez-Paloma L, Riccio R (2001) Isolation and NMR characterization of rosacelose, a novel sulfated polysaccharide from the sponge *Myxilla rosacea*. Carbohydr Res 334(1): 39–47

Cirne-Santos CC, Souza TM, Teixeira VL, Fontes CF, Rebello MA, Castello-Branco LR, Abreu CM, Tanuri A, Frugulhetti IC, Bou-Habib DC (2008) The dolabellane diterpene dolabelladienetriol is a typical noncompetitive inhibitor of HIV-1 reverse transcriptase enzyme. Antiviral Res 77(1):64–71

Comin MJ, Maier MS, Roccatagliata AJ, Pujol CA, Damonte EB (1999) Evaluation of the antiviral activity of natural sulfated polyhydroxysteroids and their synthetic derivatives and analogs. Steroids 64(5): 335–340

Cooper S, Bennett W, Andrade J, Reubinoff BE, Thomson J, Martin FP (2002) Biochemical properties of a keratin sulfate/chondroitin sulfate proteoglycan expressed in primate pluripotent stem cell. J Anat 200(Pt3):259–265

Cosmi B, Fredenburgh JC, Rischke J (1997) Effect of nonspecific binding to plasma proteins on the antithrombin activities of unfractionated heparin, low-molecular-weight heparin and dermatan sulfate. Circulation 95(1):118–124

da Frota ML Jr, Braganhol E, Canedo AD, Klamt F, Apel MA, Mothes B, Lerner C, Bettestinini AM, Henriques A, Moreira JC (2009) Brazilian marine sponge *Polymastia janeirensis* induces apoptotic cell death in human U138MG glioma cell line, but not in a normal cell culture. Invest New Drugs 27(1):13–20

de Garilhe P, de Rudder J (1964) Effect of 2 arbinose nucleosides on the multiplication of herpes virus and vaccine in cell culture. C R Hebd Seances Acad Sci 259:2725–2728

De Lira SP, Seleghim MH, Williams DE, Marion F, Hamill P, Jean F, Andersen RJ, Hajdu E, Nerlinck RGS (2007) A SARS-coronovirus 3CL protease inhibitor isolated from the marine sponge *Axinella cf. corrugata*: structure elucidation and synthesis. J Braz Chem Soc 18(2):440–443

De Souza PH, Leao-Ferreira LR, Moussatche N, Teixeira VL, Cavalcanti DN, Da Costa LJ, Diaz R, Frugulhetii IC (2005) Effects of diterpenes isolated from the Brazilian marine alga *Dictyota menstrualis* on HIV-1 reverse transcriptase. Planta Med 71(11):1019–1024

Devi P, Wahidulla S, Kamat T, D'Souza L (2011) Screening marine organisms for antimicrobial activity against clinical pathogens. Ind J Geo Mar Sci 40(3): 338–346

Dhinakaran I, Manohari V, Atchya B, Tamilselvi K, Lipton AP (2012) Antifungal and cytotoxic activities of some marine sponges collected from the South East Coast of India D. J Appl Pharm Sci 2(1):52–55

Elion GB, Furman PA, Fyfe JA, de Miranda P, Beauchamp L, Schaeffer HJ (1977) Selectivity of action of an antiherpetic agent, 9-(2-hydroxyethoxymethyl) guanine. Proc Natl Acad Sci U S A 74(12):5716–5720

Ellithey MS, Lall N, Hussein AA, Meyer D (2014) Cytotoxic and HIV-1 enzyme inhibitory activities of Red Sea marine organisms. BMC Complement Altern Med 14:77

Esteves AIS, Nicolai M, Humanes M, Goncalves J (2011) Sulfated polysaccharides in marine sponges: extraction methods and anti-HIV activity. Mar Drugs 9(1): 139–153

Falshaw R, Furneaux RH, Slim GC (1999) In: Finch P (ed) Carbohydrate sulphates carbohydrates structures, syntheses and dynamics, 1st edn. Springer science + Business media, Dordrecht, pp 107–141

Ferreira EG, Wilke DV, Jimenez PC, Portela TA, Silveira ER, Hajdu E Pessoa C, de Moraes MO, Lotufo LVC (2007) Cytotoxic activity of hydroethanolic extracts of sponges (Porifera) collected at Pedra da Risca do Meio Marine State Park, Ceara State, Brazil. In: Custodio, M.R., Lobo- Hajdu, G., Hajdu, E. & Muricy, G. (Eds.) Porifera research: biodiversity, innovation and sustainability, Serie Livros, 28, Museu Nacional, Rio de Janeiro, pp 313–318

Ford PW, Gustafson KR, McKee TC, Shigematsu N, Maurizi LK, Pannell LK, Williams DE, de Silva ED, Lassota P, Allen TM, Soest RV, Andersen RJ, Boyd MR (1999) Papuamides A–D, HIV-inhibitory and cytotoxic depsipeptides from the sponges *Theonella mirabilis* and *Theonella swinhoei* collected in Papua New Guinea. J Am Chem Soc 121(25):5899–5909

Goding JW (1993) Monoclonal antibodies: principles and practice: production and application, 2nd edn. Academic Press Limited, San Diego

Grant AC, Linhardt RJ, Fitzgerald G, Park JJ, Langer R (1984) Metachromatic activity of heparin and heparin fragments. Anal Biochem 137(1):25–32

Guerardel Y, Czeszak X, Sumanovski LT, Karamanos Y, Popescu O, Strecker G, Misevic GN (2004) Molecular fingerprinting of carbohydrate structure phenotypes of three porifera proteoglycan like glyconectins. J Biol Chem 279(15):15591–15603

Guimaraes TR, Quiroz CG, Rigotto C, de Oliveira SQ, de Almeida MTR, Bianco EM, Moritz MIG, Carraro JL, Palermo JA, Cabrera G, Schenkel EP, Reginatto FH, Simoes CMO (2013) Anti HSV-1 activity of halistanol sulfate and halistanol sulfate C isolated from Brazilian marine sponge *Petromica citrina* (Demospongiae). Mar Drugs 11(11):4176–4192

Guo Z, Boons ZJ (2009) Carbohydrate-based vaccines and immunotherapies, 1st edn. Wiley, New Jersey

Gustafson KR, Cardellina JH, Fuller RW, Weislow OS, Kiser RF, Snader KM, Patterson GM, Boyd MR (1989a) AIDS – antiviral sulfolipids from cyanobacteria (blue–green algae). J Natl Cancer Inst 81(16):1254–1258

Gustafson KR, Roman M, Fenical W (1989b) The macrolactins, a novel class of antiviral and cytotoxic macrolides from a deep-sea marine bacterium. J Am Chem Soc 111(19):7519–7524

Harlow E, Lane D (1998) Antibodies: a laboratory manual, 1st edn. Cold Spring Harbor Laboratory Press, Cold Spring Harbor

Haseley SR, Vermeer HJ, Kamerling JP, Vliegenthart JFG (2001) Carbohydrate self-recognition mediates marine sponge cellular adhesion. Proc Natl Acad Sci 98(16):9419–9424

Hasui M, Matsuda M, Okutani K, Shigeta S (1995) In vitro antiviral activities of sulfated polysaccharides from a marine microalga (*Cochlodinium polykrikoides*) against human immunodeficiency virus and other enveloped viruses. Int J Biol Macromol 17(5):293–297

Hayashi T, Hayashi K, Maed M, Kojima I (1996) Calcium spirulan, an inhibitor of enveloped virus replication, from a blue–green alga *Spirulina platensis*. J Nat Prod 59(1):83–87

Hellio C, Tsoukatou M, Maréchal JP, Aldred N, Beaupoil C, Clare AS, Vagias C, Roussis V (2005) Inhibitory effects of Mediterranean sponge extracts and metabolites on larval settlement of the barnacle *Balanus Amphitrite*. Mar Biotechnol 7(4):297–305

Hirsh J (1992) Overview of low molecular weight heparins and heparinoids: basic and clinical aspects. Aust N Z J Med 22(5):487–495

Holick MF, Judkiewicz A, Walworth N, Wang MH (1985) Recovery of heparin from fish wastes. In: Colwell RR, Pariser ER, Sinskay AJ (eds) Biotechnology of marine polysaccharides. Hemisphere Publishing Corporation, New York, pp 389–397

Hooper JNA (1995) Sponge guide. Queensland Museum, Queensland

Hooper JNA, van Soest RWM (2002) Systema Porifera: a guide to the classification of sponges. Kluwer Academic/Plenum Publishers, New York

Horwitz JP, Chua J, Noel M (1964) The monomesylates of 1-(2-deoxy-dlyxofuranosyl) thymidine. J Org Chem 29:2076–2078

Hovingh P, Linker A (1982) An unusual heparan sulfate isolated from lobsters (*Homarus americanus*). J Biol Chem 257(16):9840–9844

Hovingh P, Linker A (1998) Glycosaminoglycans in two mollusks, *Aplysia californica* and *Helix aspersa*, and in the leech, *Nephelopsis obscura*. Biochem Mol Biol 119(4):691–696

Hutagalung RA, Victor, Karjadidjaja M, Prasasty VD, Mulyono N (2014) Extraction and characterization of bioactive compounds from cultured and natural sponge, *Haliclona molitba* and *Stylotella aurantium* origin of Indonesia. Int J Biosci Biochem Bioinforma 4(1):14–18

Hwang Y, Rowley D, Rhodes D, Gertsch J, Fenical W, Bushman F (1999) Mechanism of inhibition of a poxvirus topoisomerase by the marine natural product sansalvamide A. Mol Pharmacol 55(6): 1049–1053

Iwashima M, Mori J, Ting X, Matsunaga T, Hayashi K, Shinoda D, Saito H, Sankawa U, Hayashi T (2005) Antioxidant and antiviral activities of plastoquinones from the brown alga *Sargassum micracanthum*, and a new chromene derivative converted from the plastoquinones. Biol Pharm Bull 28(2):374–377

Jackson RL, Busch SJ, Cardin AD (1991) Glycosaminoglycans: molecular properties, protein interactions, and role in physiological processes. Physiol Rev 71(2):481–539

Jarchow J, Fritz J, Anselmetti D, Calabro A, Hascall VC, Gerosa D, Burger MM, Busquets XF (2000) Supramolecular structure of a new family of circular proteoglycans mediating cell adhesion in sponges. J Struct Biol 132(2):95–105

Joseph B, Sujatha S, Jeevitha MV (2010) Screening of pesticidal activities of some marine sponge extracts against chosen pests. J Biopest 3(2):495–498

Kjellen L, Lindahl U (1991) Proteoglycans: structures and interactions. Annu Rev Biochem 60:443–475

Laurent TC, Tengblad A, Thunberg L, Hook M, Lindahl U (1978) The molecular-weight dependence of the anti-coagulant activity of heparin. Biochem J 175: 691–701

Laurienzo P (2010) Marine polysaccharides in pharmaceutical applications: an overview. Mar Drugs 8(9): 2435–2465

Lee JB, Hayashi K, Hirata M, Kuroda E, Suzuki E, Kubo Y, Hayashi T (2006) Antiviral sulfated polysaccharide from *Navicula directa*, a diatom collected from deep-sea water in Toyama Bay. Biol Pharm Bull 29(10):2135–2139

Li C, Chen J, Hua T (1998) Precambrian sponges with cellular structures. Science 279(5352):879–882

Loya S, Hizi A (1993) The interaction of illimaquinone, a selective inhibitor of the RNase H activity, with the reverse transcriptases of human immunodeficiency and murine leukemia retroviruses. J Biol Chem 268(13):9323–9328

Loya S, Rudi A, Kashman Y, Hizi A (2002) Mode of inhibition of HIV-1 reverse transcriptase by polyacetylenetriol, a novel inhibitor of RNA- and DNA-directed DNA polymerases. Biochem J 362(Pt3):685–692

Lu CX, Li J, Sun YX, Qi X, Wang QJ, Xin XL, Geng MY (2007) Sulfated polymannuroguluronate, a novel anti-AIDS drug candidate, inhibits HIV-1 Tatinduced angiogenesis in Kaposi's sarcoma cells. Biochem Pharmacol 74(9):1330–1339

Marinho PR, Muric GRS, Silva MFL, Marval MG, Laport MS (2008) Antibiotic-resistant bacteria inhibited by extracts and fractions from Brazilian marine sponges. Braz J Pharmacogn 20(2):267–275

Marinho PR, Simas NK, Kuster RM, Duarte RS, Fracalanzza SE, Ferreira DF, Romanos MT, Muricy G, Giambiagi-Demarval M, Laport MS (2012) Antibacterial activity and cytotoxicity analysis of halistanol trisulphate from marine sponge *Petromica citrina*. J Antimicrob Chemother 67(10): 2396–2400

Matsuhiro B, Conte AF, Damonte EB, Kolender AA, Matulewicz MC, Mejias EG, Pujol CA, Zuniga EA (2005) Structural analysis and antiviral activity of a sulfated galactan from the red seaweed *Schizymenia binderi* (Gigartinales, Rhodophyta). Carbohydr Res 340(15):2392–2402

Meisenberg G, Simmons WH (2006) Principles of medical biochemistry. Elsevier Health Sciences, Philadelphia, pp 243

Menozzi F, Mutombo R, Renauld G, Gantiez C, Hannah J, Leininger E, Brennan MJ, Locht C (1994) Heparin-inhibitable lectin activity of the filamentous hemagglutinin adhesion of *Bordetella pertussis*. Infect Immunol 62(3):767–788

Misevic GN, Guerardel Y, Sumanovski LT, Slomianny MC, Demarty M, Ripoll C, Karamanos Y, Maes E, Popescu O, Strecker G (2004) Molecular recognition between glyconectins as an adhesion self-assembly pathway to multicellularity. J Biol Chem 279: 15579–15590

Misevic GN, Ripoll C, Norris J, Norris V, Guerardel Y, Maes E (2007) Evolution of multicellularity in Porifera via self-assembly of glyconectin carbohydrates. In: Custódio MR, Lôbo-Hajdu G, Hajdu E, Muricy G (eds) Porifera research: biodiversity, innovation and sustainability. Museu Nacional, Rio de Janeiro

Moen LK, Clark GF (1993) A novel reverse transcriptase inhibitor from *Fucus vesiculosus*. Int Conf AIDS 9: 145–161

Monks NR, Lerner C, Henriques AT, Farias FM, Schaopoval EES, Suyenaga ES, Rocha AB, Schwartsmann G, Mothes B (2002) Anticancer, antichemotactic and antimicrobial activities of marine sponges collected off the coast of Santa Catarina, southern Brazil. J Exp Mar Biol Ecol 281(1–2):1–12

Mori T, O'Keefe BR, Sowder RC, Bringans S, Gardella R, Berg S, Cochran P, Turpin JA, Buckheit RW Jr, McMahon JB, Boyd MR (2005) Isolation and characterization of griffithsin, a novel HIV-inactivating protein, from the red alga *Griffithsia* sp. J Biol Chem 280(10):9345–9353

Muller WEG, Maidhof A, Zahn RK, Schroder HC, Gasic MJ, Heidemann D, Bernd A, Kurelec B, Eich E, Seibert G (1985) Potent antileukemic activity of the novel cytostatic agent avarone and its analogues *in vitro* and *in vivo*. Cancer Res 45:4822–4826

Muller WEG, Sobe C, Diehl-Seifert B, Maidhof A, Schroder HC (1987) Influence of the antileukemic and anti-human immunodeficiency virus agent avarol on selected immune responses *in vitro* and *in vivo*. Biochem Pharmacol 36(9):1489–1494

Muricy G, Hajdu E, Araujo FV, Hagler AN (1993) Antimicrobial activity of southwestern Atlantic shallow-water marine sponges Porifera. Sci Mar 57(4): 427–432

Nakao Y, Takada K, Matsunaga S, Fusetani N (2001) Calyceramides A–C: neuraminidase inhibitory sulfated ceramides from the marine sponge *Discodermia calyx*. Tetrahedron 57:3013–3017

Nakashima H, Kido Y, Kobayashi N, Motoki Y, Neushul M, Yamamoto N (1987a) Purification and characterization of an avian myeloblastosis and human immunodeficiency virus reverse transcriptase inhibitor, sulfated polysaccharides extracted from sea algae. Antimicrob Agents Chemother 31(10): 1524–1528

Nakashima H, Kido Y, Kobayashi N, Motoki Y, Neushul M, Yamamoto N (1987b) Antiretroviral activity in a marine red alga: reverse transcriptase inhibition by an aqueous extract of *Schizymenia pacifica*. J Cancer Res Clin Oncol 113(5):413–416

Newbold RW, Jensen PR, Fenical W, Pawlik JR (1999) Antimicrobial activity of Caribbean sponge extracts. Aquat Microb Ecol 19:279–284

Oku N, Gustafson KR, Cartner LK, Wilson JA, Shigematsu N, Hess S, Pannell LK, Boyd MR, McMahon JB (2004) Neamphamide A, a new HIV-inhibitory depsipeptide from the Papua New Guinea marine sponge *Neamphius huxleyi*. J Nat Prod 67(8):1407–1411

Pape PL, Zidane M, Abdala H, More MT (2000) A glycoprotein isolated from the sponge *Pachymatisma johnstonii*, has anti-leishmanial activity. Cell Biol Int 24(1):51–56

Parrish CR, Jakobsen KB, Coombe DR, Bacic A (1991) Isolation and characterization of cell adhesions molecules from the marine sponge *Ophlitaspongia tenius*. Biochim Biophys Acta 1073(1):56–64

Pavao MSG, Aiello KRM, Werneck CC, Silva LCF, Valente AP, Mulloy B, Colwell NS, Tollefsen DM, Mourao PAS (1998) Highly sulfated dermatan sulfates from ascidians. J Biol Chem 273(43):27848–27857

Pereira HS, Leao-Ferreira LR, Moussatche N, Teixeira VL, Cavalcanti DN, Costa LJ, Diaz R, Frugulhetti IC (2004) Antiviral activity of diterpenes isolated from the Brazilian marine alga *Dictyota menstrualis* against human immunodeficiency virus type 1 (HIV-1). Antivir Res 64(1):69–76

Plaza A, Gustchina E, Baker HL, Kelly M, Bewley CA (2007) Mirabamides A–D, depsipeptides from the sponge *Siliquariaspongia mirabilis* that inhibit HIV-1 fusion. J Nat Prod 70(11):1753–1760

Popescu O, Misevic GN (1997) Self-recognition by proteoglycans. Nature 386:231–232

Prado MP, Torres YR, Berlinck RGS, Desidera C, Sanchez MA, Craveiro MV, Hajdu E, da Rocha RM, Machado-Santelli GM (2004) Effects of marine organisms extracts on microtubule integrity and cell cycle progression in cultured cells. J Exp Mar Biol Ecol 313(1):125–137

Purushottama GB, Venkateshvaran K, Pani Prasad K, Nalini P (2009) Bioactivities of extracts from the marine sponge *Halichondria panacea*. J Venom Anim Toxins Incl Trop Dis 15(3):444–459

Putten JPDT, Cole R (1998) Entry of OpaA1 gonococci into HEp-2 cells requires concerted action of glycosaminoglycans, fibronectin and integrin receptors. Mol Microbiol 29(1):369–379

Rangel M, Sanctis B, Freitas JC, Polatto JM, Granato AC, Berlinck RG, Hajdu E (2001) Cytotoxic and neurotoxic activities in extracts of marine sponges (Porifera) from southeastern Brazilian coast. J Exp Mar Biol Ecol 262(1):31–40

Rao JV, Usman PK, Kumar JB (2008) Larvicidal and insecticidal properties of some marine sponges collected in Palk Bay and Gulf of Mannar waters. Afr J Biotechnol 7(2):109–113

Rashid MA, Gustafson KR, Cartner LK, Shigematsu N, Pannell LK, Boyd MR (2001) Microspinosamide, a new HIVinhibitory cyclic depsipeptide from the marine sponge *Sidonops microspinosa*. J Nat Prod 64(1):117–121

Reddy MV, Rao MR, Rhodes D, Hansen MS, Rubins K, Bushman FD, Venkateshwarlu Y, Faulkner DJ (1999) Lamellarin alpha 20-sulfate, an inhibitor of HIV-1 integrase active against HIV-1 virus in cell culture. J Med Chem 42(11):1901–1907

Reegan AD, Kinsalin AV, Paulraj MG, Ignacimuthu S (2013) Larvicidal, ovicidal, and repellent activities of marine sponge *Cliona celata* (Grant) extracts against *Culex quinquefasciatus* say and *Aedes aegypti* L. (Diptera: Culicidae) ISRN Entomol 2013:1–8

Reintamm T, Lopp A, Kuusksalu A, Pehk T, Kelve M (2003) ATP N-glycosidase A novel ATP-converting activity from a marine sponge *Axinella polypoides*. Eur J Biochem 270(20):4122–4132

Rodriguez MC, Merino ER, Pujol CA, Damonte EB, Cerezo AS, Matulewicz MC (2005) Galactans from cystocarpic plants of the red seaweed *Callophyllis variegata* (Kallymeniaceae, Gigartinales). Carbohydr Res 340(18):2742–2751

Rostand KSEJ (1997) Microbial adherence to and invasion through proteoglycans. Infect Immun 65(1):1–8

Rowley DC, Hansen MS, Rhodes D, Sotriffer CA, Ni H, McCammon JA, Bushman FD, Fenical W (2002) Thalassiolins A–C: new marine derived inhibitors of HIV cDNA integrase. Bioorg Med Chem 10(11): 3619–3625

Rudd TR, Skidmore MA, Guerrini M, Hricovini M, Powell AK, Siligardi G, Yates EA (2010) The conformation and structure of GAGs: recent progress and perspectives. Curr Opin Struct Biol 20(5):567–674

Rudi A, Yosief T, Loya S, Hizi A, Schleyer M, Kashman Y (2001) Clathsterol, a novel anti-HIV-1 RT sulfated sterol from the sponge *Clathria* species. J Nat Prod 64(11):1451–1453

Rutzler K (2004). Sponges on coral reefs: a community shaped by competitive cooperation. In M Pansini, R Pronzato, G Bavestrello, R Manconi, eds. Sponge science in the new millennium. Genova, Italy: Boll Mus Ist Biol Univ Genova, 68, 85–148

Saravanan R, Shanmugam A (2010) Isolation and characterization of low molecular weight glycosaminoglycans from marine mollusk *Amussium pleuronectus* (Linne) using chromatography. Appl Biochem Biotechnol 160(3):791–799

Saravanan R, Shanmugam A (2011) Is isolation and characterization of heparan sulfate from marine scallop *Amussium pleuronectus* (Linne) an alternative source of heparin?!! Carbohydr Polym 86(2):1082–1084

Seleghim MHR, Lira SP, Kossuga MH, Batista T, Berlinck RGS, Hajdu E, Muricy G, da Rocha RM, do Nascimento GGF, Silva M, Pimenta EF, Theimann OH, Oliva G, Cavalcanti BC, Pessoa C, de Moraes MO, Galetti FCS, Silva CL, de Souza AO, Peixinho S (2007) Antibiotic, cytotoxic and enzyme inhibitory activity of crude extracts from Brazilian marine invertebrates. Rev Bras Farmacogn 17(3): 287–318

Sepcic K, Kauferstein S, Mebs D, Turk T (2010) Biological activities of aqueous and organic extracts from tropical Marine sponges. Mar Drugs 8(5): 1550–1566

Silbert JE, Bernfield M, Kokenyesi R (1997) Proteoglycans: a special class of glycoproteins. In: Montrevil J, Vliegentharc JFG, Schachter J (eds) Glycoproteins II. Elsevire, Oxford, pp 1–31

Silva AC, Kratz JM, Farias FM, Henriques AT, Santos JD, Leonel RM, Lerner C, Mothes B, Barardi CR, Simoes CM (2006) In vitro antiviral acticity of marine sponges collected off Brazilian coast. Biol Pharm Bull 29(1):135–140

Sipkema D, Franssen MCR, Osinga R, Tramper J, Wijffels RH (2005) Marine sponges as pharmacy. Mar Biotechnol 7(3):142–162

Sonia AS, Lipton AP, Raj RP (2009) Lethal concentration of methanol extract of sponges to the brine shrimp, *Artemia salina* G. J Mar Biol Assoc 51(1):122–125

Stead P, Hiscox S, Robinson PS, Pike NB, Sidebottom PJ, Roberts AD, Nicholas WL, Wright AE, Pompony SA, Langley D (2000) Eryloside F, a novel penasterol disaccharide possessing potent thrombin receptor antagonist activity. Bioorg Med Chem Lett 10(7): 661–664

Sujatha S, Joseph B (2011) Effect of few marine sponges and its biological activity against *Aedes aegypti* Linn. *Musca domestica* (Linnaeus, 1758) (Diptera: Culicidae). J Fish Aquat Sci 6(2):170–177

Sun HH, Cross SS, Gunasekera M, Koehn FE (1991) Weinbersteroldisulfates A and B, antiviral steroid sulfates from the sponge *Petrosia weinbergi*. Tetrahedron 47:1185–1190

Talarico LB, Duarte ME, Zibetti RG, Noseda MD, Damonte EB (2007) An algal derived DL-galactan hybrid is an efficient preventing agent for in vitro dengue virus infection. Planta Med 73(14):1464–1468

Tersariol IL, Ferreira TM, Medeiros MG, Porcionatto MA, Moraes CT, Abreu LR, Nader HB, Dietrich CP (1994) Sequencing of heparan sulfate proteoglycans: identification of variable and constant oligosaccharide regions in eight heparan sulfate proteoglycans of different origins. Braz J Med Biol Res 27(9):2097–2102

Uzair B, Mahmood Z, Tabassum S (2011) Antiviral activity of natural products extracted from marine organisms. Bio Impacts 1(4):203–211

Van de Westerlo EM, Smetsers TF, Dennissen MA, Linhardt RJ, Veerkamp JH, van Muijen GNP, Van Kuppevelt TH (2011) Human single chain antibodies against heparin: selection, characterization and effect on coagulation. Blood 99(7):2427–2433

Van Soest RWM, Boury-Esnault N, Hooper JNA, Rutzler K, de Voogd NJ, Alvarez B, Hajdu E, Pisera AB, Manconi R, Schoenberg C, Janussen D, Tabachnick KR, Klautau M, Picton B, Kelly M, Vacelet J, Dohrmann M, Diaz MC, Cardenaz P (2008) World Porifera database http://www.marinespecies.org/porifera on 2015-05-26

Varki A, Cummings R, Esko J, Freeze H, Hart G, Marth J (1999) Essentials of glycobiology, 1st edn. Cold Spring Harbor Laboratory Press, Cold Spring Harbor

Venkateshwar Goud T, Srinivasa Reddy N, Raghavendra Swamy N, Siva Ram T, Venkateswarlu Y (2003) Anti-HIV active petrosins from the marine sponge *Petrosia similis*. Biol Pharm Bull 26(1):1498–1501

Vilela-Silva AC, Werneck CC, Valente AP, Vacquier VD, Mourão PA (2001) Embryos of sea urchin *Strongylocentrotus purpuratus* synthesize a dermatan sulfate enriched in 4-O and 6-O-disulfated galactosamine units. Glycobiol 11(6):433–440

Volpi N (1993) "Fast moving" and "slow moving" heparins, dermatan sulfate, and chondroitin sulfate: qualitative and quantitative analysis by agarose-gel electrophoresis. Carbohydr Res 2(247):263–278

Volpi N (1994) Fractionation of heparin, dermatan sulfate, and chondroitin sulfate by sequential precipitation: a method to purify a single glycosaminoglycan species from a mixture. Anal Biochem 218(2):382–391

Volpi N (1996) Electrophoresis separation of glycosaminoglycans on nitrocellulose membranes. Anal Biochem 240(1):114–118

Volpi N (2005) Occurrence and structural characterization of heparin from mollusks-review. ISJ 2:6–16

Warabi K, Hamada T, Nakao Y, Matsunaga S, Hirota H, Van Soest RWM, Fusetani N (2005) Axinelloside A, an unprecented highly sulfated lipopolysaaccharide inhibiting telomerase, from the marine sponge, *Axinella infundibula*. J Am Chem Soc 127(38): 13262–13270

Wellington KD, Cambie RC, Rutledge PS, Bergquist PR (2000) Chemistry of sponges, 19: novel bioactive metabolites from *Hamigera tarangaensis*. J Nat Prod 63(1):79–85

Witvrouw M, Este JA, Mateu QME, Reymen D, Andrei G, Snoeck R, Ikeda S, Pauwels R, Bianchini NV, Desmyter J, De Ciercq E (1994) Antiviral activity of a sulfated polysaccharide extracted from the red seaweed *Aghardhiella tenera* against human immunodeficiency virus and other enveloped viruses. Antiviral Chem Chemother 5(5):297–303

Yalcın FN (2007) Biological activities of the marine sponge *Axinella*. Hacet Univ J Fac Pharm 27(1): 47–60

Yassine M, Shabbar A (2013) Screening a Mediterranean sponge *Axinella verrucosa* for antibacterial activity in comparison to some antibiotics. J Pharmacogn Phytochem 1(6):66–75

Zidane M, Pondaven P, Roussakis C, Quemener B, More MT (1996) Pachymatismin: a novel cytotoxic factor from the marine sponge (*Pachymatisma johnstonii*). Comp Biochem Physiol 115(1):47–53

Zierer MS, Mourao PA (2000) A wide diversity of sulfated polysaccharides are synthesized by different species of marine sponges. Carbohydr Res 328(2): 209–216

Marine Sponge-Derived Antiangiogenic Compounds for Cancer Therapeutics

14

Kalimuthu Senthilkumar, Govindan Ramajayam, Jayachandran Venkatesan, Se-Kwon Kim, and Byeong-Cheol Ahn

Abstract

The biological properties of various metabolites from sponges reported recently and marine sponges are considered as a gold mine for past 50 years. Sponge-derived compounds and their metabolites have different types of biological activity such as antimicrobial, antiinflammatory, antimalarial, antioxidant, anti-HIV, and anticancer activity. Angiogenesis is the important process in tumor progression. The term "angiogenic switch" refers to a very important event during the tumor progression between pro- and antiangiogenic factors. Angiogenesis and its mechanistic pathway targeting may be useful for therapeutic approach for cancer. Recent times many compounds from marine sources have proven important role against cancer. These compounds inhibit cell proliferation and angiogenesis of cancer. In this chapter, we discuss the antiangiogenic compounds isolated from marine sponge that work against cancer.

Keywords

Marine Sponge • Tumor • Angiogenesis • Alkaloids • Macrocyclic lactone • Triterpene

K. Senthilkumar (✉) • B.-C. Ahn
Department of Nuclear Medicine, Kyungpook National University School of Medicine, Jung-Gu, Daegu 700 842, Republic of Korea
e-mail: senthilbhus@gmail.com

G. Ramajayam
Department of Neuroscience, Christian Medical College and Hospital, Vellore 632 004, India

J. Venkatesan • S-K. Kim
Specialized Graduate School Science and Technology Convergence, Department of Marine Bio Convergence Science, Pukyong National University, Busan 608-737, Republic of Korea

14.1 Introduction: Tumor Angiogenesis

Formation of vascular network is important for the proliferation and metastasis of cancer cells, which inturn depends on the adequate source of growth nutrients and oxygen (Nishida et al. 2006). Angiogenesis in tumors will occur by the formation of fresh blood vessels. Tumor mass and metastasis cannot increase

R. Pallela, H. Ehrlich (eds.), *Marine Sponges: Chemicobiological and Biomedical Applications*,
DOI 10.1007/978-81-322-2794-6_14

without formation of blood vessels. The well-characterized step for tumor progression is angiogenesis which requires vascular endothelial cells for tumor invasion. In normal circumstances, metabolic wastes were removed through the developed vascular network. Vascular network once formed becomes a stable system that could regenerate slowly. In the body conditions, angiogenesis happens principally in developing embryo, wound healing, and ovulation. The normal tissue deficiencies have significant physiological angiogenesis between endogenous pro- and antiangiogenic factors. Endothelial cells (EC) acting as major factor in angiogenesis take notable ability to separate quickly in physiological condition, such as hypoxia and inflammation for vessels of blood and lymph (Carmeliet 2003). It is also an important factor in many pathological progressions such as tumor, psoriasis, rheumatoid arthritis, and diabetic retinopathy. In mammals embryonic vasculogenesis, the developing embryo and yolk sac progress by accumulation of de novo pathway by making angioblasts into primitive vasculogenesis, which undergoes multifaceted remodeling process through migration, sprouting, pruning, and growth are main in the development of circulatory system function (angiogenesis) (Coultas et al. 2005).

Angiogenesis and lymphangiogenesis are activated by chemical signs from cells of tumor for rapid growth (Folkman et al. 1971). Development of angiogenesis largely involved with four steps: (1) existing blood vessels of basement membrane degradation, (2) migration of endothelial cells near to angiogenic stimulus, (3) endothelial cell proliferation is important for formation of endothelial cell dense sprouts in stromal space, and (4) endothelial cell organization of capillary tubes as well as vascular loops for formation of tight junctions and deposition of fresh basement membrane (Klagsbrun and Moses 1999). Endothelial cell proliferation is occurs initially in angiogenesis and then forms fresh capillary sprout elongation.

14.2 Important Factors Regulating Angiogenesis

Angiogenic phenotype depends on alteration and balance between stimulators and inhibitors of angiogenesis. Various angiogenic particles produced from tumor or stromal cells of tumor can straightly bind to these cognate receptors in endothelial cells and then initiate angiogenesis. There are a number of proteins that were recognized as angiogenic activators, with vascular endothelial growth factor (VEGF), angiogenin, basic fibroblast growth factor (bFGF), transforming growth factor TGF-β, (TGF)-α, tumor necrosis factor (TNF)-α, platelet-derived endothelial growth factor, placental growth factor, granulocyte colony-stimulating factor, hepatocyte growth factor, interleukin-8, and epidermal growth factor (Mojzis et al. 2008; Nishida et al. 2006). Among all these factors involved in angiogenesis, VEGF appears to most appropriate role. The family of VEGFs in mammals include VEGF-A, VEGF-B, VEGF-C, VEGF-D, and placental growth factor (PlGF) (Ferrara 2002, 2005). Among the different isoforms of VEGF, VEGF-A is the one that mainly regulates tumor angiogenesis. The family of VEGF and their receptors (VEGF-R) mainly involved in vascularization. Influence of stimulators like growth factors and cytokines, VEGF family seems to cancerous tissue as well adjacent stroma of neovascularization (Folkman 1990, 1995). Among the VEGF family, VEGF-A, VEGF-B, VEGF-C, and VEGF-E, binding their respective receptors leads to blood vessel proliferation, whereas VEGF-C and also VEGF-D are involved during lymphangiogenesis (Rafii and Skobe 2003; Wang et al. 2011). VEGF/vascular permeability factor (VPF), a heparin-binding glycoprotein, occurs with six isoforms, which consist of amino acids of 121, 145, 165, 183, 189, and 206; this is due to the alternative splicing of mRNA (Ferrara 1993; Stalmans et al. 2002). VEGF isoforms share with regular tyrosine kinase receptors (VEGFR1 or Flt-1, VEGFR2 or KDR/Flk-1, VEGFR3 or Flt-4). Binding of these isoforms leads to autophosphorylation and receptor dimerization of intracytoplasmic domains of specific tyrosine residues which are located in C-terminal side, which activates tyrosine kinase pathway by involving different intracellular proteins, mainly phosphatidyl-inositol-3′ kinase and extracellular signal regulated kinase (ERK)-MAPK (Qi and Claesson-Welsh 2001; Cross et al. 2003).

VEGF-A promotes differentiation, proliferation, survival, and migration of endothelial cells, also activation and mechanisms of vascular permeability and extracellular matrix degradation. A potent and specific mitogen for vascular endothelial cells is VEGF (Leung et al. 2004) and is overexpressed in cancers (Pan et al. 2013). VEGF-B has two isoforms of protein, resulting from spliced mRNA and acting specifically to VEGFR-1, and widely expressed in vascular cells, the heart, and the skeletal muscle (Olofsson et al. 1996; Yonekura et al. 1999). VEGF-B levels are increased in both during development and later birth, closely relating with cardiac angiogenesis (Bellomo et al. 2000). VEGF-C has 30 % similarity amino acid sequence than VEGF-B (Joukov et al. 1998) and also VEGF-C; this expression looks restricted during early development as well as certain pathological conditions like lymphangiogenesis and tumor angiogenesis (Mylona et al. 2007; Xu et al. 2013). The c-FOS-induced growth factor (FIGF) or VEGF-D has 61 % similarity to amino acid sequence of VEGF-C, both binds to their receptors, namely, VEGFR-2 and VEGFR-3 in human endothelial cells (Achen et al. 1998; Baldwin et al. 2001). VEGF-E is encoded with parapoxvirus or *Orf* virus (Lyttle et al. 1994), and this interaction with this receptor induces growth of endothelial (Ogawa et al. 1998) cells.

Hypoxia (hypoxiation) is a pathological form of deprived oxygen supply. Hypoxia-inducible factors (HIFs) mediate transcriptional responses and also involved in tumor progression by changing cellular metabolism (Keith and Simon 2007). Hypoxia induces VEGF and VEGF-R expression via HIF-1α (Bottaro and Liotta 2003; Lee et al. 2012). Heterodimeric transcription factors of HIF composed of alpha and beta subunits have basic helix-loop-helix family of transcription factors. Beta subunit of HIF is constitutively expressed, while alpha subunit is oxygen regulated, tightly. There are three types of HIFs, HIF-1, HIF-2, and HIF-3, each of them is encoded by different genes. In the absence of oxygen, HIF-α protein stabilizes, accumulates, and migrates to the nucleus, where they associate with beta subunits, forming the HIF-1 and HIF-2 heterodimers. HIF binding to specific HRE (hypoxia response elements) in their promoters, these heterodimers may induce the expression of various genes that regulate survival, motility, metabolism, angiogenesis, hematopoiesis, and basement membrane integrity and other functions (Hu et al. 2013).

MMPs (matrix metalloproteinases) are the family of zinc-dependent endopeptidases, capable to break extracellular matrix and fill the spaces between cells, and made of protein and polysaccharides. MMP activity is upregulated in endothelial cells during inflammation, wound healing, and tumor growth. The activity of MMPs is controlled at different levels: (1) by proteolytic activation, since they are secreted as proenzymes; (2) by their respective endogenous inhibitors, PAIs and TIMPs; and (3) their expression is overexpressed by angiogenic factors and cytokines. Migrations of endothelial cells begin to divide and migrate into the nearby tissues. They organize into hollow tubes and evolve gradually into mature network of blood vessels with adhesion factor (integrin α or β) (Mizejewski 1999; Nelson et al. 2000). The blood vessels of new vascular growth are stabilized by angiotensin-1 and angiotensin-2 and their receptor Tie-2 (Maisonpierre et al. 1997; Tournaire et al. 2004).

Angiogenesis is also regulated by angiogenesis inhibitors; many naturally produced proteins can inhibit angiogenesis, including interferon, platelet factor 4, angiostatin, prolactin 16 kDa fragment, thrombospondin, endostatin, and tissue inhibitor of metalloproteinases 1, 2, and 3 (Stack et al. 1999). Angiostatin induces apoptosis of tumor and endothelial cells and also inhibits migration and tubule formation of endothelial cells (Claesson-Welsh et al. 1998; Lucas et al. 1998). Also, angiostatin decreased the expression of bFGF (basic fibroblast growth factor) and VEGF mRNA (Kirsch et al. 1998). The 20 kDa C-terminal fragment of type XVIII collagen endogenous inhibitor, endostatin, inhibits angiogenesis (O'Reilly et al. 1997); this is a component of basement membrane and binds $\alpha5\beta1/\alpha v\beta3$ integrin. Endostatin interfere with pro-angiogenic growth factors like basic fibroblast growth factor (bFGF/FGF-2) and VEGF (Bai et al. 2013; Olsson et al. 2004).

14.3 Angiogenesis as a Therapeutic Target for Cancer

The process of tumor angiogenesis consists of a sequence of interrelated steps which are rate limiting and useful for cancer target therapy. The outcome of the process depends on intrinsic properties of tumor cells and of host response. The efficiency of antiangiogenic compounds varies from tumors. The more specific intervention is the angiogenic pathway. If angiogenic activity of tumor is initiated mainly by one or two factors, then block activity of one factor could be sufficient to inhibit growth of tumor. The tumor angiogenesis is expected that tumor masses of cells increase from the existing local vasculature in response release of angiogenic factors (Folkman et al. 1971). The benefits of targeted therapies are with reduced toxicity which is much better for cancer therapy. Some tumors also contain vasculogenic effects on solid tumors; this mimic is strongly correlated with the advanced stage of diseases and poor outcome (Wu et al. 2008).

14.4 Marine Sponge

Sponges are multicellular organisms of the phylum *Porifera*; they appear to be long lived, with very stable growth rates, and vary enormously between other groups. Although secondary metabolites produced from sponges are comparatively less in concentrations, they help to stop predators and also compete sessile species (Pawlik et al. 2002). Also, sponges undertake symbiotic relationships with other microorganisms like bacteria and fungi; they are likely supplying bioactive compounds (Thomas et al. 2010; Richelle-Maurer et al. 2003). The "gold mine" marine sponges have the diversity of their secondary metabolites which are discovered in the past years. Marine sponges have potential agents against many diseases including antiinflammatory, immunosuppressive, antitumor, antiviral, antimalarial, and antifouling. The development of marine-derived antiangiogenic compounds increased as anticancer drugs are being successfully used for cancer, including trabectedin (PharmaMar's Yondelis®), which denotes the first anticancer agent isolated from marine. There are many components that include bioactive terpenes, nucleosides, sterols, alkaloids, peroxides, fatty acids, cyclic peptides, and halogenated amino acid derivatives (Sipkema et al. 2005). Over 30 % of compounds are isolated from sponges and have natural product patent registrations increased for cancer therapy (Faulkner 2001). Greater than 10 % of screened compounds have cytotoxic activities, identified compounds with their analogues which reached clinical trials. Eribulin mesylate, an analogue of macrocyclic polyether halichondrin B, reached phase I and II cancer trials for metastatic breast cancer (Cigler and Vahdat 2010). So marine-derived compounds inhibit proliferation of cancer by targeting tubulin polymerization and topoisomerases, modulate the antiapoptotic and proapoptotic proteins, and also able to restrain cell migration and invasion of metastasis.

14.5 Marine Sponge-Derived Antiangiogenic Compounds

14.5.1 Alkaloids

Bastadin 6 (Fig. 14.1a) is a brominated tyrosine derivative of macrocyclic tetramer isolated from *Ianthella* sp. Bastadin 6 inhibits human umbilical vein endothelial cells (HUVECs) by VEGF- or bFGF-dependent proliferation ($IC_{50} = 0.052\ \mu M$).

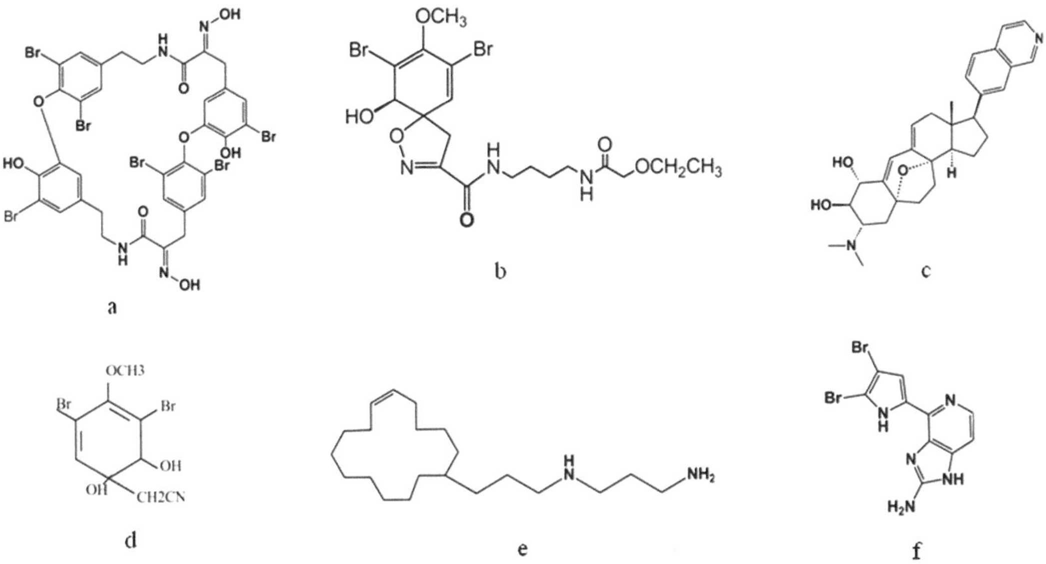

Fig. 14.1 Structure of (**a**) bastadin 6, (**b**) subereamoline A, (**c**) cortistatin A, (**d**) aeroplysinin, (**e**) motuporamine C, (**f**) and ageladine A

Bastadin 6 blocked neovascularization of mice corneal assay and suppressed growth of A431 xenograft tumor. Antiangiogenic properties of bastadin 6 are closely related to selective induction activity of endothelial cells (Aoki et al. 2006a) and also bastadins 6, 9, and 16 isolated from *Ianthella basta* which has in vitro cytotoxic activities in various mouse and human cancer cell lines and have anti-migratory effects (Mathieu et al. 2013). Subereamoline A (Fig. 14.1b) is a brominated alkaloids derived from red sea sponges *Pseudoceratina arabica*. Verongid sponge extracts, *Pseudoceratina arabica* and *Suberea mollis*, have five new alkaloids: ceratinines A–E (2–6) also with moloka'iamine (1), hydroxymo-loka'iamine, (7) and moloka'iakitamide (8). The antimigratory active fraction of the compounds subereamolline A (9), aerothionin (10), and homoaerothionin (11) was isolated from verongid sponge *Suberea mollis*. Subereamolline A is an effectively inhibited migration and invasion of MDA-MB-231 (human breast cancer cells) at nanomolar concentrations (Shaala et al. 2012). Also, four novel steroidal alkaloids, from *Corticium simplex*, a marine sponge-isolated cortistatins A–D, exhibited selective inhibition of endothelial cell proliferation (Aoki et al. 2006b). The unique 9 (10–19)-*abeo*-androstane-type steroidal alkaloids, cortistatins, have oxabicyclo [3.2.1]octene and isoquinoline units. Cortistatin A (Fig. 14.1c) showed antiproliferative and inhibited migration and tubular formation of HUVECs at 2 nM concentration (Aoki et al. 2006b). Cortistatin analogues were synthesized from estrone, and estrone-isoquinoline hybrid (EI-hybrid A) inhibits the proliferation and migration of HUVEC cells (Sato et al. 2008). The alkaloid compound aeroplysinin-1 (Fig. 14.1d) is a brominated tyrosine metabolite extracted from *Aplysina aerophoba*. Aeroplysinin-1 inhibits migration, invasion, and tube formation of endothelial cells and induces apoptosis by caspase-dependent mechanism through activation of caspases 2, 3, 8, and 9 and also cleavage of apoptotic substrates, such as poly (ADP-ribose) polymerase and lamin-A (Rodríguez-Nieto et al. 2002). The chick chorioallantoic membrane assay and in subcutaneous Matrigel implants in mice, aeroplysinin-1, inhibit angiogenesis in both assays by inhibiting endothelial cell migration and capillary tube formation and proliferation (Martínez-Poveda et al. 2012).

The macrocyclic alkaloid motuporamines, family of relatively containing spermidine-like substructure, are isolated from the sponge *Xestospongia exigua* and identified as a family of antiangiogenic alkaloids and anti-invasive type particularly motuporamines A, B, and C (Roskelley et al. 2001). Motuporamines inhibited invasion of many tumors cells including breast and prostate (MDA-231 and PC-3). Motuporamine C (Fig. 14.1e) is a very important motuporamine among these molecules, effectively induces cytoskeletal changes, delays the activation of β1-integrin, plays important role in invasion and adhesion of cancer cells, and also inhibits angiogenesis and migration (Roskelley et al. 2001). It is possible for therapeutic usefulness as an antiangiogenic or antimetastatic drug. Ageladine A (Fig. 14.1f), from *Agelas nakamurai*, inhibited the in vitro migration of endothelial cells, as well as the vascular organization model on type-I collagen gel using mouse vascular progenitor cells. Ageladine A inhibits at micromolar concentration against various matrix metalloproteinases, including MMPs 1, 2, 8, 9, 12, and 13 (Fujita et al. 2003). Some ageladine A analogues were showing more potent MMP-12 inhibitory activity (Ando and Terashima 2007; Shengule et al. 2011).

14.5.2 Triterpene

The globostellatic acid X methyl esters (1–4) have isomarabarican-type triterpenoidal skeleton of four novel compounds, and also other three related compounds (5–7) were isolated from the *Rhabdastrella globostellata*, a marine sponge species which inhibits selectively as an antiproliferative compounds against HUVECs (Aoki et al. 2007). This compound induced apoptosis in endothelial cells and antiproliferative activity but did not correlate to an inhibition of the VEGF-induced phosphorylation of ERK1/ERK2. Some structurally simplified model compounds of globostellatic acid X methyl ester have been synthesized, although their antiproliferative activity was lower than the original compound. The globostellatic acid X methyl ester (4) or compound 4 (Fig. 14.2a) inhibited VEGF-induced migration of HUVECs. HUVECs were migrated with fibronectin-coated chamber by stimulation of VEGF (20 ng/ml). 0.3 and 1 μM concentration of this compound were preincubated with HUVECs for 12 h; the number of the migrated cells decreased, and the preincubation with 1.0 μM concentration inhibited completely the migration of HUVECs. Tubular formation was partly inhibited by bFGF-induced for 12 h pretreatment with 0.1 μM concentration, although 1 or 10 μM concentrations of globostellatic acid (4) acted without inhibition of ERK1/ERK2 activation showed more than 90 % inhibition of HUVECs (Aoki et al. 2007). The structure–activity connection studies for chemically modified analogues and isolated compound proposed that unfunctionalized conjugated penta-ene side chain with 13E-geometry could significantly structure element for efficient and also selective antiproliferative activity against HUVECs (Kotoku et al. 2008). Smenospongine (Fig. 14.2b) was isolated from

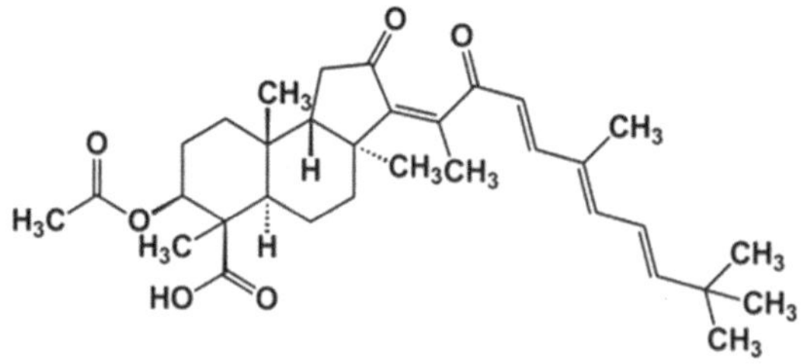

a

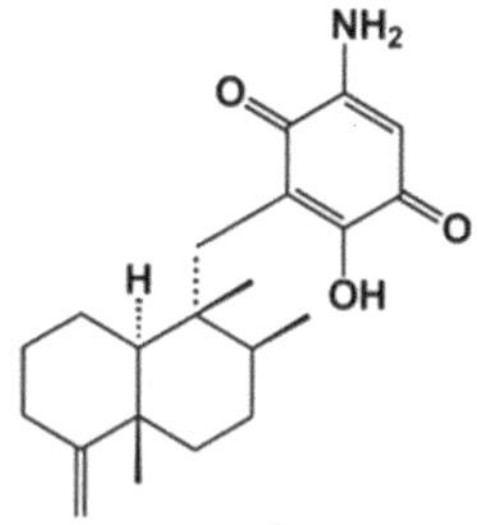

b

Fig. 14.2 Structure of (**a**) globostellatic acid 4 and (**b**) smenospongine

Dactylospongia elegans; this compound is a sesquiterpene aminoquinone from marine sponge. Smenospongine has favorably antiangiogenic activity by inhibiting tube formation, proliferation, and migration of human endothelial cells. Also this compound inhibits various cancer cell growth (Kong et al. 2011), suggested that may be useful for tumors as an anticancer drug candidate.

14.5.3 Macrocyclic Lactone

Spongistatin 1, the most cytotoxic member of the spongistatin family, a macrocyclic lactone polyether isolated from a *Spongia* species, effectively inhibits primary acute leukemic cells of patient (Schyschka et al. 2008). Spongistatin 1 inhibits proliferation of endothelial cells at 100 pM concentration, by affecting cell cycle, nor cytotoxicity or apoptosis induced. The signaling processes involved by polarization and this all process of angiogenesis depend on intracellular translocation of signaling; microtubules facilitate these translocations. PKC activity was not affected by direct inhibition with spongistatin 1; rather the translocation of PKCα from the cytosol to the membrane and related to this translocation was inhibited by spongistatin 1 (Rothmeier et al. 2009) (Fig. 14.3).

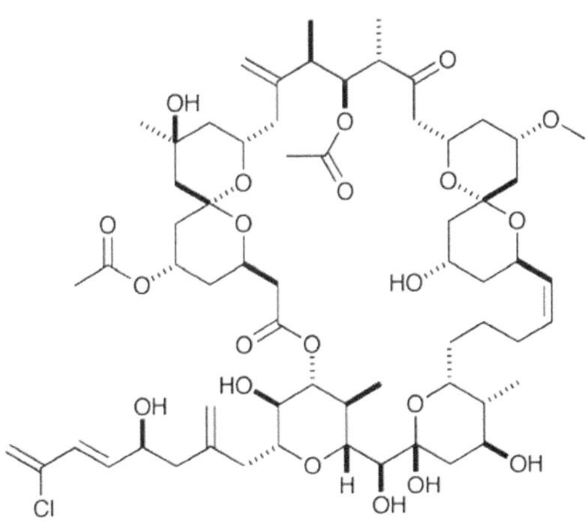

Fig. 14.3 Structure of spongistatin 1

14.5.4 Polyketide

Mycothiazole (Fig. 14.4) is a metabolite isolated from a *Petrosaspongia mycofijiensis* sponge. Nevertheless, the high neurotoxicity of this compound that suppresses selectively at the mitochondrial respiration of complex I (NADH-ubiquinone oxidoreductase) prevents its use as an anticancer drug [82]. In the same way, the potential therapeutical application of the lipophilic 2,5-disubstituted pyrroles HIF inhibitors, from the marine *Mycale* sp., of sponges is limited by the high toxicity expected from their inhibitory activity of mitochondrial respiration [83]. Heterocyclic polyketide mycothiazole belongs to structurally distinct class of mixed polyketide synthase/non-ribosomal peptide synthase (PKS-NRPS) derived from natural products that contain a thiazole ring embedded between two acyclic polyketide chains (Sonnenschein et al. 2006), inhibited hypoxic HIF-1 signaling in cancer cells at nanomolar range, with the suppression of hypoxia-stimulated VEGF secretion by angiogenesis and tumor in vitro (Morgan et al. 2010). For tumor angiogenesis hypoxia is an important stimulus. The mechanisms employed by hypoxic tumor cells to promote angiogenesis through HIF-1-dependent induction of VEGF. The compounds that inhibit VEGFs may be of clinical use for cancer (Ferrara et al. 2007). The antiangiogenic effects of mycothiazole on HUVECs based on the tube formation assay were employed in an in vitro model. Angiogenic factors (e.g., recombinant human VEGF protein) stimulated HUVEC cells to inhibit tube formation (interconnected tubelike structures). Hypoxic exposure (1 % O_2, 16 h) significantly enhanced the angiogenic activity of the T47D cancer cells. At 10 nM concentration of mycothiazole inhibits

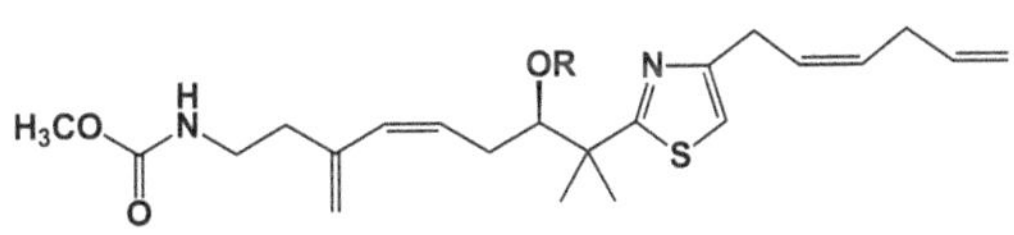

Fig. 14.4 Structure of mycothiazole

hypoxia-induced HIF-1 activation and VEGF induction (Morgan et al. 2010).

14.6 Concluding Remarks

Targeting angiogenesis is an electrifying field of biomedicine. The chemopreventive agents that selectively interfere tumor cells or those acting on the highly specialized biology of endothelial cells during neovascularization merit special consideration. Multiple factors are secreted by tumor cells and their surrounding host stromal cells and modulated by extracellular matrix and also with multiple complementary, overlapping, and independent pathways during angiogenesis. Increasing evidence recommends the use of multitargeted approaches to reach as an effective tumor angiogenesis inhibition. Increasing exploration of marine organism as a source of drug candidates has yielded a list of new natural products able to inhibit angiogenesis in vitro and in vivo. Understanding the basics of these compounds from marine sponges that inhibit angiogenesis may lead to the development of new therapeutic approaches for cancer as antiangiogenic agents.

References

Achen MG, Jeltsch M, Kukk E, Mäkinen T, Vitali A, Wilks AF, Alitalo K, Stacker SA (1998) Vascular endothelial growth factor D (VEGF-D) is a ligand for the tyrosine kinases VEGF receptor 2 (Flk1) and VEGF receptor 3 (Flt4). Proc Natl Acad Sci 95 (2):548–553

Ando N, Terashima S (2007) Synthesis and matrix metalloproteinase (MMP)-12 inhibitory activity of ageladine A and its analogs. Bioorg Med Chem Lett 17(16):4495–4499

Aoki S, Cho S-h, Ono M, Kuwano T, Nakao S, Kuwano M, Nakagawa S, Gao J-Q, Mayumi T, Shibuya M (2006a) Bastadin 6, a spongean brominated tyrosine derivative, inhibits tumor angiogenesis by inducing selective apoptosis to endothelial cells. Anticancer Drugs 17(3):269–278

Aoki S, Watanabe Y, Sanagawa M, Setiawan A, Kotoku N, Kobayashi M (2006b) Cortistatins A, B, C, and D, anti-angiogenic steroidal alkaloids, from the marine sponge Corticium simplex. J Am Chem Soci 128(10):3148–3149

Aoki S, Watanabe Y, Tanabe D, Arai M, Suna H, Miyamoto K, Tsujibo H, Tsujikawa K, Yamamoto H, Kobayashi M (2007) Structure–activity relationship and biological property of cortistatins, anti-angiogenic spongean steroidal alkaloids. Bioorg Med Chem 15(21):6758–6762

Bai Y-j, Huang L-z, Zhou A-y, Zhao M, W-z Y, Li X-x (2013) Antiangiogenesis effects of endostatin in retinal neovascularization. J Ocul Pharmacol Ther 29 (7):619–626

Baldwin ME, Catimel B, Nice EC, Roufail S, Hall NE, Stenvers KL, Karkkainen MJ, Alitalo K, Stacker SA, Achen MG (2001) The specificity of receptor binding by vascular endothelial growth factor-d is different in mouse and man. J Biol Chem 276(22):19166–19171

Bellomo D, Headrick JP, Silins GU, Paterson CA, Thomas PS, Gartside M, Mould A, Cahill MM, Tonks ID, Grimmond SM (2000) Mice lacking the vascular endothelial growth factor-B gene (Vegfb) have smaller hearts, dysfunctional coronary vasculature, and impaired recovery from cardiac ischemia. Circ Res 86(2):e29–e35

Bottaro DP, Liotta LA (2003) Cancer: out of air is not out of action. Nature a-z index 423(6940):593–595

Carmeliet P (2003) Angiogenesis in health and disease. Nat Med 9(6):653–660

Cigler T, Vahdat LT (2010) Eribulin mesylate for the treatment of breast cancer. Expert Opin Pharmacother 11(9):1587–1593

Claesson-Welsh L, Welsh M, Ito N, Anand-Apte B, Soker S, Zetter B, O'Reilly M, Folkman J (1998) Angiostatin induces endothelial cell apoptosis and activation of focal adhesion kinase independently of the integrin-binding motif RGD. Proc Natl Acad Sci 95(10):5579–5583

Coultas L, Chawengsaksophak K, Rossant J (2005) Endothelial cells and VEGF in vascular development. Nature 438(7070):937–945

Cross MJ, Dixelius J, Matsumoto T, Claesson-Welsh L (2003) VEGF-receptor signal transduction. Trends Biochem Sci 28(9):488–494

Faulkner DJ (2001) Marine natural products. Nat Prod Rep 18(1):1R–49R

Ferrara N (1993) Vascular endothelial growth factor. Trends Cardiovasc Med 3(6):244–250

Ferrara N (2002) VEGF and the quest for tumour angiogenesis factors. Nat Rev Cancer 2(10):795–803

Ferrara N (2005) VEGF as a therapeutic target in cancer. Oncology 69(Suppl. 3):11–16

Ferrara N, Mass RD, Campa C, Kim R (2007) Targeting VEGF-A to treat cancer and age-related macular degeneration. Annu Rev Med 58:491–504

Folkman J (1990) What is the evidence that tumors are angiogenesis dependent? J Natl Cancer Inst 82(1):4–7

Folkman J (1995) Angiogenesis in cancer, vascular, rheumatoid and other disease. Nat Med 1(1):27–30

Folkman J, Merler E, Abernathy C, Williams G (1971) Isolation of a tumor factor responsible for angiogenesis. J Exp Med 133(2):275–288

Fujita M, Nakao Y, Matsunaga S, Seiki M, Itoh Y, Yamashita J, van Soest RW, Fusetani N (2003) Ageladine A: an antiangiogenic matrix metalloproteinase inhibitor from the marine sponge *Agelas nakamurai* 1. J Am Chem Soci 125(51):15700–15701

Hu Y, Liu J, Huang H (2013) Recent agents targeting HIF-1α for cancer therapy. J Cell Biochem 114(3): 498–509

Joukov V, Kumar V, Sorsa T, Arighi E, Weich H, Saksela O, Alitalo K (1998) A recombinant mutant vascular endothelial growth factor-C that has lost vascular endothelial growth factor receptor-2 binding, activation, and vascular permeability activities. J Biol Chem 273(12):6599–6602

Keith B, Simon MC (2007) Hypoxia-inducible factors, stem cells, and cancer. Cell 129(3):465–472

Kirsch M, Strasser J, Allende R, Bello L, Zhang J, Black PM (1998) Angiostatin suppresses malignant glioma growth in vivo. Cancer Res 58(20):4654–4659

Klagsbrun M, Moses MA (1999) Molecular angiogenesis. Chem Biol 6(8):217–224

Kong D, Yamori T, Kobayashi M, Duan H (2011) Antiproliferative and antiangiogenic activities of smenospongine, a marine sponge sesquiterpene aminoquinone. Mar Drugs 9(2):154–161

Kotoku N, Tamada N, Hayashi A, Kobayashi M (2008) Synthesis of BC-ring model of globostellatic acid X methyl ester, an anti-angiogenic substance from marine sponge. Bioorg Med Chem Lett 18(12): 3532–3535

Lee SW, Jeong HK, Lee JY, Yang J, Lee EJ, Kim SY, Youn SW, Lee J, Kim WJ, Kim KW (2012) Hypoxic priming of mESCs accelerates vascular lineage differentiation through HIF1-mediated inverse regulation of Oct4 and VEGF. EMBO Mol Med 4(9):924–938

Leung T-W, Xue W-C, Cheung AN, Khoo U-S, Ngan H (2004) Proliferation to apoptosis ratio as a prognostic marker in adenocarcinoma of uterine cervix. Gynecol Oncol 92(3):866–872

Lucas R, Holmgren L, Garcia I, Jimenez B, Mandriota SJ, Borlat F, Sim B, Wu Z, Grau G, Shing Y (1998) Multiple forms of angiostatin induce apoptosis in endothelial cells. Blood 92(12):4730–4741

Lyttle DJ, Fraser KM, Fleming SB, Mercer AA, Robinson AJ (1994) Homologs of vascular endothelial growth factor are encoded by the poxvirus orf virus. J Virol 68(1):84–92

Maisonpierre PC, Suri C, Jones PF, Bartunkova S, Wiegand SJ, Radziejewski C, Compton D, McClain J, Aldrich TH, Papadopoulos N (1997) Angiopoietin-2, a natural antagonist for Tie2 that disrupts in vivo angiogenesis. Science 277(5322): 55–60

Martínez-Poveda B, Rodríguez-Nieto S, García-Caballero M, Medina M-Á, Quesada AR (2012) The antiangiogenic compound aeroplysinin-1 induces apoptosis in endothelial cells by activating the mitochondrial pathway. Mar Drugs 10(9):2033–2046

Mathieu V, Wauthoz N, Lefranc F, Niemann H, Amighi K, Kiss R, Proksch P (2013) Cyclic versus hemi-bastadins. Pleiotropic anti-cancer effects: from apoptosis to anti-angiogenic and anti-migratory effects. Molecules 18(3):3543–3561

Mizejewski GJ (1999) Role of integrins in cancer: survey of expression patterns. In: Proceedings of the Society for Experimental Biology and Medicine. Society for Experimental Biology and Medicine, New York. Royal Society of Medicine, pp 124–138

Mojzis J, Varinska L, Mojzisova G, Kostova I, Mirossay L (2008) Antiangiogenic effects of flavonoids and chalcones. Pharmacol Res 57(4):259–265

Morgan JB, Mahdi F, Liu Y, Coothankandaswamy V, Jekabsons MB, Johnson TA, Sashidhara KV, Crews P, Nagle DG, Zhou Y-D (2010) The marine sponge metabolite mycothiazole: a novel prototype mitochondrial complex I inhibitor. Bioorg Med Chem 18(16):5988–5994

Mylona E, Alexandrou P, Mpakali A, Giannopoulou I, Liapis G, Markaki S, Keramopoulos A, Nakopoulou L (2007) Clinicopathological and prognostic significance of vascular endothelial growth factors (VEGF)-C and-D and VEGF receptor 3 in invasive breast carcinoma. Eur J Surg Oncol (EJSO) 33(3):294–300

Nelson AR, Fingleton B, Rothenberg ML, Matrisian LM (2000) Matrix metalloproteinases: biologic activity and clinical implications. J Clin Oncol 18(5): 1135–1135

Nishida N, Yano H, Nishida T, Kamura T, Kojiro M (2006) Angiogenesis in cancer. Vasc Health Risk Manag 2(3):213–219

O'Reilly MS, Boehm T, Shing Y, Fukai N, Vasios G, Lane WS, Flynn E, Birkhead JR, Olsen BR, Folkman J (1997) Endostatin: an endogenous inhibitor of angiogenesis and tumor growth. Cell 88(2):277–285

Ogawa S, Oku A, Sawano A, Yamaguchi S, Yazaki Y, Shibuya M (1998) A novel type of vascular endothelial growth factor, VEGF-E (NZ-7 VEGF), preferentially utilizes KDR/Flk-1 receptor and carries a potent mitotic activity without heparin-binding domain. J Biol Chem 273(47):31273–31282

Olofsson B, Pajusola K, Kaipainen A, Von Euler G, Joukov V, Saksela O, Orpana A, Pettersson RF, Alitalo K, Eriksson U (1996) Vascular endothelial growth factor B, a novel growth factor for endothelial cells. Proc Natl Acad Sci 93(6):2576–2581

Olsson A-K, Johansson I, Åkerud H, Einarsson B, Christofferson R, Sasaki T, Timpl R, Claesson-Welsh L (2004) The minimal active domain of endostatin is a heparin-binding motif that mediates inhibition of tumor vascularization. Cancer Res 64(24): 9012–9017

Pan L, Baek S, Edmonds PR, Roach M, Wolkov H, Shah S, Pollack A, Hammond ME, Dicker AP (2013) Vascular endothelial growth factor (VEGF) expression in locally advanced prostate cancer: secondary analysis of radiation therapy oncology group (RTOG) 8610. Radiat Oncol 8(1):100

Pawlik JR, McFall G, Zea S (2002) Does the odor from sponges of the genus *Ircinia* protect them from fish predators? J Chem Ecol 28(6):1103–1115

Qi JH, Claesson-Welsh L (2001) VEGF-induced activation of phosphoinositide 3-kinase is dependent on focal adhesion kinase. Exp Cell Res 263(1):173–182

Rafii S, Skobe M (2003) Splitting vessels: keeping lymph apart from blood. Nature Med 9(2):166–168

Richelle-Maurer E, Gomez R, Braekman J-C, Van de Vyver G, Van Soest RW, Devijver C (2003) Primary cultures from the marine sponge Xestospongia muta (Petrosiidae, Haplosclerida). J Biotechnol 100(2): 169–176

Rodríguez-Nieto S, González-Iriarte M, Carmona R, Muñoz-Chápuli R, Medina MA, Quesada AR (2002) Antiangiogenic activity of aeroplysinin-1, a brominated compound isolated from a marine sponge. FASEB J 16(2):261–263

Roskelley CD, Williams DE, McHardy LM, Leong KG, Troussard A, Karsan A, Andersen RJ, Dedhar S, Roberge M (2001) Inhibition of tumor cell invasion and angiogenesis by motuporamines. Cancer Res 61(18):6788–6794

Rothmeier AS, Ischenko I, Joore J, Garczarczyk D, Fürst R, Bruns CJ, Vollmar AM, Zahler S (2009) Investigation of the marine compound spongistatin 1 links the inhibition of PKCα translocation to nonmitotic effects of tubulin antagonism in angiogenesis. FASEB J 23(4):1127–1137

Sato Y, Kamiyama H, Usui T, Saito T, Osada H, Kuwahara S, Kiyota H (2008) Synthesis and antiangiogenic activity of cortistatin analogs. Biosci Biotechnol Biochem 72(11):2992–2997

Schyschka L, Rudy A, Jeremias I, Barth N, Pettit G, Vollmar A (2008) Spongistatin 1: a new chemosensitizing marine compound that degrades XIAP. Leukemia 22(9):1737–1745

Shaala LA, Youssef DT, Sulaiman M, Behery FA, Foudah AI, El Sayed KA (2012) Subereamolline A as a potent breast cancer migration, invasion and proliferation inhibitor and bioactive dibrominated alkaloids from the Red Sea sponge *Pseudoceratina arabica*. Mar Drugs 10(11):2492

Shengule SR, Loa-Kum-Cheung WL, Parish CR, Blairvacq M, Meijer L, Nakao Y, Karuso P (2011) A one-pot synthesis and biological activity of Ageladine A and analogues. J Med Chem 54(7):2492–2503

Sipkema D, Franssen MC, Osinga R, Tramper J, Wijffels RH (2005) Marine sponges as pharmacy. Marine Biotechnol 7(3):142–162

Sonnenschein RN, Johnson TA, Tenney K, Valeriote FA, Crews P (2006) A reassignment of (-)-mycothiazole and the isolation of a related diol. J Nat Prod 69(1): 145–147

Stack MS, Gately S, Bafetti LM, Enghild JJ, Soff GA (1999) Angiostatin inhibits endothelial and melanoma cellular invasion by blocking matrix-enhanced plasminogen activation. Biochem J 340(Pt 1):77

Stalmans I, Ng Y-S, Rohan R, Fruttiger M, Bouché A, Ÿuce A, Fujisawa H, Hermans B, Shani M, Jansen S (2002) Arteriolar and venular patterning in retinas of mice selectively expressing VEGF isoforms. J Clin Invest 109(3):327–336

Thomas T, Rusch D, DeMaere MZ, Yung PY, Lewis M, Halpern A, Heidelberg KB, Egan S, Steinberg PD, Kjelleberg S (2010) Functional genomic signatures of sponge bacteria reveal unique and shared features of symbiosis. ISME J 4(12):1557–1567

Tournaire R, Simon M-P, Le Noble F, Eichmann A, England P, Pouysségur J (2004) A short synthetic peptide inhibits signal transduction, migration and angiogenesis mediated by Tie2 receptor. EMBO Rep 5(3):262–267

Wang TB, Chen ZG, Wei XQ, Wei B, Dong WG (2011) Serum vascular endothelial growth factor C and lymphangiogenesis are associated with the lymph node metastasis and prognosis of patients with colorectal cancer. ANZ J Surg 81(10):694–699

Wu H-C, Huang C-T, Chang D-K (2008) Anti-angiogenic therapeutic drugs for treatment of human cancer. J Cancer Mol 4(2):37–45

Xu H, Zhang T, Man GCW, May KE, Becker CM, Davis TN, Kung AL, Birsner AE, D'Amato RJ, Wong AWY (2013) Vascular endothelial growth factor C is increased in endometrium and promotes endothelial functions, vascular permeability and angiogenesis and growth of endometriosis. Angiogenesis 16(3):541–551

Yonekura H, Sakurai S, Liu X, Migita H, Wang H, Yamagishi S-i, Nomura M, Abedin MJ, Unoki H, Yamamoto Y (1999) Placenta growth factor and vascular endothelial growth factor B and C expression in microvascular endothelial cells and pericytes. J Biol Chem 274(49):35172–35178

Chronicles of Sponge Biomaterials: The Saga in Biomedicine

15

Ira Bhatnagar, Ramjee Pallela, P.V. Bramhachari, and Kranti Kiran Reddy Ealla

Abstract

Marine environment is a prolific source of natural products and biomaterials of utmost importance in disease. Sponges are one of the better-known, diverse, multicellular invertebrates and abundant members of marine benthic communities. They are among the richest known sources of biologically active secondary metabolites of pharmaceutical significance. Researchers have been trying to explore the marine sponges not only for their associated pharmaceutical potential but also for the biomaterials including chitin/chitosan, ceramic, biosilica, and collagen since sponges are an excellent source of biocompatible materials to be used in biomedicine. This chapter covers an overview of sponge biomaterials and their possible applications in biomedicine.

Keywords

Biomaterials • Sponges • Collagen • Biomedicine • Biosilica • Tissue engineering • Drug delivery

15.1 Introduction

Marine environment harbors a wide variety of molecularly diversified range of organisms when compared to terrestrial regions which is mainly attributed to their longer evolutionary history. Sponges (phylum Porifera) are the most basic of the multicelled animals that have existed for 700–800 million years and are important resources for several unique cytotoxic and anti-cancer compounds. Of the roughly 15,000 sponge species, most available in marine environments, only about 1 % of the species

I. Bhatnagar (✉)
Center for Cellular Molecular Biology, Habsiguda, Uppal Road, Hyderabad 500007, Telangana, India
e-mail: ira@ccmb.res.in; ibhatnagar@gmail.com

R. Pallela
IKP Knowledge Park, Genome Valley, Turkapally, Hyderabad 500078, Telangana, India
e-mail: rpallela@gmail.com

P.V. Bramhachari
Department of Biotechnology, Krishna University, Machilipatnam 521 001, Andhra Pradesh, India
e-mail: veerabramha@gmail.com

K.K.R. Ealla
Department of Oral Pathology, MNR Dental College and Hospital, Sangareddy, Telangana 502294, India

R. Pallela, H. Ehrlich (eds.), *Marine Sponges: Chemicobiological and Biomedical Applications*,
DOI 10.1007/978-81-322-2794-6_15

populates in freshwater (Belarbi et al. 2003). To add to this statement, members of phylum Porifera are well known, multicellular, and diverse and characterized by a pronounced plasticity in the determination of cell lineages and are conspicuous and abundant members of marine benthic groups ranging from the euryhaline, to the estuarine, to the intertidal, to the deep sea (Rao et al. 2011). Sponges can filter up to 24,000 l of seawater per kg sponge per day (Vogel 1977), and up to 60 % of their biomass can be made up of microorganisms. They are among the richest known sources of biologically active compounds and have produced more compounds than any other group of marine organisms (Blunt et al. 2012).

Marine biomaterials are an important new developing area of research with noteworthy appliance. Recently, scientists are dedicating considerable consideration to marine-sponge biomaterials for various biomedical implications. Sponges are an excellent source of biocompatible materials to be used in biomedicine and natural products to combat various diseases of mankind. It has been reported earlier that the chitin isolated from arthropods (crabs, shrimps, lobsters, crayfish, king crabs, and insects) as well as mollusks (e.g., squids) occurs in the form of granules, sheets, or powders with no evidence for the presence as 3D scaffolds. The unique chitin-based scaffolds found in sponges may therefore find applications in biomedicine, materials science, and bioengineering (Ehrlich et al. 2010).

Collagen, a fibrous protein kind, is the major constituent of the skin and bone and represents roughly 25 % of the total dry weight of mammals. Since collagen owns a major benefit in being biocompatible, being biodegradable is simply obtainable and extremely versatile; numerous innovations have occurred in the field of collagen-based biomaterials during the past decade. From injectable collagen matrices to bone regeneration scaffolds, production and cross-linking methods have evolved and improved. Collagen is now widely used in both research environments and medical applications (Parenteau-Bareil et al. 2010). The use of collagen-based biomaterials in the field of tissue engineering applications has been intensively growing over the past decades. Collagenous marine sponges were already shown to be successful templates for the formation and support of musculoskeletal tissue in vitro and in vivo (Ehrlich et al. 2010).

Apart from collagen and chitin, other sponge biomaterials of utmost importance are biosilica and hydroxyapatite. This book chapter presents an overview of sponge biomaterials and their specific use in biomedical sciences including tissue engineering.

15.2 Sponge Biomaterials

15.2.1 Sponge Biosilica and Its Biomedical Application

Silica is, in principle, a mechanical brittle material. However, siliceous organisms use silica as a composite material. The measurement of the biogenic silica content of sediments is a chemical estimate of the siliceous microfossil abundance. Briefly, sediments are leached with a weak base, usually Na_2CO_3. The aliquots are then measured for the amount of Si extracted, and a least-squares regression is made on the increase in concentration with time to separate the Si extracted from amorphous Si compounds, e.g., diatoms, sponges, etc., from that of mineral silicates (Conley and Schelske 2001). Biosilicification is an evolutionarily long-standing type of biomineralization both in multicellular and unicellular and organisms, including diatoms, sponges, choanoflagellates, radiolarians, and higher plants (Mann 2001). Marine organisms process about 6.7 gigatonnes of silicon every year to build their silica skeletons (DeMaster et al. 1995). Approx. 10,000 species of sponges are capable to form an enormous variety of biosilica structures which are species-specific and often used as systematic characters for a given species (Fig. 15.1).

Sponges are the oldest Metazoa that use silica as a biomineral to form their inorganic skeleton (Müller 1995). However, it has been reported that only two classes of sponges, the Hexactinellida

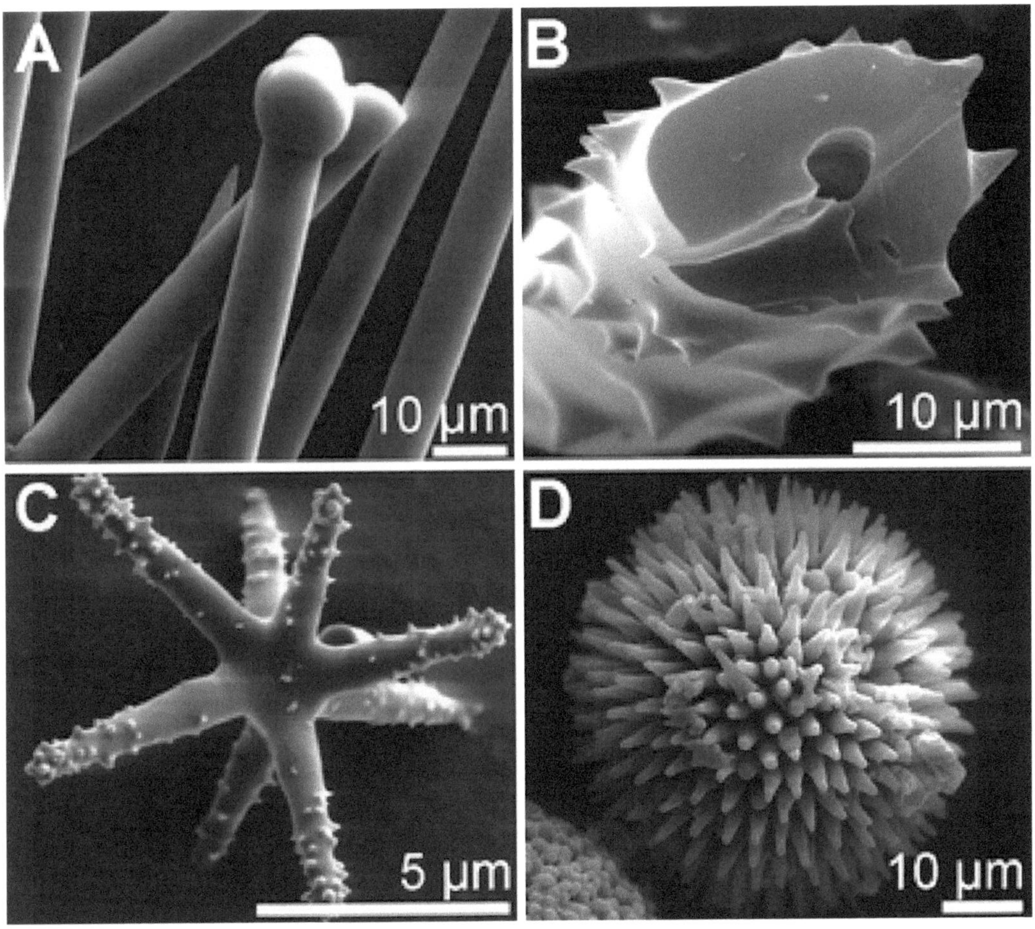

Fig. 15.1 SEM analysis of (**a**) spicules from the marine demosponge *Suberites domuncula*, (**b**) a spicule from the freshwater demosponge *Lubomirskia baicalensis*, (**c**) spheraster and (**d**) sterraster from the marine demosponge *Geodia cydonium* (Reproduced with permission from Schröder et al. 2008)

and Demospongiae, have a silica skeleton, while the evolutionary younger third class of sponges, the Calcarea, has spicules made of calcium carbonate (Simpson 1984). Sponge spicules represent the main components of the biogenic silica (Bavestrello et al. 1996). Silica formation of sponge spicule has attracted much attention in the last decade since it could provide key information to elaborate new hierarchically structured materials and nanodevices.

Following the concept of "nature as model," biosilicification is a model for development of novel fabrication material development for nanotechnology (Morse 1999). Technical production of silica commonly requires higher pressure, extreme temperature, and different pH range. But living organisms are, however, able to form silica under low temperature, ambient condition, and neutral pH (Baeuerlein 2006). SEM and AFM (scanning electron and atomic force microscopic) analyses of the annular substructure of demosponge biosilica spicules reveal that the deposited material is nanoparticulate, with the diameter range at 74 ± 13 nm (Weaver et al. 2003). These biologically produced silicas exhibit a genetically controlled precision of nanoscale architecture that, in many cases, exceeds the capabilities of present-day human engineering (Morse 1999). In the last few years, combined efforts in molecular biology, cell biology, and inorganic and analytical chemistry have allowed the first insight into the molecular mechanisms by which these organisms form an astonishing variety of siliceous structures that cannot be achieved by chemical methods (Schröder et al. 2008). Müller (2003)) has published an extensive review on biomineralization which throws light on the essentialities of sponges in this process (Müller 2003).

It has been reported that bioinspired silicification has become one of the promising ways in the field of bone-repairing biomaterials. Silica is a component of many materials used as scaffolds in tissue engineering of the bone and cartilage, including bioactive glasses and composite materials (Schröder et al. 2007). Gao et al. (2001) suggested that silica-based bioactive glasses modulate the expression of bone morphogenetic protein-2 mRNA in SaOS-2 osteoblasts and a chemical exchange of the silica gel layer forming on the surface of bioactive glasses was thought to be the principal reaction for bone–bioactive glass bonding (Gao et al. 2001). Studies have also demonstrated that biosilica, synthesized by the enzyme silicatein, induces hydroxyapatite formation in osteoblast-like SaOS-2 cells (Schröder et al. 2005). Moving a step further, Wiens et al. (2010) suggested the effect of biosilica on the expressions of osteoprotegerin [OPG] and the receptor activator for NF-kB ligand [RANKL] in the SaOS-2 cell model. They found that during the growth of SaOS-2 cells on biosiliceous matrices, hydroxyapatite formation is induced, while synthesis of cartilaginous proteoglycans and sulfated glycosaminoglycans is downregulated (Fig. 15.2) and proposed the considerable biomedical potential of biosilica for treatment and prophylaxis of osteoporotic disorders (Wiens et al. 2010).

Apart from bone tissue engineering, the potential application of biosilica and silicatein enzymes (biosilica synthesizing enzymes) lies in the surface modification of biomaterials used either as a bone replacement material, carrier for tissue engineering, or coating of metal implants. They may also be applied for the encapsulation of biomolecules such as drugs, hormones, and other bioactive molecules and the controlled release of these compounds. The applications are further extended to the synthesis of nanostructures of amorphous silica for the synthesis of semiconductors, biosensors, and catalysts with unique properties. They may be in the form of nanowires, nanotubes, or nanoparticles

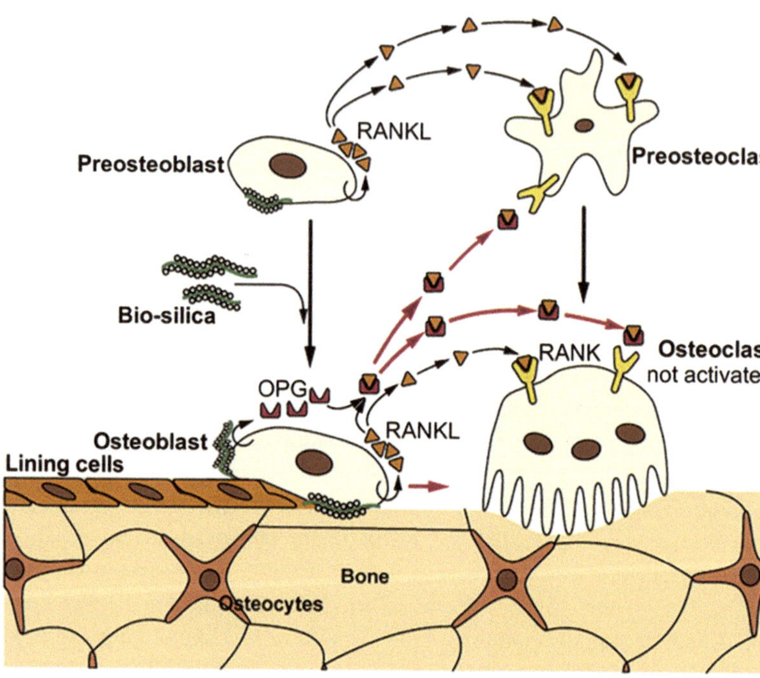

Fig. 15.2 Proposed effects of biosilica on osteoblasts, osteoclasts, and their progenitors. Biosilica enhances expression of OPG in osteoblasts. Osteoblasts have the potential to differentiate to hydroxyapatite-forming osteocytes and lining cells. OPG counteracts various effects of RANKL, a cytokine that induces pre-osteoclast maturation and osteoclast activation (Reproduced with permission from Wiens et al. 2010)

(Schröder et al. 2007). The finding that the silicateins can be harnessed to produce synthetic silsesquioxane (silicone) polymer networks in vitro suggests the possibility of adapting these biomolecular mechanisms to develop new, environmentally benign routes to the synthesis of high-performance materials (Morse 1999).

15.2.2 Sponge Collagens in Biomedicine

Collagens, the major constituent of the extracellular matrix of multicellular animals, are a family of proteins endowed with specific structural functions. Collagens are neutral macromolecules of low immunogenicity used in many pharmaceutical applications, and this property spurred intense research on a sustainable and safe source of these proteins (Pozzolini et al. 2012). Collagen is a natural material of low immunogenicity. Many pharmaceutical applications are known for collagen, e.g., shields, injectable dispersions, sponges, and microparticles (Friess 1998). The use of cattle as the main source for collagen has to be reconsidered because of the risks of BSE (bovine spongiform encephalopathy) and TSE (transmissible spongiform encephalopathy). One alternative is the use of porcine collagen or, much safer, collagen from sea animals, such as marine sponges (Swatschek et al. 2002a).

The molecular characterization of some partial collagen sequences in sponges, the most ancient phylum of the animal kingdom, is rendered essential to provide a clue to understand the evolution of these molecules (Pallela 2013).Very few reports are available on characterization of fibrillar and nonfibrillar marine-sponge collagens at the biophysical, biochemical and biomedical level (Pallela et al. 2011, 2013). Together with chitin, fibrillar collagen (Fig. 15.3) has been described as universal template for biosilicification in Hexactinellides, whose long siliceous spicules are constituted by a composite material with remarkable mechanical and physical properties potentially useful for technology (Ehrlich 2010).

Recent studies sparked a high biotechnological interest of collagenous extracts from marine sponges, as witnessed by a wide pattern of applications in biomedicine (Pallela 2013), food science, and cosmetics (Rao et al. 2011). About a decade ago, Swatschek et al. (2002b) prepared and characterized the marine-sponge collagen spherical microparticles, with a diameter of 120–300 nm and a particle size range from 126 (±2.9) to 2179 (±342) nm, targeting the dermal delivery of all-trans retinol. The surface charge was measured as a function of pH. At pH 2.8 the particles were nearly uncharged; however, at pH 9.0 the particles showed a strong negative charge of about −60 mV. The dermal penetration of retinol into the skin increased significantly by approximately twofold when

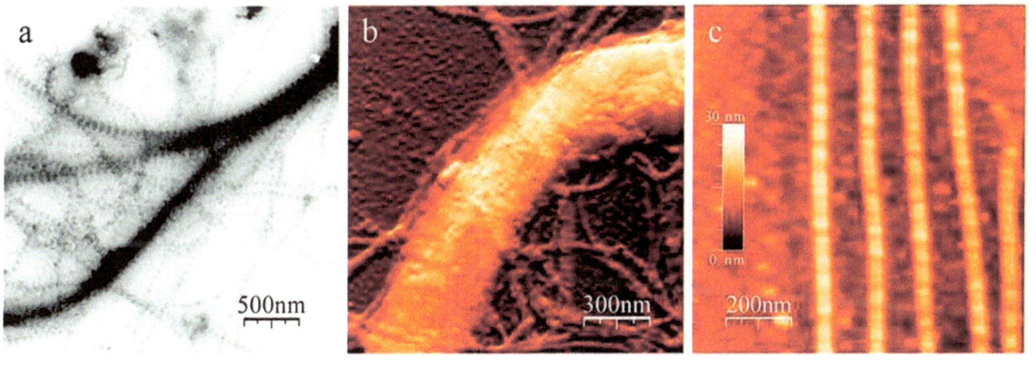

Fig. 15.3 TEM image (**a**) of positively stained *Chondrosia* collagen fibrils shows the directed lateral association of the single fibrils as well as high-resolution AFM imaging (deflection image, **b**). The height image in panel **c** shows the unique banding pattern of five slightly separated *Chondrosia* collagen fibrils (Adopted with permission from Heinemann et al. 2007). Copyright © 2007 American Chemical Society)

collagen microparticles were employed (Swatschek et al. 2002b). Yet another group moved a step forward and fabricated the collagen nanoparticles for the transdermal delivery of 17-β-estradiol-hemihydrate (Nicklas et al. 2009).

Based on a couple of reports on the ultrastructure (Heinemann et al. 2007; Pallela 2013) and biomimetic potential of marine sponges and their utilization for development of new biomaterials (Ehrlich and Worch 2007), research was initiated on the role of marine collagen in drug delivery (Sehgal and Srinivasan 2009) and bone tissue engineering. Green et al. (2003) have firstly reported the fiber skeleton of natural marine sponge and suggested its application for tissue-engineered bone (Green et al. 2003; Granito et al. 2016). The main important parameter to be considered while constructing the artificial bone is osteoconductive, mechanical strength, and osteoinductive (Bruder and Fox 1999). In search for a suitable scaffold matrix which is critical for cell-based bone tissue engineering, the role of marine-sponge collagen was explored and evaluated in vitro. Lin et al. (2011) characterized natural marine sponges as potential bioscaffolds for osteogenesis (Fig. 15.4) and found that alkaline phosphatase expression, a marker of early osteoblast differentiation, was evident at 7 days although expression decreased steadily with long-term culture. Gene expression of osteoblast markers, osteocalcin and osteopontin, was also observed at 7, 14, and 21 days of culture (Lin et al. 2011).

Marine-sponge collagen has also been used in conjunction with chitosan-grafted hydroxyapatite as a scaffold system for bone tissue engineering

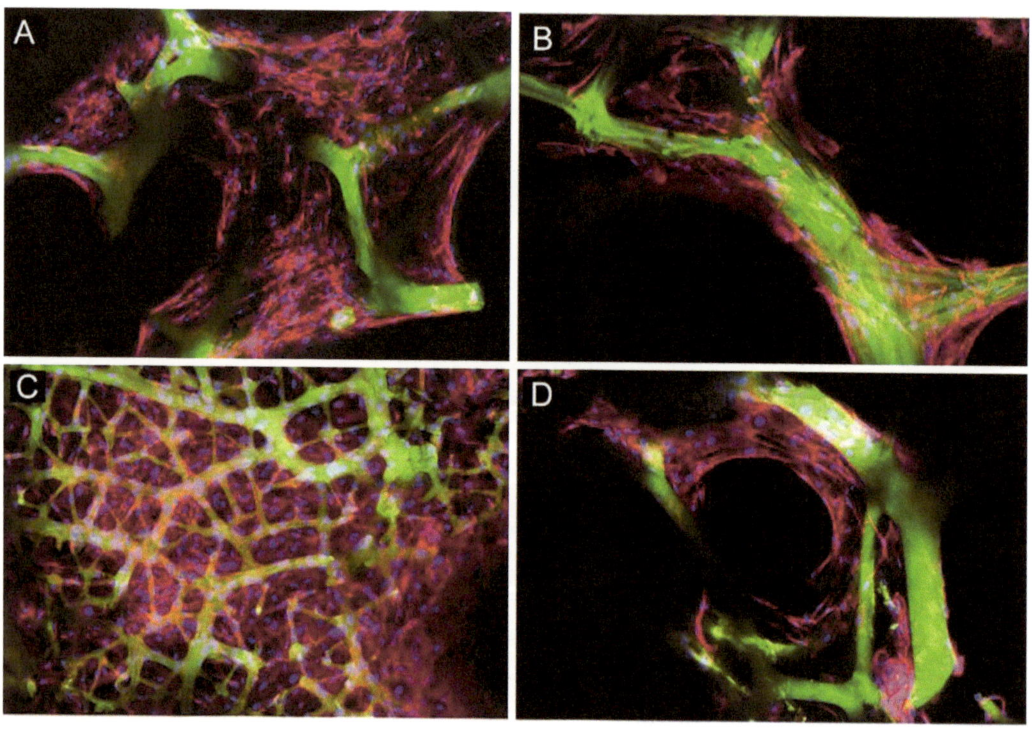

Fig. 15.4 Confocal microscopy showing F-actin fluorescent staining of cells (*red*) and nucleus (*purple*) on sponge skeletons (*green autofluorescence*) after 14-day culture. (**a**) Low magnification, cells form a mat over the ectosomal skeleton. (**b**) High magnification, cells form a thin layer over the fibers of the scaffold. (**c**) Low magnification, cells infiltrate pores of the ectosomal skeletal fibers. (**d**) High magnification showing cellular infiltration of a pore of the sponge skeleton. Scale bar = 20 μm (Adopted with permission from Lin et al. 2011)

(Pallela et al. 2012). Collagen-based biomaterials are of the utmost importance for tissue engineering and regenerative medicine. An elaborative review on collagen-based biomaterials for tissue engineering applications has recently been published (Parenteau-Bareil et al. 2010). Because of its superior biocompatibility and low immunogenicity, sponge collagen is still the protein of choice for biomaterials preparation.

15.2.3 Sponge Ceramics in Biomedicine

Marine sponges are excellent sources of natural materials such as biopolymers; however, they harbor tremendous amount of inorganic materials with significant relevance for tissue replacement and regeneration. Although ample of research work has been done in the application area, data on their properties, sources, as well as isolation, chemical modification, and purification methods are still scarce (Silva et al. 2012).

Most of the marine sponges produce mineralized spicular skeletons that consist either of silica (in Hexactinellida and Demospongiae) (Müller et al. 2006) or of calcium carbonate (almost exclusively calcite, in the Calcarea) (Uriz 2006) in order to maintain morphological rigidity of the body wall and aquiferous system. However, Sethmann and Wörheide (2008) reported that some demosponges, for example, the "Keratosa" (a polyphyletic group), only produce collagenous spongin fibers and completely lack mineralized spicules. Another polyphyletic group of sponges formerly known as "Sclerospongiae," as reported by them, produce a secondary calcareous skeleton of aragonite or Mg-calcite, in addition to their primary spicular skeleton. It has been suggested that the Calcareous sponge spicules from triactines of *Pericharax heteroraphis* may be used as precursor material for bioceramic coatings (Sethmann and Wörheide 2008).

Among the inorganic materials, calcium phosphorous compounds such as hydroxyapatite (HAp), $Ca_{10}(PO_4)_6(OH)_2$, have a special importance in the biomedical field due to its similarities with the mineral constituents of bones (Fig. 15.5). Apart from hydroxyapatite, another inorganic material in abundance is calcium carbonate ($CaCO_3$), which is not as interesting as calcium phosphates from the biomedical application point of view. However, previously published reports suggest that it can be the precursor material for obtaining different calcium phosphates, and consequently, there is a growing interest in finding new sources of this inorganic material. Interestingly, calcium carbonate (aragonite or calcite forms) can be found in many marine organisms. Certain reviews summarizing the aspects dealing with the evolution and physiology of those organisms and looking into their inorganic/organic composition and mechanical properties are available in public academic domains. Silva et al. (2012) have suggested that a good example of a marine species possessing calcium carbonates that might be used as calcium precursors, and thus further exploited in the biomedical field, is sponges (Silva et al. 2012).

Currently, some authors have proposed the use of sponges as three-dimensional biomatrices (Cunningham et al. 2010; Green 2008). The results confirmed that the three-dimensional topography, the porosity, and the surface parameters of these materials influence positively in the cell differentiation. Studies by Green et al. (2003) reported the role of natural marine-sponge fiber skeleton as a biomimetic scaffold for human osteoprogenitor cell attachment, growth, and cell differentiation (Green et al. 2003), whereas Cunningham et al. (2010) reported the development of hydroxyapatite bone substitutes through replication of natural marine sponges. The marine-sponge ceramics are sure going to be the most promising material for tissue engineering and biomedicine.

15.2.4 Sponge Biomaterials in Biomedicine

As stated in earlier sections, sponges have a proven worth in biomedicine. Thus, investigations of the compositions and the microstructures of the sponge skeletons as examples for natural structural biomaterials are of scientific importance (Ehrlich and Worch 2007). By investigating the internal proteinaceous (spongin) skeleton of two demosponges

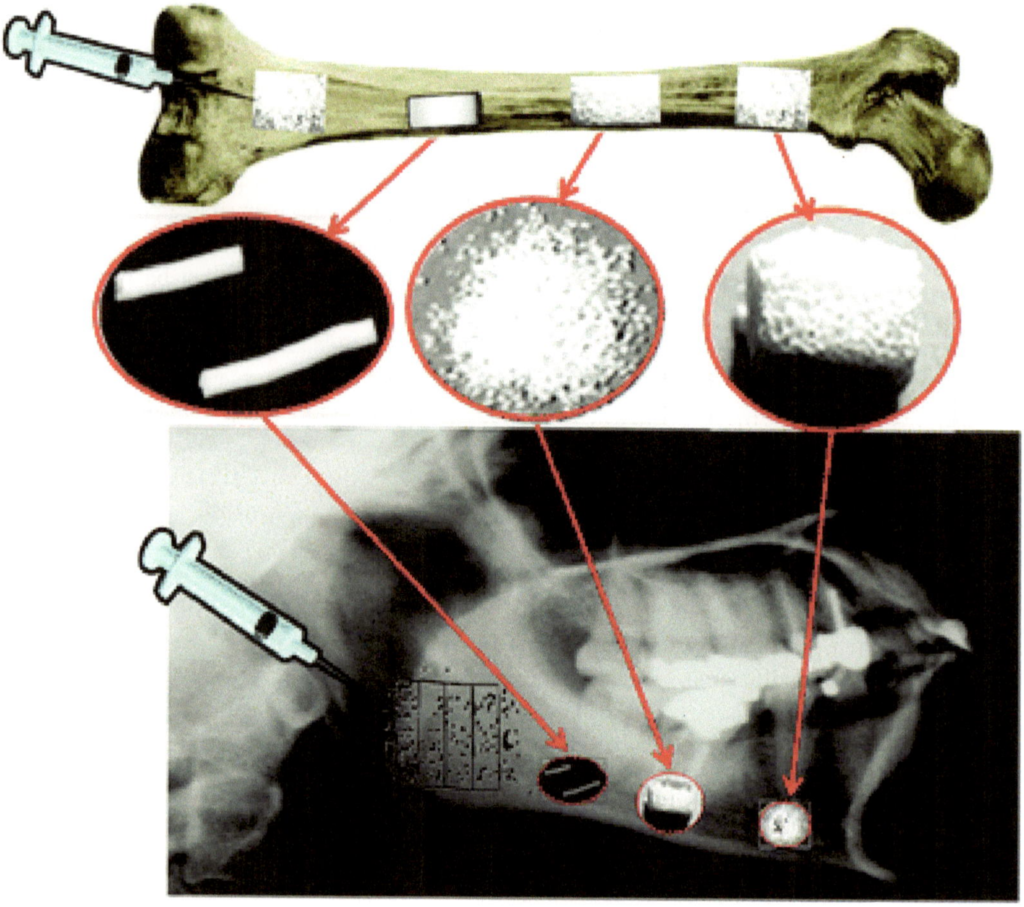

Fig. 15.5 Different uses of calcium phosphate bioceramics (Adopted with permission from Salinas et al. 2013 © Royal Society of Chemistry 2014)

(*Aplysina* sp. and *Verongula gigantea*), Ehrlich et al. (2007b) demonstrated that chitin is a component of the outermost layer (cuticle) of the skeletal fibers of these demosponges. Study consistently revealed that sponge chitin is much closer to the α-chitin known from other animals than to β-chitin (Ehrlich et al. 2007b). The same group then worked on for identifying that some glass sponges (*Farrea occa*) also possess chitin as a component of their skeletons (Ehrlich et al. 2007a). Another report suggested a unique silica–chitin composite biomaterial found in *Euplectella aspergillum* (Ehrlich and Worch 2007).

Irrespective of the source, collagen and chitin are the most investigated materials of biological origin with wide fields of applications in biomedicine because of their unique multifunctional engineering mechanical properties and biocompatibility. A comprehensive understanding of silica–chitin-based sponge skeletons with respect to chemical composition and structure may prove to be a novel model for the biomimetic synthesis of sponge-like three-dimensional chitin-based composites analogous to well-established chitosan–silica hybrid materials with specific optical and bioactive properties for different modern applications (Ehrlich et al. 2008).

From the biomaterial developmental point of view, apart from chitin in the form of biopolymers isolated from sponges, chitin-based fibrous skeletons recently isolated from some keratose sponges are also of great interest for practical use. The practical value of similar sponge skeletons is due to their large internal surface area, which enables considerable liquid absorption to take place by capillary attraction. This phenomenon is the key principle for application of 3D chitinous networks of the sponge origin to carry cells, bacteria, or yeast for biotechnological applications (Ehrlich et al. 2010).

A recent study has suggested that the chitin-based networks of sponge origin are useful for effective uranium adsorption with a higher adsorption capacity than many other chitinous sorbents. It has been proposed under the light of these observations that this renewable material may provide an alternative to more elaborate and expensive chitin-based sorbents. The advantage is that marine-sponge chitin networks are porous, mechanically stable, and flexible (can be cut or pressed into any desired form) and can be easily extracted (Schleuter et al. 2013).

15.3 Sponge Symbiotic Microorganisms in Medicine

Marine environment has been a storehouse of bioactive substances (Bhatnagar and Kim 2010a) with marine sea weeds being the most potential candidates (Kim and Bhatnagar 2011; Bhatnagar et al. 2013) and microorganisms being the second (Kim et al. 2012). Sponges have been long reported as being sessile, benthic, and most primitive filter feeders among metazoans (Amarendra et al. 2013). Microbial communities associated with marine sponges are well-known producers of novel bioactive compounds. Both marine fungi and bacteria have been reported as producers of these metabolites (Bhatnagar and Kim 2010b). Isolation of marine *Micrococcus* sp. from sponge *Tedania ignis* that produces metabolites previously ascribed to the sponge steepened the interest of sponge-associated microbes for the production of biologically active secondary metabolites (Stierle et al. 1988). A major drawback lies in procuring these natural products in limited quantity. Due to limited literature on the culturing techniques and media formulation for culturing marine microbes, pharmaceutical industries have not been able to fully utilize this enormous resource. Sponge-associated microbes are known for their tremendous activities covering a wide range of biological functions (Thomas et al. 2010). Recently, Baker et al. (2009) carried out a study aimed at isolation and identification of a diverse range of fungi from *H. simulans*. They used varieties of media for identification and determination of antimicrobial activities, if any. They isolated 19 different genotypes belonging to *Agaricomycotina*, *Mucoromycotina*, *Saccharomycotina*, and *Pezizomycotina*; some of these isolates show antimicrobial inhibition of *Escherichia coli*, *Bacillus* sp., *Staphylococcus aureus*, and *Candida glabrata* (Baker et al. 2009). The sponge-microbial association is a potential chemical and ecological phenomenon, which provides sustainable resource for developing novel pharmaceutical leads.

Studies suggest that the sponge *Dendrilla nigra* is a rich source of cultivable marine actinomycetes. It is reported that apart from other bioactive symbionts isolated from the surface of this sponge, *Streptomyces* sp. BLT7 isolated from *Dendrilla nigra* obtained from Kanyakumari (southeast coast of India) showed potential antibacterial activity in their extracellular products (Thomas et al. 2010). The species *Ircinia fasciculata*, collected from the shallow coastal habitats of the Mediterranean Sea (~15 m depth), showed antimicrobial activity in the agar media inoculated with different indicator organisms such as *Escherichia coli*, *Staphylococcus lentus*, *Candida* sp., *Bacillus subtilis*, and *Mycobacterium* sp. An antileukemic marine natural product, sorbicillactone A, was isolated from the salt water culture of the fungus *Penicillium chrysogenum* obtained from another Mediterranean specimen of *Ircinia fasciculata* (Bringmann et al. 2005). This alkaloid is also known to have antiviral and neuroprotective properties (Bringmann et al. 2007).

Numerous sponge-derived natural products exhibit promising activities against various diseases, most notably cancer (Hentschel et al. 2012). For instance, *eribulin* (Fig. 15.6), a

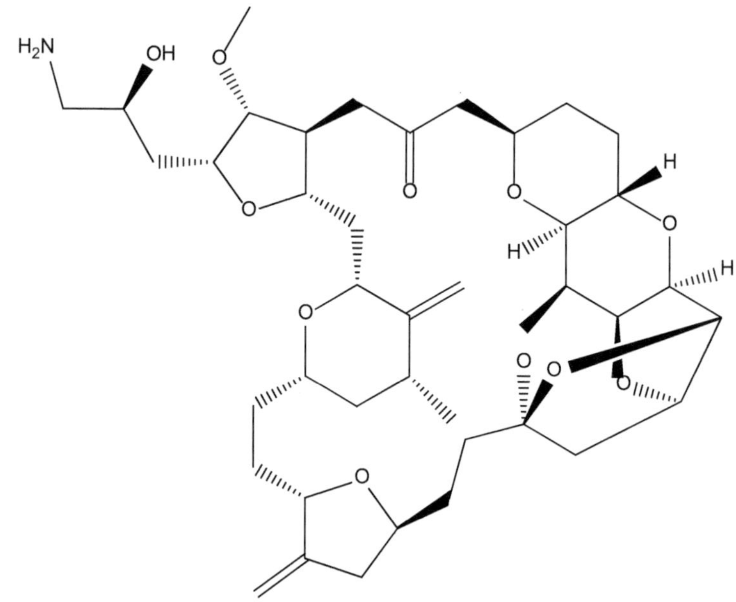

Fig. 15.6 Structure of eribulin

synthetic analogue of halichondrin B from *Halichondria* and *Lissodendoryx* spp. sponges, was recently approved as a drug for the treatment of metastatic breast cancer (Cortes et al. 2010).

15.4 Future Roadmap

Sponges can provide potential drugs against many major worldwide occurring diseases. Regardless of this fact, no sustainable production method has been developed so far to ensure an uninterrupted supply of metabolites and maintenance of ecosystem at the same time. It is important to unveil why, when, where, and how these metabolites are produced in sponges. The foremost task would be to study the factors that influence metabolite production as our knowledge pertaining to the same is minimal. Focus should be on the interrelationship of stress factors and the production of bioactive metabolites. Studies are needed of sponge nutrition and how nutrition can influence growth and metabolite production. It would be handy to develop some fruitful in vitro sponge culture system, sponge cell culture, culture methods for symbionts, or the transfer of production routes into another host (Koopmans et al. 2009). In the near future, sea-based sponge culture seems to be the best production method as the culture of sponge cells more likely primmorphs can become a future source of metabolites (Belarbi et al. 2003).

As Koopmans et al. (2009) suggested, the location of production within the sponge should be identified in order to choose between sponge cell culture and symbiont culture. Molecular biological approaches for delineating the biosynthetic pathway of metabolite production and unveiling the genes involved should be employed (Koopmans et al. 2009). An important query of sponge biomedicine production is that whether methods can be developed for culturing healthy sponge without its endosymbionts. Can endosymbiotic bacteria be cultured in the absence of live sponge tissue and cells, to produce metabolites of interest (Belarbi et al. 2003)?

Not only the research pertaining to the underlying facts of sponge physiology is important; the applicability of sponge biomaterials is equally essential. Venkatesan et al. (2012) utilized marine sponge, *Ircinia fusca*, collagen to fabricate chitosan–amylopectin/hydroxyapatite and chitosan–chondroitin sulfate/hydroxyapatite composite scaffolds for bone tissue engineering (Venkatesan et al. 2012). Ehrlich and Worch

(2007) have clearly foresighted that a comprehensive understanding of collagen- and chitin-based sponge skeletons with respect to chemical composition and structure may prove to be a novel model for biomimetic synthesis of three-dimensional collagen- and chitin-based composites with specific mechanical, optical, and bioactive properties for applications in different modern technologies, including materials science and biomedicine (Ehrlich and Worch 2007). Thus, an inclusive approach of molecular cell biology and biomedical technology is essential for exploring the sponge potential for biomaterial production and its indispensable role in biomedicine.

References

Amarendra V, Santhosh RS, Dhevendaran K (2013) Sponges: a reservoir for microorganism-derived bioactive metabolites. In: Marine microbiology: bioactive compounds and biotechnological applications. Wiley-VCH, Weinheim

Baeuerlein E (2006) Biomineralization: progress in biology, molecular biology and application. Wiley, New york

Baker PW, Kennedy J, Dobson AD, Marchesi JR (2009) Phylogenetic diversity and antimicrobial activities of fungi associated with Haliclona simulans isolated from Irish coastal waters. Mar Biotechnol 11(4): 540–547

Bavestrello G, Cattaneo-Vietti R, Cerrano C, Cerutti S, Sará M (1996) Contribution of sponge spicules to the composition of biogenic silica in the Ligurian Sea. Mar Ecol 17:41–50

Belarbi EH, Contreras Gómez A, Chisti Y, GarcÂa Camacho F, Molina Grima E (2003) Producing drugs from marine sponges. Biotechnol Adv 21(7):585–598

Bhatnagar I, Kim S-K (2010a) Immense essence of excellence: marine microbial bioactive compounds. Mar Drugs 8(10):2673–2701

Bhatnagar I, Kim S-K (2010b) Marine antitumor drugs: status, shortfalls and strategies. Mar Drugs 8(10): 2702–2720

Bhatnagar I, Thomas NV, Kim S-K (2013) Natural flora and anticancer regime: milestones and roadmap. Anti Cancer Agents Med Chem (Formerly Curr Med Chem-Anti-Cancer Agents) 13(6):910–922

Blunt JW, Copp BR, Keyzers RA, Munro MH, Prinsep MR (2012) Marine natural products. Nat Prod Rep 29(2):144–222

Bringmann G, Lang G, Gulder TA, Tsuruta H, Mühlbacher J, Maksimenka K, Steffens S, Schaumann K, Stöhr R, Wiese J (2005) The first sorbicillinoid alkaloids, the antileukemic sorbicillactones A and B, from a sponge-derived *Penicillium chrysogenum* strain. Tetrahedron 61(30):7252–7265

Bringmann G, Gulder TA, Lang G, Schmitt S, Stöhr R, Wiese J, Nagel K, Imhoff JF (2007) Large-scale biotechnological production of the antileukemic marine natural product sorbicillactone A. Mar Drugs 5(2): 23–30

Bruder SP, Fox BS (1999) Tissue engineering of bone: cell based strategies. Clin Orthop Relat Res 367:S68–S83

Conley DJ, Schelske CL (2001) Biogenic silica. In: Tracking environmental change using lake sediments. Springer, Dordrecht, pp 281–293

Cortes J, Vahdat L, Blum JL, Twelves C, Campone M, Roché H, Bachelot T, Awada A, Paridaens R, Goncalves A (2010) Phase II study of the halichondrin B analog eribulin mesylate in patients with locally advanced or metastatic breast cancer previously treated with an anthracycline, a taxane, and capecitabine. J Clin Oncol 28(25):3922–3928

Cunningham E, Dunne N, Walker G, Maggs C, Wilcox R, Buchanan F (2010) Hydroxyapatite bone substitutes developed via replication of natural marine sponges. J Mater Sci Mater Med 21(8):2255–2261

DeMaster DJ, Leynaert A, Queguiner B (1995) The silica balance in the world ocean: a reestimate. Science 268(5209):375–379

Ehrlich H (2010) Chitin and collagen as universal and alternative templates in biomineralization. Int Geol Rev 52(7–8):661–699

Ehrlich H, Worch H (2007) Sponges as natural composites: from biomimetic potential to development of new biomaterials. In: Porifera research: biodiversity, innovation & sustainability. Museu Nacional, Rio de Janeiro

Ehrlich H, Krautter M, Hanke T, Simon P, Knieb C, Heinemann S, Worch H (2007a) First evidence of the presence of chitin in skeletons of marine sponges. Part II. Glass sponges (Hexactinellida: Porifera). J Exp Zool 308(4):473–483

Ehrlich H, Maldonado M, Spindler KD, Eckert C, Hanke T, Born R, Goebel C, Simon P, Heinemann S, Worch H (2007b) First evidence of chitin as a component of the skeletal fibers of marine sponges. Part I. Verongidae (Demospongia: Porifera). J Exp Zool 308(4):347–356

Ehrlich H, Janussen D, Simon P, Bazhenov VV, Shapkin NP, Erler C, Mertig M, Born R, Heinemann S, Hanke T (2008) Nanostructural organization of naturally occurring composites-part II: silica-chitin-based biocomposites. J Nanomater 2008:54

Ehrlich H, Ilan M, Maldonado M, Muricy G, Bavestrello G, Kljajic Z, Carballo J, Schiaparelli S, Ereskovsky A, Schupp P (2010) Three-dimensional chitin-based scaffolds from Verongida sponges (Demospongiae: Porifera). Part I. Isolation and identification of chitin. Int J Biol Macromol 47(2): 132–140

Friess W (1998) Collagen–biomaterial for drug delivery. Eur J Pharm Biopharm 45(2):113–136

Gao T, Aro HT, Ylänen H, Vuorio E (2001) Silica-based bioactive glasses modulate expression of bone morphogenetic protein-2 mRNA in Saos-2 osteoblasts in vitro. Biomaterials 22(12):1475–1483

Granito RN, Custódio MR, Rennó ACM (2016) Natural marine sponges for bone tissue engineering: The state of art and future perspectives. J Biomed Mater Res Part B 2016:00B:000–000

Green DW (2008) Tissue bionics: examples in biomimetic tissue engineering. Biomed Mater 3(3):034010

Green D, Howard D, Yang X, Kelly M, Oreffo R (2003) Natural marine sponge fiber skeleton: a biomimetic scaffold for human osteoprogenitor cell attachment, growth, and differentiation. Tissue Eng 9(6): 1159–1166

Heinemann S, Ehrlich H, Douglas T, Heinemann C, Worch H, Schatton W, Hanke T (2007) Ultrastructural studies on the collagen of the marine sponge Chondrosia reniformis Nardo. Biomacromolecules 8 (11):3452–3457. doi:10.1021/bm700574y

Hentschel U, Piel J, Degnan SM, Taylor MW (2012) Genomic insights into the marine sponge microbiome. Nat Rev Microbiol 10(9):641–654

Kim S-K, Bhatnagar I (2011) Physical, chemical, and biological properties of wonder kelp—Laminaria. Adv Food Nutr Res 64:85–96

Kim S-K, Bhatnagar I, Kang K-H (2012) Development of marine probiotics: prospects and approach. Adv Food Nutr Res 65:353–362

Koopmans M, Martens D, Wijffels RH (2009) Towards commercial production of sponge medicines. Mar Drugs 7(4):787–802

Lin Z, Solomon KL, Zhang X, Pavlos NJ, Abel T, Willers C, Dai K, Xu J, Zheng Q, Zheng M (2011) In vitro evaluation of natural marine sponge collagen as a scaffold for bone tissue engineering. Int J Biol Sci 7(7):968–977

Mann S (2001) Biomineralization. Oxford University Press, Oxford

Morse DE (1999) Silicon biotechnology: harnessing biological silica production to construct new materials. Trends Biotechnol 17(6):230–232. doi: http://dx.doi.org/10.1016/S0167-7799(99)01309-8

Müller WG (1995) Molecular phylogeny of metazoa (animals): monophyletic origin. Naturwissenschaften 82(7):321–329. doi:10.1007/bf01131528

Müller WE (2003) Silicon biomineralization: biology, biochemistry, molecular biology, biotechnology. Springer, Berlin/New York

Müller WE, Belikov SI, Tremel W, Perry CC, Gieskes WW, Boreiko A, Schröder HC (2006) Siliceous spicules in marine demosponges (example *Suberites domuncula*). Micron 37(2):107–120

Nicklas M, Schatton W, Heinemann S, Hanke T, Kreuter J (2009) Preparation and characterization of marine sponge collagen nanoparticles and employment for the transdermal delivery of 17β-estradiol-hemihydrate. Drug Dev Ind Pharm 35(9):1035–1042

Pallela R (2013) Isolation and characterization of native collagens from marine sponges. Lambert Academic Publishing, Saarbrücken

Pallela R, Bojja S, Janapala VR (2011) Biochemical and biophysical characterization of collagens of marine sponge, *Ircinia fusca* (Porifera: Demospongiae: Irciniidae). Int J Biol Macromol 49(1):85–92

Pallela R, Venkatesan J, Janapala VR, Kim S-K (2012) Biophysicochemical evaluation of chitosan-hydroxyapatite-marine sponge collagen composite for bone tissue engineering. J Biomed Mater Res 100(2): 486–495

Pallela R, Venkatesan J, Bhatnagar I, Shim Y-B. and Kim S-K (2013) Application of marine collagen based scaffolds in bone tissue engineering. Marine Biomaterials: characterization, isolation, and applications. CRC Press, Boca Raton

Parenteau-Bareil R, Gauvin R, Berthod F (2010) Collagen-based biomaterials for tissue engineering applications. Materials 3(3):1863–1887

Pepi F, Barone V, Cimino P, Ricci A (2008) Contribution of sponge spicules to the composition of biogenic silica in the ligurian sea. Mar Ecol

Pozzolini M, Bruzzone F, Berilli V, Mussino F, Cerrano C, Benatti U, Giovine M (2012) Molecular characterization of a nonfibrillar collagen from the marine sponge Chondrosia reniformis Nardo 1847 and positive effects of soluble silicates on its expression. Mar Biotechnol 14(3):281–293

Rao JV, Pallela R, Prakash G (2011) Prospects of marine sponge collagen and its applications in cosmetology. In: Marine cosmeceuticals: trends and prospects. CRC Press, Boca Raton, pp 77–103

Salinas AJ, Esbrit P, Vallet-Regi M (2013) A tissue engineering approach based on the use of bioceramics for bone repair. Biomater Sci 1(1):40–51. doi:10.1039/c2bm00071g

Schleuter D, Günther A, Paasch S, Ehrlich H, Kljajić Z, Hanke T, Bernhard G, Brunner E (2013) Chitin-based renewable materials from marine sponges for uranium adsorption. Carbohydr Polym 92(1):712–718

Schröder HC, Boreiko O, Krasko A, Reiber A, Schwertner H, Müller WE (2005) Mineralization of SaOS-2 cells on enzymatically (silicatein) modified bioactive osteoblast-stimulating surfaces. J Biomed Mater Res 75(2):387–392

Schröder HC, Brandt D, Schloßmacher U, Wang X, Tahir MN, Tremel W, Belikov SI, Müller WE (2007) Enzymatic production of biosilica glass using enzymes from sponges: basic aspects and application in nanobiotechnology (material sciences and medicine). Naturwissenschaften 94(5):339–359

Schröder HC, Wang X, Tremel W, Ushijima H, Müller WE (2008) Biofabrication of biosilica-glass by living organisms. Nat Prod Rep 25(3):455–474

Sehgal PK, Srinivasan A (2009) Collagen-coated microparticles in drug delivery. Expert Opin Drug Deliv 6(7):687–695

Sethmann I, Wörheide G (2008) Structure and composition of calcareous sponge spicules: a review and comparison to structurally related biominerals. Micron 39(3):209–228

Silva TH, Alves A, Ferreira B, Oliveira J, Reys L, Ferreira R, Sousa R, Silva S, Mano J, Reis R (2012) Materials of marine origin: a review on polymers and ceramics of biomedical interest. Int Mater Rev 57(5): 276–306

Simpson TL (1984) The cell biology of sponges. Springer, New York

Stierle A, Cardellina Ii J, Singleton F (1988) A marine Micrococcus produces metabolites ascribed to the sponge Tedania ignis. Experientia 44(11–12): 1021–1021

Swatschek D, Schatton W, Kellermann J, Müller WE, Kreuter J (2002a) Marine sponge collagen: isolation, characterization and effects on the skin parameters surface-pH, moisture and sebum. Eur J Pharm Biopharm 53(1):107–113

Swatschek D, Schatton W, Müller WEG, Kreuter J (2002b) Microparticles derived from marine sponge collagen (SCMPs): preparation, characterization and suitability for dermal delivery of all-trans retinol. Eur J Pharm Biopharm 54(2):125–133. doi:http://dx.doi.org/10.1016/S0939-6411(02)00046-2

Thomas TRA, Kavlekar DP, LokaBharathi PA (2010) Marine drugs from sponge-microbe association—a review. Mar Drugs 8(4):1417–1468

Uriz M-J (2006) Mineral skeletogenesis in sponges. Can J Zool 84(2):322–356

Venkatesan J, Pallela R, Bhatnagar I, Kim S-K (2012) Chitosan–amylopectin/hydroxyapatite and chitosan–chondroitin sulphate/hydroxyapatite composite scaffolds for bone tissue engineering. Int J Biol Macromol 51(5):1033–1042

Vogel S (1977) Current-induced flow through living sponges in nature. Proc Natl Acad Sci 74(5): 2069–2071

Weaver JC, Pietrasanta LI, Hedin N, Chmelka BF, Hansma PK, Morse DE (2003) Nanostructural features of demosponge biosilica. J Struct Biol 144(3):271–281. doi: http://dx.doi.org/10.1016/j.jsb.2003.09.031

Wiens M, Wang X, Schröder HC, Kolb U, Schloßmacher U, Ushijima H, Müller WE (2010) The role of biosilica in the osteoprotegerin/RANKL ratio in human osteoblast-like cells. Biomaterials 31(30):7716–7725

16 Biomedical Potential of Marine Sponges

Sushrut Sharma, Renesha Srivastava, Ananya Srivastava, Pawan Kumar Maurya, and Pranjal Chandra

Abstract

Marine sponges, ubiquitously occurring invertebrates, are sources of diverse variety and unique metabolites that indicate their potential in therapeutics and biomedical science. They are the richest sources of pharmacologically active compounds from marine organisms. These bioactive compounds have the potential to become future drugs against important diseases such as cancer and malaria. They also have a range of biomedical applications but are yet to be commercialized to leverage the benefits for the society. In this chapter compounds or metabolites that had been isolated from marine sponges have been defined, followed by a brief account of their characteristics and numerous activities that may have potent impact on biomedical applications as they may emerge out as convincing solutions for numerous significant diseases. The products obtained show antiviral, antimicrobial, and antiprotozoal activity. These compounds can also be used as immuno- and neurosuppressors and can be implemented to an array of medical diagnostics.

Keywords

Bioactive compounds • Marine sponges • Therapeutics • Biomedical applications

S. Sharma • R. Srivastava
College of Professional Studies, Northeastern University, 360 Huntington Ave, Boston 02115, MA, USA

A. Srivastava
Department of Chemistry, Indian Institute of Technology-Delhi, Hauz Khas, New Delhi 110 016, India

P.K. Maurya
Amity Institute of Biotechnology, Amity University Uttar Pradesh, Sector 125, Noida, India

P. Chandra (✉)
Department of Biosciences and Bioengineering, Indian Institute of Technology, Guwahati 781039, Assam, India
e-mail: pranjalmicro13@gmail.com

16.1 Marine Sponges

Marine sponges belonging to the phylum Porifera are multicellular invertebrates that attach to solid substrates in benthic habitats and have existed for the past 700–800 million years resulting in enormous molecular diversity as compared to their terrestrial counterparts. Poriferans are very simple in terms of cellular

R. Pallela, H. Ehrlich (eds.), *Marine Sponges: Chemicobiological and Biomedical Applications*, DOI 10.1007/978-81-322-2794-6_16

organization having an exceptional body design as filter-feeders, devoid of any organs or specialized tissue (Vacelet and Duport 2004). Sponges are classified mainly into three classes: Calcarea, Demospongiae, and Hexactinellida (Thomas et al. 2010). Only 1 % of 15,000 known sponge species dwells in freshwater. Few sponges can grow into huge sponges. However, their growth is slow because of their biomass doubling time that ranges from months to years. Apart from living independently, they can form symbiotic associations with cyanobacteria and microalgae (Belarbi et al. 2003).

Marine sponges are sessile invertebrates that need silica to develop needlelike spicules which are a significant element for their skeletal support. Spicules are synthesized by sclerocytes by deposition of dissolved silica from water on protein filaments. Therefore, lack of silica in their environment can limit their growth (Belarbi et al. 2003). But some species have no spicules, thus lacking physical defenses. These species then produce toxins and secondary metabolites as chemical defenses that repel and deter predators. These secondary metabolites also help in competing for space apart from protection and communication (Laport et al. 2009).

Comparing all metazoan phyla, sponges provide a huge number of bioactive compounds serving as gold mine for diverse unique metabolites that may have high medicinal and therapeutic value. The metabolites produced include antifouling agents, anticancer compounds, cytotoxic, and immunomodulators accompanied with antimicrobial agents (Muller et al. 2000).

16.2 History of Marine Products for Biomedical Applications

During the eighteenth century, Russian and Polish treated patients with lung disease or rheumatism by rubbing dry powder of a freshwater sponge called Badiaga. Later in the twentieth century, it was discovered by Oficjalski (1937) that Badiaga was a mixture of several freshwater sponges and not the sponges but the high iodine concentration in these species was responsible for the effect of Badiaga. In 1951, nucleosides spongouridine and spongothymidine were discovered by Bergmann and Feeney that was the base for synthesis of the first marine-derivative anticancer agent (Sipkema et al. 2005).

Marine flora and fauna serve as the inexhaustible reserve for unique prime agents that are beneficial for biomedical approach, among which phylum Porifera provide the leading figure of bioactive compounds. Various biomedical applications of marine sponges can be seen in Table 16.1.

16.3 Anti-inflammatory Sponge Products

The first sesterterpenoid was isolated from *Luffariella variabilis* by De Silva and Scheuer in 1980 which was found to have antibiotic and analgesic properties, called as manoalide. Later studies by Bennet in 1987 suggested that manoalide showed anti-inflammatory behavior too. The mode of action was preventing enzyme phospholipase A2 from binding to membranes by irreversible inhibition of release of arachidonic acid from membrane phospholipids (Fig. 16.1). Few sponge-derived terpenoids have a different mode of action against inflammatory response by inhibiting a different enzyme (lipoxygenase) (Sipkema et al. 2005).

Five oxygenated hexylitaconic acid derivatives were isolated from sponge-derived fungus *Penicillium* sp. The ester forms (two out of five) of these hexylitaconates showed strong inhibition of pro-inflammatory mediators interleukin-6 and interleukin-1beta (Lin et al. 2011). Contignasterol isolated from marine sponge *Petrosia contignata* inhibits release of histamine from lung cells of guinea pigs and rat mast cells (Newman and Cragg 2004). Barettin, isolated from the marine sponge *Geodia barretti*, was confirmed to have antifouling properties caused by the serotonin-like structure but later it was found that it also has anti-inflammatory activity (Lind et al. 2013).

Table 16.1 Few bioactive compounds from marine sponges

Compound name	Species	Activity	References
Avarol	*Dysidea avara*	Cytotoxic, antitumor	Muller et al. (2000)
Manoalide	*Luffariella variabilis*	Antibiotic	Silva et al. (1980)
Barettin	*Geodia barretti*	Anti-inflammatory	Lind et al. (2013)
Ara-C	*Cryptotethya crypta*	Anticancer	Proksch et al. (2002)
Ara-A	*Cryptotethya crypta*	Antiviral	Proksch et al. (2002)
Contignasterol	*Petrosia contignata*	Anti-inflammatory	Newman and Cragg 2004
Manzamine A	*Haliclona* sp.	Antitumor	Kalifatidis et al. (2013)
Haliclonayclamine E	*Arenosclera brasiliensis*	Antibacterial	Lapor et al. (2009)
Aurantoside K	*Melophlus* sp.	Fungicide	Kumar et al. (2012)
1-methylisoguanosine	*Tedania digitata*	Antiallergic	Quinn et al. (1980)
Agelasine F	*Agelas* sp.	Fungicide	Gordaliza (2009)
Haplosamates A and B	*Xestospongia* sp.	Antiviral (HIV-1)	Qureshi and Faulkner (1999)
Theopalauamide	*Theonella swinhoei*	Antifungal	Thomas et al. (2010)
Agelasine J-L	*Agelas mauritiana*	Antiprotozoal	Gordaliza (2009)
Callyspongynic acid	*Callyspongia truncata*	Antiviral	Nakao et al. (2002)
Simplexides	*Plakortis simplex*	Inhibitor of T-cell proliferation	Costantino et al. (1999)
Pateamine A	*Mycale* sp.	IL-2 inhibitor	Northcote et al. (1991)
(Z)-17-methyl-13-octadecenoic acid	*Polymastia penicillus*	Antiprotozoal	Carballeira et al. (2009)
Kalihinol A	*Acanthella* sp.	Antimalarial	Sipkema et al. 2005)
Eribulin mesylate	*Halichondria okadai*	Anticancer drug	Shin et al. (2013)
Halistanol sulfate C	*Petromica citrina*	Anti-herpes	Guimarães et al. (2013)
Leucascandrolide A	*Leucascandra caveolata*	Fungicide	Sipkema et al. (2005)
S1319	*Dysidea* sp.	Antiasthmatic, uterine relaxation	Suzuki et al. (1999)

16.4 Immunosuppressive and Neurosuppressive Compounds

Northern Australian sponge *Dysidea* sp. has three polyoxygenated sterols that display selective immunosuppression. These sterols inhibit the binding of the neutrophil-attracting cytokine, interleukin 8, by noncytotoxic mechanism. Proliferation of activated T-lymphocytes is also inhibited by a group of immunosuppressive glycolipids called simplexides. Simplexides were isolated from Caribbean sponge *Plakortis simplex* (Sipkema et al. 2005).

Contignasterol isolated from marine sponge *Petrosia contignata* inhibits allergen-induced release of histamine from lung cells of guinea pigs and rat mast cells in vitro and the eosinophil activation in airways of guinea pigs (Sipkema et al. 2005). In 1990, a polyhydroxylated lactone, discodermolide, was reported by the Harbor Branch group as a new immunosuppressive agent. It was isolated from Caribbean sponge *Discodermia dissoluta* (Newman and Cragg 2004). A compound that is a serotonergic receptor antagonist, keramidine from an *Agelas* sp., is a neurosuppressive compound that blocks neural communication which is serotonin mediated (Sipkema et al. 2005).

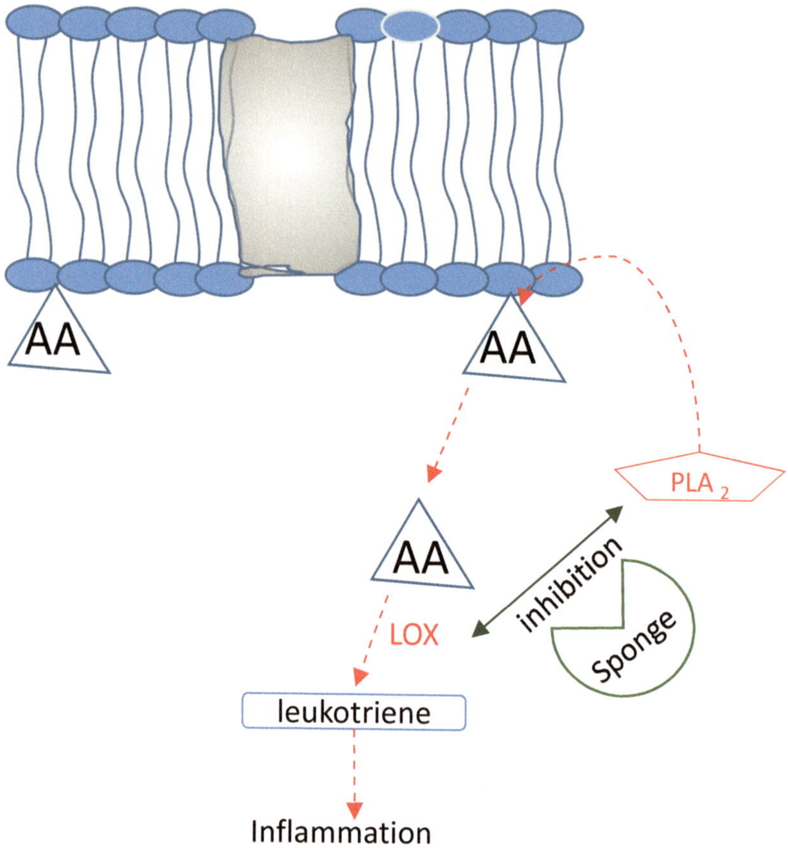

Fig. 16.1 Inflammatory response. Membrane bound arachidonic acid (AA) gets released to free arachidonic acid by action of phospholipase A2 (PLA2). Free arachidonic acid gets converted into leukotriene by lipoxygenase (LOX). Sponge-derived molecules inhibit PLA2 or LOX (Sipkema et al. 2005)

16.5 Antitumor Sponge Products

A range of bioactive compounds is available from around 11 sponge genera. Out of these, genera *Haliclona*, *Petrosia*, and *Discodermia* produce potent anticancer agents. Decritin isolated from *Dercitus* sp. is an aminoacridine alkaloid that shows cytotoxic activity and is also active against B16 melanoma cells and small cell Lewis lung carcinoma (Jha and Zi-Rong 2004).

Manzamine A is an alkaloid isolated from sponges of genera *Haliclona* sp., *Xestospongia* sp., and *Pellina* sp. Manzamine A can be used for treatment of pancreatic ductal adenocarcinoma as it blocks the autophagic pathway in pancreatic cancer cells at level of autophagosome turnover or autophagosome–lysosome fusion (Fig. 16.2). This antitumor activity can be a promising treatment option for tumors resistant to chemotherapy as autophagy is essential for pancreatic tumor growth and chemoresistance (Kallifatidis et al. 2013).

Eribulin mesylate is a simplified derivative of halichondrin B which was isolated from the marine sponge *Halichondria okadai*, is an instance for lately accepted anticancer drug for metastatic breast cancer (Shin et al. 2013). High levels of protein kinase C (PKC) are observed during pathogenesis of arthritis and tumor development. PKC inhibitors isolated from various sponges prevent carcinoma cells to bind to the endothelium. Fucose residues are important for binding of carcinoma cells to receptors in endothelium, so fucosyltransferase inhibitors are good for combating tumor growth. Few meroterpenoid and spiro-sesquiterpene aldehyde compounds, like corallidictyals and akadisulfates, have been sequestered from the extracts of the marine

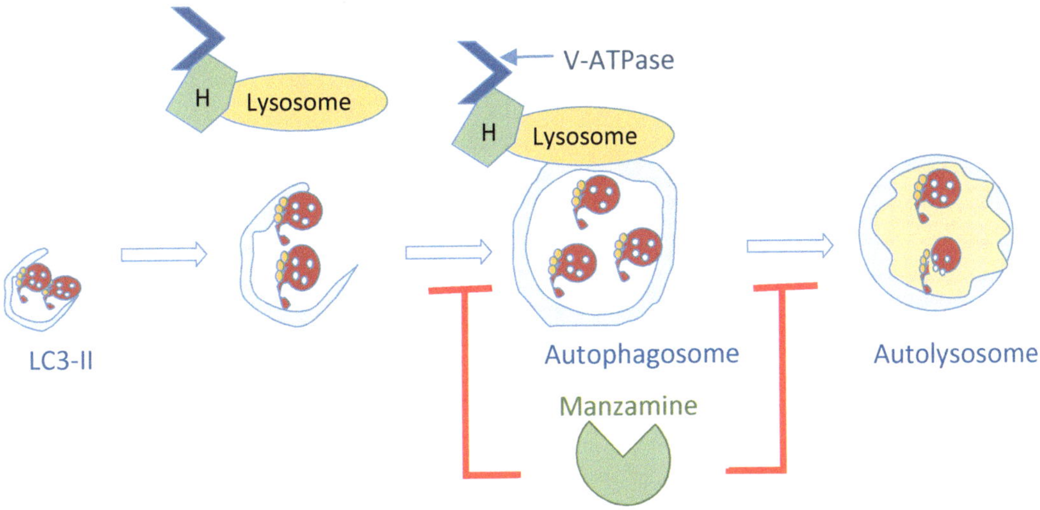

Fig. 16.2 Manzamine A's potential mode of action. Manzamine A inhibits autophagy at the level of autophagosome–lysosome fusion and/or autophagosome turnover by causing a hiatus with proton pump activity of v-ATPases in pancreatic cancer cells (Kallifatidis et al. 2013)

sponge *Siphonodictyon coralliphagum* that inhibits protein kinase C (Pandey et al. 2014).

Most abundant secondary metabolite, triterpenoids, present in marine sponges shows intense antitumor activity. Isomalabaricane-type triterpenoids stelletins A-K are a rare group of triterpenoids that can be isolated from marine sponge species of the genus *Jaspis*, *Stelleta*, and *Rhabdastrella*. Stellettin A showed toxicity to P388 leukemia cells; stelletins A-D have shown selective cytotoxicity toward p21-deficient human colon tumor (HCT-116) cells and stelletins J and K against the A2780 ovarian cancer cell line (Li et al. 2013).

Lectins can serve as specific biomarkers for tumor cell glycoconjugates and can be conjugated with carrier agents to act on malignant cells. A lectin from marine sponge *Cinachyrella apion* denominated as CaL works in human adenocarcinoma cells inducing apoptosis through the instigation (not exclusive) of the mitochondrial intrinsic pathway in both independent and dependent manner. It stimulates mitochondrial membrane permeability releasing protein cytochrome C (Rabelo et al. 2012). Nepheliosyne B and nepheliosyne A, oxygenated polyacetylenes isolated from *Niphates* sp., display temperate cytotoxicity against K562 (chronic myelogenous leukemia), U266 (myeloma), SKM1 (myelodysplastic syndrome), and Kasumi (acute myeloid leukemia) cancer cell lines (Legrave et al. 2013).

16.6 Antiviral Sponge Products

Marine sponges produce antiviral compounds such as haplosamates, papuamides C and D, and avarol (Vacelet and Duport 2004). Demosponge *Dysidea avara* is used for production of avarol by establishment of primmorphs system. Avarol displays strong cytotoxic activity (in vitro) and antitumor activity (in vivo) (Guimarães et al. 2013). It also inhibits progression of HIV infection by increased production of antibodies IgG and IgM, also interfering in post-transcriptional processes. Avarol blocks synthesis of a viral protease obligatory for its proliferation by blocking synthesis of natural UAG suppressor glutamine transfer tRNA (Vacelet and Duport 2004; Laport et al. 2009). Ara-A, a semisynthetic compound (vidarabine), isolated from *Cryptotethya crypta*, is based on arabinosyl nucleosides which inhibit viral DNA production. Mycalamides A and B are isolated from *Mycale* sp. Pure mycalamide A obstructs polio virus type 1 and herpes simplex virus in vitro but mycalamide B showed greater cytotoxicity and antiviral activity (Laport et al. 2009).

High molecular weighted sulfated polysaccharides from sponge *Erylus discophorus* displayed exceptionally potent HIV-1 inhibitory activity which seems to be species-specific. The anti-HIV activity mechanism works by inhibition of virus entry, prevention of HIV adsorption, and fusion with lymphatic cell (Esteves et al. 2011). A new *Pseudomonas* species isolated from the surface of marine sponge *Homophymia* sp. yielded five compounds. 2-undecyl-4-quinolone, which is one of the isolated compounds, showed activity against *Plasmodium falciparum* and HIV-1 (Thomas et al. 2010). The soluble ethyl acetate extract C-29EA that was prepared from sponge *Amphimedon* sp. inhibited HCV replication in a dose-dependent manner regardless of cytotoxicity (Fujimoto et al. 2012).

Few sulfated steroidal compounds, halistanol sulfate and halistanol sulfate C isolated from sponge *Petromica citrina* found on the Brazilian coast, displayed anti-herpes activity by reduction of viral infectivity and by the impairment of levels of ICP27 and glycoprotein D of HSV-1. These compounds also inhibited the entry of virus into cells (Guimarães et al. 2013).

16.7 Antibacterial Activity from Sponge Products

Marine sponges tend to show high antibacterial activity against terrestrial bacteria as compared to marine bacteria producing more than 3,300 antibiotics alone and other bioactive compounds. Psammaplin A, a symmetrical bromotyrosine-derived disulfide compound isolated from *Psammaplysilla* sponge displayed in vitro antibacterial activity against methicillin-resistant *Staphylococcus aureus* (Radic and Bratkovic 2012).

The fungus *Curvularia lunata* yields two antibacterial anthraquinones, lunatin and cytoskyrin A. This fungus is isolated from sponge *Niphates olemda*. The two antibacterial compounds showed activity against *Staphylococcus aureus*, *Escherichia coli*, and *Bacillus subtilis* (Thomas et al. 2010). Manoalide along with its anti-inflammatory properties shows bactericidal activity (Sipkema et al. 2005). The axinellamines B-D isolated from *Axinella* sp., shows bactericidal activity against Gram-negative bacterium *Helicobacter pylori*. Petro amine B isolated from *Oceanapia* species inhibits enzyme aspartyl semi-aldehyde dehydrogenase in *H. pylori* responsible for production of one-fourth of all amino acid residues (Laport et al. 2009).

Three alkaloids, hyrtioerectines D–F isolated from red sea sponge, show variable antimicrobial activity against Gram-positive *Staphylococcus aureus* (ATCC 6538) and Gram-negative *E. coli* (ATCC 8739) (Youssef et al. 2013). Marine sponges from order Haplosclerida such as *Arenosclera brasiliensis* produce alkylpiperidine alkaloids. Four types of alkylpiperidine alkaloids, haliclonacyclamine E and arenosclerins A, B, and C, that are generated after fractionation of crude extract showed inhibitory effects on numerous antibiotic resistant bacteria inclusive of *Staphylococcus aureus* and *Pseudomonas aeruginosa* (Laport et al. 2009). Alkaloids, such as clathrodin and oroidin belonging to structural class pyrrole-2-aminoimidazole of secondary metabolites, can be isolated from sponges of genus *Agelas*, *Hymeniacidon*, *Cymbaxinella*, and *Axinella*. Oroidins have low molecular mass and a simple chemical structure and show antibacterial and anti-biofilm action by disruption of bacterial cell membrane, targeting the response regulator protein BfmR and inhibiting enoyl reductases (Fig. 16.3) (Zidar et al. 2014).

Broad-spectrum antimicrobial activity is observed from 20 % of bacteria connected with marine sponge and corals in a symbiotic association in different coastal areas of the China Sea that includes bacteria isolated from marine sponges *S. tenuis*, *H. rugosa*, and *D. avara*

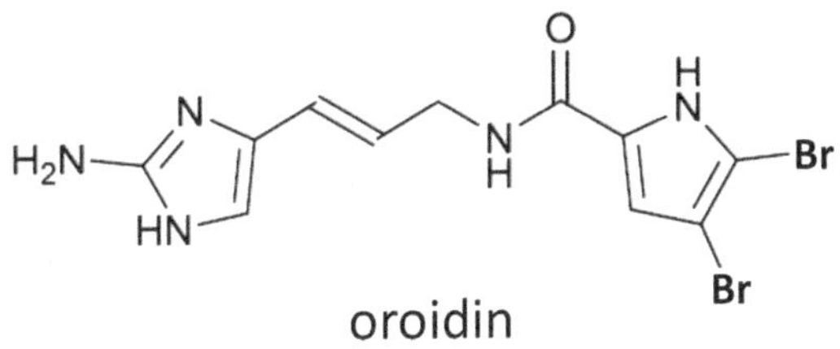

Fig. 16.3 Structure of marine alkaloid, oroidin (Zidar et al. 2014)

(Li 2009). Broad-spectrum antibacterial activity is associated with presence of terpenes, alkaloids, and tannins found in sponges *Biemna tubulosa and Stylissa* spp. (Govinden-Soulange et al. 2014).

Monocyclic diterpenes agelasine F, isolated from Pacific sea sponge *Agelas* sp., inhibits drug-resistant strains of *Mycobacterium tuberculosis* in vitro accompanied with inhibition of tuberculosis H37Rv growth, thus showing anti-tuberculosis activity. Agelasine F, agelasidine A, and ageline B are isolated from *Agelas* sp. that shows antibacterial activity against Gram-positive bacteria *S. aureus* (Gordaliza 2009). Marine sponge species, *Haliclona* sp., *P. citrina*, and *Cinachyrella* sp., displayed antibacterial activity alongside CNS strains that were isolated from bovine mastitis cases. The ethanol extract of *Cinachyrella* sp. is responsible for inhibition of the poly-resistant strain *S. chromogenes* 4606 plus the ethanolic extract of *Haliclona* sp. showed activity against *S. chromogenes* 4476, which is an oxacillin-resistant strain (Laport et al. 2011).

16.8 Antiprotozoal Activity by Marine Sponge Compounds

The most promising antimalarial compound found in various sponges is manzamine. Manzamine shows antimalarial effect with a boosted immune reaction (Sipkema et al. 2005). Agelasine J-L was isolated from orange marine sponge *Agelas mauritiana* found in the Solomon Islands. Agelasine D and its analogs K and L displayed in vitro antiprotozoal activity against *Plasmodium falciparum*, *Leishmania infantum*, *Trypanosoma brucei*, *and Trypanosoma cruzi* (Gordaliza 2009).

Plakortin and dihydroplakortin displayed antimalarial activity when evaluated in vitro. These two simple six-membered endoperoxide compounds were extracted from sponge *P. simplex* (Fattorusso et al. 2002). The unusual fatty acid (Z)-17-methyl-13-octadecenoic acid present in the 0.8 % abundant phospholipid is collected from sponge *Polymastia penicillus*. This unprecedented fatty acid is the iso methyl-branched nonadecanoic acid. This compound shows antiprotozoal activity against *Leishmania donovani*. It inhibits the DNA topoisomerase IB of *Leishmania* and also appears to be cytotoxic to the protozoa. As the DNA topoisomerase IB and the human DNA topoisomerase 1 have considerable differences, therefore there is a possibility that it may restrict the protozoan DNA topoisomerase IB without damaging the mammalian DNA topoisomerase I. (Z)-14-methyl-9-pentadecenoic acid, which hinders the human DNA topoisomerase 1 only at high concentrations (Carballeira et al. 2009).

The crude methylene chloride extract from sponge *N. nolitangere* exhibits antimalarial activity signifying the potential applications of marine sponge products in the health sector (Thompson and Gallimore 2013). Selective in vitro antiprotozoal activity is displayed against *P. falciparum* by a number of terpenoid isocyanates and isonitriles extracted from *Cymbastela hooperi*. Another compound, kalihinol A, isolated from sponge *Acanthella* sp., displays the same antimalarial activity (Sipkema et al. 2005). An array of glycosphingolipids is produced by sponge *Axinyssa djiferi* which is found in African mangrove. They contain a rare compound, namely, galactopyranose. These lipids after downstream processing yield axidjiferosides which displayed convincing antiplasmodial activity (Djiferi 2013).

16.9 Antifungal Compounds

Marine *Streptomyces* sp. DA11 which is known to be associated with sponge *Craniella australiensis* has a protein chitinase that exhibits antifungal activity. On comparing the terrestrial organisms derived chitinase with marine chitinase, the latter with higher pH and salinity tolerance may add to exceptional biomedical applications. The antifungal activity of chitinase was observed against *Aspergillus niger and Candida albicans* (Li 2009).

An antifungal compound theopalauamide, isolated from the marine sponge *Theonella swinhoei* from Palau, encompasses a cytotoxic polyketide, swinholide A, and a bicyclic glycopeptide. Interestingly, another antifungal glycopeptide, namely, theonegramide, was earlier isolated from the same marine sponge, but from the Philippines (Thomas et al. 2010). Agelasine F, agelasidine A, and ageline B are isolated from *Agelas* sp. that show antifungal activity against *Candida albicans* and *C. utilis* (Gordaliza 2009) Roridin A and roridin D are two antifungal trichothecenes produced by fungus *Myrothecium* sp. isolated from the marine sponge *Axinella* sp. that have the potential to inhibit plant pathogen *S. sclerotiorum* (Li 2009).

Fungicides used at present, on relating with the antimicrobials, are less diverse and thus their use is constrained because of their toxic effects on eukaryotic species. Macrolide leucascandrolide A, isolated from sponge *Leucascandra caveolata*, may have different characteristics and a potential of being potent fungicide and being less toxic than the fungicides that are currently used, as they are produced by the eukaryotic organism itself (Sipkema et al. 2005). Aurantoside K, a tetramic acid glycoside isolated from marine sponge *Melophlus* sp., showed promising antifungal activity against a wide spectrum of fungal pathogens (Kumar et al. 2012).

16.10 Applications of Marine Sponges in Diagnostics

Marine sponges are attention-grabbing entities for biomonitoring. The minimal tissue differentiation in these organisms characterizes an advantage of simplified usage in laboratorical procedures (Marques et al. 2007). Several biomarkers such as protein tyrosine phosphatases (PTP's) and PRL-3 associated with tumor metastasis as elevated PRL-3 mRNA level are observed in colon cancer cells, lung cancer cells, and prostate cancer cells along with PRL-3 overexpression in colorectal cancer-derived metastatic lesions (Shin et al. 2013). These biomarkers can be analyzed through use of biosensors.

Biosensors are sensitive, simple, compact, rapid analytical devices entailing a recognition element that has a biological origin such as enzymes, immunoagents, and proteins joined with a transducer which converts the response measured from the biological event into a measurable electrical signal (Costa et al. 2014; Chandra et al. 2013, 2014; Ye Zhu et al. 2012, 2013; Won et al. 2013; Noh et al. 2012). Biosensors are classified on the basis of receptors/recognition elements into biocatalytic and bio-affinity receptors (Davis et al. 1995; Thévenot et al. 2001; Phadke 1992).

The current biosensor technology is based on various nanomaterials, bio-affinity receptors (Fig. 16.4) and chemical modifiers is very much successful. However, the real clinical implications in implantable biosensors suffer due to the toxic or immunogenic response of transducer element. Therefore, it would be interesting to attempt usage of substrates which may

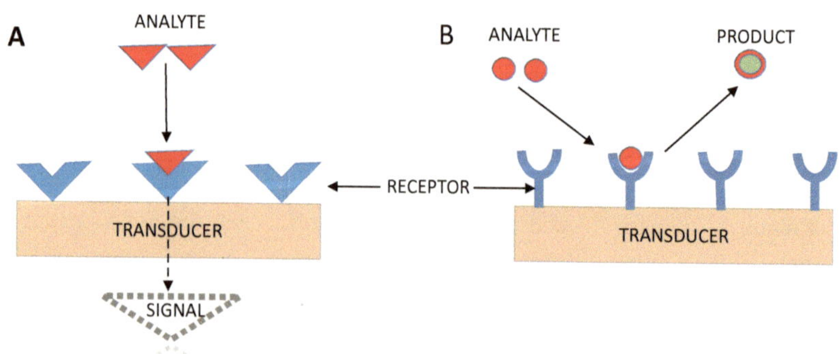

Fig. 16.4 Graphical illustration of (**a**) bio-affinity receptors (**b**) biocatalytic receptors

be more biocompatible, nontoxic, and do not elicit any immunogenic reaction. In this regard the biomaterials based on marine sponges will be interesting substrates. In recent years some biomedical devices have been developed to detect biological molecules such as glucose (Koschwanez and Reichert 2007), phenolic metabolites (Diaconu et al. 2010), cholesterol (Khan et al. 2008), ethanol (Secchi et al. 2013), etc.

For diagnostics and other biomedical applications, long-time reliability and biocompatibility should be the pinnacle (Reina et al. 2014). These implantable devices developed are very effective and were developed using commercially available substrates but biomolecules from marine sponges were rarely used for fabrication of subcutaneous or implantable devices. It is possible that the devices that are developed based on the biomaterials of marine sponges will be more biocompatible, nontoxic, and may not elicit an immunogenic response. Also the devices are composed of some metal ions such as zinc (Khan et al. 2008), platinum (Luz et al. 2013), and titanium (Chamberlain et al. 2011) which can catalyze some biological reaction such as electrocatalysis and often show enhanced biocompatibility.

For proper functioning of a biosensor, it must have an implantable electrode, generally needle shaped, a reference, and working electrode. In case of subcutaneously implanted biosensors, few elements of the sensor can be immunogenic and elicit a response that can affect the device's functioning and performance by forming fibrotic tissue which prevents movement of metabolites (Yoo and Lee 2010). The implantable electrode is coated with bio-inert materials to shun any immunogenic response. Though it may still cause an immunogenic reaction, the sensor may function properly for few days (Wang 2001). Thus for proper functioning and extended life of sensor the implantable material must be coated with natural polymer such as alginate and chitin (Wang et al. 2007). Hydrogels having natural polymers are challenging to alter but they show enhanced biocompatibility and do not elicit inflammatory reaction. Polymers such as collagen and chitin are found in humans and marine sponges, respectively. Chitosan, a deacetylated form of chitin, is nontoxic, biodegradable, and non-immunogenic and can be isolated from marine sponges (Muzzarelli 2000).

Chitin has wide array of importance. It is used as a support in various biosensors, for instance, in implantable glucose sensors; the supporting material used is crystalline beta chitin. Also chitin is coupled with acetylcholine esterase in acetylcholine estimating biosensor to form rapid and biocompatible device (Felse and Panda 1999). The commercial synthesis of chitin is highly expensive and technologically burdensome. A cheap alternative is isolation of chitin and related biomaterials from sponge scaffolds from various marine sponges such as *I. basta*. (Brunner et al. 2009).

16.11 Marine Sponges in Tissue Engineering

Tissue engineering is now developing as a convincing method for wound healing. Scaffold materials of biological origin having extracellular matrix (ECM) assist rebuilding of diverse tissues. Collagen shows high biocompatibility and high biological affinity (Kirk et al. 2013). Using biomaterials for matrix development from marine sponges may assure efficient biocompatibility and may provide scaffold for cell adherence and cell proliferation.

For bone and cartilage regeneration, human osteoblasts derivative of human multipotent stromal cells (hMSC) are convincing candidates because of their osteogenic differentiation ability and their delivery is done by embedding in a platelet lysate to desired location which is moderately challenging. Scaffold established on alginates which are supplemented with biosilica fabricates the hydrogel delivering an active scaffold. Biosilica can be obtained from sponges as this naturally occurring inorganic polymer is involved in spicule formation of Poriferans (Wang et al. 2014). Silicatein assists the development of biosilica lamellae which on immobilization to matrix can synthesize nanoparticulate biosilica from precursors (Müller et al. 2009).

For monitoring and controlling wound infections, chitosan-based wound dressing entailing silver sulfadiazine is efficacious (Phaechamud and Charoenteeraboon 2008). Chitin as observed in fungal cells surges production of fibroblasts and serves as a matrix for fibroblast adherence leading to granulation phase of healing cycle. Chitin and chitosan can be used as wound dressing material because of their high biocompatibility and biodegradability (Felse and Panda 1999). Chitin and chitosan that can be isolated from marine demosponges as required (Brunner et al. 2009) have numerous functions such as scaffold for nerve regeneration (Mullen et al. 2010), wound dressing material (Phaechamud and Charoenteeraboon 2008), and supporting aid for biosensors (Felse and Panda 1999) etc.

16.12 Conclusion

Marine sponges produce copious compounds and metabolites that have a diverse array of biomedical applications with high commercial value. The compounds produced belong to colossal array of antitumor, antiviral, antiprotozoal, antimicrobial, anti-inflammatory, and neurosuppressors that may affect pathogenesis of various diseases by targeting different components in different mechanisms. In spite of their potential uses, they are yet to be leveraged and commercialized as advancement in this field requires multidisciplinary methodology. The biomaterials from sponges can be used to fabricate biocompatible implantable biosensors which till now use synthetic, commercially produced biomaterials. In the future, metabolites obtained from sponges can be a promising solution to numerous diseases and can be utilized in medical diagnostics with other biomedical applications.

Acknowledgments Dr. Pranjal Chandra is acknowledge, Indian Institute of Technology-Guwahati, India for providing adequate facility for completion of this chapter.

References

Axinyssa djiferi (2013) Antimalarial activity of axidjiferosides, new β-galactosylceramides from the African Sponge. Mar Drugs 11(4):1304–1315

Belarbi EH et al (2003) Producing drugs from marine sponges. Elsevier J Biotech Adv 21:585–598

Brunner et al (2009) Chitin-based scaffolds are an integral part of the skeleton of the marine demosponge *Ianthella basta*. J Struct Biol 168(3):539–547

Carballeira NM et al (2009) First total synthesis and antiprotozoal activity of (Z)-17-methyl-13-octadecenoic acid, a new marine fatty acid from the sponge *Polymastia penicillus*. Chem Phys Lipids 161 (1):38–43

Chamberlain L et al (2011) Macrophage inflammatory response to TiO_2 nanotube surfaces. J Biomat Nanobiotech 2(3):293

Chandra P, Noh H-B, Shim Y-B (2013) Cancer cell detection based on the interaction between an anticancer drug and cell membrane components. Chem Commun 49(19):1873–1972

Chandra P, Suman P et al (2014) Prospects and advancements in C-reactive protein detection. World J Methodol 4(1):1–5

Costa C et al (2014) Biosensors for the detection of circulating tumour cells. Sensors 14:4856–4875

Costantino V et al (1999) A new cytotoxic diterpene with the dolabellane skeleton from the marine sponge *Sigmosceptrella quadrilobata*. Eur J Org Chem 227–230

Davis J, Vaughan DH, Cardosi MF (1995) Elements of biosensor construction. Enzyme Microb Technol 17:1030–1035

Diaconu M, Litescu SC, Radu GL (2010) Laccase–MWCNT–chitosan biosensor—a new tool for total polyphenolic content evaluation from in vitro cultivated plants. Sensors Actuators 145:800–806

Esteves AIS et al (2011) Sulfated polysaccharides in marine sponges: extraction methods and anti-HIV activity. Mar Drugs 9:139–153

Fattorusso E et al (2002) Activity against Plasmodium falciparum of cycloperoxide compounds obtained from the sponge *Plakortis simplex*. J Antimicrob Chemother 50:883–888

Felse PA, Panda T (1999) Studies on applications of chitin and its derivatives. Bioprocess Eng 20:505–512

Fujimoto Y, Salam KA, Furuta A, Matsuda Y, Fujita O et al (2012) Inhibition of both protease and helicase activities of hepatitis C virus NS3 by an ethyl acetate extract of marine sponge *Amphimedon* sp. PLoS ONE 7(11):e48685

Gordaliza M (2009) Terpenyl-purines from the sea. Mar Drugs 7:833–849

Govinden-Soulange J et al (2014) Antibacterial properties of marine sponges from Mauritius waters. Trop J Pharm Res 13(2):249–254

Guimarães T d R et al (2013) Anti HSV-1 activity of halistanol sulfate and halistanol sulfate C isolated from Brazilian marine sponge *Petromica citrina* (Demospongiae). Mar Drugs 11:4176–4192

Jha RK, Zi-Rong X (2004) Biomedical compounds from marine organisms. Mar Drugs 2:123–146

Kallifatidis G et al (2013) The marine natural product Manzamine A targets vacuolar ATPases and inhibits autophagy in pancreatic cancer cells. Mar Drugs 11:3500–3516

Khan R, Kaushik A, Solanki PR, Ansari AA, Pandey MK, Malhotra BD (2008) Zinc oxide nanoparticles-chitosan composite film for cholesterol biosensor. Anal Chim Acta 616:207–213

Kirk JF, Ritter G, Finger I, Sankar D, Reddy JD, Talton JD, Nataraj C, Narisawa S, Millan JL, Cobb RR (2013) Mechanical and biocompatible characterization of a cross-linked collagen-hyaluronic acid wound dressing. Biomatter 34:e25633

Koschwanez HE, Reichert WM (2007) In vitro, in vivo and post explantation testing of glucose- detecting biosensors: current methods and recommendations. Biomaterials 28(25):3687–3703

Kumar R et al (2012) Aurantoside K, a new antifungal tetramic acid glycoside from a Fijian marine sponge of the Genus Melophlus. Mar Drugs 10:200–208

Lapor MS et al (2009) Marine sponges: potential sources of new antimicrobial drugs. Curr Pharm Biotechnol 10:86–105

Laport MS et al (2011) Antimicrobial activity of marine sponges against coagulase-negative staphylococci isolated from bovine mastitis. Vet Microbiol 155 (2–4):362–368

Legrave N et al (2013) Nepheliosyne B, a new polyacetylenic acid from the new caledonian marine sponge *Niphates* sp. Mar Drugs 11:2282–2292

Li Z (2009) Advances in marine microbial symbionts in the China Sea and related pharmaceutical metabolites. Mar Drugs 7:113–129

Li JL et al (2011) Oxygenated hexylitaconates from a marine sponge-derived fungus *Penicillium* sp. Chem Pharm Bull 59(1):120–123

Li Y-X et al (2013) Triterpenoids of marine origin as anti-cancer agents. Molecules 18(7):7886–7909

Lind KF et al (2013) Antioxidant and anti-inflammatory activities of barettin. Mar Drugs 11:2655–2666

Luz RAS, Iost RM, Crespilho FN (2013) Nanomaterials for biosensors and implantable biodevices: nanobioelectrochem. Springer, Berlin

Marques D, et al (2007) Biomarkers in marine sponges: acetylcholinesterase in the sponge Cliona celata. In: Porifera research: biodiversity, innovation and sustainability. Museu Nacional Rio de Janeiro, Rio de Janeiro, pp 427–432

Mullen et al. (2010) Binding and release characteristics of insulin-like growth factor-1 from a collagen–glycosaminoglycan scaffold. Tissue Eng Mary Ann Liebert 16(6):1439–1448

Muller et al (2000) Application of cell culture for the production of bioactive compounds from sponges: synthesis of avarol by primmorphs from *Dysidea avara*. J Nat Prod 63(8):1077–1081

Müller WEG, Wang X, Cui F-Z, Jochum KP, Tremel W, Bill J, Schröder HC, Natalio F, Schloßmacher U, Wiens M (2009) Sponge spicules as blueprints for the biofabrication of inorganic–organic composites and biomaterials. Appl Microbiol Biotechnol 83:397–413

Muzzarelli R (2000) Chitin and chitosan hydrogels. In: Handbook of hydrocolloids. CRC Press, Boca Raton, pp 849–888

Nakao Y et al (2002) Callyspongynic acid, a polyacetylenic acid which inhibits α-glucosidase, from the marine sponge *Callyspongia truncate*. J Nat Prod 65(6):922–924

Newman DJ, Cragg GM (2004) Marine natural products and related compounds in clinical and advanced preclinical trials. J Nat Prod 67:1216–1238

Noh H-B, Chandra P, Moon JO, Shim Y-B (2012) In vivo detection of glutathione disulfide and oxidative stress monitoring using a biosensor. Biomaterials 33:2600–2607

Northcote PT et al (1991) Pateamine: a potent cytotoxin from the New Zealand marine sponge, *Mycale* sp. Tetrahedron Lett 32:6411–6414

Pandey et al (2014) A marine sponge associated strain of Bacillus subtilis and other marine bacteria can produce anticholinesterase compounds. Microb Cell Factories 13:24

Phadke RS (1992) Biosensors and enzyme immobilized electrodes. Biosystems 27:203–206

Phaechamud T, Charoenteeraboon J (2008) Antibacterial activity and drug release of chitosan sponge containing Doxycycline hyclate. AAPS PharmSciTech 9(3):829–835

Proksch P et al (2002) Drugs from the seas – current status and microbiological implications. Appl Microbiol Biotechnol 59:125–134

Quinn RJ et al (1980) Isolation and synthesis of 1-methylisoguanosine. a potent pharmacologically active constituent from the marine sponge *Tedania digitata*. Tetrahedron Lett 21:567–568

Qureshi A, Faulkner J (1999) Haplosamates A and B: new steroidal sulfamate esters from two haplosclerid sponges. Tetrahedron 55(28):8323–8330

Rabelo L et al (2012) A lactose-binding lectin from the marine sponge *Cinachyrella apion* (Cal) induces cell death in human cervical adenocarcinoma cells. Mar Drugs 10:727–743

Radic N, Bratkovic T (2012) Future antibiotic agents: turning to nature for inspiration: In book: Antimicrobial agents, Chapter 2: Future antibiotic agents: turning to nature for inspiration: INtech, chapters

Reina G, Tamburri E, Orlanducci S, Gay S, Matassa R, Guglielmotti V, Lavecchia T, Terranova ML, Ross M (2014) Nanocarbon surfaces for biomedicine. Landes Biosci Biomatter 4:e28537

Secchi O, Zinellu M, Spissu Y, Pirisinu M, Bazzu G, Mighel R, Desole MS, O'Neil RD, Serra PA, Rocchitta G (2013) Further in-vitro characterization of an implantable biosensor for ethanol monitoring in the brain. Sensors 13:9522–9535

Shin Y et al (2013) Antimetastatic effect of halichondramide, a trisoxazole macrolide from the marine sponge *Chondrosia corticata*, on human prostate cancer cells via modulation of epithelial-to-mesenchymal transition. Mar Drugs 11:2472–2485

Sipkema D, et al. (2005) Marine sponges in pharmacy. Mar Biotechnol (NY) 7(3):142–162

Suzuki H et al (1999) S1319: A novel ß2-adrenoceptor agonist from a marine sponge *Dysidea* sp. Bioorg Med Chem Lett 9:1361–1364

Thévenot DR, Toth K, Durst RA et al (2001) Electrochemical biosensors: recommended definitions and classification. Biosens Bioelectron 16:121–131

Thomas TRA et al (2010) Marine drugs from Sponge-Microbe Association—a review. Mar Drugs 8:1417–1468

Thompson MN, Gallimore W (2013) Antileishmanial, antimalarial and antimicrobial activity of the Jamaican 'Touch-me-not' sponge *Neofibularia nolitangere*. J Appl Pharm Sci 3(08):080–083

Vacelet J, Duport E (2004) Prey capture and digestion in the carnivorous sponge *Asbestopluma hypogea* (Porifera: Demospongiae). Zoomorphology 123 (4):179–190

Wang J (2001) Glucose biosensors: 40 years of advances and challenges. Electroanalysis 13(12):983–988

Wang X, Wenk E, Hu X, Castro G, Lorenz M, Wang X, Kaplan D (2007) Silk coatings on PLGA and alginate microspheres for protein delivery. Biomaterials 28:4141–4169

Wang X, Schröder HC, Grebenjuk V, Diehl-Seifert B, Mailänder V, Steffen R, Schloßmacher U, Müller WEG (2014) The marine sponge-derived inorganic polymers, biosilica and polyphosphate, as morphogenetically active matrices/scaffolds for the differentiation of human multipotent stromal cells: potential application in 3D printing and distraction osteogenesis. Mar Drugs 12:1131–1147

Won S-Y, Chandra P, Hee TS, Shim Y-B (2013) Simultaneous detection of antibacterial sulfonamides in a microfluidic device with amperometry. Biosens Bioelectron 39:204–209

Yoo E, Lee S (2010) Glucose biosensors: an overview of use in clinical practice. Sensors 10:4558–4576

Youssef DTA et al (2013) Bioactive compounds from the Red Sea marine sponge *Hyrtios species*. Mar Drugs 11:1061–1070

Zhu Y, Chandra P, Song K-M, Ban C, Shim Y-B (2012) Label-free detection of kanamycin based on the aptamer-functionalized conducting polymer/gold nanocomposite. Biosens Bioelectron 36:29–34

Zhu Y, Chandra P, Shim Y-B (2013) Ultrasensitive and selective electrochemical diagnosis of breast cancer based on a hydrazine–Au nanoparticle–aptamer bioconjugate. Anal Chem 85:1058–1064

Zidar N et al (2014) Antimicrobial activity of the marine alkaloids, clathrodin and oroidin, and their synthetic analogues. Mar Drugs 12:940–963

Sponge Biomass for the Development of Biomedical Products and Their Applications

17

Naveen Kumar Mekala, Rama Raju Baadhe, and Sreenivasa Rao Parcha

Abstract

Advancements in the areas of natural biomaterials are not a new branch of science, and it has existed from many decades. The application of natural materials as biomaterials is recently undergoing a revival in the biomedical engineering. The major limitations of natural biomaterials are due to the uncontrolled degradation that can occur following in vivo implantation and a lot of variabilities in molecular structure associated with animal sourcing. The main applications of natural biomaterials as materials in medicine are, namely, tissue engineering, wound management products, and drug delivery systems. In recent times, a significant number of biomaterials with biomedical significance have been discovered from the sponges. This book chapter is going to enlighten few important outcomes and their applications in the field of biomedical sciences.

Keywords

Biomaterials • Extracellular matrix • Natural biopolymers • Chitosan polysaccharide

17.1 Introduction

The marine coral reef is a wealthy environment with a variety of organisms, which could yield a variety of biomaterials. Few of the biomaterials like polysaccharides are already commercialized in food and chemical industries. However, very little is known about marine biomaterial application in tissue engineering and regenerative medicine (Radjasa et al. 2011). The definite evolution of novel chemicals/materials from marine organisms into substances of biomedical importance started from the early 1950s at a very slow pace. The developments in this area happened in the form of several strategic publications. One remarkable illustration was the 1976 article entitled "Drugs from the Sea" (Ruggieri 1976). Also during the 1970s the emphasis on marine

N.K. Mekala
Clinical Research Facility, Center for Cellular and Molecular Biology (CSIR), Hyderabad, India
e-mail: mekalanaveenkumar@gmail.com

R.R. Baadhe • S.R. Parcha (✉)
Department of Biotechnology, National Institute of Technology, Warangal, India
e-mail: parcha@nitw.ac.in

R. Pallela, H. Ehrlich (eds.), *Marine Sponges: Chemicobiological and Biomedical Applications*,
DOI 10.1007/978-81-322-2794-6_17

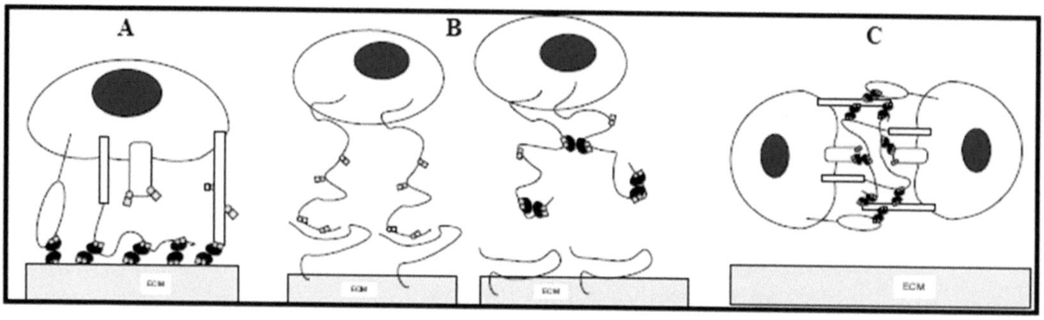

Fig. 17.1 Role of galectin in the system of ECM. (**a**) Galectin may promote or inhibit cell adhesion to extracellular matrix. (**b**) Aggregates of cell surface glycoproteins may concentrate on one side of the cell, resulting in polarization of cell surface glycoproteins. This polarization may impart directionality for migrating lymphocytes or tumor cells. (**c**) Galectin secreted by a tumor cell may favor tumor cell aggregation over adhesion to ECM and promote tumor embolization as suggested by Allen and colleagues. By this mechanism, galectin-1 secreted by tumors may promote tumor cell metastasis (Adapted from Perillo et al. 1998)

invertebrates like sponges, soft corals, and other macroorganisms clearly emerged due to their abundance in near shore and they were easy to collect (Kijjoa and Sawangwong 2004).

When we look for the composition of sponge extracellular matrix (ECM), majority of the polypeptide components include galectin, collagen, fibronectin, etc. These polypeptides are biocompatible and support the cell adhesion to ECM through integrins (Schütze et al. 2001). Here in this book chapter, we discuss the above polypeptides and their significance in the biomedical studies.

17.2 Galectin

Galectin is a family of proteins distinct by their binding efficacy with β-galactoside sugars, like N-acetyllactosamine (Galβ1-3GlcNAc or Galβ1-4GlcNAc), which can bind to other proteins by either O-linked or N-linked glycosylation (Barondes et al. 1994; Iacobini et al. 2003; Leffler et al. 2002). Members of these soluble galectin proteins were discovered in various amphibians, fish, mammals, sponges, and some fungi (Liu and Rabinovich 2010). These galectins are a large family of proteins with relatively broad functionalities including mediation of cell-matrix adhesion, cell-cell interactions, and transmembrane signaling.

Galectins due to their variations can both inhibit and promote integrin-mediated adhesion. To improve integrin-mediated adhesion, they cross-link between two glycans of different cells, which brings the cells closer, and mutually so integrin binding occurs (Perillo et al. 1998; Yang et al. 2008). At the same time these galectins can also hinder intercellular adhesion by binding to two glycans on the same cell, which chunk the integrin-binding site. Galectin-8 is specific for the glycans bound to integrin and has a direct role in adhesion as well as activating integrin-specific signaling cascades (Zick et al. 2002) (Fig. 17.1).

17.3 Collagen

Collagen is the most abundant protein in the connective tissue of most vertebrates and constitutes about 30 % of total proteins. Collagen is the most widely used natural biomaterial in the last 20 years (In Jeong et al. 2007). Other than various well-studied collagen sources, collagen from marine origin is gaining much importance toward development of various cosmetic, pharmaceutical, biomaterials, etc. Marine organisms like sponges, octopus, fish, and jellyfish are best available sources for collagen (Song et al. 2006). Cell walls of sponges belonging to Demospongiae are typically made up of collagen/spongin. The silica- and calcium-abundant marine environments make these sponges biocompatible and make them suitable for various biomedical applications. Also various mollusks contain unique muscular arrangement made up of different collagen types and collagen-like proteins. On the other hand various echinoderms also possess collagen fibrils to make their functional tissues (Szulgit 2007; Pallela et al. 2011).

As discussed above collagen is the dominant constituent of the skeletal matrix of the sea sponges, and it is the chief protein constituent of the extracellular matrix of human bone. The use of collagen-made sponges as three-dimensional (3D) scaffolds in tissue engineering has a number of advantages (Exposito et al. 2002; Datta et al. 2005). Since collagen is a "native" constituent, it is well recognized as a cell-matrix adhesion molecule to support cell adhesion. Furthermore, the netted orientation of collagen fibers within the sponge skeleton is structurally similar to the lattice work of fibers in human trabecular bone (Blumbach et al. 1998). This study demonstrated that the collagen fibers of the marine sponge skeleton indeed provide a suitable framework for the attachment, migration, and proliferation of osteoblasts. The aggregation of osteoblastic cells on spongin fibers may be attributed by the collagenous composition of the sponge fibers together with the presentation of matrix moieties at the skeleton surface (Green et al. 2003; Lin et al. 2011).

In addition, the chemical properties of collagen are also beneficial to its use in tissue engineering scaffolds (Fromont 2003). The collagen skeleton is also analogous to the connective tissue of more complex life forms and is analogous to collagen type XIII; therefore, collagen provides a natural setting for cellular attachment and aggregation (Fromont et al. 2006). Lists of few marine species with better collagen yields were listed in Table 17.1. These collagens are widely used in the industry but less for research and clinical usage. Various collagen sources are worth investigating considering that properties of collagen vary from one animal source to another.

The application of collagen-based 3D biomaterials, prepared either from acellular matrix or extracted pure collagen, has a vast range of applications both in vitro and in vivo (Noah et al. 2002). Biomedical engineers use these collagen scaffolds to study cell proliferation, migration, as well as differentiation and phenotype expression. Additionally, primary studies on cell behavior in complex environments depend on the ability of cells to grow in vitro in a 3D tissue scaffold. Hydrogels from collagen sources are also pretty convenient scaffolds when the access to cell membrane is desirable, for example, in electrophysiological protocols (Ma et al. 2004; Parenteau-Bareil et al. 2010). Other collagen-based scaffolds are used in nervous system to visualize motor neuron myelinization by Schwann cells (Gingras et al. 2008). Currently, 3D collagen scaffolds were also widely used for cancer studies. In this way, the persistent quality of cancer cells and interaction between cancer cells and with the healthy

Table 17.1 Marine sponges explored for collagen molecules

Name of the animal	Classification	References
Spongia species	Demospongiae (Porifera)	Green et al. (2003)
Glass sponges	Hexactinellida (Porifera)	Ehrlich and Worch (2007)
Hyalonema sieboldi	Hyalonematidae (Porifera)	
Monorhaphis chuni	Monorhaphididae (Porifera)	Heinemann et al. (2007)
Farrea occa	Farreidae (Porifera)	
Euplectella aspera	Euplectellidae (Porifera)	Walter et al. (2007)
Euplectella aspergillum		
Several species of *Ircinia*	Demospongiae (Porifera)	Pallela et al. (2012)

Table 17.2 Marine sponges explored for fibronectin and fibronectin-like peptides

Name of sponge	Family	Reference
Ephydatia fluviatilis	Spongillidae	Labat-Robert et al. (1981)
Geodia cydonium	Geodiidae	Conrad et al. (1982)
Tethya aurantia	Tethyidae	Labat-Robert et al. (1981)
Halisarca dujardini	Halisarcidae	Ereskovsky (2010)
Ircinia oros	Irciniidae	Ereskovsky (2010)
Hydra vulgaris	Hydridae	Sarras Jr et al. (1991)

tissues in a 3D environment can be analyzed (Shanmugasundaram et al. 2001). Collagen scaffold can also be used as a 3D environment to test anticancer drugs, in vitro. As per immunology perspective, in vitro experiments can also be done to evaluate 3D T cell migration studies. Moreover, collagen-based biomaterials could serve as anchorage material to cultivate organs ex vivo or as 3D models for diseases like osteoarthritis (Stachowiak and Irvine 2008; Spencer et al. 2008).

Collagen-based implants are also necessary when osteochondral defects reach critical illness or when autografts have to be evaded due to practical or pathological reasons (Ma 2008; Athanasiou et al. 2009). In the above cases hardness of bone tissue engineering scaffolds depends on the mineralization of collagen scaffolds with calcium phosphate and/or on cross-linking with other substances like hydroxyapatite and TCP (Du et al. 2000). During cartilage regeneration, collagen biomaterials are made to be more flexible and ideally built with type II collagen in contrast to the majority of other collagen-based scaffolds, which are produced using type I collagen (Liao et al. 2009). However, some studies exhibit that small amounts of autologous chondrocytes can grow in dynamic culture on type I or II collagen scaffolds without any notable differences.

17.4 Fibronectin

Fibronectin (FN) is a high molecular weight glycoprotein that consists of three types of repeating amino acid units, namely, type I, type II, and type III repeats. FN is the most extensively utilized ECM protein, with existing data about its structure and functions in vertebrates (Paz et al. 2002). This large dimeric glycoprotein is made up of two nearly identical polypeptide subunits (220–250 kDa) connected by disulfide chains. Heterogeneity among FN subunits isolated from different sources arises, at least in part, from alternative splicing of the primary FN transcript. The FN molecule contains several domains with different binding activities specific for collagen, heparin, as well as a cell domain (DeSimone et al. 1985; Matranga et al. 1995). FN exists in soluble form in body fluids, in insoluble form in ECM, and in basal membranes.

Fibronectin-like peptides are predominantly found in invertebrates like sponges, insects, mollusks, and in sea urchins. It is practically proven that these fibronectin-like polypeptides play a role in supporting the cell adherence and cell spreading over the biomedical devices (Li et al. 2006). In general in marine sponges, fibronectin does not show fibrillar structure but show more diffused patterns when compared to vertebrates. The list of various sponges with FN contents was listed in Table 17.2.

Other than peptide biomaterials, sponges are very good sources for polysaccharides. These polysaccharides were natural, inexpensive, and easily biodegradable. Most of these polysaccharides can be utilized as materials in medicine including tissue engineering, drug delivery systems, hemostatic devices, and wound management.

17.5 Chitin

Chitin is a nitrogen-enclosed polysaccharide, chemically resemble to cellulose, and it is not soluble in most of the solvents due to its specific structure which is based on hydrogen bonding

Table 17.3 Marine sponges explored for chitin and chitosan polysaccharide

Name of sponge	Family	References
Aplysina fistularis	Aplysinidae	Ehrlich et al. (2010)
Verongula gigantea	Aplysinidae	Schleuter et al. (2013)
Ianthella basta	Ianthellidae	Ehrlich et al. (2007)
Aplysina bathyphila	Aplysinidae	Ehrlich et al. (2010)
Aiolochroia crassa	Aplysinidae	Attaway and Zaborsky (1993)

among acetamide groups, hydroxyl groups, and carbonyl groups (Aribo 2012; Khor 2005). Controlled deacetylation is used to produce chitosan which is a derivative with approximately 50 % free amine. Chitin is well known as the second most abundant natural polysaccharide after cellulose, and there are three possible sources of chitin as raw materials including traditional shellfish sources like shrimps and crabs; fungal mycelia from bioreactor processes, a classic example is mushrooms; and production from monomeric/dimeric units using chemical and enzymatic strategies (Dutta et al. 2004; Naznin 2005).

Recently, for large-scale chitin production, studies have shifted from the traditional to exciting and novel sources like marine sponges (Ehrlich et al. 2010). Most commonly in sponges, chitin and collagen serve as 3D supporting matrix with crystalline or amorphous inorganic deposits. In case of marine sponges, chitin distribution is found abundant at ectodermal origin and helps in exoskeleton synthesis (Maldonado 2006). Biochemical and physiological studies also revealed that chitin from marine animal and fungal origin are the same, and members of Verongida sponges have apparently higher chitin content in them. Table 17.3 will give us the list of sponges with better chitin yields.

Chitins exhibit different functional roles including immunity enhancing, hemostatic, antithrombogenic, wound healing, etc. (Shelma et al. 2008). It has proven that chitin and its polymer chitosan are nonallergenic and nonhazardous, so our immune system does not reject these polysaccharides as foreign particles (Honary et al. 2009). When compared to cellulose, chitin and its derivatives have better biocompatibility, biodegradability, and adsorption properties.

In the broad areas of regenerative medicine, chitosan derivatives have been effectively used in drug delivery systems, wound dressings, and as 3D scaffolds in tissue engineering. Chitosan has been reported to be a promising material as implant material for engineering human tissue like the cartilage, bone, and skin due to its resorbability (Aribo 2012). Chitosan implants can also be used as an intraocular lens because of its oxygen permeability, and it has also been found to accelerate blood coagulation.

17.6 Conclusion

Natural biomaterials from various marine sponges are found to have numerous applications in the areas of biomedical engineering. The purpose of this book chapter is to highlight peptide and polysaccharide biomolecules as well as to emphasize the applications of these natural biomaterials. After detailed review of available literature, here we advise that future work should be focused on various procedures to minimize the major drawbacks like cytocompatibility, immunogenic responses, and the technological processing methods. Also we should look for better physical and chemical processing methods by which we can alter the surface of the biomaterials for superior bioavailability.

References

Aribo S (2012) Natural products: a minefield of biomaterials. ISRN Mater Sci 983062:1–20

Athanasiou KA, Darling EM, Hu JC (2009) Articular cartilage tissue engineering. Synth Lect Tissue Eng 1(1):1–182

Attaway DH, Zaborsky OR (1993) Marine biotechnology: pharmaceutical and bioactive natural products, vol 1. Springer, New York, p 500

Barondes SH, Cooper DN, Gitt MA, Leffler H (1994) Galectins. Structure and function of a large family of animal lectins. J Biol Chem 269:20807–20807

Blumbach B, Pancer Z, Diehl-Seifert B, Steffen R, Munkner J, Muller I, Muller WE (1998) The putative sponge aggregation receptor. Isolation and characterization of a molecule composed of scavenger receptor cysteine-rich domains and short consensus repeats. J Cell Sci 111(17):2635–2644

Conrad J, Diehl-Seifert B, Zahn RK, Uhlenbruck G, Zimmermann E, Muller WEG (1982) Fibronectin is apparently not involved in species-specific reaggregation of cells from the marine sponge *Geodia cydonium*. J Cell Biochip 19(4):395–404

Datta N, Holtorf HL, Sikavitsas VI, Jansen JA, Mikos AG (2005) Effect of bone extracellular matrix synthesized *in vitro* on the osteoblastic differentiation of marrow stromal cells. Biomaterials 26(9):971–977

DeSimone DW, Spiegel E, Spiegel M (1985) The biochemical identification of fibronectin in the sea urchin embryo. Biochem Bioph Res Co 133(1):183–188

Du C, Cui FZ, Zhang W, Feng QL, Zhu XD, De Groot K (2000) Formation of calcium phosphate/collagen composites through mineralization of collagen matrix. J Biomed Mater Res 50(4):518–527

Dutta PK, Dutta J, Tripathi VS (2004) Chitin and chitosan: chemistry, properties and applications. J Sci Ind Res India 63(1):20–31

Ehrlich H, Worch H (2007) Sponges as natural composites: from biomimetic potential to development of new biomaterials. In: Custodio MR, Lobo-Hajdu G, Hajdu E, Muricy G (eds) Porifera research: biodiversity, innovation & sustainability. Museu Nacional, Rio de Janeiro, pp 303–312

Ehrlich H, Maldonado M, Spindler KD, Eckert C, Hanke T, Born R, Worch H (2007) First evidence of chitin as a component of the skeletal fibers of marine sponges. Part I. Verongidae (Demospongia: Porifera). J Exp Zool Part B 308(4):347–356

Ehrlich H, Ilan M, Maldonado M, Muricy G, Bavestrello G, Kljajic Z, Brunner E (2010) Three-dimensional chitin-based scaffolds from Verongida sponges (Demospongiae: Porifera). Part I. Isolation and identification of chitin. Int J Biol Macromol 47(2):132–140

Ereskovsky AV (2010) The comparative embryology of sponges. Springer, Dordrecht, p 323

Exposito JY, Cluzel C, Garrone R, Lethias C (2002) Evolution of collagens. Anat Rec 268(3):302–316

Fromont J (2003) Porifera (sponges) in the Dampier Archipelago: taxonomic affinities and biogeography. In: The marine flora and fauna of Dampier, Western Australia. Western Australian Museum, Perth, pp 405–417

Fromont J, Vanderklift MA, Kendrick GA (2006) Marine sponges of the Dampier Archipelago, Western Australia: patterns of species distributions, abundance and diversity. Biodiversity Conserv 15 (11):3731–3750

Gingras M, Beaulieu MM, Gagnon V, Durham HD, Berthod F (2008) In vitro study of axonal migration and myelination of motor neurons in a three-dimensional tissue-engineered model. Glia 56(3):354–364

Green D, Howard D, Yang X, Kelly M, Oreffo ROC (2003) Natural marine sponge fiber skeleton: a biomimetic scaffold for human osteoprogenitor cell attachment, growth, and differentiation. Tissue Eng 9(6): 1159–1166

Heinemann S, Heinemann C, Ehrlich H, Meyer M, Baltzer H, Worch H, Hanke T (2007) A novel biomimetic hybrid material made of silicified collagen: perspectives for bone replacement. Adv Eng Mater 9(12):1061–1068

Honary S, Maleki M, Karami M (2009) The effect of chitosan molecular weight on the properties of alginate/chitosan microparticles containing prednisolone. Trop J Pharm Res 8(1):53–61

Iacobini C, Amadio L, Oddi G, Ricci C, Barsotti P, Missori S, Pugliese G (2003) Role of galectin-3 in diabetic nephropathy. J Am Soc Nephrol 14(suppl 3):S264–S270

In Jeong S, Kim SY, Cho SK, Chong MS, Kim KS, Kim H, Lee YM (2007) Tissue-engineered vascular grafts composed of marine collagen and PLGA fibers using pulsatile perfusion bioreactors. Biomaterials 28(6):1115–1122

Khor E (2005) Chitin and chitosan as biomaterials: going forward based on lessons learnt. J Met Mater Min 15(1):69–72

Kijjoa A, Sawangwong P (2004) Drugs and cosmetics from the sea. Mar Drugs 2(2):73–82

Labat-Robert J, Robert L, Auger C, Lethias C, Garrone R (1981) Fibronectin-like protein in Porifera: its role in cell aggregation. Proc Natl Acad Sci U S A 78(10): 6261–6265

Leffler H, Carlsson S, Hedlund M, Qian Y, Poirier F (2002) Introduction to galectins. Glycoconj J 19(7–9):433–440

Li WJ, Cooper JA Jr, Mauck RL, Tuan RS (2006) Fabrication and characterization of six electrospun poly (α-hydroxy ester)-based fibrous scaffolds for tissue engineering applications. Acta Biomater 2(4): 377–385

Liao S, Ngiam M, Chan CK, Ramakrishna S (2009) Fabrication of nano-hydroxyapatite/collagen/osteonectin composites for bone graft applications. Biomed Mater 4(2):025019

Lin Z, Solomon KL, Zhang X, Pavlos NJ, Abel T, Willers C, Zheng M (2011) *In vitro* evaluation of natural marine sponge collagen as a scaffold for bone tissue engineering. Int J Biol Sci 7(7):968–977

Liu FT, Rabinovich GA (2010) Galectins: regulators of acute and chronic inflammation. Ann NY Acad Sci 1183(1):158–182

Ma PX (2008) Biomimetic materials for tissue engineering. Adv Drug Deliv Rev 60(2):184–198

Ma W, Fitzgerald W, Liu QY, O'shaughnessy TJ, Maric D, Lin HJ, Barker JL (2004) CNS stem and progenitor cell differentiation into functional neuronal circuits in three-dimensional collagen gels. Exp Neurol 190(2):276–288

Maldonado M (2006) The ecology of the sponge larva. Can J Zoolog 84(2):175–194

Matranga V, Zito F, Tesoro V, Yokota Y, Nakano E (1995) Biochemical and immunological relationships among fibronectin-like proteins from different sea urchin species. Roux Arch Dev Biol 204(7–8):413–417

Naznin R (2005) Extraction of chitin and chitosan from shrimp (*Metapenaeus monoceros*) shell by chemical method. Pak J Biol Sci 8(7):1051–1054

Noah EM, Chen J, Jiao X, Heschel I, Pallua N (2002) Impact of sterilization on the porous design and cell behavior in collagen sponges prepared for tissue engineering. Biomaterials 23(14):2855–2861

Pallela R, Bojja S, Janapala VR (2011) Biochemical and biophysical characterization of collagens of marine sponge, *Ircinia fusca* (Porifera: Demospongiae: Irciniidae). Int J Biol Macromol 49(1):85–92

Pallela R, Venkatesan J, Janapala VR, Kim S-K (2012) Biophysicochemical evaluation of chitosan-hydroxyapatite-marine sponge collagen composite for bone tissue engineering. J Biomed Mater Res A 100(2): 486–495

Parenteau-Bareil R, Gauvin R, Berthod F (2010) Collagen-based biomaterials for tissue engineering applications. Materials 3(3):1863–1887

Paz M, Ruiz MF, Sánchez L, Mikhailov A (2002) Identification of a fibronectin-like molecule from a marine bivalve *Pecten maximus* (L., 1758) and its hyperaccumulation in the female compartment of the gonad. Bol Inst Esp Oceanogr 18(1–4):393–400

Perillo NL, Marcus ME, Baum LG (1998) Galectins: versatile modulators of cell adhesion, cell proliferation, and cell death. J Mol Med 76(6):402–412

Radjasa OK, Vaske YM, Navarro G, Vervoort HC, Tenney K, Linington RG, Crews P (2011) Highlights of marine invertebrate-derived biosynthetic products: their biomedical potential and possible production by microbial associants. Bioorgan Med Chem 19(22): 6658–6674

Ruggieri GD (1976) Drugs from the sea. Science 194 (4264):491–497

Sarras MP Jr, Madden ME, Zhang X, Gunwar S, Huff JK, Hudson BG (1991) Extracellular matrix (mesoglea) of *Hydra vulgaris*: I. Isolation and characterization. Dev Biol 148(2):481–494

Schleuter D, Gunther A, Paasch S, Ehrlich H, Kljajic Z, Hanke T, Brunner E (2013) Chitin-based renewable materials from marine sponges for uranium adsorption. Carbohydr Polym 92(1):712–718

Schütze J, Skorokhod A, Muller IM, Muller WE (2001) Molecular evolution of the metazoan extracellular matrix: cloning and expression of structural proteins from the demosponges *Suberites domuncula* and *Geodia cydonium*. J Mol Evol 53(4–5):402–415

Shanmugasundaram N, Ravichandran P, Neelakanta Reddy P, Ramamurty N, Pal S, Panduranga Rao K (2001) Collagen–chitosan polymeric scaffolds for the in vitro culture of human epidermoid carcinoma cells. Biomaterials 22(14):1943–1951

Shelma R, Paul W, Sharma CP (2008) Chitin nanofibre reinforced thin chitosan films for wound healing application. Trends Biomater Artif Organs 22:111–115

Song E, Yeon Kim S, Chun T, Byun HJ, Lee YM (2006) Collagen scaffolds derived from a marine source and their biocompatibility. Biomaterials 27(15): 2951–2961

Spencer NJ, Cotanche DA, Klapperich CM (2008) Peptide-and collagen-based hydrogel substrates for *in vitro* culture of chick cochleae. Biomaterials 29(8):1028–1042

Stachowiak AN, Irvine DJ (2008) Inverse opal hydrogel-collagen composite scaffolds as a supportive microenvironment for immune cell migration. J Biomed Mater Res A 85(3):815–828

Szulgit G (2007) The echinoderm collagen fibril: a hero in the connective tissue research of the 1990s. Bioessays 29(7):645–653

Walter SL, Flinn BD, Mayer G (2007) Mechanisms of toughening of a natural rigid composite. Mater Sci Eng C Mater Biol Appl 27(3):570–574

Yang RY, Rabinovich GA, Liu FT (2008) Galectins: structure, function and therapeutic potential. Expert Rev Mol Med 10, e17

Zick Y, Eisenstein M, Goren RA, Hadari YR, Levy Y, Ronen D (2002) Role of galectin-8 as a modulator of cell adhesion and cell growth. Glycoconj J 19(7–9): 517–526

18 Marine Sponges as Future Biomedical Models

Jayachandran Venkatesan, Sukumaran Anil, Elna P. Chalisserry, and Se-Kwon Kim

Abstract

Marine sponges are animals of the phylum Porifera and also excellent source of various biomaterials and organic compounds. In recent years, significant developments on marine sponges derived biomaterials have been explored for various biological and biomedical applications (tissue engineering, drug delivery and biosensor). Biosilica and collagen of marine sponge are important constituents and has huge potential application in regenerative medicine. In the present chapter, we have discussed about isolation procedure of biosilica and collagen from marine sponge. Furthermore, tissue engineering of biogenic silica toward bone tissue engineering is explained in details. Finally, sponge-derived compounds and its use in regenerative medicine and collagen in drug delivery are discussed. As a conclusion, marine sponges are promising source of future biomaterials for various biological and biomedical applications.

Keywords

Sponge • Marine biomaterials • Bone tissue engineering • Collagen and biosilica

18.1 Introduction

The marine source has huge amount of mineralized organisms with excellent pore structures, which are being used for biomedical applications (tissue engineering and drug delivery) and others that are in initial level of development (Clarke et al. 2011). Marine sponges are simple and multicellular organisms of the phylum Porifera and usually made up of spongin (a modified type of collagen protein) and spicules of calcium

J. Venkatesan • S.-K. Kim (✉)
Department of Marine-Bio Convergence Science and Marine Bioprocess Research Center, Pukyong National University, Busan, South Korea
e-mail: venkatjchem@gmail.com; sknkim@pknu.ac.kr

S. Anil
College of Dentistry, Prince Sattam Bin Abdulaziz University, AlKharj, Riyadh, Saudi Arabia

E.P. Chalisserry
Department of Maxillofacial Surgery & Diagnostic Sciences, College of Dentistry, Jazan University, Jazan, Saudi Arabia

R. Pallela, H. Ehrlich (eds.), *Marine Sponges: Chemicobiological and Biomedical Applications*, DOI 10.1007/978-81-322-2794-6_18

carbonate or silica. It was estimated that approximately 15,000 sponge species are available in natural habitat (Hooper et al. 2002). From the ancient days, sponges have been widely used in bathing, painting, cleaning, and some medical purpose (Gresswell 1922). The application can be varied according to the sponge structural and chemical features of skeletons (Ehrlich 2010; Blunt et al. 2005). The Porifera has been divided to three different classes as follows (Bergquist 1978):

- Hexactinellida
- Calcarea
- Demospongiae

Siliceous skeleton are commonly called spicules and mainly present in Hexactinellida and Demospongiae (Müller et al. 2007). Demosponges are the main group and commonly living in deeps seas (Hooper and Van Soest 2002). The sponge skeletons consist of tetraxonic or monaxonic spicules with spongin. The molecular biology, biochemistry, and the applications of demosponges well explained (Wang et al. 2012a). The underwater images of marine sponges are shown in Fig. 18.1.

18.2 Isolation of Biosilica from Marine Sponges

Spicules are structural element that is commonly present in almost all the sponges. It usually provides the structural support to sponges. Sponge spicules represent the main components of the biogenic silica (Bavestrello et al. 1996). Silica formation of sponge spicules provides the key information to develop novel nanostructure materials for several applications. Identification of biomolecules is important in silicon development organism to know the secret of silicon development (Lopez et al. 2005). Siliceous spicules of marine sponge morphology are shown in Fig. 18.2.

18.3 Biomedical Application of Marine Sponge and Its Biosilica

The biomineralization process inspired the researchers to develop novel hierarchy materials with "Nature as model" toward tissue engineering application. Marine sponges have significant structures, with pores which can be used as tissue engineering scaffolds. Recently, much development has been focused on to developing the biomaterials to cure the bone-related diseases (Wang et al. 2011a). Marine sponges and its biosilica have become popular in biomedical application such as tissue engineering and drug delivery. In recent years, significant development has been achieved in marine-derived biomaterials for tissue regeneration (Kim 2013; Silva et al. 2012). Silica and its biocomposite are becoming familiar in making the artificial scaffold for bone tissue engineering (Sowjanya et al. 2013). Sowjanya et al. prepared the blends of chitosan and alginate with nanosilica composite scaffold for bone regeneration. The presence of nanosilica ($nSiO_2$) in the composite increased the protein adsorption and controlled the swelling. The addition of $nSiO_2$ in the composite scaffold significantly increases the mineral deposition on the composite scaffolds. The developed scaffold is biocompatible with osteo-lineage cells. In another study, Ravichandran et al. presented the fabrication procedure of gelatin with mesoporous silica fibers for bone tissue regeneration. The bioactivity of the developed scaffolds initiates the viability of MG63 cells and also increases the alkaline phosphatase activity. The expression of important genes such as osteocalcin, osteopontin, bone sialoprotein, collagen I, and alkaline phosphatase are also increased. (Ravichandran et al. 2014).

Muller et al. found that biosilica positively affects the SaOS-2 cells growth and mineralization; further it significantly increases the osteoprotegerin (OPG) expression. Owing to this excellent property of biosilica, it is extensively used in the biomedical application, especially

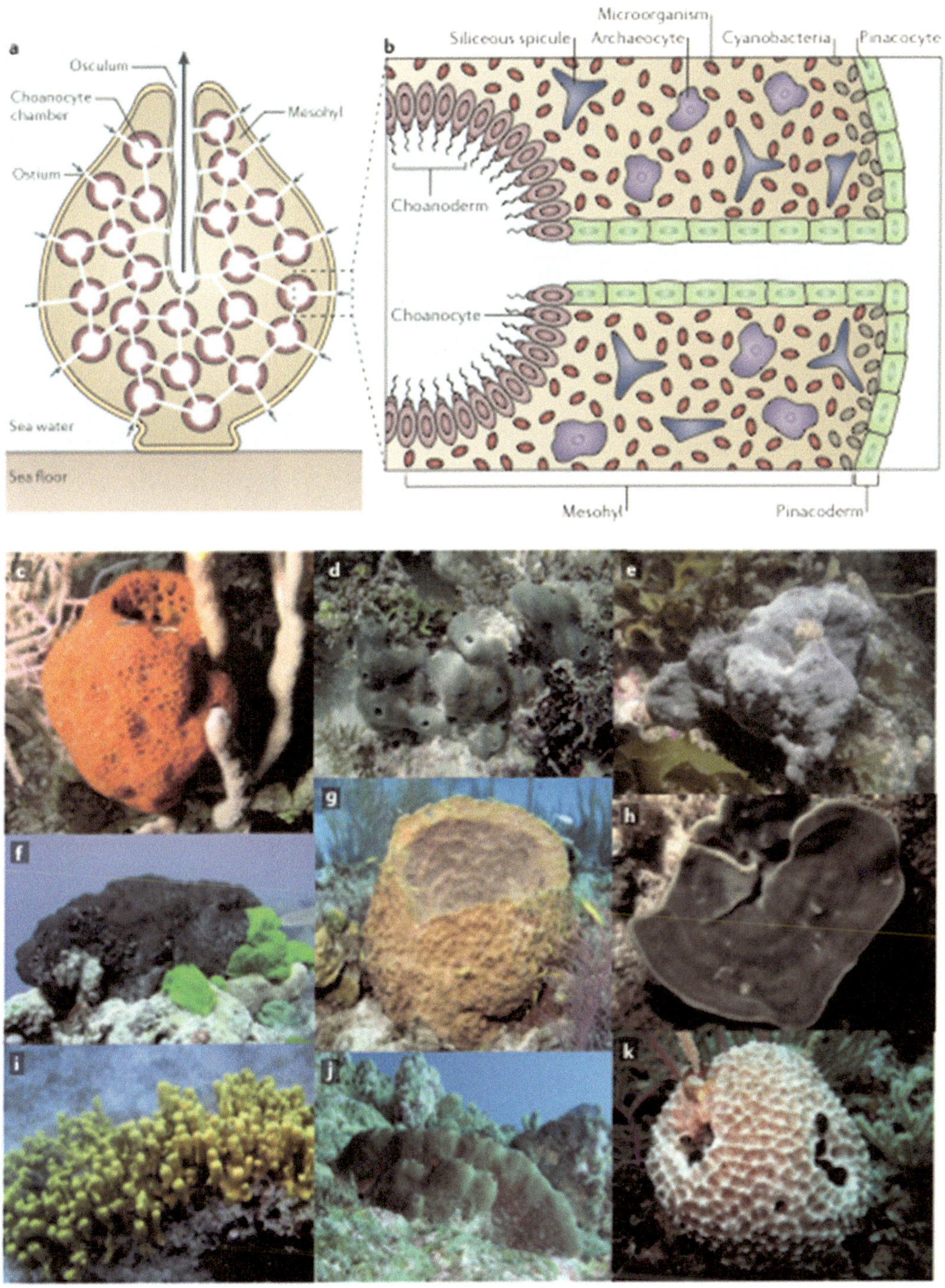

Fig. 18.1 Body plan and underwater images of marine sponges. (**a**) A schematic overview of a typical demosponge. (**b**) An enlargement of the internal structure of a typical demosponge. (**c–k**) Underwater photography of important model sponge species: *Mycale laxissima* (**c**), *Amphimedon queenslandica* (**d**), *Ancorina alata* (**e**), *Rhopaloeides odorabile* (**f**), *Xestospongia muta* (**g**), *Cymbastela concentrica* (**h**), *Aplysina aerophoba* (**i**), *Theonella swinhoei* (**j**), and *Ircinia felix* (**k**) (Reproduced with permission from Hentschel et al. 2012)

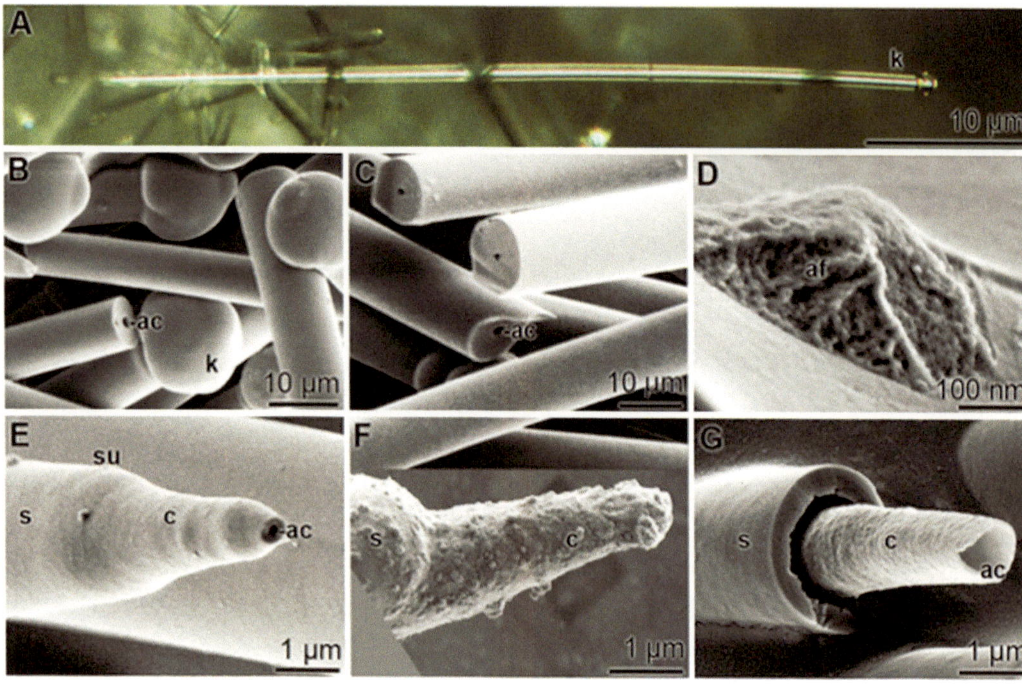

Fig. 18.2 Different microscopic techniques reveal the morphology of *S. domuncula* spicules: (**a**) light microscopy, (**b–g**) scanning electron microscopy (Reproduced with permission from Wang et al. 2011b)

bone-related diseases (Müller et al. 2009) (Fig. 18.3).

Wang et al. reported that biosilica have shown morphogenetic effect and differentiation effect on osteoblast cells and stem cells. Biosilica have a capacity to induce the and also found to increase substantial gene expression such as bone morphogenetic protein-2 (BMP-2) and ALP in osteogenic cells. It was suggested that biosilica are morphogenetically active additives for several composites biomaterials (Wang et al. 2014). Biosilica can be used for osteoporosis treatment, as mentioned earlier, its huge capacity to induce the BMP-2 directly and also inhibiting the function of osteoclast. So, it will be an excellent candidate to treat the bone-related diseases (Wang et al. 2012b).

Marine sponges show the interesting property such as higher degree of swelling ability, and its structure is mainly composed of collagen and silica materials. From the research study, several marine sponges (*Dysidea avara*, *Axinella damicornis*, *Chondrosia reniformis*, *Petrosia ficiformis*, *Sarcotragus spinosulus*, *Agelas oroides*, and *Psammocinia* sp.) were developed as natural scaffold for tissue engineering application using supercritical fluid system. The scaffold was developed in the presence of ethanol at 40 °C and 200 bar and for 6 h. Saos-2 cells were grown on the scaffold and it was checked by SEM; further, cell viability was checked. The in vitro results show excellent biocompatibility (Duarte et al. 2012).

18.4 Marine Sponge Apatite for Bone Tissue Engineering

Different kinds of sponge species – elephant ear (*Spongia agaricina*), Dalmata Fina (*Spongia officinalis Linnaeus*), and Fina Silk (*Spongia zimocca*) – were used to develop the scaffold materials. First, the sponge scaffold was infiltrated and sintered. It produces excellent scaffold with proper pore structure with interconnected. The scaffold developed from *Spongia Agaricina* was promising with overall porosity of 56–61 %

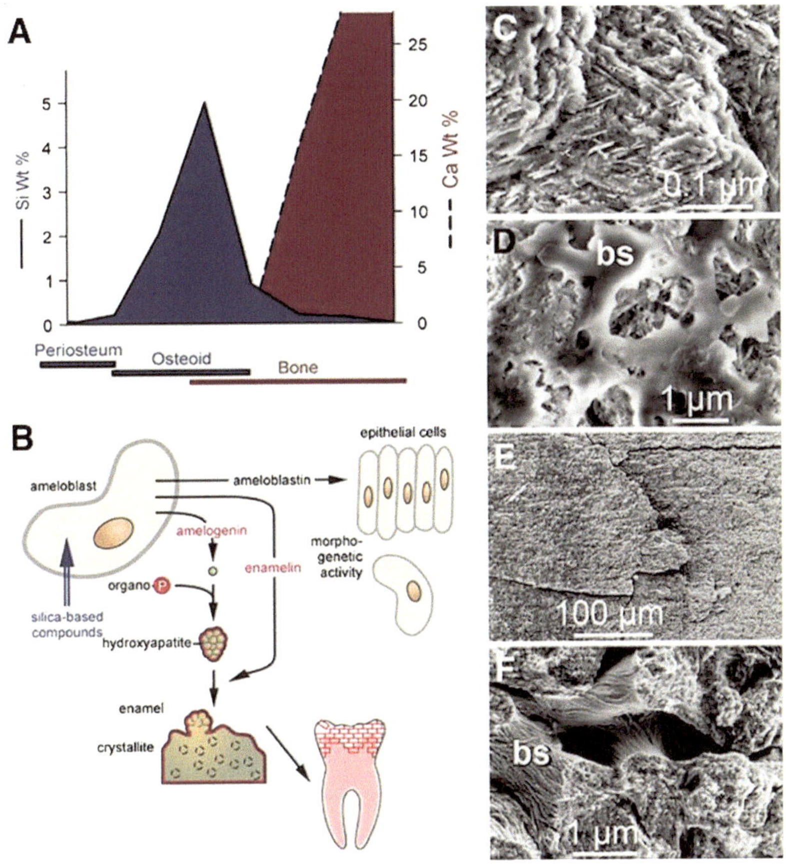

Fig. 18.3 (**A**) Spatial relationship between silicon accumulation and calcium composition during early stages of boneformation in rats. Biomedical application of biosilica and silicatein. (**B**) Schematic representation of the effect of silica-based components on the expression of the three marker genes (amelogenin, ameloblastin, enamelin) in ameloblasts. (**C, D**) Formation of biosilica layers on pig molars. (**D**) Of recombinant silicatein (4 µg/ml PBS) for 12 h at 20 °C. (**E, F**) In parallel, biosilica formation on femur bone samples was examined: untreated control (**E**) or silicatein treated (**F**) (Reproduced with permission from Müller et al. 2009)

and interconnectivity of 99.92 % (Cunningham et al. 2010).

The properties and composition of the marine sponge (*Verongula gigantea*) and octocorals (*Isidella* sp.) were investigated. It was shown that the demosponge *V. gigantea* has much potential as a biomaterial due to the multilayered structure of its rigid fibrous skeletons. Nanocrystalline aragonite was isolated and identified in *V. gigantea*, a sponge usually described as lacking a mineral skeleton (Born et al. 2010). Sponges as natural composites: from biomimetic potential to development of new biomaterials also explained by Ehrlich (Ehrlich and Worch 2007).

The main aspect of tissue engineering is the use of scaffold metrics in proper way (Vats et al. 2003). Scanning electron microscopy was used to check *Callyspongiidae* sponge morphology and checked its bioactivity with osteoblast cells. The morphology of the sponge skeleton is interconnected with a pore size 100–300 um in diameter. At 21 days of cell culture on the marine sponge scaffolds, mineralization was seen. The important gene expressions OCN and OPN were also observed at different days (Lin et al. 2011).

18.5 Marine-Sponge-Derived Compounds

In the last five decades, marine-sponge-related compounds are promising in terms of drugs against various diseases, including cancer, viral, inflammatory diseases, and malaria. It has the capacity to produce different kinds of chemical compounds with widely varied carbon skeletons (Sipkema et al. 2005). Around 7,000 sponge species are alive both in marine and freshwater environment. Interesting fact is that sponge can live

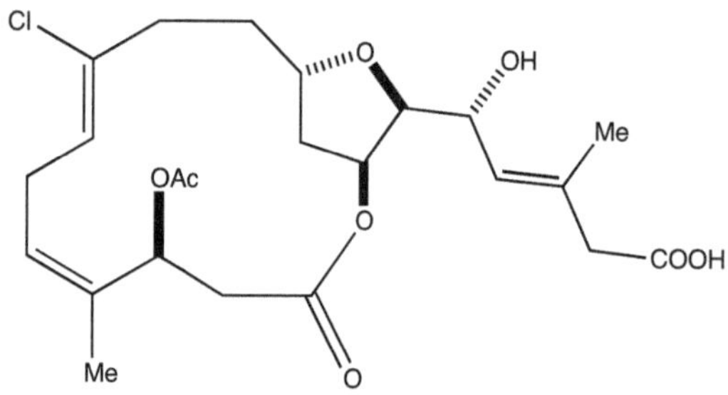

Fig. 18.4 Structure of haterumalides

all around the world at any region. Around 99 % of sponges are present in marine environment. The chemical constituent varies depending on the region. Different bioactivity compounds have been isolated from marine sponge source and checked for antimicrobial test, and results are promising to develop the drugs (Fusetani et al. 1981). The isolation, structure, biological activities, and synthetic studies of marine secondary metabolites, symbioimine and haterumalide, are explored well and inhibited the differentiation of RAW 264.7 cells into osteoclasts. Haterumalides (Fig. 18.4), 14-membered cytotoxic macrolides from the Okinawan sponge *Ircinia* sp., show potent cytotoxicity (Kita et al. 2006).

18.6 Role of Marine Sponge Biomaterials in Tissue Engineering

The abundance and structural variety of natural marine sponge frameworks and their potential as multifunctional, cell conductive and inductive frameworks indicate a promising new source of scaffold for tissue regeneration (Green et al. 2003). Marine sponge (*Ircinia fusca*)-produced collagen with chitosan/hydroxyapatite has been developed for bone tissue engineering in vitro. Cell proliferation in chitosan/hydroxyapatite/marine sponge collagen scaffolds is higher than pure chitosan scaffold (Pallela et al. 2012). The physicochemical characterization has been done with various marine sponges such as (*Verongula gigantea*) and octocorals (*Isidella* sp.) suggests huge resource for bone replacement (Born et al. 2010).

Marine-sponge-derived collagenous fiber framework provides a suitable architecture as bioscaffold for tissue regeneration, as it supports the proliferation, migration, and adhesion of osteoblasts in vitro (Zheng et al. 2007). Sulfate poly-N-acetyl glucosamine seed on sponge shows promising bone healing of both bone and cartilage when compared to the control group (Kang et al. 2005). The collagenous structure of marine sponge is shown as an excellent candidate to support the wide range of cells for tissue regeneration, specifically bone tissue regeneration (Green et al. 2003; Lin et al. 2011). The collagen fiber in the marine sponges is fully bonded (Pallela et al. 2011; Ehrlich 2010; Heinemann et al. 2007a). Zheng et al. marine sponges from the genus, *Hippospongia* used for tissue regeneration in terms of osteoblast proliferation and cell adhesion (Zheng et al. 2007). There is another way to develop tissue engineering scaffolds using marine-sponge-derived silica and collagen (Heinemann et al. 2007b; Ehrlich et al. 2010; Green et al. 2014).

18.7 Marine Sponges for Drug Delivery

Biogenic inorganic ceramic materials are extensively used for drug delivery purpose, for example, biogenic apatite and biogenic silica. Biogenic silica from sponges has advantages over the synthetic silica materials in drug delivery purpose (Roveri et al. 2008). Spicules are constituted of concentric layers of hydrated silica. They exhibit low elastic modulus, which leads to flexibilities of material for which there is no equivalent man-made synthetic silica-based material. Spicules are multifunctional materials and carry light-showing optical properties very close to those of modern optics fiber. A growing spicule is embedded in a membrane called silicalemma, and silica polycondensation takes place on poly protein fibers which catalyze the hydrolysis, orienting the tetraethoxysilane polymerization parallel to the proteases' lineament (Cha et al. 1999, 2000; Levi et al. 1989; Aizenberg et al. 2005; Sundar et al. 2003; Meyers et al. 2008; Shimizu et al. 1998; Krasko et al. 2000).

Nicklas et al. prepared the nanoparticles of *Chondrosia reniformis* sponge collagen in hormone therapy. Estradiol hemihydrate was used as a drug model. The loading efficiency of the estradiol hemihydrate in nanoparticles is 13.1 %. Comparative study was performed between marine-sponge-derived nanoparticles with hydrogel with commercial gel without drugs (Nicklas et al. 2009).

Collagen microparticles were developed using marine sponge. Retinol was used as a drug model and incorporated into the nanoparticle, and drug stability was also investigated. The dermal penetration of drug into the skin increases approximately twofold (Swatschek et al. 2002). Release properties from marine sponge collagen coated formulations are also studied (Pergament et al. 2011).

18.8 Conclusion

Marine-sponge-derived biomaterials show promising direction in artificial organ development. Marine sponge biogenic silica aids in differentiating stem cells into osteogenic cell. This will be a promising approach to develop bone tissue construction. Furthermore, marine sponge collagen can be used for nano-biotechnological applications.

Acknowledgments This paper was supported by research funds of Pukyong National University in 2015 and grant from the Marine Bioprocess Research Center of the Marine Biotechnology Program, funded by the Ministry of Oceans and Fisheries, Republic of Korea.

References

Aizenberg J, Weaver JC, Thanawala MS, Sundar VC, Morse DE, Fratzl P (2005) Skeleton of *Euplectella sp.*: structural hierarchy from the nanoscale to the macroscale. Science 309(5732):275–278

Bavestrello G, Cattaneo-Vietti R, Cerrano C, Cerutti S, Sará M (1996) Contribution of sponge spicules to the composition of biogenic silica in the Ligurian Sea. Mar Ecol 17(1–3):41–50. doi:10.1111/j.1439-0485.1996.tb00488.x

Bergquist PR (1978) Sponges. University of California Press, Los Angeles

Blunt JW, Copp BR, Munro MHG, Northcote PT, Prinsep MR (2005) Marine natural products. Nat Prod Rep 22(1):15–61. doi:10.1039/B415080P

Born R, Ehrlich H, Bazhenov V, Shapkin NP (2010) Investigation of nanoorganized biomaterials of marine origin. Arab J Chem 3(1):27–32

Cha JN, Shimizu K, Zhou Y, Christiansen SC, Chmelka BF, Stucky GD, Morse DE (1999) Silicatein filaments and subunits from a marine sponge direct the polymerization of silica and silicones in vitro. Proc Natl Acad Sci 96(2):361–365

Cha JN, Stucky GD, Morse DE, Deming TJ (2000) Biomimetic synthesis of ordered silica structures mediated by block copolypeptides. Nature 403(6767):289–292

Clarke S, Walsh P, Maggs C, Buchanan F (2011) Designs from the deep: marine organisms for bone tissue engineering. Biotechnol Adv 29(6):610–617

Cunningham E, Dunne N, Walker G, Maggs C, Wilcox R, Buchanan F (2010) Hydroxyapatite bone substitutes developed via replication of natural marine sponges. J Mater Sci Mater Med 21(8):2255–2261

Duarte ARC, Silva JM, Silva TH, Osinga R, Ilan M, Marques A, Mano J, Reis R (2012) Marine sponges as natural scaffolds: decellularization by supercritical fluid technology and cellularization with osteoblasts for tissue engineering applications. J Tissue Eng Regent Med 6(1):171–172

Ehrlich H (2010) Biological materials of marine origin. Springer, Heidelberg

Ehrlich H, Worch H (2007) Sponges as natural composites: from biomimetic potential to development of new biomaterials. In: Porifera research: biodiversity, innovation & sustainability. Museu Nacional, Rio de Janeiro

Ehrlich H, Ilan M, Maldonado M, Muricy G, Bavestrello G, Kljajic Z, Carballo J, Schiaparelli S, Ereskovsky A, Schupp P (2010) Three-dimensional chitin-based scaffolds from Verongida sponges (Demospongiae: Porifera). Part I. Isolation and identification of chitin. Int J Biol Macromol 47(2):132–140

Fusetani N, Matsunaga S, Konosu S (1981) Bioactive marine metabolites II. Halistanol sulfate, an antimicrobial novel steroid sulfate from the marine sponge *Halichondria cf. moorei Bergquist*. Tetrahedron Lett 22(21):1985–1988

Green D, Howard D, Yang X, Kelly M, Oreffo R (2003) Natural marine sponge fiber skeleton: a biomimetic scaffold for human osteoprogenitor cell attachment, growth, and differentiation. Tissue Eng 9(6):1159–1166

Green DW, Lai W-F, Jung H-S (2014) Evolving marine biomimetics for regenerative dentistry. Mar Drugs 12(5):2877–2912

Gresswell EJ (1922) Sponges: their nature, history, modes of fishing, varieties, cultivation, etc. Sir I. Pitman & Sons, ltd

Heinemann S, Ehrlich H, Douglas T, Heinemann C, Worch H, Schatton W, Hanke T (2007a) Ultrastructural studies on the collagen of the marine sponge *Chondrosia reniformis Nardo*. Biomacromolecules 8(11):3452–3457

Heinemann S, Ehrlich H, Knieb C, Hanke T (2007b) Biomimetically inspired hybrid materials based on silicified collagen. Int J Mater Res 98(7):603–608

Hentschel U, Piel J, Degnan SM, Taylor MW (2012) Genomic insights into the marine sponge microbiome. Nat Rev Microbiol 10(9):641–654

Hooper JN, Van Soest RW (2002) Systema Porifera. A guide to the classification of sponges. Springer, Berlin

Hooper JN, Kennedy JA, Quinn RJ (2002) Biodiversity 'hotspots', patterns of richness and endemism, and taxonomic affinities of tropical Australian sponges (Porifera). Biodivers Conserv 11(5):851–885

Kang QK, Hill CM, Demcheva MV, Vournakis JN, An YH (2005) Poly-N-acetyl glucosamine-SO4 for repairing osteochondral defect in rabbits. Key Eng Mater 288–289: 83–86

Kim S-K (2013) Marine biomaterials: characterization, isolation and applications. CRC Press, New York

Kita M, Sakai E, Uemura D (2006) Pursuit of novel bioactive marine metabolites. J Synth Org Chem 64(5):471–480

Krasko A, Lorenz B, Batel R, Schröder HC, Müller IM, Müller WE (2000) Expression of silicatein and collagen genes in the marine sponge Suberites domuncula is controlled by silicate and myotrophin. Eur J Biochem 267(15):4878–4887

Levi C, Barton J, Guillemet C, Bras E, Lehuede P (1989) A remarkably strong natural glassy rod: the anchoring spicule of the Monorhaphis sponge. J Mater Sci Lett 8(3):337–339

Lin Z, Solomon KL, Zhang X, Pavlos NJ, Abel T, Willers C, Dai K, Xu J, Zheng Q, Zheng M (2011) In vitro evaluation of natural marine sponge collagen as a scaffold for bone tissue engineering. Int J Biol Sci 7(7):968

Lopez PJ, Gautier C, Livage J, Coradin T (2005) Mimicking biogenic silica nanostructures formation. Curr Nanosci 1(1):73–83. doi:10.2174/1573413052953156

Meyers MA, Chen P-Y, Lin AY-M, Seki Y (2008) Biological materials: structure and mechanical properties. Prog Mater Sci 53(1):1–206

Müller W, Li J, Schröder H, Qiao L, Wang X (2007) The unique skeleton of siliceous sponges (Porifera; Hexactinellida and Demospongiae) that evolved first from the Urmetazoa during the Proterozoic: a review. Biogeosciences 4(1):219–232

Müller WEG, Wang X, Cui FZ, Jochum KP, Tremel W, Bill J, Schröder HC, Natalio F, Schloßmacher U, Wiens M (2009) Sponge spicules as blueprints for the biofabrication of inorganic-organic composites and biomaterials. Appl Microbiol Biotechnol 83(3): 397–413

Nicklas M, Schatton W, Heinemann S, Hanke T, Kreuter J (2009) Preparation and characterization of marine sponge collagen nanoparticles and employment for the transdermal delivery of 17β-estradiol-hemihydrate SCNPs for dermal delivery of estradiol. Drug Dev Ind Pharm 35(9):1035–1042

Pallela R, Bojja S, Janapala VR (2011) Biochemical and biophysical characterization of collagens of marine sponge, *Ircinia fusca* (Porifera: Demospongiae: Irciniidae). Int J Biol Macromol 49(1):85–92

Pallela R, Venkatesan J, Janapala VR, Kim SK (2012) Biophysicochemical evaluation of chitosan-hydroxyapatite-marine sponge collagen composite for bone tissue engineering. J Biomed Mater Res 100 A(2): 486–495

Pergament V, Schatton W, Fotaki N (2011) Study of release properties from marine sponge collagen coated formulations. AAPS J 13 (S2)

Ravichandran R, Sundaramurthi D, Gandhi S, Sethuraman S, Krishnan UM (2014) Bioinspired hybrid mesoporous silica–gelatin sandwich construct for bone tissue engineering. Microporous Mesoporous Mater 187(0):53–62. doi:http://dx.doi.org/10.1016/j.micromeso.2013.12.018

Roveri N, Palazzo B, Iafisco M (2008) The role of biomimetism in developing nanostructured inorganic matrices for drug delivery. Expert Opin Drug Deliv 5(8):861–877. doi:10.1517/17425247.5.8.861

Shimizu K, Cha J, Stucky GD, Morse DE (1998) Silicatein α: cathepsin L-like protein in sponge biosilica. Proc Natl Acad Sci 95(11):6234–6238

Silva TH, Alves A, Ferreira B, Oliveira J, Reys L, Ferreira R, Sousa R, Silva S, Mano J, Reis R (2012) Materials of marine origin: a review on polymers and ceramics of biomedical interest. Int Mater Rev 57(5): 276–306

Sipkema D, Franssen MC, Osinga R, Tramper J, Wijffels RH (2005) Marine sponges as pharmacy. Mar Biotechnol 7(3):142–162

Sowjanya J, Singh J, Mohita T, Sarvanan S, Moorthi A, Srinivasan N, Selvamurugan N (2013) Biocomposite scaffolds containing chitosan/alginate/nano-silica for bone tissue engineering. Colloids Surf B 109:294–300

Sundar VC, Yablon AD, Grazul JL, Ilan M, Aizenberg J (2003) Fibre-optical features of a glass sponge. Nature 424(6951):899–900

Swatschek D, Schatton W, Müller WEG, Kreuter J (2002) Microparticles derived from marine sponge collagen (SCMPs): preparation, characterization and suitability for dermal delivery of all-trans retinol. Eur J Pharm Biopharm 54(2):125–133

Vats A, Tolley N, Polak J, Gough J (2003) Scaffolds and biomaterials for tissue engineering: a review of clinical applications. Clin Otolaryngol Allied Sci 28(3): 165–172

Wang S-F, Wang X-H, Gan L, Wiens M, Schröder H, Müller W (2011a) Biosilica-glass formation using enzymes from sponges [silicatein]: basic aspects and application in biomedicine [bone reconstitution material and osteoporosis]. Front Mater Sci 5(3):266–281. doi:10.1007/s11706-011-0145-1

Wang X, Wiens M, Schröder HC, Schloßmacher U, Pisignano D, Jochum KP, Müller WE (2011b) Evagination of cells controls bio-silica formation and maturation during spicule formation in sponges. PLoS ONE 6(6):e20523

Wang X, Schröder HC, Wiens M, Schloβmacher U, Müller WE (2012a) 5 biosilica: molecular biology, biochemistry and function in demosponges as well as its applied aspects for tissue engineering. Adv Mar Biol 62:231

Wang X, Schröder HC, Wiens M, Ushijima H, Müller WE (2012b) Bio-silica and bio-polyphosphate: applications in biomedicine (bone formation). Curr Opin Biotechnol 23(4):570–578

Wang X, Schröder HC, Grebenjuk V, Diehl-Seifert B, Mailänder V, Steffen R, Schloßmacher U, Müller WEG (2014) The marine sponge-derived inorganic polymers, biosilica and polyphosphate, as morphogenetically active matrices/scaffolds for the differentiation of human multipotent stromal cells: potential application in 3D printing and distraction osteogenesis. Mar Drugs 12(2):1131–1147

Zheng MH, Hinterkeuser K, Solomon K, Kunert V, Pavlos N, Xu J (2007) Collagen-derived biomaterials in bone and cartilage repair. In: Macromolecular symposia, vol 1. Wiley Online Library, pp 179–185

Chemistry and Biology of Marine Sponge Collagens

19

Kota Sobha and Devarai Santhosh Kumar

Abstract

Collagens are the proteins found in the extracellular matrix of multicellular organisms, from primitive sponges (parazoans) to highly advanced mammals (metazoans). These proteins, classified as the "collagen superfamily," comprise about 28 members each with at least one triple-helical domain. Collagens deposited in the extracellular matrix (ECM) and connective tissues form supramolecular assemblies and function as structural proteins contributing to mechanical properties, organization, and shape of tissues. They regulate cell proliferation, migration, and differentiation by binding to cognate receptors on the cell surface and triggering signal transduction cascades. Collagens with restricted tissue distribution perform specific biological functions. This chapter envisages the structural and functional characteristics of both invertebrate and vertebrate collagens with a special account of sponge collagens and their significance in tissue engineering.

Keywords

Collagen structure • Tissue engineering • Marine sponges • Types of collagen

19.1 Introduction

The term collagen derives from the Greek word *glue* and is defined as that constituent of connective tissue which yields gelatin on boiling. With the advancements in protein sequence determination, collagens are defined as the proteins that assemble into fibrous supramolecular aggregates in the extracellular space and which comprise three polypeptide chains (Figs. 19.1 and 19.2)

K. Sobha
Department of Biotechnology, RVR & JC College of Engineering, Guntur 522 019, Andhra Pradesh, India

D.S. Kumar (✉)
Industrial Bioprocess and BioProspecting Laboratory, Department of Chemical Engineering, Indian Institute of Technology Hyderabad, 2nd Floor, Academic Block 'E', Kandi Campus, Hyderabad 502285, Telangana, India
e-mail: devarai@iith.ac.in

R. Pallela, H. Ehrlich (eds.), *Marine Sponges: Chemicobiological and Biomedical Applications*,
DOI 10.1007/978-81-322-2794-6_19

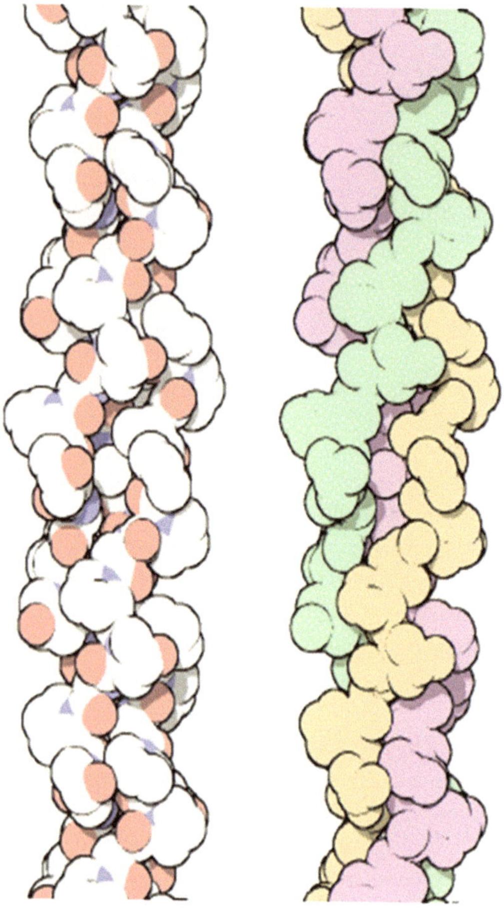

Fig. 19.1 Collagen triple-helical structure

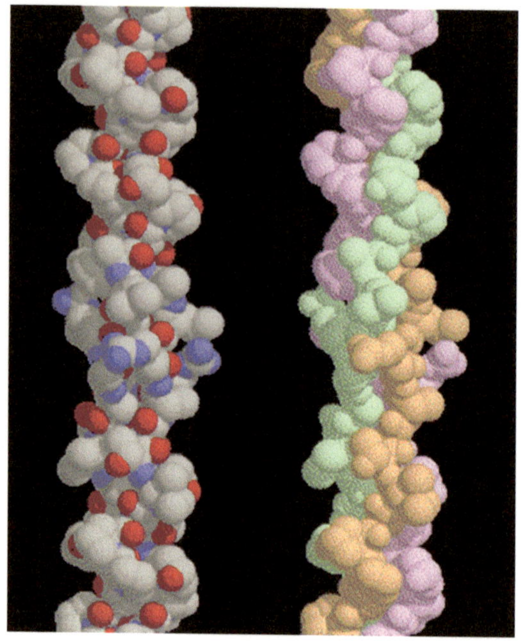

Fig. 19.2 A segment of human collagen

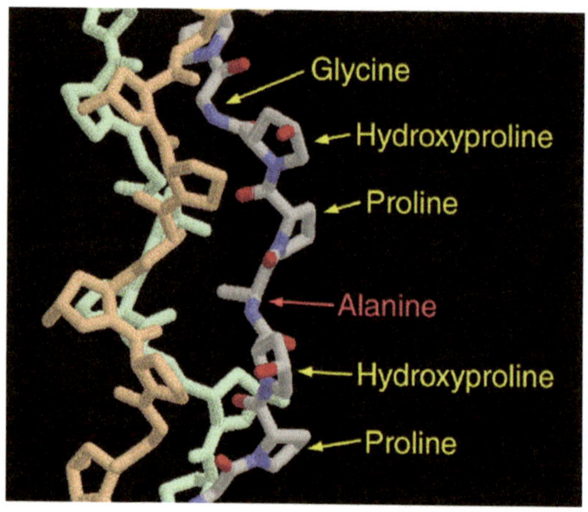

Fig. 19.3 A special amino acid sequence

with a large number of repeat sequences Gly-X-Y where X is often proline and Y is often hydroxyproline (Fig. 19.3). Hydroxyproline is derived from proline during posttranslational modifications by a specific enzyme called prolyl-hydroxylase. Hydroxyproline is rarely found in other proteins. Each polypeptide chain of collagen contains about 1000 amino acid residues and the entire triple chain molecule is about 3000 Å (Branden and Tooze 1999).

Collagen is a long chain fibrous protein and serves as structural material. The individual long chain molecules obtain bulk properties by cross-linking, interleaving, and intertwining the proper combination of individual chain molecules and these properties enable multiple different functions of the molecule. Depending on the secondary structure of the individual molecules, fibrous proteins are classified as coiled coil α-helices (e.g., keratin and myosin), the triple helix (e.g., collagen), and β sheet (e.g., amyloid fibers and silks) (Branden and Tooze 1999). Collagen fibers are strong, resistant to stretching, and relatively rigid. Fibrous proteins like collagen often form protofilaments or protofibrils that assemble into structurally specific, higher-ordered filaments and fibrils. These filaments cannot be crystallized because they can be ordered only in two dimensions. However, ordered fibers give two-dimensional diffraction patterns which could be used for overall structure determination.

19.2 Synthesis of Collagens

Collagens are synthesized as longer precursors called procollagens with globular extensions called propeptides (about 200 residues) at the ends. When the procollagen polypeptides get transported from ribosomes into the lumen of the endoplasmic reticulum, they undergo hydroxylation and other chemical modifications before assembling into triple chain molecules. The function of the terminal propeptides is to form proper triple chains in register through interchain disulfide bonds. Propeptides are cleaved off only after the precursor proteins get released by exocytosis. Excision of both propeptides allows the triple chain molecules to polymerize into fibrils several micrometers long and 50–200 nm in diameter. The fibrils then pack side by side into parallel bundles, the collagen fibers, which are stronger than steel of same size. When mature collagen devoid of propeptides is denatured, the polypeptides associate at many different places other than their ends, forming triple chains that are out of register. These "out of register" triple chains polymerize to form a gel called gelatin (Branden and Tooze 1999).

From fiber diffraction studies of collagen by Linus Pauling, Francis Crick, and others, it is established that each of the three polypeptide chains is folded into an extended left-handed helix with 3.3 residues per turn and a rise per residue along the helical axis of 2.9 Å. This is in contrast to the normal right-handed α helix with 3.6 residues per turn and a rise per residue of 1.5 Å. Accordingly, the rise per turn in the collagen helix is 9.6 Å, compared with 5.4 Å for the α helix and this gives such an extended chain that it must aggregate to form a stable structure. Synthetic polymers of proline or glycine fold into similar extended, left-handed helices and so the helix is called a polyproline type II helix. The three polyproline type II helices in collagen form a trimeric molecule by coiling about a central axis to form a right-handed superhelix with a repeat distance of about 100 Å. The side chain of every third residue in the super helix is very close to the central axis that no space is available for a side chain and hence every third residue must be a glycine. If this glycine is substituted by any other residue, the super helix undergoes deformation (Branden and Tooze 1999). Certain genetic diseases of connective tissue occur due to mutations in codons for these glycine residues. This sequence requirement is a hall mark of triple helix collagen-like domains and is used in sequence analyses of proteins of unknown structure. Varieties of supramolecular aggregates contain the triple-helical domains ranging from the collagen fibrils in tendons and cartilage to reticulate forms of basement membranes and to the parallel clusters of short triple helices seen in the complement component C1q of blood and in the sugar-binding collectins. Some collagens have interruptions and imperfections in the triple helix and the resulting conformational changes are visualized in crystal structures of model peptides (Bella et al. 2006).

Helen Berman et al. at Rutgers University determined the crystal structure of collagen triple helix to 1.9 Å resolution of a synthetic collagen-like peptide (Pro-hydroxyproline-Gly)$_{10}$ with one glycine substituted by alanine (Bella et al. 1994). This collagen-like peptide formed single crystals but not fibers. When the details of the regular collagen triple helix structures were compared with the structure obtained by the effect of mutation at a glycine position, the importance of direct as well as water-mediated hydrogen bonds in stabilizing the triple helix structure became obvious. It further showed that the alanine side chain can be accommodated inside the triple helix by a local small change of the helix geometry which in turn allows the incorporation of interstitial water molecules to link the chains. These kinds of conformational shifts help to accommodate the sequence variations that deviate from the consensus. In the regular triple helix structure of collagen, the three chains are held close together by direct hydrogen bonds between proline $C = O$ groups of one chain and the glycine N-H groups of another. In the region around the alanine residues, the three polypeptide chains are forced apart by the alanine side chains and four water molecules are incorporated between the chains; this allows the direct hydrogen bonds to be replaced by water-mediated hydrogen bonds.

All side chains as well as the C=O group of glycines in all three chains are on outside of the triple helix molecule and in contact with water molecules. These water molecules mediate hydrogen bonds between the hydroxyl groups of hydroxyproline and the peptide C=O and N-H groups both within each chain and between different chains. These water-mediated hydrogen bonds are essential for the stability of the triple helix and are presumably the reason for the presence of hydroxyproline in collagen.

19.3 Extracellular Matrix (ECM) and Collagen

Extracellular matrix (ECM) is the dynamic structural environment of cells in tissues and organs and is constantly remodeled for maintaining tissue integrity and mechanical properties. ECM is essential for maintaining tissue homeostasis, morphogenesis, and differentiation through specific interactions with cells. The various macromolecular components of ECM are collagens, proteoglycans, elastic proteins, and non-collagenous adhesive glycoproteins (Lu et al. 2011). Collagens are distinguishable from other ECM components by their relative abundance and their capacity to self-assemble into supramolecular organized structures. Thus, collagen is a common and major component of the extracellular matrices of all multicellular life beginning from Parazoa (sponges) to connective tissues of invertebrates and vertebrates (bone and tendon) (Silver 2009). The metazoan development is closely correlated with the evolution of collagens and only two types of collagen have been conserved from the whole of multicellular animals. These are the fibrillar and the basement membrane collagens. There are more than one lakh research papers on collagens, and of late, collagen research has renewed due to the identification of transmembrane collagens on the surfaces of a wide variety of cells and the collagens that are precursors of bioactive peptides with paracrine functions.

19.4 Collagen Superfamily

The superfamily comprises of 28 members with considerable complexity and diversity in structure, assembly, and function (Table 19.1). However, there are certain common features shared by all of them like modular nature with collagenous domains flanked by non-collagenous domains (linker regions), homo- or heterotrimeric nature with a characteristic amino acid signature of Gly-X-Y, assembly into supramolecular aggregates, posttranslational modifications including proteolytic processing, fibril formation, reticulation, production of functional domains, and shedding of transmembrane collagens (Ricard-Blum and Ruggiero 2005).

19.5 Polymorphism in Collagens

There is a large degree of polymorphism in collagens as they form a variety of different structures. In humans, in addition to the normal type I collagen, 27 different other collagenous polypeptides of varied ultrastructure are known with different functions, and for some of them, the functions are yet to be explored. Each collagen type consists of a triple-stranded subunit called tropocollagen made of either identical (homotrimeric) or genetically diverse (heterotrimeric) alpha chains. However, all alpha chains contain at least one collagenous domain comprising of a repeating "signature" triplet, Gly-X-Y, in which X and Y are often the amino acids proline and hydroxyproline, respectively. Fibrillar collagens display specific lateral assembly of the units induced by specific conditions, sometimes showing alternating regions of diverse density (D-period) along the longitudinal axis.

Table 19.1 Types of collagens based on their structure and supramolecular organization

Collagen type	Collagen subfamily	Function	Characteristics
I, II, III, V, XI, XXIV, and XVII	Fibril-forming collagens	Assemble into organized fibrils	Long central COL domain with about 1000 amino acids (330 G-x-y tripeptide repeats), flanked by small terminal globular extensions (NC domains)
IV, VIII, X and dogfish egg case collagen	Network-forming collagens	Reticulation-intertwining of different collagen molecules	Help reduce the stiffness of the collagen molecule, allowing more spatial freedom for the molecule, and also promoting supercoiling
IX, XII, XIV, XVI, XIX, XX, XI, and XXII	Fibril-associated collagen with interrupted triple helix (FACIT) collagens	Mediate protein-protein interactions	NC domains are predominant
XV and XVIII	Basement membrane multiplexin (multiple triple helix domains and interruptions)	Cell adhesion, growth and differentiation, tissue repair, molecular ultrafiltration, cancer cell invasion, and metastasis	COL domains are shorter and/or contain interruptions
XXV (neuronal collagen), XIII, XVII, XXIII	Transmembrane collagens	Dynamic bidirectional links between the extracellular matrix and the cytoskeleton	Pass information across the cell membranes to regulate extracellular matrix assembly, cell proliferation, differentiation, and death
VII	Anchoring fibrils	Extend from the basal lamina of epithelial cells and attach to the lamina reticularis	Essential to the functional integrity of the dermoepidermal junction
VI	Ubiquitous collagen	Form extensive microfibrillar arrays and often seen in association with hyaluronan	Growth and remodeling of connective tissue

Modified from Myllyharju and Kivirikko (2001)

Fibrillar collagens are present from sponges to humans and are involved in the formation of the well-known striated fibrils. Among the 28 different types identified in vertebrates (Heino 2007; Söderhäll et al. 2007), basement membrane type IV and the fibrillar collagens are the only ones to have been hitherto described from sponges to humans (Boute et al. 1996; Exposito and Garrone 1990). Types of fibrillar collagens are summarized in Table 19.1.

Most collagen types are recognized by cognate receptors like ECM integrin receptors, collagen-specific discoidin domain receptors (DDR), and the transmembrane proteoglycan syndecans. Collagen receptor binding induces appropriate cellular pathways and regulates migration, proliferation, and differentiation of cells. Certain collagens like cartilage collagen II can also bind to growth factors and control their bioavailability by acting as reservoirs.

19.6 Basement Membranes

These are sheet-like complexes of extracellular matrix structures found beneath the epithelial and endothelial tissues and surrounding cells like those of muscle tissue, peripheral nerves, and adipocytes. They serve as selective barriers for macromolecules and scaffold support for cells and in cell behavior (Erickson and Couchman 2000). For example, basement membrane found in kidney glomerulus helps in molecular filtration.

One of the major collagen constituents of basement membranes is type IV collagen. Each

Fig. 19.4 Type IV collagen of basement membrane

molecule has a globular head at one end and an extra tail at the other. Four such collagen molecules associate through their tails and form an X-shaped complex while their heads bind strongly together in a "head-on" pattern (Fig. 19.4). In humans, six units of type IV collagen chains (α1–α6) have been identified. These are involved in the formation of heterotrimeric molecules with $(\alpha1)_2\,\alpha2$ being the most abundant and ubiquitous isoform (Hudson et al. 1993). Each type IV chain contains a long triple-helical or "collagenous domain" of approximately 1400 amino acids flanked by the 7S region at N-terminus and non-collagenous (NC1) domain at the C-terminus. The hexameric network assembly of type IV collagen molecules is mediated through NC1 domain. This domain plays a vital role in the selection and association of the three type IV α chains and the initiation of triple helix formation (Borza et al. 2001; Boutaud et al. 2000; Khoshnoodi et al. 2006; Söder and Pöschl 2004). Triple-helical type IV molecules, also called "protomers," assemble into a complex network, mediated through the NC1 regions from two protomers associating to form dimmers and 7S domains aiding in the formation of tetramers (Timpl et al. 1981). X-ray structures revealed a characteristic 3D fold in the NC1 monomer composed predominantly of β-sheets, which interact through a domain swapping mechanism. Both hydrophobic and hydrophilic interactions at the interface favor the association of two NC1 protomers and, in addition, the association is stabilized by a covalent cross-link, termed S-hydroxylysyl methionine. This link is made by methionine and lysine residues contributed by both NC1 trimers. NC1 monomers have attracted the attention of researchers as they are the targets of pathogenic antibodies in Goodpasture's syndrome and Alport's syndrome (Hudson et al. 2003). Further, research findings suggest that NC1 proteolytic fragments from type IV collagen chains have potent anti-angiogenic and antitumor activities in vivo (Hamano and Kalluri 2005; Ortega and Werb 2002).

Type IV collagen which is characteristic of vertebrates is also widely distributed in invertebrates and more particularly in sponges. Homoscleromorpha, earlier included in the class Demospongiae, is now one of the four main sponge taxa, and the common characteristic between homoscleromorpha and eumetazoa is the presence of a basal membrane with type IV collagen. Other types of collagen are reported in Demospongiae sps. A family of collagens comprising a collagenous domain of about 120 Gly-Xaa-Yaa triplets and a carboxy-terminal region sharing some similarities with cuticular and fibril-associated collagens of nematodes and vertebrates, respectively, with interrupted triple helices, has been reported in the sponge *Microciona prolifera* (Aho et al. 1993). In addition, a fibrillar collagen chain and a short-chain collagen family have been described in the freshwater sponge *Ephydatia muelleri* (Figs. 19.5 and 19.6) (Exposito and Garrone 1990; Exposito et al. 1991). High expression levels of the genes encoding these two collagen families were reported during the early development of sponges from the asexual buds. During the developmental process, striated fibrils and the fibrillar collagens are found to be involved in the formation of striated fibrils while spongins are made by the short-term collagens. Genes encoding the sponge short-chain collagens are highly expressed in cells located in the epithelial layer and around the inorganic skeleton. At molecular level, the short-chain collagens of spongin

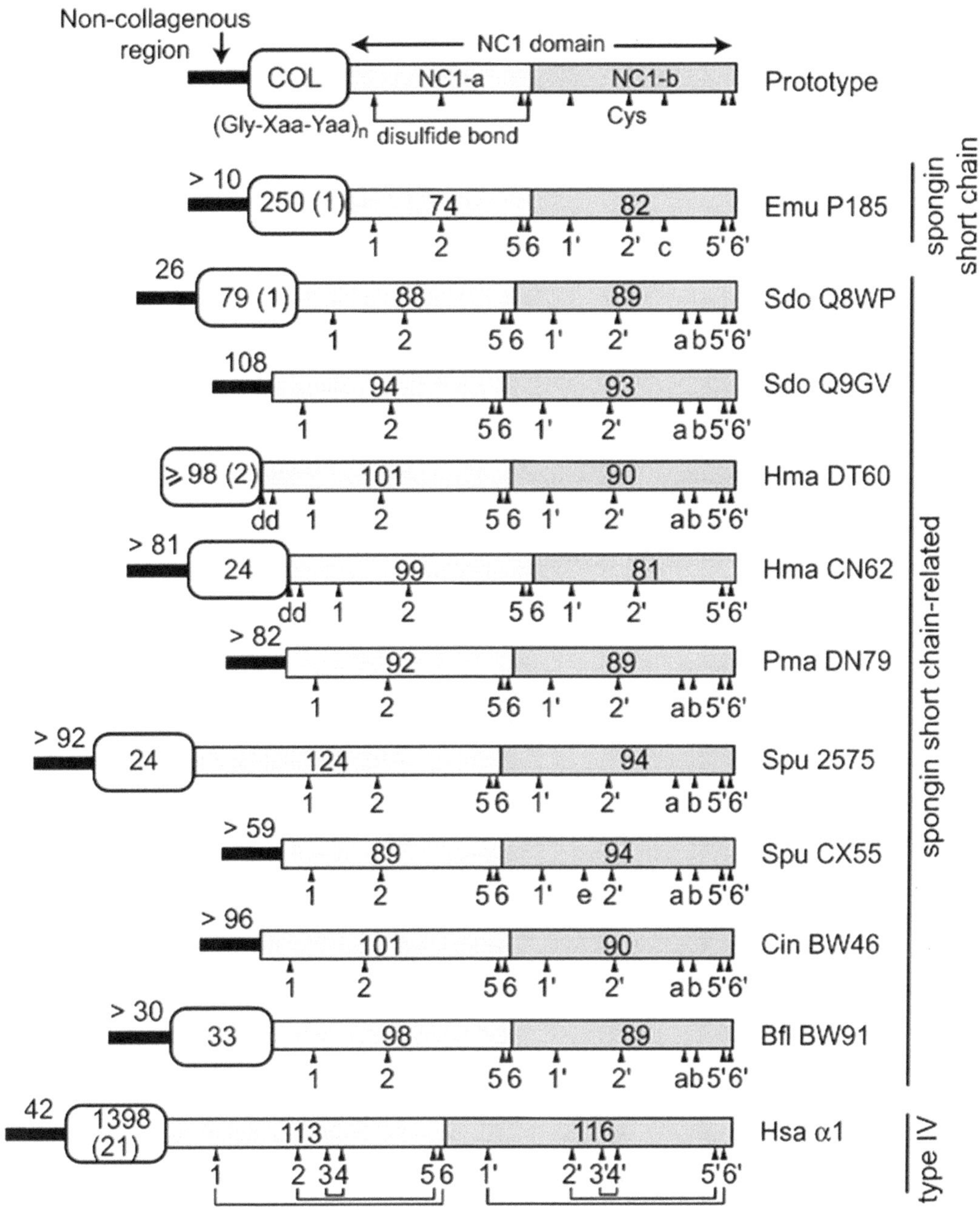

Fig. 19.5 A prototypal protein consisting of different regions and structural motifs is depicted at the top. *Arabic numbers in bold* represent the length in amino acids of different regions. *Arabic numbers in parentheses* indicate the number of interruptions within the collagenous domain. The critical cysteine residues are represented by *black triangles* and labeled according to their position in multiple sequence alignments (Reproduced with permission from Aouacheria et al. 2006)

contain two collagenous domains encompassing 79 Gly-Xaa-Yaa triplets and three non-collagenous domains. The presence of non-collagenous C-terminal domain in two proteins of the sponge *Suberites domuncula* with one of them including a short collagenous

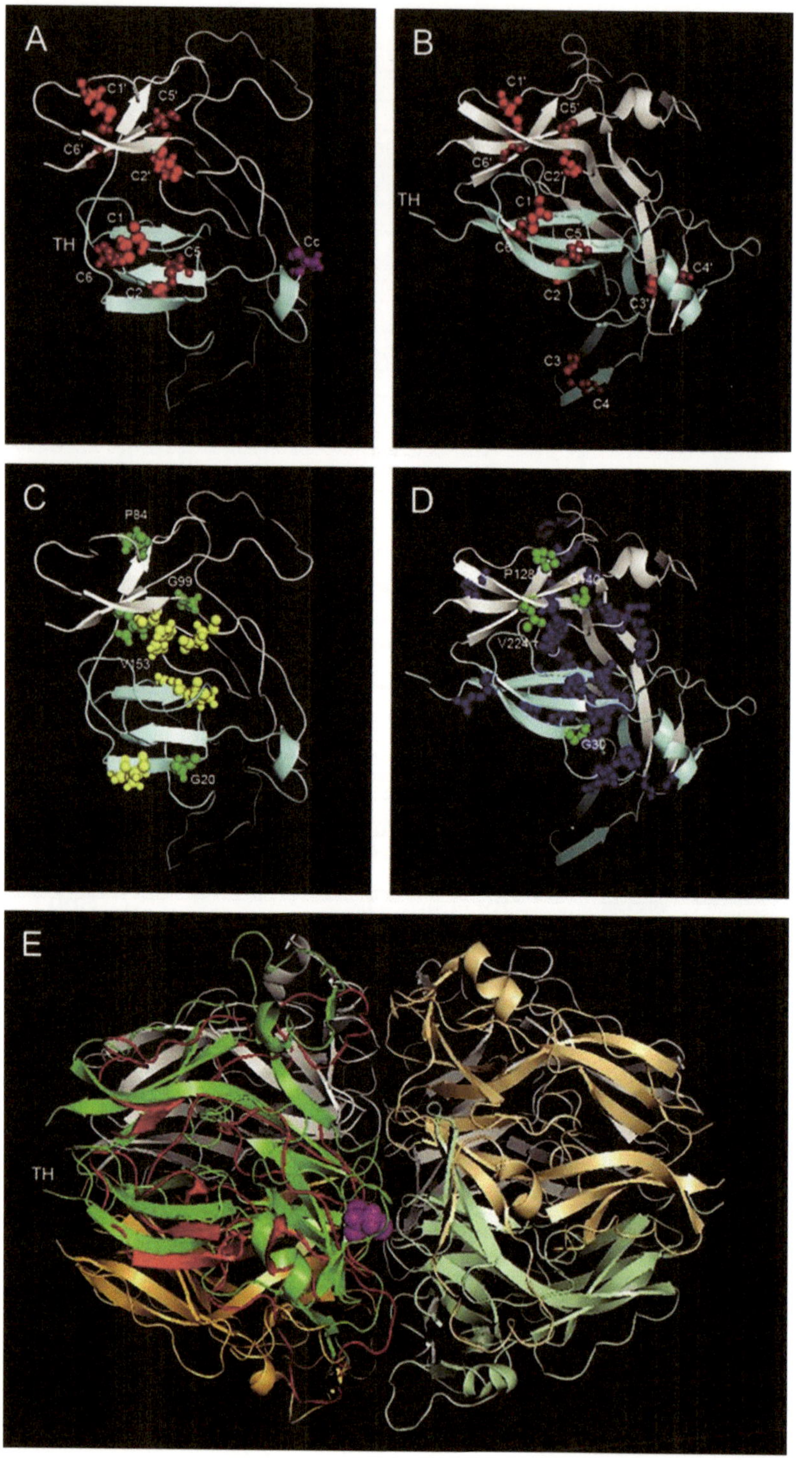

Fig. 19.6 Homology derived model of *Ephydatia muelleri* spongin short-chain collagen NC1 domain. (**a**, **c**) Represent *E. muelleri* spongin short-chain collagen. (**b**, **d**) Represent human α1 collagen NC1 domains: conserved residues within the spongin short-chain collagen-related family are indicated as *yellow balls* (**c**), and *blue balls* in type IV collagen NC1 domains (**d**). Conserved residues common for both the families are marked as *green balls*. (**e**) Constructed using Pymol depicts the type IV collagen NC1 hexamer down the twofold pseudo-exact axis (Reproduced with permission from Aouacheria et al. 2006)

Table 19.2 Summarized data on sponge collagens as taken from NCBI

S. no.	Source	Phylum/class	No. of amino acids	Accession number	Publication
1	*Aphrocallistes vastus* (cloud sponge)	Porifera; Hexactinellida	784	CAL69616	Müller et al. (2007)
2	*Clathria prolifera*	Porifera; Demospongiae	380	AAA29291	Aho et al. (1993)
3	*Chondrosia reniformis* (nonfibrillar collagen)	Porifera; Demospongiae	743	ABI79457	Biologia, University of Genova (source)
4	*Ephydatia muelleri* (partial short chain)	Porifera; Demospongiae;	366	CAA36831	Exposito et al. (1990)
5	*Amphimedon queenslandica* (fibrillar collagen COL1alpha, partial)	Porifera; Demospongiae	111	CAQ63559	Exposito et al. (2008)
6	*Amphimedon queenslandica*	Porifera; Demospongiae	114	CAQ63560	Exposito et al. (2008)
7	*Amphimedon queenslandica* (fibrillar collagen COL7alpha, partial)	Porifera; Demospongiae	268	CAQ63563	Exposito et al. (2008)
8	*Amphimedon queenslandica* (fibrillar collagen COL6alpha, partial)	Porifera; Demospongiae	268	CAQ63562	Exposito et al. (2008)
9	*Amphimedon queenslandica* (fibrillar collagen COL5alpha, partial)	Porifera; Demospongiae	211	CAQ63561	Exposito et al. (2008)
10	*Suberites domuncula*	Porifera; Demospongiae	295	CAC03736	Krasko et al. (2000)
11	*Suberites domuncula*	Porifera; Demospongiae	282	CAC81019	Unpublished
12	*Suberites domuncula*	Porifera; Demospongiae	120	CAC38782	Schröder et al. (2000)
13	*Suberites domuncula*	Porifera; Demospongiae	330	CAC03737	Krasko et al. (2000)

domain of 24 Gly-Xaa-Yaa triplets is of significance.

19.7 Sponge Collagen

Sponges, belonging to Parazoa, have the simplest anatomical organization with their body containing an internal tissue called mesohyl surrounded by an outer layer of pinacoderm cells. Mesohyl is composed of cells and a structured matrix of collagen fibrils. In horny sponges (Demospongiae), the siliceous spicules are glued together by collagenous microfibrillar cement. In glass sponges (Hexactinnelidae), fibrillar collagen is also present within the spicules and acts like a template for biosilicification process. The occurrence of fibrillar collagen in glass sponges which evolved during the Cambrian period of the Palaeozoic era reflects the significance of collagen in the evolution of the earliest metazoans and their skeletons. Collagen-controlled highly structured silica networks of the primitive metazoans must have evolved over ages to form the skeletal elements (bones and teeth) of higher vertebrates. Data on sponge collagens, available at NCBI, is summarized in Table 19.2.

19.8 Collagen Immunogenicity

Natural collagens isolated from calf skin have applications in the fields of medicine, surgery, and cosmetics. They are used as shields, injectable dispersions, sponges, and microparticles. However, bovine collagens are reported to elicit immunogenicity as is the case in bovine spongiform encephalopathy (BSE) and transmissible spongiform encephalopathy (TSE) and hence sponge collagens may be considered as alternatives. In such a case, thorough investigations on the ultrastructure and biochemical properties of sponge collagens need to be carried out. The major drawbacks that hamper these studies are the insolubility and the mineralization of the sponge collagens. Swatschek et al. (2002) demonstrated the marine demosponge, *Chondrosia reniformis* Nardo, as a potential candidate and developed a standard protocol for the isolation of collagen. With the study, it became evident that the conventional collagen can be substituted by the collagen of marine origin. Pioneering investigations on the fine structure and physicochemical properties of the collagen of the marine sponge *Chondrosia reniformis* Nardo were made by Garrone et al. (1975). The amino acid composition of sponge collagen was similar to that of vertebrate collagen, but the infrared spectra obtained from the whole cortex of the sponge showed variations in some typical peaks when compared to pure collagen. X-ray investigation data indicated the classical helical structure of the *Chondrosia* sponge collagen with an apparent period of about 22 nm and a diameter of about 20 nm. Transmission electron microscopy of stained fibrils exposed a periodic banding pattern of one dark and two light segments alternating with one another. For the first time, topographical details of the segments were elucidated by using atomic force microscopy (AFM) and an advanced model of the collagen's ultrastructure and organization was provided by Heinemann et al. (2007). The cortex of *Chondrosia reniformis* is demonstrated to be composed of interlacing collagen fibers. Studies of Garrone et al. (1975) via TEM showed that single fibrils are intertwined into bundles by minute filamentous connections. However, the collagen isolated by Heinemann et al. (2007) by acid supported solubilization occurred as thick and long bundles. On treatment with acid, fibrils of soluble collagen type I degrade into the monomer tropocollagen but neutral buffers induce the self-assembly of the tropocollagen into fibrils. Corroborative results from the studies of Imhoff and Garrone (1983) and Heinemann et al. (2007) suggest that the strong fibrillar cross-links in the native sponge collagen are functionally similar to the covalent cross-links that stabilize the fibrillar aggregates in vertebrates; yet the chemistry of the cross-links remains to be established. The isolated *Chondrosia reniformis* sponge collagen showed no separation of fibers from the bundles even after prolonged suspension with stirring for 7 days in acetic acid medium. On the contrary, the use of neutral buffer solutions like 0.1M Tris/Hcl (pH 7.4) facilitated formation of homogeneous milk-like suspensions within a few hours of stirring. After 7 days, the solvent accomplished the plain separation of the fibers into single collagen fibrils. Unlike collagen type I, these fibrils were quite flexible and exhibited no breaks or kinks. Hence, it is assumed that the solvent unstitches the interfibrillar filaments, responsible for the aggregation of the fibrils to form fibers. The observed length of single fibrils runs up to several hundred micrometers, and considering the measured average diameter of about 20 nm, aspect ratios of about 1:5000 have been reported and these observations correlate with the reported data of other invertebrate collagen fibrils as exemplified in the cases of spine ligaments of *Eucidaris tribuloides* (1:2500), sea urchin (1:5000), and dermis of sea cucumber *Cucumaria frondosa* (1:2000). Extraction of collagen from connective tissue is normally done by alkali treatment. This type of collagen I is generally found to lose the ability to form fibrils at neutral pH although the triple-helical conformation of the collagen molecule was maintained through the period of treatment. In the case of *Chondrosia* collagen treated in 0.1M NaOH, hydrolysis led to the formation of clear solutions in a few days and the AFM imaging showed unequal fragments ranging from nanoparticles to short fibrils or huge aggregates. A comparison of the amino acid compositions of *Chondrosia* collagen with that of bovine type I collagen and the collagen of *Hyalonema sieboldi* reveals certain differences like reduced values for glycine and hydroxyproline in *Chondrosia* collagens, and this could be attributed to either the impurities of glycoproteins or non-triple-helical portions in the analyzed collagens. However, the deviations are restricted to only certain amino acids and hence the overall composition remains more or less the same.

Based on the results of solubilization experiments, single fibrils were favored for further purification and analysis. Preprocessing procedure includes suspension of fibrils in 0.1M Tris/HCl buffer pH 7.4 solution for 7 days at 4 °C. Then the suspensions filtered to remove insoluble constituents before lyophilization. Dialysis and lyophilization are done several times and the final lyophilizate was analyzed by Fourier transform-infrared reflection absorption spectroscopy (FT-IRAS). This is a method used to study changes in the secondary structure of collagen and collagen denaturation, cross-linking, thermal self-assembly, and comparison with gelatin. A comparison of the FT-IRAS spectra of the *Chondrosia* collagen with that of calf skin collagen indicated that the two collagens are very similar. The wide scan infrared spectra showed the typical bands such as N-H stretching for the amide A (3330 cm^{-1}), C-H stretching for the amide B (3070 cm^{-1}), C=O stretching for the amide I (1660 cm^{-1}, 1634 cm^{-1}), N-H deformation for the amide II (1555 cm^{-1}, 1537 cm^{-1}), and N-H deformation for the amide III (1341 cm^{-1}, 1281 cm^{-1}, 1239 cm^{-1}, 1205 cm^{-1}). The clear correlation of nearly all peak positions and band intensities is suggestive of homology despite phylogenetic differences. A clear distinction, hitherto, has not been established between sponge collagen and bovine type I collagen.

Transversal and longitudinal height profiles of AFM images were measured to understand the morphological details, especially the banding pattern of the separated *Chondrosia reniformis* sponge collagen fibrils. Accordingly, two groups of peaks were identified corresponding to the course of the observed height level. Along the fibril, one characteristically thick protrusion-like segment of about 28 nm in diameter is followed by two equal thinner and closer conjoined inter-band segments of about 20 nm in diameter, respectively. The average distance between the protrusion is about 67–69 nm. Between two following peaks of the inter-band regions or between a protrusion and adjoined inter-band region about 21–23 nm was measured and the average step height between the protrusions and the inter-band regions was calculated to be about 4 nm. Thus the new model proposed by Heinemann et al. (2007) envisages the combination of the prime topographical data obtained by AFM and the results of TEM investigations and hence stands as a refined model of what has been proposed by Garrone et al. (1975).

19.9 Sponge Collagen Nanoparticles and Drug Delivery

The use of nanoparticulate drug delivery systems confers many advantages such as sustained release, improved bioavailability, reduced side effects, and drug protection against enzymatic and chemical degradation. Collagen, a biodegradable biomaterial, has till date been used for drug delivery, as shields in ophthalmology, injectable dispersions for local tumor treatment, as a scaffold, or for transdermal drug delivery (Aishwarya et al. 2008; Friess 1998; Kleinmann et al. 2007; Rössler et al. 1994; Swatschek et al. 2002; Takezawa et al. 2007). Collagen isolated from the marine sponge *Chondrosia reniformis* Nardo does not bear the risk of bovine spongiform encephalopathy (BSE) and transmissible spongiform encephalopathy (TSE).

Transdermal administration of estradiol is found to be advantageous over oral formulations and earlier studies established that the collagen from the marine sponge *Chondrosia reniformis* Nardo (Demospongiae, Hadromerida, Chondrosiidae) could be used for the dermal delivery of trans retinol. This prompted the preparation and characterization of marine sponge collagen nanoparticles and their employment for the transdermal delivery of 17-β-estradiol-hemihydrate by Nicklas et al. (2009). The results of the study showed that these particles could be used as penetration enhancers in hormone replacement therapy. For the study, collagen nanoparticles were prepared by controlled alkaline hydrolysis and characterized using atomic force microscopy and photon correlation spectroscopy.

One of the key aspects of tissue engineering is the use of scaffold matrices with properties that closely match the properties of the tissue it would replace. An ideal scaffold for tissue engineering, viz., bone, must possess suitable biocompatibility and osteoconductive and osteoinductive capacities in addition to the structure that mimics the trabecular network of bone tissue. Studies on synthetic and natural biomaterials suggest that natural scaffolds display highly optimized structures and comprise extracellular matrix components that offer a foundation for cell attachment, migration, and proliferation. Marine sponges belonging to phylum Porifera are important components of the ecosystem with economical and scientific importance (Lin et al. 2011). More recently marine sponges are identified as potential sources of therapeutic drugs and antibiotic substances. They are shown to display a structure which is very similar to the architecture of bone tissue. The complex canal system in sponges creates a porous environment which is ideal for cellular integration when combined with cells for tissue engineering.

The sponges collected from the Fremantle coast of Western Australia are a member of the sponge family Callyspongiidae (belonging to the order Haplosleridia). The sponge is irregular in shape and cushion-like and scattered with oscules at regular intervals over its surface. The microstructure of the sponge, as examined by SEM, is composed of choanosomal skeleton and a rectangular meshed network, made of regularly interconnected spongin fibers which differentiate into the primary, secondary, and occasional tertiary fibers. Primary fibers are generally ramified to form secondary and tertiary fibers. The spongin fibers range from 30 to 50 μm in diameter and create pores ranging from 100 to 300 μm in diameter. The fibers of the scaffold contain occasional siliceous spicules which are attached or embedded within the fibrous network in a random orientation. The spicules are 20–40 μm in length and align longitudinally along the fiber axis (Lin et al. 2011).

Collagens are considered to represent the most important molecular innovations in the metazoan evolution. All eumetazoa contain basement membrane type IV collagen and was also found in the sponge group Homoscleromorpha with a well-organized epithelium. In contrast, spongin appears to be a demosponge-specific collagenous protein which can effectively substitute an inorganic skeleton, as known in bath sponge. A family of short-chain collagens that are likely to be the main components of spongins were characterized from the freshwater sponge *Ephydatia muelleri* (Aouacheria et al. 2006). Further, evidence of remote homology between the carboxy-terminal non-collagenous NC1 domain of spongin short-chain collagens and type IV collagen was presented using a combination of sequence and structure homology methods. Retrieval of spongin short-chain collagen-related proteins from non-sponge animals suggests that a sponging-related family constitutes an evolutionary sister to the type IV collagen family.

19.10 Conclusions

From the extensive research over the past four decades, 28 types of collagens' molecular and functional properties were deciphered and this knowledge paved way for understanding both the genetic and the nongenetic collagen-associated diseases and their treatment by cell therapy. Spongin short-chain and type IV collagens are the members of the oldest modular proteins that were revealed to be unique to metazoa. Both spongin and collagen give flexibility and support to the tissues. Research studies indicate that the natural marine sponge skeleton could be favored as a bioscaffold for the repair of bone defects and hence could be effectively employed in bone tissue engineering.

Acknowledgements DSK thanks the director of the Indian Institute of Technology Hyderabad for providing facilities to accomplish the present book chapter. KS expresses her sincere thanks to the principal and the management of RVR & JC College of Engineering for their continued support and encouragement.

References

Aho S, Turakainen H, Onnela M-L, Boedtker H (1993) Characterization of an intronless collagen gene family in the marine sponge *Microciona prolifera*. PNAS 90(15):7288–7292. doi:10.2307/2362698

Aishwarya S, Mahalakshmi S, Sehgal PK (2008) Collagen-coated polycaprolactone microparticles as a controlled drug delivery system. J Microencapsul 25(5):298–306. doi:10.1080/02652040801972004

Aouacheria A, Geourjon C, Aghajari N, Navratil V, Deléage G, Lethias C, Exposito J-Y (2006) Insights into early extracellular matrix evolution: spongin short chain collagen-related proteins are homologous to basement membrane type IV collagens and form a novel family widely distributed in invertebrates. Mol Biol Evol 23(12):2288–2302. doi:10.1093/molbev/msl100

Bella J, Eaton M, Brodsky B, Berman H (1994) Crystal and molecular structure of a collagen-like peptide at 1.9 Å resolution. Science 266:75–81

Bella J, Liu J, Kramer R, Brodsky B, Berman HM (2006) Conformational effects of Gly–X–Gly interruptions in the collagen triple helix. J Mol Biol 362(2):298–311. doi:http://dx.doi.org/10.1016/j.jmb.2006.07.014

Borza D-B, Bondar O, Ninomiya Y, Sado Y, Naito I, Todd P, Hudson BG (2001) The NC1 domain of collagen IV encodes a novel network composed of the α1, α2, α5, and α6 chains in smooth muscle basement membranes. J Biol Chem 276(30):28532–28540. doi:10.1074/jbc.M103690200

Boutaud A, Borza D-B, Bondar O, Gunwar S, Netzer K-O, Singh N, Ninomiya Y, Sado Y, Noelken ME, Hudson BG (2000) Type IV collagen of the glomerular basement membrane: evidence that the chain specificity of network assembly is encoded by the noncollagenous nc1 domains. J Biol Chem 275(39):30716–30724. doi:10.1074/jbc.M004569200

Boute N, Exposito J-Y, Boury-Esnault N, Vacelet J, Noro N, Miyazaki K, Yoshizato K, Garrone R (1996) Type IV collagen in sponges, the missing link in basement membrane ubiquity. Biol Cell 88(1–2):37–44. doi:10.1016/S0248-4900(97)86829-3

Branden C, Tooze J (1999) Introduction to protein structure, 2nd edn. Garland Publishing, Taylor and Francis group, New York

Erickson AC, Couchman JR (2000) Still more complexity in mammalian basement membranes. J Histochem Cytochem 48(10):1291–1306. doi:10.1177/002215540004801001

Exposito JY, Garrone R (1990) Characterization of a fibrillar collagen gene in sponges reveals the early evolutionary appearance of two collagen gene families. PNAS 87(17):6669–6673

Exposito J-Y, Ouazana R, Garrone R (1990) Cloning and sequencing of a porifera partial cDNA coding for a short-chain collagen. Eur J Biochem 190(2):401–406. doi:10.1111/j.1432-1033.1990.tb15589.x

Exposito JY, Le Guellec D, Lu Q, Garrone R (1991) Short chain collagens in sponges are encoded by a family of closely related genes. J Biol Chem 266(32):21923–21928

Exposito J-Y, Larroux C, Cluzel C, Valcourt U, Lethias C, Degnan BM (2008) Demosponge and sea anemone fibrillar collagen diversity reveals the early emergence of A/C clades and the maintenance of the modular structure of type V/XI collagens from sponge to human. J Biol Chem 283(42):28226–28235. doi:10.1074/jbc.M804573200

Friess W (1998) Collagen – biomaterial for drug delivery. Eur J Pharm Biopharm 45(2):113–136. doi:http://dx.doi.org/10.1016/S0939-6411(98)00017-4

Garrone R, Huc A, Junqua S (1975) Fine structure and physicochemical studies on the collagen of the marine sponge *Chondrosia reniformis* Nardo. J Ultrastruct Res 52(2):261–275. doi:http://dx.doi.org/10.1016/S0022-5320(75)80117-1

Hamano Y, Kalluri R (2005) Tumstatin, the NC1 domain of α3 chain of type IV collagen, is an endogenous inhibitor of pathological angiogenesis and suppresses tumor growth. Biochem Biophys Res Commun 333(2):292–298. doi:http://dx.doi.org/10.1016/j.bbrc.2005.05.130

Heinemann S, Ehrlich H, Douglas T, Heinemann C, Worch H, Schatton W, Hanke T (2007) Ultrastructural studies on the collagen of the marine sponge *Chondrosia reniformis* Nardo. Biomacromolecules 8(11):3452–3457. doi:10.1021/bm700574y

Heino J (2007) The collagen family members as cell adhesion proteins. BioEssays 29(10):1001–1010. doi:10.1002/bies.20636

Hudson BG, Reeders ST, Tryggvason K (1993) Type IV collagen: structure, gene organization, and role in human diseases. Molecular basis of Goodpasture and Alport syndromes and diffuse leiomyomatosis. J Biol Chem 268(35):26033–26036

Hudson BG, Tryggvason K, Sundaramoorthy M, Neilson EG (2003) Alport's syndrome, Goodpasture's syndrome, and type IV collagen. N Engl J Med 348(25):2543–2556. doi:10.1056/NEJMra022296

Imhoff JM, Garrone R (1983) Solubilization and characterization of *Chondrosia reniformis* sponge collagen. Connect Tissue Res 11(2–3):193–197. doi:10.3109/03008208309004855

Khoshnoodi J, Sigmundsson K, Cartailler J-P, Bondar O, Sundaramoorthy M, Hudson BG (2006) Mechanism of chain selection in the assembly of collagen IV: a prominent role for the α2 chain. J Biol Chem 281(9):6058–6069. doi:10.1074/jbc.M506555200

Kleinmann G, Larson S, Hunter B, Stevens S, Mamalis N, Olson RJ (2007) Collagen shields as a drug delivery system for the fourth-generation fluoroquinolones. Ophthalmologica 221(1):51–56

Krasko A, Lorenz B, Batel R, Schröder HC, Müller IM, Müller WEG (2000) Expression of silicatein and collagen genes in the marine sponge *Suberites domuncula* is controlled by silicate and myotrophin. Eur J

Biochem 267(15):4878–4887. doi:10.1046/j.1432-1327.2000.01547.x

Lin Z, Solomon KL, Zhang X, Pavlos NJ, Abel T, Willers C, Dai K, Xu J, Zheng Q, Zheng M (2011) In vitro evaluation of natural marine sponge collagen as a scaffold for bone tissue engineering. Int J Biol Sci 7(7):968–977. doi:10.7150/ijbs.7.968

Lu P, Takai K, Weaver VM, Werb Z (2011) Extracellular matrix degradation and remodeling in development and disease. Cold Spring Harb Perspect Biol 3 (12):1–24. doi:10.1101/cshperspect.a005058

Müller WG, Eckert C, Kropf K, Wang X, Schloßmacher U, Seckert C, Wolf S, Tremel W, Schröder H (2007) Formation of giant spicules in the deep-sea hexactinellid *Monorhaphis chuni* (Schulze 1904): electron-microscopic and biochemical studies. Cell Tissue Res 329(2):363–378. doi:10.1007/s00441-007-0402-x

Myllyharju J, Kivirikko KI (2001) Collagens and collagen-related diseases. Ann Med 33(1):7–21. doi:10.3109/07853890109002055

Nicklas M, Schatton W, Heinemann S, Hanke T, Kreuter J (2009) Preparation and characterization of marine sponge collagen nanoparticles and employment for the transdermal delivery of 17β-estradiol-hemihydrate. Drugs Dev Ind Pharm 35(9):1035–1042. doi:10.1080/03639040902755213

Ortega N, Werb Z (2002) New functional roles for non-collagenous domains of basement membrane collagens. J Cell Sci 115(22):4201–4214. doi:10.1242/jcs.00106

Ricard-Blum S, Ruggiero F (2005) The collagen superfamily: from the extracellular matrix to the cell membrane. Pathol Biol 53(7):430–442. doi:http://dx.doi.org/10.1016/j.patbio.2004.12.024

Rössler B, Kreuter J, Ross G (1994) Effect of collagen microparticles on the stability of retinol and its absorption into hairless mouse skin in vitro. Pharmazie 49(2–3):175–179

Schröder HC, Krasko A, Batel R, Skorokhod A, Pahler S, Kruse M, Müller IM, Müller WEG (2000) Stimulation of protein (collagen) synthesis in sponge cells by a cardiac myotrophin-related molecule from *Suberites domuncula*. FASEB J 14(13):2022–2031. doi:10.1096/fj.00-0043com

Silver FH (2009) The importance of collagen fibers in vertebrate biology. J Eng Fiber Fabr 4(2):9–17

Söder S, Pöschl E (2004) The NC1 domain of human collagen IV is necessary to initiate triple helix formation. Biochem Biophys Res Commun 325(1):276–280. doi:http://dx.doi.org/10.1016/j.bbrc.2004.10.034

Söderhäll C, Marenholz I, Kerscher T, Rüschendorf F, Esparza-Gordillo J, Worm M, Gruber C, Mayr G, Albrecht M, Rohde K, Schulz H, Wahn U, Hubner N, Lee Y-A (2007) Variants in a novel epidermal collagen gene *COL29A1* are associated with atopic dermatitis. PLoS Biol 5(9):e242. doi:10.1371/journal.pbio.0050242

Swatschek D, Schatton W, Kellermann J, Müller WEG, Kreuter J (2002) Marine sponge collagen: isolation, characterization and effects on the skin parameters surface-pH, moisture and sebum. Eur J Pharm Biopharm 53(1):107–113. doi:http://dx.doi.org/10.1016/S0939-6411(01)00192-8

Takezawa T, Takeuchi T, Nitani A, Takayama Y, Kinooka M, Taya M, Enosawa S (2007) Collagen vitrigel membrane useful for paracrine assays *in vitro* and drug delivery systems *in vivo*. J Biotechnol 131(1):76–83. doi:http://dx.doi.org/10.1016/j.jbiotec.2007.05.033

Timpl R, Wiedemann H, Van Delden V, Furthmayr H, KÜHn K (1981) A network model for the organization of type IV collagen molecules in basement membranes. Eur J Biochem 120(2):203–211. doi:10.1111/j.1432-1033.1981.tb05690.x

Biomedical Applications of Marine Sponge Collagens

20

Ramjee Pallela, Hermann Ehrlich, and Ira Bhatnagar

Abstract

Collagens are the most abundant protein present in vertebrates and invertebrate organisms and play major roles in their structural organizations, body flexibility, and elastic properties. The major sources of collagen production are porcine and bovine origin, which are widely used for nutraceutical, pharmaceutical, and cosmeceutical developments for humankind. However, problems exist in transmissible diseases from bovine and porcine source like bovine spongiform encephalopathy disease. Thus, there is need to search some alternative source for collagen, likely fish and sponges. Marine sponge is a significant and unexplored source for collagen productions until now. Only few reports have been suggested that marine sponge collagen can be used for biological and biomedical applications (tissue engineering and drug delivery). Marine sponge collagens have a capacity for cell adhesion, ability to form pores, and capability to induce the osteogenic differentiation, which makes marine sponge collagen suitable for tissue engineering purpose. In the present chapter, we have discussed about isolation and use of marine sponge collagen in the form of bioscaffolds and nanoparticles in the areas of regenerative medicine and drug delivery, respectively. Recent technological methods have been provided for marine sponge collagen with biomedical applications, which makes marine sponge collagen an alternative source for industrial developments for the commercial usage.

Keywords

Marine sponge collagen • Tissue engineering and drug delivery

R. Pallela (✉)
IKP Knowledge Park, Genome Valley, Turkapally, Hyderabad 500078, Telangana, India
e-mail: rpallela@gmail.com

H. Ehrlich
Institute of Experimental Physics, TU Bergakademie Freiberg, 09599 Freiberg, Germany
e-mail: Hermann.Ehrlich@physik.tu-freiberg.de

I. Bhatnagar
Center for Cellular Molecular Biology, Habsiguda, Uppal Road, Hyderabad 500007, Telangana, India
e-mail: ira@ccmb.res.in; ibhatnagar@gmail.com

R. Pallela, H. Ehrlich (eds.), *Marine Sponges: Chemicobiological and Biomedical Applications*,
DOI 10.1007/978-81-322-2794-6_20

20.1 Introduction

In the previous chapters, extensive information on the chemicobiological and biomedical properties of molecules/materials from marine sponges and their symbionts has been covered (Sipkema et al. 2005; Rao et al. 2011). Past two to three decades of marine sponge research, although presumably, focused on the importance of sponge-derived molecules in biological and pharmaceutical fields and also in biomedicine with prior information on the real applications of these molecules and materials. Still, even with that vast treasure chest of natural compounds, researchers say finding something promising is rare. According to David Newman, who heads the Natural Products branch at the National Cancer Institute (NCI, Maryland), "*Mother Nature has been doing her chemistry for three billion years, while chemists have been at work for less than 300*" (Shapiro 2014). Although sponges are uniquely referred for their vast diversity in their species level and chemical molecules, the hallmark protein, collagen, should not be avoided when referring these marvelous organisms.

Among the sponges, Demospongiae is the largest and most diverse class, which unites sponges with siliceous spicules (either monaxonic or tetraxonic, never triaxonic) and/or with a skeleton of organic fibers or fibrillar collagen (Van Soest et al. 2012; Heinemann et al. 2007). Although numerous collagen families have been phylogenetically characterized in metazoans, only two of them are present from marine sponges corresponding to human collagens, i.e., the fibrillar (type I) and the basement membrane (type IV) collagens (Aouacheria et al. 2006; Pallela et al. 2011; Addad et al. 2011). In earlier days of sponge collagen research, Gross and colleagues isolated two distinct forms of collagens, named as spongin A and spongin B. The collagen lineages discovered in these species became a landmark for identifying potential sponge candidates for their possible implications in current biomedical applications. Exposito and Garrone have characterized cDNA and genomic clone coding for a sponge collagen. Their previous studies showed the existence of a nonfibrillar collagen in the same sponge species, which demonstrates that at least two collagen gene families are represented in the most primitive metazoan (Exposito and Garrone 1990). However, studies on the comparative significance of collagen genes from sponge with humans remain far behind for expediting the sponge collagen research towards the notion of applying them in biology and biomedicine.

Five different types of collagen are available in animal tissues. Type I, type II, type III, type IV, and type V are the most abundant ones present in invertebrates and vertebrates (Deyl et al. 2003). Collagens are formed by polypeptide chain mainly based on Gly-X-Y, X, and Y designated as proline and hydroxyproline amino acids (Gómez-Guillén et al. 2002; Kołodziejska et al. 1999; Mendis et al. 2005). But, in the case of sponges, particularly *H. sieboldi* glass sponge, -X-Y position can engage with isomer of hydroxyproline (3-hydroxyaproline and 4-hydroxyaproline), which is responsible for biomineralization process (Ehrlich et al. 2010; Ehrlich 2010; Silva et al. 2014). Complete cDNA sequence of a nonfibrillar collagen has been isolated from the marine sponge *Chondrosia reniformis*, Nardo 1847, using a PCR approach. The phylogenetic analysis on the deduced amino acid sequence of C-terminal end shows that the isolated sequence belongs to the short-chain spongin-like collagen subfamily, a nonfibrillar group of invertebrate collagens similar to type IV collagen (Pozzolini et al. 2012).

Collagen has advantages of biodegradability, low toxicity, and immunogenicity. Collagen can be derived from several sources including bovine (Francis and Thomas 1975; Quereshi et al. 2010), but often collagen from bovine origin associates the risk of bovine spongiform encephalopathy (BSE) disease. Therefore, collagen from other sources is an attractive alternative, especially jellyfish (Kimura et al. 1983; Miura and Kimura 1985; Song et al. 2006; Nagai et al. 2000) and sponges (Garrone 1985; Heinemann et al. 2007; Swatschek et al. 2002a; Gruner et al. 1993). Jellyfish-derived collagen scaffold has been prepared using 1-ethyl-(3-3-dimethylaminopropyl)carbodiimide hydrochloride/N-hydroxysuccinimide for tissue engineering purpose. These are highly porous non-cytotoxic with higher cell proliferation than other collagen materials (bovine) (Song et al. 2006).

Collagens from invertebrate sources (sponges) are considered for biocompatible studies comparably with vertebrate collagens. Till now very few reports of sponge research highlight the ultrastructure, biochemical properties, and applications of collagen derived from the marine sponges (Heinemann et al. 2007; Pallela et al. 2011, 2012). State-of-the-art description of spongins and collagens has well been presented in the recent narrations of monograph by Prof. Hermann Ehrlich to describe the chemico-biological and biophysical analyses of these biomedically important materials (Ehrlich 2010). Because of difficulties in isolation and purification of marine sponge collagens, the real application of these collagens did not gain extensive pace of implication in biological and biomedical applications. Few decades of earlier research on sponge collagen purification though formed a basis of extracting pure collagens; the limited availability of sponge biomass restricted researchers not to move further to launch real marine sponge collagen products in mass. Hence, developing sustainable sponge cultivation methodologies is very necessary to meet the requirements of producing sponge materials like collagen to implement them in biology and medicine. KliniPharm scientists, through the sponsorship of European Community, screened hundreds of marine sponges from the deep waters of the Aegean Sea to identify biologically important marine sponges. As part of their study, KliniPharm cultivates sponges in underwater farms to generate sufficient raw material (potential production methods: mariculture, ex situ culture, and cell culture) for the pharmacologically as well as biomedically important products without disturbing the marine habitat (Sipkema et al. 2005). Collagen can be extracted from the marine sponge *Chondrosia reniformis* using different extraction methods as follows:

1. 0.5 M acetic acid with 10 % pepsin
2. 50 mM Tris-HCl with 1 M NaCl
3. 100 mM Tris-HCl, 10 mM EDTA, 8 M urea and 100 mM 2-mercaptoethanol

Collagen extracted from *C. reniformis* is not cytotoxic and promotes proliferation. *C. reniformis* collagen has been characterized and identified as mainly of type IV, and thus has promising application in epidermal regeneration strategies (Moreira-Silva et al. 2013).

In the current chapter, a brief application of marine sponge collagens in biomedicine and the prospects of implementing these collagens for various allied biological and biomedical fields are discussed.

20.2 Biomedical Applications of Sponge Collagens

20.2.1 Tissue Engineering

Tissue engineering is an interdisciplinary and emerging field of research to construct and/or prepare the artificial organs with the use of materials, cells, and growth factors (Nerem and Sambanis 1995; Langer and Vacanti 1993). The basic principle of tissue engineering is shown in the Fig. 20.1.

The availability of suitable scaffolds for the treatment of bone-related defects is limited. Scaffolds play a major role in the construction of artificial organs; scaffolds are often made from synthetic and natural materials. Marine sponge collagens possess a unique structure that mimics the cancellous architecture of bone tissue. The complex canal system within the sponges creates a porous environment ideal for infiltration, attachment, and growth of cells, thereby facilitating these structures suitable for tissue engineering (Lin et al. 2011). Marine sponge collagen has effective structure to support wide range of cell and tissues (Green et al. 2003). According to Aouacheria et al. (2006), collagen provides flexibility and support in sponge, whereas spicules and extracellular matrix both integrate cells into 3D structures (Aouacheria et al. 2006). Identifying the potential of marine sponges as a bioscaffold

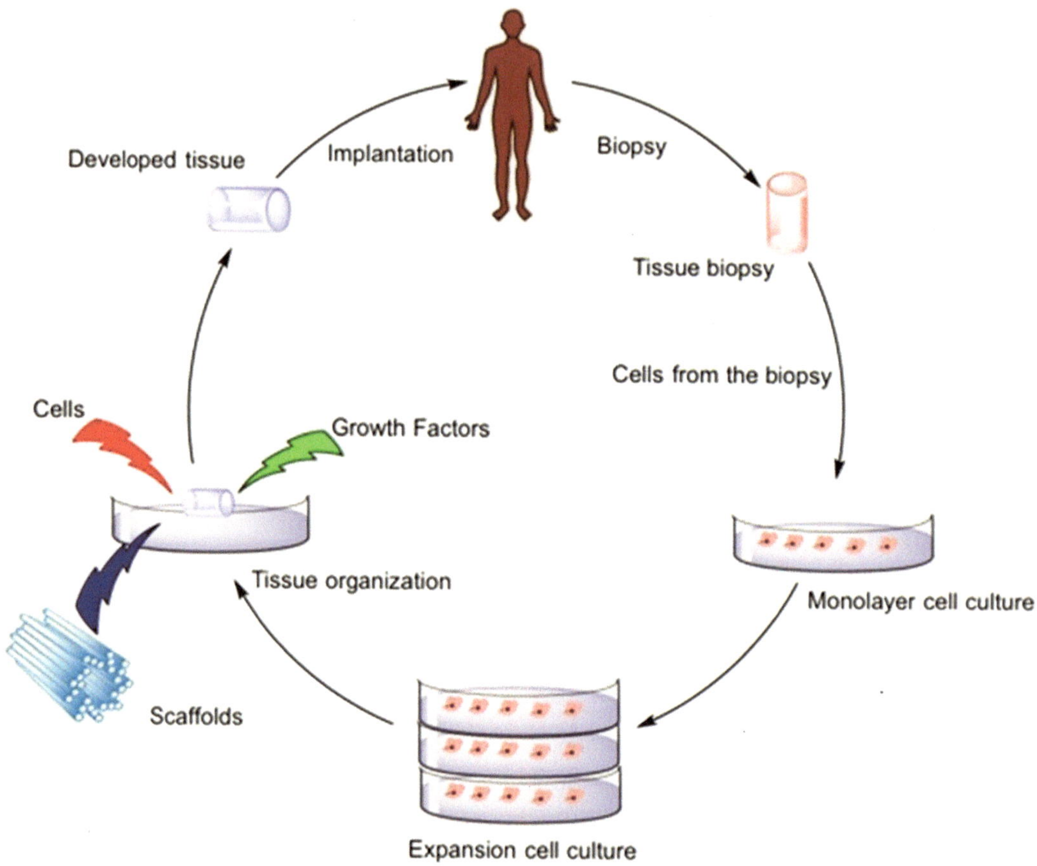

Fig. 20.1 The basic principle of tissue engineering

is very challenging, however, promoting these sponges for osteogenesis has a great advantage in tissue engineering. Silva et al. reported marine-based collagens and their potential applications towards tissue regeneration. Recently, much attention is being paid on marine-based collagen when compared to normal collagen; as shown by Silva et al. (2014) in his publication (Silva et al. 2014). Marine sponge collagen has 3D structure which consists of micrometer pore structure that is suitable for tissue growth and nutrient supplement; optimum pore size for cell migration is around 80–500 um (O'Brien et al. 2005; Griffith 2002).

Green et al. (2003) reported a natural marine sponge skeleton as a potential scaffold for tissue regeneration. Marine sponge skeleton often consists of collagen fiber (Pallela et al. 2011); commonly called spongin, this skeleton has been used for human osteoprogenitor cell attachment, growth, and differentiation. The important gene parameter for bone tissue engineering, alkaline phosphatase in sponge skeleton with osteoprogenitor cells, was significantly greater than in control (cell culture plastic) (Green et al. 2003; Granito et al. 2016).

Zheng et al. (2007) presented the marine sponge collagen bioscaffolds for bone and cartilage tissue regeneration. Marine sponge skeletons (*Hippospongia* (1), the genus *Callyspongia* (3), and the family *Chalinidae* (1)) were used as scaffolds for bone tissue repair on the basis of their collagen fiber extracellular matrix, interconnecting canal systems forming porosity, ability to hydrate to a high degree, and the diverse skeletal architecture within the phylum *Porifera* (Zheng et al. 2007) (Fig. 20.2).

On the other hand, apart from collagen, some other marine skeletal proteins have also been used for bone tissue regeneration (Green et al. 2013). Green et al. (2014) reported the usage of

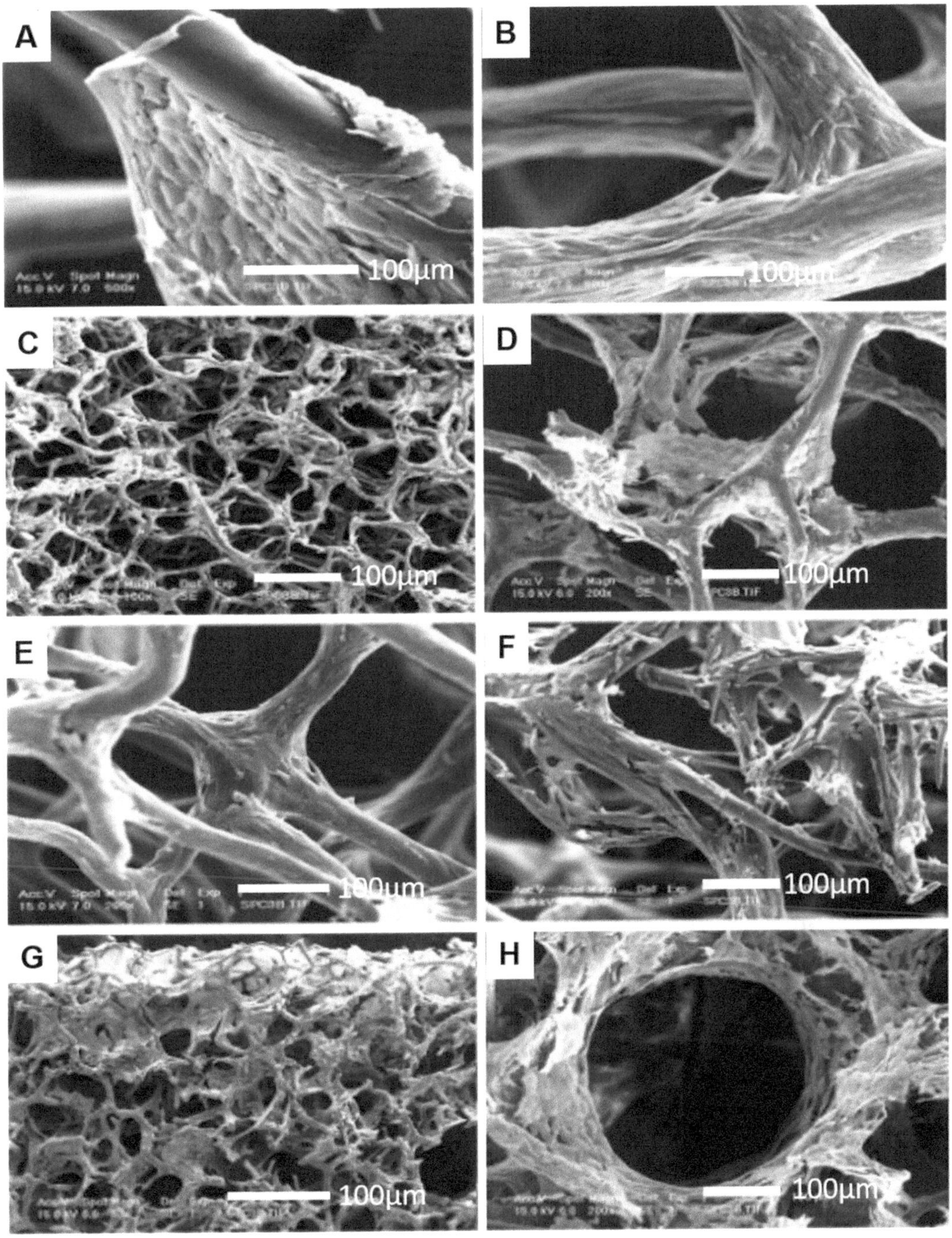

Fig. 20.2 SEM micrographs of sponge-cell constructs after 7 days of culture. (**a**, **b**) Species 1, showing cells with globular and spindle-like morphology; Species 2 at low (**c**) and high (**d**) magnification showing heterologous cell coverage; (**e**) Species 3, showing cells forming layers over the fibers of the sponge skeleton; (**f**) Species 4, showing sporadic cell growth over fibers and spicules; Species 5 at low (**g**) and high (**h**) magnification, showing limited filling-in of large pores by cells (Reproduced with permission from Zheng et al. 2007 © 2010 by the authors. This article is an open-access article distributed under the terms and conditions of the Creative Commons Attribution license (http://creativecommons.org/licenses/by/3.0/))

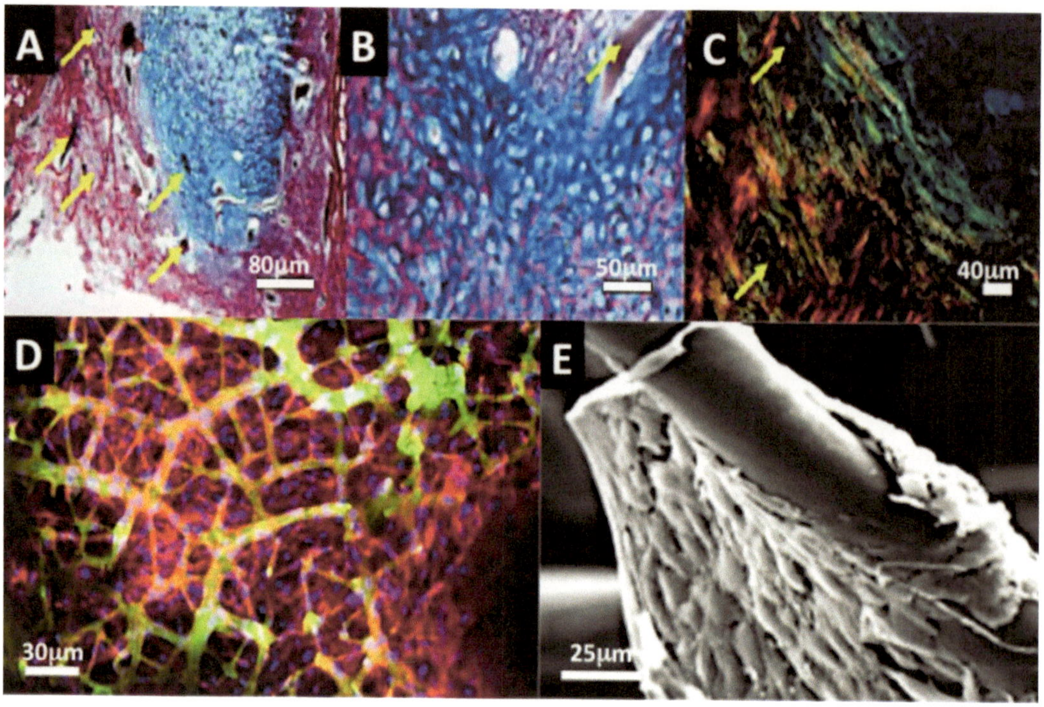

Fig. 20.3 Collagenous marine sponge comprises of a fibrous framework of bonded fibers and this could be an ideal substitute for a periodontal ligament and bone tissue. (**a–d**) Confocal fluorescence image of osteoblast cell sheets attached and suspended in marine sponge framework at 14 days. (**e**) SEM image of osteoblast cell aggregation on *Hippospongia* fiber

marine sponge collagen in dentistry (Green et al. 2014). The two marine inorganic polymers, biosilica and poly phosphate, have been shown to display a morphogenetic effect on osteoblasts (Wang et al. 2014) (Fig. 20.3).

20.2.2 Drug Delivery

Marine sponge collagen can be used as drug delivery vehicles due to its adsorbing capacity, nonantigenic nature, nontoxicity, and biocompatibility (Chak et al. 2013). After years of research and development, many of the compounds from the explored marine sponge species by KliniPharm have found their way into the product market including the eleana™ skin care. Collagen for dermatological applications and drug delivery technologies has been initiated by KliniPharm scientists, who have engineered a complex precipitation process to create pharmaceutically important collagen nanoparticles from organic marine sponge collagen. The developed collagen nanoparticles are capable of penetrating the skin's deeper tissues that show tremendous anti-aging effects (Swatschek et al. 2002a). Sterilization of the marine sponge collagens is very important to be considered for their application in biomedicine and health care (Palmer et al. 2012).

Swatschek et al. (2002b) prepared microparticles derived from marine sponge collagen for dermal delivery of all-trans retinol. Scanning electron microscopy results confirmed spherical size of 120–130 nm, and photon correlation spectroscopic measurement indicated particle size from 126 to 2,179 nm. Retinol-loaded sponge collagen microparticles were investigated into hairless mice skin and it was observed that the dermal penetration of retinol into the skin increases significantly (Swatschek et al. 2002b). Nicklas et al. (2009b) developed nanoparticles of *Chondrosia reniformis* sponge collagen as penetration enhancers for the transdermal drug delivery of 17β-estradiol-hemihydrate in hormone replacement therapy. Traditional alkaline hydrolysis method was used to prepare collagen nanoparticles and subsequently characterized by several spectroscopic techniques. Drug loading up to 13.1 % of sponge collagen particle mass was found (Nicklas et al. 2009b). No absorbance was noticed in samples with sponge collagen nanoparticles without drug. At estradiol-hemihydrate concentrations between 1.25 and 5 mg/mL, an increasing drug loading up to 13.1 % of sponge collagen particle mass was found (Fig. 20.4).

The hydrogel with estradiol-loaded collagen nanoparticles enabled a prolonged estradiol release compared to a commercial gel and yielded a considerably enhanced estradiol absorption. Consequently, sponge collagen nanoparticles represent promising carriers for transdermal drug delivery (Nicklas et al. 2009b). The novel coating based on the marine sponge collagen (using 12.9 mg/cm^2 coating material) complied with the requirements of Ph. Eur. for gastroresistant tablets. This coating material also meets the regulatory requirements for dietary supplements (Nicklas et al. 2009a).

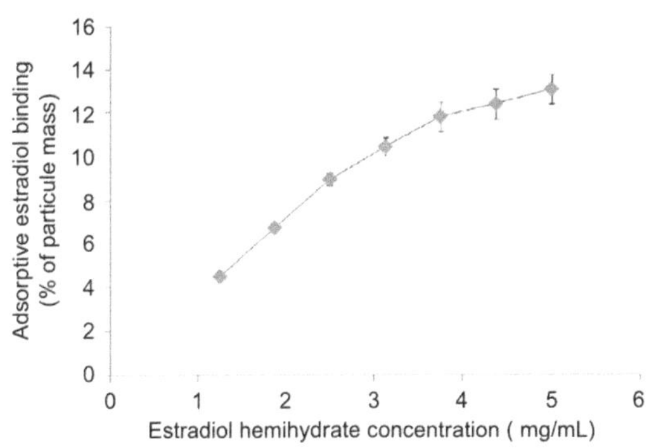

Fig. 20.4 Adsorption isotherm (20 °C) of estradiol to 10 mg SCNPs

20.3 Conclusion and Future Perspectives

Collagen has several advantages as a biomaterial in tissue engineering and drug delivery. The usage of marine sponge collagen has been described in this chapter briefly. Marine sponge collagen architecture has been found usefull for cell attachment and proliferation, which renders marine collagen biomaterials suitable for tissue construction. In addition, the tools for genetic characterization have not been considered for quick assessment and analysis of sponge genome. The next-generation sequencing technologies have to be put forward for the application in sponge collagen gene targeting to express such important collagen protein in microbes and plants in order to meet the human demands rather than destroying sponge habitat to isolate bulk proteins or other biomedical molecules.

References

Addad S, Exposito J-Y, Faye C et al. (2011) Isolation, characterization and biological evaluation of jellyfish collagen for use in biomedical applications. Mar Drugs 9(6):967–983

Aouacheria A, Geourjon C, Aghajari N et al. (2006) Insights into early extracellular matrix evolution: spongin short chain collagen-related proteins are homologous to basement membrane type IV collagens and form a novel family widely distributed in invertebrates. Mol Biol Evol 23(12):2288–2302

Chak V, Kumar D, Visht S (2013) A review on collagen based drug delivery systems. Int J Pharm Teach Pract 4(4):811–820

Deyl Z, Mikšík I, Eckhardt A (2003) Preparative procedures and purity assessment of collagen proteins. J Chromatogr B 790(1):245–275

Ehrlich H (2010) Biological materials of marine origin. Springer, Dordrecht

Ehrlich H, Deutzmann R, Brunner E et al. (2010) Mineralization of the metre-long biosilica structures of glass sponges is templated on hydroxylated collagen. Nat Chem 2(12):1084–1088

Exposito JY, Garrone R (1990) Characterization of a fibrillar collagen gene in sponges reveals the early evolutionary appearance of two collagen gene families. Proc Natl Acad Sci U S A 87(17):6669–6673

Francis G, Thomas J (1975) Isolation and chemical characterization of collagen in bovine pulmonary tissues. Biochem J 145(2):287–297

Garrone R (1985) The collagen of the Porifera. In: Biology of invertebrate and lower vertebrate collagens. Springer, Boston, pp 157–175

Gómez-Guillén M, Turnay J, Fernández-Dıaz M et al. (2002) Structural and physical properties of gelatin extracted from different marine species: a comparative study. Food Hydrocoll 16(1):25–34

Granito RN, Custódio MR, Rennó ACM (2016) Natural marine sponges for bone tissue engineering: the state of art and future perspectives. J Biomed Mater Res Part B 00B:000–000

Green D, Howard D, Yang X et al. (2003) Natural marine sponge fiber skeleton: a biomimetic scaffold for human osteoprogenitor cell attachment, growth, and differentiation. Tissue Eng 9(6):1159–1166

Green DW, Padula M, Santos J et al. (2013) A new role for marine skeletal proteins in regenerative orthopaedics. Key Eng Mater 529–530:654–659

Green DW, Lai W-F, Jung H-S (2014) Evolving marine biomimetics for regenerative dentistry. Mar Drugs 12 (5):2877–2912

Griffith LG (2002) Emerging design principles in biomaterials and scaffolds for tissue engineering. Ann N Y Acad Sci 961(1):83–95. doi:10.1111/j.1749-6632.2002.tb03056.x

Gruner H, Moritz M, Dunger W (1993) Lehrbuch der Speziellen Zoologie, Band I: Wirbellose Tiere, 4. Teil: Arthropoda (ohne Insekten), Gustav-Fischer-Verlag

Heinemann S, Ehrlich H, Douglas T et al. (2007) Ultrastructural studies on the collagen of the marine sponge *Chondrosia reniformis* Nardo. Biomacromolecules 8(11):3452–3457

Kimura S, Miura S, Park YH (1983) Collagen as the major edible component of jellyfish (*Stomolophus nomural*). J Food Sci 48(6):1758–1760

Kołodziejska I, Sikorski ZE, Niecikowska C (1999) Parameters affecting the isolation of collagen from squid (*Illex argentinus*) skins. Food Chem 66(2): 153–157

Langer R, Vacanti J (1993) Tissue engineering. Science 260(5110):920–926. doi:10.1126/science.8493529

Lin Z, Solomon KL, Zhang X et al. (2011) In vitro evaluation of natural marine sponge collagen as a scaffold for bone tissue engineering. Int J Biol Sci 7(7): 968–977

Mendis E, Rajapakse N, Byun H-G, Kim S-K (2005) Investigation of jumbo squid (*Dosidicus gigas*) skin gelatin peptides for their in vitro antioxidant effects. Life Sci 77(17):2166–2178

Miura S, Kimura S (1985) Jellyfish mesogloea collagen. Characterization of molecules as alpha 1 alpha 2 alpha 3 heterotrimers. J Biol Chem 260(28):15352–15356

Moreira-Silva J, Silva TH, Prata MB et al. (2013) Potential of marine sponge collagen coatings for skin regeneration strategies. J Tissue Eng Regen Med 7: 6–52. doi:10.1002/term.1822

Nagai T, Worawattanamateekul W, Suzuki N et al. (2000) Isolation and characterization of collagen from rhizostomous jellyfish (*Rhopilema asamushi*). Food Chem 70(2):205–208. doi:http://dx.doi.org/10.1016/S0308-8146(00)00081-9

Nerem RM, Sambanis A (1995) Tissue engineering: from biology to biological substitutes. Tissue Eng 1(1): 3–13

Nicklas M, Schatton W, Heinemann S et al. (2009a) Enteric coating derived from marine sponge collagen. Drug Dev Ind Pharm 35(11):1384–1388

Nicklas M, Schatton W, Heinemann S et al. (2009b) Preparation and characterization of marine sponge collagen nanoparticles and employment for the transdermal delivery of 17β-estradiol-hemihydrate SCNPs for dermal delivery of estradiol. Drug Dev Ind Pharm 35(9):1035–1042

O'Brien FJ, Harley BA, Yannas IV et al. (2005) The effect of pore size on cell adhesion in collagen-GAG scaffolds. Biomaterials 26(4):433–441. doi:http://dx.doi.org/10.1016/j.biomaterials.2004.02.052

Pallela R, Bojja S, Janapala VR (2011) Biochemical and biophysical characterization of collagens of marine sponge, Ircinia fusca (Porifera: Demospongiae: Irciniidae). Int J Biol Macromol 49(1):85–92. doi:10.1016/j.ijbiomac.2011.03.019

Pallela R, Venkatesan J, Janapala VR et al. (2012) Biophysicochemical evaluation of chitosan-hydroxyapatite-marine sponge collagen composite for bone tissue engineering. J Biomed Mater Res 100A(2):486–495. doi:10.1002/jbm.a.33292

Palmer I, Clarke SA, Nelson J et al. (2012) Identification of a suitable sterilisation method for collagen derived from a marine Demosponge. Inter J Nano Biomater 4(2):148–163

Pozzolini M, Bruzzone F, Berilli V et al. (2012) Molecular characterization of a nonfibrillar collagen from the marine sponge chondrosia reniformis nardo 1847 and positive effects of soluble silicates on its expression. Mar Biotechnol 14(3):281–293

Quereshi S, Mhaske A, Raut D, Singh R, Mani A, Patel J (2010) Extraction and partial characterization of collagen from different animal skins. Recent Res Sci Technol 2(9):28–31

Rao JV, Pallela R, Prakash G (2011) Prospects of marine sponge collagen and its applications in cosmetology. Marine cosmeceuticals: trends and prospects. CRC Press, Boca Raton, pp 77–103

Shapiro AD (2014) Scientists search Palau's coral reefs for new anti-cancer drugs. Public Radio International, the World. http://www.pri.org/stories/2014-02-24/scientists-search-palaus-coral-reefs-new-anti-cancer-drugs

Silva TH, Moreira-Silva J, Marques AL et al. (2014) Marine origin collagens and its potential applications. Mar Drugs 12(12):5881–5901

Sipkema D, Osinga R, Schatton W et al. (2005) Large-scale production of pharmaceuticals by marine sponges: sea, cell, or synthesis? Biotechnol Bioeng 90(2):201–222. doi:10.1002/bit.20404

Song E, Yeon Kim S, Chun T et al. (2006) Collagen scaffolds derived from a marine source and their biocompatibility. Biomaterials 27(15):2951–2961

Swatschek D, Schatton W, Kellermann J et al. (2002a) Marine sponge collagen: isolation, characterization and effects on the skin parameters surface-pH, moisture and sebum. Eur J Pharm Biopharm 53(1): 107–113

Swatschek D, Schatton W, Müller WEG et al. (2002b) Microparticles derived from marine sponge collagen (SCMPs): preparation, characterization and suitability for dermal delivery of all-trans retinol. Eur J Pharm Biopharm 54(2):125–133

Van Soest RW, Boury-Esnault N, Vacelet J et al. (2012) Global diversity of sponges (Porifera). PLoS One 7(4): e35105

Wang X, Schröder HC, Grebenjuk V et al. (2014) The marine sponge-derived inorganic polymers, biosilica and polyphosphate, as morphogenetically active matrices/scaffolds for the differentiation of human multipotent stromal cells: potential application in 3D printing and distraction osteogenesis. Mar Drugs 12(2):1131–1147

Zheng MH, Hinterkeuser K, Solomon K et al. (2007) Collagen-derived biomaterials in bone and cartilage repair. Macromol Symp 253:179–185